DUXBURY

Statistical Ideas and Methods

JESSICA M. UTTS
University of California, Davis

ROBERT F. HECKARD
Pennsylvania State University

THOMSON
BROOKS/COLE

Australia • Canada • Mexico • Singapore • Spain • United Kingdom • United States

Statistical Ideas and Methods
Jessica M. Utts and Robert F. Heckard

Senior Acquisitions Editor: Carolyn Crockett
Development Editor: Danielle Derbenti and Carol Pritchard-Martinez
Assistant Editor: Ann Day
Editorial Assistant: Fiona Chong
Technology Project Manager: Burke Taft
Marketing Manager: Stacy Best
Marketing Assistant: Jessica Bothwell
Advertising Project Manager: Nathaniel Bergson-Michelson
Project Manager, Editorial Production: Belinda Krohmer
Print/Media Buyer: Barbara Britton

Permissions Editor: Joohee Lee
Production Service: Martha Emry
Text Designer: Terri Wright, Andrew Ogus
Photo Researcher: Pat Quest
Copy Editor: Constance Day
Cover Designer: Bill Reuter
Cover Image: © 2004 PictureQuest/Brand X Picture
Cover Printer: Phoenix Color Corp
Compositor: G&S Typesetters, Inc.
Printer: R.R. Donnelly/Willard

Text printed in the United States of America
1 2 3 4 5 6 7 09 08 07 06 05

For more information about our products, contact us at:
Thomson Learning Academic Resource Center
1-800-423-0563

For permission to use material from this text or product, submit a request online at:
http://www.thomsonrights.com

Any additional questions about permissions can be submitted by email to **thomsonrights@thomson.com.**

Library of Congress Control Number: 2004116850

Student Edition: ISBN 0-534-40284-4

Instructor's Edition: ISBN 0-495-01595-4

Thomson Higher Education
10 Davis Drive
Belmont, CA 94002-3098
USA

Asia (including India)
Thomson Learning
5 Shenton Way
#01-01 UIC Building
Singapore 068808

Australia/New Zealand
Thomson Learning Australia
102 Dodds Street
Southbank, Victoria 3006
Australia

Canada
Thomson Nelson
1120 Birchmount Road
Toronto, Ontario M1K 5G4
Canada

UK/Europe/Middle East/Africa
Thomson Learning
High Holborn House
50/51 Bedford Row
London WC1R 4LR
United Kingdom

Latin America
Thomson Learning
Seneca, 53
Colonia Polanco
11560 Mexico
D.F. Mexico

Spain (includes Portugal)
Thomson Paraninfo
Calle Magallanes, 25
28015 Madrid, Spain

Brief Contents

Contents

CHAPTER 3   Gathering Useful Data   58

CHAPTER 4   Sampling: Surveys and How to Ask Questions   88

CHAPTER 5   Relationships Between Quantitative Variables   130

CHAPTER 12   More About Confidence Intervals   390

CHAPTER 13   More About Significance Tests   436

CHAPTER 14   More About Regression   492

CHAPTER 15   More About Categorical Variables   526

CHAPTER 16   Analysis of Variance   560

Preface

Before you continue, think about how you would answer the question in the first bullet, and read the statement in the second bullet. We will return to them a little later in this Preface.

- What do you *really know* is true, and how do you know it?
- The diameter of the moon is about 2160 miles.

WHAT IS STATISTICS AND WHO SHOULD CARE?

Because people are curious about many things, chances are that your interests include topics to which statistics has made a useful contribution. As written in Chapter 22, "information developed through the use of statistics has enhanced our understanding of how life works, helped us learn about each other, allowed control over some societal issues, and helped individuals make informed decisions. There is almost no area of knowledge that has not been advanced by statistical studies."

Statistical methods have contributed to our understanding of health, psychology, ecology, politics, music, lifestyle choices, and dozens of other topics. A quick look through this book, especially Chapters 1 and 22, should convince you of this. Watch for the influences of statistics in your daily life as you learn this material.

Although statistics courses are often offered through a mathematics department, statistics is not a branch of mathematics. Mathematics is to statistics as wood, hammer, and nails are to building a house: a partial set of materials and tools. Statistics also draws materials and tools from philosophy, graphics, computing, psychology, and language.

HOW IS THIS BOOK DIFFERENT?
TWO BASIC PREMISES OF LEARNING

We wrote this book because we were tired of being told that what statisticians do is boring and difficult. We think statistics is useful and not difficult to learn, and yet the majority of college graduates we've met seemed to have had a negative experience taking a statistics class in college. We hope this book will help to overcome these misguided stereotypes.

Let's return to the two bullets at the beginning of this Preface. Without looking, do you remember the diameter of the moon? Unless you already had a pretty good idea, or have an excellent memory for numbers, you probably don't remember. One premise of this book is that new material is much easier to learn and remember if it is related to something interesting or previously known. The diameter of the moon is about the same as the air distance between Atlanta and Los Angeles, San Francisco and Chicago, London and Cairo, or Moscow and Madrid. Picture the moon sitting between any of those pairs of cities, and you are not likely to forget the size of the moon again. Throughout this book, new material is presented in the context of interesting and useful examples. The first and last chapters (1 and 22) are exclusively devoted to examples and case studies, which illustrate the wisdom that can be generated through statistical studies.

Now answer the question asked in the first bullet: What do you really know is true and how do you know it? If you are like most people, you know because it's something you have experienced or verified for yourself. It is not likely to be something you were told or heard in a lecture. The second premise of this book is that new material is easier to learn if you actively ask questions and answer them for yourself. *Statistical Ideas and Methods* is designed to help you learn statistical ideas by actively thinking about them. Throughout most of the chapters there are boxes entitled *Turn On Your Mind*. Thinking about the questions in those boxes will help you to discover and verify important ideas for yourself. We encourage you to think and question, rather than simply read and listen.

TOOLS FOR EXPANDED LEARNING

There are a number of tools provided in this book and beyond to enhance your learning of statistics.

Turn On Your Mind boxes appear throughout each chapter to help you develop your statistical reasoning and intuition. ***Hints*** are provided to help you develop these skills.

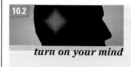

10.2 | *turn on your mind*

Explain in your own words what it means to say that we have 95% confidence in the interval estimate. Then give an example of something you do in your life that illustrates the same concept—you follow the same procedure each time, and it either works (most of the time) or does not work to produce the desired result. What confidence level would you assign to the procedure in your example; i.e., what percent of the time do you think it produces your desired result?*

*HINT: What is a task you frequently perform that you can't always do successfully?

Interactive Video Skillbuilder CD-ROM, in the back of every new text, contains hours of video instruction presented by Lisa M. Sullivan of Boston University. Watch as she walks you through key examples from the text, step by step—giving you a foundation in the skills that you need to know. Each example found on the CD is identified by the video icon located in the margin.

EXAMPLE 3.3
What Confounding Variables Lurk Behind Lower Blood Pressure?

In Case Study 1.5, the people who attended church regularly had lower blood pressure than people who stayed home and watched religious services on television. But it may be that those who attended church had lower blood pressure as a result of a strong social support network rather than from attending church regularly. If this is the case, the "amount of social support" is a *confounding* variable. Notice that two conditions must be met for "amount of social support" to qualify as a confounding variable:

- Amount of social support (confounding variable) *affects* blood pressure (the response variable).
- Amount of social support (confounding variable) *is related to* attending church regularly (the explanatory variable) because the same people who attend church regularly are likely to have a strong social support network.

Health status is another possible confounding variable, since those who are not well may not be able to attend church regularly. There are numerous other possibilities, like age, attitude toward life, and so on, that could affect the observed relationship between religious activity and blood pressure. ◆

Turn On Your Computer features provide additional opportunities for independent, hands-on exploration of key statistical concepts.

Random Sampling in Action

It should be obvious to you by now that each time you take a random sample from a population you are likely to get a different group of individuals, even though the population remains fixed. The **Sampling** applet on the CD for this text allows us to watch what happens when we repeatedly take random samples from the same population. Figure 4.1 illustrates how the applet appears when it starts. Each stick figure represents one member of a population of 100 college students: 55 females (blue) and 45 males (red). Their heights have been recorded as well, and if you look carefully you will see that the little stick figures are all of slightly different heights. The mean height for the population is 68 inches.

FIGURE 4.1 **The sampling applet starting point**

Key Terms at the end of each chapter, organized by section, can be used as a "quick-finder" and as a review tool.

KEY TERMS

Section 9.1
statistic, 294, 296
parameter, 294, 296
sampling distribution, 294–295, 296

Section 9.2
sample proportion, 296
normal curve approximation rule for sample proportions, 297, 298
sampling distribution of $\hat{p}$, 298
standard deviation of $\hat{p}$, 300
standard error of $\hat{p}$, 300

Section 9.3
normal curve approximation rule for sample means, 302–303, 304

sampling distribution of $\bar{x}$, 304
sampling distribution of the mean, 304
standard deviation of the sample mean, 304
standard error of the mean, 305
law of large numbers, 306

Section 9.4
central limit theorem, 309

Section 9.6
standardized statistic, 313–314
standardized z-statistic, 313–314, 315

Section 9.7
Student's t-distribution, 314
t-distribution, 314
degrees of freedom, (df), 315
standardized t-statistic, 315

Section 9.8
statistical inference, 316
confidence interval, 317
hypothesis testing, 317
significance testing, 317
statistical significance, 317

Special *Tech Note* boxes provide additional technical discussion as well as details about using MINITAB® statistical software* and Microsoft® Excel®.

Additional Notes on Creating Histograms

tech note ◆ Consider using intervals that make the range and width of each interval convenient. For instance, to create a histogram of ages at death for First Ladies, it would be convenient to use ten-year periods—died in her 30s, died in her 40s, and so on, up to 90s. This would create seven intervals.

◆ To show relative frequency, you can use either the proportion or the percent that are in an interval.

Basics *Exercises* focus on practice and review; these exercises, color-coded in blue and appearing at the beginning of each exercise section, complement the conceptual and data-analysis exercises.

EXERCISES   Blue = Basics   Bold, **Bold** = Answer in back

1.1 A five-number summary for the heights in inches of the women who participated in the survey in Case Study 1.1 is

Female Heights (Inches)		
Median	65	
Quartiles	63.5	67.5
Extremes	59	71

a. What is the median height for these women?
b. What is the range of heights, that is, the difference in heights between the shortest and the tallest women?
c. What is the interval of heights containing the shortest 1/4 of the women?
d. What is the interval of heights containing the middle 1/2 of the women?

1.2 In the year 2000, Vietnamese American women had the highest rate of cervical cancer in the country. Suppose that among 200,000 Vietnamese American women, 86 developed cervical cancer.

1.8 Suppose that an observational study showed that students who get at least 7 hours of sleep performed better on exams than students who didn't. Which of the following are possible confounding variables, and which are not? Explain why in each case.
a. Number of courses the student took that term.
b. Weight of the student.
c. Number of hours the student spent partying in a typical week.

1.9 Explain the distinction between statistical significance and practical significance. Can the result of a study be statistically significant but not practically significant?

1.10 A headline in a major newspaper read, "Breast-Fed Youth Found to Do Better in School."
a. Do you think this statement was based on an observational study or a randomized experiment? Explain.
b. Given your answer in part (a), which of these two alternative headlines do you think would be preferable: "Breast-feeding Leads to Better School Performance" or "Link Found Between Breast-feeding and School Performance"? Explain.

A *Student's Suite CD* provided with the book includes the *Turn On Your Computer* applets, tutorial quizzes, InfoTrac® College Edition and Internet exercises, Microsoft PowerPoint® presentation slides, links to both the Companion Website and *Internet Companion for Statistics* by Michael Larsen, datasets formatted for MINITAB, Microsoft Excel, SPSS®, SAS®, JMP®, and ASCII, and technology manuals designed for use with MINITAB, SPSS, Excel, and the TI-83.

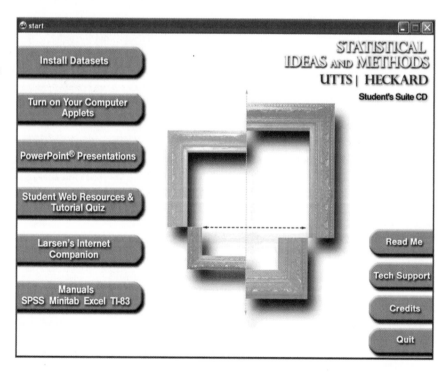

*MINITAB is a trademark of Minitab, Inc. and is used herein with the owner's permission (www.minitab.com).

Case Studies apply statistical ideas to intriguing news stories. As the *Case Studies* are developed, they model the statistical reasoning process.

CASE STUDY 6.1
Drinking, Driving, and the Supreme Court

In the early 1970s a young man challenged an Oklahoma State law that prohibited the sale of 3.2% beer to males under 21 but allowed its sale to females in the same age group. The case (*Craig* v. *Boren*, 429 U.S. 190, 1976) was ultimately heard by the U.S. Supreme Court.

Laws are allowed to use gender-based differences as long as they "serve important governmental objectives" and "are substantially related to the achievement of these objectives" (Gastwirth, 1988, p. 524). The defense argued that traffic safety was an important governmental objective and that data clearly show that young males are more likely to have alcohol-related accidents than young females.

The Supreme Court examined evidence from a "random roadside survey" that measured information on age, gender, and whether or not the driver had been drinking alcohol in the previous two hours. Although the survey was called a "random" survey of drivers, it probably was not. In roadside surveys, police tend to stop all drivers at certain locations at the time of the survey. This procedure does not really provide a random sampling of drivers in an area, but we'll treat it as though it does. Table 6.8 gives the results of the roadside survey for the drivers under 20 years of age.

TABLE 6.8 ■ **Results of Roadside Survey for Young Drivers**

	Drank Alcohol in Last Two Hours?			
	Yes	No	Total	Percent Who Drank
Males	77	404	481	16.0%
Females	16	122	138	11.6%
Total	93	526	619	15.0%

Source: Gastwirth, 1988, p. 526.

Notice that the percentage of young men who had been drinking alcohol is slightly higher than the percentage of young women. The difference is 16% − 11.6% = 4.4%. However, we cannot rule out chance as a reasonable explanation for this difference. In other words, if there really is no difference between the percents of young male and female drivers in the population who drink and drive, we could possibly see a difference as large as the one observed in a sample of this size.

In Figure 6.6, we present the results of asking the Minitab program to compute the chi-square statistic for this example. The chi-square summary statistic is 1.637, and the *p*-value for this statistic is 0.201. This *p*-value tells us that if there's really no association in the population (the null hypothesis), there's about a 20% chance that the sample would have a chi-square statistic as large as 1.637, or larger. In other words, the observed relationship could easily have occurred even if there is no relationship in the population represented by the sample.

The Supreme Court overturned the law, concluding that "the showing offered by the appellees does not satisfy us that sex represents a legitimate, accurate proxy for the regulation of drinking and driving" (Gastwirth, 1988, p. 527). Based on the chi-square analysis, you can see why the Supreme Court was reluctant to conclude that the difference in the sample represented sufficient evidence for a real difference in the population.

Minitab v. 14 update on CD

Expected counts are printed below observed counts

	Drank in Last 2 Hours?		
	Yes	No	Total
Males	77	404	481
	72.27	408.73	
Females	16	122	138
	20.73	117.27	
Total	93	526	619

Chi-Sq = 0.310 + 0.055 +
                1.081 + 0.191 = 1.637
DF = 1, P-Value = 0.201

FIGURE 6.6   Minitab output for Case Study 6.1

In Summary boxes appear at appropriate points to enhance key concepts and calculations.

in summary   **Possible Reasons for Outliers and Reasonable Actions**

♦ *A mistake was made while taking a measurement or entering it into the computer.* If this can be verified, the values should be discarded or corrected.

♦ *The individual in question belongs to a different group than the bulk of individuals measured.* Values may be discarded if a summary is desired and reported for the majority group only.

♦ *The outlier is a legitimate data value and represents natural variability for the group and variable(s) measured.* Values may not be discarded in this case—they provide important information about location and spread.

Many *Exercises* (indicated by bold numbers) have complete or partial solutions, found in the **Answers to Selected Exercises** at the back of the book, to check your answers on those exercises and guide your thinking on similar exercises.

Answers to Selected Exercises

Chapter 1

1.2 a. .00043
1.5 a. 400
1.7 c. Randomized experiment.
     d. Observational study.
1.11 189/11,034, or about 17/1000, based on placebo group.
1.15 a. 150 mph.
     b. 55 mph.
     c. 80 mph.
     d. 1/2
     e. 51
1.19 No.
1.22 The base rate for that type of cancer.
1.26 a. 212/1525 = .139.
     b. $1/\sqrt{1525}$ = .026.
     c. .113 to .165.
1.28 a. Self-selected or volunteer.
     b. No. Readers with strong opinions will respond.

Chapter 2

2.1 a. Categorical.
     b. Quantitative.
     c. Quantitative.
     d. Categorical.
2.2 a. Categorical.
     b. Ordinal.
2.4 a. Not continuous.
2.5 a. Explanatory is amount of walking or running; response is lung function.
2.6 a. Population.
     b. Sample.
2.8 a. Support ban or not; categorical.
     b. Gain on verbal and math SATs after program; quantitative.

c. Smoker or not and Alzheimer sufferer or not; both categorical.
2.9 a. Question 1a.
2.11 Example: Letter grades (A, B, etc.) converted to GPA.
2.14 a. 1427/2530 = .564, or 56.4%.
     b. 100% − 56.4% = 43.6%.
2.15 a. 1700/2470 = .688, or 67.8%.
     b. 1056/1700 = .621, or 62.1%.
2.16 c. Overweight = 26.57%; about right = 69.23%; underweight = 4.20%.
2.18 a. Explanatory is smoked or not; response is developed Alzheimer's disease or not.
2.21 Either could be justified as being more informative.
2.23 a. Skewed to the right.
     b. 13 ear pierces may be an outlier.
     c. 2 ear pierces; about 45 women had this number.
     d. About 32 or so.
2.25 a. Roughly symmetric.
     b. Highest = 92.
     c. Lowest = 64.
     d. 5/20 = .25, or 25%
2.29 Example: The age of a person who is 80 years old would be an outlier at a traditional college, but not at a retirement home.
2.31 Whether it's the male author (then not an outlier) or the female author (then an outlier).
2.34 Yes. Values inconsistent with the bulk of the data will be obvious.
2.36 a. This is personal preference; some may prefer a very large family.
     b. An outlier (in the high direction).
2.37 a. Median = (72 + 76)/2 = 74; mean = 74.33.
     b. Median = 7; mean = 25.
2.39 a. Range = 225 − 123 = 102.
     b. IQR = 35.
     c. 50%

HOW TO USE THIS TEXT

Chapters 1–13 cover the basics. Chapters 14 through 16 can be covered in any order after that. Links between Chapters 17 through 22 and earlier chapters are as follows:

◆ Chapter 17, Additional Discrete Random Variables, can be covered any time after Section 8.4 is completed. The three sections of Chapter 17 do not depend upon each other, so any of the sections can be included or omitted.

◆ Chapter 18, Nonparametric Tests of Hypotheses, can be covered at any time after sections 13.1–13.3 have been covered. The four sections of Chapter 18 do not depend upon each other, so any of the sections can be included or omitted.

◆ Chapter 19, Multiple Regression, can be covered after Chapter 14 has been completed.

◆ Chapter 20, Two-Way Analysis of Variance, can be covered after the completion of Chapter 16.

◆ Chapter 21, Ethics, can be covered at any time after Chapter 13 has been completed, although there is a reference in Section 21.3 to the Mann–

Whitney two-sample rank-sum test. It may be desirable to cover Section 18.2 either before or in conjunction with Chapter 21.

♦ Chapter 22, Turning Information into Wisdom, can be covered any time after Chapter 13 has been completed.

A Note to Instructors

The Instructor's Suite CD includes everything featured on the Student's Suite CD, plus a Test Bank in Microsoft Word® format, sample course outlines and syllabi, Web resources for examples and case studies, complete solutions to the exercises found in the text, suggested answers for the *Turn On Your Mind* questions, recommended in-class noncomputer-oriented projects, and information on how to correlate text material with the AP Statistics curriculum. Additional Instructor's Resources are available through your Duxbury/Thomson representative.

ACKNOWLEDGMENTS

We thank William Harkness, Professor of Statistics at Penn State University, for continued support and feedback throughout our careers and during the writing of this book, and for his remarkable dedication to undergraduate statistics education. Preliminary editions of *Mind on Statistics,* the basis for this text, were used at Penn State, the University of California at Davis, and Texas A & M University, and we thank the many students who provided comments and suggestions on those and on subsequent editions. Thanks to Dr. Melvin Morse (Valley Children's Clinic and University of Washington) for suggesting the title for Chapter 22 and to Deb Niemeier, University of California, Davis, for suggesting that we add a chapter on ethics. At Penn State, Dave Hunter, Steve Arnold, and Tom Hettmansperger have provided many helpful insights. At the University of California at Davis, Rodney Wong has provided insights as well as material for some exercises and the Test Bank. The following reviewers offered valuable suggestions: Patti B. Collings, Brigham Young University; James Curl, Modesto Junior College; Donald Harden, Georgia State University; Rosemary Hirschfelder, University of Puget Sound; Sue Holt, Cabrillo Community College; Tom Johnson, North Carolina State University; Andre Mack, Austin Community College; D'Arcy Mays, Virginia Commonwealth; Mary Murphy, Texas A & M University; N. Thomas Rogness, Grand Valley State University; Heather Sasinouska, Clemson University; Robert Alan Wolf, University of San Francisco; Elizabeth Clarkson, Wichita State University; Megan Meece, University of Florida; Thomas Nygren, Ohio State University; Wade Ellis, West Valley College; David Robinson, St. Cloud University; Jay Gregg, Colorado State University; Mark Johnson, University of Central Florida; and Nancy Pfenning, University of Pittsburgh.

Our sincere appreciation and gratitude goes to Carolyn Crockett and the staff of Duxbury, without whom this book could not have been written, and to Martha Emry who always keeps us on track throughout the editing and production process. Finally, for their support, patience, and numerous prepared dinners, we thank our families and friends, especially Candace Heckard, Molly Heckard, Wes Johnson, Claudia Utts-Smith, and Dennis Smith.

Jessica Utts
Robert Heckard

Statistics Success Stories and Cautionary Tales

If they could drive faster, would they? See Case Study 1.1.

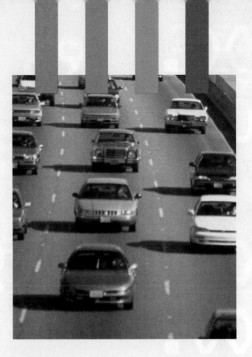

The seven stories in this chapter are meant to bring life to the term *statistics*. When you are finished reading these stories, if you still think the subject of statistics is lifeless or gruesome, check your pulse!

et's face it. You're a busy person. Why should you spend your time learning about a subject that sounds as dull as statistics? In this chapter we give seven examples of situations in which statistics either provided enlightenment or misinformation. With these examples, we hope to convince you that learning about this subject will be interesting and useful.

Each of the stories in this chapter illustrates one or more concepts that will be developed throughout the book. These concepts are given as "the moral of the story" after a case is presented. Definitions of some terms used in the story also are provided following each case. By the time you read all of these stories, you already will have an overview of what statistics is all about. ◆

1.1 | WHAT IS STATISTICS?

When you hear the word *statistics* you probably think of lifeless or gruesome numbers, like the population of your state or the number of violent crimes committed in your city last year. The word *statistics,* however, actually is used to mean two different things. The better-known definition is that statistics are numbers measured for some purpose. A more complete definition, and the one that forms the substance of this book, is the following:

> *definition* **Statistics** is a collection of procedures and principles for gathering data and analyzing information in order to help people make decisions when faced with uncertainty.

The stories in this chapter are meant to bring life to this definition. When you are finished reading them, if you still think the subject of statistics is lifeless or gruesome, check your pulse!

1.2 | SEVEN STATISTICAL STORIES WITH MORALS

The best way to gain an understanding of some of the ideas and methods used in statistical studies is to see them in action. Each of the seven stories presented in this chapter includes interesting lessons about how to gain information from data. The methods and ideas will be expanded throughout the book, but these seven stories will give you an excellent overview of why it is useful to study statistics. To help you understand some basic statistical principles, each case study is accompanied by a "moral of the story" and by some definitions. All of the ideas and definitions will be discussed in greater detail in subsequent chapters.

CASE STUDY 1.1
Who Are Those Speedy Drivers?

A survey taken in a large statistics class at Penn State University contained the question "What's the fastest you have ever driven a car? _____mph." The *data* provided by the 87 males and 102 females who responded are listed here.

Males: 110 109 90 140 105 150 120 110 110 90 115 95 145 140 110 105 85 95 100 115 124 95 100 125 140 85 120 115 105 125 102 85 120 110 120 115 94 125 80 85 140 120 92 130 125 110 90 110 110 95 95 110 105 80 100 110 130 105 105 120 90 100 105 100 120 100 100 80 100 120 105 60 125 120 100 115 95 110 101 80 112 120 110 115 125 55 90

Females: 80 75 83 80 100 100 90 75 95 85 90 85 90 90 120 85 100 120 75 85 80 70 85 110 85 75 105 95 75 70 90 70 82 85 100 90 75 90 110 80 80 110 110 95 75 130 95 110 110 80 90 105 90 110 75 100 90 110 85 90 80 80 85 50 80 100 80 80 80 95 100 90 100 95 80 80 50 88 90 90 85 70 90 30 85 85 87 85 90 85 75 90 102 80 100 95 110 80 95 90 80 90

From these numbers, can you tell which sex tends to have driven faster, and by how much? Notice how difficult it is to make sense of the *data* when you are simply presented with a list. Even if the numbers had been presented in numerical order, it would be difficult to compare the two sexes.

Your first lesson in statistics is how to formulate a simple summary of a long list of numbers. The *dotplot* shown in Figure 1.1 helps us see the pattern in the data. In the plot, each dot represents the response of an individual student. We can see that the men tend to claim a higher "fastest ever driven" speed than do the women.

The graph shows us a lot, and calculating some statistics that summarize the data will provide additional insight. There are a variety of ways to do so, but for this example we examine a *five-number summary* of the data for males and females. The five numbers are the lowest value, the cutoff points for 1/4, 1/2, and 3/4 of the data, and the highest value. The three middle values of the summary (the cutoff points for 1/4, 1/2, and 3/4 of the data)

FIGURE 1.1   **Responses to "What's the fastest you've ever driven?"**

are called the *lower quartile, median,* and *upper quartile.* Five-number summaries can be represented like this:

	Males (87 Students)		**Females (102 Students)**	
Median		110		89
Quartiles	95	120	80	95
Extremes	55	150	30	130

Some interesting facts become immediately obvious from these summaries. By looking at the medians, you see that half of the men have driven 110 miles per hour or more, whereas the halfway point for the women is only 89 miles per hour. In fact, 3/4 of the men have driven 95 miles per hour or more, but only 1/4 of the women have done so. These facts were not at all obvious from the original lists of numbers.

Moral of the Story: *Simple summaries of data can tell an interesting story and are easier to digest than long lists.*

Definitions: Data is a plural word referring to numbers or non-numerical labels (such as male/female) collected from a set of entities (people, cities, and so on). The **median** of a numerical list of data is the value in the middle when the numbers are put in order. For an even number of entities, the median is the average of the middle two values. The **lower quartile** and **upper quartile** are (roughly) the medians of the lower and upper halves of the data.

CASE STUDY 1.2
Safety in the Skies?

Air travelers already have enough concerns, with the potential for delays, lost baggage, and so on. So if you do a lot of air travel, or know anyone who does, you may have been disturbed by the headline in *USA Today* that read, "Planes get closer in midair as traffic control errors rise" (Levin, 1999). You may have been even more disturbed by the details: "Errors by air traffic controllers climbed from 746 in fiscal 1997 to 878 in fiscal 1998, an 18% increase." Don't cancel your next vacation yet. A look at the statistics indicates that the news is actually pretty good! And, there is some reassurance when we are told that "most [errors] involve planes passing so far apart that there is no danger of collision."

The headline and details do sound ominous — all those errors and an increase of almost 20%! If things continue at that rate, won't your next flight be quite likely to suffer from air traffic controller error? The answer is a resounding "no," which becomes obvious when we are told the *base rate* or *baseline risk* for errors. "The errors per million flights handled by controllers climbed from 4.8 to 5.5." So, the original *rate* of errors in 1997, from which the 18% increase was calculated, was less than 5 errors per million flights.

Fortunately, the rates for the two years were provided in the story. This is not always the case in news reports of increases in rates. For instance, an article may say that the rate of a certain type of cancer is doubled if you eat a certain unhealthful food. But what good is that information unless you know the actual risk? Doubling your chance of getting cancer from 1 in a million to 2 in a million is trivial, but doubling your chance from 1 in 50 to 2 in 50 is not.

Moral of the Story: *When discussing the change in the rate or risk of occurrence of something, make sure you also include the base rate or baseline risk.*

Definitions: The **rate** at which something occurs is simply the number of times it occurs per number of opportunities for it to occur. In fiscal year 1998, the rate of errors was 5.5 per million flights. The **risk** of a bad outcome in the future can be estimated using the past rate for that outcome, if it is assumed the future is like the past. Based on 1998 data, the estimated risk of an error for any given flight in 1999 would be 5.5/1,000,000, or .0000055. The **base rate** or **baseline risk** is the rate or risk at a beginning time period or under specific conditions. For instance, the base rate of air traffic controller errors was 4.8 per million flights in fiscal year 1997.

CASE STUDY 1.3
Did Anyone Ask Whom You've Been Dating?

"According to a new *USA Today*/Gallup Poll of teenagers across the country, 57 percent of teens who go out on dates say they've been out with someone of another race or ethnic group" (Peterson, 1997). That's over half of the dating teenagers, so of course it was natural for the headline in the *Sacramento Bee* to read, "Interracial dates common among today's teenagers." The article contained other information as well, such as: "In most cases, parents aren't a major obstacle. Sixty-four percent of teens say their parents don't mind that they date interracially, or wouldn't mind if they did."

There are millions of teenagers in the United States whose experiences are being reflected in this story. How could the poll-takers manage to ask so many teenagers these questions? The answer is that they didn't. The article states that "the results of the new poll of 602 teens, conducted Oct. 13–20, reflect the ubiquity of interracial dating today. . . ." They asked only 602 teens? Could such a small sample possibly tell us anything about the millions of teenagers in the United States? The answer is "yes" if those teens constituted a *random sample* from the *population* of interest.

The featured statistic of the article is that "57 percent of teens who go out on dates say they've been out with someone of another race or ethnic group." Only 496 of the 602 teens in the poll said they date, so the 57% value is actually a percent based on 496 responses. In other words, the pollsters were using information from only 496 teenagers to estimate something about all teenagers. Figure 1.2 illustrates this situation.

How accurate could this sample possibly be? The answer may surprise you. The results of this poll are accurate to within a *margin of error* of about 4.5%. As surprising as it may seem, the true percentage of all teens in the United States who date interracially is reasonably likely to be within 4.5% of the reported percentage that's based only on the 496 teens asked! We'll be conservative and round the 4.5% margin of error up to 5%. The percent of all teenagers in the United States who date that would say they have dated interracially is likely to be in the range 57% $\pm$ 5%, or between 52% and 62%. (The symbol $\pm$ is read "plus and minus" and means that the value on the right should be added to and subtracted from the value on the left, to create an interval.)

FIGURE 1.2   Population and sample for the survey

Polls and *sample surveys* are frequently used to assess public opinion and to estimate population characteristics like the percent of teens who have dated interracially. Many sophisticated methods have been developed that allow pollsters to gain the information they need from a very small number of individuals. The trick is to know how to select those individuals. In Chapter 4 we examine a number of other strategies used to ensure that sample surveys provide reliable information about populations.

Moral of the Story: *A representative sample of only a few thousand, or perhaps even a few hundred, can give reasonably accurate information about a population of many millions.*

Definitions: A **population** is a collection of all individuals about which information is desired. The "individuals" are usually people, but could also be schools, cities, pet dogs, agricultural fields, and so on. A **random sample** is a subset of the population selected so that every individual has a specified probability of being part of the sample. In a **sample survey,** the investigators gather opinions or other information from each individual included in the sample. The **margin of error,** for a properly conducted survey, is a number that is added to and subtracted from the sample information to produce an interval that is 95% certain to contain the truth about the population. In the most common types of sample surveys, the margin of error is approximately equal to 1 divided by the square root of the number of individuals in the sample. Hence, a sample of 496 teenagers who have dated produces a margin of error of about $1/\sqrt{496} = .045$, or about 4.5%.

CASE STUDY 1.4
Who Are Those Angry Women?

A well-conducted survey can be very informative, but a poorly conducted one can be a complete disaster. As an extreme example, Moore (1997, p. 11) reports that for her highly publicized book *Women and Love,* Shere Hite sent questionnaires to 100,000 women asking about love, sex, and relationships. Only 4.5% of the women responded, and Hite used those responses to write her book. As Moore notes, "The women who responded were fed up with men and eager to fight them. For example, 91% of those who were divorced said that they had initiated the divorce. The anger of women toward men became the theme of the book." Do you think that women who were angry with men would be likely to answer questions about love relationships in the same way as the general population of women?

The Hite sample exemplifies one of the most common problems with surveys—the sample data may not represent the population. Extensive *nonresponse* from a random sample, or the use of a *self-selected* (i.e., *all-volunteer*) *sample,* will probably produce biased results. Those who voluntarily respond to surveys tend to care about the issue and thus have stronger and different opinions than those who do not respond.

Moral of the Story: *An unrepresentative sample, even a large one, tells you almost nothing about the population.*

Definitions: Nonresponse bias can occur when many people who are selected for the sample either do not respond at all or do not respond to some of the key survey questions. This may occur even when an appropriate random sample is selected and contacted. The survey is then based on a nonrepresentative sample, usually those who feel strongly about the issues. Some surveys don't even attempt to contact a random sample, but instead ask anyone who wishes to respond to do so. Magazines, television stations, and Internet websites routinely conduct this kind of poll, and those who respond are called a **self-selected** **sample** or a **volunteer sample.** This kind of sample tells you nothing about the larger population at all; it only tells you about those who responded.

CASE STUDY 1.5
Does Prayer Lower Blood Pressure?

Newspaper headlines are notorious for making one of the most common mistakes in the interpretation of statistical studies— jumping to unwarranted conclusions. A headline in *USA Today* read, "Prayer can lower blood pressure" (Davis, 1998). The story that followed continued the possible fallacy it began by stating, "Attending religious services lowers blood pressure more than tuning into religious TV or radio, a new study says." The words "attending religious services lowers blood pressure" imply a direct cause-and-effect relationship. This is a strong statement, but it's not a statement that is justified by the research project described in the article.

The article was based on an *observational study* conducted by the U.S. National Institutes of Health, which followed 2391 people aged 65 or older for six years. The article described one of the study's principal findings: "People who attended a religious service once a week and prayed or studied the Bible once a day were 40% less likely to have high blood pressure than those who don't go to church every week and prayed and studied the Bible less" (Davis, 1998). So, the researchers did observe a relationship, but it's a mistake to think that this justifies the conclusion that prayer actually *causes* lower blood pressure.

When groups are compared in an observational study, the groups usually differ in many important ways that may contribute to the observed relationship. In this example, people who attended church and prayed regularly may have been less likely than the others to smoke or to drink alcohol. These could affect the results because smoking and alcohol use are both believed to affect blood pressure. The regular church attendees may have had a better social network, a factor that could lead to reduced stress, which in turn could reduce blood pressure. People who were generally somewhat ill may not have been as willing or able to go out to church. We're sure you can think of other possibilities for *confounding variables* that may have contributed to the observed relationship between prayer and lower blood pressure.

Moral of the Story: *Cause-and-effect conclusions cannot generally be made based on an observational study.*

Definitions: An **observational study** is one in which participants are merely observed and measured. Comparisons based on observational studies are comparisons of naturally occurring groups. A **variable** is a characteristic that differs from one individual to the next. It may be numerical, like blood pressure, or it may be categorical, like whether or not someone attends church regularly. A **confounding variable** is a variable that is not the main concern of the study, but may be partially responsible for the observed results.

CASE STUDY 1.6
Does Aspirin Reduce Heart Attack Rates?

In 1988, the Steering Committee of the Physicians' Health Study Research Group released the results of a five-year *randomized experiment* conducted using 22,071 male physicians between the ages of 40 and 84. The purpose of the experiment was to determine if taking aspirin reduces the risk of a heart attack. The physicians had been *randomly assigned* to one of the two *treatment* groups. One group took an ordinary aspirin tablet every other day, while the other group took a *placebo.* None of the physicians knew whether he was taking the actual aspirin or the placebo.

TABLE 1.1 ■ **The Effect of Aspirin on Heart Attacks**

Treatment	Heart Attacks	Doctors in Group	Attacks Per 1000 Doctors
Aspirin	104	11,037	9.42
Placebo	189	11,034	17.13

The results, shown in Table 1.1, support the conclusion that taking aspirin does indeed help reduce the risk of having a heart attack. The rate of heart attacks in the group taking aspirin was only about half the rate of heart attacks in the placebo group. In the aspirin group, there were 9.42 heart attacks per 1000 participating doctors, while in the placebo group, there were 17.13 heart attacks per 1000 participants.

Because the men in this experiment were randomly assigned to the two conditions, other important risk factors like age, amount of exercise, and dietary habits should have been similar for both groups. The only important difference between the two groups should have been whether they took aspirin or placebo. This makes it possible to conclude that taking aspirin actually *caused* the lower rate of heart attacks for that group. In a later chapter, you will learn how to determine that the difference seen in this sample is *statistically significant*. In other words, the observed sample difference probably reflects a true difference within the population.

To what population does the conclusion of this study apply? The participants were all male physicians, so the conclusion that aspirin reduces the risk of a heart attack may not hold for the general population of men. No women were included, so the conclusion may not apply to women at all. More recent evidence, however, has provided additional support for the benefit of aspirin in broader populations.

Moral of the Story: *Unlike with observational studies, cause-and-effect conclusions can generally be made on the basis of randomized experiments.*

Definitions: A **randomized experiment** is a study in which treatments are randomly assigned to participants. A **treatment** is a specific regimen or procedure assigned to participants by the experimenter. A **random assignment** is one in which each participant has a specified probability of being assigned to each treatment. A **placebo** is a pill or treatment designed to look just like the active treatment but with no active ingredients. A **statistically significant** relationship or difference is one that is large enough to be unlikely to have occurred in the sample if there was no relationship or difference in the population.

CASE STUDY 1.7

Does the Internet Increase Loneliness and Depression?

It was big news. Researchers at Carnegie Mellon University had found that "greater use of the Internet was associated with declines in participants' communication with family members in the household, declines in size of their social circle, and increases in their depression and loneliness" (Kraut et al., 1998, p. 1017). An article in the *New York Times* reporting on this study was entitled "Sad, lonely world discovered in cyberspace" (Harmon, 1998). The study included 169 individuals in 73 households in Pittsburgh, Pennsylvania who were given free computers and Internet service in 1995. The participants answered a series of questions at the beginning of the study and either one or two years later, measuring social contacts, stress, loneliness, and depression. The *New York Times* reported:

> In the first concentrated study of the social and psychological effects of Internet use at home, researchers at Carnegie Mellon University have found that people who spend even a few hours a week online have higher levels of depression and loneliness than they would if they used the computer network less frequently . . . it raises troubling questions about the nature of "virtual" communication and the disembodied relationships that are often formed in cyberspace. (Harmon, 1998)

Given these dire reports, one would think that using the Internet for a few hours a week is devastating to your mental health. But a closer look at the findings reveals that the changes were actually quite small, although statistically significant. Internet use averaged 2.43 hours per week for participants. The number of people in the participants' "local social network" decreased from an average of 23.94 people to an average of 22.90 people, hardly a noticeable loss. On a scale from 1 to 5, self-reported loneliness decreased from an average of 1.99 to 1.89 (lower scores indicate greater loneliness). And on a scale from 0 to 3, self-reported depression dropped from an average of 0.73 to an average of 0.62 (lower scores indicate higher depression).

The *New York Times* did report the magnitude of some of the changes, noting for instance, that "one hour a week on the Internet was associated, on average, with an increase of 0.03, or 1 percent on the depression scale." But the attention the research received masked the fact that the impact of Internet use on depression, loneliness, and social contact was actually quite small.

As a follow-up to this study, in July 2001 *USA Today* (Elias, 2001) reported that in continued research, the bad effects had mostly disappeared. The article, titled "Web use not always a downer: Study disputes link to depression," began with the statement "Using the Internet at home doesn't make people more depressed and lonely after all." However, the article noted that the lead researcher, Robert Kraut of Carnegie Mellon University, believes that the earlier findings were correct, but that "the Net has become a more social place since the study began in 1995." His explanation for the change in findings is that "either the Internet has changed, or people have learned to use it more constructively, or both." Whether the link ever existed will never be known, but it is not surprising, given the small magnitude of the original finding, that it subsequently disappeared.

Moral of the Story: *A "statistically significant" finding does not necessarily have practical importance. When a study reports a statistically significant finding, find out the magnitude of the relationship or difference. A secondary moral to this story is that the implied direction of cause and effect may be* wrong. In this case, it could be that people who were more lonely and depressed were more prone to use the Internet. And, as the follow-up research makes clear, remember that "truth" doesn't necessarily remain fixed across time. Any study should be viewed in the context of society at the time it was done.

1.3 | THE COMMON ELEMENTS IN THE SEVEN STORIES

The seven stories were meant to bring life to our definition of statistics. Let's consider that definition again:

> **Statistics** is a collection of procedures and principles for gathering data and analyzing information in order to help people make decisions when faced with uncertainty.

Think back over the seven stories. In every story, *data are used to make a judgment about a situation.* This common theme is what statistics is all about.

The Discovery of Knowledge

Each story illustrates part of the process of discovery of new knowledge, for which statistical methods can be very useful. The basic steps in this process are

1. *Asking the right question(s).*
2. *Collecting useful data*, which includes deciding how much is needed.
3. *Summarizing and analyzing data*, with the goal of answering the questions.
4. *Making decisions and generalizations* based on the observed data.
5. *Turning the data and subsequent decisions into new knowledge.*

We'll explore these five steps throughout the book, concluding with a chapter on "Turning Information into Wisdom." We're confident that your active participation in this exploration will benefit you in your everyday life and in your future professional career.

In a practical sense, almost all decisions in life are based on knowledge obtained by gathering and assimilating data. Sometimes the data are quantitative, as when an instructor must decide what grades to give based on a collection of homework and exam scores. Sometimes the information is more qualitative and the process of assimilating it is informal, such as when you decide what you are going to wear to a party. In either case, the principles in this book will help you understand how to be a better decision maker.

1.1

turn on your mind

Think about a decision that you recently had to make. What "data" did you use to help you make the decision? Did you have as much information as you would have liked? If you could freely use them, how would you use the principles in this chapter to help you gain more useful information?*

HINT: As an example, how did you decide to live where you are living? What additional data, if any, would have been helpful?

in summary Some Important Statistical Principles

The "moral of the story" items for the case studies presented in this chapter give a good overview of many of the important ideas covered in this book. Here is a summary:

- Simple summaries of data can tell an interesting story and are easier to digest than long lists.

- When discussing the change in the rate or risk of occurrence of something, make sure you also include the base rate or baseline risk.

- A representative sample of only a few thousand, or perhaps even a few hundred, can give reasonably accurate information about a population of many millions.

- An unrepresentative sample, even a large one, tells you almost nothing about the population.

- Cause-and-effect conclusions cannot generally be made based on an observational study.

- Cause-and-effect conclusions can generally be made on the basis of randomized experiments.

- A statistically significant finding does not necessarily have practical importance. When a study reports a statistically significant finding, find out the magnitude of the relationship or difference.

KEY TERMS

Every term in this chapter is discussed more extensively in later chapters, so don't worry if you don't understand all of the terminology that's been introduced here. The following list indicates the page number(s) where the important terms in this chapter are introduced and defined.

EXERCISES   Blue = Basics   Bold, **Bold** = Answer in back

1.1   A five-number summary for the heights in inches of the women who participated in the survey in Case Study 1.1 is

Female Heights (Inches)		
Median	65	
Quartiles	63.5	67.5
Extremes	59	71

 a. What is the median height for these women?
 b. What is the range of heights, that is, the difference in heights between the shortest and the tallest women?
 c. What is the interval of heights containing the shortest 1/4 of the women?
 d. What is the interval of heights containing the middle 1/2 of the women?

1.2   In the year 2000, Vietnamese American women had the highest rate of cervical cancer in the country. Suppose that among 200,000 Vietnamese American women, 86 developed cervical cancer.
 a. Calculate the rate of cervical cancer for these women.
 b. What is the estimated risk of developing cervical cancer for Vietnamese American women in the next year?
 c. Explain the conceptual difference between the rate and the risk, in the context of this example.

1.3   Explain the difference between a population and a sample.

1.4   For a survey based on a random sample of 2000 adults, what is the approximate margin of error?

1.5   What sample size produces each of the following values as the approximate margin of error?
 a. Margin of error = .05 or 5%.
 b. Margin of error = .30 or 30%.

1.6   A proposed study design is to leave 100 questionnaires by the checkout line in a student cafeteria. The questionnaire could be picked up by any student and returned to the cashier. Explain why this volunteer sample is a poor study design.

1.7   For each of the examples given here, decide whether the study was an observational study or a randomized experiment.
 a. A group of 100 students was randomly divided, with 50 assigned to receive vitamin C and the remaining 50 to receive a placebo, to see if vitamin C helps prevent colds.
 b. All patients who received a hip transplant operation at Stanford University Hospital during 1990 to 2000 will be followed for 10 years after their operation to determine the success (or failure) of the transplant.
 c. A group of students who were enrolled in an introductory statistics course were randomly assigned to take a Web-based course or to take a traditional lecture course. The two methods were compared by giving the same final examination in both courses.
 d. A group of smokers and another group of nonsmokers who visited a particular clinic were asked to come in for a physical exam every 5 years for the rest of their lives in order to monitor and compare their health status.

1.8   Suppose that an observational study showed that students who get at least 7 hours of sleep performed better on exams than students who didn't. Which of the following are possible confounding variables, and which are not? Explain why in each case.
 a. Number of courses the student took that term.
 b. Weight of the student.
 c. Number of hours the student spent partying in a typical week.

1.9   Explain the distinction between statistical significance and practical significance. Can the result of a study be statistically significant but not practically significant?

1.10   A headline in a major newspaper read, "Breast-Fed Youth Found to Do Better in School."
 a. Do you think this statement was based on an observational study or a randomized experiment? Explain.
 b. Given your answer in part (a), which of these two alternative headlines do you think would be preferable: "Breast-feeding Leads to Better School Performance" or "Link Found Between Breast-feeding and School Performance"? Explain.

1.11   Refer to Case Study 1.6, in which the relationship between aspirin and heart attack rates was examined. Using the results of this experiment, what do you think is the base rate of heart attacks for men like the ones in this study? Explain.

1.12   A random sample of 1001 University of California faculty members taken in December 1995 was asked, "Do you favor or oppose using race, religion, sex, color, ethnicity, or national origin as a criterion for admission to the University of California?" (Roper Center, 1996). Fifty-two percent responded "favor."
 a. What is the population for this survey?
 b. What is the approximate margin of error for the survey?
 c. Based on the results of the survey, could it be concluded that a majority (over 50%) of *all* University of California faculty members favor using these criteria? Explain.

1.13   In this chapter you learned that cause and effect can be concluded from randomized experiments but generally not from observational studies. Why don't researchers simply conduct all studies as randomized experiments rather than observational studies?

1.14   Give an example of a question you would like to have answered, such as whether eating chocolate helps prevent depression. Then explain how a randomized experiment or an observational study could be done to study this question.

1.15   Refer to the data and five-number summaries given in Case Study 1.1. Give a numerical value for each of the following:
 a. The fastest speed driven by anyone in the class.
 b. The slowest speed driven by a male.
 c. The speed for which 1/4 of the women had driven at that speed or faster.
 d. The proportion of females who had driven 89 mph or faster.
 e. The number of females who had driven 89 mph or more.

1.16   Why was the study described in Case Study 1.5 conducted as an observational study instead of an experiment?

1.17 Students in a statistics class at Penn State were asked, "About how many minutes do you typically exercise in a week?" Responses from the *women* in the class were

60, 240, 0, 360, 450, 200, 100, 70, 240, 0, 60, 360, 180, 300, 0, 270

Responses from the *men* in the class were

180, 300, 60, 480, 0, 90, 300, 14, 600, 360, 120, 0, 240

a. Compare the women to the men using a dotplot. What does your plot show you about the difference between the men and the women?
b. For each gender, determine the median response.
c. Do you think there's a "significant" difference between the weekly amount that men and women exercise? Explain.

1.18 Refer to the previous exercise.
a. Create a five-number summary for the men's responses. Show how you found your answer.
b. Use your five-number summary to describe in words the exercise behavior of this group of students.

1.19 Case Study 1.6 reported that the use of aspirin reduces the risk of heart attack and that the relationship was found to be "statistically significant." Does either of the cautions in the "moral of the story" for Case Study 1.7 apply to this result? Explain.

1.20 An article in the magazine *Science* (Service, 1994) discussed a study comparing the health of 6000 vegetarians and a similar number of their friends and relatives who were not vegetarians. The vegetarians had a 28% lower death rate from heart attacks and a 39% lower death rate from cancer, even after the researchers accounted for differences in smoking, weight, and social class. In other words, the reported percentages were the remaining differences after adjusting for differences in death rates due to those factors.
a. Is this an observational study or a randomized experiment? Explain.
b. On the basis of this information, can we conclude that a vegetarian diet causes lower death rates from heart attacks and cancer? Explain.
c. Give an example of a potential confounding variable and explain what it means to say that this is a confounding variable.

1.21 Refer to the previous exercise, comparing vegetarians and nonvegetarians for two causes of death. Were base rates given for the two causes of death? If so, what were they? If not, explain what a base rate would be for this study.

1.22 An article in the *Sacramento Bee* (March 8, 1984, p. A1) reported on a study finding that "men who drank 500 ounces or more of beer a month (about 16 ounces a day) were three times more likely to develop cancer of the rectum than nondrinkers." In other words, the rate of cancer was triple in the beer-drinking group when compared with the nonbeer drinkers in this study. What important numerical information is missing from this report?

1.23 Dr. Richard Hurt and his colleagues (Hurt et. al., 1994) randomly assigned volunteers wanting to quit smoking to wear either a nicotine patch or a placebo patch to determine whether wearing a nicotine patch improves the chance of quitting. After 8 weeks of use, 46% of those wearing the nicotine patch but only 20% of those wearing the placebo patch had quit smoking.
a. Was this a randomized experiment or an observational study?
b. The difference in the percent of participants who quit (20% versus 46%) was statistically significant. What conclusion can be made on the basis of this study?
c. Why was it advisable to assign some of the participants to wear a placebo patch?

1.24 Refer to the study in the previous exercise, in which there was a statistically significant difference in the percent of smokers who quit using a nicotine patch and a placebo patch. Now read the two cautions in the "moral of the story" for Case Study 1.7. Discuss each of them in the context of this study.

1.25 Refer to the study in the previous two exercises, comparing the percent of smokers who quit using a nicotine patch and a placebo patch. Refer to the definition of statistics given on page 1, and explain how it applies to this study.

1.26 The Roper Organization conducted a poll in 1992 (Roper, 1992) in which one of the questions asked was whether or not the respondent had ever seen a ghost. Of the 1525 people in the 18 to 29-year-old age group, 212 said "yes."
a. What is the risk of someone in this age group seeing a ghost?
b. What is the approximate margin of error that accompanies the proportion in (a)?
c. What is the interval that is 95% certain to contain the actual proportion of people in this age group who have seen a ghost?

1.27 Refer to the previous exercise. The Roper Organization selected a random sample of adults in the United States for this poll. Suppose listeners to a late-night radio talk show were asked to call and report whether or not they had ever seen a ghost.
a. What is this type of sample called?
b. Do you think the proportion reporting that they had seen a ghost for the radio poll would be higher or lower than the proportion for the Roper poll? Explain.

1.28 A popular Sunday newspaper magazine often includes a yes or no survey question such as "Do you think there is too much violence on television?" or "Do you think parents should use physical discipline?" Readers are asked to mail their answers to the magazine and the results are reported in a subsequent issue.
a. What is this type of sample called?
b. Do you think the results of these polls represent the opinions of all readers of the magazine? Explain.

1.29 Refer to Case Study 1.6. Go through the five steps listed for "The Discovery of Knowledge" in Section 1.3, and show how each step was addressed in this study.

1.30 Refer to Case Study 1.5. Explain what mistakes were made in the implementation of steps 4 and 5 of "The Discovery of Knowledge" when *USA Today* reported the results of this study.

REFERENCES

Davis, Robert (1998). "Prayer can lower blood pressure," *USA Today,* August 11, 1998, p. 1D.

Elias, Marilyn (2001). "Web use not always a downer. Study disputes link to depression," *USA Today,* July 23, 2001, p. 1D.

Harmon, Amy (1998). "Sad, lonely world discovered in cyberspace," *New York Times,* August 30, 1998, p. A3.

Hite, Shere (1987). *Women and Love,* New York: Knopf.

Hurt, R., L. Dale, P. Fredrickson, C. Caldwell, G. Lee, K. Offord, G. Lauger, Z. Maruisic, L. Neese, and T. Lundberg (1994). "Nicotine patch therapy for smoking cessation combined with physician advice and nurse follow-up," *Journal of the American Medical Association,* Feb. 23, 1994, Vol. 271, No. 8, pp. 595–600.

Kraut, R., V. Lundmark, M. Patterson, S. Kiesler, T. Mukopadhyay, and W. Scherlis (1998). "Internet paradox: A social technology that reduces social involvement and psychological well-being?" *American Psychologist,* Vol. 53, No. 9, pp. 1017–1031.

Levin, Alan (1999). "Planes get closer in midair as traffic control errors rise," *USA Today,* February 24, 1999, p. 1A.

Moore, David S. (1997). *Statistics: Concepts and Controversies,* 4th ed., New York: W. H. Freeman.

Peterson, Karen S. (1997). "Interracial dates common among today's teenagers," *Sacramento Bee,* November 3, 1997, p. A6.

Roper Center (1996). "University of California Faculty Opinion Survey on Affirmative Action Policies," Roper Center for Public Opinion Research, Press Release, January 15, 1996.

The Roper Organization (1992). *Unusual personal experiences: An analysis of the data from three national surveys,* Las Vegas: Bigelow Holding Corp.

Service, R. F. (1994). "Dropping cholesterol—safely," *Science,* Nov 25, 1994, Vol. 266, No. 5189, p. 1323.

The Steering Committee of the Physicians' Health Study Research Group (1988). "Preliminary report: Findings from the aspirin component of the ongoing Physicians' Health Study," *New England Journal of Medicine,* Vol. 318, No. 4, Jan. 28,1988, pp. 262–264.

Turning Data Into Information

The average weight of the men in this boat might be lower than you think. See Example 2.7.

Bob Daemmrich/Stock Boston/PictureQuest

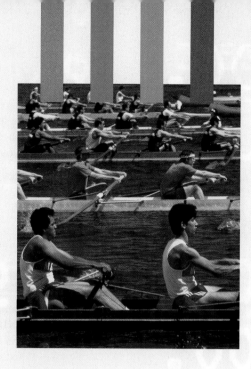

Looking at a long list of numbers is about the same as looking at a scrambled set of letters. To get information from numerical data, you have to organize it in ways that allow you to answer questions of concern to you.

I n Case Study 1.1, we analyzed the responses that 189 college students gave to the question "What's the fastest you've ever driven a car?" The "moral of the story" for that case study was that *simple summaries of data can tell an interesting story and are easier to digest than long lists.* In this chapter, you will learn how to create simple summaries and pictures from various kinds of raw data. ◆

2.1 | RAW DATA

Raw data is a term used for numbers and category labels that have been collected but have not yet been processed in any way. For example, here is a list of questions that were asked in a large statistics class and the "raw data" given by one of the students:

1. What is your sex (m = male, f = female)?	Raw data: m
2. How many hours did you sleep last night?	Raw data: 5 hours
3. Randomly pick a letter—*S* or *Q*.	Raw data: *S*
4. What is your height in inches?	Raw data: 67 inches
5. Randomly pick a number between 1 and 10.	Raw data: 3
6. What's the fastest you've ever driven a car (mph)?	Raw data: 110 mph
7. What is your right handspan in centimeters?	Raw data: 21.5 cm
8. What is your left handspan in centimeters?	Raw data: 21.5 cm

For questions 7 and 8, a centimeter ruler was provided on the survey form and handspan was defined as the distance covered on the ruler by a stretched hand, from the tip of the thumb to the tip of the small finger. You may be wondering why questions 3 and 5 were asked. Your curiosity will be satisfied as you keep reading this chapter!

2.1

turn on your mind

There were almost 200 students who answered these survey questions. Formulate four interesting questions you would like to answer using the data from these students. What kind of summary information would help you answer your questions?*

Data from Populations and Samples

tech note   In Chapter 1, Case Study 1.3, we learned that data from a representative sample can provide information about the larger population represented by the data. When measurements are taken from a subset of a population, they represent **sample data.** When all individuals in a population are measured, the measurements represent **population data.** Sometimes the reason for collecting the data is important in making this distinction. For instance, data measured in a statistics class is population data if just that class is of interest but is sample data if it is to be used to represent a larger collection of students.

The use of sample data to make inferences about populations is a major theme of this book, and it is important to determine whether raw data were obtained from a sample or from the complete population. However, the material in this chapter explains how to summarize data, and these descriptive methods are the same for both sample and population data. Therefore, in this chapter we will distinguish between sample and population data only when the notation differs for the two situations.

The generic names used for summary measures from sample and population data also differ. A summary measure computed from sample data is called a **statistic,** while a summary measure for an entire population is called a **parameter.** However, this distinction is often overlooked when numerical summaries are the only information of interest from either a population or a sample. In that case, the summary numbers for either a population or a sample are simply called **descriptive statistics.** We will begin to emphasize the distinction between populations and samples in Chapter 9.

2.2 | TYPES OF DATA

Different types of summaries are appropriate for different types of data. It makes sense, for example, to calculate the average number of hours of sleep last night for the members of a group, but it doesn't make sense to calculate the average sex (male, female) for the group. For gender data, it makes more sense to determine the proportion of the group that's male and the proportion that's female.

A **variable** is a characteristic that differs from one individual to the next. A variable may be a *categorical* characteristic, like a person's sex, or a *numerical* characteristic, like hours of sleep last night. To determine what type of summary

*HINT: An example is, "What was the average amount of sleep for these students?" Case Study 1.1 could be utilized to generate another example.

might provide meaningful information, you first have to recognize which type of variable you want to summarize.

The raw data from a **categorical variable** consists of group or category names that don't necessarily have any logical ordering. Each individual falls into one and only one category. For a categorical variable, the most fundamental summaries are how many individuals and what percent of the group fall into each category. Here are a few examples of categorical variables and their possible categories:

Variable	Possible Categories
Dominant hand	Left-handed, Right-handed
Regular church attendance	Yes, No
Opinion about marijuana legalization	Yes, No, Not sure
Eye color	Brown, Blue, Green, Hazel

Raw data from **quantitative variables** are recorded as numerical values, and the data are either measurements or counts taken on each individual. All individuals can be meaningfully ordered according to these values, and averaging and other arithmetic operations make sense for these data. Some examples of quantitative variables are height, weight, hours of sleep last night, years of education, and number of classes missed this week.

Quantitative variable, **measurement variable,** and **numerical variable** are all similar terms for quantitative data. The term **continuous variable** can also be used for quantitative data when every value within some interval is a possible result. For example, height is a continuous quantitative variable because any height within some range is possible. The limitations of measuring tapes, however, don't allow us to measure heights accurately enough to find that a person's actual height is 66.5382617 inches. Even if we could measure that accurately, we would usually prefer to round such a height to 66.5 inches.

Not all numbers fit the definition of a quantitative variable. For instance, Social Security numbers or student identification numbers may carry some information (like region of the country where the Social Security number was obtained), but it is not generally meaningful to put them into numerical order or to determine the average Social Security number.

A data type can be a function of how something is measured. For instance, household income is a numerical value with two digits after the decimal place, and if it's recorded this way, it is a quantitative variable. Researchers, however, often collect household income data using ordered categories like 1 = less than $25,000; 2 = $25,000 to $50,000; 3 = $50,000 to $75,000; and so on. When household income is recorded using categories like these, it becomes a categorical variable. In some situations, household income could be categorized very broadly, as when it is used to determine whether or not someone qualifies for a loan. In that case, income may either be "high enough" to qualify for the loan or "not high enough."

Sometimes the distinction between categorical and quantitative data is ambiguous or depends on the purpose. For example, "shoe size" could be used as a categorical variable if the Army needed to know how many of each size to manufacture for new soldiers, but used as a quantitative variable as a measure of foot size. As another example, suppose that you are asked to rate your driving skills compared to the skills of other drivers, using the codes 1 = better than average,

2 = average, and 3 = worse than average. The response is a categorical variable with arbitrary numbers attached to the category labels. It could, however, still be informative to find the average of the responses to see if it is close to 2, which it should be if everyone gives an honest appraisal. On the other hand, it would not make sense to talk about "average household income" using the numerical codes attached to the income categories described in the previous paragraph (1 = less than $25,000; 2 = $25,000 to $50,000; and so on).

When a categorical variable has ordered categories, the term **ordinal variable** is sometimes used to describe the data. Both the rating of driver skill variable and the categorized household income variable just described could be called ordinal variables. Another example of an ordinal variable is the response to the instruction "On a scale of 1 to 7 where 1 = poor and 7 = excellent, rate your teacher."

---

in summary   Types of Variables

- ◆ Raw data from *categorical variables* consist of group or category names that don't necessarily have a logical ordering. Examples: eye color, country of residence.
- ◆ Categorical variables for which the categories have a logical ordering are called *ordinal variables*. Examples: highest educational degree earned, tee shirt size (S, M, L, XL).
- ◆ Raw data from *quantitative variables* consist of numerical values taken on each individual. Examples: height, number of siblings.

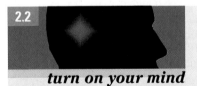

2.2

turn on your mind

Review the data collected in the statistics class, listed in Section 2.1, and identify a type for each variable. The only one that is ambiguous is question 5. That question asks for a numerical response, but as we will see later in this chapter, it is more interesting to summarize the responses as if they are categorical.*

Asking the Right Questions

As with most situations in life, the information you get when you summarize a data set depends on how careful you are about asking for what you want. Here is a brief summary of the questions that are most commonly of interest for various kinds of variables and combinations, with an example given in each case.

One Categorical Variable

Question 1a: How many and what percentage of individuals fall into each category?

---

*HINT: For each variable, consider whether the raw data are meaningful quantities or category names.

Example: What percentage of college students favor the legalization of marijuana, and what percentage of college students oppose legalization of marijuana?

Question 1b: Are individuals equally divided across categories, or do the percentages across categories follow some other interesting pattern?

Example: When individuals are asked to choose a number from 1 to 10, are all numbers equally likely to be chosen?

Two Categorical Variables

Question 2a: Is there a relationship between the two variables, so that the category into which individuals fall for one variable seems to depend on which category they are in for the other variable?

Example: In Case Study 1.6, we asked if the risk of having a heart attack was different for the physicians who took aspirin than for those who took a placebo.

Question 2b: Do some combinations of categories stand out because they provide information that is not found by examining the categories separately?

Example: The relationship between smoking and lung cancer was detected, in part, because someone noticed that the *combination* of being a non-smoker and having lung cancer is unusual.

One Quantitative Variable

Question 3a: What are the interesting summary measures, like the average or the range of values, that help us understand the collection of individuals who were measured?

Example: What is the average handspan measurement, and how much variability is there in handspan measurements?

Question 3b: Are there individual data values that provide interesting information because they are unique or stand out in some way?

Example: What is the oldest recorded age of death for a human? Are there many people who have lived nearly that long, or is the oldest recorded age a unique case?

One Categorical and One Quantitative Variable

Question 4a: Are the measurements similar across categories?

Example: Do men and women drive at the same "fastest speeds" on average?

Question 4b: When the categories have a natural ordering (an ordinal variable), does the measurement variable increase or decrease, on average, in that same order?

Example: Do high school dropouts, high school graduates, college dropouts, and college graduates have increasingly higher average incomes?

Two Quantitative Variables

Question 5a: If the measurement on one variable is high (or low), does the other one also tend to be high (or low)?

Example: Do taller people also tend to have larger handspans?

Question 5b: Are there individuals whose combination of data values provides interesting information because that combination is unusual?

Example: An individual who has a very low IQ score but can perform complicated arithmetic operations very quickly may shed light on how the brain works. Neither the IQ nor the arithmetic ability may stand out as uniquely low or high, but it is the combination that is interesting.

Sometimes it's the summary or aggregate information that is most interesting, and sometimes it is the uniqueness of certain individuals. Part of the fun and creativity of analyzing data comes from deciding which questions to ask in each context. When you encounter a dataset, return to these questions and others we will add as we encounter new situations to guide your exploration.

Explanatory and Response Variables

Many of the questions just listed were questions about the relationship between two variables (questions 2a, 2b, 4a, 4b, 5a, and 5b). In these instances, it is useful to identify one variable as the **explanatory variable** and the other variable as the **response variable.** In general, the value of the explanatory variable for an individual is thought to partially explain the value of the response variable for that individual. For example, in examining the relationship between smoking and lung cancer, whether or not an individual smokes is the explanatory variable, and whether or not he or she develops lung cancer is the response variable. If we note that people with higher education levels generally have higher incomes, education level is the explanatory variable and income is the response variable. The identification of one variable as "explanatory" and the other as "response" does not imply that there is a *causal* relationship. It simply implies that knowledge of the value of the explanatory variable may help provide knowledge of the value of the response variable for an individual.

2.3 | SUMMARIZING ONE OR TWO CATEGORICAL VARIABLES

Numerical Summaries

The first step in summarizing a categorical variable is to simply count how many individuals fall into each category. Because percents are usually more informative than counts, the second step is to calculate the percent in each category. This simple method can also be used to summarize a combination of two categorical variables.

EXAMPLE 2.1
The Importance of Order

In the sample survey described in Section 2.1, there were 92 students who responded to question 3, "Randomly pick a letter—*S* or *Q*." The results are displayed in Table 2.1.

Notice that about 66% picked the first choice they were given, *S,* and only 34% picked the second choice, *Q.* This result raises an interesting question that might be relevant when you read the results of a survey or look at the results of an election. When people are given a set of options, are they more likely to pick the first choice?

A second possible explanation of the result is that the students liked the letter *S* better than the letter *Q.* For the 92 students described above, the letter *S* was always the first letter presented, so we can't tell if there was a general preference for the first possible choice or for the letter *S.* To clarify this confusion, the teacher actually used two different forms of the question, randomly distributed among the class. The second form asked students to "Randomly pick a letter—*Q* or *S.*" The order of presenting the two letters has been reversed.

TABLE 2.1 ■ **Letter Picked When Asked to Pick Either *S* or *Q***

Letter Picked	Frequency	Percent
S	61	66%
Q	31	34%
Total	92	100%

TABLE 2.2 ■ **Order of Letters on Form and Choice of Letter**

	S Picked	*Q* Picked	Total
S Listed First	61 (66%)	31 (34%)	92
Q Listed First	45 (46%)	53 (54%)	98
Total	106 (56%)	84 (44%)	190

The results are displayed in Table 2.2. There was a clear propensity for picking the letter that was presented first. This tendency was stronger when *S* was listed first. In that case, 66% of the students picked the first choice (*S*). With *Q* listed first, 54% picked the first choice (*Q*). In all, (61 + 53)/190 = 60% of the 190 respondents picked the first letter.

One more explanation for the results should be considered. Perhaps everyone really did pick randomly, and the results were this unbalanced just by chance. In Chapter 6, we will learn how to determine that the chance such unbalanced results would occur if everyone picked randomly is less than 1 in 200. This isn't much of a chance, so a "chance luck" explanation seems unreasonable. ◆

EXAMPLE 2.2
Lighting the Way to Nearsightedness

A survey of 479 children found that those who had slept with a nightlight or in a fully lit room before the age of 2 had a higher incidence of nearsightedness (myopia) later in childhood (*Sacramento Bee,* May 13, 1999, pp. A1, A18). The raw data for each child consisted of two categorical variables, each with three categories. Table 2.3 gives the categories and the number of children falling into each combination of them. The pattern in Table 2.3 is striking. As the amount of sleeptime light increases, the incidence of myopia also increases. However, this study does not prove that sleeping with light actually *caused* myopia in more children. There are other possible explanations. For example, myopia has a genetic component, so those children whose parents have myopia are more likely to suffer from it themselves. Maybe nearsighted parents are more likely to provide light while their children are sleeping. ◆

TABLE 2.3 ■ **Nighttime Lighting in Infancy and Eyesight**

Slept with:	No Myopia	Myopia	High Myopia	Total
Darkness	155 (90%)	15 (9%)	2 (1%)	172
Nightlight	153 (66%)	72 (31%)	7 (3%)	232
Full Light	34 (45%)	36 (48%)	5 (7%)	75
Total	342 (71%)	123 (26%)	14 (3%)	479

2.3

turn on your mind

Can you think of possible explanations for the observed relationship between use of nightlights and myopia, other than direct cause and effect? What additional information might help provide an explanation?*

*HINT: Reread Example 2.2 where one possible explanation is mentioned. What data would we need to investigate the possible explanation mentioned there?

Explanatory and Response Variables for Categorical Variables

In both Example 2.1 and Example 2.2, percentages were given across rows. For instance, Table 2.3 in Example 2.2 shows that 90% ($100\% \times 155/172$) of those who slept in darkness were free of myopia, but only 45% of those who slept in full light were free of myopia. The table does not provide "column percentages" such as the fact that 34/342 or about 10% of those who were free of myopia had slept in full light.

When summarizing two categorical variables, it may be possible to identify one variable as a response variable (or **outcome variable**), and the other as an explanatory variable. In Example 2.1, the outcome variable is the letter chosen and the explanatory variable is the question form the student received. In Example 2.2, the outcome variable is degree of myopia and the explanatory variable is the amount of sleeptime lighting.

Notice that in both examples, the categories of the explanatory variable defined the rows of the table and the categories of the response variable defined the columns. Tables often are formed this way, and when they are, the row percentages are of more interest than the column percentages. No matter how the table is constructed, you should determine if one variable is a response variable and the other is an explanatory variable. Within each explanatory group, we are interested in knowing what percent fell into each response or outcome.

Minitab Tip

Minitab v. 14 update on CD

 tech note **Numerically Describing One or Two Categorical Variables**

- ◆ To determine how many and what percent fall into the categories of a categorical variable, use **Stat>Tables>Tally.** In the dialog box, specify a column containing the raw data for a categorical variable, and click on any desired options for counts and percents.

- ◆ To create a two-way table for two categorical variables, use **Stat>Tables>Cross Tabulation.** In the dialog box, specify the two columns containing the raw data for the variables, and then choose any desired percents (row and/or column and/or total).

Visual Summaries for Categorical Variables

There are two simple visual summaries used for categorical data:

- ◆ **Pie charts** are useful for summarizing a single categorical variable if there are not too many categories.

- ◆ **Bar graphs** are useful for summarizing one or two categorical variables and are particularly useful for making comparisons when there are two categorical variables.

Both of these simple graphical displays are easy to construct and interpret, as the following examples demonstrate.

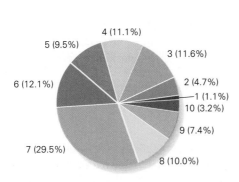

FIGURE 2.1   **Pie chart of numbers picked**

FIGURE 2.2   **Bar graph of numbers picked**

EXAMPLE 2.3
Humans Are Not Good Randomizers

Question 5 in the class survey described in Section 2.1 asked students to "Randomly pick a number between 1 and 10." The pie chart shown in Figure 2.1 illustrates that the results are not even close to being evenly distributed across the numbers. Notice that almost 30% of the students chose 7 while only just over 1% chose the number 1.

Figure 2.2 illustrates the same results with a bar graph. This bar graph shows the actual frequencies of responses on the vertical axis. The display makes it even more obvious that the number of students who chose 7 was more than double that of the next most popular choice. We also see that very few students chose either 1 or 10. ◆

EXAMPLE 2.4
Revisiting Nightlights and Nearsightedness

Figure 2.3 illustrates the data presented in Example 2.2 with a bar chart showing, for each lighting group, the percent that ultimately had each level of myopia. This bar chart differs from the one in Figure 2.2 in two respects. First, it is used to present data for two categorical variables instead of just one. Second, the vertical axis represents percents instead of counts, with the percents for myopia status computed separately within each lighting category. ◆

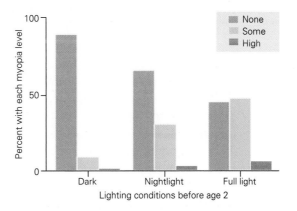

FIGURE 2.3   **Bar chart for myopia and nighttime lighting in infancy**

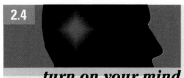

2.4

turn on your mind

Redo the bar graph in Figure 2.3 using counts instead of percents. The necessary data are given in Table 2.3. Would the comparison of frequency of myopia across the categories of lighting be as easy to make using the bar graph with counts? Generalize your conclusion to provide guidance about what should be done in similar situations.*

---

in summary   Bar Graphs for Categorical Variables

In a bar graph for *one categorical variable*, you can choose one of the following to display as the height of a bar for each category, indicated by labeling the vertical axis:

- ◆ Frequency or count
- ◆ Relative frequency = number in category/overall number
- ◆ Percent = relative frequency × 100%

In a bar graph for *two categorical variables*, if an explanatory and response variable can be identified it is most common to:

- ◆ Draw a separate group of bars for each category of the explanatory variable.
- ◆ Within each group of bars, draw one bar for each category of the response variable.
- ◆ Label the vertical axis with percents and make the heights of the bars for the response categories sum to 100% within each explanatory category group. It can sometimes be useful to make the heights of the bars equal the counts in the category groups instead of percents.

Minitab Tip

Minitab v. 14 update on CD

tech note   **Graphically Describing a Categorical Variable**

- ◆ To draw a *bar graph* of a categorical variable, use **Graph> Chart**. In the dialog box, specify a column containing the raw data for a categorical variable as the X variable. Do not make any other specifications.
- ◆ To draw a *pie chart* of a categorical variable, use **Graph>Pie Chart**.

*****HINT:** Which graph makes it easier to compare the percent with myopia for the three graphs? What could be learned from the graph of counts that isn't apparent from the graph of percents?

2.4 | FINDING INFORMATION IN QUANTITATIVE DATA

Looking at a long list of numbers is about the same as looking at a scrambled set of letters. To get information from numerical data, you have to organize it in ways that allow you to answer questions of concern to you. For instance, do you know if, compared to others of your sex, your hands are large, small, or average? Table 2.4 displays the raw data for the right handspan measurements (in centimeters) made by the students in the survey described in Section 2.1.

Find a ruler with a centimeter scale and use the ruler to measure your own handspan. Stretch out your hand and measure from the tip of your thumb to the tip of your small finger. Can you see how your measurement compares to the students of your sex in Table 2.4? That's probably hard to see because the list of data values is disorganized.

Five-Number Summaries

In the "fastest ever driven" example of Case Study 1.1, you learned how to display a **five-number summary** of a list of data values by finding the extremes (high, low), the median, and the quartiles (roughly, the medians of the lower and upper halves of the values). A five-number summary provides a quick overview of the data values and information about the center, spread, and shape of the data as well. Specific instructions for finding the median and quartiles are given in Section 2.6. In this section, the focus is on interpreting these summary values.

EXAMPLE 2.5
Right Handspans

Here are five-number summaries for male and female responses to the survey question asking students to measure the span of their right hands:

	Males (87 Students)		Females (103 Students)	
Median		22.5		20.0
Quartiles	21.5	23.5	19.0	21.0
Extremes	18.0	26.0	12.5	23.25

Remember that the five-number summary divides the dataset into approximate quarters. For example, about 25% of the handspans of females are between

TABLE 2.4 ■ **Stretched Right Handspans (cm) of 190 College Students**

Males (87 students):
21.5, 22.5, 23.5, 23, 24.5, 23, 26, 23, 21.5, 21.5, 24.5, 23.5, 22, 23.5, 22, 22, 24.5, 23, 22.5, 19.5, 22.5, 22, 23, 22.5, 20.5, 21.5, 23, 22.5, 21.5, 25, 24, 21.5, 21.5, 18, 20, 22, 24, 22, 23, 22, 22, 23, 22.5, 25.5, 24, 23.5, 21, 25.5, 23, 22.5, 24, 21.5, 22, 22.5, 23, 18.5, 21, 24, 23.5, 24.5, 23, 22, 23, 23, 24, 24.5, 20.5, 24, 22, 23, 21, 22.5, 21.5, 24.5, 22, 22, 21, 23, 22.5, 24, 22.5, 23, 23, 23, 21.5, 19, 21.5

Females (103 Students):
20, 19, 20.5, 20.5, 20.25, 20, 18, 20.5, 22, 20, 21.5, 17, 16, 22, 22, 20, 20, 20, 20, 21.7, 22, 20, 21, 21, 19, 21, 20.25, 21, 22, 18, 20, 21, 19, 22.5, 21, 20, 19, 21, 20.5, 21, 22, 20, 20, 18, 21, 22.5, 22.5, 19, 19, 19, 22.5, 20, 13, 20, 22.5, 19.5, 18.5, 19, 17.5, 18, 21, 19.5, 20, 19, 21.5, 18, 19, 19.5, 20, 22.5, 21, 18, 22, 18.5, 19, 22, 17, 12.5, 18, 20.5, 19, 20, 21, 19, 19, 21, 18.5, 19, 21.5, 21.5, 23, 23.25, 20, 18.8, 21, 21, 20, 20.5, 20, 19.5, 21, 21, 20

12.5 and 19.0 centimeters, about 25% are between 19 and 20 cm, about 25% are between 20 and 21 cm, and about 25% are between 21 and 23.25 cm. The five-number summary should give you a sense of where your own handspan measurement fits into the distribution of handspans for your gender. ◆

Interesting Summary Features of Quantitative Variables

For most quantitative variables, there are three summary characteristics of the overall distribution of the data that tend to be of the most interest. These are the **location** (center, average), the **spread** (variability), and the **shape** of the data. From a five-number summary, you can learn about location and spread. The median is an estimate of the *location* or *center* of the data. The difference between the two extremes and the difference between the two quartiles tell us about the *spread* of the data. Later in this section, you'll learn how to characterize the *shape* of a distribution.

Outliers

Another interesting feature to consider is whether any individual values are **outliers.** There is no precise definition for an outlier, but in general, an outlier is a data point that is not consistent with the bulk of the data. For a single variable, an outlier is a value that is unusually high or low. When two variables are considered, an outlier is an unusual combination of values. For instance, if you look back at Example 2.5, you will see that a female with a handspan of 24.5 cm would be an outlier because this handspan is well past the largest of the measurements made by the 103 females. On the other hand, a male with a handspan of 24.5 cm is not an outlier because this measurement is clearly within the normal range for males.

The next example illustrates that sometimes the extreme points are the most interesting features of a dataset.

EXAMPLE 2.6
Ages of Death of U.S. First Ladies

Much has been written about ages of U.S. Presidents when elected and at death, but what about their wives? Do these women tend to live short lives or long lives? Table 2.5 lists the First Ladies of the United States and their approximate ages at death (to within one year) as listed in late 2002 at the White House web-site.[1] It is not completely accurate to label all of these women "First Ladies" if the strict definition is "the wife of a President while in office." For example, Harriet Lane served socially as "First Lady" to President James Buchanan, but he was unmarried and she was his niece. A few of the women listed died before their husband's term in office. Nonetheless, we will use the data as provided by the White House and summarize the ages at death for these women. Here is a five-number summary for these ages:

	First Ladies' Ages at Death	
Median		70
Quartiles	60	81.5
Extremes	34	97

[1]http://www.whitehouse.gov/WH/glimpse/firstladies/html/firstladies.html

TABLE 2.5 ■ The First Ladies of the United States of America

Name	Born–Died	Age at Death
Martha Dandridge Custis Washington	1731–1802	71
Abigail Smith Adams	1744–1818	74
Martha Wayles Skelton Jefferson	1748–1782	34
Dolley Payne Todd Madison	1768–1849	81
Elizabeth Kortright Monroe	1768–1830	62
Louisa Catherine Johnson Adams	1775–1852	77
Rachel Donelson Jackson	1767–1828	61
Hannah Hoes Van Buren	1783–1819	36
Anna Tuthill Symmes Harrison	1775–1864	89
Letitia Christian Tyler	1790–1842	52
Julia Gardiner Tyler	1820–1889	69
Sarah Childress Polk	1803–1891	88
Margaret Mackall Smith Taylor	1788–1852	64
Abigail Powers Fillmore	1798–1853	55
Jane Means Appleton Pierce	1806–1863	57
Harriet Lane	1830–1903	73
Mary Todd Lincoln	1818–1882	64
Eliza McCardle Johnson	1810–1876	66
Julia Dent Grant	1826–1902	76
Lucy Ware Webb Hayes	1831–1889	58
Lucretia Rudolph Garfield	1832–1918	86
Ellen Lewis Herndon Arthur	1837–1880	43
Frances Folsom Cleveland	1864–1947	83
Caroline Lavinia Scott Harrison	1832–1892	60
Ida Saxton McKinley	1847–1907	60
Edith Kermit Carow Roosevelt	1861–1948	87
Helen Herron Taft	1861–1943	82
Ellen Louise Axson Wilson	1860–1914	54
Edith Bolling Galt Wilson	1872–1961	89
Florence Kling Harding	1860–1924	64
Grace Anna Goodhue Coolidge	1879–1957	78
Lou Henry Hoover	1874–1944	70
Anna Eleanor Roosevelt Roosevelt	1884–1962	78
Elizabeth Virginia Wallace Truman	1885–1982	97
Mamie Geneva Doud Eisenhower	1896–1979	83
Jacqueline Lee Bouvier Kennedy Onassis	1929–1994	65
Claudia Taylor Johnson	1912–	
Patricia Ryan Nixon	1912–1993	81
Elizabeth Bloomer Ford	1918–	
Rosalynn Smith Carter	1927–	
Nancy Davis Reagan	1923–	
Barbara Pierce Bush	1925–	
Hillary Rodham Clinton	1947–	
Laura Welch Bush	1946–	

If you are at all interested in history, this summary will make you curious about the extreme points. Who died at 34? Who lived to be 97? The extremes are more interesting features of this dataset than the summary of ages in the middle, which tend to match what we would expect for ages at death. From Table 2.5 you can see that Thomas Jefferson's wife Martha died in 1782 at age 34, almost 20 years before he entered office. He reportedly was devastated and he never re-married, although historians believe that he may have had other children in his relationship with Sally Hemmings. At the other extreme, Bess Truman died in 1982 at age 97; her husband Harry preceded her in death by ten years, but he too lived a long life—he died at age 88. ◆

How to Handle Outliers

Should we attach the label "outlier" to either of the most extreme points in the list of ages at death for the First Ladies? To study this issue, we have to examine all of the data to see if the two extremes clearly stand apart from the other values. If you look over Table 2.5, you may be able to form an opinion about whether Martha Jefferson and Bess Truman should be called outliers. Making sense of a list of numbers, however, is difficult. The most effective way to look for outliers is to graph the data, which we will learn to do later in this chapter.

Outliers need special attention because they can have a big influence on conclusions drawn from a dataset, and because they can lead to erroneous conclusions if they are not treated appropriately. Remember that outliers are data values that are not consistent with the bulk of the data, for either a single variable or a combination of variables.

Outliers can also cause complications in some statistical analysis procedures, as you will learn throughout this book. As a result, some researchers wrongly discard them rather than treat them as legitimate data. Outliers should never be discarded without justification. The first step in deciding what to do with outliers is determining why they exist, so that appropriate action can be taken.

> ### *in summary*   Possible Reasons for Outliers and Reasonable Actions
>
> - *A mistake was made while taking a measurement or entering it into the computer.* If this can be verified, the values should be discarded or corrected.
> - *The individual in question belongs to a different group than the bulk of individuals measured.* Values may be discarded if a summary is desired and reported for the majority group only.
> - *The outlier is a legitimate data value and represents natural variability for the group and variable(s) measured.* Values may not be discarded in this case—they provide important information about location and spread.

Let's consider each of these possible reasons for outliers, and what to do about them.

A Mistake Was Made While Taking a Measurement or Entering It into the Computer

Sometimes faulty equipment or unclear instructions can cause an outlier. For example, a data value of 8 would be recorded for a woman's right handspan measurement if she had a span of 20 cm but misunderstood and reported the measurement in inches instead of centimeters. A faulty thermometer or a clock with a dying battery can result in erroneous measurements of temperature and time. Misuse of equipment or inexperienced technicians can distort measurements. More blatant mistakes can occur when data values are recorded or typed incorrectly, such as when a height of 68 inches is recorded as 86 inches. Fortunately, outliers caused by these kinds of problems are often easy to identify and should be removed from the dataset or, if possible, corrected and retained.

The Individual in Question Belongs to a Different Group Than the Bulk of Individuals Measured

Sometimes a collection of measurements is taken on a group that contains a few individuals who are not like everyone else. For instance, a college class may include mostly traditional-age students (perhaps 18 to 22) but a few retirees. Measurements taken on variables related to age, such as total dollar value of assets owned, may result in outliers for the retirees. Or consider the relationship in real estate sales between the size of a house (square feet) and sales price for the house and property. If the home sales in an area are mostly suburban dwellings but a few farms are included, the sales prices for the farms will be much higher than comparable houses of the same size.

When outliers result from the inclusion of a few individuals who are different from the others for an obvious reason, the decision about whether to retain the outliers must be made based on the question of interest. For example, in measuring assets owned by college students, if interest is in the value of assets owned by traditional-age students, then the measurements taken on students who are not in that age range should be discarded before the dataset is analyzed. If the question of interest concerns all college students, then all measurements should be retained. In that case, the relationship between age and assets may be of primary interest.

EXAMPLE 2.7
Tiny Boatsmen

Here are the weights (in pounds) of 18 men who were on the crew teams at Oxford and Cambridge universities (*The Independent,* March 31, 1992, also Hand, D. J. et al., 1994, p. 337.):

> *Cambridge:* 188.5, 183.0, 194.5, 185.0, 214.0, 203.5, 186.0, 178.5, 109.0
> *Oxford:* 186.0, 184.5, 204.0, 184.5, 195.5, 202.5, 174.0, 183.0, 109.5

Read over the list. Do you notice anything unusual? The last weight given in each list is very different from the others. In fact, those two men were the coxswains for their teams, while the other men were the rowers. What is the mean weight for the crew team members? If all members are included, it is 181 pounds. If only the rowers are included, it is 190 pounds. Different questions, different answers. ◆

The Outlier Is a Legitimate Data Value and Represents Natural Variability for the Group and Variable(s) Measured

Characterizing the natural variability in a collection of measurements is one of the most important themes in statistics and data analysis. Outliers that are inherent in measurements of a particular type should never be discarded unless the goal is to study only a partial range of the possible values. As we will see in the next section, the two smallest female handspan measurements, 12.5 cm and 13.0 cm, were well below the remainder of the measurements. But they provide important information about the spectrum of possibilities. Discarding them for no reason would result in an erroneous depiction of these measurements, with a mean that was too high and measure of spread that was too low.

2.5 | PICTURES FOR QUANTITATIVE DATA

There are three similar types of pictures that are used to represent quantitative variables, all of which are valuable for assessing center, spread, shape, and outliers. **Histograms** are similar to bar graphs and can be used for any number of

FIGURE 2.4   **Histogram of women's right handspans**

```
12 | 5
13 | 0
14 |
15 |
16 | 0
17 | 005
18 | 00000005558
19 | 0000000000000005555
20 | 000000000000000000022555555
21 | 00000000000000055557
22 | 00000000555555
23 | 02
```

Example: | 12 | 5 = 12.5

FIGURE 2.5   **Stem-and-leaf plot of women's right handspans**

data values. **Stem-and-leaf plots** and **dotplots** present all individual values, so for very large datasets, they are not as useful as histograms. Another kind of picture, called a **boxplot** or **box-and-whisker plot,** is useful for summarizing quantitative data and for comparing summary measures for two or more groups. Boxplots visually display the type of information given in a five-number summary and will be discussed in the next section along with some other methods for summarizing quantitative variables.

Figures 2.4 to 2.6 illustrate a histogram, stem-and-leaf plot, and dotplot, respectively, for the females' right handspans displayed in Table 2.4. Examine the three figures. Notice that if the stem-and-leaf plot were turned on its side, all three pictures would look similar.

Interpreting Histograms, Stem-and-Leaf Plots, and Dotplots

Each of these pictures is useful for assessing the location, spread, and shape of a distribution, and each is also useful for detecting outliers. For the data presented in Figures 2.4 to 2.6, notice that the values are "*centered*" around 20 cm, which we learned in Example 2.5 is indeed the median value. There are two possible *outlier* values that are low compared to the bulk of the data. These are identifiable in the stem-and-leaf plot as 12.5 and 13.0 cm, but are evident in the other two pictures as well. Except for those values, the handspans *range* from about 16 to 23 cm, and tend to be "clumped" around 20 and taper off toward 16 and 23.

Describing Shape

The *shape* of a dataset is usually described as either **symmetric** or **skewed.** A skewed dataset is either *skewed to the left* or *skewed to the right*. The shape in

FIGURE 2.6   **Dotplot of women's right handspans**

Figures 2.4 to 2.6 is called "skewed to the left" because the values trail off to the left. Were it not for the two low values, the shape would be relatively "symmetric," meaning that it is similar on both sides of the center. A symmetric dataset can either be **bell-shaped** or not. Without the two low values, the data set in Figures 2.4 to 2.6 would be considered bell-shaped because the pictures would look somewhat like the shape of a bell. We will learn much more about bell-shaped curves in Section 2.7.

There are many computer programs that you can use to create these pictures. Figures 2.4, 2.5, and 2.6, for instance, are slight modifications of pictures created using Minitab. We'll go through the steps for creating each type of picture by hand, but keep in mind that statistical software like Minitab automates most of the process.

Creating a Histogram

A histogram is a bar chart of a quantitative variable that shows how many values are in various intervals of the data. The steps in creating a histogram are

Step 1: Decide how many *equally spaced* intervals to use for the horizontal axis. The experience of many researchers is that somewhere between 6 and 15 intervals is a good number. The data in Figure 2.4 ranged from 12.5 to 23.25, and Minitab chose to use 11 intervals, with each interval *centered* on an integer (12.5 to 13.5 is centered on 13, 13.5 to 14.5 is centered on 14, and so on). The intervals *must* all be the same width, so there will often be gaps like we see in Figure 2.4.

Step 2: Decide whether to use *frequencies* or *relative frequencies* on the vertical axis. Frequencies simply represent the actual number of individuals who fall into each interval. Relative frequencies represent the proportion. For instance, 2 of the 36 First Ladies died in their 30s. A relative frequency histogram would show a bar with a height of 2/36 or about .056 (or 5.6%) for the interval 30–40. It is useful to use relative frequencies when you want to compare histograms for groups with different numbers of individuals.

Step 3: Draw the appropriate number of equally spaced intervals on the horizontal axis, making sure you cover the entire range of the data. Determine the frequency or relative frequency of data values in each interval and draw a bar with corresponding height. You have to decide what rule to use for values that fall on the border between two intervals. For Figure 2.4, Minitab assigned individuals with handspans of exactly 13.5 to the interval 13.5 to 14.5, those with exactly 14.5 to the interval 14.5 to 15.5, and so on.

Additional Notes on Creating Histograms

 tech note ◆ Consider using intervals that make the range and width of each interval convenient. For instance, to create a histogram of ages at death for First Ladies, it would be convenient to use ten-year periods—died in her 30s, died in her 40s, and so on, up to 90s. This would create seven intervals.

◆ To show relative frequency, you can use either the proportion or the percent that are in an interval.

Creating a Dotplot

The dotplot in Figure 2.6 was created using the Minitab Statistical Package, and the online help feature in Minitab gives a good description of how it's done: "A dotplot displays a dot for each observation along a number line. If there are multiple occurrences of an observation, or if observations are too close together, then dots will be stacked vertically. If there are too many points to fit vertically in the graph, then each dot may represent more than one point" (Minitab, Release 12.1, 1998).

Creating a Stem-and-Leaf Plot

A stem-and-leaf plot is created much like a histogram, except every individual data value is shown. This type of plot is a quick way to summarize small datasets and is also a useful tool for ordering from lowest to highest. The basic structure of the plot is that the "stem" contains all but the last digit of a number, and the "leaf" is the last digit of the number, regardless of whether it falls before or after a decimal point.

To simplify the work, data values sometimes are truncated or rounded off. To truncate a value, digits are simply dropped. The number 23.58 is truncated to 23.5, but it is rounded off to 23.6. In Figure 2.5 the handspan values are truncated to one digit after the decimal point. The largest handspan is 23.25 cm, which is truncated to 23.2. This number is shown on the "23" stem along with the value of 23.0 cm. These two handspans are displayed as |23| 02. In other words, this stem has two leaves, each representing a different individual.

Here are the steps required to create a stem-and-leaf plot:

Step 1: Determine the stem values. Remember that the "stem" contains all but the last of the displayed digits of a number. As with histograms, it is reasonable to have between 6 and 15 stems (each stem defines an interval of values). The stems should define equally spaced intervals.

Step 2: For each individual, attach a "leaf" to the appropriate stem. A "leaf" is the last of the displayed digits of a number. It is standard, but not mandatory, to put the leaves in increasing order at each stem value.

Note: There will be more than one way to define equally spaced stems. For example, the ages at death of First Ladies range from 34 to 97. We could have stem values representing the decades (3, 4, . . . , 9) for a total of seven stems, *or* we could allow two stem values for each decade, for a total of 14 stem values. With two stems for each decade, the first instance of each stem value would receive leaves of 0 to 4 and the second would receive leaves of 5 to 9. So, the two deaths in the 30s, at ages 34 and 36, could be represented in two different ways:

$$|3|\ 46 \quad\quad \text{or} \quad\quad \begin{array}{c|c} 3 & 4 \\ 3 & 6 \end{array}$$

The first method, with fewer stem values, is generally preferable for small datasets, while the second, with more stem values, is preferable for larger datasets.

2.5

turn on your mind

For the First Ladies data, could you use three stems for each decade? Why or why not? Could you use five stems for each decade? Why or why not?*

EXAMPLE 2.8
Big Music Collections

About how many music CDs do you own? Responses to this question for 24 students in a senior-level statistics course in 1999 at Penn State University were

220, 20, 50, 450, 300, 30, 20, 50, 200, 35, 25, 50,
250, 100, 0, 100, 20, 13, 200, 2, 125, 150, 90, 60

Here is a stem-and-leaf plot of these data:

Reported music CDs owned for *n* = 24 Penn State students (stem unit = 100s, leaf unit = 10s)

```
0 | 001222233
0 | 55569
1 | 002
1 | 5
2 | 002
2 | 5
3 | 0
3 |
4 |
4 | 5
```

(Source: Class data collected by Robert Heckard.)

Notice that the numbers range from 0 to about 450. The stem unit is "100s" and the leaf unit within a row is "10s." The final digit of the response is truncated. For instance, the row |1| 002 represents the data values 100, 100, and 125. There is a clear outlier, with someone reporting that they own 450 CDs. Even without the outlier of 450, the data are skewed to the right (turn the picture on its side to see this), with the bulk of the values in the 0 to 90 range. The median number of CDs owned is 55. The arithmetic average or mean is about 105. The outlier of 450 is a legitimate part of the data, but (along with the skewness) causes a big difference between the mean and the median. ◆

Minitab Tip
Minitab v. 14 update on CD

tech note **Visually Describing a Quantitative Variable**

- To draw a *histogram* of a quantitative variable, use **Graph>Histogram.** In the dialog box, specify under X a column containing the raw data for a quantitative variable. Use the **Options** button if you want to display percents rather than counts or if you want to control how many intervals are created.

- To draw a *dotplot* of a quantitative variable, use **Graph>Dotplot.**

- To draw a *stemplot* of a quantitative variable, use **Graph>Stem-and-Leaf.**

*HINT: Is the number of possibilities for the second digit of age evenly divisible by 3?

2.6 | NUMERICAL SUMMARIES OF QUANTITATIVE VARIABLES

We have discussed the interesting features of a quantitative dataset and learned how to look for them with pictures. The final stage of the process is to compute numerical summaries of these features.

Notation for Raw Data

formula n = the number of individuals in a dataset

$x_1, x_2, x_3, \ldots, x_n$ represent the individual raw data values

Example: A dataset consists of handspan values in centimeters for six females; the values are 21, 19, 20, 20, 22, and 19. Then,

$n = 6$

$x_1 = 21, x_2 = 19, x_3 = 20, x_4 = 20, x_5 = 22,$ and $x_6 = 19$

Describing the Location of a Dataset

The words *location* and *center* often are used as synonyms for the middle of a dataset. There are two common ways to describe this feature. The **mean** is simply the numerical average, calculated as the sum of the data values divided by the number of values. The **median** is either the middle value (for an odd number of individuals) or the average of the middle two values (for an even number of individuals).

In a perfectly symmetric dataset, the mean and the median are equal, but in a skewed data set they differ. When the data are skewed to the left, the mean will tend to be smaller than the median. When the data are skewed to the right, the mean will tend to be larger than the median. Example 2.8 illustrates an extreme case with data skewed to the right, and the mean of about 105 CDs is much larger than the median of 55.

Determining the Mean and Median

formula   **The Mean**

The symbol $\bar{x}$ is nearly always used to represent the mean of a sample. Notation for calculating the sample mean of a list of values is

$$\bar{x} = \frac{\Sigma x_i}{n}$$

The capital Greek letter sigma, written as Σ, is the universal symbol meaning "add up whatever follows." Therefore, for a dataset with individual values $x_1, x_2, x_3, \ldots, x_n$, the notation Σx_i is the same as saying "add together all the values."

The Median

It would require more notation than is convenient to write a formula for the median, so we simply write the rule:

- If n is odd, the median M is the middle of the ordered values. Find M by counting $(n + 1)/2$ up from the bottom or down from the top of the ordered list.

- If n is even, the median M is the average of the middle two of the ordered values. Find M by averaging the values that are $(n/2)$ and $(n/2) + 1$ from the top or the bottom of the ordered list.

Note: If you're determining a median "by hand," your first step should be to put the data in order from lowest to highest!

EXAMPLE 2.8—*CONTINUED*
Median Number
of CDs Owned

In Example 2.8, we gave data for the number of music CDs owned by $n = 24$ college students in a statistics class, and also reported that the median number owned was 55 CDs. To find the median of a data set, first write the values in order from lowest to highest. For the music CDs owned data, the ordered list is

$$0 \ \ 2 \ \ 13 \ \ 20 \ \ 20 \ \ 20 \ \ 25 \ \ 30 \ \ 35 \ \ 50 \ \ 50 \ \ 50$$
$$60 \ \ 90 \ \ 100 \ \ 100 \ \ 125 \ \ 150 \ \ 200 \ \ 200 \ \ 220 \ \ 250 \ \ 300 \ \ 450$$

From the display of the ordered data, it's easy to see that the median is located between 50 and 60, the 12th- and 13th-highest values in the dataset, which are the two middle values among the 24 observations. The median is found by averaging these two middle values, which gives $M = (50 + 60)/2 = 55$. ◆

EXAMPLE 2.9
Will "Normal" Rainfall
Get Rid of Those Odors?

A company (that will remain unnamed) located near Davis, California, was having an odor problem in its wastewater facility. Blame it on the weather:

> [According to a company official] "Last year's severe odor problems were due in part to the 'extreme weather conditions' created in the Woodland area by El Niño." She said Woodland saw 170 to 180 percent of its normal rainfall. "Excessive rain means the water in the holding ponds takes longer to exit for irrigation, giving it more time to develop an odor." (Amy Goldwitz, *The Davis Enterprise,* March 4, 1998, p. A1)

This wording is typical of weather-related stories in which it is often remarked that rainfall is vastly "above normal" or "below normal." In fact, these stories occur so frequently that one wonders if there is *ever* a normal year. The rainfall (in inches) for Davis, California, for 47 years is shown in Table 2.6. A histogram is shown in Figure 2.7.

What is the "normal" annual rainfall in Davis? The mean in this case is 18.69 inches and the median is 16.72 inches. For the year under discussion in the article (1997–98), rainfall was 29.69 inches, hence the comment that it was "170 to 180

TABLE 2.6 ■ **Annual Rainfall for Davis, California (July 1–June 30), in Inches**

Year	Rainfall	Year	Rainfall	Year	Rainfall	Year	Rainfall
1951	20.66	1964	18.56	1977	27.69	1990	13.84
1952	16.72	1965	11.41	1978	17.25	1991	17.46
1953	13.51	1966	27.64	1979	25.06	1992	29.84
1954	14.1	1967	11.49	1980	12.03	1993	11.86
1955	25.37	1968	24.67	1981	31.29	1994	31.22
1956	12.05	1969	17.04	1982	37.42	1995	24.5
1957	28.74	1970	16.34	1983	16.67	1996	19.52
1958	10.98	1971	8.6	1984	15.74	1997	29.69
1959	12.55	1972	27.69	1985	27.47		
1960	12.75	1973	20.87	1986	10.81		
1961	14.99	1974	16.88	1987	16.3		
1962	27.1	1975	6.14	1988	11.38		
1963	11.2	1976	7.69	1989	15.79		

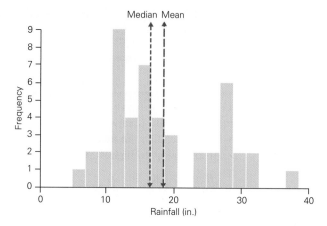

FIGURE 2.7   **Annual rainfall in Davis, California**

percent of normal." But, 29.69 inches of rain is within the range of rainfall values over this 47-year period, and more rain occurred in 4 of the other 46 years. The next time you hear about weather conditions that are not "normal," pay attention to whether they are truly outliers or just different from the *average* value. In this case, the company's excuse for the problem just doesn't hold water. ◆

The Influence of Outliers on the Mean and Median

Outliers have a larger influence on the mean than the median. Outliers at the high end will increase the mean, while outliers at the low end will decrease it. For instance, suppose there had been only five First Ladies, with ages at death of 70, 72, 74, 76, and 78. Then the median and the mean age of death would both be 74. But now suppose the youngest age had been 35 instead of 70. The median would *still* be 74 years old. But the mean would be

$$\bar{x} = \frac{35 + 72 + 74 + 76 + 78}{5} = 67 \text{ years}$$

You should remember this when you hear statistics about "average life expectancy." Those values are calculated based on averaging the anticipated age at death for all babies born in a given time period. The majority of individuals will live to be older than this "average," but those who die in infancy are the outliers that pull down the average. Datasets that are skewed to the right or that have large outliers at the high end generally have higher means than medians. Examples include annual incomes of executives and sales prices of homes for a large area.

Describing Spread: Range and Interquartile Range

Three summary measures that describe the *spread* or *variability* of a dataset are

- ◆ **Range** = high value − low value
- ◆ **Interquartile range** = upper quartile − lower quartile. The notation **IQR** is often used to represent the interquartile range.
- ◆ **Standard deviation**

The standard deviation is easiest to interpret in the context of bell-shaped data, so we postpone a description of it until Section 2.7, where bell-shaped distributions are discussed. In this section, we describe how to calculate the range and the interquartile range.

EXAMPLE 2.10

Range and Interquartile Range for Fastest Speeds Ever Driven

In Case Study 1.1 we summarized responses to the question "What's the fastest you've ever driven a car?" The five-number summary for the 87 males surveyed is

	Males (87 Students)	
Median		110
Quartiles	95	120
Extremes	55	150

This summary provides substantial information about the location, spread, and possible outliers. Remember that the median, 110 mph in this case, measures the *center* or *location* of the data. The other four numbers in the five-number summary can be used to describe how *spread out* or *variable* the responses are.

- The two *extremes* describe the spread over 100% of the data. Here, the responses are spread from 55 mph to 150 mph.
- The two *quartiles* describe the spread over approximately the middle 50% of the data. About 50% of the men gave responses between 95 and 120 mph.

Given the values of the extremes and the quartiles, it's simple to calculate the range and the interquartile range (IQR). For the fastest speed reportedly driven by males, values for these two measures of variability are

- Range = high − low = 150 − 55 = 95 mph
- IQR = upper quartile − lower quartile = 120 − 95 = 25 mph

While the range from the smallest to the largest data point is 95 mph, the middle 50% of the data fall in a relatively narrow range of only 25 mph. In other words, the responses are more densely clumped near the center of the data and are more spread out toward the extremes. ◆

Finding Quartiles

tech note Q_1 = **lower quartile** = median of lower half of the ordered values

Q_3 = **upper quartile** = median of upper half of the ordered values

A simple way to find the quartiles is to split the ordered values into the half that is below the median and the half that is above the median.

- The lower quartile (Q_1) is the median of the data values that are below the median.
- The upper quartile (Q_3) is the median of the data values that are above the median.

These values are called **quartiles** because, along with the median, they divide the ordered data into quarters.

Note: The Minitab program uses a different procedure, so its quartile estimates could occasionally differ slightly from the estimates determined by the procedure described here.

EXAMPLE 2.10—CONTINUED
***Fastest Driving
Speeds for Men***

Here are the 87 males' responses to the question about how fast they have driven a car, as given in Case Study 1.1, except now the data are in numerical order. To make it easier to count, the data are arranged in rows of ten numbers:

55 60 80 80 80 80 85 85 85 85
90 90 90 90 90 92 94 95 95 95
95 **95** 95 100 100 100 100 100 100 100
100 100 101 102 105 105 105 105 105 105
105 105 109 **110** 110 110 110 110 110 110
110 110 110 110 110 112 115 115 115 115
115 115 120 120 120 ***120*** 120 120 120 120
120 120 124 125 125 125 125 125 125 130
130 140 140 140 140 145 150

The median is the middle value in an ordered list, so for 87 values, the median is the $(87 + 1)/2 = 88/2 = 44$th value in the list. The 44th value is 110, and this value is shown in bold in the data list. Aside from the middle value of 110, there were 43 values at or below 110, and another 43 values at or above 110. Notice that there are many responses of 110, which is why we are careful to say that 43 of the values are *at or above* the median.

There are 43 values on either side of the median. To find the *quartiles,* simply find the median of each of those sets of 43 values. The lower quartile is the $(43 + 1)/2 = 22$nd value from the bottom of the data, and the upper quartile is the 22nd value from the top. These values are in bold and italics in the data list; $Q_1 = 95$ and $Q_3 = 120$. Notice that the 87 values have been partitioned as follows:

21 values • Q_1 • 21 values • M • 21 values • Q_3 • 21 values
|←——————Interquartile Range——————→|
|←——————————————Range——————————————→|

The median and quartiles divide the data into equal numbers of values but do not necessarily divide the data into equally wide intervals. For example, the lowest 1/4 of the males had responses ranging over the 40-mph interval from 55 mph to 95 mph, while the next 1/4 had responses ranging over only a 15-mph interval, from 95 to 110. Similarly, the third quarter had responses in only a 10-mph interval (110 to 120), while the top 1/4 had responses in a 30-mph interval (120 to 150). It is common to see the majority of values clumped in the middle and the remainder tapering off into a wider range. ◆

2.6

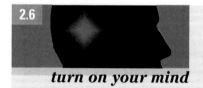

turn on your mind

A **resistant statistic** is a numerical summary of the data that is "resistant" to the influence of outliers. In other words, an outlier is not likely to have a major influence on its numerical value. Two of the summary measures from the list *mean, median, range,* and *interquartile range* are *resistant,* while the other two are not. Explain which two are resistant and which two are not.*

Percentiles

The quartiles and the median are special cases of **percentiles** for a dataset. In general, the *k*th *percentile* is a number that has *k*% of the data values at or below it and $(100 - k)$% of the data values at or above it. The lower quartile, median, and upper quartile are also the 25th percentile, 50th percentile, and 75th percentile, respectively. If you are told that you scored at the 90th percentile on a standardized test (like the SAT), it indicates that 90% of the scores were at or below your score, while 10% were at or above your score.

Using Excel to Find Summary Measures

tech note   Suppose the dataset has been stored in a range of cells, which we represent by the word "list" in what follows. For instance, if the dataset is in column A, rows 1 to 30, then "list" is A1:A30. You can also "list" the actual numerical values themselves, rather than the range of cells containing them. All of these commands are part of the "statistical functions" provided by Excel. You can insert them directly into a cell by preceding the command with the symbol @. Some values have multiple options. For instance, there are many commands that give the minimum value.

Average(list) = mean

Quartile(list, 0) = *Min(list)* = minimum value

Quartile(list, 1) = lower quartile

Quartile(list, 2) = *Median(list)* = median

Quartile(list, 3) = upper quartile

Quartile(list, 4) = *Max(list)* = maximum value

Small(list,k) gives the kth-smallest value, for example, *small(list,1)* = minimum

Large(list,k) gives the kth-largest value, for example, *large(list,n)* = minimum

Percentile(list,p) gives the kth percentile where $p = k/100$. In other words, you must express the desired percentile as a proportion rather than a percent. For instance, to find the 90th percentile, use *percentile(list,.9)*.

Count(list) = n, the number of values in the dataset.

Note: As with most computer programs, Excel uses a more precise algorithm to find the upper and lower quartiles than the one we recommend using if you are finding them "by hand," so your values may differ slightly.

Minitab Tip

tech note   **Numerical Summaries of a Quantitative Variable**

◆ To determine summary statistics for a quantitative variable, use **Stat>Basic Statistics>Display Descriptive Statistics.** In the dialog box, specify a column(s) containing the raw data for a quantitative variable(s). The **Graphs** button, of the dialog box provides options for several different graphs.

◆ If you wish to compare numerical summaries of a quantitative variable across categories (for example, compare the handspans of men and women), check "By Variable" and specify the variable that defines the categories.

Note: The output from **Stat>Basic Statistics>Display Descriptive Statistics** includes the mean, the median, the elements of the five-number summary, and the standard deviation, a statistic that we'll discuss in the next section.

Picturing Location and Spread with Boxplots

We have already seen how to create a simple and informative five-number summary using the extremes, the quartiles, and the median. A simple way to picture that information is to use a **boxplot.** This type of graph is particularly useful for comparing two or more groups and also is an effective tool for identifying outliers.

Figure 2.8 uses boxplots to compare the spans of the right hands of males and females. For each group, the box covers the middle 50% of the data, and the line within a box marks the median value. With the exception of possible outliers, the lines extending from a box reach to the minimum and maximum data values. Possible outliers are marked with an asterisk. In Figure 2.8, the vertical axis is used for the quantitative variable (handspan), but the graph could also be drawn so that the horizontal axis is used for the quantitative variable.

How to Draw a Boxplot and Identify Outliers

Step 1: Label either a vertical axis or a horizontal axis with numbers from the minimum to the maximum of the data.

Step 2: Draw a box with the lower end of the box at Q_1 and the upper end at Q_3.

Step 3: Draw a line through the box at the median M.

Step 4: Draw a line that extends from the Q_1 end of the box to the smallest data value that is not further than $1.5 \times$ IQR from Q_1. Also draw a line that extends from the Q_3 end of the box to the largest data value that is not further than $1.5 \times$ IQR from Q_3.

Step 5: Mark data points further than $1.5 \times$ IQR from either edge of the box with an asterisk. *Points represented with asterisks are considered to be outliers.*

In Figure 2.8, we see that several features of each group are immediately obvious. The comparison between the two groups is simplified as well. The only feature of a dataset that is not obvious from a boxplot is the shape, although there is information about whether the values tend to be clumped in the middle or tend to stretch more toward one extreme than the other.

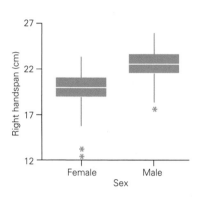

FIGURE 2.8   Boxplots for right handspans of men and women

EXAMPLE 2.7—CONTINUED
Five-number Summary and Outlier Detection for the Cambridge Boatsmen

In Example 2.7, the weights (in pounds) given for nine men on the Cambridge crew team were

188.5, 183.0, 194.5, 185.0, 214.0, 203.5, 186.0, 178.5, 109.0

The first step in finding a five-number summary is to write the data in order from smallest to largest. Here, the ordered values are

109.0   178.5   183.0   185.0   **186.0**   188.5   194.5   203.5   214.0

The median, 186.0, is underlined and in bold typeface in the ordered list. Because the number of data values is odd ($n = 9$), the median value is the middle of the ordered values. The lower quartile is the median of the four values below the median (186.0), which are 109.0, 178.5, 183.0, and 185.0. Thus, the lower quartile is $Q_1 = (178.5 + 183)/2 = 180.75$. The upper quartile is the median of the four values above the median. These are 188.5, 194.5, 203.5, and 214.0, so the value of the upper quartile is $Q_3 = (194.5 + 203.5)/2 = 199.0$.

The value 109.0, the weight of the coxswain, is an obvious outlier. We'll use this value to illustrate the criterion for identifying outliers on a boxplot. A value below the lower quartile is marked as an outlier if it is more than $1.5 \times$ IQR below Q_1. The interquartile range for these data is IQR $= Q_3 - Q_1 = 199.0 - 180.75 = 18.25$, and $1.5 \times$ IQR $= 1.5 \times 18.25 = 27.375$. Any value smaller than $Q_1 - 1.5 \times$ IQR $= 180.75 - 27.375 = 153.375$ would be marked as an outlier on a boxplot, the case for the weight 109.0. No other values would be identified as outliers in this example. For instance, a value above the upper quartile would have to be larger than $Q_3 + 1.5 \times$ IQR $= 199.0 + 27.375 = 226.375$ to be identified as an outlier. None of the crew team members was that heavy. ◆

2.7

turn on your mind

Using the boxplots in Figure 2.8, what can you say about the respective handspans for males and females? Are there any surprises, or do you see what you would expect?*

Minitab Tip

Minitab v. 14 update on CD

tech note **Drawing a Boxplot**

◆ To draw a *boxplot* of a quantitative variable, use **Graph>Boxplot.** In the dialog box, specify a column containing the raw data for a quantitative variable under Y. It is not necessary to enter anything under X.

◆ To use **Graph>Boxplot** to create a boxplot that compares a quantitative response variable across categories of an explanatory variable, specify the response variable under Y and specify the categorical explanatory variable under X. Figure 2.8 is an example of this type of plot.

2.7 | BELL-SHAPED DISTRIBUTIONS OF NUMBERS

Nature seems to follow a predictable pattern for many kinds of measurements. Most individuals are clumped around the center, and the greater the distance a value is from the center, the fewer individuals have that value. Except for the two outliers at the lower end, that pattern is evident in the females' right handspan measurements in Figures 2.4 to 2.6. If we were to draw a smooth curve connecting the tops of the bars on a histogram with this shape, the smooth curve would resemble the shape of a bell.

Numerical variables that follow this pattern are said to follow a **bell-shaped curve,** or to be "bell-shaped." A special case of this distribution of measurements is so common it is also called a **normal distribution** or **normal curve.** There is a precise mathematical formula for this smooth curve, which we will study in more depth in Chapter 8, but in this chapter we will limit ourselves to a few convenient descriptive features for data of this type. Most variables with a bell shape do not fit the mathematical formula for a normal distribution exactly, but they come close enough that the results in this section can be applied to them to provide useful information.

*HINT: Is there a difference in the location of the data for the two graphs? Is either group more spread out than the other?

EXAMPLE 2.11
The Shape of
British Women's Heights

A representative sample of 199 married British couples, taken in 1980, provided information on five variables: height of each spouse (in millimeters), age of each spouse, and husband's age at the time they were married (Hand et al., 1994). Figure 2.9 displays a histogram of the wives' heights, with a normal curve superimposed. The particular normal curve shown in Figure 2.9 was generated using Minitab statistical software, and of all the possible curves of this type, it was chosen because it is the best match for the histogram. The mean height for these women is 1602 millimeters, and the median at 1600 millimeters is very close. Although it is difficult to tell precisely in Figure 2.9, the normal curve is centered at the mean of 1602. For bell-shaped curves, there is a useful measure of spread called the *standard deviation,* which we describe next. ◆

Minitab Tip
Minitab v. 14 update on CD

 tech note   **Superimposing a Normal Curve onto a Histogram**

- ◆ To draw a normal curve onto the histogram of a quantitative variable, use **Stat**>**Basic Statistics**>**Display Statistics.** Specify the column of interest, click the ***Graphs*** button, and choose the *Histogram of data, with Normal Curve* option.

Describing Spread with Standard Deviation

Because normal curves are so common in nature, a whole set of descriptive features has been developed that apply mostly to variables with that shape. In fact, two summary features uniquely determine a normal curve, so that if you know those two summary numbers, you can draw the curve precisely. The first summary number is the mean, and the bell shape is centered on that number. The second summary number is called the **standard deviation,** and it is a measure of the spread of the values.

The Concept of Standard Deviation

You can think of the **standard deviation** as *roughly the average distance values fall from the mean.* Put another way, it measures variability by summarizing how

FIGURE 2.9   **Histogram of wife's height and normal curve**

far individual data values are from the mean. Consider, for instance, the standard deviations for the following two sets of numbers, both with a mean of 100:

Set	Numbers	Mean	Standard Deviation
1	100, 100, 100, 100, 100	100	0
2	90, 90, 100, 110, 110	100	10

In the first set of numbers, all values equal the mean value so there is no variability or spread at all. For this set, the standard deviation is 0, as will *always* be the case for such a set of numbers. In the second set of numbers, one number equals the mean, while the other four numbers are each 10 points away from the mean, so the average distance away from the mean is about 10, the standard deviation for this set of data.

Calculating the Standard Deviation

The formula for calculating the standard deviation is a bit more involved than the conceptual interpretation just discussed. This is the first instance of a summary measure that differs based on whether the data represent a sample or an entire population. The version given here is appropriate when the dataset is considered to represent a sample from a larger population. This distinction will become clear later in the book.

The Formula for the Standard Deviation

formula   The formula for the (sample) standard deviation is

$$s = \sqrt{\frac{\Sigma(x_i - \bar{x})^2}{n - 1}}$$

The value of s^2 is called the (sample) **variance.**

An equivalent formula for the (sample) standard deviation that is easier to compute on a calculator is

$$s = \sqrt{\frac{\Sigma x_i^2 - n\bar{x}^2}{n - 1}}$$

In practice, statistical software like Minitab or a spreadsheet program like Excel typically is used to find the standard deviation for a dataset. For situations where you have to calculate the standard deviation by hand, here is a step-by-step guide to the steps involved:

Step 1: Calculate $\bar{x}$, the sample mean.

Step 2: For each observation, calculate the difference between the data value and the mean.

Step 3: Square each difference calculated in step 2.

Step 4: Sum the squared differences calculated in step 3, and then divide this sum by $n - 1$. The answer for this step is called the **variance.**

Step 5: Take the square root of the variance calculated in step 4.

VIDEO

EXAMPLE 2.12

Calculating a Standard Deviation

Calculate the standard deviation of the four pulse rates 62, 68, 74, 76.

Step 1: The sample mean is

$$\bar{x} = \frac{62 + 68 + 74 + 76}{4} = \frac{280}{4} = 70$$

Steps 2 and 3: For each observation, calculate the difference between the data value and the mean. Then, square this difference. The results of these two steps are shown here:

Data Value	Step 2 Value − Mean	Step 3 (Value − Mean)²
62	62 − 70 = −8	(−8)² = 64
68	68 − 70 = −2	(−2)² = 4
74	74 − 70 = 4	4² = 16
76	76 − 70 = 6	6² = 36

Step 4: The sum of step 3 quantities is $64 + 4 + 16 + 36 = 120$. Divide this sum by $n - 1 = 4 - 1$ to get the variance:

$$s^2 = \frac{120}{4 - 1} = \frac{120}{3} = 40$$

Step 5: Take the square root of the variance computed in step 4:

$$s = \sqrt{40} = 6.3 \quad \blacklozenge$$

Population Mean and Standard Deviation

tech note For reasons that will become clear later in this book, datasets are commonly treated as if they represent a sample from a larger population. However, in situations where the dataset includes measurements for an entire population, the notations for the mean and standard deviation are different, and the formula for the standard deviation is also slightly different. A **population mean** is represented by the symbol μ ("mu"), and

$$\text{population standard deviation} = \sigma = \sqrt{\frac{\Sigma(x_i - \mu)^2}{n}}$$

Notice that the difference between this formula and the sample version is that the denominator is now n instead of $n - 1$. Also, the appropriate notation for the mean (population) is used.

Excel Tip

tech note The **Excel** commands for the standard deviation and variance are

Stdev(list) = sample standard deviation

Stdevp(list) = population standard deviation

Var(list) = sample variance

Varp(list) = population variance

Interpreting the Standard Deviation
for Bell-Shaped Curves: The Empirical Rule

Once you know the mean and standard deviation for a bell-shaped curve, you can also determine the approximate proportion of the data that will fall into any specified interval. We will learn much more about how to do this in Chapter 8, but for now, here are some useful benchmarks.

> *definition* The **Empirical Rule** states that for any bell-shaped curve, approximately
>
> ♦ 68%   of the values fall within 1 standard deviation of the mean in either direction
>
> ♦ 95%   of the values fall within 2 standard deviations of the mean in either direction
>
> ♦ 99.7% of the values fall within 3 standard deviations of the mean in either direction

Combining the Empirical Rule with knowledge that bell-shaped variables are symmetric allows the "tail" ranges to be specified as well. The first statement of the Empirical Rule implies that about 16% of the values fall more than 1 standard deviation *below* the mean, and 16% fall more than 1 standard deviation *above* the mean. Similarly, about 2.5% fall more than 2 standard deviations below the mean, and so on.

The Empirical Rule, the Standard Deviation, and the Range

The Empirical Rule implies that the range from the minimum to maximum data values equals about 4 to 6 standard deviations for data with an approximate bell shape. *You can get a rough idea of the value of the standard deviation by dividing the range of the data values by 6.* In other words, the standard deviation of bell-shaped data can be approximated as

$$s \approx \frac{\text{Range}}{6}$$

Generally, if the rough estimate of standard deviation based on dividing the range by 6 is decidedly different from the actual standard deviation, the dataset might contain outliers or could have a skewed shape. In this case, the Empirical Rule won't work well.

EXAMPLE 2.11—
CONTINUED
Women's Heights and
the Empirical Rule

The mean for the 199 British women's heights is 1602 millimeters, and the standard deviation is 62.4 millimeters. Figure 2.10 illustrates how the Empirical Rule would apply if the distribution exactly followed a normal curve.

For instance, about 68% of the 199 heights would fall into the range 1602 ± 62.4, or 1539.6 to 1664.4 mm. (The symbol "±" is read "plus or minus" and indicates that you form an interval by first subtracting and then adding the value that follows the symbol from the value that precedes it.) About 95% of the heights would fall into the interval 1602 ± (2 × 62.4), or 1477.2 to 1726.8 mm. And about 99.7% of the heights would be in the interval 1602 ± (3 × 62.4), or 1414.8

FIGURE 2.10   **The Empirical Rule applied to British women's heights**

to 1789.2 mm. In fact, these intervals work well for the actual data. Here is a summary of how well the Empirical Rule compares with the actual numbers and percents of heights falling within 1, 2, and 3 standard deviations (s.d.) of the mean:

Interval	Numerical Interval	Empirical Rule % and Number	Actual Number	Actual Percent
Mean ± 1 s.d.	1539.6 to 1664.4	68% of 199 = 135	140	140/199 or 70%
Mean ± 2 s.d.	1477.2 to 1726.8	95% of 199 = 189	189	189/199 or 95%
Mean ± 3 s.d.	1414.8 to 1789.2	99.7% of 199 = 198	198	198/199 or 99.5%

Notice that the women's heights, although not perfectly "bell-shaped," follow the Empirical Rule quite well. The minimum height was 1410 mm, while the maximum was 1760, for a range of $1760 - 1410 = 350$ mm. Therefore, a reasonable guess for the standard deviation is Range/6 = 350/6 = 58.3 mm. This is indeed close to the actual standard deviation of 62.4 mm. ◆

Standardized *z*-Scores

The standard deviation is also useful as a "yardstick" for measuring how far an individual value falls from the mean. Suppose you were told that scores on your last statistics exam were bell-shaped (they often are) and that your test score was 2 standard deviations above the mean for the class. Without even knowing your score or the mean score, you would know that only about 2.5% of the students had scores exceeding yours. From the Empirical Rule, we know that the scores for about 95% of the class are within 2 standard deviations of the mean. Of the remaining 5% of the scores, about half, or 2.5%, will be more than 2 standard deviations above the mean.

The **standardized score** or **z-score** is a useful measure of the relative value of any observation in a dataset. The formula for this score is simple:

$$z = \frac{\text{Observed value} - \text{Mean}}{\text{Standard deviation}}$$

Notice that a *z*-score is simply the distance between the observed value and the mean, measured in terms of number of standard deviations. Data values below the mean have negative *z*-scores, and values above the mean have positive *z*-scores.

FIGURE 2.11   **The Empirical Rule applied to standardized scores (z-scores)**

As an example, suppose that the mean resting pulse rate for adult men is 70 beats per minute, the standard deviation is 8 beats per minute, and we calculate the standardized score for a resting pulse rate of 80 beats per minute. The calculation is

$$z = \frac{80 - 70}{8} = 1.25$$

The value $z = 1.25$ indicates that a pulse rate of 80 is 1.25 standard deviations above the mean pulse rate for adult men.

---

definition The **Empirical Rule** for bell-shaped data can be restated as follows:

♦ About 68% of values have z-scores between -1 and $+1$
♦ About 95% of values have z-scores between -2 and $+2$
♦ About 99.7% of values have z-scores between -3 and $+3$

Figure 2.11 illustrates this version of the Empirical Rule.

Many computer programs and calculators will find the approximate proportion of a bell-shaped variable falling below any z-score you specify. For instance, the function NORMSDIST(z) in Excel does this. Remember that a common special case of a bell-shaped distribution is called the "normal distribution," and it is this special case that is used by Excel. As an example, NORMSDIST(-1) = 0.158655, or 15.8655%, corresponding to the information from the Empirical Rule that about 16% of values fall more than 1 standard deviation below the mean. In Chapter 8 you will learn a more precise interpretation for z-scores.

2.8

turn on your mind

Why do you think measurements with a bell-shaped distribution are so common in nature? For example, why do you think women's heights are distributed in this way, rather than, for instance, being equally spread out from about 5 feet tall to 6 feet tall?*

---

*HINT: What factors contribute to persons' adult height? Considering these factors, why would it be more likely that heights are close to the mean than far from the mean?

The Empirical Rule in Action

The **Empirical** applet on the CD accompanying this book can be used to explore how well the Empirical Rule works for each of eight variables, some with bell-shaped distributions and some with skewed distributions. The data are from the **UCDavis1** and **pennstate1** datasets on the CD for this book. A general description of the eight variables is given at the top of the Web page that includes the applet. For each variable, the applet will display a histogram along with information about the intervals *mean*±*s* and *mean*±*2s*. The percent of the sample contained in each interval is reported. When the Empirical Rule applies, these two percents should be about 68% and 95%, respectively.

What to Do

Open the **Empirical** applet. Figure 2.12 shows the initial applet display, a summary of the hours of sleep the previous night for *n* = 173 students in a UC Davis statistics class. A histogram of the hours of sleep data is displayed with superimposed vertical lines indicating the intervals *mean*±*s* and *mean*±*2s*. Notice that the histogram has approximately a bell shape. Below the histogram, we see that the sample mean = 6.935 hours and the standard deviation is *s* = 1.705. Notice also that the interval *mean*±*s* = (5.23, 8.64) contains 114 of the 173 data values, which is 65.9%. The interval *mean*±*2s* = (3.525, 10.35) contains 166/173 = 95.95% of the data values. These percents are consistent with the Empirical Rule—not surprising since the distribution is approximately bell-shaped.

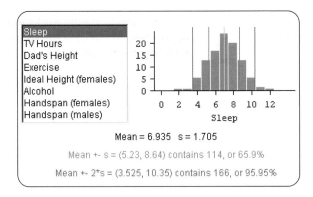

FIGURE 2.12   **The Empirical applet summary of hours of sleep reported by *n* = 173 students**

After examining the results for the hours of sleep variable, click on *TV Hours,* the second variable in the menu at the left of the applet display. Figure 2.13 displays the result, a summary of self-reported weekly hours of watching television for the same 173 students in the hours of sleep example. Notice that the distribution is skewed to the right, and there's an extreme outlier at 100 hours, so the Empirical Rule won't work well. Here, the interval *mean*±*s*, given as −1.484 to 19.26 hours, contains about 89% of the dataset, much more than the (approximate) 68% that would be in this interval if the Empirical Rule applied. Another difficulty is that the lower value of the interval is negative, an impossible value for weekly hours of watching television. This also is the case for the interval **mean±2s** = (−11.86, 29.63).

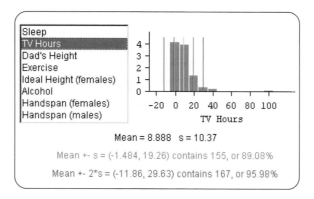

FIGURE 2.13   **The Empirical applet display of weekly hours watching television reported by** *n* = **173 students**

Click on each of the other six variables, to further explore the connection between the shape of the histogram and the applicability of the Empirical Rule. For each variable, judge the shape of the distribution and take note of the percents in the two intervals given. Which of the variables are well described by the Empirical Rule? Typically, what is the approximate shape of the distributions of these variables? Which variables are not well described by the Empirical Rule? What shape do the distributions of these variables typically have?

Lessons Learned

You'll see that the Empirical Rule works well when the distribution is more or less bell-shaped. But, when the distribution is skewed or an extreme outlier is present, you'll see that the interval **mean±s** tends to include noticeably more than 68% of the dataset, and the values in the interval **mean ± 2s** generally don't match the characteristics of the actual data.

KEY TERMS

Section 2.1
raw data, 13
sample data, 14
population data, 14
statistic, 14
parameter, 14
descriptive statistics, 14

Section 2.2
variable, 14
categorical variable, 15
quantitative variable, 15, 16

measurement variable, 15
numerical variable, 15
continuous variable, 15
ordinal variable, 16
questions for variable types, 16–18
explanatory variable, 18, 20
response variable, 18, 20

Section 2.3
outcome variable, 20
pie chart, 20–21
bar graph, 20–22

Section 2.4
five-number summary, 23
location, 24
spread, 24, 34
shape, 24, 28
outlier, 24, 26–27, 34, 38–39

Section 2.5
histogram, 27, 29
stem-and-leaf plot, 28, 30
dotplot, 28, 30
boxplot, 28, 38

EXERCISES   Blue = Basics   **Bold, Bold** = Answer in back

Sections 2.1 and 2.2

2.1 For each of the following variables, indicate whether the variable is categorical or quantitative:
  a. Importance of religion to respondent (very, somewhat, or not very important).
  b. Hours of sleep last night.
  c. Weights of adult women, measured in pounds.
  d. Favorite color for an automobile.

2.2 For each of the following, indicate whether the variable is ordinal or not. If the variable is not ordinal, indicate its variable type.
  a. Opinion about a new tax law (favor or oppose).
  b. Letter grade in a statistics course (A, B, and so on).
  c. Whether the person believes in love at first sight.
  d. Student rating of teacher effectiveness on a 7-point scale where 1 = not at all effective and 7 = extremely effective.

2.3 For each of the following characteristics of an individual, indicate whether the variable is categorical or quantitative:
  a. Length of forearm from elbow to wrist (in cm).
  b. Whether the person has ever been the victim of a crime.
  c. Number of music CDs owned.
  d. Feeling about own weight (overweight, about right, underweight).

2.4 For each of the following quantitative variables, explain whether the variable is continuous or not:
  a. Number of classes a student misses in a week.
  b. Head circumference (cm).
  c. Time it takes students to walk from their dorm to a classroom.
  d. Number of coins presently in someone's pockets and/or purse.

2.5 For each pair of variables, specify which variable is the explanatory variable and which is the response variable in the relationship between them.
  a. Amount person walks or runs per day, and performance on a test of lung function.
  b. Feeling about importance of religion, and age of respondent.
  c. Score on the final exam, and the final course grade in a psychology course.
  d. Opinion about the death penalty (favor or oppose), and gender (male or female).

2.6 In each situation, explain whether it would be more appropriate to treat the observed data as a sample from a larger population or as data from the whole population.
  a. An instructor surveys all students in her class to determine whether students would prefer a take-home exam or an in-class exam.
  b. The Gallup organization polls 1000 individuals to estimate the percent of American adults who approve of the President's job performance.
  c. A historian summarizes the ages at death for all past Presidents of the United States.
  d. A nutritionist wants to determine which of two weight-loss programs is more effective. He assigns 25 volunteers to each program, and records the weight loss after two months for each participant.

2.7 Refer to the previous exercise. In each part, identify the variable of interest, and specify whether that variable is a categorical variable or a quantitative variable.

2.8 For each of the following situations reported in the news, specify what variable(s) were measured on each individual and whether they are best described as categorical, ordinal, or quantitative.
  a. A *Los Angeles Times* survey found that 60% of the 1515 adult Californians polled supported a recent state law banning smoking in bars (*Sacramento Bee,* May 28, 1998, p. A3).
  b. According to the *College Board News* (Dec. 1998, p. 1), "Students using either one of two major coaching programs [for the SAT] were likely to experience an average gain of 5 to 19 points on verbal and 5 to 38 points on math."
  c. According to the *Associated Press* (June 19, 1998), "Smokers are twice as likely as lifetime nonsmokers to develop Alzheimer's disease and other forms of dementia . . . [according to a study that] followed 6,870 men and women ages 55 and older."

2.9 Review the questions listed in Section 2.2 labeled "Asking the Right Questions." For each of the situations in exercise 2.8, determine which of those questions is being answered.

2.10 Give an example of an ordinal variable that is likely to be treated as a categorical variable because numerical summaries like the average would not make much sense.

2.11 Give an example of an ordinal variable for which a numerical summary like the average would make sense.

2.12 To answer the following questions, researchers would need to measure two variables for each individual unit measured or observed in the study. In each case, specify the two variables, and whether each one would most likely be categorical, ordinal, or quantitative. Then, specify which of the two variables you would call the explanatory variable and which you would designate as the response variable.

a. Is the average IQ of left-handed people higher than the average IQ of right-handed people?

b. For married couples, is there a relationship between owning a pet and whether or not they get divorced?

c. For college students, is there a relationship between grade point average (GPA) and number of hours spent studying each week?

d. Individuals in the United States fall into one of a small number of tax brackets based on level of income. Is there a relationship between a person's tax bracket and the percent of income donated to charities?

2.13 Find an example of a study in a magazine, newspaper, or website. Determine what variables were measured, and, for each variable, determine its type. Which of the questions listed under "Asking the Right Questions" were addressed in this study? Describe the question(s) in the context of the study, then explain what answer was found.

Section 2.3

2.14 The Youth Risk Behavior Surveillance System (YRBSS) is a biennial survey of American high school students that is sponsored by the U.S. Centers for Disease Control. In the 2001 survey, students were asked how often they wear a seatbelt when they drive. A summary of the responses given by 2530 students in the 12th grade who said they drive is as follows:

Response	Frequency
Never	105
Rarely	248
Sometimes	286
Most times	464
Always	1427

Source: http://www.cdc.gov/nccdphp/dash/yrbs/.

a. What percent of the 12th-grade students who drive said they always wear a seatbelt when driving?

b. What percent of the 12th-grade students who drive do not always wear a seatbelt when they drive?

c. Find the percent in each of the five response categories.

d. Draw a bar graph of the percents found in part (c).

2.15 Refer to the previous exercise. Students also were asked what grades they usually get in school. For 12th-grade students who responded to this question and the question about how often they wear seatbelts when driving, a summary of frequency counts for combinations of responses to the two questions is as follows:

	Usual School Grades			
Wears Seatbelt	A and B	C	D and F	Total
Never	52	32	18	102
Rarely	128	93	22	243
Sometimes	166	104	8	278
Most times	298	128	24	450
Always	1056	300	41	1397
Total	1700	657	113	2470

a. The total number of students in the table is 2470. What percent of these 2470 students said they usually get A's and B's in school?

b. What percent of the 1700 students who said they usually get A's and B's said they always wear a seatbelt when driving?

c. What percent of the 657 students who said they usually get C's said they always wear a seatbelt when driving?

d. What percent of the 113 students who said they usually get D's and F's said they always wear a seatbelt when driving?

2.16 A sample of college students is asked how they feel about their weight. Of the 143 women in the sample who responded, 38 women said they felt overweight, 99 felt their weight was about right, and 6 felt they were underweight. Of the 78 men in the sample, 18 men felt they were overweight, 35 felt their weight was about right, and 25 felt they were underweight. (*Data source:* **pennstate3** dataset on the CD for this book.)

a. In the relationship between feelings about weight and gender, which variable is the explanatory variable and which is the response variable?

b. Summarize the observed counts by creating a table similar to Table 2.3.

c. For the 143 women, find the percent responding in each category for how they feel about their weight.

d. For the 78 men, find the percent responding in each category for how they feel about their weight.

e. Using the percents found in parts (c) and (d), summarize how the women and men differed in how they felt about their weight.

2.17 Refer to the previous exercise about gender and feelings about weight. To compare the men and women, draw a bar graph of the percents found in parts (c) and (d). Use Figure 2.3 for guidance.

2.18 For each of the following situations, which is the explanatory variable and which is the response variable?

a. The two variables are whether or not someone smoked and whether or not they developed Alzheimer's disease.

b. The two variables are whether or not somebody voted in the last election and their political party (Democrat, Republican, Independent, or Other).

c. The two variables are income level and whether or not the person has ever been subjected to a tax audit.

2.19 Refer to the data in Example 2.1 and Table 2.2.

a. Reconstruct the table using the two categorical variables "letter listed first (*S* or *Q*)" and "ordering of letter chosen (listed first or second)."

b. Draw an appropriate picture to accompany your numerical summary.

c. Explain whether you think the variables used in Example 2.1 or the variables used in this exercise were more appropriate for illustrating the point of this dataset.

2.20 Each of the following quotes is taken from an article entitled "Education seems to help in selecting husbands" (*Sacramento Bee,* Dec. 4, 1998, p. A21), which reported on new data in the *Statistical Abstract of the United States.* Draw an appropriate graph to represent each situation.

a. "The data show that 3.8 percent of women who didn't complete high school had four or more husbands. For high school graduates, the share with four or more partners drops to 3 percent. Among those who attended college 2 percent had four or more husbands, and that fell to 1 percent for those with college degrees."

b. "From 1997 on, 5.5 percent of children lived with their grandparents, a share that has been rising steadily. It was only 3.6 percent in 1980, and by 1990 it was 4.9 percent."

c. "The center said 20.1 percent of Americans took part in some regular activity—21.5 percent of men and 18.9 percent of women."

2.21 According to Krantz (1992, p. 190) the percentage of women having their first child at various ages is as follows (for all women who have children):

Under 20	20–24	25–29	30–34	35 and Over
25%	33%	25%	12.5%	4%

(The numbers do not sum to 100% due to rounding them off.) Draw a pie chart and a bar graph to represent the data. Explain which picture you think is more informative.

Sections 2.4 and 2.5

2.22 Hand et al. (1994, p. 148) provide data on the number of words in each of 600 randomly selected sentences from the book *Shorter History of England* by G. K. Chesterton. They summarized the data as follows:

Number of Words	Frequency
1–5	3
6–10	27
11–15	71
16–20	113
21–25	107
26–30	109
31–35	68
36–40	41
41–45	28
46–50	18
51–55	12
56–60	3

a. Create a histogram for the number of words in sentences in the *Shorter History of England.*

b. Provide a summary of the dataset based on your histogram.

c. Explain why you could not create a stem-and-leaf plot for this dataset.

d. Count the number of words in the first 20 sentences in Chapter 1 of this book (not including headings), and create a histogram of sentence lengths. Compare the sentence lengths to those in the *Shorter History of England.*

2.23 The figure for this exercise is a histogram summarizing the responses given by 137 college women to a question asking how many ear pierces they have. (*Data source:* **pennstate2** dataset on the CD for this book.)

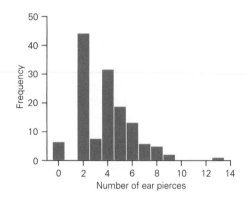

EXERCISE 2.23

a. Describe the shape of the dataset. Explain whether it is symmetric or skewed.

b. Are there any outliers? For any outlier, give a value for the number of ear pierces, and explain why you think the value is an outlier.

c. What number of ear pierces was the most frequently reported value? Roughly, how many women said they have this number of ear pierces?

d. Roughly, how many women said they have four ear pierces?

2.24 The figure for this exercise (see next page) is a histogram summarizing the responses given by 116 college students to a question asking how much they had slept the previous night. (*Data source:* **sleepstudy** dataset on the CD for this book.)

a. Describe the shape of the dataset. Explain whether it is symmetric or skewed.

b. Are there any outliers? For any outlier, give an approximate value for the amount of sleep, and explain why you think the value is an outlier.

c. What was the most frequently reported value (approximately) for the amount of sleep the previous night?

d. Roughly, how many students said they slept eight hours the previous night?

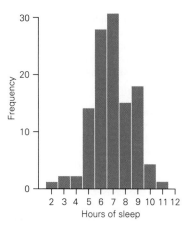

EXERCISE 2.24

2.25 The following stem-and-leaf plot is for the mean August temperatures (Fahrenheit) in 20 U.S. cities. The "stem" (row label) gives the first digit of a temperature, while the "leaf" gives the second digit. (*Data source:* **temperature** dataset on the CD for the book.)

```
6 | 44
6 | 89
7 | 01124
7 | 56667
8 | 1223
8 | 5
9 | 2
```

  a. Describe the shape of the dataset. Is it skewed or is it symmetric?
  b. What is the highest temperature in the dataset?
  c. What is the lowest temperature in the dataset?
  d. What percent of the 20 cities have a mean August temperature in the 80s?

2.26 A set of fifteen exam scores is

75, 84, 68, 95, 87, 93, 56, 87, 83, 82, 80, 62, 91, 84, 75

  a. Draw a stem-and-leaf plot of the scores.
  b. Draw a dotplot of the scores.

2.27 Case Study 1.1 presented data on the fastest speed men and women had driven a car, and dotplots were shown for each sex. Data for the men are also on the **pennstate1M** dataset on the CD for the text.[1]
  a. Create a stem-and-leaf plot for the male speeds.
  b. Create a histogram for the male speeds.
  c. Compare the pictures created in (a) and (b) and the dotplot in Case Study 1.1. Comment on which is more informative, if any of them are, and comment on any other differences you think are important.
  d. How would you describe the shape of this dataset?

2.28 In the data discussed in Section 2.1, one student reported having slept 16 hours the previous night.

[1] A small CD next to an exercise indicates that the data used in the exercise are on the CD and that it may be useful to use statistical software to answer the question.

  a. What additional information do you need in order to determine whether or not this value is an outlier?
  b. If you do determine that the data value of 16 hours of sleep is an outlier, what additional information do you need to decide whether or not to discard it before summarizing the data?

2.29 A male whose height is 78 inches might be considered to be an outlier among males in a statistics class but not among males who are professional basketball players. Give another example in which the same measurement taken on the same individual would be considered to be an outlier in one dataset but not in another dataset.

2.30 Annual rainfall for Davis, California, for 1951 to 1997 is given in Table 2.6 in Section 2.6 and in the **rainfall** dataset on the CD for the text. A histogram is shown in Figure 2.7.
  a. Create a stem-and-leaf plot for the rainfall data, rounded (not truncated) to the nearest inch.
  b. Create a dotplot for the rainfall data, rounded to the nearest inch.
  c. Describe the shape of the rainfall data.
  d. Discuss whether there are any outliers, and if so, whether to discard them or not.

2.31 One of the authors of this book (one male, one female) has a right handspan measurement of 23.5 cm. Would you consider this value to be an outlier? What additional information do you need to make your decision?

2.32 Here are the ages, arranged in order, for the CEOs of the 60 top-ranked small companies in America in 1993 (*Forbes,* "America's Best Small Companies," Nov. 8, 1993; also at http://lib.stat.cmu.edu/DASL/Datafiles/ceodat.html). These data are part of the **ceodata** dataset on the CD for the text.

32, 33, 36, 37, 38, 40, 41, 43, 43, 44, 44, 45, 45, 45, 45, 46, 46, 47, 47, 47, 48, 48, 48, 48, 49, 50, 50, 50, 50, 50, 50, 51, 51, 52, 53, 53, 53, 55, 55, 55, 56, 56, 56, 56, 57, 57, 58, 58, 59, 60, 61, 61, 61, 62, 62, 63, 69, 69, 70, 74

  a. Create a histogram for these ages.
  b. Create a stem-and-leaf plot for these ages.
  c. Create a dotplot for these ages.
  d. Describe the shape of this dataset.
  e. Are there any outliers in this dataset?
  f. In general, would outliers be more likely to occur in the salaries of heads of companies or in the ages of heads of companies? Explain.

2.33 Does a five-number summary provide sufficient information to determine whether a dataset contains an outlier? Explain.

2.34 Does a stem-and-leaf plot provide sufficient information to determine whether a dataset contains an outlier? Explain.

2.35 For the following situations, would you be most interested in knowing the average value, the spread, or the maximum value for each dataset? Explain. If you think it would be equally useful to know more than one of these summaries, explain that as well. (Answers may differ for different individuals. It is your reasoning that is important.)
  a. A dataset with the annual salaries for all employees in a large company that has offered you a job.

b. You need to decide from which of two statistics instructors you will take a class. You have two datasets, with previous final exam scores given by each of the two instructors.

c. A dataset with ages at death for 20 of your relatives who died of natural causes.

2.36 For each of the following situations, would you prefer your value to be average, a low outlier, or a high outlier? Explain.

a. Number of children you have.

b. Your annual salary.

c. Gas mileage for your car.

d. Crime rate in the city or town where you live.

Section 2.6

2.37 Find the mean and the median for each list of values:
a. 64, 68, 72, 76, 80, 86
b. 10, 6, 2, 7, 100
c. 30, 10, 40, 30

2.38 Refer to part (b) of the previous exercise. Explain why there is such a large difference between the mean and median values.

2.39 Sixty-three college men were asked what they thought was their ideal weight. A five- number summary of the responses (in pounds) is

Median		175
Quartiles	155	190
Extremes	123	225

(*Data source:* **idealwtmen** dataset on the CD for the text)

a. Find the value of the range for these data.

b. Find the value of the interquartile range (IQR).

c. About what percent of the men gave a response that falls in the interval 155 to 190 pounds?

2.40 Students in a statistics class wrote as many letters of the alphabet as they could in 15 seconds using their nondominant hand. The figure for this exercise is a boxplot that compares the number of letters written by males and females in the sample. (*Data source:* **letters** dataset on the CD for this book.)

a. What is the median number of letters written by females?

b. What is the median for males?

EXERCISE 2.40

c. Explain whether the interquartile range is larger for males or for females.

d. Find the value of the range for males.

e. Find the value of the range for females.

2.41 A set of eight systolic blood pressures is

110, 123, 132, 150, 127, 118, 102, 122

a. Find the median value for the dataset.

b. Find the values of the lower and upper quartiles.

c. Find the value of the interquartile range (IQR).

d. Identify any outliers in the dataset. Use the criterion that a value is an outlier if it is either more than $1.5 \times$ IQR above Q_3 or more than $1.5 \times$ IQR below Q_1.

e. Draw a boxplot of the dataset.

2.42 Create side-by-side boxplots for the "fastest ever driven a car" data from Case Study 1.1, one for males and one for females. Compare the two sexes using the information displayed in the boxplots. The data are in the **pennstate1** dataset on the CD for the text.

2.43 Refer to "Asking the Right Questions" in Section 2.2, and notice the two questions that are asked for one quantitative variable (questions 3a and 3b). Use those questions as a basis for describing the age of First Ladies' death data in Table 2.5. Compute whatever numerical summaries you think are appropriate, then write a narrative summary based on the computed information. Include pictures if appropriate.

2.44 Refer to the previous exercise. Repeat that exercise to describe the rainfall data in Table 2.6.

2.45 Refer to Example 2.9, Table 2.6, and Figure 2.7 for the rainfall data. Specify whether the shape is skewed to the left or to the right, and explain whether the relationship between the mean and the median (which one is higher) is what you typically expect for data with that shape.

2.46 Create a five-number summary for the rainfall data in Example 2.9, Table 2.6. Write a few sentences describing the dataset. The data are in the **rainfall** dataset on the CD.

2.47 Refer to the sentence-length dataset in exercise 2.22. Notice that you cannot compute exact summary values. Provide as much information as you can about the median, interquartile range, and range for sentence lengths in the *Shorter History of England*.

For Exercises 2.48 to 2.50 use the data for the number of CDs owned by students, listed in Example 2.8, and in the **Music-CDs** *dataset on the CD for this book.*

2.48 Create a five-number summary for the number of music CDs owned by the students listed in Example 2.8. Write a few sentences describing the dataset.

2.49 Create a boxplot for the number of music CDs owned by the students listed in Example 2.8.

2.50 a. Find the mean and median for the number of music CDs owned by the students listed in Example 2.8.

b. Compare the two measures and discuss which one is a more representative "average" for these values.

c. Is the relationship between the mean and median what you typically expect for data with the shape of this dataset? Explain.

For Exercises 2.51 to 2.53 use the data for the ages of CEOs, listed in Exercise 2.32 and in the **ceodata** *dataset on the CD for this book.*

2.51 Create a five-number summary for the ages of CEOs of small companies listed in exercise 2.32. Write a few sentences describing the dataset.

2.52 Create a boxplot for the ages of CEOs of small companies listed in exercise 2.32.

2.53 Find the mean and median for the ages of CEOs of small companies listed in exercise 2.32. Is the relationship between them what you would typically expect for data with the shape of this dataset? Explain.

Section 2.7

2.54 The typical amount of sleep per night for college students has a bell-shaped distribution with a mean of 7 hours and a standard deviation equal to 1.7 hours. Use the Empirical Rule to complete each sentence:
 a. About 68% of college students typically sleep between ____ and ____ hours per night.
 b. About 95% of college students typically sleep between ____ and ____ hours per night.
 c. About 99.7% of college students typically sleep between ____ and ____ hours per night.

2.55 Refer to the previous exercise about hours of sleep per night for college students. Draw a picture of the distribution. Indicate the locations of the three intervals found in the previous exercise. Use Figure 2.10 for guidance.

2.56 Suppose that the mean weight for men 18 to 24 years old is 170 pounds, and the standard deviation is 20 pounds. In each part, find the value of the standardized score (z-score) for the given weight:
 a. 200 pounds
 b. 140 pounds
 c. 170 pounds
 d. 230 pounds

2.57 Find the mean and standard deviation for each sample of values:
 a. 18, 19, 20, 21, 22
 b. 20, 20, 20, 20, 20
 c. 1, 5, 7, 8, 79

2.58 The scores on the final exam in a course have approximately a bell-shaped distribution. The mean score was 70, the highest score was 98, and the lowest score was 41.
 a. Find the value of the range for the exam scores.
 b. Refer to the previous part. Use the value of the range to estimate the value of the standard deviation for the exam scores.

2.59 A sample of $n = 500$ individuals is asked how many hours they typically spend using a computer in a week. The mean response is $\bar{x} = 8.3$ hours and the standard deviation is $s = 7.2$ hours. Find values for the interval $\bar{x} \pm 2s$, and explain why the result is evidence that the distribution of weekly hours spent using the computer is not bell-shaped.

2.60 The mean for the women's right handspans is about 20 cm, with a standard deviation of about 1.8 cm. Using the stem-and-leaf plot in Figure 2.5, determine how well this set of measurements fits with the Empirical Rule.

2.61 Do you think the "ages at death" in Table 2.5 are likely to fit a bell-shaped curve? Explain why or why not.

2.62 a. Would the "ages at death" in Table 2.5 be considered a population of measurements or a sample from some larger population? Explain.
 b. Find the appropriate standard deviation (sample or population) for the "ages at death" data in Table 2.5.

2.63 The interquartile range and the standard deviation are two different measures of spread. Which measure do you think is more affected by outliers? Explain.

2.64 Explain why women's heights are likely to have a bell shape but their ages at marriage do not.

2.65 Can a categorical variable have a bell-shaped distribution? Explain.

2.66 The data for 103 women's right handspans are shown in Figures 2.4 to 2.6, and a five-number summary is given in Example 2.5.
 a. Examine Figures 2.4 to 2.6 and comment on whether or not the Empirical Rule should hold.
 b. The mean and standard deviation for these measurements are 20.0 cm and 1.8 cm, respectively. Determine whether the range of the data (found from the five-number summary) is about what would be expected using the Empirical Rule.

2.67 Refer to the women's right handspan data. As you can see in Figures 2.4 to 2.6, there are two apparent outliers, at 12.5 cm and 13.0 cm.
 a. If these values are removed, do you think the mean will increase, decrease, or remain the same? What about the standard deviation? Explain.
 b. With the outliers removed, the mean and standard deviation for the remaining 101 values are 20.2 cm and 1.45 cm, respectively. The range is 7.25, from 16.0 to 23.25. Determine whether the Empirical Rule for mean $\pm$ 3 standard deviations appears to hold for these values.
 c. Refer to Figures 2.4 to 2.6. Based on those figures, do you think the Empirical Rule should hold when the outliers are removed? How about when the outliers have not been removed? Explain.
 d. Is there any justification for removing the two outliers? Explain.

2.68 Heights for traditional college-age students in the United States have means and standard deviations of approximately 70 inches and 3 inches for males and 65 inches and 2.5 inches for females. Specify your own height and find the z-score for a college-age student of your sex and height.

2.69 a. If a data value has a z-score of 0, the value equals one of the summary measures discussed in this chapter. Which summary is that?
 b. Verify that a data value having a z-score of 1.0 is equal to the mean plus 1 standard deviation.

2.70 Using a computer or calculator that provides proportions falling below a specified z-score, determine the approximate proportion for each of the following situations. In each case, assume the values are approximately bell-shaped.
 a. The proportion of SAT scores falling below 450 for a group with a mean of 500 and a standard deviation of 100.
 b. The proportion of boys with heights below 36.5 inches for a group with mean height of 34 inches and standard deviation of 1 inch.
 c. The proportion of a large class that scored below you

on a test for which the mean was 75, the standard deviation was 8, and your score was 79.

d. The proportion of a large class that scored below you on a test for which the mean was 75, the standard deviation was 4, and your score was 79.

2.71 If you learn that your score on an exam was 80 and the mean was 70, would you be more satisfied if the standard deviation was 5 or if it was 15? Explain.

2.72 Write a set of seven numbers with a mean of 50 and a standard deviation of 0. Is there more than one possible set of numbers? Explain.

2.73 Remember that a resistant statistic is a numerical summary whose value is not unduly influenced by an outlier of any magnitude. Is the standard deviation a resistant statistic? Justify your answer by giving an example of a small dataset, then adding a very large outlier and noting how the standard deviation is affected.

2.74 For a bell-shaped dataset with a large number of values, approximately what z-score would correspond to a data value equaling each of the following?
a. The median.
b. The lowest value.
c. The highest value.
d. The mean.

2.75 Refer to exercise 2.32, in which the ages for the 60 CEOs of America's best small companies of 1993 were given. These data are in the **ceodata** dataset.
a. Find the mean and standard deviation for these ages.
b. Recall that the range should be equivalent to 4 to 6 standard deviations for bell-shaped data. Determine whether that relationship holds for these ages.
c. Find the z-scores for the youngest and oldest CEO. Are they about what you would expect? Explain.

Chapter Exercises

2.76 In an experiment conducted by one of this book's authors, 26 students were asked to estimate (in millions) the population of Canada, which was about 30 million at that time. Before they made their estimates, ten of the students (group 1) were told that the population of the United States was about 290 million at that time. Nine of the students (group 2) were told that the population of Australia was roughly 20 million at that time. The other seven students (group 3) were not given any information. The estimates (millions) made by the students in these three groups were

Group 1: 2, 30, 35, 70, 100, 120, 135, 150, 190, 200
Group 2: 8, 12, 16, 29, 35, 40, 45, 46, 95
Group 3: 5, 23, 50, 55, 150, 150, 300

a. Find the median estimate in group 1.
b. Find the median estimate in group 2.
c. Find the median estimate in group 3.
d. Compare the values of the range for the three groups. Which group had the largest range? Which group had the smallest range?

2.77 Refer to the previous exercise.
a. Find values of the lower and upper quartiles for the data in group 1.

b. Find values of the lower and upper quartiles for the data in group 2.
c. Find values of the lower and upper quartiles for the data in group 3.
d. Draw a boxplot that compares the three groups. Refer to Figure 2.8 for guidance.

2.78 Specify the type (categorical, ordinal, quantitative) for each of the following variables recorded in a survey of telephone usage in student households:
a. Telephone exchange (first three numbers after area code).
b. Number of telephones in the household.
c. Dollar amount of last month's phone bill.
d. Long-distance phone company used.

2.79 In the same survey for which wives' heights are given in Example 2.11, husbands' heights were also recorded. A five-number summary of husbands' heights (mm) is

	Husbands' Heights ($n = 199$)	
Median		1725
Quartiles	1691	1774
Extremes	1559	1949

a. Construct a boxplot for the husbands' heights.
b. Use the range to approximate the standard deviation for these heights.
c. What assumption did you need to make in part (b) to make the approximation appropriate?
d. The mean and standard deviation for these heights are 1732.5 mm and 68.8 mm. Use the Empirical Rule to construct an interval that should cover 99.7% of the data, and compare your interval to the extremes. Does the interval cover both extremes?

2.80 Can a variable be *both* of the following types? If so, give an example.
a. An explanatory variable and a categorical variable.
b. A continuous variable and an ordinal variable.
c. A quantitative variable and a response variable.
d. A bell-shaped variable and a categorical variable.
e. A bell-shaped variable and a response variable.

2.81 Reach into your wallet, pocket, or wherever you can find at least ten coins, and sort all of the coins you have by type.
a. Count how many of each kind of coin you have (pennies, nickels, and so on, or the equivalent for your country). Draw a pie chart illustrating the distribution of your coins.
b. In part (a), "kind of coin" was the variable of interest for each coin. Is that a categorical, ordinal, or quantitative variable?
c. Now consider the total monetary value of all of your coins as a single data value. What type of variable is it?
d. Suppose you had similar data for all of the students in your statistics class. Write a question for which the variable "kind of coin" is the data of interest for answering the question. Then write a question for which the variable "total monetary value of the coins" is the data of interest for answering the question.

2.82 Look around your living space or current surroundings

and find a categorical variable for which there are at least three categories and for which you can collect at least 20 observations (example: color of the shirts in your closet). Collect the data.

a. Draw a pie chart for your data.

b. Draw a bar graph for your data.

c. Is one of the pictures more informative than the other? Explain. (Your answer may depend on the variable you chose.)

2.83 Look around your living space or current surroundings and find a quantitative variable for which you can collect at least 20 observations (example: monetary amounts of the last 25 checks you wrote). List the data with your response.

a. Create a five-number summary.

b. Draw a boxplot.

c. Draw your choice of a histogram, stem-and-leaf plot, or dotplot.

d. Refer to your picture in part (c) and comment on the shape and presence or absence of outliers.

e. Compute the mean and compare it to the median. Explain whether the relationship between them is what you would expect based on the information you discussed in part (d).

2.84 For each of the following two sets of data, explain which one is likely to have a larger standard deviation:

a. Set 1: Heights of the children in a kindergarten class.
   Set 2: Heights of all of the children in an elementary school.

b. Set 1: Systolic blood pressure for a single individual taken daily for 30 days.
   Set 2: Systolic blood pressure for 30 people who visit a health clinic in 1 day.

c. Set 1: SAT scores (which range from 200 to 800) for the students in an honors class.
   Set 2: Final examination scores (which range from 0 to 100) for all of the students in the English classes at a high school.

2.85 For each of the following datasets, explain whether you would expect the mean or the median of the observations to be higher:

a. In a rural farming community, for each household the number of children is measured.

b. For all households in a large city, yearly household income is measured.

c. For all students in a high school (not just those who were employed), income earned in a job outside the home in the past month is measured.

d. For the coins in someone's pocket that has 1/3 pennies, 1/3 nickels, and 1/3 quarters, the monetary value of each coin is recorded.

2.86 For each of the following research questions, explain what two variables would have to be measured for each individual in the study.

a. Is there a relationship between the amount of beer people drink and their systolic blood pressure?

b. Is there a relationship between calories of protein consumed per day and incidence of colon cancer?

c. Is there a relationship between eye color and whether or not corrective lenses are needed by age 18?

d. For women who are HIV positive when they get pregnant, is there a relationship between whether or not the HIV is transmitted to the infant and the length of time the woman had been infected prior to getting pregnant?

2.87 Refer to the previous exercise. In each case, specify which of the two variables is the explanatory variable and which is the response variable. If it is ambiguous, explain why.

Exercises 2.88 to 2.91 each describe one or two variables and the individuals for whom they were measured. Determine which question listed under "Asking the Right Questions" in Section 2.2 is of interest, and state the question in the context of the situation, as is done for the examples listed in the text with the questions.

2.88 Individuals are all of the kindergarten children in a school district.
   One variable: Adult(s) with whom the child lives (both parents, mother only, father only, one or both grandparents, other).

2.89 Individuals are all mathematics majors at a college.
   Two variables: Grade point average and hours spent studying last week.

2.90 Individuals are a representative sample of adults in a large city.
   Two variables: Ounces of coffee consumed per day and marital status (currently married or not).

2.91 Individuals are a representative sample of college students.
   Two variables: Male or female and whether the person dreams in color (yes or no).

Dataset Exercises

2.92 The data for this exercise are in the **GSS-93** dataset. The variable ***gunlaw*** is whether a respondent favors or opposes stronger gun control laws.

a. Determine the percent of respondents who favor stronger gun control laws and the percent of respondents who oppose stronger gun control laws. (*Note:* Not all survey participants were asked the question about gun laws so the sample size for ***gunlaw*** is smaller than the overall sample size.)

b. Draw a graphical summary of the ***gunlaw*** variable.

c. Create a two-way table of counts that shows the relationship between gender (variable name ***sex***) and opinion about stronger gun control laws. From looking at this table of counts, are you able to judge whether the two variables are related? Briefly explain.

d. What percent of females favors stronger gun control laws? What percent of males favors stronger gun control laws?

e. Based on the percents found in part (d), do you think that gender and opinion about gun control are related? Briefly explain.

2.93 Use the **pennstate1** dataset for this exercise. The data for the variable ***HrsSleep*** are responses by *n* = 190 students to the question "How many hours did you sleep last night?"

a. Draw a histogram of the data for the ***HrsSleep*** variable. Describe the shape of this histogram, and comment on any other interesting features of the data.

b. Determine the five-number summary for these data.

c. What is the range of the data? What is the interquartile range?

2.94 Use the **pennstate2** dataset for this exercise. The variable **CDs** is the approximate number of music CDs owned by a student.
a. Draw a stem-and-leaf plot for the **CDs** variable.
b. Draw a histogram for the **CDs** variable.
c. Draw a dotplot for the **CDs** variable.
d. Describe the shape of the data for the **CDs** variable, and comment on any other interesting features of the data.
e. Calculate the mean number and the median number of CDs. Compare these two values.
f. For these data, do you think the mean or the median is a better description of the location of the data? Briefly explain.

2.95 For this exercise, use the **GSS-93** dataset. The variable **cappun** is the respondent's opinion about the death penalty for persons convicted of murder, and the variable **polparty** is the respondent's political party preference (Democrat, Republican, Independent, Other).
a. In this dataset, what percent favors the death penalty? What percent opposes the death penalty?
b. Create a table that displays the relationship between political party and opinion about the death penalty. Calculate an appropriate set of conditional percentages for describing the relationship.
c. Are the variables **polparty** and **cappun** related? Explain.

2.96 Use the **cholest** dataset for this exercise. The dataset contains cholesterol levels for 30 "control" patients and 28 heart attack patients at a medical facility. For the heart attack patients, cholesterol levels were measured 2 days, 4 days, and 14 days after the heart attack.
a. Calculate the mean, the standard deviation, and the five-number summary for the control patients.
b. Calculate the mean, the standard deviation, and the five-number summary for the heart attack patients' cholesterol levels 2 days after their attacks.
c. Generally, which group has the higher cholesterol levels? How much difference is there in the *location* of the cholesterol levels of the two groups?
d. Which group of measurements has a larger *spread*? Compare the groups with regard to all three measures of spread introduced in Sections 2.6 and 2.7.
e. Compare the controls and the heart attack patients using a comparative dotplot (as in Case Study 1.1). Briefly explain what this plot indicates about the difference between the two groups.

2.97 Use the **pennstate1** dataset.
a. Draw a histogram of the **height** variable.
b. What is the shape of this histogram? Why do you think it is not a bell shape?
c. Draw a boxplot of the height variable.
d. Which graph, the histogram or the boxplot, is more informative about this dataset? Briefly explain.

2.98 Use the **GSS-93** dataset. The variable **degree** indicates the highest educational degree achieved by a respondent.
a. Is the **degree** variable quantitative, categorical, or ordinal? Explain.
b. Determine the number and percent falling into each degree category.

c. What percent of the sample has a degree that is beyond a high school degree?
d. The variable **tvhours** is the self-reported number of hours of watching television in a typical day. Find the mean number of television-watching hours for each of the five degree groups.
e. Is there a relationship between self-reported hours of watching television and educational degree? Explain.
f. Draw any visual summary of the variable **tvhours** (for the whole sample). What are the interesting features of your graph?

Turn On Your Computer

*For Exercises 2.99 to 2.103, use the **Empirical** applet described in Section 2.7. It is on the CD for this text.*

2.99 Examine the results given by the applet for each of the eight variables.
a. Among the eight variables, which variables are best described by the Empirical Rule, and which are not well described by the Empirical Rule?
b. Generally, what is the shape of the histogram for the variables that are well described by the Empirical Rule? What is the shape of the histogram for the variables that are not well described by the Empirical Rule?

2.100 The parts of this exercise concern the variable **Dad's Height**, which is data for father's height as reported by $n = 167$ college students.
a. What is the shape of the histogram given for the variable **Dad's Height**? Are there any outliers?
b. Refer to part (a). Based on the shape of the histogram, explain whether the Empirical Rule will apply or not.
c. What numerical values are given by the applet for the interval *Mean*$\pm s$? What percent of the data values are in this interval? Compare this percent to the percent that would be expected if the Empirical Rule applies.
d. What numerical values are given by the applet for the interval *Mean*$\pm 2s$? What percent of the data values are in this interval? Compare this percent to the percent that would be expected if the Empirical Rule applies.

2.101 The parts of this exercise concern the variable **Alcohol**, which is data for number of alcoholic beverages consumed in a typical week as reported by $n = 167$ college students.
a. What is the shape of the histogram given for **Alcohol**? Are there any outliers?
b. Refer to part (a). Based on the shape of the histogram, explain whether the Empirical Rule will apply or not.
c. What numerical values are given by the applet for the interval *Mean*$\pm s$? What percent of the data values are in this interval? Compare this percent to the percent that would be expected if the Empirical Rule applies.
d. What numerical values are given by the applet for the interval *Mean*$\pm 2s$? Explain why this interval is not a good description of possible values for the variable **Alcohol**.

2.102 The parts of this exercise concern the variable **Ideal Height**, which is the respondent's desired height as reported by $n = 149$ college women.
a. Use information given by the applet to explain whether the Empirical Rule applies for **Ideal Height**.

b. Assuming the Empirical Rule applies, give numerical values for the interval that will contain about 99.7% of the data values.

c. Compare the interval found in part (b) to the histogram of the data. Explain whether or not the interval is a reasonable description of the data.

2.103 The data for each of the following variables includes one or more outliers. In each case, identify the outlier(s).

Then explain whether the Empirical Rule would apply to the remaining data if the outlier(s) were removed from the dataset.

a. *TV Hours*.

b. *Handspan (females)*.

c. *Dad's Height*.

REFERENCES

Hand, D. J., F. Daly, A. D. Lunn, K. J. McConway, and E. Ostrowski (1994). *A Handbook of Small Data Sets,* London: Chapman and Hall.

Krantz, Les (1992). *What the Odds Are,* New York: Harper-Perennial.

CHAPTER 3

Gathering Useful Data

What can you tell about a book by its cover? See Example 3.4.

Tom Stewart /CORBIS

Probably the biggest misinterpretation made when reporting studies is to imply that a *cause-and-effect* relationship can be concluded on the basis of an observational study. Groups that naturally differ for the explanatory variable of interest are almost certain to differ in other ways, any of which may contribute to differences in the responses.

An old saying that we should retire immediately is that "the data speaks for itself." In fact, data should never be allowed to speak without a good translator. Part of the intent of this book is to develop your translation skills. The knowledge of how the data were generated is one of the key ingredients for translating data intelligently. In this chapter, and the next, we focus on how to collect useful data. ◆

3.1 | DESCRIPTION OR DECISION? USING DATA WISELY

There are two major categories of statistical techniques that can be applied to data. The first category is **descriptive statistics,** in which we use numerical and graphical summaries to characterize a dataset. We partially covered descriptive statistics in Chapter 2, and we will introduce additional descriptive techniques in Chapters 5 and 6. Descriptive statistics such as the mean, interquartile range, and so on, as well as graphical displays, can be applied to any set of data. They are generally not difficult to interpret.

The second important category of statistical techniques is **inferential statistics,** in which we use sample information to make conclusions about a broader range of individuals than just those who are observed. For example, based on the 22,071 physicians who participated in the study described in Case Study 1.6, we can *infer* that taking aspirin helps prevent heart attacks for all men similar to the participants.

Inference methods can only be applied when it is reasonable to assume that the data in hand are *representative* for the question being considered about a larger group. In Chapters 3 and 4 you will learn how to collect representative data and the extent to which it can be used to generalize to a larger group. The key concept requires you to think about *both* the source of the data *and* the question(s) of interest. A dataset may contain representative information for certain questions but not for others.

> *definition* The **Fundamental Rule for Using Data for Inference** is that available data can be used to make inferences about a much larger group *if the data can be considered to be representative with regard to the question(s) of interest.*

EXAMPLE 3.1
Do First Ladies Represent Other Women?

If we were interested in how old American women live to be, it would not make sense to use the ages of death given in Chapter 2 for the First Ladies. Even if you ignore the fact that most of the First Ladies did not live in modern times with modern health care, you can probably think of many reasons why those women do not represent the full population of American women. In fact, there is probably no larger group for which these women are representative. Past First Ladies are not even likely to be representative of future First Ladies on the question of age at death, since medical, social, and political conditions keep changing in ways that may affect their health. ◆

EXAMPLE 3.2
Do Penn State Students Represent Other College Students?

One dataset in Chapter 2 was collected at Penn State University. Do Penn State students represent all college students? This probably depends on the question. If we wanted to know the average handspan of female college students in the United States, we could convincingly argue that the data collected in statistics classes at Penn State University are indeed representative. We could even argue that the Penn State women are representative of all females in the same age group, not just college students. On the other hand, Penn State students may not be representative of all college students when questioned about how fast they have ever driven a car. Penn State is in rural Pennsylvania, where there are many country roads and little traffic. Students in larger cities, even nearby ones like New York or Philadelphia, may not have access to the same open spaces and may not have had the opportunity to drive a car as fast as students in rural areas. Many students in larger cities may not even drive cars at all. ◆

As you can see from these examples, it may require knowledge of the subject matter to determine whether a sample is representative of the larger group for the question of interest. There are, however, a small number of common strategies that researchers use to improve the likelihood that the data they collect can be used to make inferences about a larger population. We will examine some of those strategies in the remainder of this chapter and the next. Also, we will discuss the strengths of those strategies and some problems that may arise when they are used.

Populations, Samples, and Simple Random Samples

In most statistical studies, the objective is to use a small group of units to make an inference about a larger group. The larger group of units about which inferences are to be made is called the **population.** The smaller group of units actually measured is called the **sample.** Sometimes measurements *are* taken on the whole group of interest, in which case *these measurements comprise the whole population.* Occasionally you will see someone make the mistake of trying to make inferences to some hypothetical "larger group" when there isn't one.

Remember the fundamental rule for making valid inferences about the group represented by the sample for which the data were measured: *The data must be representative of the larger group with respect to the question of interest.* The principal way to guarantee that sample data will be representative of the population is to use a **simple random sample** from the population.

> ◄ *definition* With a **simple random sample,** every conceivable group of units of the required size from the population has the same chance to be the selected sample.

Although obtaining a simple random sample of the population of interest is ideal, for most research studies it is not possible for both practical and ethical reasons. For instance, suppose researchers wanted to study the effect of using marijuana for the reduction of pain in cancer patients. It would not be practical or ethical to select a random sample of all cancer patients to use in the study. Instead, researchers would use volunteers who wanted to participate and would hope that they were representative of the larger population of all cancer patients.

Simple random samples and related sampling methods *are* typically used for one type of statistical study: sample surveys or polls. Remember from Chapter 1 that in a *sample survey* the investigators gather opinions or other information from each individual included in the sample. Because this gathering of information is usually not time-consuming or invasive, it is often both practical and ethical to contact a large random sample from the population of interest. In Chapter 4 we will learn more about how to select simple random samples and how to conduct sample surveys.

Most research studies must rely on volunteers or on samples that are readily available rather than on random samples. Therefore, when you read the results of a study it is important to pay attention to how the participants were chosen. You can then decide for yourself to what extent you think the results can be extended to a larger group. You will learn much more about how to critically assess statistical studies and surveys in the remainder of this chapter and in Chapter 4.

3.2 | SPEAKING THE LANGUAGE OF RESEARCH STUDIES

Although there are a number of different strategies for collecting meaningful data, there is common terminology used in most of them. Statisticians tend to borrow words from common usage and apply a slightly different meaning, so be sure you are familiar with the special usage of a word in a statistical context.

Types of Research Studies

There are two basic types of statistical research studies:

   ◆ *observational studies*

   ◆ *experiments*

In an **observational study,** the researchers simply observe or question the participants about opinions, behaviors, or outcomes. Participants are not asked to do anything differently. For example, Case Study 1.5 described an observational study in which blood pressure and frequency of certain types of religious activity (like prayer and church attendance) were measured. The goal was to see if people with higher frequency of religious activity had lower blood pressure. Two special cases of observational studies are *sample surveys* and *case-control* *studies.* We will examine case-control studies later in this chapter and take an in-depth look at sample surveys in Chapter 4.

In an **experiment,** researchers manipulate something and measure the effect of the manipulation on some outcome of interest. **Randomized experiments** are experiments in which the participants are *randomly assigned* to participate in one condition or another. The different "conditions" are called **treatments.**

Case Study 1.6 described a randomized experiment in which physicians were randomly assigned to take an aspirin or a placebo every other day. The goal was to see if regular intake of aspirin resulted in lower risk of heart attacks than that experienced by the group taking only a placebo. The two possible treatments were taking aspirin or taking a placebo. The random assignment should have ensured that the groups were similar in all other respects, so any difference in heart attack rates could actually be attributed to the difference in the pill they took (aspirin or placebo).

In many cases, there are practical and ethical issues that prevent researchers from conducting an experiment. For instance, it would not be reasonable to randomly assign some people to attend church and pray and others not to do so. Or, in an investigation of the effect of oral contraceptive usage on blood pressure, it would certainly not be reasonable to randomly assign some women to use oral contraceptives and others to use a placebo.

Researchers often recruit volunteers to participate in randomized experiments. These volunteers agree to receive whichever treatment is assigned to them, and they often don't know which treatment it is until the experiment is concluded. For example, volunteers in a study of the effectiveness of a new vaccine may be randomly assigned to receive either the vaccine or a placebo. Because the volunteers sign an "informed consent" to participate, it is deemed ethically acceptable to ask them to accept whichever treatment is assigned to them.

Observational Study or Randomized Experiment?

In some cases, researchers can choose between the strategies for collecting data and must weigh the advantages and disadvantages of each strategy. For instance, suppose researchers want to study whether people lose more weight by exercising or by limiting their fat intake. They could conduct an *observational study,* in which they question people about their exercise and eating habits, as well as their history of weight loss, and compare people who exercise with those who eat limited fat. Or, they could conduct a *randomized experiment* in which they randomly assign people to either exercise or eat a specific diet, and then measure and compare weight loss. As we will see later in this chapter, there are advantages and disadvantages to each type of study.

3.1

turn on your mind

Many randomized experiments recruit volunteers who agree to accept whichever treatment is randomly assigned to them. Why do you think this strategy cannot always be used, thus requiring observational studies to be used instead?*

*HINT: See the discussion under "Types of Research Studies."

Who Is Measured: Units, Subjects, Participants

Researchers use a variety of terms to describe who is measured. The generic term **unit** is used to indicate a single individual or object being measured. In experiments, the most basic entity (person, plant, and so on) to which different treatments can be assigned is called an **experimental unit.** For instance, in Case Study 1.6 each physician was an experimental unit, since each one was randomly assigned to take aspirin or a placebo. When the experimental units are people, they often are called **subjects.** In both experiments and observational studies, the subjects may also be called **participants.**

Roles Played by Variables — Measured or Not

Most experiments and observational studies are conducted because researchers want to learn about the relationship between two or more variables. For instance, in the observational study in Case Study 1.5, the researchers wanted to know if religious activity and blood pressure were related. In the randomized experiment in Case Study 1.6, they wanted to know if aspirin consumption and risk of heart attack were related.

It is useful to assign generic names to the different roles played by the variables of interest. An **explanatory variable** is one that may explain or may cause differences in a **response variable** (sometimes called an **outcome variable**). For example, in Case Study 1.5, frequency of religious activity is the explanatory variable and blood pressure is the response variable. We distinguished between explanatory variables and response variables in Chapter 2 when creating summaries for categorical variables. The data for a response variable or an explanatory variable may be either categorical or quantitative.

Another term used for a response variable is **dependent variable** because the values are sometimes thought to *depend* on the values of the explanatory variable(s). In contrast, and completely separate from the common usage of the word *independent,* explanatory variables are sometimes called **independent variables.**

Most relationships are complex, and it is unlikely that any explanatory variable is the direct and sole explanation for the values of the response variable. In many studies, particularly in observational studies, *confounding* variables may be present. For example, suppose that an observational study finds that people who take at least 500 mg of vitamin C every day get fewer colds than other people do. But, perhaps people who take vitamins also eat more healthful foods and the better diet reduces the risk of a cold. Or, perhaps the extra glass of water used to take the vitamin pill helped to prevent colds.

> *definition* A **confounding variable** is a variable that both *affects the response variable* and also *is related to the explanatory variable.* The effect of a confounding variable on the response variable cannot be separated from the effect of the explanatory variable.

Confounding might cause the appearance of a nonexistent meaningful relationship between the explanatory and response variables, or it can even mask an actual relationship. The term **lurking variable** is sometimes used to describe a potential confounding variable that is not measured and is not considered in the interpretation of a study. Notice that whether or not the researchers measure confounding variables, such variables are an inherent property of the nature of the relationship between explanatory and response variables. When interpreting the results of an observational study, always consider the possibility that confounding variables may exist.

EXAMPLE 3.3
What Confounding Variables Lurk Behind Lower Blood Pressure?

In Case Study 1.5, the people who attended church regularly had lower blood pressure than people who stayed home and watched religious services on television. But it may be that those who attended church had lower blood pressure as a result of a strong social support network rather than from attending church regularly. If this is the case, the "amount of social support" is a *confounding* variable. Notice that two conditions must be met for "amount of social support" to qualify as a confounding variable:

◆ Amount of social support (confounding variable) *affects* blood pressure (the response variable).

◆ Amount of social support (confounding variable) *is related to* attending church regularly (the explanatory variable) because the same people who attend church regularly are likely to have a strong social support network.

Health status is another possible confounding variable, since those who are not well may not be able to attend church regularly. There are numerous other possibilities, like age, attitude toward life, and so on, that could affect the observed relationship between religious activity and blood pressure. ◆

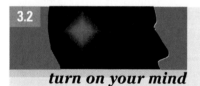

3.2

turn on your mind

Choose a possible confounding variable for this situation (Case Study 1.5), and explain how it meets the two conditions necessary to qualify as a confounding variable.*

EXAMPLE 3.4
The Fewer the Pages, the More Valuable the Book?

If you peruse the bookshelves of a typical college professor, you will find a variety of books, ranging from textbooks to esoteric technical publications to paperback novels. In order to determine whether or not there is a relationship between the price of a book and the number of pages it contains, a college professor recorded the number of pages and price for 15 books on one shelf.

The numbers are shown in Table 3.1. The books are ordered from fewest to most number of pages. Do prices increase accordingly, with the books with the fewest number of pages costing the least? No, the trend appears to be the reverse—many of the books with fewer pages seem to be more expensive than many of the books with a larger number of pages!

HINT: Think of something that would probably be different for people who attend church compared to those who don't and that might affect blood pressure.

TABLE 3.1 ■ **Pages Versus Price for the Books on a Professor's Shelf**

Book #	Pages	Price	Book #	Pages	Price
1	104	32.95	9	417	4.95
2	188	24.95	10	417	39.75
3	220	49.95	11	436	5.95
4	264	79.95	12	458	60.00
5	336	4.50	13	466	49.95
6	342	49.95	14	469	5.99
7	378	4.95	15	585	5.95
8	385	5.99			

Why does the relationship seem to be the opposite of what we might expect? The answer is that there is a confounding variable, the type of book. The books on the professor's shelf consisted of paperback novels, which tend to be long but inexpensive, and hardcover technical books, which tend to be shorter but expensive. As illustrated in Figure 3.1, when we look at the relationship between price (the response variable) and pages (the explanatory variable) for each type of book separately, we see that the price does tend to increase with number of pages, especially for the technical books. Notice that type of book (confounding variable) *affects* price (response variable) and type of book (confounding variable) *is related to* number of pages (explanatory variable), because hardcover technical books tend to have fewer pages than paperback novels. ◆

As these examples illustrate, it is unlikely that any explanatory variable is the direct and sole explanation for the values of the response variable. There are almost always confounding variables. These confounding variables might be measured and accounted for in the analysis of the data, or they could be unmeasured lurking variables. In either case, always think about the possible effect of confounding variables when you consider the results of statistical studies. Confounding variables can be especially problematic when interpreting the results of observational studies. Randomized experiments are designed to help control the influence of confounding variables.

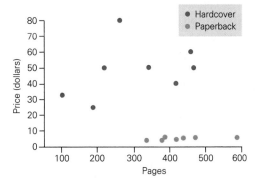

FIGURE 3.1 **Price versus pages for two types of books**

CASE STUDY 3.1
Lead Exposure and Bad Teeth

The following article appeared on the *USA Today* website (www.usatoday.com/life/health/). As you read the article, identify the explanatory and response variables and any possible lurking or confounding variables mentioned.

Lead exposure linked to bad teeth in children

Children exposed to lead are more likely to suffer tooth decay, and vitamin C might help lower blood lead levels, say two studies out Thursday. In the first of the reports in the *Journal of the American Medical Association,* researchers calculate that lead exposure could account for tooth decay in 2.7 million children. "Other people may debate that, but that's our position," says head researcher Mark Moss of the University of Rochester (N.Y.) School of Medicine and Dentistry. Prior studies showed that lead exposure can depress a child's IQ. "There are a lot worse things that lead can do to you than hurt your teeth," Moss says. He notes, however, that one of the key questions in dentistry is why low-income people experience more tooth decay than higher-income people. "This study suggests lead might be one of the reasons," he says.

The study involved 24,901 children ages 2 and older. It showed that the greater the child's exposure to lead, the more decayed or missing teeth. "The risk of getting tooth decay increased as the amount of lead went up," Moss says.

Thomas Matte of the Centers for Disease Control and Prevention says that nearly 1 million U.S. children younger than age 6 still have too much lead in their blood, particularly poor children living in older housing. An estimated 57 million residences still have lead paint on the walls, Matte says. He cautions in an editorial, however, that the problems with tooth decay might be associated with something other than lead, perhaps a sugar-rich diet or poor access to fluoridated water. "Those are good points," Moss says. "But I doubt it. We controlled for income level, the proportion of diet due to carbohydrates, calcium in the diet and the number of days since the last dental visit. We still saw this consistent pattern between dental decay and blood lead level—whether they were baby teeth or adult teeth."

In the second study, Joel Simon and his colleagues at the University of California at San Francisco studied 19,578 people who had no history of excess lead exposure. They found that the higher a person's intake of vitamin C, the lower his blood lead level. (Steve Sternberg, *USA Today*, June 23, 1999, Internet.)

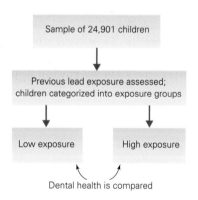

FIGURE 3.2 **An observational study in Case Study 3.1: Researchers assess prior lead exposure and compare dental health of low and high exposure groups**

Discussion of Case Study 3.1

The *USA Today* article is based on two *observational studies,* the first of which is described in the most detail.

Figure 3.2 illustrates the process the researchers used in the first study. Because it is an observational study, there is no randomization step in the process. The steps for this study include identifying the participants, measuring the explanatory and response variables on each participant, and making appropriate comparisons.

The researchers are interested in studying the relationship between exposure to lead and tooth decay and loss. The *explanatory variable* is level of exposure to lead, and the *response variable* is the extent to which the child has missing or decayed teeth. The article mentions a number of possible *confounding variables.* Remember that confounding variables are variables whose effect on the response variable cannot be separated from the effect of the explanatory variable. In this example, these are factors that may differ for children exposed to high and low amounts of lead and that may also affect tooth decay. The researchers measured some of these and tried to account for their impact, including "income level, the proportion of diet due to carbohydrates, calcium in the diet and the number of days since the last dental visit." They did not measure or account for other possible *lurking variables* that could also be confounding variables, such as amount of fluoride in the water and amount of general health care available to these children, which may differ for those exposed to high and low amounts of lead.

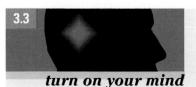

3.3

turn on your mind

For most randomized experiments, such as medical studies comparing a new treatment with a placebo, it is unrealistic to recruit a simple random sample of people to participate. Why is this the case? What can be done instead to make sure the Fundamental Rule for Using Data for Inference is not violated?*

3.3 | DESIGNING A GOOD EXPERIMENT

An *experiment* measures the effect of manipulating the environment of the participants in some way. With human participants, the manipulation may include receiving a drug or medical treatment, going through a training program, agreeing to a special diet, and so on. Most experiments on humans use volunteers because you can't very well force someone to accept a manipulation. Experiments are also done on other kinds of experimental units, such as when different growing conditions are compared for their effect on plant yield, or different paints are applied on highways to see which ones last longer. The idea is to measure the effect of the feature being manipulated, the explanatory variable, on the response variable.

Randomized experiments are particularly important because, unlike most other studies, they often allow us to determine **cause and effect.** The participants in randomized experiments are usually randomly assigned to either receive a specific treatment or to take part in a control group. The purpose of the random assignment is to make the groups approximately equal in all respects except for the explanatory variable, which is purposely manipulated. Differences in the response variable between the groups, if large enough to rule out natural chance variability, can then be attributed to the manipulation of the explanatory variable.

CASE STUDY 3.2
Kids and Weight Lifting

Is weight training good for children? If so, is it better for them to lift heavy weights for a few repetitions or moderate weights a larger number of times? Researchers at the University of Massachusetts set out to answer this question with a randomized experiment using 43 young volunteers between the ages of 5.2 and 11.8 years old. The children, recruited from a YMCA after-school program, were randomly assigned to one of three groups. Group 1 performed 6 to 8 repetitions with a heavy load, group 2 performed 13 to 15 repetitions with a moderate load, and group 3 was a control group.

Figure 3.3 illustrates the process used in this study. The steps for this randomized experiment include recruiting volunteers to participate, randomly assigning participants to the three groups, carrying out the weight-lifting treatments (the explanatory variable), and comparing muscular strength and endurance (the response variable) for the three groups.

*HINT: As an example, in a study to test a new medication to treat people with athlete's foot, would it be possible to identify the population of those afflicted, get a random sample, and get their cooperation? How might a "representative" sample be found instead?

FIGURE 3.3 **The randomized experiment in Case Study 3.2: Children were randomly assigned to one of three different groups**

The exercises were performed twice a week for eight weeks. According to the research report, the higher-repetition, moderate-load group (group 2) showed the best results. Here is a sample of the results:

> Leg extension strength significantly increased in both exercise groups compared with that in the control subjects. Increases of 31.0% and 40.9%, respectively, for the low repetition–heavy load and high repetition–moderate load groups were observed. (Faigenbaum et al., 1999, p. e5)

Because the children were randomly assigned to the exercise groups, they should have been similar, on average, before the weight-training intervention. Therefore, it is reasonable to conclude that the different training programs actually *caused* the difference in results.

Who Participates in Randomized Experiments?

In most cases, the participants in randomized experiments are **volunteers.** Sometimes they are passive volunteers, such as when all patients treated at a particular medical facility are asked to sign a consent form agreeing to participate in a study. Often, researchers recruit volunteers through the newspaper. For instance, a "Well-Being" feature page in *The Sacramento Bee* had an article titled "Volunteers for Study." It said:

> People who suffer from anxiety disorders are needed for a study being conducted by the University of California, Davis, Department of Psychiatry. Researchers are examining the effectiveness of a new medication on moderate-to-severe symptoms of anxiety. The study requires volunteers who are at least 18 and in generally good health and who are not taking oral medications for their anxiety. The study lasts 14 weeks and requires routine visits to the Department of Psychiatry. All study procedures, visits and medications are free. To enroll, call . . . (*Sacramento Bee,* Stephanie McKinnon McDade, July 7, 1999, p. G5. Copyright, *The Sacramento Bee,* 2004.)

When researchers recruit volunteers for studies, they may not be able to extend the results to a larger population. For instance, investigators often recruit volunteers by offering a small payment or free medical care as compensation for participation. Those who respond are more likely to be from lower socioeconomic backgrounds, and the results may not apply to the general population.

Remember the Fundamental Rule for Using Data for Inference: Available data can be used to make inferences about a much larger group *if the data can be considered to be representative with regard to the question(s) of interest.* In many randomized experiments, the question of interest is whether the treatment and control groups differ on the outcome variable(s). Common sense should enable you to figure out if the volunteers in the study are representative of a larger population for that question.

Randomization: The Crucial Element

Researchers do experiments to reduce the likelihood that the results will be affected by confounding variables and other sources of bias that often are present in observational studies. The simple principle of **randomization** is the key to doing this.

In experimentation, we want to make sure that the experimenters do not have the flexibility to choose which of the units (people, animals, etc.) receive each of the potential treatments, or in some cases, the order in which each unit receives multiple treatments. For instance, in Case Study 3.2 it was important for the researchers to randomly assign the kids to different weight-lifting routines. If they handpicked which kids received which routine, they could have chosen the strongest kids to receive the treatment they wanted to show was superior.

In many experiments, each experimental unit receives or participates in only a single treatment, such as in Case Study 1.6, when the physicians were given either aspirin or a placebo. In these experiments, the treatments are randomly assigned to the experimental units, or equivalently, the experimental units are assigned to the treatment groups. In other experiments, each experimental unit receives or participates in all of the treatments, such as when people are asked to taste and compare three brands of soda. In this case, the order in which the treatments are presented must be randomized. In Chapter 4 we will learn how both of these types of randomization are actually accomplished. For now, we discuss why randomization is necessary.

Randomizing the Type of Treatment

Randomly assigning the treatments to the experimental units keeps the researchers from making assignments favorable to their hypotheses and also helps protect against hidden or unknown biases. For example, suppose that in the experiment in Case Study 1.6, approximately the first 11,000 physicians who enrolled were given aspirin and the remaining physicians were given placebos. It could be that the healthier, more energetic physicians enrolled first, thus giving aspirin an unfair advantage.

In statistics, "random" is not synonymous with "haphazard," despite what your thesaurus might say. As we have seen, random assignments may not be possible or ethical under some circumstances. But in situations where randomization is feasible, it is usually not difficult to accomplish. It can be done with a table of random digits, a computer, or even, if done carefully, by physical means like flipping a coin or drawing numbers from a hat. The important feature, ensured by proper randomization, is that the chances of being assigned to each condition are the same for each participant.

Randomizing the Order of Treatments

In some experiments, all treatments are applied to each unit. In that case, randomization should be used to determine the *order* in which they are applied. For example, suppose an experiment is conducted to determine the extent to which drinking alcohol or smoking marijuana impairs driving ability. Because drivers are all different, it makes sense to test the same drivers under all three conditions (alcohol, marijuana, and sober) rather than using different drivers for each condition. But if everyone were tested under alcohol first, then marijuana, then sober, the performances could improve on the second and third times just from having learned something about the test.

A better method would be to randomly assign some drivers to each of the possible orderings, so the learning effect would average out over the three treatments. Notice that it is important that the assignments be made *randomly*. If we let the experimenter decide which drivers to assign to which ordering, assignments could be made to give an unfair advantage to one of the treatments.

Control Groups, Placebos, and Blinding

Control Groups

In order to determine whether or not a drug, weight-lifting routine, meditation technique, and so on has an effect, we need to know what would have happened to the response variable if the treatment had not been applied. To find that out, experimenters create **control groups,** which are treated identically in all respects except that they don't receive the active treatment. Occasionally, a control group will receive a standard or existing treatment against which a new treatment is to be compared.

Placebos

A special kind of control group is often used in studies of the effectiveness of drugs. A substantial body of research shows that people respond not only to active drugs but also to **placebos.** A placebo looks like the real drug but has no active ingredients. Placebos can be amazingly effective; studies have shown that they can help up to 62% of headache sufferers, 58% of those suffering from seasickness, and 39% of those with postoperative wound pain.

Because the **placebo effect** is so strong, drug research is generally conducted by randomly assigning half of the volunteers to receive the drug and the other half to receive a placebo, without telling them which they are receiving. The placebo looks just like the real thing, so the participants will not be able to distinguish between it and the actual drug and thus will not be influenced by belief biases.

Blinding

It isn't only the patient who can be affected by knowing whether or not he or she has received an active drug. If the researcher who is measuring the reaction of the patients were to know which group they were in, the measurements could be taken in a biased fashion.

To avoid these biases, good experiments use **double-blind** procedures. A double-blind experiment is one in which neither the participant nor the researcher taking the measurements knows who had which treatment. A **single-blind** experiment is one in which the participants do not know which treatment they have been assigned. An experiment would also be called *single-blind* if the participants knew the treatments but the researcher was kept blind.

Although double-blind experiments are preferable, they are not always possible. For example, to test the effect of meditation on blood pressure, the subjects would obviously know if they were in the meditation group or the control group. Only a single-blind study is possible, in which the person taking the blood pressure measurement would not know who was in which group.

Double Dummy

Sometimes in an experiment to compare two treatments it is impossible to mask which treatment is which, so each group receives one active treatment and one placebo treatment. For instance, suppose researchers wanted to compare the effects of nicotine patches to nicotine gum for helping people quit smoking. Obviously, participants would know if they were wearing a patch or chewing gum. So participants are asked to both wear a patch and to chew gum. One of the two will have the active ingredient (nicotine), but the other will be a placebo. In other words, the treatments would be

Group One: Nicotine patch, placebo gum
Group Two: Placebo patch, nicotine gum

This method of giving each participant two "treatments" to ensure that the experiment is blind is called **double dummy.** Sometimes a third group is used and given placebos of both methods to compare the results from each treatment method to what would happen with no active treatment at all. In our example, a third group would receive a placebo patch and placebo gum.

Pairing and Blocking

It is sometimes easier and more efficient to have each person in a study serve as his or her own control. That way, differences that inherently exist among individuals don't obscure the treatment effects. We encountered this idea when we discussed how to compare driving ability under the influence of alcohol and marijuana and when sober.

Sometimes instead of using the same individual for the treatments, researchers will match people on traits that are likely to be related to the outcome, such as age, IQ, or weight. They will then randomly assign each of the treatments to one member of each matched pair or grouping. For example, in a study comparing chemotherapy to surgery to treat cancer, patients might be matched by sex, age, and level of severity of the illness. One from each pair would then be randomly chosen to receive the chemotherapy and the other to receive surgery. (Of course, such a study would only be ethically feasible if there were no prior knowledge that one treatment was superior to the other. Patients in such cases are always required to sign an informed consent.)

Matched-Pair Designs

Experimental designs that use either two matched individuals or the same individual to receive each of two treatments are called **matched-pair designs.** The important feature of these designs is that randomization is used to assign the order of the two treatments. Of course, it is still important to try to conduct the experiment in a double-blind fashion, so that neither the participant nor the researcher knows which *order* was used.

Block Designs

The matched-pair design is a special case of a **block design.** Experimental units are divided into homogeneous groups called **blocks,** and each treatment is randomly assigned to one or more units in each block. Sometimes the "blocks" are individuals, and the "experimental units" are the repeated time periods in which they receive the varying treatments. The method discussed for comparing drivers under three conditions was a block design. Each driver is a block, and the experimental units are the three time periods during which they drove sober, with alcohol, and with marijuana. This somewhat peculiar terminology results from the fact that these ideas were first used in agricultural experiments, where the experimental units were plots of land that had been subdivided into homogeneous blocks. In the social sciences, designs such as these, in which the same participants are measured repeatedly under differing conditions, are referred to as **repeated-measures designs.**

Block designs are sometimes used when the experimental units are different enough on an important variable that experimenters group them on that variable, then assign all treatments within each group. For instance, an experiment to compare two methods for increasing memorization skills might first block participants into age groups (young, middle-aged, elderly) and then randomly assign the two methods to half of the individuals in each age group. Comparison of the two

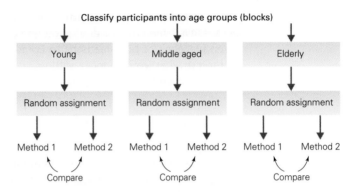

Classify participants into age groups (blocks)

| Young | Middle aged | Elderly |

Random assignment   Random assignment   Random assignment

Method 1   Method 2   Method 1   Method 2   Method 1   Method 2

Compare   Compare   Compare

FIGURE 3.4   **A blocked experiment to compare two memorization methods. Within each of three age groups (blocks), individuals are randomly allocated to the two methods.**

treatments would be done separately in each age group so that differences in memorization capacity with age would not mask the differences in the two methods. Figure 3.4 illustrates the steps that would be taken in such an experiment.

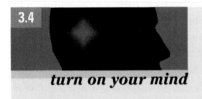

3.4

turn on your mind

Students are sometimes confused by the reasons for blocking and for randomization. One method is used to control *known* sources of variability among the experimental units, and the other is used to control *unknown* sources of variability. Explain which is which, and provide examples illustrating these ideas.*

Design Terminology and Examples

If the treatments in an experiment are randomly assigned to experimental units without using matched pairs or blocks, the experiment has what is called a **completely randomized design.** When matched-pairs are used, the experiment has a **matched-pairs design,** and when blocks are used, the experiment has a **randomized block design.** When researchers design an experiment, they must decide whether to use matched-pairs, blocks, or neither. Blocks or matched-pairs are used to reduce known sources of variability in the response variable across experimental units and should be used when such sources are known and can be determined for each experimental unit. For example, if you wanted to study the relationship between exercise and weight loss, it would make sense to separate people based on whether they were initially underweight, overweight, or about right. Initial weight would be used to create matched-pairs or blocks.

Notice that there is some overlap in the terminology used for these designs because a matched-pairs design could also be considered to be a randomized block design, with each pair representing one block. Therefore, matched-pairs designs are a subset of randomized block designs, in which there are only two treatments, each assigned only once within each block.

To help you distinguish the types of experimental design, here are two examples of experiments that might be done using those designs and how each design would be accomplished.

*HINT: Consider the example in Figure 3.4. Two reasons for differing ability to memorize might be age and individual brain structure. Which one can be measured and used to assign people to blocks?

Experiment 1: The Effect of Caffeine on Swimming Speed

Suppose a researcher wants to know if taking caffeine an hour before swimming affects the time that it takes swimmers to complete a one-mile swim and that 50 volunteers are available for the study.

◆ *Completely randomized design.* Randomly assign 25 volunteers to take a caffeine pill and the remaining 25 to take a placebo, have them all swim a mile, and compare the times for the two groups as the analysis.

◆ *Matched-pairs design.* There are two ways this could be done.

  Method 1: Have each of the 50 swimmers take both treatments, with adequate time, perhaps a week, between them. In week one, randomly assign swimmers to take either a caffeine pill or a placebo. Have them swim a mile and record the time. The next week, give each swimmer the treatment he or she did not have the previous week, have them swim a mile, and record the time. For each swimmer, take the difference in the two times and use those measurements for the analysis.

  Method 2: Create 25 pairs of swimmers based on speed. Before assigning treatments, measure how long it takes each one to swim a mile. Match the two fastest swimmers as one pair, the next two as the second pair, and so on. A week later, randomly assign one swimmer in each pair to take a caffeine pill and the other to take a placebo. Measure their times to swim a mile. For each pair, take the difference in the two times and use those measurements for the analysis.

◆ *Randomized block design.* This would be most appropriate if there were different types of swimmers. For instance, suppose that 10 of the volunteers were males who swim competitively, 20 were males who don't swim competitively, 8 were females who swim competitively, and the remaining 12 were females who don't swim competitively. Then each of these four groups would constitute a block, and half of each block would be randomly assigned to take a caffeine pill while the other half would take a placebo. Each person would swim a mile. The times for caffeine versus placebo would be compared within each block as well as across the four blocks combined.

Experiment 2: Advertising Strategies for Selling Coffee Near Colleges

Suppose that a national chain of coffee cafes wants to compare three methods for increasing sales at their cafes near college campuses. Each method involves placing a weekly advertisement in the campus newspaper. With method one, a coupon for a free cup of coffee is included in the ad. With method two, a coupon for a free cookie with any coffee drink is included in the ad. The third method is to place the ad without any coupons. They place the ads for one full semester and measure how much sales change from the same semester the previous year.

◆ *Completely randomized design.* For this design, they would randomly choose one-third of the campuses in the study to receive each of the treatments, and the analysis would compare the change in sales for the three treatment groups.

◆ *Matched-pairs design.* This design would not be possible if they want to compare all three treatments.

◆ *Randomized block design.* Decide what known factors contribute to differing sales across campuses. Group or "block" the campuses based on those factors. For instance, they might categorize the campuses based on size of the student body (small or large) and population of the town or city (small, medium, large) for a total of six blocks. Once the campuses have been blocked, randomly assign one-third of the campuses within each block to receive each treatment. Compare change in sales within each block, then across blocks. Other blocking variables might include initial sales, distance from the campus, and whether or not there is a competing cafe in the neighborhood.

Debate About How Many Blocks To Use

tech note   Some statisticians argue that the number of experimental units in a block should only equal the number of treatments, so that each treatment is assigned only once in each block. That allows as many sources of known variability as possible to be controlled. For the example of the effect of caffeine on swimming speed there are two treatments, so blocks of size two (matched pairs) would be created. This could be accomplished by matching people on known variables such as sex, initial swim speed, usual caffeine consumption, age, and so on.

CASE STUDY 3.3
Quitting Smoking with Nicotine Patches

There is no longer any doubt that smoking cigarettes is hazardous to your health and to those around you. Yet, for someone addicted to smoking, quitting is no simple matter. One promising technique is to apply a patch to the skin that dispenses nicotine into the blood. In fact, these nicotine patches have become one of the most frequently prescribed medications in the United States.

In order to test the effectiveness of these patches on the cessation of smoking, Dr. Richard Hurt (1994) and his colleagues recruited 240 smokers at Mayo Clinics in Rochester, Minnesota, Jacksonville, Florida, and Scottsdale, Arizona. Volunteers were required to be between the ages of 20 and 65, have an expired carbon monoxide level of 10 ppm or greater (showing that they were indeed smokers), be in good health, have a history of smoking at least 20 cigarettes per day for the past year, and be motivated to quit.

Volunteers were randomly assigned to receive either 22-mg nicotine patches or placebo patches for eight weeks. They were also provided with an intervention program recommended by the National Cancer Institute, in which they received counseling before, during, and for many months after the eight-week period of wearing the patches.

After the eight-week period of patch use, almost half (46%) of the nicotine group had quit smoking, while only one-fifth (20%) of the placebo group had. Having quit was defined as "self-reported abstinence (not even a puff) since the last visit and an expired air carbon monoxide level of 8 ppm or less" (p. 596). After a year, rates in both groups had declined, but the group that had received the nicotine patch still had a higher percentage who successfully quit than did the placebo group, 27.5% versus 14.2%.

The study was double-blind, so neither the participants nor the nurses taking the measurements knew who had received the active nicotine patches. The study was funded by a grant from Lederle Laboratories and was published in the *Journal of the American Medical Association*. Because this was a well-designed randomized experiment, it is likely that the nicotine patches actually *caused* the additional percentage who had quit in the nicotine patch group compared with the control group.

3.4 | DESIGNING A GOOD OBSERVATIONAL STUDY

Observational studies start with a distinct *disadvantage* compared to experiments when trying to establish causal links. The researchers observe, but cannot control, the explanatory variables. However, they have the *advantage* that they are more likely to measure participants in their natural setting. Before looking at some complications that can arise, let's look at an example of a well-designed observational study.

CASE STUDY 3.4
Baldness and Heart Attacks

On March 8, 1993, *Newsweek* announced, "A really bad hair day: Researchers link baldness and heart attacks" (p. 62). The article reported that "men with typical male pattern baldness . . . are anywhere from 30 to 300 percent more likely to suffer a heart attack than men with little or no hair loss at all." (Pattern baldness is the type affecting the crown or vertex and not the same thing as a receding hairline; it affects approximately one-third of middle-aged men.)

The report was based on an observational study conducted by researchers at Boston University School of Medicine. They compared 665 men who had been admitted to the hospital with their first heart attack to 772 men in the same age group (21 to 54 years old) who had been admitted to the same hospitals for other reasons. There were 35 hospitals involved, all in eastern Massachusetts and Rhode Island. Lesko, Rosenberg, and Shapiro reported the full results in the *Journal of the American Medical Association* (1993).

The study found that the percentage of men who showed some degree of pattern baldness was substantially higher for those who had had a heart attack (42%) than for those who had not (34%). Further, when they used sophisticated statistical tests to ask the question in the reverse direction, they found that there was an increased risk of heart attack for men with any degree of pattern baldness. The analysis methods included adjustments for age and other heart attack risk factors. The increase in risk was more severe with increasing severity of baldness, after adjusting for age and other risk factors.

The authors of the study speculated that a third variable, perhaps a male hormone, might simultaneously increase the risk of heart attacks and the propensity for baldness. With an observational study such as this one, scientists can establish a connection, and they can then look for causal mechanisms in future investigations.

Special Types of Observational Studies

Retrospective or Prospective Studies

Observational studies can be classified according to whether they are **retrospective,** in which participants are asked to recall past events, or **prospective,** in which participants are followed into the future and events are recorded. The prospective approach is a better procedure, because people often do not remember past events accurately.

Case-Control Studies

In a **case-control study,** "cases" who have a particular attribute or condition are compared to "controls" who do not. The idea is to compare the cases and controls to see how they differ on an explanatory variable of interest. In medical settings, the cases usually are individuals who have been diagnosed with a particular disease. Researchers then identify a group of controls who are as similar as possible to the cases, except that they don't have the disease. In order to achieve this similarity, researchers often use patients hospitalized for other causes as the controls.

Case Study 3.4 is an example of a case-control study. In this example, men who had been admitted to the hospital with a heart attack were the cases, and men who had been admitted for other reasons were the controls. The cases and controls were compared on a variable of interest, which in Case Study 3.4 was the degree of baldness. In essence, the explanatory variable is whether a participant is a case or a control (heart attack or not), and the response variable is

Extent of baldness compared

FIGURE 3.5   **The case-control design in Case Study 3.4: Samples of male heart attack patients (cases) and other male hospital patients (controls) were compared for the extent of baldness**

the measurement of interest (degree of baldness). Figure 3.5 illustrates the steps taken in this study.

Sometimes cases are matched with controls on an individual basis. That type of design is similar to a matched-pair experimental design. The analysis proceeds by first comparing the pair, then summarizing over all pairs. Unlike a matched-pair experiment, the researcher does not randomly assign treatments within pairs, but is restricted to how they occur naturally. For example, to identify whether or not left-handed people die at a younger age, researchers might match each left-handed case with a right-handed sibling as a control and compare their ages at death. Handedness could obviously not be randomly assigned to the two individuals, so confounding factors (like the fact that most tools are made for right-handed people) might be responsible for any observed differences.

Advantages of Case-Control Studies

Case-control studies have become increasingly popular in medical research, with good reason. They do not suffer from the ethical considerations inherent in the random assignment of potentially harmful or beneficial treatments. In addition, case-control studies can be efficient in terms of time, money, and the inclusion of enough people with the disease.

Efficiency

In an exercise given later in this book, researchers were interested in whether or not owning a pet bird is related to the incidence of lung cancer. Imagine trying to design an experiment to study this problem. You would randomly assign people to either own a bird or not and then wait to see how many in each group contracted lung cancer. The problem is that you would have to wait a long time, and even then, you would observe very few cases of lung cancer in either group. In the end, you may not have enough cases for a valid comparison.

A case-control study, in contrast, would identify a large group of people who had just been diagnosed with lung cancer, then ask them whether or not they had owned a pet bird. A similar control group would be identified and asked the same question. A comparison would then be made between the proportion of cases (lung-cancer patients) who had birds and the proportion of controls who had birds.

Reducing Potential Confounding Variables

Another advantage of case-control studies over other observational studies is that the controls are chosen to try to reduce potential confounding variables. For example, in Case Study 3.4, suppose it was simply the case that balding men were less healthy than other men and were therefore more likely to get sick in some way. An observational study that recorded only whether or not someone had baldness and whether or not he had experienced a heart attack would not be able to control for that fact. By using other hospitalized patients as the controls, the researchers were able to partially account for general health as a potential confounding factor.

You can see that careful thought is needed to choose controls that reduce potential confounding factors and do not introduce new ones. For example, suppose we want to know if heavy exercise induces heart attacks, and as cases we use people recently admitted to the hospital with a heart attack. We would certainly not want to use other newly admitted patients as controls. People who are sick enough to enter the hospital (for anything other than sudden emergencies) would probably not have been recently engaging in heavy exercise. When you read the results of a case-control study, you should pay attention to how the controls were selected.

3.5 | DIFFICULTIES AND DISASTERS IN EXPERIMENTS AND OBSERVATIONAL STUDIES

There are a number of complications that can arise if researchers are not careful in conducting and reporting experiments and observational studies. Some apply only to observational studies, some only to experiments, and some to both settings. The first step in reading the results of a study is to determine if it's an experiment or an observational study. The conclusions you can make differ for the two types. We first list the possible problems, then describe each one in detail.

Difficulties and Disasters in Experiments and Observational Studies

- Confounding variables and the implication of causation in observational studies
- Extending results inappropriately
- Interacting variables
- Placebo, Hawthorne, and experimenter effects
- Ecological validity and generalizability
- Relying on memory or secondhand sources

Confounding Variables and the Implication of Causation in Observational Studies

Probably the biggest single misinterpretation made when reporting studies is to imply that a *cause-and-effect* relationship can be concluded on the basis of an observational study. There is simply no way to separate the role of confounding variables from the role of explanatory variables in producing the outcome variable if randomization has not been used.

A partial solution is for researchers to measure possible confounding variables and include them in the analysis to see if they are related to the outcome variable. Another partial solution can be achieved in case-control studies by choosing the controls to be as similar as possible to the cases. In Chapter 22 we discuss some nonstatistical considerations that can help you determine whether a causal link is plausible based on observational studies. The other part of the solution is up to you as a consumer of information: Don't be misled into thinking a causal relationship exists when it doesn't.

EXAMPLE 3.5
*Will Preventing Artery
Clog Prevent Memory Loss?*

An observational study based on a random sample of over 6000 older people (average age 70 when the study began) found that 70% of the participants did not lose cognitive functioning over time. One important finding, however, was that "those who have diabetes or high levels of atherosclerosis in combination with a gene for Alzheimer's disease are eight times more likely to show a decline in cognitive function." This finding led one of the researchers to comment that "that has implications for prevention, which is good news. If we can prevent atherosclerosis, we can prevent memory loss over time, and we know how to do that with behavior changes—low-fat diets, weight control, exercise, not smoking, and drug treatments." (*Source:* "Study: Age doesn't sap memory," *Sacramento Bee,* Kathryn Doré Perkins, July 7, 1999. pp. A1 and A10. Copyright, *The Sacramento Bee,* 2004.)

But notice that this conclusion assumes that the memory loss is *caused* by atherosclerosis. This conclusion is not justified on the basis of an observational study. Perhaps certain people have a genetic predisposition to artherosclerosis and to memory loss, in which case preventing the former condition with lifestyle changes will not reduce the memory loss. Or perhaps people who suffer from depression tend to eat poorly, not exercise well, and also suffer from memory loss. There are simply too many possible confounding factors that could be related to both artherosclerosis and memory loss to conclude that preventing artherosclerosis would prevent memory loss. ◆

Extending Results Inappropriately

In interpreting both randomized experiments and observational studies, keep in mind the Fundamental Rule for Using Data for Inference: Available data can be used to make inferences about a much larger group *if the data can be considered to be representative with regard to the question(s) of interest.* Some observational studies are based on random samples, in which case the results can readily be extended to the population from which the sample was drawn. Example 3.5 is one such example; the study on aging and memory loss was part of the large Cardiovascular Health Study, sponsored by the National Heart, Lung and Blood Institute, part of the National Institutes of Health.

Many observational studies and almost all randomized experiments use convenience samples or volunteers. It is up to you to determine whether or not the results can be extended to any larger group for the question of interest. For example, the weight-lifting experiment in Case Study 3.2 used 11 girls and 32 boys who volunteered to participate in the study. There is no reason to suspect that these volunteers would differ from other children in terms of the differential effects of no training, moderate lifting, and heavy lifting measured in the study. Thus, although the sample consisted solely of volunteers, the results can probably be extended to other children in this age group.

Interacting Variables

Sometimes a second variable *interacts* with the explanatory variable in its relationship with the outcome variable. This can be a problem if the results of a study are reported without taking the interaction into account. Although the reported results may be true on average, they may not hold for specific subgroups.

For example, in Case Study 3.3, there was an interaction between the treatment (nicotine or placebo patch) and whether there were other smokers at

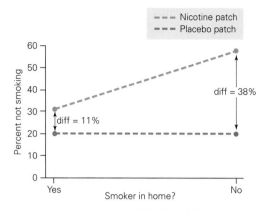

FIGURE 3.6   **Interaction in Case Study 3.3: The difference between the nicotine and placebo patches is greater when there are no smokers in the home than when there are smokers in the home**

home. The researchers measured and reported this interaction, which is illustrated in Figure 3.6. When there were no other smokers in the participant's home, the eight-week success rate was actually 58% for the nicotine patch users compared to 20% for the placebo patch group. When there were other smokers in the home, the eight-week success rate for the nicotine patch was only 31%, but the success rate for the placebo users remained at about 20%. In other words, the amount by which the nicotine and placebo groups differed was affected by whether other smokers were present in the home. Therefore, it would be misleading to merely report that 46% of the nicotine recipients had quit without also providing the information about the interaction.

Hawthorne and Experimenter Effects

We have already discussed the fact that a placebo can have a very strong effect on experimental outcomes, since the power of suggestion is somehow able to affect the result. A similar difficulty is that participants in an experiment respond differently than they otherwise would, just because they are in the experiment. This is called the **Hawthorne effect** because it was first detected in 1924 when studying factory workers at the Hawthorne, Illinois plant of the Western Electric Company. (The phrase was not actually coined until much later; see French, 1953.)

The Hawthorne effect is a common problem in medical research. Many treatments have been observed to have a higher success rate in clinical trials than they do in actual practice. This may occur because patients and researchers are highly motivated to correctly carry out a treatment protocol in a clinical trial.

Related to these effects are numerous ways in which the experimenter can bias the results. These **experimenter effects** include recording the data erroneously to match the desired outcome, treating subjects differently based on which condition they are receiving, or subtly making the subjects aware of the desired outcome. Most of these problems can be overcome by using double-blind designs and by including a placebo group or a control group that gets identical handling except for the active part of the treatment. Other problems, such as incorrect data recording, should be addressed by allowing data to be automatically entered into a computer when it is collected, if possible. Depending on the experiment, there may still be subtle ways in which experimenter effects can sneak into the results. You should be aware of these possibilities when you read the results of a study.

EXAMPLE 3.6
Dull Rats

In a classic experiment designed to test whether or not the expectations of the experimenter could really influence the results, Rosenthal and Fode (1963) deliberately conned 12 experimenters. They gave each one 5 rats that had been taught to run a maze. They told 6 of the experimenters that the rats had been bred to do well (i.e., they were "maze bright") and told the other 6 that their rats were "maze dull" and should not be expected to do well. Sure enough, the experimenters who had been told they had bright rats found learning rates far superior to those found by the experimenters who had been told they had dull rats. Hundreds of other studies have since confirmed the experimenter effect. ◆

Ecological Validity and Generalizability

Suppose you want to compare three assertiveness-training methods to see which is more effective in teaching people how to say "no" to unwanted requests for their time. Would it be realistic to give participants the training, then measure the results by asking them to role-play in situations in which they should say "no"? Probably not, since everyone involved would know that they were only role-playing. The usual social pressures to say "yes" would not be as present.

This is an example of an experiment with little **ecological validity.** The variables have been removed from their natural setting and are measured in the laboratory or in some other artificial setting. Thus, the results do not accurately reflect the impact of the variables in the real world or in everyday life. Whenever you read the results of a study you should question its ecological validity and its generalizability.

EXAMPLE 3.7
Real Smokers with a Desire to Quit

The researchers in Case Study 3.3, the nicotine patch study, did many things to help ensure ecological validity and generalizability. First, they used a standard intervention program available from and recommended by the National Cancer Institute, instead of inventing their own, so that other physicians could follow the same program. Next, they used participants at three different locations around the country rather than in one community only, and they involved a wide range of ages (20 to 65). They included individuals who lived in households with other smokers as well as those who did not. Finally, they recorded numerous other variables (sex, race, education, marital status, psychological health, etc.) and checked to make sure these were not related to the response variable or the patch assignment. ◆

Relying on Memory or Secondhand Sources

Retrospective observational studies can be particularly unreliable because they ask people to recall past behavior. Some studies, in which the response variable is whether or not someone has died, can be even worse because they rely on the memories of relatives and friends rather than the actual participants to measure explanatory and other variables. Retrospective studies also suffer from the fact that variables that confounded things in the past may no longer be similar to those that would currently be confounding variables, and researchers may not think to measure them.

If at all possible, prospective studies should be used. That's not always possible. For example, researchers who first considered the potential causes of AIDS or toxic-shock syndrome had to start with those who were ill and try to find common factors from their pasts. If possible, retrospective studies should use authoritative sources such as medical records rather than rely on memory.

EXAMPLE 3.8
Do Left-Handers
Die Young?

A few years ago, a highly publicized study pronounced that left-handed people did not live as long as right-handed people (Coren and Halpern, 1991). In one part of the study, the researchers had sent letters to next of kin for a random sample of recently deceased individuals, asking which hand the deceased had used for writing, drawing, and throwing a ball. They found that the average age of death for those who had been left-handed was 66, while for those who had been right-handed it was 75.

What the researchers failed to take into account was that in the early part of the 20th century many children were forced to write with their right hands, even if their natural inclination was to be left-handed. Therefore, people who died in their 70s and 80s during the time of this study were more likely to be right-handed than those who died in their 50s and 60s. The confounding factor of how long ago one learned to write was not taken into account. A better study would be a prospective one, following current left- and right-handers to see which group survived longer, but you can imagine the practical difficulties of conducting such a study. The participants could well outlive the researchers. ◆

KEY TERMS

Section 3.1
descriptive statistics, 59
inferential statistics, 59
fundamental rule for using data for
   inference, 60
population, 60
sample, 60
simple random sample, 61

Section 3.2
observational study, 61
experiment, 62
randomized experiment, 62, 67
treatments, 62
units, 63
experimental units, 63
subjects, 63
participants, 63
explanatory variable, 63
response variable, 63

outcome variable, 63
dependent variable, 63
independent variable, 63
confounding variable, 63
lurking variable, 63–64

Section 3.3
cause and effect, 67, 77
volunteers, 68
randomization, 69
control groups, 70
placebo, 70
placebo effect, 70
blinding, 70
double-blind, 70
single-blind, 70
double dummy, 70–71
matched-pair design, 71, 72
block design, 71–72
blocks, 71

repeated-measures design, 71
completely randomized design, 72
randomized block design, 72

Section 3.4
retrospective study, 75
prospective study, 75
case-control study, 75–76

Section 3.5
confounding variable and causation, 77
extending results inappropriately, 78
interacting variable, 78–79
Hawthorne effect, 79
experimenter effect, 79–80
ecological validity, 80
memory and secondhand sources,
   80–81

Exercises
effect modifier, 85

EXERCISES   Blue = Basics   Bold, **Bold** = Answer in back

Sections 3.1 and 3.2

3.1 The median income for a random sample of households in a school district was found to be $36,300. From this information, the school board decided that the median income for all households in the district was probably between about $32,000 and $40,000. Describe the difference between descriptive statistics and inferential statistics in the context of this example.

3.2 According to the Fundamental Rule for Using Data for Inference, when can available data be used to make inferences about a much larger group?

3.3 In a class of 20 students (John, Maria, Inez, Bill, etc.), a simple random sample of two people will be selected from the class. Use the definition of a simple random sample to compare the chance that John and Maria will be the group selected with the chance that Maria and Bill will be the group selected.

3.4 Suppose a statistics teacher wants to know whether the number of hours students spend studying in a group affects the final course grade. In each part, explain whether the research method described is a randomized experiment or an observational study.
   a. Each student keeps a log of the hours he or she spends studying in a group, and reports the total after the course is completed.
   b. Students are randomly assigned to study groups. The teacher tells each group how often to meet. This varies from one hour the day before each exam to two hours per week.
   c. Students voluntarily join groups based on how often the groups will meet. The groups are designated as meeting weekly, meeting only before exams, or meeting whenever enough members feel it is necessary.

3.5 To estimate the percent of households in the United States that use a DVD player, a researcher surveys a randomly selected sample of 500 households and asks about DVD player usage.
   a. What is the unit being measured in this survey?
   b. What is the population of interest in this survey?
   c. What is the sample in this survey?

3.6 In each part, identify the response variable and the explanatory variable in the relationship between the two given variables.
   a. Daughter's height and mother's height.
   b. Age and weight for children between the ages of 3 and 10 years old.
   c. Opinion about the death penalty for persons convicted of murder and political party membership (Democrat, Republican, etc.)
   d. Smoking habits and performance on a test of lung function.

3.7 Remember that a confounding variable affects the response variable and is related to the explanatory variable. For each of the following situations, explain how the given confounding variable meets these criteria.
   a. *Response variable* = Math skills of children aged 6 to 12 years old; *explanatory variable* = shoe size; *confounding variable* = age.

   b. *Response variable* = Whether or not a student gets a cold during a school term; *explanatory variable* = whether or not the student procrastinates in doing assignments; *confounding variable* = number of hours per week the student spends socializing.
   c. *Response variable* = Child's IQ at age 10; *explanatory variable* = whether mother smokes or not; *confounding variable* = mother's educational level.
   d. *Response variable* = Number of missing or decayed teeth a child has; *explanatory variable* = amount of past exposure to lead; *confounding variable* = number of visits to a dentist child has made in the past three years.

3.8 For each of the following situations explain whether or not the Fundamental Rule for Using Data for Inference holds.
   a. *Available Data:* Salaries for a random sample of male and female professional basketball players.
   Research Question: Are women paid less than men who are in equivalent jobs?
   b. *Available Data:* Pulse rates for smokers and nonsmokers in a large statistics class at a major university.
   Research Question: Do college-age smokers have higher pulse rates than college-age nonsmokers?
   c. *Available Data:* Opinions on whether or not the legal drinking age should be lowered to 19 years old, collected from a random sample of 1000 adults in the state.
   Research Question: Does a majority of adults in the state support lowering the drinking age to 19?
   d. *Available Data:* Opinions on whether or not the legal drinking age should be lowered to 19 years old, collected from a random sample of parents of high school students in the state.
   Research Question: Does a majority of adults in the state support lowering the drinking age to 19?

3.9 For each of the following research scenarios, explain whether a randomized experiment could be used.
   a. To study the relationship between long-term practice of meditation and blood pressure.
   b. To determine whether a special training program improves scores on a standard college admissions test.
   c. To compare two programs to reduce the number of commuters who drive to work: providing discount coupons for the bus, or providing shuttle service for people who need to run errands during the workday.
   d. To study the relationship between age and opinion about whether marijuana should be legal.

3.10 Refer to exercise 3.9. In each case, state the explanatory variable and the response variable.

3.11 Refer to exercise 3.9. In each case, give an explanation of a possible confounding variable that could be present if an observational study were done.

3.12 Give an example of an observational study, and explain the difference between a confounding variable and a lurking variable in the context of your example.

3.13 Explain whether a variable can be both
   a. A confounding variable and a lurking variable.
   b. A response variable and a confounding variable.
   c. An explanatory variable and a dependent variable.

3.14 Refer to the first observational study described in Case Study 3.1. Give an example of a possible confounding variable in addition to those described in the text. Explain why it could be a confounding variable.

3.15 A friend has recommended the work of a musician who has recorded five CDs, each containing ten selections. You decide to visit a music store and listen to a simple random sample of four songs. Explain how you could select the four songs.

3.16 Refer to the previous exercise. Suppose you randomly chose four of the CDs, then randomly chose one song from each one. Would this be considered a simple random sample of the musician's songs? Explain.

3.17 A news article on the *Reuters Health* website, Dec. 18, 1998 reported that

> A study of over 1200 people over age 65 showed that the "owls"—those who go to sleep after 11 p.m. and rise after 8 a.m.—tend to be as healthy and intelligent as "larks"—those who go to bed before 11 p.m. and rise before 8 a.m.—according to a report in the December 19/26 issue of the *British Medical Journal* [Vol. 317, pp. 1675–1677]. The researchers also found that owls tend to have higher average income than the early birds.

a. Do you think this is based on an observational study or a randomized experiment? Explain.

b. It is mentioned that the "owls" have higher average income than the "larks." Explain whether or not income is a possible confounding variable in this study.

c. Give an example of a possible confounding variable that is not mentioned.

d. Draw a figure similar to Figure 3.2, illustrating the steps in this study.

3.18 A news article by *Reuters* on October 6, 1998 reported:

> In a new study, Dr. Matti Uhari and colleagues at the University of Oulu in Finland, randomly gave 857 healthy children in daycare centers xylitol in syrup, gum or a lozenge form, or a placebo gum or syrup in five doses per day for 3 months. According to a report in the October issue of the journal *Pediatrics* [Vol. 102, pp. 879–884, 971–972, 974–975], the incidence of ear infections were reduced by 40% in children given xylitol chewing gum, 30% in those given syrup and 20% in those given lozenges when compared to children given a placebo.

a. Is this an observational study or a randomized experiment? Explain.

b. What are the explanatory and response variables?

c. Are confounding variables likely to be a problem in this study? Explain.

Sections 3.3 and 3.4

3.19 Twenty students agreed to participate in a study on colds. Ten were randomly assigned to receive vitamin C, and the remaining 10 received a tablet that looked and tasted like vitamin C but in fact contained only sugar and flavoring. The students did not know whether they were taking vitamin C or not, but the investigators did. The students were followed for two months to see who came down with a cold and who didn't. Explain whether each of the following terms applies to this study.

a. Observational study.

b. Randomized experiment.

c. Placebo.

d. Single-blind.

e. Double-blind.

f Matched pairs.

g. Repeated measures.

3.20 A school district has 20 elementary schools and each school has 12 classes that can be used for a study.

a. Describe a randomized block design to compare three teaching methods, using schools as blocks.

b. Explain why schools should be used as blocks.

3.21 Twenty volunteers aged 40 to 65 years old will participate in an experiment done to compare two methods of memorizing information. The age variation in the participants concerns the researchers because memorization skills decrease with age.

a. Describe how the study could be done using a matched-pair design.

b. Explain why a matched-pair design is a good idea in this situation.

3.22 Refer to Case Study 3.2, "Kids and Weight Lifting." Did the experiment described use a completely randomized design, a matched-pair design, or a randomized block design? Explain.

3.23 Students who had meningitis were matched with other students without meningitis on gender, undergraduate (or graduate) status, and college. The purpose of the study was to discover risk factors for meningitis. Explain whether each of the following terms applies to this observational study.

a. Prospective study.

b. Retrospective study.

c. Case-control study.

3.24 Students whose undergraduate major was economics were followed for 10 years after graduation to study the number of different jobs they took. Explain whether each of the following terms applies to this observational study.

a. Prospective study.

b. Retrospective study.

c. Case-control study.

3.25 What type of variable is being described by, "a variable that affects the response variable and is also related to the explanatory variable?"

3.26 Refer to Case Study 3.3, "Quitting Smoking with Nicotine Patches," which used a completely randomized design. Explain how the experiment could have been done using a randomized block design instead.

3.27 Explain whether each of the following experiments was single-blind, double-blind, or neither.

a. An electric company wanted to know if residential customers would use less electricity during peak hours if they were charged more during those hours. One hundred customers were randomly selected to receive the special time-of-day rates, and they were matched with 100 customers who were similar in terms of past electrical use and household size. Special meters were installed for all 200 customers, and at the end of three months the usage during peak hours was compared for each pair. The technician who read the meters did not know who had which rate plan.

b. To test an herbal treatment for depression, 100 volunteers who suffered from mild depression were randomly divided into two groups. Each person was given a month's supply of tea bags. For one group, the tea contained the herb mixed with a spice tea, whereas for the other group, the bags contained only the spice tea. Participants were not told which type of tea they had and were asked to drink one cup of the tea per day for a month. At the end of the month, a psychologist evaluated them to determine if their mood had improved. The psychologist did not know who had the tea with the herbal ingredient added.

c. To compare three packaging designs for a new food, a manufacturer randomly selected three grocery stores in each of 50 cities. The three designs were randomly assigned to the three stores in each city, and sales were compared for a two-month period.

3.28 Refer to exercise 3.27. In each case, explain whether the experiment was a matched-pair design, a block design, or neither.

3.29 Refer to exercise 3.27. In each case, designate the explanatory variable and the response variable.

3.30 Refer to exercise 3.27. In each case, explain whether a control group was used and whether a placebo treatment was used.

3.31 Refer to exercise 3.27. In each case, explain the extent to which you think the results from the sample in the experiment could be extended to a larger population.

3.32 Explain why confounding variables are more of a problem in observational studies than in randomized experiments. Give an example.

3.33 Echinacea is an herb that may help prevent colds and flu. Explain how researchers could conduct a double-blind, double-dummy experiment to compare echinacea tea with chewable vitamin C pills for effectiveness in preventing colds.

3.34 Suppose researchers were interested in determining the relationship, if any, between the use of cellular telephones and the incidence of brain cancer. Would it be better to use a randomized experiment, a case-control study, or an observational study that did not use cases and controls? Explain.

3.35 Find an example of an observational study in the news. Determine whether or not it was a case-control study and if it was prospective or retrospective. Specify the explanatory and response variables, and explain whether confounding variables were likely to be a major problem in interpreting the results. Be sure to include the news article with your response.

3.36 Reuters (June 24, 1997) reported on a study published in the *Journal of the American Medical Association* (1997, Vol. 277, pp. 1940–1944) in which researchers recruited 276 volunteers (aged 18 to 55) and used nose drops to infect them with a cold virus. The volunteers were then quarantined and observed to see if they came down with a cold or infection. They also answered questions about their social lives, including the number of different types of contacts they had (family, work, community, religious groups, and so on). *Reuters* reported that "those with one to three types of social contacts were 4.2 times as likely to come down with cold symptoms and signs of infection compared to those with six or more contacts."

a. Was this an observational study or a randomized experiment? Explain.

b. What were the explanatory and response variables of interest to the researchers?

c. The researchers in this study took an unusual step to help reduce the impact of confounding variables. Explain what they did and why it might have helped to reduce the impact of confounding variables.

3.37 A story at ABCNEWS.com ("Pet Contact" by Rita Rubin, March 17, 1998) reported: Karen Allen, a researcher at the University of Buffalo, found that couples who own cats or dogs have more satisfying marriages and are less stressed-out than those who don't own pets. Pet owners also have more contact with each other and with other people. Allen compared 50 couples who owned either cats or dogs with 50 pet-free couples. The volunteers completed a standard questionnaire assessing their relationship and how attached they were to their pets. They also kept track of their social contacts over a two-week period. To see how the couples responded to stress, Allen monitored their heart rates and blood pressure while they discussed sore subjects. Pet-owning couples started out with lower blood pressure readings than the others, and their numbers didn't rise as much when they argued.

a. Was this a case-control study, a randomized experiment, or an observational study that was not case-controlled?

b. What are the explanatory and response variables in this study?

c. Give an example of a possible confounding variable.

d. Draw a figure illustrating the steps in this study, similar to Figures 3.2 through 3.5.

Section 3.5

3.38 For each of the following, explain whether the study summary describes an interacting variable or a confounding variable.

a. A comparison of the mean grade point averages of male and female students finds more difference for fraternity and sorority members than for students living in dorms and off-campus.

b. A study finds that wine drinkers tend to have better health than beer drinkers, but also finds that wine drinkers tend to smoke less than beer drinkers.

c. It is found that drawing a happy face on a restaurant bill increases mean tip percent for female servers but does not increase mean tip percent for male servers.

3.39 Researchers have found that women who take oral contraceptives (birth control pills) are at higher risk of having a heart attack or stroke and that the risk is substantially higher if a woman smokes. In investigating the relationship between taking oral contraceptives (the explanatory variable) and having a heart attack or stroke (the response variable), would smoking be called a confounding variable or an interacting variable? Explain.

3.40 In a study on worker productivity done in 1927 in Cicero, Illinois, it was observed that productivity improved when the lighting was increased. However, it was observed that productivity also improved when the lighting was decreased. What term is used to describe the fact that productivity increased simply because the subjects knew they were in an experiment?

3.41 A new medication is observed to cause a weight loss in women but a weight gain in men. Identify what the response variable, explanatory variable, and interacting variable are in this situation.

3.42 An investigator believes that a new medication will help reduce anxiety and that a placebo will not. In a single-blind study (patients are blind), she finds that the group taking the new medication had reduced anxiety while the placebo group did not. Explain how the experimenter effect may have played a role in the results.

3.43 In a sample of 500 college students aged 18 to 21, it was found that offering discount coupons for restaurant meals substantially increased the likelihood that they would eat at those restaurants. Explain how "extending results inappropriately" may be a problem in the interpretation of this study.

3.44 An experiment on elephants in captivity was done to see if breeding would occur more often when the elephants were kept together in a herd or when a male was isolated with a female. Explain how ecological validity affects the ability to generalize these results to wild elephants.

3.45 A retrospective study of 517 veterans who never smoked was done to see if there was an association between lung disease and exposure to workplace gases, dust, and fumes (Clawson, *Stanford Report*, November 1, 2000). In this study, subjects were asked whether or not they remembered being exposed to gases, dust, or fumes in the workplace. A concern expressed by the investigator was that "patients who have histories of lung diseases may be more likely to recall a history of exposure—whether accurate or not—than subjects reporting good health." Discuss whether each of the following is likely to be a problem in this study.
   a. Relying on memory.
   b. Confounding variable.
   c. Hawthorne effect.

3.46 Refer to exercise 3.37, describing a study about pet ownership and marriage. Explain whether each of the following is likely to be a problem for that study:
   a. Confounding variables and the implication of causation in observational studies.
   b. Placebo, Hawthorne, and experimenter effects.
   c. Ecological validity and generalizability.

3.47 Refer to exercise 3.27. For each of the experiments described, pick one of the "Difficulties and Disasters" described in Section 3.5, and explain how it might be a problem in that experiment.

3.48 Is the experimenter effect more likely to be a problem in a study that is double-blind, single-blind, or not blind at all? Explain.

3.49 Pick the two "Difficulties and Disasters" most likely to be a problem in each of the following studies, and explain why they would be a problem:
   a. To compare marketing methods, a marketing professor randomly divides a large class into three groups and randomly assigns each group to watch one of three television commercials. The students are then asked to rate how likely they would be to buy the advertised product.
   b. Volunteers are recruited through a newspaper article to participate in a study of a new vaccine for hepatitis C, which is transmitted by contact with infected blood and is a particular problem for intravenous drug users. The article specifically requests both in-

travenous drug users and medical workers to volunteer, assuring them confidentiality and offering them free medical care for the life of the study. Volunteers are then randomly assigned to receive either a placebo or the vaccine, and a year later they are tested to see if they have the disease.
   c. A researcher at a large medical clinic wants to determine if a high-fat diet in childhood is more likely to result in heart disease in later life. As a case-control study, she gives a questionnaire to 500 patients with heart disease and 500 other clinic patients matched by age and sex. The questionnaire asks about childhood diet.

3.50 Refer to part (c) of the previous exercise, describing a case-control study relating childhood diet and heart disease. Comment on the extent to which each of the following is likely to be a problem.
   a. Confounding variables and the implication of causation in observational studies.
   b. Extending results inappropriately.
   c. Interacting variables.
   d. Placebo, Hawthorne, and experimenter effects.
   e. Ecological validity and generalizability.
   f. Relying on memory or secondhand sources.

3.51 A categorical interacting variable defines subgroups for which the effect of the explanatory variable on the outcome variable differs. For instance, as explained in the text, for the nicotine patch experiment described in Case Study 3.3, one interacting variable was whether or not there were other smokers at home. When the difference in subgroups is thought to hold for the population, the subgroup variable is sometimes called an **effect modifier**. Explain why this term makes sense, using the context of the smokers at home in the nicotine experiment. For instance, what "effect" is being modified?

Chapter Exercises

Exercises 3.52 to 3.60 refer to a study published in the Journal of the American Medical Association *and reported in the* Sacramento Bee *on Jan. 26, 2000 ("Study ties hormone replacement, cancer" by Shari Roan, p. A1). A sample of 46,355 postmenopausal women were studied for 15 years and asked about their use of hormone therapy and whether or not they had breast cancer; 2082 developed the disease. The article reported that "for each year of combined [estrogen and progestin] therapy, a woman's risk of breast cancer was found to increase by 8 percent compared to a 1 percent increase in women taking estrogen alone.*

3.52 Do you think this research was an observational study or a randomized experiment? Explain.

3.53 Discuss whether this research is the following:
   a. A case-control study.
   b. A retrospective study or a prospective study.

3.54 What are the explanatory and response variables for this study? Are they quantitative or categorical? If they are categorical, what are the categories?

3.55 Refer to the questions listed under "Asking the Right Questions" in Section 2.2 of Chapter 2. Which of those questions is being answered in this research? Give the general question, and the specific question in the context of this study.

3.56 To whom are the women taking combined therapy and estrogen alone being compared when the increased risks of 8% and 1% are computed?

3.57 Assuming the women and their physicians made the decision about which treatment to pursue (combined hormones, estrogen alone, or no hormones), discuss the possibility of confounding variables in this study.

3.58 Near the end of the *Sacramento Bee* article, it was noted that "the NCI researchers analyzed their results by the woman's weight. . . . There was no increased breast cancer risk in heavier women." Which of the "Difficulties and Disasters" in Section 3.5 is represented by this statement? Explain.

3.59 One drawback of the study quoted in the *Sacramento Bee* was that "people who develop breast cancer are much more likely to remember whether they took hormones compared to women who don't develop the disease." Which of the "Difficulties and Disasters" in Section 3.5 is represented by this statement? Explain.

3.60 Another drawback of the study quoted in the *Sacramento Bee* was that "the women in the study may not resemble women today, because many women now take lower doses of progestin than was used during the follow-up years of the study (1979–95)." Which of the "Difficulties and Disasters" in Section 3.5 is represented by this statement? Explain.

Exercises 3.61 to 3.66 refer to a study described in a Sacramento Bee *article titled "Much ado over those bad hairdos" (Jan. 26, 2000, p. A21). Here is part of the article:*

> Researchers surveyed 60 men and 60 women from 17 to 30, most of them Yale students. They were separated into three groups. One group was questioned about times in their lives when they had bad hair. The second group was told to think about bad product packaging, like leaky containers, to get them in a negative mind-set. The third group was not asked to think about anything negative. The groups were tested for self-esteem and self-judgment, and the bad hair group scored lower than the others.

3.61 Was this an observational study or a randomized experiment? Explain.

3.62 What were the explanatory and response variables for this study? Were they categorical variables or quantitative variables?

3.63 Refer to the questions listed under "Asking the Right Questions" in Chapter 2, Section 2.2. Which of those questions is being answered in this research? Give the general question, and the specific question in the context of this study.

3.64 Discuss whether this study could have been blind or double-blind, and whether you think it was blind or double-blind.

3.65 Discuss each of the following "Difficulties and Disasters" in the context of this research.
  a. Ecological validity.
  b. Extending results inappropriately.
  c. Experimenter effect.

3.66 The article also reported that "contrary to popular belief men's self-esteem may take a greater licking than women's when their hair just won't behave. Men were likely to feel less smart and less capable when their hair stuck out, was badly cut or otherwise mussed." In the context of this quote, is male/female a confounding variable, an interacting variable, or neither?

3.67 For each of the following examples from the text, explain whether there would be more interest in descriptive statistics (for the acquired data set only) or in inferential statistics (extending results to a larger population).
  a. Ages at death of First Ladies.
  b. The relationship between nicotine patch use and cessation of smoking.
  c. Annual rainfall data for Davis, California.
  d. The relationship between nightlights and myopia.

3.68 Refer to the study reported in Example 2.2, relating the use of nightlights in childhood and the incidence of subsequent myopia.
  a. Was the research based on an observational study or a randomized experiment? Explain.
  b. Which one of the "Difficulties and Disasters" in Section 3.5 do you think is most likely to be a problem in this research? Explain.

3.69 Find an example of an observational study in the news. Answer the following questions about it. Be sure to include the news article with your response.
  a. What are the explanatory and response variables, or is this distinction not possible?
  b. Briefly describe how the study was done. For instance, was it a case-control study, a prospective or retrospective study?
  c. Discuss possible confounding variables. For instance, were any possible confounding variables included in the study and news report? Were other lurking confounding variables mentioned in the report? Can you think of possible confounding variables that were not mentioned?
  d. Pick one of the "Difficulties and Disasters" in Section 3.5 and discuss how it applies to the study.

3.70 Find an example of a randomized experiment in the news. Answer the following questions about it. Be sure to include the news article with your response.
  a. What are the explanatory and response variables? What relationship was found, if any?
  b. What treatments were assigned? Was a control group or placebo used?
  c. Was the study a matched-pair design, a block design, or neither? Explain.
  d. Was the study single-blind, double-blind, or neither? Explain.

3.71 Researchers want to design a study to compare topical cream (applied to the skin) and a drug taken in capsule form for the treatment of a certain type of skin rash. Sixty volunteers who have the rash agree to participate in the study.
  a. If each volunteer is given only one treatment can the study be double-blind? Explain.
  b. If each volunteer is given only one treatment can the study be single-blind? Explain.
  c. Could a double-blind, double-dummy study be done? If not, explain why not. If so, explain how it would be done.

3.72 Specify what an individual "unit" is in each of the following studies. Then specify what two variables were measured on each unit.
  a. A study found that tomato plants raised in full sunlight produced more tomatoes than tomato plants raised in partial shade.
  b. A study found that gas mileage was higher for automo-

biles when the tires were inflated to their maximum possible pressure than when the tires were underinflated.

c. Ten randomly selected classrooms in a school district were assigned to have a morning fruit snack break, and another ten classrooms of the same grade level in similar schools were measured as a control group. The number of children who performed better than average on standardized tests was measured for each classroom.

3.73 Refer to the previous exercise and answer these questions.

a. For each study, based on the information given, is it clearly a randomized experiment? If not, explain what additional information would make it clear that the study was a randomized experiment rather than an observational study.

b. For each study, discuss whether or not matched pairs were used. If it is not clear, explain what additional information would make it clear.

c. Discuss whether any of the studies would suffer from each of the following: the Hawthorne effect, interacting variables, ecological validity.

3.74 Is it possible for each of the following to be used in the same study (on the same units)? Explain or give an example of such a study.

a. A placebo and a double-blind procedure.

b. A matched-pair design and a retrospective study.

c. A case-control study and random assignment of treatments.

3.75 Refer to the nicotine patch study described in Case Study 3.3. Draw a figure similar to those in Figures 3.2 through 3.5, illustrating the steps for that study.

3.76 Refer to the study in Example 3.8, comparing ages at death for left-handed and right-handed people. Draw a figure similar to those in Figures 3.2 through 3.5, illustrating the steps for that study.

3.77 Refer to Case Study 1.6 in which physicians were randomly assigned to take aspirin or a placebo and heart attack rates were compared. Draw a figure similar to Figures 3.2 through 3.5, illustrating the steps for that study.

3.78 A study was done (fictional) to compare the proportion of children who developed myopia after sleeping with and without a nightlight. The study found that the results differed based on whether at least one parent suffered from myopia by age 20. The percents of children suffering from myopia were as follows:

	Parent(s) Had Myopia	Parent(s) Did Not Have Myopia
Slept with:	% with Myopia	% with Myopia
No Light	10%	10%
Some Light	50%	30%

a. Identify the explanatory variable, the response variable, and an interacting variable.

b. Draw a figure similar to Figure 3.6, illustrating the interaction in this study.

c. Write a few sentences that would be understood by someone with no training in statistics explaining the concept of interaction in this study.

3.79 A fast-food chain sells its burgers alone or as part of a "value meal," which includes fries and a drink. They know some customers are health-conscious. They want to do an experiment to see if the proportion of customers choosing the meal would increase if they offered baby carrots with the meal as an alternative to fries. They will compare three treatments: (1) status quo, (2) offering carrots as an alternative but with no advertising, and (3) offering carrots as an alternative and advertising this option in the local area. They have restaurants in three types of areas—cities, suburban areas, and along major highways.

a. Explain how they would do this experiment using a completely randomized design.

b. Explain how they would do this experiment using a randomized block design.

c. Explain why it makes sense to use the type of area in which a restaurant is located as a block determinant in this experiment.

d. Explain why a matched-pair design could not be used for this experiment.

3.80 Refer to Example 3.6, "Dull Rats," which was done using a completely randomized design.

a. What were the treatments in this experiment?

b. Were the experimental units the 60 individual rats or the 12 individual experimenters? Explain.

c. Explain how the experiment could have been done using a matched-pair design instead.

REFERENCES

Coren, S., and D. Halpern (1991). "Left-handedness: A marker for decreased survival fitness," *Psychological Bulletin,* Vol. 109, No. 1, pp. 90–106.

Faigenbaum, A. D., W. L. Westcott, R. LaRosa Loud, and C. Long (1999). "The effects of different resistance training protocols on muscular strength and endurance development in children," *Pediatrics Electronic Article,* Vol. 104, No. 1 (July), p. e5.

French, J. R. P (1953). "Experiments in field settings," in L. Festinger and D. Katz (eds.), *Research Method in the Behavioral Sciences,* New York: Holt, pp. 98–135.

Hurt, R., L. Dale, P. Fredrickson, C. Caldwell, G. Lee, K. Offord, G. Lauger, Z. Maruisic, L. Neese, and T. Lundberg (1994).

"Nicotine patch therapy for smoking cessation combined with physician advice and nurse follow-up," *Journal of the American Medical Association,* Vol. 271, No. 8, (Feb. 23), pp. 595–600.

Lesko, S. M., L. Rosenberg, and S. Shapiro (1993). "A case-control study of baldness in relation to myocardial infarction in men," *Journal of the American Medical Association,* Vol. 269, No. 8, (Feb. 24), pp. 998–1003.

Rosenthal, R., and K. L. Fode (1963). "The effect of experimenter bias on the performance of the albino rat," *Behavioral Science,* Vol. 8, pp. 183–189.

Sampling: Surveys and How to Ask Questions

Mark Scott./GettyImages

Do they have your number? See Example 4.7.

Here is a fact that may astound you. If you use proper methods to sample 1500 people from a population of millions and millions, you can almost certainly gauge the percent of the entire population who have a certain trait or opinion to within 3%. The tricky part is that you have to use a proper sampling method.

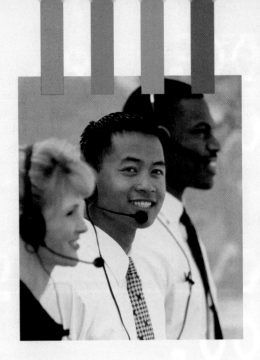

People are curious about each other. When a newsworthy event occurs, people want to know what others think. In the midst of election campaigning, people want to know which candidates are in the lead. We devour news stories about the behavior and opinions of others. For this reason, news organizations like *USA Today* and CNN often conduct sample surveys. Many recent surveys can be found on the Web. In this chapter we learn how to conduct surveys, how to make sure they are representative, and what can go wrong. ◆

4.1 | THE BEAUTY OF SAMPLING

In a **sample survey,** a subgroup of a large population is questioned on a set of topics. The investigator simply asks the participants to answer some questions. There is no manipulation of a respondent's behavior in this type of research. In other words, sample surveys are a special type of observational study.

The results from a sample survey are used as if they are representative of the larger population, which they will be if the sample is chosen correctly and if those selected cooperate in responding. National polling organizations like the Gallup organization often question only slightly over 1000 individuals. How can a survey of such a small number of individuals possibly obtain accurate information about the opinions of millions?

Here is some information that may astound you. If you use commonly accepted methods to sample 1500 people from an entire population of millions and millions, you can almost certainly gauge the percent of the entire population who have a certain trait or opinion to within 3%. Even more amazing is the fact that this result doesn't depend on how big the population is. It only depends on how many are in the sample! A sample of 1500 would do about equally well for a population of 4 billion as it would for a population of 10 million. Of course, the tricky part is that you have to use a proper sampling method, a topic that we examine in this chapter.

You can see why researchers are content to rely on public opinion polls rather than trying to ask everyone for their opinions. It is much cheaper to ask 1500 people than several million, especially when you can get an answer that is almost as accurate. It also takes less time to conduct a sample survey than it does to conduct a **census,** in which everyone in the population is measured. And because fewer interviewers are needed for a survey than for a census, there may be better quality control of interviewing procedures when only a small sample of individuals is measured.

The Margin of Error: The Accuracy of Sample Surveys

Sample surveys often are used to estimate the *proportion* or *percent* of people who have a certain trait or opinion. For example, the Nielsen ratings, used to determine the percent of American television sets tuned to a particular show, are based on a sample of a few thousand households. Newspapers and magazines routinely survey only one or two thousand people to determine public opinion on current topics of interest. As we have said, if properly conducted, these surveys are amazingly accurate.

When a survey is used to find a proportion based on a sample of only a few thousand individuals, the obvious question is how close that proportion comes to the truth for the entire population. This measure of accuracy in sample surveys is a number called the **margin of error.** The *sample proportion* and the *population proportion* with a certain opinion or trait differ by *less than* the margin of error *more than* 95% of the time, or in at least 19 in 20 surveys. In other words, the margin of error provides an upper limit on the amount by which the sample proportion differs from the true population proportion, and this upper limit holds for at least 95% of all samples. To express results in terms of percents instead of proportions, simply multiply everything by 100.

definition The amount by which the sample proportion differs from the true population proportion is less than the quantity $1/\sqrt{n}$ in at least 95% of all random samples. The letter n represents the number of people in the sample. Survey results reported in the media are usually expressed as percents and are accompanied by a margin of error, nearly always calculated as

$$\frac{1}{\sqrt{n}} \times 100\%$$

We call this a **conservative margin of error.**

In Chapter 10 we will learn a more complicated and precise formula for the margin of error that depends on the actual sample proportion. It will never give an answer larger than the formula given here, so we call this formula a *conservative* margin of error. For simplicity, many media reports use the conservative formula. For example, with a sample of 1600 people, we will usually get an estimate that is accurate to within $1/\sqrt{1600} = 1/40 = 0.025 = 2.5\%$ of the truth. You might see results such as, "Fifty-five percent of respondents support the President's economic plan. The margin of error for this survey is plus or minus 2.5 percentage points." This means that it is almost certain that between 52.5% and 57.5% of the entire population supports the plan. In other words, add and subtract the margin of error to the sample value (55% in this example), to create an interval. If you were to follow this method every time you read the results of a properly conducted survey, the interval would miss covering the truth only about 1 in 20 times (5%) and would cover the truth the remaining 95% of the time.

in summary Margin of Error and Confidence Interval

(Conservative) margin of error = $1/\sqrt{n}$ (or $100\% \times 1/\sqrt{n}$ for a percent), where n = number of individuals in the sample. For about 95% of all sample proportions found in properly conducted sample surveys,

$$|\text{population proportion} - \text{sample proportion}| \leq \frac{1}{\sqrt{n}}$$

Simply stated, for about 95% of properly conducted sample surveys, the interval

$$\textit{sample proportion} - \frac{1}{\sqrt{n}} \text{ to } \textit{sample proportion} + \frac{1}{\sqrt{n}}$$

will contain the actual population proportion. This interval is called an approximate "95% **confidence interval**" for the population proportion. In Chapter 10, a more precise formula will be given.

Another way of writing the approximate 95% confidence interval is

$$\textit{sample proportion} \pm \frac{1}{\sqrt{n}}$$

To report the results in percents instead of proportions, multiply everything by 100%.

Here are three examples of polls, with the margin of error and an approximate 95% confidence interval for each one. In each case the poll is based on a nationwide random sample of Americans aged 18 and older.

EXAMPLE 4.1
The Importance of Religion for Adult Americans

For a CNN/*USA Today*/Gallup Poll conducted on September 2 to 4, 2002, a random sample of $n = 1003$ adult Americans was asked, "How important would you say religion is in your own life: very important, fairly important, or not very important?" (http://www.pollingreport.com/religion.htm, September 17, 2002). The percents that selected each response were

Very important	65%
Fairly important	23%
Not very important	12%
No opinion	0%

Conservative margin of error: $\dfrac{1}{\sqrt{1003}} = .03$ or $.03 \times 100\% = 3\%$

Approximate 95% confidence interval for the proportion and percent of *all* adult Americans who would say religion is very important:

Proportion: $.65 \pm .03$ or $[.65 - .03$ to $.65 + .03]$ or $.62$ to $.68$
Percent: $65\% \pm 3\%$ or $[65\% - 3\%$ to $65\% + 3\%]$, or 62% to 68% ◆

EXAMPLE 4.2
Would You Eat Those Modified Tomatoes?

An ABC News.com poll conducted by TNS Intersearch on June 13 to 17, 2001 posed the following question to a random sample of $n = 1024$ adult Americans: "Scientists can change the genes in some food crops and farm animals to make them grow faster or bigger and be more resistant to bugs, weeds and disease. Do you think this genetically modified food, also known as bio-engineered food, is or is not safe to eat?" (http://www.pollingreport.com/science.htm, September 17, 2002). The percents that selected each response were

Is safe	35%
Is not safe	52%
No opinion	13%

Conservative margin of error: $\dfrac{1}{\sqrt{1024}} = .03$ or $.03 \times 100\% = 3\%$

Approximate 95% confidence interval for the proportion and percent of *all* adult Americans who would say genetically modified food is *not* safe to eat:

Proportion: .52 ± .03 or [.52 − .03 to .52 + .03] or .49 to .55
Percent: 52% ± 3% or [52% − 3% to 52% + 3%], or 49% to 55% ◆

EXAMPLE 4.3
Cloning Human Beings

A Pew Research Center for the People & the Press and Pew Forum on Religion & Public Life survey conducted by Princeton Survey Research Associates between February 25 and March 10, 2002 asked a random sample of $n = 2002$ adult Americans, "Do you favor or oppose scientific experimentation on the cloning of human beings?" (http://www.pollingreport.com/science.htm, September 17, 2002). The percents that selected each response were

Favor	17%
Oppose	77%
Don't know/Refused	6%

Conservative margin of error: $\dfrac{1}{\sqrt{2002}} = .022$ or $.022 \times 100\% = 2.2\%$

Approximate 95% confidence interval for the proportion and percent of *all* adult Americans who would say they *favor* this experimentation:

Proportion: .17 ± .022 or [.17 − .022 to .17 + .022] or .148 to .192
Percent: 17% ± 2.2% or [17% − 2.2% to 17% + 2.2%], or 14.8% to 19.2% ◆

Interpreting the Confidence Intervals in Examples 4.1 to 4.3

In each of the three examples of surveys, an approximate 95% confidence interval was found. There is no way to know whether all, some, or none of these intervals actually covers the population value of interest. For instance, the interval from 62% to 68% may or may not capture the percent of adult Americans who considered religion to be very important in their lives in September of 2002. But, in the long run, this procedure will produce intervals that capture the unknown population values about 95% of the time, as long as it is used with properly conducted surveys. This long-run performance is usually expressed after an interval is computed by saying that we are 95% confident that the population value is covered by the interval.

> ***tech note***   Because the margin of error used in this chapter is "conserva-
> tive," the procedure actually works slightly more than 95% of the
> time. Details will be covered in Chapters 10 and 12.

Advantages of a Sample Survey over a Census

When a Census Isn't Possible

Suppose you need a laboratory test to see whether or not your blood has too high a concentration of a certain substance. Would you prefer that the lab measure the entire population of your blood, or would you prefer to give them a sample? Similarly, suppose a manufacturer of firecrackers wants to know what percent of its products are duds. It would not make much of a profit if it tested them all, but it could get a reasonable estimate of the desired percentage by testing a properly selected sample. As these examples illustrate, there are situations where measurements destroy the units being tested and thus a census is not feasible.

Speed

Another advantage of a sample survey over a census is speed. For instance, it takes several years to successfully plan and execute a census of the entire population of the United States, something done every ten years. Getting figures like monthly unemployment rates would be impossible with a census; they would be quite out-of-date by the time they were released. It is much faster to collect a sample than a census if the population is large.

Accuracy

A final advantage to a sample survey is that you can devote your resources to getting the most accurate information possible from the sample you have selected. It is easier to train a small group of interviewers than a large one, and it is easier to track down a small group of nonrespondents than the larger one that would inevitably result from trying to conduct a census.

Bias: How Surveys Can Go Wrong

Not all surveys produce trustworthy results. The purpose of most surveys is to use a characteristic of the sample (such as the proportion with a certain opinion) as an estimate of the corresponding characteristic of the population from which the sample was selected. While it is unlikely that the sample value will equal the population value precisely, the goal of a good survey is to get an **unbiased** sample value. Results based on a survey are **biased** if the method used to obtain those results would consistently produce values that are either too high or too low.

There are three common types of bias that might occur in surveys:

♦ **Selection bias** occurs if the method for selecting the participants produces a sample that does not represent the population of interest. For instance, if shoppers at a mall are surveyed to determine attitudes about raising the sales tax, the results are not likely to represent all area residents.

♦ **Nonresponse bias** occurs when a representative sample is chosen for a survey, but a subset cannot be contacted or does not respond. For instance, a survey conducted by phoning people at home in the evening would omit people who work during that time. Their opinions would not be represented, so if they were much different, the sample results would be biased.

◆ **Response bias** occurs when participants respond differently from how they truly feel. The way questions are worded, the way the interviewer behaves, as well as many other factors might lead an individual to provide false information. For instance, surveys about socially unacceptable behavior such as heavy smoking or drinking, abusive behavior, and so on must be worded and conducted carefully to minimize the possibility of response bias.

4.1

The terms *nonresponse bias* and *response bias* sound very similar. Do they refer to two aspects of the same kind of problem? Explain.*

turn on your mind

4.2 | SIMPLE RANDOM SAMPLING AND RANDOMIZATION

The ability of a relatively small sample to accurately reflect a huge population does not happen haphazardly. It only happens if proper sampling methods are used. The basic idea is that everyone in the population must have a specified chance of making it into the sample, and methods with this characteristic are called **probability sampling plans.** Let's look at how to implement the most basic method of accomplishing this goal, which is to use a **simple random sample.** Remember (from Chapter 3) that with a simple random sample, every conceivable group of units of the required size has the same chance of being the selected sample.

Choosing a Simple Random Sample

Choosing a simple random sample is somewhat like choosing the winning numbers in many state lotteries. For instance, in the former California Super Lotto, six numbers were randomly selected from the choices 1, 2, . . . , 51. Every possible set of six numbers was equally likely to be the winning set. There are actually 18,009,460 different possible sets, which is why the odds of any specific individual guessing the winning set are so small! Similarly, the chances of any *particular* group of units getting selected to be the random sample from a large population is quite small, but whatever group is selected is likely to be *representative.* For instance, in a simple random sample of 1000 people in your state or country, it is extremely unlikely that you and your next-door neighbor would be selected. But it is extremely likely that someone in the sample will be representative of each of you, having similar opinions to yours.

To actually produce a simple random sample, you need only two things. First, you need a list of the units in the population. For instance, in drawing the winning lottery numbers, the list of units is the numbers 1, 2, ..., 51. In selecting a simple random sample of students from your school, the list of units is all students in the school (which the registrar can usually produce).

Second, you need a source of **random numbers.** Random numbers can be found in tables called **tables of random digits,** or they can also be generated using either a computer or the right calculator. If the population isn't too large, physical methods can be used, such as in the lottery, where the numbers were written on small hollow plastic balls and six of them physically selected.

Table 4.1 illustrates a portion of a table of random digits. It is organized into numbered rows to make it easier to find specific sections of the Table. There are only 10 rows in Table 4.1, so they are labeled using the single digits 0 to 9. A larger table would have to use longer identifying numbers for the rows. The digits are grouped into columns of 5 for easier reading. These tables are generated by the equivalent of writing the digits from 0 to 9 on slips of paper, mixing them well, choosing one, and then repeating this process over and over again. No single digit, pair of consecutive digits, triplet of digits, and so on are any more or less likely to occur than any others.

Using Table 4.1 to Choose a Simple Random Sample
Here are the steps for selecting a simple random sample using Table 4.1.

1. Number the units in the population using the same number of digits for each one.
   Example: Suppose there are 270 students in a class and the teacher wants to choose a simple random sample of 10 of them to call on in class. Number the students from 001 to 270.
2. Choose a starting point in Table 4.1. You can close your eyes and point, or use any other method as long as you haven't studied the table and then chosen numbers that give a favorable sample for your purpose.
   Example: Start in row 3, column 2 (10484 . . .)
3. From the starting point, read across the row to get numbers with the correct number of digits to identify a unit. Continue with consecutive rows. For instance, if the units are numbered 001 to 270, read three-digit numbers.
   Example: Reading three-digit numbers starting with row 3, column 2 results in 104, 842, 461, 613, 466, 416, 180, 855, 118, 314, 577, 002, 896, and so on.
4. If a number is in the range of the unit numbers, select that unit number. Otherwise, continue along the row, choosing more potential unit numbers

TABLE 4.1 ■ A Table of Random Digits

Row							
0	00157	37071	79553	31062	42411	79371	25506	69135
1	38354	03533	95514	03091	75324	40182	17302	64224
2	59785	46030	63753	53067	79710	52555	72307	10223
3	27475	10484	24616	13466	41618	08551	18314	57700
4	28966	35427	09495	11567	56534	60365	02736	32700
5	98879	34072	04189	31672	33357	53191	09807	85796
6	50735	87442	16057	02883	22656	44133	90599	91793
7	16332	40139	64701	46355	62340	22011	47257	74877
8	83845	41159	67120	56273	67519	93389	83590	12944
9	12522	20743	28607	63013	60346	71005	90348	86615

until you have a sample of the desired size. (But see step 5 for a more efficient method.) Because each unit can only be used once, if a unit number occurs that has already been selected, simply ignore that number and continue.

Example: Unit numbers can only be 001 to 270, so most of the numbers chosen are simply ignored. Select units 104, 180, 118, 002, then keep going until the required 10 students have been selected.

5. Step 4 is very inefficient. To make it more efficient, reassign some of the higher numbers onto the range of unit numbers in a way that still ensures that each unit number has an equal chance of selection. For instance, suppose the units are numbered 001 to 270. If a number between 301 and 570 is chosen, subtract 300 and use it. If a number between 601 and 870 is chosen, subtract 600 and use it. As in step 4, if a unit number occurs more than once, simply ignore subsequent occurrences after the first one. For instance, in the scenario just given, if unit 017 has already been selected, then any subsequent occurrence of 017, 317, or 617 would be discarded.

Example: Now the string generated earlier is more useful. Let's see what decision follows each three-digit entry:

Three-digit Number	Using Step 4, Choose Unit	Using Step 5, Choose Unit
104	104	104
842	Discard	242 (subtracted 600)
461	Discard	161 (subtracted 300)
613	Discard	013 (subtracted 600)
466	Discard	166 (subtracted 300)
416	Discard	116 (subtracted 300)
180	180	180
855	Discard	255 (subtracted 600)
118	118	118
314	Discard	014 (subtracted 300)
577	Discard	Not needed, 10 already selected
002	002	Not needed

Using the method in step 4, the first 4 students selected would be those numbered 104, 180, 118, and 002, and the process would continue until six more were selected. Using the method in step 5, the sample of 10 units would include 104, 242, 161, 013, 166, 116, 180, 255, 118, and 014, and unit 002 would not be needed.

Minitab Tip

tech note   **Picking a Random Sample**

♦ To create a column of ID numbers, use **Calc>Make Patterned Data>Simple Set of Numbers.** In the dialog box, specify a column for storing the ID numbers, and specify the first and last possible ID number for the population.

♦ To sample values from a column, use **Calc>Random Data> Sample from Columns.** In the dialog box, specify how many items (rows) will be selected from a particular column, and specify a column where the sample will be stored.

Note: Items can be randomly selected from a column of names or data values, so it may not be necessary to assign ID numbers to the units in the population in order to select a sample.

EXAMPLE 4.4
*Representing
the Heights of
British Women*

In Example 2.11 of Chapter 2, we examined data on the heights of 199 British women. Suppose you had a list of these 199 women and wanted to choose 10 of them to test drive a sporty, but small, automobile model and give their opinions about its comfort. The heights of the women in the sample should be representative of the range of heights in the larger group. You would not want your sample of 10 to include only short women or only tall women. Here's how you could choose a simple random sample:

1. Assign an ID number from 001 to 199 to each woman.
2. Use a table of random digits, a computer, or a calculator to randomly select ten numbers between 001 and 199, and sample the heights of the women with those ID numbers.

We used the statistical package Minitab to choose a simple random sample of heights, then used Table 4.1 to choose another one. The samples are listed next along with their sample means and the list of ten random numbers between 001 and 199 that generated them (with leading zeros dropped). In Chapter 2 the heights were given in millimeters, but here they have been converted to inches.

Sample 1 (Minitab)

> *ID numbers of the women selected:* 176, 10, 1, 40, 85, 162, 46, 69, 77, 154
> *Heights:* 60.6, 63.4, 62.6, 65.7, 69.3, 68.7, 61.8, 64.6, 60.8, 59.9; mean = 63.7 inches

Sample 2 (Table 4.1)

> *ID numbers of the women selected:* 41, 93, 167, 33, 157, 131, 110, 180, 185, 196
> *Heights:* 59.4, 66.5, 63.8, 62.6, 65.0, 60.2, 67.3, 59.8, 67.7, 61.8; mean = 63.4 inches

Sample 2 was selected by starting with the third set of five digits in the row labeled 5 (04189). Numbers from 001 to 199 can be used directly; for numbers between 201 and 399, subtract 200; for numbers between 401 and 599, subtract 400; and so on. The only numbers that would need to be discarded using this method are 000, 200, 400, 600, and 800. You can test your understanding of the use of Table 4.1 by starting at 04189 (third set in row 5), selecting 10 consecutive sets of three digits, and determining if you correctly identify the 10 women in sample 2.

As you can see, each sample is different but each should be representative of the whole collection of 199 women. Within each sample, the range of heights is from about 60 inches to about 69 inches. The sample means are both close to the mean height for the larger group, which was given in Chapter 2 as 1602 mm, or about 63 inches. ◆

Using a Table of Random Digits in a Randomized Experiment

In Chapter 3 we learned that randomization plays a key role in designing experiments to compare treatments. Remember that for a *completely randomized design*, experimental units are randomly assigned to the treatment conditions. For a *matched-pair design* or a *randomized block design*, the order in which treatments are assigned within a pair or block is randomized. In all of these designs, the randomization can be viewed as a special case of simple random sampling. As with choosing a simple random sample, a table of random digits, a computer program, or a physical method of randomization can be used. In the next example, Table 4.1 is used to randomize the treatments in a completely randomized design.

EXAMPLE 4.5
Assigning Children to Lift Weights

Case Study 3.2 involves a randomized experiment in which 43 children were assigned to one of three treatment groups. Children in group 1 were asked to perform weight-lifting repetitions with a heavy load, group 2 performed more repetitions but with a moderate load, and group 3 served as a control group and did not lift weights. According to the study report (Faigenbaum et al., 1999, p. e5), there were 15 children assigned to group 1, 16 to group 2, and the remaining 12 to group 3. Assuming no blocks or pairs were used, how would you use Table 4.1 to assign the children to the treatment groups?

First, choose the 15 children to participate in group 1. Do this by labeling the 43 children from 01 to 43, then choosing a simple random sample of 15. This can be done using Table 4.1 by choosing two-digit numbers. If the table entry is 01 to 43, choose the child with that number. For 51 to 93, subtract 50 then choose the child with that number. Discard numbers 00, 44 to 50, and 94 to 99. Let's start at the beginning of the row labeled 7. The resulting sample of 15 children for group 1 is

> 16, 33, 24, 01, 39, 14 (64 − 50), 20 (70 − 50), (14 has already been used so skip), 13 (63 − 50), 05 (55 − 50), 12 (62 − 50), 34, 02, (20 has already been used, so skip), 11, (discard 47), 25, 27 (77 − 50)

In numerical order, the children assigned to group 1 are those numbered 1, 2, 5, 11, 12, 13, 14, 16, 20, 24, 25, 27, 33, 34, and 39.

There are now 43 − 15 = 28 remaining children, of whom 16 are to be assigned to group 2 and 12 to the control group. Because 12 is the smaller number, it will be faster to choose a random sample of 12 from the 28 remaining children to assign to the control group. The 16 remaining children will constitute group 2.

There are other methods of making the random assignments that would also work. For instance, you could write the names of the children on identical slips of paper, put them in a bag, and mix well. Reach in and select 15 of them for group 1, another 16 for group 2, and the remaining 12 would constitute the control group. ◆

EXAMPLE 4.6
Redesigning Case Study 3.2

In the original research report about Case Study 3.2 (Faigenbaum et al., 1999), it was noted that the 43 children were not all available at the start of the study. The 31 children in the two weight-lifting groups were available and were randomly assigned to the two groups in a method similar to that described in Example 4.5. According to the article, "twelve children who enrolled in this study after the recruitment of the experimental groups acted as nontraining control subjects." In other words, the control group enrolled after the children in the two treatment groups.

The late enrollment may help explain another characteristic described in the research report, which may confound the results. The average weights of the children in the three groups were 35.4 kg for group 1, 39.8 kg for group 2, and only 27.7 kg for the control group. Because change in muscle strength was the response variable of interest, it would be better if the three groups were matched more closely on initial average weight.

One way to make sure that happened would be to use a randomized block design instead of a completely randomized design. Recall from Chapter 3 that a block is a collection of similar experimental units. In this situation, measure the initial weight of each child and create one block consisting of the heaviest three children, one consisting of the next three heaviest, and so on. Within each block, the children would be of similar weight. Number the children in each block 1, 2, and 3. Randomly assign the three children in each block to the three treatment groups. The assignment can be done in a variety of ways using Table 4.1. The important feature is that each child in the block has an equal chance of receiving each treatment. For instance, choose a starting point and get three unique digits. Label the three children in a block with those digits. The child with the lowest digit gets treatment 1, the one with the middle digit gets treatment 2, and the one with the highest digit goes in the control group. For instance, if the three digits are 843, then treatment assignments for children 1, 2, and 3 are control group, treatment 2, and treatment 1, respectively. ◆

Random Sampling in Action

turn on your computer

It should be obvious to you by now that each time you take a random sample from a population you are likely to get a different group of individuals, even though the population remains fixed. The **Sampling** applet on the CD for this text allows us to watch what happens when we repeatedly take random samples from the same population. Figure 4.1 illustrates how the applet appears when it starts. Each stick figure represents one member of a population of 100 college students: 55 females (blue) and 45 males (red). Their heights have been recorded as well, and if you look carefully you will see that the little stick figures are all of slightly different heights. The mean height for the population is 68 inches.

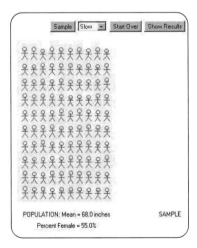

FIGURE 4.1 **The sampling applet starting point**

If we repeatedly take simple random samples of 10 people from this population, how much do you think the proportion of females in the samples will vary? Will it almost always be 50% or 60%, given that it must be a multiple of 10 and the population is 55% female? Will the sample ever consist of all males or all females? Pick one of the stick figures to represent yourself. How often do you think you would end up in the sample? Will the mean height of the sample always be close to the mean height of all 100 students, at 68 inches? If the mean height were to be 70 inches, would that indicate a possible problem in the selection of the sample, or is such a large value to be expected in some samples? In Chapters 7 to 9 you will learn the mathematical answers to these questions, but playing with the applet can give you a good intuitive idea of what happens.

What Happens

The applet always chooses a simple random sample of 10 individuals when you press the button labeled *Sample.* You can control the speed of the sampling to be slow, fast, or batch. With slow and fast sampling, you get to watch each individual move from the population into the sample, which is illustrated to the right of the population. With batch sampling, the whole sample appears as soon as you press the *Sample* button. Figures 4.2 and 4.3 illustrate two examples of the results after pressing the *Sample* button. Notice that in Figure 4.2 only 3 females were selected, so 30% of the sample is female. Not surprisingly, given that the sample has a higher percentage of men than the population, the mean height for the sample (69.6 inches) is higher than the 68-inch mean height for the population. The reverse happens in Figure 4.3, in which females make up 60% of the sample and the mean height is 67.2 inches.

FIGURE 4.2   **Sampling applet results**

FIGURE 4.3   **Additional sampling applet results**

What to Do

Open the *Sampling* applet on the CD, and then repeatedly press the *Sample* button to choose many samples. You can decide if you want to select a sample slowly, so you can carefully watch each person being taken from the population and placed in the sample; fast (you can still watch); or in a batch. This decision won't change the results. Do *not* press the *Start Over* button.

   Notice how the percent of females changes for each sample and how the mean height changes. Pick one of the stick figures to represent yourself, and notice how often you make it into the sample. At any time, you can see a summary of the samples you have selected so far by pushing the button labeled ***Show Results.*** When you press the *Start Over* button, all past results are cleared.

   Take dozens of samples, then press ***Show Results.*** Are the percent female values and the mean height values for the samples more, or less, spread out than you would have anticipated? The number of females must be 0 or 1 or 2, etc., up to 10, so the "percent female" can only be one of 11 possibilities: 0%, 10%, 20%, 30%, etc., up to 100%. Count how often each of the 11 possibilities occurred in your repeated sampling. Does the result surprise you or is it what you expected? Draw a histogram of the means from the samples you chose. What do you notice? Do mean values far away from the "truth" of 68 inches occur as often as mean values close to 68 inches? Would you suspect a problem with the sample if a sample had a mean of 70 inches? How about 72 inches?

Note for Minitab Users: The results in the "Show Results" window can be copied and then pasted into a Minitab worksheet. Minitab can then be used to analyze the results. For instance, **Graph**>**Histogram** could be used to draw a histogram of the means of the chosen samples.

Lessons Learned

This applet lets you see simple random sampling in action. Much of the remaining material in this book follows from the idea that although samples vary, the amount by which they vary over the long run can be quantified. Remember that a simple random sample is supposed to be representative of the population. However, it should be obvious to you that with only 10 individuals per sample, both the percent of females and the mean height vary considerably from sample to sample. In fact, the conservative margin of error for a sample of $n = 10$ is over 30%! Therefore, you should expect to see a wide range in the percent of females across samples.

4.3 | OTHER SAMPLING METHODS

For large populations it may not be practical to take a simple random sample because it may be difficult to get a numbered list of the units. For instance, if a polling organization wanted to take a simple random sample of all voters or all adults in a country or region, they would need to get a numbered list of them, which is simply an impossible task in most cases. Instead, they rely on more complicated sampling methods, all of which are good substitutes for simple random sampling in most situations. In fact, they often have advantages over simple random sampling. For instance, one of these other methods, stratified random sampling, can be used to increase the chance that the sample represents important subgroups within the population.

To visually illustrate the various sampling plans discussed in this chapter, let's suppose a college administration would like to survey a sample of students living in dormitories. The college has undergraduate and graduate dormitories. The undergraduate dormitories have three floors each with 12 rooms per floor. The graduate dormitories have five floors each with 8 rooms per floor.

Figure 4.4 illustrates a simple random sample of 30 rooms in the dormitories. Any collection of 30 rooms has an equal chance of being the selected sample. Notice that for the sample illustrated there are 12 undergraduate rooms and 18 graduate rooms in the sample.

Stratified Random Sampling

Sometimes the population of units falls into natural groups, called **strata.** For example, public opinion pollsters often take separate samples from each region of the country so that they can spot regional differences as well as measure

FIGURE 4.4   **A simple random sample of 30 dorm rooms**

national trends. Political pollsters may sample separately from each political party in order to compare opinions by party.

A **stratified random sample** is collected by first dividing the population of units into groups (strata) and then taking a simple random sample from each one. For example, the strata might be regions of the country or political parties. You can often recognize this type of sampling when you read the results of a survey because the results will be listed separately for each of the strata.

Figure 4.5 illustrates a stratified sample for the college survey. There are two strata, the undergraduate and graduate dorms. A random sample of 15 rooms is taken from each of the two strata. Each collection of 15 undergraduate rooms has an equal chance of being the selected sample for the undergraduate dorms, and each collection of 15 graduate rooms has an equal chance of being the selected sample for the graduate dorms. But, the total of 15 rooms within each strata is fixed before the sample is selected.

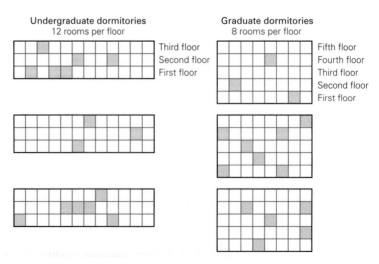

FIGURE 4.5   **A stratified sample of 15 undergraduate and 15 graduate dorm rooms**

Stratified sampling has other advantages in addition to the fact that results are available separately by strata. One is that different interviewers may work best with different people. For example, people from separate regions of the country (South, Northeast, etc.) may feel more comfortable with interviewers from the same region. It may also be more convenient to stratify before sampling. If we were interested in opinions of college students across the country, it would probably be easier to train interviewers at each college rather than to send the same interviewer to all campuses.

So far we have been focusing on measuring categorical variables, such as opinions or traits people might have. Surveys are also used to measure quantitative variables, such as age at first intercourse or number of cigarettes smoked per day. We are often interested in the population average for such measurements.

The accuracy with which we can estimate the average depends on the natural variability among the measurements. The less variable they are, the more precisely we can assess the population average on the basis of the sample values. For instance, if everyone in a relatively large sample reports that the age at first intercourse was between 16 years 3 months and 16 years 4 months, then we can be relatively sure that the average age in the population is close to that. On the other hand, if the reported ages range from 13 years to 25 years, then we cannot pinpoint the average age for the population nearly as accurately.

Stratified sampling can help to solve this problem. Suppose we could figure out how to stratify in a way that there is little natural variability in the answers within each strata. We could then get an accurate estimate for each stratum and combine them to get a much more precise answer for the group than if we measured everyone together. For example, if we wanted to estimate the average weight loss for participants in a new exercise program, we could do so more accurately by dividing the participants into groups based on their initial weight. Heavy people are more likely to lose weight during the first few days, but light individuals might even gain weight as they build muscles.

Cluster Sampling

Cluster sampling is often confused with stratified sampling, but actually it is a radically different concept and can be much easier to accomplish. The population units are again divided into groups, called **clusters,** but rather than sampling within each group, *we select a random sample of clusters and measure only those clusters.* In most applications of stratified sampling, the population is divided into a few *large* strata, such as regions of the country, and a small subset within each one is randomly sampled. In most applications of cluster sampling, the population is divided into *small* clusters, such as city blocks, a large number of clusters are randomly sampled, and everyone in those clusters is measured.

Figure 4.6 illustrates a cluster sampling plan for the college survey. Each floor of each dormitory is a cluster. A random sample of five floors is selected, and all students on those floors are surveyed. Any collection of five floors has an equal chance of being the selected sample, but once those floors are selected, all rooms on them are surveyed.

One obvious advantage of cluster sampling is that you need only a list of clusters instead of a list of all individual units. City blocks are commonly used as clusters in surveys that require door-to-door interviews. To measure customer satisfaction, airlines sometimes randomly sample a set of flights, then distribute a survey to everyone on those flights. Each flight is a cluster. It is clearly much

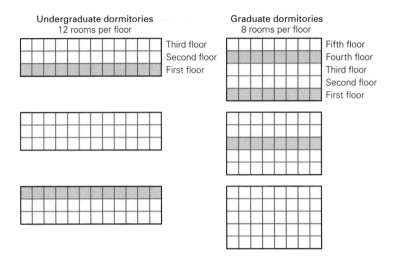

FIGURE 4.6 **A cluster sample in which five floors (clusters) are randomly selected**

easier for the airline to choose a random sample of flights than it would be to identify and locate a random sample of individual passengers to whom to distribute surveys.

If cluster sampling is used, the analysis must proceed differently because there may be similarities among the members of the clusters that must be taken into account. Numerous books are available that describe proper analysis methods based on which sampling plan was employed. (See, for example, *Sampling: Design and Analysis* by Sharon Lohr, Duxbury Press, 1999.)

Systematic Sampling

Suppose you have a list of 5000 names and telephone numbers from which you want to select a sample of 100 individuals. That means you would want to select 1 out of every 50 people on the list. You would not want to simply choose the first 100 names because, depending on how the list was ordered, they might not constitute a representative sample.

An idea that would work in most cases is to choose every 50th name on the list. If you did so, you would be using a **systematic sampling plan.** With this plan, you divide the list into as many consecutive segments as you need, randomly choose a starting point in the first segment, then sample at that same point in each segment. In our example, you would randomly choose a starting point in the first 50 names, then sample every 50th name after that. When you were finished, you would have selected one person from each of 100 segments, equally spaced throughout the list. As another example, suppose a company wanted to assess the quality of items produced on an assembly line. They might use a systematic sampling plan by selecting every 200th unit that was produced and testing its quality. This would be a much simpler plan than trying to make a list of all units produced and selecting a simple random sample of them.

Systematic sampling is often a good alternative to random sampling. In a few instances it can lead to a biased sample, and common sense must be used to avoid those. As an example, suppose you were doing a survey of potential noise problems in a high-rise college dormitory. Further, suppose a list of residents was provided, arranged by room number, with 20 rooms per floor and two people

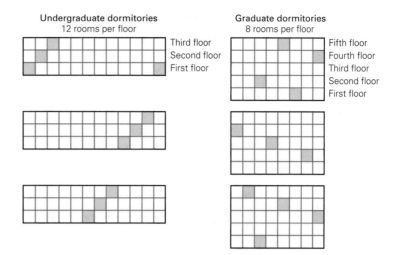

FIGURE 4.7 **A systematic sample of every eleventh room, randomly starting with the third room**

per room. If you were to take a systematic sample of, say, every 40th person on the list, you would get people who lived in the same location on every floor and thus a biased sampling of opinions about noise problems.

Figure 4.7 illustrates a systematic sample of the college dormitory rooms that were shown in Figures 4.4 through 4.6. Consider an ordered list of rooms starting with the top floor of the first undergraduate dorm, numbering those rooms 1 to 12. The second-floor rooms are numbered 13 to 24 and so on, continuing with the undergraduate dorms, then moving to the fifth floor of the first graduate dorm. In the figure, room 3 was randomly selected as a starting point chosen from the first 11 rooms. After that, every eleventh room is sampled. Notice that there is some pattern to the rooms selected, which may or may not lead to biased results, depending on the questions asked in the survey.

Random-Digit Dialing

Most of the national polling organizations and many of the government surveys in the United States now use a method of sampling called **random-digit dialing.** This method results in a sample that approximates a simple random sample of all households in the United States that have telephones.

The method proceeds as follows. First, the polling organization makes a list of all possible telephone *exchanges,* where the exchange consists of the area code and the next three digits. Then, a computer is used to generate a sample of exchanges. Using numbers listed in the white pages, the pollsters can approximate the proportion of all households in the country that have a specific exchange, and this proportion governs the chance that the exchange is sampled. Next, the same method is used to randomly sample *banks* within each exchange, where a bank consists of the next two numbers. Phone companies assign numbers using banks, so that certain banks are mainly assigned to businesses, certain ones are held for future neighborhoods, and so on. Finally, to complete the number, the computer randomly generates the last two digits from 00 to 99.

Once a phone number has been determined, the pollsters should make multiple attempts to reach someone at that household. Sometimes they will ask to speak to a male because females are more likely to answer the phone and would thus be overrepresented if the researchers always survey the first individual who answers the phone.

Multistage Sampling

Many large surveys use a combination of the methods we have discussed. The survey designers might stratify the population by region of the country, then stratify by urban, suburban, and rural, and then choose a random sample of communities within those strata. They would then divide those communities into city blocks or fixed areas, as clusters, and sample some of those. Everyone on the block or within the fixed area may then be sampled. This is called a **multistage sampling plan.**

> ### *in summary*   Getting a Representative Sample
>
> All of the sampling plans described in this section will provide samples that are representative of the population from which they are drawn, except for special problems where noted (such as improper use of a systematic sample). It might be possible to get a representative sample without using one of these plans, but caution must be used. As we will learn in Section 4.4, there are sampling methods that are complete disasters. For instance, relying on volunteers to respond to a survey in a magazine will not provide a sample that represents any population. Results of such a survey cannot be extended beyond those people who actually respond. When you read the results of a survey, make sure you learn how the sample was obtained.

EXAMPLE 4.7
The Nationwide Personal Transportation Survey

Every five years, the U.S. Department of Transportation conducts The Nationwide Personal Transportation Survey to learn about the amount and type of personal travel by people in the United States. The 1995 survey included 21,000 households nationally. Interviews were conducted by telephone, using a **computer-assisted telephone interviewing (CATI)** system. This system helps interviewers keep track of which additional questions to ask based on specific answers, when to skip entire sections, and so on. Here is a description of who was interviewed:

> Interviews are conducted for all persons age 5 and older in the household. Persons 14 and older are interviewed directly and a household adult is asked to report for children age 5–13. A list-assisted Random Digit Dialing sample framework is used. The sample is stratified by region of the country, size of metropolitan area and presence or absence of a subway system. (Liss, 1997, p. 15)

Notice that this is a multistage sample. U.S. households are *stratified* by region of the country, size of metropolitan area, and whether there is a subway system. Households are then selected by *random-digit dialing*. Finally, everyone in a selected household is included, so each household is a *cluster.* ◆

EXAMPLE 4.8
A Los Angeles Times
National Poll
on the Millennium

Most polling agencies now include information on recent polls on their websites, including a brief description of how the poll was conducted. For example, a poll reported by the *Los Angeles Times,* taken in early 1999, reported these results:

> A full quarter of Americans surveyed in a recent Los Angeles Times poll said they believe the onset of a new millennium heralds the second coming of Jesus Christ. While half of Americans polled said they view Jan. 1, 2000, as "just another New Year's Day," considerable numbers say they expect an increase in natural disasters (26%) or civil unrest (30%). About one in 10 report that they are stockpiling goods. (Teresa Watanabe, Mar. 31, 1999)

A description of how the poll was conducted was also given, as follows:

> The Times Poll contacted 1,249 adults nationwide by telephone February 27–28. Telephone numbers were chosen from a list of all exchanges in the nation. Random-digit dialing techniques were used so that listed and non-listed numbers could be contacted.

In other words, the poll was taken over a two-day period in February 1999, by telephone, using a simple random-digit dialing sample. ◆

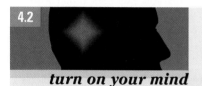

4.2

turn on your mind

In Chapter 3, the Fundamental Rule for Using Data for Inference stated that "available data can be used to make inferences about a much larger group *if the data can be considered to be representative with regard to the question(s) of interest.*" Read the description of how the *Los Angeles Times* poll in Example 4.8 was conducted. Do you think the results of the poll can be extended to a larger group than the 1249 people in the sample? If so, to what group can the results be extended and why?*

4.4 | DIFFICULTIES AND DISASTERS IN SAMPLING

In theory, designing a good sampling plan is easy and straightforward. However, the real world rarely cooperates with well-designed plans, and trying to collect a proper sample is no exception. Difficulties that can occur in practice need to be considered when you evaluate a study. If a proper sampling plan is never implemented, this can lead to very misleading and inaccurate conclusions. Some problems can occur even when a sampling plan has been well designed. Here is a list of possible problems:

- ◆ Using the wrong sampling frame
- ◆ Not reaching the individuals selected
- ◆ Nonresponse or volunteer response
- ◆ Self-selected sample
- ◆ Convenience or haphazard sample

*HINT: Read the section on random-digit dialing in this chapter.

Using the Wrong Sampling Frame

> *definition* The **sampling frame** is the list of units from which the sample is selected. This list may or may not be the same as the list of all units in the desired "target" population.

Sometimes a sampling frame either will include unwanted units or exclude desired units. For example, if we sample from a list of registered voters in order to predict election outcomes, we will include individuals who are not likely to vote as well as those who are likely to do so. Using a telephone directory to survey the general population excludes those who move often, those with unlisted home numbers (like many physicians and teachers), and those who cannot afford a telephone. Using the wrong sampling frame is one way to create selection bias. Figure 4.8 illustrates this situation.

Common sense can often lead to a solution for this problem. In the example of registered voters, interviewers may try to first ascertain the voting history of the people contacted by asking them where they vote and then continuing the interview only if the person knows the answer. Instead of using a telephone directory, surveys use random-digit dialing. This still excludes those without phones, but not those who didn't happen to be in the last printed directory.

Not Reaching the Individuals Selected

Even if a proper sample of units is selected, the desired units may not be reached. For example, *Consumer Reports* magazine mails a lengthy survey to its subscribers in order to obtain information on the reliability of various products. If you were to receive such a survey and you had a close friend who had been having trouble with a highly rated automobile, you may very well decide to pass the questionnaire on to your friend to answer. That way, he would get to register his complaints about the car, but *Consumer Reports* would not have reached the intended recipient.

Telephone surveys tend to reach a disproportionate number of women since they are more likely to answer the phone. Researchers sometimes ask to speak to the oldest adult male at home to try to counter that problem. Surveys are also likely to have trouble contacting people who work long hours and are rarely

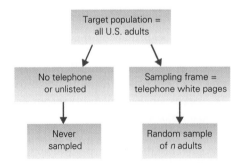

FIGURE 4.8   **The sampling frame doesn't cover the target population, leading to selection bias**

home or those who tend to travel extensively. Some people screen their calls or simply refuse to cooperate even when they do answer the phone.

In recent years, there has been pressure on news organizations to produce surveys of public opinions quickly. When a controversial story breaks, people want to know how others feel about it. This pressure results in what *Wall Street Journal* reporter Cynthia Crossen calls "**quickie polls.**" As she notes, these are "most likely to be wrong because questions are hastily drawn and poorly pretested, and it is almost impossible to get a random sample in one night" (Crossen, 1994, p. 102). Even if the computer can randomly generate phone numbers for the sample, there will be groups of people who are not likely to be home that night, and they may have different opinions from those who are likely to be home. Most responsible reports about polls include information about the dates during which the poll was conducted. For instance, the millennium poll in Example 4.8 was conducted on a Saturday and Sunday, so people who frequently travel on weekends may be underrepresented.

It is very important once a sample has been selected that those individuals are the ones who are actually measured. It is better to put resources into getting a smaller sample than to get one that has been biased by moving on to the next person on the list when someone is initially unavailable. Failing to contact or measure the individuals who were selected in the sampling plan leads to nonresponse bias.

Nonresponse or Volunteer Response

Even the best surveys are not able to contact everyone on the list, and not everyone contacted will respond. The General Social Survey (GSS), run by the prestigious National Opinion Research Center (NORC) at the University of Chicago, noted in its September 1993 *GSS News:*

> In 1993 the GSS achieved its highest response rate ever, 82.4%. This is five percentage points higher than our average over the last four years. Given the long length of the GSS (90 minutes), the high response rates of the GSS are testimony to the extraordinary skill and dedication of the NORC field staff.

The GSS is now conducted biennially, and the response rates for 1994, 1996, and 1998 have been 78%, 76%, and 76%, respectively, according to the NORC website (http://www.norc.uchicago.edu/gss/homepage.htm). Beyond having a dedicated staff, not much can be done about getting everyone in the sample to respond. Response rates should be reported in research summaries.

As a reader, remember that the lower the response rate, the less the results can be generalized to the population as a whole. Responding to a survey (or not) is voluntary, and those who respond are likely to have stronger opinions than those who do not respond. As mentioned earlier, this type of **nonresponse bias** can lead to systematically over- or underestimating the truth about a population.

With mail surveys, it may be possible to compare those who respond immediately with those who need a second prodding, and in telephone surveys you could compare those who are home on the first try with those who require numerous call-backs. If those groups differ on the measurement of interest, then those who were never reached are probably different as well.

In a mail survey, it is best not to rely solely on **volunteer response.** In other words, researchers should not just accept that those who did not respond the first time can't be cajoled into responding. Often, sending a reminder with a brightly colored stamp or following up with a personal phone call will produce the desired effect. Surveys that simply use those who respond voluntarily are sure to be biased in favor of those with strong opinions or with time on their hands.

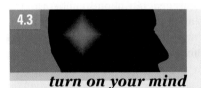

4.3

turn on your mind

Suppose you want to know how students at your school feel about the computer services that are offered. You are able to obtain the list of e-mail addresses for all students taking statistics classes, so you send a survey to a simple random sample of 100 of those students and 65 respond. Using the difficulties discussed so far in this section, explain to whom you could extend the results of your survey and why.*

EXAMPLE 4.9
Which Scientists Trashed the Public?

According to a poll taken among scientists and reported in the prestigious journal *Science* (Mervis, 1998), scientists don't have much faith in either the public or the media. The article reported that based on the results of a "recent survey of 1400 professionals" in science and in journalism, 82% of scientists "strongly or somewhat agree" with the statement, "The U.S. public is gullible and believes in miracle cures or easy solutions." Eighty percent agreed that "the public doesn't understand the importance of federal funding for research." Eighty-two percent also trashed the media, agreeing with the statement, "The media do not understand statistics well enough to explain new findings." It isn't until the end of the article that we learn who responded: "The study reported a 34% response rate among scientists, and the typical respondent was a white, male physical scientist over the age of 50 doing basic research." Remember that those who feel strongly about the issues in a survey are the most likely to respond. With only about a third of those contacted responding, it is inappropriate to generalize these findings and conclude that most scientists have so little faith in the public and the media. This is especially true since we were told that the respondents represented only a narrow subset of scientists. ◆

Disasters in Sampling

Basing a sample survey on a **self-selected sample** (also called a **volunteer sample**) or a **convenience sample** is usually so problematic that the results cannot be extended to anyone beyond the sample. This warning applies only to sample surveys and to some observational studies. As we learned in Chapter 3, randomized experiments often make use of volunteers and convenience samples.

Self-Selected Sample

While relying on volunteer responses from a random sample presents somewhat of a difficulty in determining the extent to which surveys can be generalized, relying on a self-selected sample consisting of only those who volunteer to participate is usually a complete waste of time. If a television or radio station, magazine, or Internet site presents a survey and asks any readers or viewers who are interested to respond, the results reflect the opinions of only those listeners or readers who decide to take the trouble to respond. And people who weren't watching or listening, or reading the magazine or Internet site, would not even know the poll existed.

As noted earlier, those who have a strong opinion about the question are more likely to respond than those who do not have a strong opinion. Thus, the self-selected responding group is simply not representative of any larger group. Most media outlets now acknowledge that such polls are "unscientific" when they report the results, but most readers are not likely to understand how misleading the results can be. The next example illustrates the contradiction that can result between a scientific poll and one relying solely on a self-selected sample.

EXAMPLE 4.10
A Meaningless Poll

On the 18th of February, 1993, shortly after Bill Clinton became President of the United States, a television station in Sacramento, California asked viewers to respond to the question, "Do you support the President's economic plan?" The next day the results of a properly conducted study that asked the same question were published in the newspaper. Here are the results from these two different surveys:

	Television Poll	Survey
Yes (support plan)	42%	75%
No (don't support plan)	58%	18%
Not sure	0%	7%

As you can see, those who were dissatisfied with the President's plan were much more likely to respond to the television poll than those who supported it, and no one who was "Not sure" called the television station, since they were not invited to do so. Trying to extend those results to the general population is misleading. It is irresponsible to publicize such studies, especially without a warning that they result from an unscientific survey and are not representative of general public opinion. You should never interpret such polls as anything other than a count of who bothered to respond. ◆

Convenience or Haphazard Sample

Another worthless sampling technique for surveys is to use the most convenient group available or to decide haphazardly on the spot who to sample. It is almost always the case that these types of samples break the Fundamental Rule of Using Data for Inference given in Section 3.1. Rarely do the responses from a **convenience sample** or **haphazard sample** represent any larger population for the question of interest.

EXAMPLE 4.11
Haphazard Sampling

A few years ago, the student newspaper at a California university announced as a front-page headline "Students ignorant, survey says." The article explained that a "random survey" indicated that American students were less aware of current events than international students were.

The article quoted the undergraduate researchers, who were international students themselves, as saying "the students were randomly sampled on the quad." The quad is a wide expanse of lawn where students relax, eat lunch, and so on. There is simply no proper way to collect a random sample of students by selecting them in an area like that. In such situations, the researchers are likely to approach people whom they think will support the results they intended for their survey. Or, they are likely to approach friendly looking people who look like they will easily cooperate. This is called a haphazard sample, and it cannot be expected to be representative at all. ◆

You have seen the proper way to collect a sample and have been warned about the many difficulties and dangers inherent in the process. Here is a famous example that helped researchers learn some of these pitfalls.

CASE STUDY 4.1
The Infamous Literary Digest *Poll of 1936*

Before the election of 1936, a contest between Democratic incumbent Franklin Delano Roosevelt and Republican Alf Landon, the magazine *Literary Digest* had been extremely successful in predicting the results in U.S. presidential elections. But 1936 turned out to be the year of its downfall, when it predicted a 3-to-2 victory for Landon. To add insult to injury, young pollster George Gallup, who had just founded the American Institute of Public Opinion in 1935, not only correctly predicted Roosevelt as the winner of the election, he also predicted that the *Literary Digest* would get it wrong. He did this before they even conducted their poll! And Gallup surveyed only 50,000 people, while the *Literary Digest* sent questionnaires to 10 million people (Freedman, Pisani, Purves, and Adhikari, 1991, p. 307).

The *Literary Digest* made two classic mistakes. First, the lists of people to whom they mailed the 10 million questionnaires were taken from magazine subscribers, car owners, telephone directories, and, in a few instances, lists of registered voters. In 1936, those who owned telephones or cars, or subscribed to magazines, were more likely to be wealthy individuals who were not happy with the Democratic incumbent.

Despite what many accounts of this famous story conclude, the bias produced by the more affluent list was not likely to have been as severe as the second problem (Bryson, 1976). *The main problem was volunteer response.* They received 2.3 million responses, a response rate of only 23%. Those who felt strongly about the outcome of the election were most likely to respond. And that included a majority of those who wanted a change, the Landon supporters. Those who were happy with the incumbent were less likely to bother to respond.

Gallup, on the other hand, knew the value of random sampling. He was not only able to predict the election but he also predicted what the results of the *Literary Digest* poll would be to within 1%. How did he do this? According to Freedman et al. (1991, p. 308), "he just chose 3,000 people at random from the same lists the *Digest* was going to use, and mailed them all a postcard asking them how they planned to vote."

This example illustrates the beauty of random sampling and the idiocy of trying to base conclusions on nonrandom and biased samples. The *Literary Digest* went bankrupt the following year, so it never had a chance to revise its methods. The organization founded by George Gallup has flourished, although it has made a few sampling blunders as well. (See exercise 4.46.)

4.5 | HOW TO ASK SURVEY QUESTIONS

You may be surprised at how much the answers to questions can change based on simple changes in wording. Here is one example. Loftus and Palmer (1974, quoted in Plous, 1993, p. 32) showed college students films of an automobile accident, after which they asked them a series of questions. One group was asked the question, "About how fast were the cars going when they contacted each other?" The average response was 31.8 miles per hour. Another group was asked, "About how fast were the cars going when they collided with each other?" In that group, the average response was 40.8 miles per hour. Simply changing from the word *contacted* to the word *collided* increased the estimates of speed by 9 miles per hour, or 28%, even though the respondents had witnessed the same film.

The wording and presentation of questions can significantly influence the results of a survey. The seven pitfalls discussed in this section are all possible sources of response bias in surveys.

Possible Sources of Response Bias in Surveys

Many pitfalls can be encountered when asking questions in a survey or experiment. Here are some of them, each of which will be discussed in turn:

- Deliberate bias
- Unintentional bias
- Desire to please
- Asking the uninformed
- Unnecessary complexity
- Ordering of questions
- Confidentiality and anonymity

Deliberate Bias

Sometimes, if a survey is being conducted to support a certain cause, questions are deliberately worded in a biased manner. Be careful about survey questions that begin with phrases like "Do you agree that . . . ?" Most people want to be agreeable and will be inclined to answer "yes" unless they have strong feelings the other way.

For example, suppose an antiabortion group and a prochoice group each wanted to conduct a survey in which it would find the best possible agreement with its position. Here are two questions that would each produce an estimate of the proportion of people who think abortion should be completely illegal. The questions are almost certain to produce different estimates:

1. Do you agree that abortion, the murder of innocent beings, should be outlawed?
2. Do you agree that there are circumstances under which abortion should be legal, to protect the rights of the mother?

The wording of a question should not indicate a desired answer. For instance, a Gallup Poll conducted in June 1998 contained the question, "Do you think it was a good thing or a bad thing that the atomic bomb was developed?" Notice that the question does not indicate which answer is preferable (61% of the respondents said "bad," while 36% said "good" and 3% were undecided).

Unintentional Bias

Sometimes questions are worded in such a way that the meaning is misinterpreted by a large percentage of the respondents. For example, if you were to ask people whether or not they use drugs, you would need to specify if you mean prescription drugs, illegal drugs, over-the-counter drugs, or common substances like caffeine. If you were to ask people to recall the most important date in their life, you would need to clarify if you meant the most important calendar date or the most important social engagement with a potential partner. (It is unlikely that anyone would mistake the question as being about the shriveled fruit, but you can see that the same word can have multiple meanings.)

EXAMPLE 4.12
Laid Off or Fired?

Plewes (1994) described a number of changes implemented at the beginning of 1994 to the Current Population Survey (CPS), which is a monthly survey used to determine government unemployment figures and other government statistics in the United States. One change involved clarification and expansion of questions about being laid off. Plewes explained that people were answering that they had been laid off, when in reality they had been fired: "The CPS concept of 'on layoff' differs from the everyday usage by respondents. CPS defines 'on layoff' as embodying an expectation of recall to the job; common usage considers it a euphemism for 'fired'" (p. 38). ◆

Desire to Please

Most survey respondents have a desire to please the person who is asking the question. They tend to understate their responses about undesirable social habits and opinions, and vice versa. For example, in recent years estimates of the prevalence of cigarette smoking based on surveys do not match those based on cigarette sales. Either people are not being completely truthful, or lots of cigarettes are ending up in the garbage.

EXAMPLE 4.13
Most Voters Don't Lie but Some Liars Don't Vote

Abelson, Loftus, and Greenwald (1992) discuss how the National Election Surveys always find that more people claim to vote than actually did. They explain that researchers have tried to make it more socially acceptable to admit to not voting by adding a preamble to questions about voting. The preamble reads, "In talking to people about elections, we often find that a lot of people were not able to vote because they weren't registered, they were sick, or they just didn't have time. How about you—did you vote in the elections this November?" (p. 139). In a survey six to seven months after the November 1986 election, the authors of the report surveyed people for whom they knew whether or not they had actually voted. The question started with the preamble and finished with "did you vote in the 1986 elections for United States Congress?" (p. 141). They found that of the 211 respondents who had actually voted, 203 (96%) accurately reported that they did. But of the 98 respondents who had *not* voted, 39 (40%) claimed that they *did* vote. ◆

Asking the Uninformed

People do not like to admit that they don't know what you are talking about when you ask them a question. Crossen (1994, p. 24) gives an example: "When the American Jewish Committee studied Americans' attitudes toward various ethnic groups, almost 30% of the respondents had an opinion about the fictional Wisians, rating them in social standing above a half-dozen other real groups, including Mexicans, Vietnamese and African blacks."

Political pollsters, who are interested in surveying only those who will actually vote, learned long ago that it is useless to simply ask people if they plan to vote. Most of them will say "yes." Instead, they ask questions to establish a history of voting, such as "Where did you go to vote in the last election?"

Unnecessary Complexity

If questions are to be understood, they must be kept simple. A question like "Shouldn't former drug dealers not be allowed to work in hospitals after they are released from prison?" is sure to lead to confusion. Does a "yes" answer mean they should or should not be allowed to work in hospitals? It would take a few readings to figure that out.

Another way in which a question can be unnecessarily complex is by actually asking more than one question at once. An example would be a question like "Do you support the president's health care plan, since it would ensure that all Americans receive health coverage?" If you agree with the idea that all Americans should receive health coverage but disagree with the remainder of the plan, do you answer "yes" or "no"? Or what if you support the president's plan, but not for that reason?

EXAMPLE 4.14
Why Weren't You at Work Last Week?

In the work leading to the revision of the Current Population Survey discussed in Example 4.13, researchers discovered that some questions were too complex to be understood by many respondents. For instance, as Plewes (1994, p. 38) described, one such question on the old version was "Did you have a job from which you were temporarily absent or on layoff LAST WEEK?" In the revised version of the questionnaire, people are first asked if they were absent in the last week, and if they were, further questions about temporary absence are asked. A new and separate series of questions asks about layoffs. ◆

Ordering of Questions

If one question requires respondents to think about something that they may not have otherwise considered, then the order in which questions are presented can change the results. For example, suppose a survey were to ask, "To what extent do you think teenagers today worry about peer pressure related to drinking alcohol?" and then ask, "Name the top five pressures you think face teenagers today." It is quite likely that respondents would use the idea they had just been given and name peer pressure related to drinking alcohol as one of the five choices.

EXAMPLE 4.15
Is Happiness Related to Dating?

Clark and Schober (1992, p. 41) report on a survey that asked the following two questions:

1. How happy are you with life in general?
2. How often do you normally go out on a date? about ____ times a month

When the questions were asked in this order, there was almost no relationship between the two answers. But when question 2 was asked first, the answers were highly related. Clark and Schober speculate that in that case, respondents interpreted question 1 to mean "Now, considering what you just told me about dating, how happy are you with life in general?" ◆

Confidentiality and Anonymity

People will often answer questions differently based on the degree to which they believe they are anonymous. Because researchers often need to perform follow-up surveys, it is easier to try to ensure confidentiality than anonymity. In ensuring **confidentiality,** the researcher promises not to release identifying information about respondents. In an **anonymous** survey, the researcher does not know the identity of the respondents.

Questions on issues like sexual behavior and income are particularly difficult because people consider those to be private matters. A variety of techniques have been developed to help ensure confidentiality, but surveys on such issues are hard to conduct accurately.

The problems discussed in this section are well known to professional pollsters, and most reputable polling agencies go to great lengths to overcome them to the extent possible. The survey questions in Examples 4.1 to 4.3 in Section 4.1 are excellent illustrations of how questions should be worded. For instance, the question in Example 4.3, "Do you favor or oppose scientific experimentation on the cloning of human beings?" is worded so as not to suggest which answer the interviewer would prefer. Here are two examples that illustrate attempts to overcome bias but also illustrate how difficult that can be.

EXAMPLE 4.16

When Will Adolescent Males Report Risky Behavior?

It is particularly difficult to measure information on behavior that is sensitive, risky, or illegal. In an article in *Science*, Turner et al. (1998) describe the use of audio, computer-assisted self-interviews as part of the 1995 National Survey of Adolescent Males (aged 15 to 19). The respondents were randomly assigned to answer the survey using the traditional paper form ($n = 368$) or using a laptop computer ($n = 1361$). The paper questionnaire was filled out and sealed in an envelope to return to the surveyor but was accompanied by identifying code numbers. The computer method included listening to questions through headphones, then recording the answers on a laptop computer. The authors believed this method would allow respondents to feel that their responses were more private. In particular, respondents who could not read would need to have the paper version read to them, whereas the audio-computer method allowed complete self-administration of the survey.

Results indicated that the additional perception of privacy increased the reported incidence of certain behaviors, while decreasing others. The behaviors with increased reported incidence for the audio-computer version tended to be less socially acceptable, while those with higher incidence for the paper version tended to be more socially acceptable (for the adolescent males in this survey). Here are some examples reported in the article:

Question:	Estimated Prevalence per 100 Males in Population	
Have you ever...	Paper Version	Audio-Computer Version
Had sex with a prostitute	0.7	2.5
Had intercourse with a female	68.1	63.9
Made a girl pregnant	7.9	6.5
Had sex with another male	1.5	5.5
Taken street drugs using a needle	1.4	5.2

◆

EXAMPLE 4.17

Politics Is All in the Wording

Crossen (1994, p. 112) described three very different sets of responses to a question about whether the president of the United States should be able to veto specific items in the federal budget. The first set came from a questionnaire published in *TV Guide* in which founder of the Reform Party and presidential candidate Ross Perot asked, "Should the president have the line item veto to eliminate waste?" Respondents were volunteers who mailed in the survey. Of those who responded, 97% said "yes." The second set of responses came from asking a properly selected random sample the same question, and 71% said "yes." But clearly the question is written to elicit a "yes" answer. When the question was reworded to read, "Should the president have the line item veto, or not?" and a proper random sample was queried, only 57% answered yes.

Cohn and Cope (2001, pp. 145–146) report that Perot also asked the question "Should laws be passed to eliminate all possibilities of special interests giving huge sums of money to candidates?" Again the "yes" responses were an overwhelming 99%. When a properly selected random sample was asked the same, biased version of the question, 80% said "yes." But when a properly selected random sample was asked, "Should laws be passed to prohibit interested groups from contributing to campaigns, or do groups have a right to contribute to the candidate they support?" only 40% said that such laws should be passed. ◆

4.4

turn on your mind

You have now learned that survey results have to be interpreted in the context of *who* responded and *to what questions* they responded. When you read the results of a survey, for which of these two areas do you think it would be easier for you to recognize and assess possible biases? Why?*

Be Sure You Understand What Was Measured

Sometimes words mean different things to different people. When you read about survey results, you should get a precise definition of what was actually asked or measured. Here are two examples that illustrate that even common terminology may mean different things to different people.

EXAMPLE 4.18
Teenage Sex

A letter to advice columnist Ann Landers stated, "According to a report from the University of California at San Francisco . . . sexual activity among adolescents is on the rise. There is no indication that this trend is slowing down or reversing itself." The letter went on to explain that these results were based on a national survey (*Davis Enterprise,* Feb. 19, 1990, p. B-4). On the same day, in the same newspaper, an article entitled "Survey: Americans Conservative with Sex" reported that "teenage boys are not living up to their reputations. [A study by the Urban Institute in Washington] found that adolescents seem to be having sex less often, with fewer girls and at a later age than teenagers did a decade ago" (*Ibid.,* p. A-9).

Here we have two apparently conflicting reports on adolescent sexuality, both reported on the same day in the same newspaper. One indicated that teenage sex was on the rise, while the other indicated that it was on the decline. Although neither report specified exactly what was measured, the letter to Ann Landers proceeded to note that "national statistics show the average age of first intercourse is 17.2 for females and 16.5 for males." The article stating that adolescent sex was on the decline measured it in terms of *frequency.* It was based on interviews with 1880 boys between the ages of 15 and 19, in which "the boys said they had had six sex partners, compared with seven a decade earlier. They reported having had sex an average of three times during the previous month, compared with almost five times in the earlier survey."

Thus it is not enough to know that both surveys were measuring adolescent or teenage sexual behavior. In one case the author was, at least partially, discussing the *age* of first intercourse, whereas in the other case the author was discussing the *frequency*. ◆

EXAMPLE 4.19
The Unemployed

Ask people whether or not they know anyone who is unemployed and invariably they will say that they do. Most people, however, don't realize that to be officially unemployed and included in the unemployment statistics given by the U.S. government, you must meet very stringent criteria. The government excludes "discouraged workers" who are "identified as persons who, though not working or seeking work, have a current desire for a job, have looked for one within the past year [but not within the past four weeks], and are currently available for work" (U.S. Dept. of Labor, 1992, p. 10). If you knew someone who fit that definition, you would undoubtedly think of him or her as unemployed. But they are not included in the official statistics. You can see that the true number of people who are not working is higher than the government statistics would lead you to believe. ◆

Some Concepts Are Hard to Precisely Define

Sometimes it is not the language but the concept itself that is ill-defined. For example, there is still not universal agreement on what should be measured with intelligence, or IQ, tests. The tests were originated at the beginning of the 20th century in order to determine the mental level of schoolchildren. The intelligence quotient (IQ) of a child was found by dividing the child's "mental level" by his or her chronological age. The "mental level" was determined by comparing the child's performance on the test with that of a large group of "normal" children to find the age group the individual's performance matched. Thus, if an 8-year-old child performed as well on the test as a "normal" group of 10-year-old children, he or she would have an IQ of $100 \times (10/8) = 125$.

IQ tests have been expanded and refined since the early days, but they continue to be surrounded by controversy. One reason is that it is very difficult to define what is meant by intelligence. It is hard to measure something if you can't even agree on what it is you are trying to measure. If you are interested in knowing more about these tests and the surrounding controversies, you can find numerous books on the subject. Anastasi (1988) provides a detailed discussion of a large variety of psychological tests, including IQ tests.

Measuring Attitudes and Emotions

Similar problems exist with trying to measure attitudes and emotions like self-esteem and happiness. The most common method for trying to measure such things is to have respondents read statements and determine the extent to which they agree with the statement. For example, a test for measuring happiness might ask respondents to indicate their level of agreement, from "strongly disagree" to "strongly agree," with statements such as "I generally feel optimistic when I get up in the morning." In order to produce agreement on what is meant by characteristics such as "introversion," psychologists have developed standardized tests that claim to measure those attributes.

Open or Closed Questions: Should Choices Be Given?

An **open question** is one in which respondents are allowed to answer in their own words, while a **closed question** is one in which they are given a list of alternatives from which to choose their answer. Usually the latter form offers a choice of "other" in which the respondent is allowed to fill in the blank.

Problems with Closed Questions

To show the limitation of closed questions, Schuman and Scott (1987) asked about "the most important problem facing this country today." Half of the sample, 171 people, were given this as an open question. The most common responses were

> Unemployment (17%)
> General economic problems (17%)
> Threat of nuclear war (12%)
> Foreign affairs (10%)

In other words, one of these four choices was volunteered by over half of the respondents.

The other half of the sample was given the question as a closed question, with the list of choices and the percent who chose them being

> The energy shortage (5.6%)
> The quality of public schools (32.0%)
> Legalized abortion (8.4%)
> Pollution (14.0%)

These choices combined were mentioned by only 2.4% of respondents in the open question survey, yet they were selected by 60% when they were the only specific choices given. Don't think respondents had no choice; in addition to the list of four, they were told, "If you prefer, you may name a different problem as most important." On the basis of the closed-form questionnaire, policymakers would have been seriously misled about what is important to the public. It is possible to avoid this kind of astounding discrepancy. If closed questions are preferred, they first should be presented as open questions to a test sample before the real survey, and the most common responses should then be included in the list of choices for the closed question. This kind of exercise is usually done as part of what's called a "pilot survey" in which various aspects of a study design can be tried before it's too late to change them.

Problems with Open Questions

The biggest problem with open questions is that the results can be difficult to summarize. If a survey includes thousands of respondents, it can be a major chore to categorize their responses. Another problem, found by Schuman and Scott (1987), is that the wording of the question might unintentionally exclude answers that would have been appealing had they been included in a list of choices (such as in a closed question). To test this, they asked 347 people to "name one or two of the most important national or world event(s) or change(s) during the past 50 years." The most common choices and the percent who mentioned them were

> World War II (14.1%)
> Exploration of space (6.9%)
> Assassination of John F. Kennedy (4.6%)
> The Vietnam war (10.1%)
> Don't know (10.6%)
> All other responses (53.7%)

The same question was then repeated in closed form to a new group of 354 people. Five choices were given, the preceding ones and "invention of the computer." Of the 354 respondents, the percent who selected each choice were

World War II (22.9%)
Exploration of space (15.8%)
Assassination of John F Kennedy (11.6%)
The Vietnam war (14.1%)
Invention of the computer (29.9%)
Don't know (0.3%)
All other responses (5.4%)

The most frequent response was "invention of the computer," which had been mentioned by only 1.4% of respondents in the open question. Clearly, the wording of the question led respondents to focus on "events" rather than "changes" and the invention of the computer did not readily come to mind. When it was presented as an option, however, people realized that it was indeed one of the most important events or changes during the past 50 years.

In summary, there are advantages and disadvantages to both approaches. One compromise is to ask a small test sample to list the first several answers that come to mind, then use the most common of those. These could be supplemented with additional answers, like "invention of the computer," that may not readily come to mind.

Remember that, as the reader, you have an important role in interpreting the results. You should always be informed as to whether questions were asked in open or closed form, and if the latter, you should be told what the choices were. You should also be told whether or not "don't know" or "no opinion" was offered as a choice in either case.

CASE STUDY 4.2
No Opinion of Your Own? Let Politics Decide

This is an excellent example of how people will respond to survey questions when they do not know about the issues, and how the wording of questions can influence responses. In 1995, the *Washington Post* decided to expand on a 1978 poll taken in Cincinnati, Ohio in which people were asked whether they "favored or opposed repealing the 1975 Public Affairs Act" (Morin, 1995, p. 36). There was no such act, but about one-third of the respondents expressed an opinion about it.

In February 1995, the *Washington Post* added this fictitious question to its weekly poll of 1000 randomly selected respondents: "Some people say the 1975 Public Affairs Act should be repealed. Do you agree or disagree that it should be repealed?" Almost half (43%) of the sample expressed an opinion, with 24% agreeing that it should be repealed and 19% disagreeing!

The *Post* then tried another trick that produced even more disturbing results. This time, they polled two separate groups of 500 randomly selected adults. The first group was asked: "President Clinton [a Democrat] said that the 1975 Public Affairs Act should be repealed. Do you agree or disagree?" The second group was asked: "The Republicans in Congress said that the 1975 Public Affairs Act should be repealed. Do you agree or disagree?" Respondents were also asked about their party affiliation.

Overall, 53% of the respondents expressed an opinion about repealing this fictional act! The results by party affiliation were striking: For the Clinton version, 36% of the Democrats but only 16% of the Republicans agreed that the act should be repealed. For the "Republicans in Congress" version, 36% of the Republicans but only 19% of the Democrats agreed that the act should be repealed.

KEY TERMS

EXERCISES   Blue = Basics   Bold, **Bold** = Answer in back

Section 4.1

4.1 Briefly explain the difference between a sample survey and a census.

4.2 A survey is planned to estimate the proportion of voters in a community who plan to vote for candidate Y. Calculate the conservative margin of error, as a proportion and as a percent, for each of the following possible sample sizes:
a. $n = 100$.
b. $n = 400$.
c. $n = 900$.

4.3 The Nielsen survey produces the rankings of television shows. The survey is based on a random sample of about 5000 households with TV sets. Calculate a conservative margin of error for this survey as a proportion and as a percent.

4.4 In a random sample of 90 students at a university, 72 students (80% or .80 of the sample) say they use a laptop computer.
a. Calculate the conservative margin of error for the survey.
b. Compute an approximate 95% confidence interval for the population *proportion* that uses a laptop computer.

4.5 Briefly explain what it means to say that a survey method is biased.

4.6 For each definition, identify the correct term for the type of bias being defined. Possible answers are selection bias, nonresponse bias, response bias.
a. Participants respond differently from how they truly feel.
b. The method for selecting the participants produces a sample that does not represent the population of interest.

c. A representative sample is chosen, but a subset of the sample cannot be contacted or does not respond.

4.7 In each of the following situations, indicate whether the potential bias is a selection bias, a nonresponse bias, or a response bias.
a. A survey question asked of unmarried men was, "What is the most important feature you consider when deciding whether to date somebody?" The results were found to depend upon whether the interviewer was male or female.
b. In a study of women's opinions about community issues, investigators randomly selected a sample of households and interviewed a woman from each selected household. When no woman was present in a selected household, a next-door neighbor was interviewed instead. The survey was done during daytime hours, so working women might have been disproportionately missed.
c. A telephone survey of 500 residences is conducted. People refused to talk to the interviewer in 200 of the residences.

4.8 In a CNN/*Time* poll conducted December 17–18, 1998, a sample of $n = 1031$ adults in the United States was asked, "Do you think the police should or should not be allowed to collect DNA information from suspected criminals, similar to how they take fingerprints?" Of those sampled, 66% answered "should" (http://www.pollingreport.com/crime.htm).
a. Calculate the conservative margin of error for the survey, as a percent.
b. Compute an approximate 95% confidence interval for the percent of all American adults who think police should be allowed to collect DNA information from suspected criminals.

4.9 In an ABC News Poll conducted between January 21st and 26th, 2000 a random sample of $n = 1006$ adult Americans was asked, "Compared to buying things by mail order or in a store, do you think that buying things over the Internet poses more of a threat to your personal privacy, less of a threat, or about the same?" (http://www.pollingreport.com/computer.htm, September 17, 2002). The percents that selected each response were

More of a threat	40%
Less of a threat	7%
About the same	47%
No opinion	6%

a. What is the margin of error for this poll? Give your answer as both a proportion and a percent.
b. What is a conservative 95% confidence interval for the population proportion that would have responded "More of a threat?" Give your answer for the population percent as well.
c. Write a sentence or two interpreting the interval you found in part (b).
d. Based on the interval you found in part (b), could you conclude that in January 2000 fewer than half of all adult Americans perceived Internet shopping as posing more of a threat than buying things by mail order or in a store? Explain.

4.10 Suppose a national polling agency conducted 100 polls in a year, using proper random sampling, and reported a 95% confidence interval for each poll. About how many of those confidence intervals would be wrong, i.e., would not cover the true population value?

4.11 The question about the importance of religion in Example 4.1 was asked in a survey of elementary statistics students at Penn State University in the fall of 2001 and spring of 2002. The results were as follows:

Very important	460	= 27%
Fairly important	703	= 41%
Not very important	543	= 32%
Total	1706	= 100%

Because these students were not a random sample from a larger population, and because statistics students may not represent other students on questions of religion, treat these 1706 students as a population. Therefore, we know that for the population of elementary statistics students at Penn State in the 2001–2002 academic year, 27% felt that religion was very important in their lives.
a. Based on the results in Example 4.1, can we conclude that this percent differs from the percent of the population of all adult Americans who felt religion was very important in their lives in September 2002? Explain.
b. Based on the preceding results and those in Example 4.1, answer the question in part (a) for the percent who felt religion was fairly important and for the percent who felt it was not very important.
c. Write a short news article comparing the two surveys as if you were writing it for the student newspaper.
d. Suppose we believe the Penn State sample is representative of all students in the academic programs that require the elementary statistics course. Calculate an approximate 95% confidence interval for the popula-

tion proportion that would say religion is very important in their lives.

4.12 Give an example of a situation where a sample must be used because a census is not possible.

4.13 The U.S. government gathers numerous statistics based on random samples, but every ten years it conducts a census of the U.S. population. What can it learn from a census that cannot be learned from a sample?

4.14 An Internet report on a 1999 Gallup Poll (August 2, 1999, http://www.gallup.com/poll/index.asp) included the statement: "The results below are based on telephone interviews with a randomly selected national sample of 1,021 adults, 18 years and older, conducted July 22–25, 1999. For results based on this sample, one can say with 95 percent confidence that the maximum error attributable to sampling and other random effects is plus or minus 3 percentage points." What is the margin of error for this survey? Give two separate ways in which you know this is the answer.

4.15 Refer to the previous exercise. One of the questions asked was "Do you think there will or will not come a time when Israel and the Arab nations will be able to settle their differences and live in peace?" The choices and percent choosing them were "Yes, will be a time" (49%), "No, will not" (47%), "No opinion" (4%).
a. Give the interval of values that is likely to cover the true proportion of the population who would answer, "Yes, will be a time."
b. Give the interval of values that is likely to cover the true proportion of the population who would answer, "No, will not."
c. Compare your answers in parts (a) and (b). Is there a clear majority of the population for either opinion? Explain.

4.16 A poll commissioned by the Student Aid Alliance sampled 1022 American adults in May 1999 and was reported to have a margin of error of 3.1%. Respondents were asked, "How important do you think it is to increase the current level of federal funding for each of the following things?" The programs and the percent of the sample who answered "Very important" or "Somewhat important" were "Health care" (90%), "Social Security" (89%), "Financial aid for college students" (87%), "Defense" (77%), "Welfare" (56%), and "Foreign aid" (47%). (*Source:* NASULGC[1] *Newsline,* July–August 1999, p. 7.) Write a short news article explaining these results. Address the question of which program(s) the highest percentage of *all* adult Americans think it is at least somewhat important to fund at increased levels.

4.17 Refer to the three types of bias given in Section 4.1. Which type of bias do you think would be introduced in each of the following situations?
a. A list of registered automobile owners is used to select a random sample for a survey about whether people think homeowners should pay a surtax to support public parks.
b. A survey is mailed to a random sample of residents in a city asking if they think the current mayor is doing an acceptable job.

[1]NASULGC is the National Association of State Universities and Land Grant Colleges.

c. In a college town, college students are hired to conduct door-to-door interviews, based on a multistage cluster sample, to determine if city residents think there should be a law forbidding loud music at parties.

d. A magazine sends a survey to a random sample of its subscribers asking them if they would like the frequency of publication reduced from biweekly to monthly or would prefer that it remain the same.

e. A random sample of registered voters is contacted by phone and asked whether or not they are going to vote in the upcoming presidential election.

Section 4.2

4.18 Define each of the following terms:
a. Probability sampling plan.
b. Simple random sample.

4.19 There are 8000 items in a population, and these items are labeled using four-digit numbers ranging from 0000 to 7999. Use the following stream of random digits to select four items from the population. Explain how you determined your answer.

Random Digits: 76429 69730 23395 12694 43387

4.20 Ten grocery stores, numbered from 0 to 9, will participate in an experiment to compare the effectiveness of two methods for displaying a product. Five of these stores will be randomly selected to use the first method, while the other five stores will use the second method. Use the following stream of random digits to select the five stores that will use the first display method.

Random Digits: 68538 70621 32205 23593 61396

4.21 Refer to Example 4.5, in which 43 children were randomly assigned to three treatment groups. The example demonstrated how the children for group 1 were selected. Use Table 4.1 to select the 12 children (from the remaining 28) to go into the control group. Number the children from 1 to 28. Start at the beginning of the row labeled 4, explain your process, and list the unit numbers of the children selected.

4.22 A lottery game is played by choosing six whole numbers between 1 and 49. The grand prize is won if all six numbers chosen match the winning numbers drawn.
a. Use Table 4.1 to draw six numbers. Start at the beginning of the table, explain your process, and write down the six numbers chosen.
b. George plays this game weekly. Assume the winning numbers are drawn fairly, so that any number has the same chance of occurrence. Would George have a better chance of winning if he chose the same numbers every week or chose different numbers every week, or doesn't it matter?

4.23 One hundred volunteers agree to participate in an experiment. There are four treatments, each of which will be assigned to 25 participants. Explain how you could use Table 4.1 to assign the treatments.

4.24 Refer to the previous exercise. Suppose the participants were grouped into 25 blocks of 4 each, with each treatment assigned once per block. Explain how you could use Table 4.1 to assign the treatments.

4.25 A radio station has a daily contest in which they randomly select one birthday (month and day, not year) and announce it on the air. The first person with that birthday who calls the station wins a prize. Each day the station selects a new birthday, and it doesn't mind if the same birthday is selected multiple times over the course of the contest. Explain how it could use Table 4.1 to select the birthdays. Start at the beginning of the row labeled 9 and write down the first five birthdays the station would use.

4.26 Refer to Figure 4.2, showing the results of taking a simple random sample of 10 stick figures from Figure 4.1. Label the stick figures in Figure 4.1 from 00 to 99, going across rows. Suppose a table of random digits had been used to select the sample in Figure 4.2. Write down a stream of random digits that would have led to the selection of the sample shown.

4.27 Refer to the previous exercise and write down a stream of random digits that would have led to the selection of the sample shown in Figure 4.3.

4.28 An exercise in Chapter 3 described the following scenario: To compare three packaging designs for a new food, a manufacturer randomly selected three grocery stores in each of 50 cities. The three designs were randomly assigned to the three stores in each city, and sales were compared for a two-month period.
a. Explain how you would use Table 4.1 to choose the three grocery stores in each city.
b. Explain how you would use Table 4.1 to randomly assign the three packaging designs to the three stores in each city.

4.29 The right handspan measurements for 103 female college students in Table 2.4 in Chapter 2 are given again here and in the **pennstate1F** dataset on the CD.

Females (103 Students): 20, 19, 20.5, 20.5, 20.25, 20, 18, 20.5, 22, 20, 21.5, 17, 16, 22, 22, 20, 20, 20, 20, 21.7, 22, 20, 21, 21, 19, 21, 20.25, 21, 22, 18, 20, 21, 19, 22.5, 21, 20, 19, 21, 20.5, 21, 22, 20, 20, 18, 21, 22.5, 22.5, 19, 19, 19, 22.5, 20, 13, 20, 22.5, 19.5, 18.5, 19, 17.5, 18, 21, 19.5, 20, 19, 21.5, 18, 19, 19.5, 20, 22.5, 21, 18, 22, 18.5, 19, 22, 17, 12.5, 18, 20.5, 19, 20, 21, 19, 19, 21, 18.5, 19, 21.5, 21.5, 23, 23.25, 20, 18.8, 21, 21, 20, 20.5, 20, 19.5, 21, 21, 20

a. Draw three simple random samples of ten measurements each from this data set (an individual can be in more than one of your samples). Explain how you chose the samples and list the ten handspan measurements in each sample.
b. Find the median for each of your three samples. Compare them to the median for the full data set (20.0 cm).
c. Find the mean for each of your three samples. Compare them to the mean for the full data set (also 20.0 cm).
d. There were 21 women whose recorded handspan measurements were 20.0 cm. Would it be possible to have a random sample of ten measurements with a mean of 20.0 and a standard deviation of 0? Explain.

Section 4.3

4.30 A class of 200 students is numbered from 1 to 200, and a table of random digits is used to choose 60 students from the class. Is the group of students selected a simple random sample, a stratified random sample, a cluster sample, or a systematic sample?

4.31 In each part, identify whether the sample is a stratified random sample or a cluster sample.
   a. A class of 200 students is seated in 20 rows of 10 students per row. Three students are randomly selected from every row.
   b. An airline company randomly chooses one flight from a list of all flights taking place that day. All passengers on that selected flight are asked to fill out a survey on meal satisfaction.

4.32 In a factory producing television sets, every 100th set produced is inspected. Is the collection of sets inspected a simple random sample, a stratified random sample, a cluster sample, or a systematic sample?

4.33 The **Sampling** applet described in Section 4.2 drew a simple random sample of 10 individuals from a population of 45 men and 55 women and measured their heights. The average height for the sample could be used as an estimate of the average height in the population. Another possibility for estimating the average height would be to use stratified sampling, where men and women are the two strata, then estimate the heights of men and women separately and use the knowledge that 45% of the population is male to combine the two estimates. Do you think stratified sampling would be a more accurate or a less accurate method than simple random sampling for estimating the mean height in the population? Explain.

4.34 Suppose a state has 10 universities, 25 four-year colleges, and 50 community colleges, each of which offer multiple sections of an introductory statistics class each year. Researchers want to conduct a survey of students taking introductory statistics in the state. Explain a method for collecting each of the following types of samples:
   a. A stratified sample.
   b. A cluster sample.
   c. A simple random sample.

4.35 Refer to the previous question. Give one advantage of each sampling method in the context of the problem.

4.36 Find an example of a survey routinely conducted by the U.S. government. (The Internet is a good source, for instance, try http://www.fedstats.gov.) Explain how the survey is conducted and identify the type(s) of sampling used.

4.37 Is a sample found using random-digit dialing more like a stratified sample or a cluster sample? Explain.

Sections 4.4 and 4.5

4.38 A group of biologists wants to estimate the abundance of barrel cactus in a desert. They divide the desert into a grid of 100 rectangular areas, but exclude 10 of those areas because they are difficult to access. The biologists then measure the density of cactus in a randomly selected sample of 40 of the 90 accessible areas.
   a. What was the sampling frame in this study, and how did it differ from the population of interest?
   b. Explain why "using the wrong sampling frame" might lead to a biased estimate of the abundance of cactus in the desert.

4.39 A local government wants to determine whether taxpayers support increasing local taxes in order to provide more public funding to schools. They randomly select 500 schoolchildren from a list of all children enrolled in local schools and then survey the parents of these children about possible tax increases.
   a. What is the population of interest for the local government?
   b. What sampling frame did the government use for the survey?
   c. Explain why "using the wrong sampling frame" might lead to a biased estimate of taxpayer support for increasing taxes.

4.40 In each part, indicate whether the sample should be called a self-selected sample or a convenience sample.
   a. To assess passenger satisfaction, an airline distributed questionnaires to 100 passengers in the airline's frequent flyer lounge. All 100 individuals responded, and 95 respondents said they had a high degree of satisfaction with the airline.
   b. A magazine contains a survey about sexual behavior. Readers are asked to mail in their answers, and $n = 2000$ readers do so.
   c. A political scientist surveys the 80 people in a class he teaches in order to evaluate student political views.

4.41 A survey question will be asked to determine whether people think smoking should be banned on all airline flights.
   a. Write a version of the question that is as neutral and unbiased as possible.
   b. Write a version of the question likely to get people to respond that smoking should be forbidden on all flights.
   c. Write a version of the question likely to get people to respond that smoking should be allowed on airline flights.

4.42 An example of an unnecessarily complex survey question is "Shouldn't former drug dealers not be allowed to work in hospitals after they are released from prison?" Restate this question so that it is clearer.

4.43 In a planned survey about moviegoing, two questions that will be asked are:
   ◆ How many times per month do you go to the movies?
   ◆ Do you consider yourself to be well-informed about recent movies, or not?
   The questions could be asked in either order. Briefly explain how you think the question order for asking them might affect how people answer them.

4.44 In presidential election years in the United States, a Gallup Poll is conducted in which the first survey question asks which presidential candidate the voter prefers. Subsequent questions concern other political, social, and election issues. Explain why this question order might be the best order for estimating how people may vote in the presidential election.

4.45 Medical tests, such as those for detecting HIV, sometimes concern such sensitive information that people do not want to give their names when they take the test. In some instances, a person taking such a medical test is given a number or code that they can use later to learn the test result by phone. Explain whether this procedure is an example of anonymous testing or confidential testing.

4.46 Despite his success in 1936, George Gallup failed miserably in trying to predict the winner of the 1948 U.S. presidential election. His organization, as well as two others, predicted that Thomas Dewey would beat incumbent Harry Truman. All three used what is called **quota sampling.** The interviewers were told to find a certain number, or quota, of each of several types of people. For example, they might have been told to interview six women under age 40, one of whom was black and the other five of whom were white. Imagine that you are one of their interviewers, trying to follow these instructions. Who would you ask? Now explain why you think these polls failed to predict the true winner and why quota sampling is not a good method.

4.47 Explain why the main problem with the *Literary Digest* poll is described as "volunteer response" and not "volunteer sample."

4.48 Gastwirth (1988, p. 507) describes a court case in which Bristol Myers was ordered by the Federal Trade Commission to stop advertising that "twice as many dentists use Ipana as any other dentifrice" and that more dentists recommended it than any other dentifrice. Bristol Myers had based its claim on a survey of 10,000 randomly selected dentists from a list of 66,000 subscribers to two dental magazines. They received 1983 responses, with 621 saying they used Ipana and only 258 reporting that they used the second most popular brand. As for the recommendations, 461 respondents recommended Ipana, compared with 195 for the second most popular choice.

  a. Specify the sampling frame for this survey, and explain whether or not you think "using the wrong sampling frame" was a difficulty here, based on what Bristol Myers was trying to conclude.

  b. Of the remaining four "Difficulties and Disasters in Sampling" listed in Section 4.4 (other than "using the wrong sampling frame"), which do you think was the most serious in this case? Explain.

  c. What could Bristol Myers have done to improve the validity of the results after it had mailed the 10,000 surveys and received 1983 of them back? Assume it kept track of who had responded and who had not.

4.49 An Internet poll sponsored by a site called About.com asked Internet users to pick one of two choices in response to the question, "Should jurors opposed to gun control laws refuse to convict defendants even if they have clearly broken gun laws?" The two choices and the number and percent choosing them were

  ◆ Yes, that's an effective way to defeat unjust laws (16,864, 23%).

  ◆ No, that undermines the legal system (55,519, 77%). Discuss this poll, including whether or not you think the results are representative of all adults and whether you think the wording is appropriate.

4.50 A Gallup Poll released on July 9, 1999 included a series of questions about possible religious activities in public schools. The poll was based on telephone interviews with a randomly selected sample of 1016 U.S. adults conducted June 25–27, 1999. The questions, asked in random order, included the following:

  ◆ Do you favor or oppose teaching creationism ALONG WITH evolution in public schools? Favor—68%, Oppose—29%, No opinion—3%.

  ◆ Do you favor or oppose teaching creationism INSTEAD

OF evolution in public schools? Favor—40%, Oppose—55%, No opinion—5%.

  a. What is the margin of error for this poll?

  b. Write a few sentences describing the results of the survey, being truthful but biasing the story in favor of teaching creationism.

  c. Repeat part (b), but bias the story in favor of not teaching creationism.

  d. Repeat part (b), but write an unbiased story.

  e. Explain which of the pitfalls in Section 4.5 is illustrated by this example.

4.51 Suppose a community group wants to convince city officials to put more trash containers on the city streets. They decide to conduct a survey, but rather than a scientifically valid result, they want their results to show that as many citizens as possible want additional trash containers. Give an example of how they could use each of the following to their advantage:

  a. Self-selected sample.

  b. Deliberate bias.

  c. Ordering of questions.

  d. Desire to please.

4.52 Find an example of a poll based on a self-selected sample. Report the wording of the questions and the number who chose each response. Comment on whether or not you think the results can be extended to any population.

4.53 Give an example of two survey questions for which you think the results would be substantially different depending on which order they were asked.

4.54 Explain which of three methods—a door-to-door interview, a telephone interview, or a mail survey—would be *most* likely and which would be *least* likely to suffer from each of the following problems:

  a. Bias due to desire to please the interviewer.

  b. Volunteer response.

  c. Bias due to perceived lack of confidentiality.

4.55 Find an example of a poll conducted by a reputable scientific polling organization. Write a few paragraphs describing how the poll was conducted, what was asked, and the results. Explain whether or not any of the problems discussed in Sections 4.4 and 4.5 were likely to have biased the results of the poll.

4.56 Refer to Example 4.16, "When Will Adolescent Males Report Risky Behavior?" Explain which two of the seven "Possible Sources of Response Bias in Surveys" are illustrated by this example.

4.57 An advertiser of a certain brand of aspirin (let's call it Brand B) claims that it is the preferred painkiller for headaches, based on the results of a survey of headache sufferers. But further investigation reveals that the choices given to respondents were: Tylenol, Extra-Strength Tylenol, Brand B aspirin, Advil.

  a. Is this an open- or closed-form question? Explain.

  b. Comment on the choices given to respondents.

  c. Comment on the advertiser's claim.

  d. What choices do you think should have been given? Explain.

4.58 An NBC News/*Wall Street Journal* poll conducted at the end of the 20th century (September 9–12, 1999) asked a random sample of $n = 1010$ adult Americans, "Which one of the following do you consider to be the best American movie of the 20th century?" The choices given and the percent who chose them were: *Gone With the Wind* 28%, *Schindler's List* 18%, *Titanic* 11%, *Star Wars* 11%, *Casablanca* 8%, *The Godfather* 7%, *Citizen Kane* 6%, *The Graduate* 1%, Not sure 4%. In addition, 2% volunteered the answer "all of them," and 4% volunteered an answer not listed (http://www.pollingreport.com/hollywoo.htm, September 29, 2002).

a. Was this asked as an open-form or closed-form question?

b. Do you think the form used was appropriate for trying to ascertain what movies adult Americans thought were the best of the 20th century? Explain.

c. Give one advantage and one disadvantage each for open-form and closed-form questions in this situation.

4.59 Rock singer Elvis Presley died on August 16, 1977 and his life received substantial publicity as the 25th anniversary of his death approached on August 16, 2002. On August 7–11, 2002 an ABC News poll asked a random sample of $n = 1023$ adult Americans, "Who do you think is the greatest rock 'n' roll star of all time?" There were 128 different stars volunteered by respondents. The top responses and the percent they received were: Elvis Presley 38%, Jimi Hendrix 4%, John Lennon 2%, Mick Jagger 2%, Bruce Springsteen 2%, Paul McCartney 2%, Eric Clapton 2%, Michael Jackson 2%, Other 25%, None 7%, No opinion 15% (http://abcnews.go.com/sections/us/DailyNews/elvis_poll.html, Sept 29, 2002).

a. Was this an open- or closed-form question?

b. Do you think the percent who responded "Elvis Presley" was influenced by the timing of the poll? Explain.

c. Do you think the percent who responded "Elvis Presley" would have been higher, lower, or about the same if the question had given specific choices?

4.60 Refer to the previous exercise. Another question asked in the poll was "Do you consider yourself a fan of Elvis Presley, or not?" The responses were almost equally split, with 49% saying "yes" and 51% saying "no." Do you think the order in which this question and the one in the previous exercise were asked would influence the results to either question? Explain (http://www.pollingreport.com/music.htm, September 29, 2002).

Chapter Exercises

4.61 Refer to Example 4.17, "Politics Is All in the Wording."

a. Explain which one of the "Difficulties and Disasters" listed in Section 4.4 is illustrated by this example.

b. Explain which one of the seven "Possible Sources of Response Bias in Surveys" is illustrated by this example.

4.62 Refer to Example 4.3 in which a survey question was "Do you favor or oppose scientific experimentation on the cloning of human beings?" Suppose an agency that advocated cloning of human beings wanted to conduct a survey that would show support for their cause. Explain how they might use each of the following to do so:

a. Selection bias.

b. Deliberate bias.

c. Desire to please.

d. Ordering of questions.

4.63 Suppose you wanted to estimate the proportion of adults who write with their left hands and decide to watch a sample of n people signing credit card receipts at a mall.

a. Do you think this sample would be representative for the question of interest? Explain.

b. Assuming the sample was representative, how many people would you have to include to estimate the proportion with a margin of error of 5%?

c. Suppose you observed 500 people and 60 of them signed with their left hand. What interval of values would you give as being likely to contain the true proportion of people in the larger population who would sign with their left hand?

d. Is the sampling plan used a probability sampling plan? Explain.

4.64 The faculty senate at a large university wanted to know what proportion of the students thought a foreign language should be required for everyone. The statistics department offered to cooperate in conducting a survey, and a simple random sample of 500 students was selected from all students enrolled in statistics classes. A survey form was sent by e-mail to these 500 students. Discuss the extent to which each of the three types of bias would be likely to occur in this survey:

a. Selection bias.

b. Nonresponse bias.

c. Response bias.

d. Which of these three types of bias do you think would be the most serious? Explain.

4.65 Refer to the previous exercise.

a. What is the population of interest to the faculty senate?

b. What is the sampling frame?

c. What is the sample?

d. Is the sample representative of the population of interest? Explain.

4.66 A large medical professional organization with membership consisting of doctors, nurses, and other medical employees wanted to know how its members felt about HMOs (health maintenance organizations). Name the type of sampling plan they used in each of the following scenarios:

a. They randomly selected 500 members from each of the lists of all doctors, all nurses, and all other employees and surveyed those 1500 members.

b. They randomly selected ten cities from all cities in which its members lived, then surveyed all members in those cities.

c. They randomly choose a starting point from the first 50 names in an alphabetical list of members, then chose every 50th member in the list starting at that point.

4.67 An article in the *Sacramento Bee* was headlined "Drop found in risky behavior among teens over last decade" (June 7, 2000, p. A6; reprinted from the *Los Angeles Times*). The article reported that "the findings show that for the general population [not a specific ethnic group], the share of teens abstaining from all 10 risky behaviors jumped from 20 percent to 25 percent [from the beginning to the end of the 1990s]." The article did not say who was measured, but was obviously based on *samples* taken at each end of the decade. What additional information would you need to know to determine whether or not this drop represents a real drop in the percent of the *population* of teens abstaining? Explain.

4.68 Case Study 3.3 discussed an experiment in which smokers wore a nicotine patch or a placebo patch. Of the 120 participants wearing a nicotine patch, 46% had quit smoking after eight weeks. Assume this group is equivalent to a random sample of smokers desiring to quit.
   a. What is the margin of error accompanying this percent?
   b. What is the approximate 95% confidence interval accompanying this percent?
   c. Suppose the manufacturer of the nicotine patches was to claim that "evidence shows that a majority of smokers with a desire to quit may be helped by nicotine patches." Comment on this statement, referring to your answer in part (b).

4.69 One evening around 6 P.M. authorities in a major metropolitan area received hundreds of phone calls from people reporting that they had just seen an unidentified flying object (UFO). A local television station would like to report the story on their 10 P.M. news and wants to include an estimate of the proportion of area residents who witnessed the UFO. Explain how each of the following methods of obtaining this information would be biased, if at all:
   a. A "quickie poll" taken between 7 P.M. and 9 P.M. that evening, using random-digit dialing for residents of the area.
   b. During commercial breaks between 7 P.M. and 9 P.M. that evening, providing two numbers for viewers to phone—one if they had seen the UFO, and one if they had not.
   c. A door-to-door survey taken between 7 P.M. and 9 P.M., based on a multistage sampling plan, in which ten city blocks (or equivalent area) were randomly selected and interviewers knocked on the doors of all homes in those areas and surveyed those who answered the door.
   d. An e-mail survey sent at 7 P.M. to all e-mail addresses for local residents served by a local Internet Service Provider in the area, using responses received by 9 P.M.

4.70 Refer to the previous exercise. Acknowledging that some bias is inevitable when information is required on such short notice, which of the four methods would you recommend that the television station use to get the information it desires? Explain.

4.71 Refer to the previous two exercises. Suppose that a local newspaper conducted a survey during the next week, based on random-digit dialing of 1400 residents of the area, and found that 20% of them reported having seen a UFO on the evening in question.
   a. What is the margin of error for the survey?
   b. Give an interval of values that probably covers the true percentage of the population that saw the UFO.

4.72 Refer to the survey described in Example 4.9, "Which Scientists Trashed the Public?" Explain which type of bias (selection, nonresponse, or response) was the most problematic in that survey.

4.73 Suppose an (unscrupulous!) organization wanted to conduct a survey in which the results supported its position that drinking coffee in public should be illegal except in designated coffee bars. Explain how they could use each of the following sources of bias to help produce the results they wanted:

   a. Selection bias.
   b. Nonresponse bias.
   c. Response bias.

4.74 Suppose a survey reported that 55% of respondents favored gun control, with a margin of error of ±3 percentage points.
   a. What was the approximate size of the sample?
   b. What is an approximate 95% confidence interval for the percent of the corresponding population who favor gun control?
   c. Based on the material in this chapter, write a statement interpreting the interval you found in part (b) that could be understood by someone with no training in statistics. (In Chapter 10 you will learn a more precise way to interpret this type of interval.)

4.75 Refer to the *Literary Digest* poll in Case Study 4.1. Discuss the extent to which each type of bias (selection, nonresponse, and response) played a role in producing the disastrous results, if at all.

4.76 Use the data in Case Study 1.1 (p. 2) or in the file **pennstate1** on the CD accompanying this book for this exercise.
   a. Select a random sample of ten males and a random sample of ten females and write down their answers to the question "What's the fastest you've ever driven a car?"
   b. Compare the males' and females' responses using appropriate numerical and/or graphical summary information.
   c. Recall from Chapters 1 and 2 that there was an obvious difference between male and female responses for the set of all students. Is this difference obvious in your samples? Explain.
   d. It is true in statistics that the larger the actual difference between two groups is in the population, the smaller the sample sizes needed to detect the difference. Discuss your answer to part (c) in the context of this statement.

Exercises 4.77 to 4.80 are based on a report of a Field Poll printed in the Sacramento Bee *on July 7, 2000 (Dan Smith, p. A4) with the headline "Ratings up for Boxer, Feinstein in new poll." The results were reported with a graphical display containing the information that "results are based on a statewide survey of 750 adults conducted June 9–18. The sample includes 642 voters considered likely to vote in the Nov. 7 general election. The margin of error is 3.8 percentage points."*

4.77 The text of the article said that the 750 adults surveyed were actually 750 *registered voters*. Based on the information about who was surveyed, if the poll results are to be used to predict the proportion of voters likely to vote for Senator Feinstein in November, should the results of all 750 voters be used? Explain, using the Fundamental Rule for Using Data for Inference in your explanation.

4.78 The report stated that the margin of error was 3.8 percentage points. This value is based on the 642 voters considered likely to vote and uses the precise formula you will learn in Chapter 10.
   a. Compute the conservative margin of error for this poll based on those likely to vote, and compare it to the precise margin of error presented.

b. For this sample, 57% reported that they approve of Dianne Feinstein's job performance. What is an approximate 95% confidence interval for the percent of all likely voters who approve of her performance?

4.79 Read the headline that accompanied the article, which stated that ratings were "up" for both senators. For the June 2000 poll, 57% approved of Feinstein's performance, compared with 56% and 53%, respectively, for similar polls taken in February 2000 and October 1999. Presumably, the headline refers to all registered voters in the state, and not just those in the samples. Considering the margin of error accompanying each poll, do you agree with the headline? Explain.

4.80 The article also reported that "among GOP voters, 39 percent give Feinstein high marks." Would the margin of error accompanying this result be larger, smaller, or the same as the margin of error for the entire sample? Explain.

Exercises 4.81 to 4.83 refer to a survey of University of California faculty members on affirmative action policies for the university (Roper Center, 1996). The survey was based on telephone interviews with a random sample of 1001 faculty members conducted in December 1995.

4.81 One of the questions asked was "Do you favor or oppose using race, religion, sex, color, ethnicity, or national origin as a criterion for admission to the University of California?" Of the 804 male respondents, 47% said "Favor," and of the 197 female respondents, 73% said "Favor."
   a. Find the conservative margin of error for the males, and use it to compute an approximate 95% confidence interval for the percent of males in the population who favor this policy.
   b. Repeat the analysis in part (a) for females.
   c. Compare the approximate 95% confidence intervals you found in parts (a) and (b), and comment on whether you think male and female faculty members in the population differed on this question.

4.82 Refer to the previous exercise. Of the 166 arts and humanities faculty members, 66% favored the policy, while of the 229 engineering, mathematical, and physical sciences faculty, 38% favored it.
   a. Find the conservative margin of error for the arts and humanities faculty, and use it to compute an approximate 95% confidence interval for the percent of them in the population who favor this policy.
   b. Repeat the analysis in part (a) for the engineering and sciences faculty.
   c. Compare the approximate 95% confidence intervals you found in parts (a) and (b), and comment on whether you think faculty members in these disciplines in the population differed on this question.

4.83 One question asked was as follows: "The term affirmative action has different meanings to different people. Please tell me which statement best describes what you mean by the term: First, affirmative action means granting preferences to women and certain racial and ethnic groups. Second, affirmative action means promoting equal opportunities for all individuals without regard to their race, sex, or ethnicity." The options "Neither" or "Both" were not given and were recorded only if the respondent offered them. The first statement was chosen by 37%, the second one by 43%, "Neither" by 13%, and "Both" by 2%. The remainder

said, "Don't know." The organization that commissioned the poll, the California Association of Scholars, issued a press release that said in reference to this question, "These findings demolish the claim that the faculty at UC wants preferential policies." Refer to the "Possible Sources of Response Bias" in Section 4.5, and discuss them in the context of this question and resulting press release.

Turn On Your Computer

*For Exercises 4.84 to 4.88, use the **Sampling** applet described in Section 4.2. The applet is on the CD accompanying this text.*

4.84 Choose one stick figure to represent yourself and identify which one it was in your answers to this exercise by giving the row number (1 to 10) and column number (1 to 10) where you reside.
   a. Suppose you are in a class of 100 individuals, represented by these stick figures. Your teacher randomly selects 10 people each class period and asks them to answer a question about the reading assignment. What are your chances of being in the sample each day?
   b. Repeat the sample selection process 20 times, each time noting whether you made it into the sample. How many times did you make it into the sample? Does that surprise you or is it about what you expected?

4.85 Repeat the sampling process 50 times without pressing the **Start Over** button. Use **Show Results** to display the results of the 50 samples. Cut and paste the results into Minitab or another program.
   a. Display a histogram of the 50 means of the chosen samples.
   b. Describe the histogram in part (a), including shape, center, and spread.
   c. Display a histogram of the 50 values of "percent female."
   d. Describe the histogram in part (c).

4.86 Repeat the sampling process 20 times without pressing the **Start Over** button. Use **Show Results** to display the results of the 20 samples. Write down the results, or cut and paste them into a computer program.
   a. What is the median of the 20 sample means?
   b. What is the range of the 20 sample means?
   c. Display a boxplot of the 20 sample means.
   d. We know that the population has a mean height of 68 inches. Based on your results in parts (a), (b), and (c), discuss how useful the mean of a sample of size 10 is for estimating the population mean in this situation.

4.87 Repeat the sampling process 20 times without pressing the **Start Over** button. Use **Show Results** to display the results of the 20 samples. Write down the results, or cut and paste the results into a computer program. Find the mean of the 20 sample means. This value represents the mean of all 200 sample values. Is this combined mean a better estimate of the population mean of 68 than most of the individual sample means? Is that what you would expect, or not? Explain.

4.88 Generate one random sample and print the display showing the applet and your result. Show what stream of numbers from a table of random digits would have produced your sample, numbering the stick figures from 00 to 99, starting at the top and going across rows.

REFERENCES

Abelson, R. P., E. F. Loftus, and A. G. Greenwald (1992). "Attempts to improve the accuracy of self-reports of voting," in *Questions about Questions*, J. M. Tanur (ed.), New York: Russell Sage Foundation, pp. 138–153.

Anastasi, Anne (1988). *Psychological Testing*, 6th ed., New York: Macmillan.

Bryson, M. C. (1976). "The *Literary Digest* poll: Making of a statistical myth," *American Statistician,* Vol. 30, pp. 184–185.

Clark, H. H., and M. F. Schober (1992). "Asking questions and influencing answers," in *Questions about Questions*, J. M. Tanur (ed.), New York: Russell Sage Foundation, pp. 15–48.

Cohn, Victor, and Lewis Cope (2001). *News and Numbers: A Guide to Reporting Statistical Claims and Controversies in Health and Other Fields*, 2nd ed., Ames, IA: Iowa State University Press.

Crossen, Cynthia (1994). *Tainted Truth,* New York: Simon and Schuster.

Freedman, D., R. Pisani, R. Purves, and A. Adhikari (1991). *Statistics,* 2nd ed., New York: W. W. Norton.

Gastwirth, Joseph L. (1988). *Statistical Reasoning in Law and Public Policy,* Vol. 2: *Tort Law, Evidence and Health,* Boston: Academic Press.

Liss, Susan (1997). "The Effects of Survey Methodology Changes in the NPTS," Technical Report, Office of Highway Policy Information, Federal Highway Administration.

Loftus, E. F., and J. C. Palmer (1974). "Reconstruction of automobile destruction: An example of the interaction between language and memory," *Journal of Verbal Learning and Verbal Behavior,* Vol. 13, pp. 585–589.

Mervis, Jeffrey (1998). "Report deplores science-media gap," *Science,* Vol. 279, p. 2036.

Morin, Richard (1995). "What informed public opinion?" *The Washington Post National Weekly Edition,* April 10–16, 1995, p. 36.

Plewes, Thomas J. (1994). "Federal agencies introduce redesigned Current Population Survey," *Chance,* Vol. 7, No. 1, pp. 35–41.

Plous, Scott (1993). *The Psychology of Judgment and Decision Making,* New York: McGraw-Hill.

Roper Center (1996). "University of California Faculty Opinion Survey on Affirmative Action Policies," Roper Center for Public Opinion Research, Press Release, January 15, 1996.

Schuman, H. and J. Scott (1987). "Problems in the use of survey questions to measure public opinion," *Science,* Vol. 236, May 22, 1987, pp. 957–959.

Turner, C. F., L. Ku, S. M. Rogers, L. D. Lindberg, J. H. Pleck, F. L. Sonenstein (1998). "Adolescent sexual behavior, drug use, and violence: Increased reporting with computer survey technology," *Science,* Vol. 280, pp. 867–873.

U.S. Department of Labor, Bureau of Labor Statistics (1992). *BLS Handbook of Methods,* Sept. 1992, Bulletin 2414.

CHAPTER 5

Relationships Between Quantitative Variables

Do college students give up study time to sleep, or vice versa? See Example 5.9.

A *statistical relationship* is different from a *deterministic relationship*, for which the value of one variable can be determined exactly from the value of the other variable. In a statistical relationship, there is variation from the average pattern. Our ability to predict what happens for an individual depends on the amount of natural variability from that pattern.

The description and confirmation of relationships between variables is so important in research that entire courses are devoted to the topic. You have already seen several examples that involved a potential relationship. For instance, in the observational study in Case Study 3.4, the investigators wanted to know if male pattern baldness and the risk of a heart attack were related. In the randomized experiment in Case Study 1.6, the researchers wanted to know if aspirin consumption and risk of heart attack were related.

In this chapter, we will learn how to describe the relationship between two *quantitative variables.* Remember (from Chapter 2) that the terms *quantitative variable* and *measurement variable* are synonyms for data that can be recorded as numerical values and then ordered according to those values. The relationship between weight and height is an example of a relationship between two quantitative variables.

The questions we ask about the relationship between two variables often concern specific numerical features of the association. For example, we may want to know how much weight will increase on average for each 1-inch increase in height. Or, we may want to estimate what the college grade point average will be for a student whose high school grade point average was 3.5.

We will use three tools to describe, picture, and quantify the relationship between two quantitative variables:

- **Scatterplot,** a two-dimensional graph of data values.
- **Correlation,** a statistic that measures the *strength and direction* of a linear relationship between two quantitative variables.
- **Regression equation,** an equation that describes the average relationship between a quantitative response variable and an explanatory variable. ◆

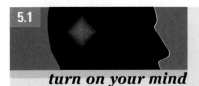

5.1

turn on your mind

For adults, there is a *positive association* between weight and height. For used cars, there is a *negative association* between the age of the car and the selling price. Explain what it means for two variables to have a positive association. Explain what it means when two variables have a negative association. What is an example of two variables that would have *no association*?*

5.1 | LOOKING FOR PATTERNS WITH SCATTERPLOTS

A **scatterplot** is a two-dimensional graph of the measurements for two numerical variables. A point on the graph represents the combination of measurements for an individual observation. The vertical axis, which is called the *y axis,* is used to locate the value of one of the variables. The horizontal axis, called the *x axis,* is used to locate the value of the other variable.

As we learned in Chapter 2, when looking at relationships we can often identify one of the variables as an **explanatory variable** that may explain or cause differences in the **response variable.** The term **dependent variable** is used as a synonym for *response variable.* In a scatterplot, the response variable is plotted on the vertical axis (the *y* axis), so it may also be called the **y variable.** The explanatory variable is plotted along the horizontal axis (the *x* axis) and may be called the **x variable.**

Questions to Ask About a Scatterplot

- What is the *average* pattern? Does it look like a straight line or is it curved?
- What is the direction of the pattern?
- How much do individual points vary from the average pattern?
- Are there any unusual data points?

EXAMPLE 5.1
Height and Handspan

TABLE 5.1 ■ **Handspans and Height**

Height (in.)	Span (cm)
71	23.5
69	22.0
66	18.5
64	20.5
71	21.0
72	24.0
67	19.5
65	20.5
76	24.5
67	20.0
70	23.0
62	17.0

and so on, for *n* = 167 observations.

Table 5.1 displays the first 12 observations of a dataset that includes the heights (in inches) and fully stretched handspans (in centimeters) of 167 college students. The data values for all 167 students are the raw data for studying the connection between height and handspan. Imagine how difficult it would be to see the pattern in the data if all 167 observations were shown in Table 5.1. Even when we just look at the data for 12 students, it takes a while to confirm that there does seem to be a tendency for taller people to have larger handspans.

Figure 5.1 is a scatterplot that displays the handspan and height measurements for all 167 students. The handspan measurements are plotted along the vertical axis (*y*), and the height measurements are plotted along the horizontal axis (*x*). Each point represents the two measurements for an individual.

We see that taller people tend to have greater handspan measurements than shorter people do. When two variables tend to increase together, as they do in Figure 5.1, we say that they have a **positive association.** Another noteworthy characteristic of the graph is that we can describe the general pattern of this re-

*HINT: Average weight increases as height increases. The selling price decreases as a car's age increases. Use these patterns to define positive and negative association more generally.

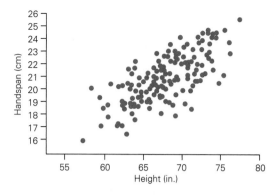

FIGURE 5.1   **Handspans and height**

lationship with a straight line. In other words, the handspan and height measurements may have a **linear relationship.** ◆

> *definition* ◆ Two variables have a **positive association** when the values of one variable tend to increase as the values of the other variable increase.
>
> ◆ Two variables have a **negative association** when the values of one variable tend to decrease as the values of the other variable increase.

EXAMPLE 5.2
Driver Age and the Maximum Legibility Distance of Highway Signs

In a study of the legibility and visibility of highway signs, a Pennsylvania research firm determined the maximum distance at which each of 30 drivers could read a newly designed sign. The 30 participants in the study ranged in age from 18 to 82 years old. The government agency that funded the research hoped to improve highway safety for older drivers and wanted to examine the relationship between age and the sign legibility distance.

Table 5.2 lists the data and Figure 5.2 shows a scatterplot of the ages and distances. The sign legibility distance is the response variable, so that variable is plotted on the *y* axis (the vertical axis). The maximum reading distance tends to

TABLE 5.2 ■ **Data Values for Example 5.2**

Age	Distance	Age	Distance	Age	Distance
18	510	37	420	68	300
20	590	41	460	70	390
22	560	46	450	71	320
23	510	49	380	72	370
23	460	53	460	73	280
25	490	55	420	74	420
27	560	63	350	75	460
28	510	65	420	77	360
29	460	66	300	79	310
32	410	67	410	82	360

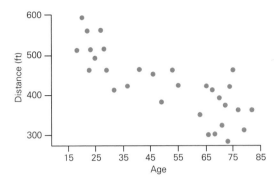

FIGURE 5.2   Driver age and the maximum distance at which highway sign is read
(Source: Adapted from data collected by Last Resource, Inc., Bellefonte, PA.)

decrease as age increases, so there is a negative association between distance and age. This is not a surprising result. As a person gets older, his or her eyesight tends to get worse, so we would expect the distances to decrease with age.

The researchers collected the data to determine numerical estimates for two questions about the relationship:

- How much does the distance decrease when age is increased?
- For drivers of any specific age, what is the average distance at which the sign can be read?

We'll examine these questions in the next section. For now, we simply point out that the pattern in the graph looks *linear,* so a straight-line equation that links distance to age will help us answer these questions. ◆

Curvilinear Patterns

A linear pattern is common, but it is not the only type of relationship. Sometimes, a curve describes the pattern of a scatterplot better than a line does, and when that's the case the relationship is called **nonlinear,** or **curvilinear.**

EXAMPLE 5.3
The Development
of Musical Preferences

Will the music that you like now always be your favorite music? If you are about 20 years old, the answer is likely to be "yes" according to research reported in a 1989 issue of the *Journal of Consumer Research* (Holbrook and Schindler, 1989). The research finding was that popular music preferences are acquired in late adolescence and early adulthood.

The researchers asked 108 participants from 16 to 86 years old to listen to 28 top hits from the years 1932 to 1986. Each song had been on *Billboard*'s Top 10 list for popular music. Respondents rated the 28 songs using a 10-point scale on which the value "1" corresponded to the statement "I dislike it a lot" and the value "10" corresponded to "I like it a lot." After collecting the data, the researchers adjusted the music-preference scores so that the mean score for each individual was 0. On this scale, a positive score indicates that a song rating was above average for the respondent, and a negative score indicates a below-average rating for the respondent. This was done to adjust for person-to-person

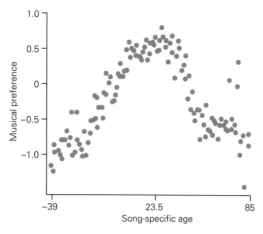

FIGURE 5.3 **Scatterplot of music-preference score versus song-specific age**
(Source: Figure adapted from Holbrook and Schindler, 1989, p. 122.)

variation in the rating levels because some respondents gave all songs high ratings while others assigned low ratings to all songs.

The researchers also created an adjusted age variable that they called the song-specific age of the respondent. This was the respondent's "age" in the year the song was popular. If the song was popular before the respondent was born, the song-specific age is negative. For example, if a song was on the *Billboard* Top 10 list in 1934 and the respondent was born in 1954, the song-specific age is −20, because the song was popular 20 years before the respondent was born.

Figure 5.3 shows the relationship between the preference rating and song-specific age. The overall pattern looks somewhat like an inverted U, and the maximum preference ratings occur when the song-specific age is in the late teens and early twenties. With this pattern, we can't characterize the relationship as being either generally positive or negative. Also, a straight line does not describe the overall pattern, so the association is nonlinear (or curvilinear). ◆

Indicating Groups Within the Data on Scatterplots

When we examined the connection between height and handspan in Example 5.1, you may have wondered whether we should be concerned about student gender. Both height and handspan tend to be greater for men than for women, so we should consider the possibility that gender differences might be completely responsible for the observed relationship.

It's easy to indicate subgroups on a scatterplot. We just use different symbols or different colors to represent the different groups. Figure 5.4 is the same as Figure 5.1, but now different symbols are used for males and females. Notice that the positive association between handspan and height appears to hold within each sex. For both men and women, handspan tends to increase as height increases. It's not always the case that the pattern in each subgroup is consistent with the pattern in the whole group. Later in this chapter, we will see that when we combine subgroups inappropriately, the relationship for the combined group can misrepresent the relationship that we see in each subgroup.

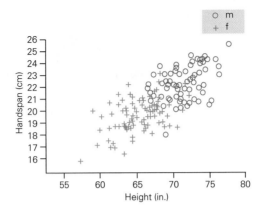

FIGURE 5.4   **Handspan and height by gender**

Look for Outliers

Outliers can have a big impact when we quantify a relationship, as we will see in more detail later in this chapter. When we consider two variables, an **outlier** is a point that has an unusual combination of data values. For instance, a man 6′7″ tall who weighs 145 pounds would probably be an outlier in a scatterplot of weights and heights because this is an unusual *combination* of weight and height measurements. Outliers can occur because there are interesting and unusual data points, or they may occur because mistakes were made when the data were recorded or entered into the computer.

EXAMPLE 5.4
Heights and Foot Lengths of College Women

Figure 5.5 shows the relationship between foot length (in centimeters) and height (in inches) for a sample of 41 college women. The two shortest women have much greater foot lengths than we would predict based on the rest of the data. On the other side of the height scale, the tallest woman (74 inches) has a much shorter foot length than we might expect. Fortunately, the students submitted their measurements on a paper form, and a look at those forms revealed that the heights of the three "unusual" women were incorrectly entered into the computer. The woman who appears to be 74 inches tall is actually 64 inches tall.

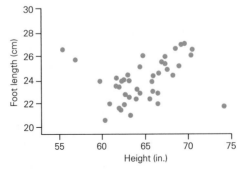

FIGURE 5.5   **Outliers in the relationship between the heights and foot lengths of women**
(Data Source: Collected in class by one of the authors.)

The women with heights of 55 inches and 57 inches on the plot were actually 65 and 67 inches tall. ◆

Minitab Tip

Minitab v. 14 update on CD

 tech note   **Graphing the Relationship Between Two Quantitative Variables**

◆ To draw a scatterplot, use **Graph>Plot.** In the dialog box, specify the columns containing the raw data for Y and X.

◆ To mark different subgroups with different symbols, use the "Data display:" area of the Plot dialog box, put the word Group under "For each," and specify the column that defines the groups under "Group Variables."

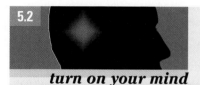

5.2

turn on your mind

Suppose you were to make a scatterplot of (adult) daughters' heights versus mothers' heights by collecting data on both variables from several of your female friends. You would now like to predict how tall your infant niece will be when she grows up. How would you use your scatterplot to help you make this prediction? What other variables, aside from her mother's height, might be useful for improving your prediction? How could you use these variables in conjunction with the mother's height?*

5.2 | DESCRIBING LINEAR PATTERNS WITH A REGRESSION LINE

Scatterplots show us a lot about a relationship, but we often want more specific numerical descriptions of how the response and explanatory variables are related. Imagine, for example, that we are examining the weights and heights of a sample of college women. We might want to know what the increase in average weight is for each 1-inch increase in height. Or, we might want to estimate the average weight for women with a specific height, like 5′10″.

Regression analysis is the area of statistics used to examine the relationship between a quantitative response variable and one or more explanatory variables. A key element of regression analysis is the *estimation of an equation* that describes how, on average, the response variable is related to the explanatory variables. This **regression equation** can be used to answer the types of questions that we just asked about the weights and heights of college women.

A regression equation can also be used to make **predictions.** For instance, it might be useful for colleges to have an equation for the connection between verbal SAT score and college grade point average (GPA). They could use that equation to predict the potential GPAs of future students, based on their verbal SAT scores. Some colleges actually do this kind of prediction to decide whom to admit, but they use a collection of variables to predict GPA. The prediction equa-

*HINT: Perhaps use the approximate average of daughters' heights for mothers who are about the same height as your niece's mother.

tion for GPA usually includes high school GPA, high school rank, verbal and math SAT scores, and possibly other factors such as a rating of the student's high school or the quality of an application essay.

There are many types of relationships and many types of regression equations. *The simplest kind of relationship between two variables is a straight line,* and that's the only type we will discuss here. Straight-line relationships occur frequently in practice, so this is a useful and important type of regression equation. Before we use a straight-line regression model, however, we should always examine a scatterplot to verify that the pattern actually is linear. We remind you of the music preference and age example where a straight line definitely does not describe the pattern of the data.

> **definition** When the best equation for describing the relationship between *x* and *y* is a straight line, the resulting equation is called the **regression line.** This line is used for two purposes:
>
> - to *estimate the average value* of *y* at any specified value of *x*
> - to *predict the value* of *y* for an individual, given that individual's *x* value

EXAMPLE 5.1—CONTINUED
Describing Height and Handspan with a Regression Line

In Figure 5.1, we saw that the relationship between handspan and height has a straight-line pattern. Figure 5.6 displays the same scatterplot as Figure 5.1, but now a line is shown that describes the average relationship between the two variables. To determine this line, we used statistical software (Minitab) to find the "best" line for this set of measurements. We'll discuss the criterion for "best" later. Presently, let's focus on what the line tells us about the data.

The line drawn through the scatterplot is the regression line, and it describes how average handspan is linked to height. For example, when the height is 60 inches, the vertical position of the line is at about 18 centimeters. To see this, locate 60 inches along the horizontal axis (*x* axis), look up to the line, and then read the vertical axis to determine the handspan value. The result is that we can estimate that people 60 inches tall have an *average* handspan of about 18 centimeters (roughly 7 inches). We can also use the line to *predict* the handspan for an individual whose height is known. For instance, someone 60 inches tall is *predicted* to have a handspan of about 18 centimeters. ◆

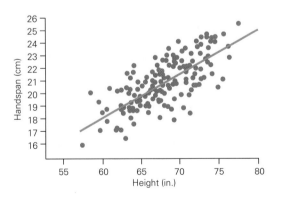

FIGURE 5.6   **Regression line describing handspan and height**

If we estimate the average handspan at a different height, we can determine how much handspan changes, on average, when height is varied. Let's use the line to estimate the average handspan for people who are 70 inches tall. We see that the vertical location of the regression line is somewhere between 21 and 22 centimeters, perhaps about 21.5 centimeters (roughly 8.5 inches). So, when height is increased from 60 inches to 70 inches, average handspan increases from about 18 centimeters to about 21.5 centimeters.

The average handspan increased by 3.5 centimeters (about 1.5 inches) when the height was increased by 10 inches. This is a rate of 3.5/10 = 0.35 centimeter per one inch increase in height, which is the **slope** of the line. For each 1-inch difference in height, there is about a 0.35-centimeter average difference in handspan.

Algebra Reminder

formula The equation for a straight line relating y and x is

$$y = b_0 + b_1 x$$

where b_0 is the "y **intercept**" and b_1 is the slope. When $x = 0$, y is equal to y intercept.

The **slope** of a line can be determined by picking any two points on the line, and then calculating

$$\text{Slope} = \frac{\text{Difference between } y \text{ values}}{\text{Difference between } x \text{ values}} = \frac{y_2 - y_1}{x_2 - x_1}$$

The letter y represents the vertical direction, and x represents the horizontal direction. The *slope* tells us how much the y variable changes for each increase of one unit in the x variable.

We ordinarily don't have to "read" the regression line as we just did. Statistical software will tell us the *regression equation,* the specific equation used to draw the line. For the handspan and height relationship, the regression equation determined by statistical software is

Handspan = −3 + 0.35 Height

When emphasis is on using the equation to estimate the average handspans for specific heights, we may write

Average handspan = −3 + 0.35 Height

When emphasis is on using the equation to predict an individual handspan, we might instead write

Predicted handspan = −3 + 0.35 Height

In most situations, the correct statistical interpretation of a regression equation is that it estimates the average value of a response variable (y) for individuals with a specific value of the explanatory variable (x). The equation Handspan = −3 + 0.35 (Height) tells us how to draw the line, but not all individuals follow this pattern exactly. Look again at Figure 5.6, in which we see that the line describes the overall pattern, but we also see substantial individual deviation from this line.

Let's use the regression equation to estimate the average handspans for some specific heights.

For height = 60, average handspan = −3 + 0.35(60) = −3 + 21 = 18 cm.
For height = 70, average handspan = −3 + 0.35(70) = −3 + 24.5 = 21.5 cm.

In the equation, the value 0.35 multiplies the height. This value is the *slope* of the straight line that links handspan and height. Consistent with our previous estimates, the slope in this example tells us that handspan increases by 0.35 centimeter, on average, for each increase of 1 inch in height. We can use the slope to estimate the average difference in handspan for any difference in height. If we consider two heights that differ by 7 inches, our estimate of the difference in handspans would be 7 × 0.35 = 2.45 centimeters, or approximately 1 inch.

Statistical Relationships Versus Deterministic Relationships

A **statistical relationship** is different from a **deterministic relationship.** In a *deterministic relationship,* if we know the value of one variable we can exactly determine the value of the other variable. For example, the relationship between the volume and weight of water is deterministic. Every pint of water weighs 1.04 pounds, so we can determine exactly the weight of any number of pints of water.

You can see from Figure 5.6 that the regression line does not predict exactly what will happen for each individual. Most individuals do not have a handspan exactly equal to −3 + 0.35 (Height), the handspan that would be predicted from the regression equation. In a *statistical relationship,* there is variation from the average pattern.

A regression equation is most useful for describing the pattern of the *average* relationship for a group. Our ability to predict what happens for an individual depends on the amount of natural variability from the overall pattern. If most measurements are close to the regression line, we may be able to accurately predict what will happen for an individual. When there is substantial variation from the line, we will not be able to accurately predict what will happen for an individual.

The Equation for the Regression Line

All straight lines can be expressed by the same formula in which y is the variable on the vertical axis and x is the variable on the horizontal axis. The equation for a regression line is

$$\hat{y} = b_0 + b_1 x$$

In any given situation, the sample is used to determine numbers that replace b_0 and b_1.

- ◆ $\hat{y}$ is spoken as "y-hat," and it is also referred to either as **predicted y** or **estimated y.**

- ◆ b_0 is the *intercept* of the straight line. The intercept is the value of y when $x = 0$.

- ◆ b_1 is the *slope* of the straight line. The slope tells us how much of an increase (or decrease) there is for the y variable when the x variable increases by one unit. The sign of the slope tells us whether y increases or decreases when x increases.

in summary   Interpreting a Regression Line

◆ $\hat{y}$ estimates the average y for a specific value of x. It also can be used as a prediction of the value of y for an individual with a specific value of x.

◆ The slope of the line estimates the average increase in y for each one unit increase in x.

◆ The intercept of the line is the value of y when $x = 0$. Note that interpreting the intercept in the context of statistical data only makes sense if $x = 0$ is included in the range of observed x values.

EXAMPLE 5.2 —
CONTINUED
Driver Age and
the Maximum Legibility
Distance of Highway Signs

The regression line $\hat{y} = 577 - 3x$ describes how the maximum sign legibility distance (the y variable) is related to driver age (the x variable). Statistical software was used to calculate this equation and to create the graph shown in Figure 5.7. Earlier, we asked these two questions about distance and age:

◆ How much does the distance decrease when age is increased?

◆ For drivers of any specific age, what is the average distance at which the sign can be read?

The slope of the equation can be used to answer the first question. Remember that the slope is the number that multiplies the x variable and the sign of the slope indicates the direction of the association. Here, the slope tells us that, on average, the legibility distance decreases 3 feet when age increases by one year. This information can be used to estimate the average change in distance for any difference in ages. For an age *increase* of 30 years, the estimated *decrease* in legibility distance is 90 feet because the slope is −3 feet per year.

The question about estimating the average legibility distances for a specific age is answered by using the specific age as the x value in the regression equation. To emphasize this use of the regression line, we write it as

Average distance = 577 − 3 Age

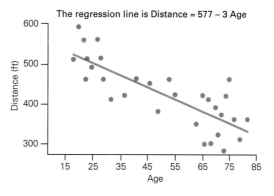

FIGURE 5.7   **Regression line for driver age and sign legibility distance**

Here are the results for three different ages:

Age	Average Distance
20	$577 - 3(20) = 517$ feet
50	$577 - 3(50) = 427$ feet
80	$577 - 3(80) = 337$ feet

The equation can also be used to predict the distance measurement for an *individual* driver with a specific age. To emphasize this use of the regression line, we write the equation as

$$\text{Predicted distance} = 577 - 3\,\text{Age}$$

For example, we can predict that the legibility distance for a 20-year-old will be 517 feet, and for an 80-year-old it will be 337 feet. ◆

Extrapolation

It is risky to use a regression equation to predict values outside the range of the observed data, a process called **extrapolation.** There is no guarantee that the relationship will continue beyond the range for which we have observed data. Suppose that a sample of adult men is used to estimate a regression equation that relates weight to height, and the equation of the line is Weight $= -180 + 5$ (Height). This equation should work well in the range of heights that we see in adult men, but it won't describe the weights of children. If we use the equation to estimate the weight of a boy 36 inches tall, the answer is $-180 + 5(36) = 0$ pounds. The straight-line equation developed for adult men doesn't accurately describe the connection between the weights and heights of children.

Extrapolation also is an issue when regression methods are used to predict future values of a variable. For instance, a straight line describes the relationship between $y =$ winning time in Olympic women's 100-meter backstroke swim and $x =$ Olympic year. This straight line could be used to predict the winning time in the near future, but it should not be used to predict the time in the year 3000.

Prediction Errors and Residuals

For any given line, we can calculate the predicted value $\hat{y}$ for each point in the observed data. To do this for any particular point, we use the observed x value in the regression equation. The **prediction error** for an observation is the difference between the observed y value and the predicted value $\hat{y}$; the formula is $error = (y - \hat{y})$. The terminology "error" is somewhat misleading, since the amount by which an individual differs from the line is usually due to natural variation rather than "errors" in the measurements. A more neutral term for the difference $(y - \hat{y})$ is that it is the **residual** for that individual.

EXAMPLE 5.2—*CONTINUED*
Prediction Errors for
the Highway Sign Data

Example 5.2 described a study in which $y =$ maximum distance at which a person can read a highway sign was related to $x =$ age. The regression equation for these data is $\hat{y} = 577 - 3x$. To calculate $\hat{y}$ for an individual, substitute his or her age for x in the equation. For individuals in the sample, an observed value of y is available and the residual $(y - \hat{y})$ can then be found. For the first three individuals shown in Table 5.2, the *residuals*, or *prediction errors*, are calculated as follows:

$x =$ Age	$y =$ Distance	$\hat{y} = 577 - 3x$	Residual $= y - \hat{y}$
18	510	$577 - 3(18) = 523$	$510 - 523 = -13$
20	590	$577 - 3(20) = 517$	$590 - 517 = 73$
22	516	$577 - 3(22) = 511$	$516 - 511 = 5$

This process could be carried out for any of the 30 observations in the dataset. The seventh individual in Table 5.2, for instance, has age = 27 years and distance = 560 feet. The predicted distance for this person is $\hat{y} = 577 - 3(27) = 496$ feet, so the residual is $(y - \hat{y}) = 560 - 496 = 64$ feet. A positive residual indicates that the individual had an observed value that was higher than what would be predicted for someone of that age. In this case, the 27-year-old in the study could see the sign at a distance 64 feet longer than would be predicted for someone of that age. Figure 5.8 illustrates this residual by showing that the residual is the vertical distance from a data point to the regression line. ◆

The Least Squares Criterion

We can use statistical software to estimate the regression line, but how does the computer find the best-fitting equation for a set of data? A mathematical criterion called **least squares** is nearly always the basis for estimating the equation of a regression line. The term *least squares* is a shortened version of "least sum of squared errors." This criterion focuses on the prediction errors $(y - \hat{y})$, the differences between the values of the response variable (y) and the predicted values. The response variable is emphasized because we often use the equation to predict that variable for specific values of the explanatory variable x. Therefore, we should minimize how far predictions will be off in that direction.

A **least squares line** has the property that the sum of squared differences between the observed values of y and the predicted values is smaller for that line than it is for any other line. Put more simply, the least squares line minimizes the sum of squared prediction errors for the observed data set. The notation **SSE,** which stands for sum of squared errors, is used to represent the sum of squared prediction errors. Thus, we can also say that that the least squares line (the regression line) has a smaller SSE than any other regression line that might be used to predict the response variable.

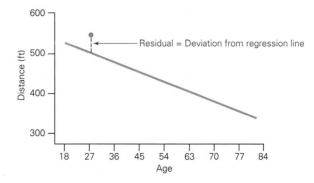

FIGURE 5.8   **Residual from regression line for 27-year-old who saw sign at a distance of 560 feet. The residual, also called the prediction error, is the difference between observed $y = 560$ feet and $\hat{y} = 496$ feet.**

There is a mathematical solution that produces general formulas for computing the slope and intercept of the least squares line. These formulas are used by all statistical software, spreadsheet programs, and statistical calculators. To be complete, we include the formulas. In practice, however, regression analysis is done using a computer so we don't include an example showing how to calculate the slope and intercept for the least squares line "by hand."

Formulas for the Slope and Intercept of the Least Squares Line

formula b_1 is the slope and b_0 is the y intercept.

$$b_1 = \frac{\sum_i (x_i - \bar{x})(y_i - \bar{y})}{\sum_i (x_i - \bar{x})^2}$$

$$b_0 = \bar{y} - b_1 \bar{x}$$

x_i represents the x measurement for the ith observation.

y_i represents the y measurement for the ith observation.

$\bar{x}$ represents the mean of the x measurements.

$\bar{y}$ represents the mean of the y measurements.

EXAMPLE 5.5
Calculating the Sum of Squared Errors

Suppose that x = score on exam 1 in a course and y = score on exam 2, and that the first two rows in Table 5.3 (shown below) give x values and y values for $n = 6$ students. For these data, the least squares regression line is $\hat{y} = 20 + 0.8x$ (found using Minitab). Values of $\hat{y}$ for all observations are given in the third row of Table 5.3, and the fourth row gives the corresponding values of the prediction errors $(y - \hat{y})$. For instance, $x = 70$ and $y = 75$ for the first observation shown in Table 5.3, so $\hat{y} = 20 + 0.8(70) = 76$ and $(y - \hat{y}) = 75 - 76 = -1$. The sum of the squared prediction errors for the regression line is

$$\begin{aligned} \text{SSE} &= (-1)^2 + (2)^2 + (-4)^2 + (2)^2 + (2)^2 + (-1)^2 \\ &= 1 + 4 + 16 + 4 + 4 + 1 = 30 \end{aligned}$$

The line $\hat{y} = 20 + 0.8x$ is the least squares line, so any other line will have a sum of squared errors greater than 30. As an example, if the line $\hat{y} = 4 + x$ were used to predict the values of y, the sum of squared values of $(y - \hat{y})$ would be

$$\begin{aligned} &(75 - 74)^2 + (82 - 79)^2 + (80 - 84)^2 + (86 - 84)^2 + (90 - 89)^2 \\ &+ (91 - 94)^2 = 40 \; \blacklozenge \end{aligned}$$

Why Regression Is Called Regression

You may wonder why the word *regression* is used to describe the study of statistical relationships. Most of the vocabulary used by statisticians has at least

TABLE 5.3 ■ Values of x, y, $\hat{y}$, and $(y - \hat{y})$ for Example 5.5

x = Exam 1 score	70	75	80	80	85	90
y = Exam 2 score	75	82	80	86	90	91
$\hat{y} = 20 + 0.8x$	76	80	84	84	88	92
$(y - \hat{y})$	−1	2	−4	2	2	−1

some connection to the common usage of the words, but this doesn't seem to be true for *regression*. The statistical use of the word *regression* dates back to Francis Galton, who studied heredity in the late 1800s. (See Stigler, 1986 or 1989, for a detailed historical account.) One of Galton's interests was whether or not a man's height as an adult could be predicted by his parents' heights. He discovered that it could, but the relationship was such that very tall parents tended to have children who were shorter than they were, and very short parents tended to have children taller than themselves. He initially described this phenomenon by saying that there was "reversion to mediocrity" but later changed to the terminology "regression to mediocrity." Thereafter the technique of determining such relationships was called *regression*.

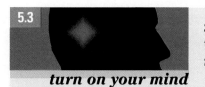

5.3

turn on your mind

Suppose the statistics community is having a contest to rename "regression" to something more descriptive of what it actually does. What would you suggest as a name for the whole procedure? As a name for the regression line?*

5.3 | MEASURING STRENGTH AND DIRECTION WITH CORRELATION

The linear pattern is so common that a statistic was created to characterize this type of relationship. The statistical **correlation** between two quantitative variables is a number that *indicates the strength and the direction of a straight-line relationship*.

♦ The *strength* of the relationship is determined by the *closeness of the points to a straight line*.

♦ The *direction* is determined by whether one variable generally increases or generally decreases when the other variable increases.

As used in statistics, the meaning of the word *correlation* is much more specific than it is in everyday life. A statistical correlation only describes *linear relationships*. Whenever a correlation is calculated, a straight line is used as the frame of reference for evaluating the relationship. When the pattern is nonlinear, as it was for the music preference data shown in Figure 5.3, a correlation is not an appropriate way to measure the strength of the relationship.

Correlation is represented by the letter *r*. Sometimes this measure is called the "Pearson product moment correlation" or the "correlation coefficient." It doesn't matter which of the two variables is called the *x* variable and which is called the *y* variable. The value of the correlation is the same either way. For instance, the correlation between height and foot length is the same regardless of whether you use height as the *y* variable or use foot length as the *y* variable. Another useful feature of the correlation coefficient is that its value doesn't change when the measurement units are changed for either or both of the variables. For instance, the correlation between weight and height is the same whether the measurements are in pounds and inches, or in kilograms and centimeters.

The formula for calculating the correlation coefficient looks complicated, although it can be described rather simply in terms of standardized scores (introduced in Section 2.7). Approximately, the correlation value is the average product of standardized scores for variables x and y. Calculating a correlation value by hand, however, generally involves much labor, so all statistical software programs and many calculators provide a way to easily calculate this statistic. In this section, we focus on how to interpret the correlation coefficient rather than how to calculate it.

A Formula for Correlation

formula

$$r = \frac{1}{n-1}\sum_i \left(\frac{x_i - \bar{x}}{s_x}\right)\left(\frac{y_i - \bar{y}}{s_y}\right)$$

n is the sample size.

x_i is the x measurement for the ith observation.

$\bar{x}$ is the mean of the x measurements.

s_x is the standard deviation of the x measurements.

y_i is the y measurement for the ith observation.

$\bar{y}$ is the mean of the y measurements.

s_y is the standard deviation of the y measurements.

Interpreting the Correlation Coefficient

Some specific features of the correlation coefficient are

- Correlation coefficients are always between -1 and $+1$.
- The magnitude of the correlation indicates the strength of the relationship, which is the overall closeness of the points to a straight line. The sign of the correlation does not tell us about the strength of the linear relationship.
- A correlation of either $+1$ or -1 indicates that there is a perfect linear relationship and all data points fall on the same straight line.
- The sign of the correlation indicates the direction of the relationship. A *positive* correlation indicates that the two variables tend to increase together (a positive association). A *negative* correlation indicates that when one variable increases the other is likely to decrease (a negative association).
- A correlation of 0 indicates that the best straight line through the data is exactly horizontal, so that knowing the value of x does not change the predicted value of y.

EXAMPLE 5.1—*CONTINUED*
The Correlation Between Handspan and Height

The relationship between handspan and height appears to be linear, so a correlation is useful for characterizing the strength of the relationship. For these data, the correlation is $r = +0.74$, a value that indicates a somewhat strong positive relationship. A look back at Figure 5.1 shows us that average handspan definitely increases when height increases, but within any specific height there is some natural variation among individual handspans. ◆

EXAMPLE 5.2—*CONTINUED*
The Correlation Between Age and Sign Legibility Distance

The correlation for the data shown in Figure 5.2, relating driver age and sign legibility distance, is $r = -0.8$. This value indicates a somewhat strong negative association between the variables. ◆

FIGURE 5.9   **Left handspan versus right handspan**

FIGURE 5.10   **Verbal SATs and grade point average**

EXAMPLE 5.6
Left and Right Handspans

If you know the span of a person's right hand, do you think you could accurately estimate his or her left handspan? Figure 5.9 displays the relationship between the right and left handspans (in centimeters) of the 190 college students in the dataset of Chapter 2. In the plot, the points nearly fall into a straight line. The correlation coefficient for this strong positive association is +0.95. ◆

EXAMPLE 5.7
Verbal SAT and GPA

The scatterplot in Figure 5.10 shows the grade point averages (GPAs) and verbal SAT scores for a sample of 100 students at a university in the northeastern United States. The correlation for the data in the scatterplot is $r = 0.485$, a value that indicates only a moderately strong relationship. ◆

EXAMPLE 5.8
Age and Hours of Television Viewing per Day

On a typical day, how many hours do you spend watching television? The National Opinion Research Center asks this question in its General Social Survey. In Figure 5.11, we see the relationship between respondent age and hours of daily television viewing for 1913 respondents in the 1996 survey. There does not seem to be much of a relationship between age and television hours, and the correlation of only +0.12 confirms this weak connection between the variables. We also see some odd responses. A few respondents claim to watch television more than 20 hours per day! ◆

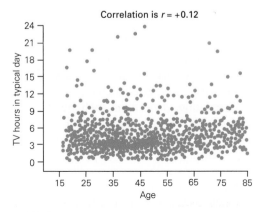

FIGURE 5.11   **Age and hours of television per day**

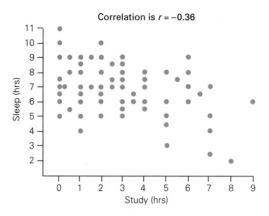

Correlation is $r = -0.36$

FIGURE 5.12   **Hours of sleep versus hours of study**
(Source: Class data collected by one of the authors.)

EXAMPLE 5.9
***Hours of Sleep
and Hours of Study***

Figure 5.12 displays, for a sample of 116 college students, the relationship between the reported hours of sleep during the previous 24 hours and the reported hours of study during the same period. The correlation value for this scatterplot is $r = -0.36$, indicating a negative association that is not particularly strong. On average, the hours of sleep decrease as hours of study increase, but there is substantial variation in the hours of sleep for any specific hours of study. ◆

turn on your computer

Exploring Correlation

The **Correlation** applet on the CD accompanying this book can be used to explore how the correlation coefficient, *r*, is sensitive to the strength and direction of the relationship between two quantitative variables. Remember that the strength of a linear relationship is measured by the absolute value of the correlation coefficient. The sign of the correlation value indicates whether the two variables have a positive association or a negative association.

What Happens

Your goal is to create a scatterplot so that the correlation value for the points on the graph is close to a "goal" value. You place points onto a graph by using the mouse to click on locations in a graph. The applet recalculates and displays the correlation value after each point is added. The applet will declare "Goal Reached!" when the correlation is within ±0.05 of the goal value after 15 or more points have been added to the plot.

What to Do

Open the **Correlation** applet. You'll see three different scatterplot regions, each with a different goal value for the correlation. Figure 5.13 shows the first of these regions before any points have been added. For that region, the goal is $r = 0.5$,

FIGURE 5.13   **Starting point for the Correlation applet. The goal is** $r = 0.5$ **and at least 15 points should be placed on the graph.**

FIGURE 5.14   **Appearance of Correlation applet after 5 points have been added. For these 5 points, the correlation is** $r = 0.6087$**, a value shown at the top of the scatter plot.**

so your task is to create a graph with at least 15 data points for which the correlation is anywhere between 0.45 and 0.55.

Begin adding points to the graph by using the mouse to click on desired locations. An example of how the applet might look after 5 data points have been added is displayed in Figure 5.14. Notice that for the 5 points shown in the figure, the correlation is $r = 0.6087$. Continue adding points until the goal is reached. You can delete points by clicking on the ***Delete*** radio button and then clicking on the point you wish to delete. The ***Clear*** button can be used to remove all points.

The correlation goals for the second and third scatterplots on the applet page are $r = -0.8$ and $r = 0$, respectively. Create scatterplots that achieve those targets. In each instance, first try achieving the correlation by making your x values range from about 10 to 90, and not including any outliers. Remember that at least 15 points should be placed on a graph. Figures 5.2, 5.10, and 5.11 in this chapter provide models for the three target correlations in the applet.

After reaching your goals, clear the plots and then explore how an outlier affects a correlation value. For instance, try reaching the $r = 0.5$ goal for the first scatterplot by adding 14 points for which the correlation is above 0.8, then adding an outlier that brings the correlation down to about 0.5. Or, where the goal is $r = -0.8$, put 14 points in the upper left portion of the graph so the correlation for those points is between -0.2 and $+0.2$, then add an outlier that makes the correlation become about -0.8.

Lessons Learned

The three target correlations exemplify how correlation is related to the strength and direction of a relationship. The algebraic sign of the correlation value gives the direction of the relationship. For linear relationships with no outliers present in the data, the absolute value of the correlation measures the strength of a relationship, which has to do with the overall closeness of points to a line. By using the applet to put outliers on the plots, you will see that an outlier can either increase or decrease a correlation.

The Interpretation of r^2, the Squared Correlation

The squared value of the correlation is frequently used to describe the strength of a relationship. A **squared correlation, r^2,** always has a value between 0 and 1, although some computer programs will express its value as a percent between 0 and 100%. By squaring the correlation, we retain information about the strength of the relationship, but we lose information about the direction.

Researchers typically use the phrase "**proportion of variation explained by x**" in conjunction with the squared correlation, r^2. For example, if $r^2 = 0.60$ (or 60%), the researcher may write that the explanatory variable explains 60% of the variation in the response variable. If $r^2 = 0.10$ (or 10%), the explanatory variable only explains 10% of the variation in the response variable. This interpretation stems from the use of the least squares line as a prediction tool.

In Example 5.6, the correlation between left and right handspans is 0.95, so r^2 is 0.90, or about 90%. This indicates that the span of one hand is very predictable if we know the span of the other hand (see Figure 5.9). In Example 5.8, the correlation between television viewing hours and age is only $r = 0.12$. The squared correlation is about $r^2 = 0.014$, only about 1.4%. As we can see from the scatterplot in Figure 5.11, knowing a person's age doesn't help us much in predicting how much television he or she watches per day.

Let's consider Example 5.1 again. For that example, the regression equation is Handspan = $-3.5 + 0.35$ (Height). The correlation is 0.74 and $r^2 = (0.74)^2 = 0.55$ (or 55%). We can say that height explains 55% of the variation in handspan, but what does it mean to say this?

Suppose that we ignore the height information when we make predictions of handspan. In other words, suppose we don't use the least squares line to predict handspan. For the 167 students in the dataset, $\bar{y}$, the average handspan, is about 20.8 centimeters. If we ignore the least squares line, we could use this value to predict the handspan for any individual, regardless of his or her height. Our prediction "equation" is simply Handspan = 20.8. For both this equation and the least squares equation involving height, we can compute the sum of squared differences between the actual handspan values and the predicted values.

- The sum of squared differences between observed y values and the sample mean $\bar{y}$ is called the *total variation in y* or **sum of squares total** and is denoted by **SSTO.**

- The sum of squared differences between observed y values and the predicted values based on the least squares line is called the **sum of squared errors** and is denoted by **SSE.** Remember that errors are sometimes called residuals, and a synonym for sum of squared errors is **residual sum of squares.**

Whenever the correlation is not 0, the least squares line will produce generally better predictions than the sample mean, so SSE will be smaller than SSTO. The squared correlation expresses the reduction in squared prediction error as a fraction of the total variation. This leads to the formula

$$r^2 = \frac{\text{SSTO} - \text{SSE}}{\text{SSTO}}$$

It can be shown (using algebra) that this quantity is exactly equal to the square of the correlation.

EXAMPLE 5.5—*CONTINUED*
Calculation of r^2

Example 5.5 involved the following six observations for x = score on exam 1 and y = score on exam 2:

x = Exam 1 score	70	75	80	80	85	90
y = Exam 2 score	75	82	80	86	90	91

The regression equation is $\hat{y} = 20 + 0.8x$, and we found in Example 5.5 that the sum of squared errors is SSE = 30. For the y values, the mean is $\bar{y} = 84$. Values of SSTO and r^2 are

$$\text{SSTO} = (75 - 84)^2 + (82 - 84)^2 + (80 - 84)^2 + (86 - 84)^2 + (90 - 84)^2 + (91 - 84)^2 = 190$$

$$r^2 = \frac{\text{SSTO} - \text{SSE}}{\text{SSTO}} = \frac{190 - 30}{190} = 0.842$$

We also can find that the correlation between x and y in this example is $r = \sqrt{0.842} = 0.918$. ◆

EXAMPLE 5.6—*CONTINUED*
Getting to Know the Left Hand Based on the Right Hand

Many statistical computer packages are available that will do all of the regression calculations for you. The box below illustrates the basic results of using the statistical package Minitab for the data in Figure 5.9. The explanatory variable is right handspan, and the response variable is left handspan.

```
The regression equation is
LftSpan = 1.46 + 0.938 RtSpan

Predictor        Coef       StDev        T         P
Constant       1.4635      0.4792      3.05      0.003
RtSpan        0.93830     0.02252     41.67      0.000

S = 0.6386      R-Sq = 90.2%      R-Sq(adj) = 90.2%

Analysis of Variance

Source           DF        SS        MS         F        P
Regression        1     708.15    708.15    1736.38    0.000
Residual Error  188      76.67      0.41
  Total         189     784.82
```

In Chapter 14, we will revisit regression and the meaning of all of the computer results will become clear. For now, you should be able to recognize the following features:

- Regression equation: $\hat{y} = b_0 + b_1 x = 1.46 + 0.938\ x$, where x = right handspan
- Slope: $b_1 = 0.938$, expanded in another part of the display to 0.93830
- Intercept: $b_0 = 1.46$, expanded in another part of the display to 1.4635
- $r^2 = 0.902$, or 90.2%
- SSTO = 784.82
- SSE = 76.67 ◆

Minitab Tip

tech note   **Determining the Regression Line and Calculating a Correlation**

♦ To find a simple regression equation, use **Stat**>**Regression**>
Regression. In the dialog box, specify the column containing
the raw data for the response variable (Y) as the "Response,"
and specify the column containing the data for the explanatory variable (X) as a "Predictor."

♦ To find a regression line and also have Minitab draw this line
onto a scatterplot of the data, use **Stat**>**Regression**>**Fitted
Line Plot.** Specify the response variable (Y) and the predictor
(X) in the dialog box.

♦ To calculate a correlation coefficient, use **Stat**>**Basic Statistics**>
Correlation. Specify two (or more) columns as the Variables.

Excel Tip

tech note   **Finding the Correlation and Regression Equation**
First, enter the y values and x values into separate columns. We refer to the range of cells as *y-range* and *x-range* in what follows:

CORREL(y-range,x-range) = the correlation.

RSQ(y-range,x-range) = squared correlation or proportion of variation
explained by x.

INTERCEPT(y-range,x-range) = intercept of the regression equation.

SLOPE(y-range,x-range) = slope of the regression equation.

For the correlation and squared correlation, the ranges can be entered in either order, but for the intercept and slope, the y-range must be listed first.

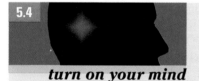

5.4

turn on your mind

Sometimes the main purpose of a regression analysis is to determine the nature of the relationship between two variables, and sometimes the main purpose is to use the equation in the future to predict a y value when the x value
is known. Explain which purpose is likely to be the main reason for a regression analysis between

x = percent fat consumed in diet, y = blood pressure
x = SAT score, y = college grade point average
x = height at age 4, y = height at age 21
x = hours of sleep per night, y = score on IQ test*

5.4 | WHY THE ANSWERS MAY NOT MAKE SENSE

As with all numerical summaries, a number of situations can cause misleading
correlation and regression results. Each of the following actions will produce an
illegitimate description of a relationship:

*value is more likely in that instance.

HINT: Colleges use SAT scores to predict GPAs of applicants, so predicting a future y

♦ Allowing outliers to overly influence the results

♦ Combining groups inappropriately

♦ Using correlation and a straight-line equation to describe curvilinear data

The Influence of Outliers

Outliers can have an impact on correlation and regression results. This is particularly true for small samples. In Example 5.4, we learned that sometimes outliers occur because mistakes are made when the data are recorded or are entered into the computer. In these cases, we may be able to make the necessary corrections. When outliers are legitimate data, we have to carefully consider their effect on the analysis. We may exclude outliers from some analyses, but we shouldn't forget about them. As in everyday life, the unusual data is often the most interesting data.

Outliers with extreme *x* values have the most *influence* on correlation and regression and are called **influential observations.** Depending on whether these points "line up" with the rest of the data, they can either deflate or inflate a correlation. An outlier at an extreme *x* value can also have a big effect on the slope of the regression line.

EXAMPLE 5.4—*CONTINUED***
*Height and Foot Length
of College Women***

Figure 5.5 displayed a scatterplot of the foot lengths and heights of 41 college women. We saw three outliers in that plot, all of which occurred because heights were incorrectly entered into the computer. If we don't correct these mistakes, the correlation between the foot lengths and heights in Figure 5.5 is only $r = 0.28$. For the corrected data set, the correlation is $r = 0.69$, a markedly higher value. The outliers also have a big effect on the equation of the *least squares* line.

♦ For the uncorrected data set, foot length = 15.4 + 0.13 height

♦ For the corrected data set, foot length = −3.2 + 0.42 height

The slope of the correct line is more than three times the size of the slope of the incorrect data. This is a big difference. For instance, let's consider a 12-inch difference in heights. The correct estimate of the associated difference in average foot lengths is $12 \times 0.42 \approx 5$ centimeters (about 2 inches). If we use the incorrect data, the estimated difference in average foot lengths is only $12 \times 0.13 \approx 1.6$ centimeters (about 5/8 of an inch). ♦

**EXAMPLE 5.10
*Earthquakes in the
Continental United States***

Table 5.4 lists the major earthquakes that occurred in the continental United States between 1850 and 1992. The correlation between deaths and magnitude for these six earthquakes is 0.73, showing a relatively strong association. It implies that, on average, higher death tolls accompany stronger earthquakes.

TABLE 5.4 ■ **Major Earthquakes in the Continental United States, 1850–1992**

Date	Location	Deaths	Magnitude
August 31, 1886	Charleston, SC	60	6.6
April 18–19,1906	San Francisco	503	8.3
March 10, 1933	Long Beach, CA	115	6.2
February 9, 1971	San Fernando Valley, CA	65	6.6
October 17, 1989	San Francisco area	62	6.9
June 28, 1992	Yuca Valley, CA	1	7.4

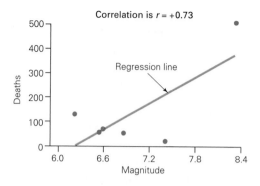

FIGURE 5.15   **Deaths and earthquake magnitude**

However, if you examine the scatterplot of the data shown in Figure 5.15, you will notice that the correlation is entirely due to the famous San Francisco earthquake of 1906. In fact, for the remaining earthquakes, the trend is actually *reversed*. Without the 1906 quake, the correlation for these five earthquakes is strongly *negative,* at −0.96, indicating that fewer deaths are associated with greater magnitudes.

Clearly, trying to interpret the correlation between magnitude and death toll for this small group of earthquakes is a misuse of statistics. The largest earthquake, in 1906, occurred before earthquake building codes were enforced. The next largest quake, with magnitude 7.4, killed only one person but occurred in a very sparsely populated area. ◆

The next example demonstrates another situation that leads to illegitimate results. We'll give you a hint—outliers don't cause the difficulty.

EXAMPLE 5.11
Does It Make Sense?
Height and Lead Feet

For a sample of college students, a scatterplot of their heights and their responses to the question "What is the fastest you have ever driven a car?" is displayed in Figure 5.16. Height is the x variable and the fastest-speed response is the y variable. We see that the fastest-speed response tends to increase as height increases. The correlation is +0.39, and the *least squares* line that describes the

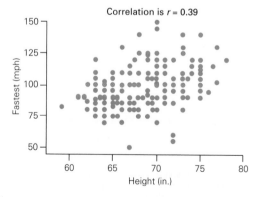

FIGURE 5.16   **Height and the fastest that college students have ever driven**
(Data Source: Class data collected in 1998 by one of the authors.)

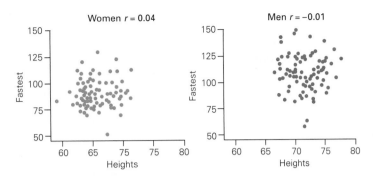

FIGURE 5.17   **Fastest speed and height separately for men and women**

average pattern is $\hat{y} = -20 + 1.7x$. The slope of the equation tells us that for every 1-inch increase in height, there is an average increase of 1.7 mph for the fastest-speed response. This means that for a 12-inch difference in heights, we would estimate the difference in the fastest-speed response to be $12 \times 1.7 \approx 20$ mph. The newspaper headline might read "Height and Lead Foot Go Together." Is this a sensible conclusion? Why might these results be misleading?

We know that men tend to be taller than women. Those of you with good memories may recall Case Study 1.1, which indicated that men tend to claim a higher fastest speed than women do. These gender differences could be causing the positive association that we see in Figure 5.16. One way to examine this possibility is to look separately at men and women. Figure 5.17 shows two scatterplots, one for each gender. For men, the correlation between height and fastest speed is -0.01, basically 0. For women, the correlation is 0.04, also basically 0. In other words, there is no relationship between height and fastest speed within either sex. The observed association in the combined data occurs only because men tend to have higher values than women do for both variables. ◆

Inappropriately Combining Groups

The height and fastest-speed example demonstrates a common mistake that can lead to an illegitimate correlation, which is combining two or more groups when the groups should be considered separately. Example 3.4 of Chapter 3 described a relationship between the cost of a book and the number of pages in the book. Surprisingly, there seemed to be a negative relationship. As the number of pages increased, the cost tended to decrease. The data included hardcover and softcover books, and this explains the unexpected negative association. The hardcover books generally had fewer pages than the softcover books, but their cost was higher. When the book types are considered separately, the association between pages and cost is positive for each type.

Once again, we should think about the association between handspan and height described in Example 5.1. In Figure 5.4, we saw that the positive association holds for each gender. The correlation for the combined group is 0.74, but within each gender the correlation is about 0.6. Combining males and females does inflate the correlation somewhat, but gender differences are not the sole reason for the observed association.

The next example shows us that it's important to look at a scatterplot before we calculate a regression line.

EXAMPLE 5.12
Does It Make Sense?
Don't Predict without
a Plot

Table 5.5 lists the population of the United States (in millions) for each census year between 1790 and 1990. There is, of course, a positive association between y = population size and x = year, because the population size has been steadily increasing through the years. The correlation between population size and year is $r = +0.96$, indicating a very strong relationship. The *least squares* line for these data has the equation $\hat{y} = -2218 + 1.218$ (Year). If we use this equation to predict the population in 2030, our estimate is about 255 million. Does this estimate make sense? What's wrong with this analysis?

Your first thought may be that we shouldn't extrapolate a prediction to 2030 because that's too far past the end of the data. That's a good thought, but the extrapolation issue is not the biggest problem here. You'll realize that something is wrong if you compare the estimate for 2030 to the 1990 population. The year 2030 estimate of 255 million is only 6 million higher than the 1990 population size. The population increased by about 20 million between 1980 and 1990, so it doesn't make sense that the increase will only be 8 million between 1990 and 2030. Also, on September 16, 1999, the United States population was already estimated to be about 273 million (*source:* http://www.census.gov).

Why does the regression line produce such a poor estimate for 2030? The reason is that the pattern of population growth is actually *curved,* so a straight-line equation isn't the right type of equation to use. Figure 5.18 shows the situation. The *least squares* line is shown as well as a curve that traces the actual data pattern. The curve is extended to the year 2030 to provide a projection of the population in that year. It looks like the population might be around 350 million at that point in time. By the way, the United States Census Bureau projects a population size of 347 million in the year 2030 (http://www.census.gov). ◆

TABLE 5.5 ■ **United States Population (millions) in Census Years Since 1790**

Year	Pop.	Year	Pop.
1790	3.9	1900	76.2
1800	5.3	1910	92.2
1810	7.2	1920	106.0
1820	9.6	1930	123.2
1830	12.9	1940	132.2
1840	17.1	1950	151.3
1850	23.2	1960	179.3
1860	31.4	1970	203.3
1870	35.6	1980	228.5
1880	50.2	1990	248.8
1890	63.0		

Source: United States Census Bureau, http://www.census.gov.

Curvilinear Data

The United States population example illustrates that we can make big mistakes if we use a straight line to describe curved data. We should describe a curvilinear pattern with an equation for a curve. This is simpler said than done. There are many different types of equations that describe curves, and it's hard to judge which type we should use just by looking at the scatterplot.

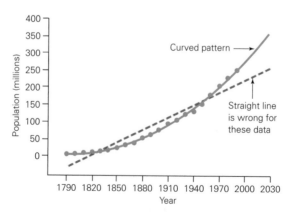

FIGURE 5.18   **Estimating the United States population in 2030**

The music-preference and age relationship in Example 5.3 illustrates another way that correlation can be misleading when the pattern is curvilinear. For the data shown in Figure 5.3, the correlation will be around 0. This value could make us believe that there is no relationship, but in fact there is. Remember that the frame of reference for calculating a correlation is a straight line. For the inverted U pattern in Figure 5.3, the best straight line will be nearly horizontal, so the correlation will be around 0. Clearly, a straight line shouldn't be used to describe these data, so the correlation value is meaningless.

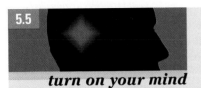

5.5

Sketch a scatterplot with an outlier that would inflate the correlation between the two variables. Sketch a scatterplot with an outlier that would deflate the correlation between the two variables.*

turn on your mind

5.5 | CORRELATION DOES NOT PROVE CAUSATION

Many spirited disagreements about the interpretation of research results revolve around whether a change in one variable actually *causes* a change in another variable. The saying "correlation does not imply causation" is used so frequently in these arguments that you may already have encountered it in everyday life or in another academic course. It is easy to construct silly, obvious examples of observed associations that don't have a causal connection. For example, weekly flu medication sales and weekly sweater sales for an area with extreme seasons would exhibit a positive association because both tend to go up in the winter and down in the summer.

The problem is that the explanation for an observed relationship usually isn't so obvious as it is in the flu medication and sweater sales example. Suppose the finding in an observational study is that people who use vitamin supplements get fewer colds than people who don't use vitamin supplements. It may be that the use of vitamin supplements causes a reduced risk of a cold, but it is also easy to think of other explanations for the observed association. Perhaps those who use supplements also sleep more and it's actually the sleep difference that causes the difference in the frequency of colds. Or, the users of vitamin supplements may be worried about good health and will take several different actions designed to reduce the risk of getting a cold.

Interpretations of an Observed Association

There are at least four possible interpretations of an observed relationship between an explanatory variable and a response variable.

1. There is causation. The explanatory variable is indeed causing a change in the response variable.
2. There may be causation, but confounding factors contribute as well and make this causation difficult to prove.

3. There is no causation. The association is explained by how the explanatory and response variables are both affected by other variables.
4. The response variable is causing a change in the explanatory variable.

Interpretation 1—Causation

It isn't easy to prove that a cause-and-effect relationship exists. Ideally, we would be able to keep everything in the environment constant except the explanatory variable. If we change only one variable, we can learn whether these changes cause changes in the outcome variable.

The use of a designed experiment is the most legitimate way to establish a causal connection statistically. As we learned in Chapter 3, in designed experiments we try to rule out confounding variables through random assignment of the experimental units to specific values of the explanatory variable. If we have a large sample, and if we use proper randomization, we can assume that the levels of confounding variables will be about equal in the different treatment groups. This reduces the chances that an observed association is due to confounding variables, even those that we have neglected to measure.

Interpretation 2—There May Be Causation, but Confounding Factors Contribute as Well and Make This Causation Difficult to Prove

The data from an observational study, in the absence of any other evidence, simply cannot be used to establish causation. It is nearly impossible to separate the effect of *confounding variables* from the effect of the explanatory variable. For example, even if we observe that smokers tend to have higher blood pressure than nonsmokers, we can't definitively say that smoking causes high blood pressure because there may be alternative explanations. Perhaps smokers are more stressed than nonsmokers and this causes the higher blood pressure.

A partial solution is to measure possible confounding variables and include them in the analysis to see how they are related to the outcome variable. Some other considerations can be taken into account as well. Each of the following items provides evidence for causation:

- There is a reasonable explanation of cause and effect.
- The connection occurs under varying circumstances.
- Potential confounding variables are ruled out.

The second item was instrumental in establishing that smoking causes the risk of lung cancer to increase. Numerous observational studies have related cigarette smoking and lung cancer. Further, the studies have shown that the higher the number of cigarettes smoked, the greater the chances of developing lung cancer; similarly, a connection has been established with the age at which smoking began. These facts make it more plausible that smoking actually causes lung cancer.

Interpretation 3—The Association Is Explained by How Other Variables Affect the Explanatory and Response Variables

The association between height and fastest speed in Example 5.11 had this interpretation. As another example of an observed association that doesn't imply causation, recall the association between verbal SAT scores and college GPAs exhibited in Example 5.7. We would certainly not conclude that higher SAT scores caused higher grades in college, except perhaps for a slight benefit of boosted self-esteem. However, we could probably agree that the causes responsible for one variable being high (or low) are the same as those responsible for the other being high (or low). Those causes would include things like intelligence, motivation, and ability to perform well on tests.

Interpretation 4—The Response Variable Is Causing a Change in the Explanatory Variable

Sometimes the causal connection is the opposite of what might be expected or claimed. If an observational study finds, for men and women over 60 years old, that regular church attendance is associated with better health, should we believe that church attendance causes good health? An alternative explanation is that the causation may be in the opposite way. Healthy people are more able to attend church, so good health could increase the likelihood of church attendance.

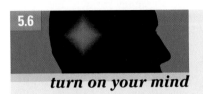

5.6

turn on your mind

An article in the *Centre Daily Times* (April 19, 1997, p. 8A) included data from the United States and several European countries that indicated a negative correlation between the cost of cigarettes and annual per capita cigarette consumption. Does this result mean that if the United States increased its cigarette tax in order to increase the price of cigarettes, the result would be that people would smoke less? What are some other explanations for the negative correlation between cigarette price and annual cigarette consumption?*

CASE STUDY 5.1
A Weighty Issue

In a large statistics class, students (119 females and 63 males) were asked to report their actual and their ideal weights. It's well known that males and females differ with regard to actual weights and their views of their weight, so the two genders should be separated for the analysis. Table 5.6 displays the mean actual and ideal weights for men and women. For women, the mean ideal is 10.7 pounds less than the mean actual, while for men, the mean ideal is only about 2.5 pounds less than mean actual.

We can use a scatterplot and regression to learn more about the connection between actual and ideal weight. Figure 5.19 shows a scatterplot of the two variables for the females, and Figure 5.20 is the same plot for the males. Each point represents one student (or multiple students with the same values), whose ideal weight can be read on the vertical axis and actual weight can be read on the horizontal axis.

If everyone had responded that his or her ideal weight was the same as his or her actual weight, all points would fall on a line with the equation

Ideal = Actual

TABLE 5.6 ■ **Mean Actual and Ideal Weights by Gender (in pounds)**

	Actual	Ideal	Difference
Females (*n* = 119)	132.8	122.1	10.7
Males (*n* = 63)	176.1	173.6	2.5

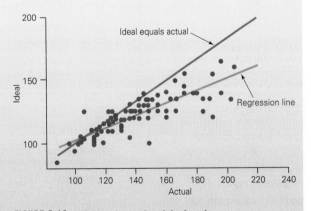

FIGURE 5.19   **Ideal and actual weight: females**

FIGURE 5.20   **Ideal and actual weight: males**

That line is drawn in each figure. Most of the women fall below that line, indicating that their ideal weight is *below* their actual weight. The situation is not as clear for the men, but a pattern is still evident. The majority of men weighing less than 175 pounds would prefer to weigh the same or more than they do, and they fall on or above the line. The majority of men weighing over 175 pounds fall on or below the line and would prefer to weigh the same or less than they do.

The *least squares regression line* is also shown on each scatterplot. The approximate regression equations are

Women: Average ideal = 44 + 0.6 Actual

Men: Average ideal = 53 + 0.7 Actual

The regression equations tell us the "average pattern" of the connection between actual and ideal weight. By substituting some different actual weights into the equations, we can explore how the ideal weight is associated with the actual weight for each gender. Table 5.7 shows regression calculations for students 15 pounds below the mean of the actual weights for their sex and for students 15 pounds above the mean of the actual weights for their sex.

TABLE 5.7 ■ **Regression Estimates of Ideal Weight**

Students with Actual Weight 15 Pounds Below the Mean for Their Sex:			
Gender	Actual	Ideal Based on Regression	Average Preference
Female	118	44 + 0.6(118) ≈ 115	Lose 3 pounds
Male	161	53 + 0.7(161) ≈ 166	Gain 5 pounds

Students with Actual Weight 15 Pounds Above the Mean for Their Sex:			
Gender	Actual	Ideal Based on Regression	Average Preference
Female	148	44 + 0.6(148) ≈ 133	Lose 15 pounds
Male	191	53 + 0.7(191) ≈ 187	Lose 4 pounds

The results in Table 5.7 reveal interesting gender differences. For instance, consider women who weigh 118 pounds, which is about 15 pounds less than the mean weight for women. On average, their ideal weight is about 115 pounds, which is 3 pounds less than their actual weight. On the other hand, men who weigh 15 pounds less than the mean for men would, on average, like to gain about 5 pounds. Women who weigh 15 pounds more than the mean for women would like to lose about 15 pounds. Men who weigh 15 pounds more than the mean for men would like to lose only about 4 pounds.

KEY TERMS

Section 5.1

scatterplot, 131, 132
explanatory variable, 132
response variable, 132
dependent variable, 132
y variable, 132
x variable, 132
positive association, 132, 133
linear relationship, 133
negative association, 133
nonlinear relationship, 134

curvilinear relationship, 134
outliers in regression, 136

Section 5.2

regression analysis, 137
regression equation, 137
prediction, 137
regression line, 138
slope of a straight line, 139, 140
intercept of a straight line, 139, 140
statistical relationship, 140

deterministic relationship, 140
predicted y, 140
estimated y, 140
extrapolation, 142
prediction error, 142
residual, 142
least squares, 143
least squares line, 143
SSE, 143
sum of squared errors (SSE), 143

EXERCISES   Blue = Basics   **Bold, Bold** = Answer in back

Section 5.1

5.1 For each of the following pairs of variables, is there likely to be a positive association, a negative association, or no association? Briefly explain your reasoning.
  a. Amount of alcohol consumed and performance on a test of coordination.
  b. Height and grade point average for college students.
  c. Miles of running per week and time for a 5-kilometer run.
  d. Forearm length and foot length.

5.2 The figure for this exercise is a scatterplot of y = average math SAT score in 1998 versus x = percent of graduating seniors who took the test that year for the 50 states and the District of Columbia The data are from the **sats98** dataset on the CD for this text.
  a. Does the plot show a positive association, a negative association, or no association between the two variables? Explain.
  b. Explain whether you think the pattern of the plot is linear or curvilinear.
  c. About what was the highest average math SAT for the 50 states and District of Columbia? Approximately, what percent of graduates took the test in that state?
  d. About what was the lowest average math SAT for the 50 states and District of Columbia? Approximately, what percent of graduates took the test in that state?

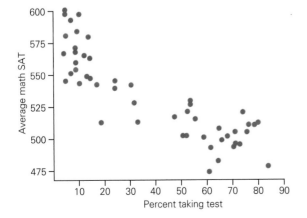

EXERCISE 5.2

5.3 Identify whether a scatterplot would or would not be an appropriate visual summary of the relationship between the following variables. In each case, explain your reasoning.
  a. Blood pressure and age.
  b. Region of country and opinion about stronger gun control laws.
  c. Verbal SAT score and math SAT score.
  d. Handspan and gender (male or female).

5.4 The figure for this exercise is a scatterplot of y = head circumference (centimeters) versus x = height (inches) for the thirty females in the **physical** dataset on the CD for this text.

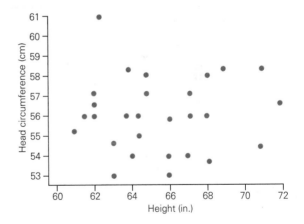

EXERCISE 5.4

  a. Does the plot show a positive association, a negative association, or no association between the two variables? Explain.
  b. One data point appears to be an outlier. What are the approximate values of height and head circumference for that point?

5.5 The figure for this exercise is a scatterplot of y = pulse rate after marching in place for one minute versus x = resting pulse rate measure before marching in place. (The data are in the **pulsemarch** dataset on the CD for this text.)

a. Does the plot show a positive association, a negative association, or no association between the two variables? Explain.
b. Explain whether you think the pattern of the plot is linear or curvilinear.
c. Explain whether there are any obvious outliers. If there are outliers, describe where they are located on the plot.

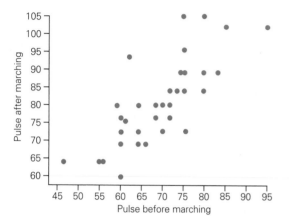

EXERCISE 5.5

5.6   The data in the following tables are the geographic latitudes and the average August temperatures (Fahrenheit) for 20 cities in the United States. The cities are listed in south to north geographic order. (These data are part of the **temperature** dataset on the CD for the text.)
a. Draw a scatterplot of y = August temperature versus x = latitude.
b. Is the pattern linear or curvilinear? What is the direction of the association?
c. Are there any outliers? If so, which cities are outliers?

Geographic Latitude and Mean August Temperature

City	Latitude	Aug Temp
Miami FL	26	83
Houston TX	30	82
Mobile AL	31	82
Phoenix AZ	33	92
Dallas TX	33	85
Los Angeles CA	34	75
Memphis TN	35	81
Norfolk VA	37	77
San Francisco CA	38	64
Baltimore MD	39	76
Kansas City MO	39	76
Washington DC	39	74
Pittsburgh PA	40	71
Cleveland OH	41	70
New York NY	41	76
Boston MA	42	72

Geographic Latitude and Mean August Temperature (continued)

City	Latitude	Aug Temp
Syracuse NY	43	68
Minneapolis MN	45	71
Portland OR	46	69
Duluth MN	47	64

Data Source: *The World Almanac and Book of Facts,* 1999.

5.7   The following data show the relationship between the speed of a car (mph) and the average stopping distance (feet) after the brakes are applied:

Speed (mph)	0	10	20	30	40	50	60	70
Distance (ft)	0	20	50	95	150	220	300	400

Source: *Defensive Driving: Managing Time and Space,* American Automobile Association, Pamphlet #3389, 1991.

a. In the relationship between these two variables, which is the response variable (y) and which is the explanatory variable (x)?
b. Draw a scatterplot of the data. Characterize the relationship between stopping distance and speed.

5.8   The following table shows sex, height (inches), and mid-parent height (inches) for a sample of 18 college students. The variable mid-parent height is the average of mother's height and father's height. (These data are in the dataset **UCDchap5**; they are sampled from the larger dataset **UCDavis2**.)

Sex, Height, and Mid-Parent Height for 18 College Students

Sex	Height	Mid-Parent Height
M	71	64.0
F	60	63.5
F	66	67.0
M	70	64.5
F	65	65.5
F	66	69.5
M	74	72.5
F	67	67.5
F	63	65.5
M	67	64.0
F	69	70.0
M	65	63.0
M	72	69.0
M	68	67.0
F	63	63.0
F	61	63.0
M	74	69.5
F	65	67.5

a. In the relationship between height and mid-parent height, which variable is the response variable (y) and which is the explanatory variable (x)?

b. Draw a scatterplot of the data for the y and x variables defined in part (a). Use different symbols for males and females.

c. Briefly interpret the scatterplot. Does the association appear to be linear? What are the differences between the males and females? Which points, if any, are outliers?

d. Calculate the difference between height and mid-parent height for each student, and draw a scatterplot of y = difference versus x = mid-parent height. Use different symbols for males and females. What does this graph reveal about the connection between height and mid-parent height?

Section 5.2

5.9   Suppose that a regression equation for the relationship between y = weight (pounds) and x = height (inches) for men aged 18 to 29 years old is

$$\text{Average weight} = -250 + 6\ \text{Height}$$

a. Estimate the average weight for men in this age group who are 70 inches tall.

b. What is the slope of the regression line for average weight and height? Write a sentence that interprets this slope in terms of how much average weight changes when height is increased by one inch.

5.10   Refer to the previous exercise in which a regression equation is given that relates average weight and height for men in the 18-to-29-year-old age group.

a. Suppose a man in this age group is 72 inches tall. Use the regression equation given in the previous exercise to predict the weight of this man.

b. Suppose this man, who is 72 inches tall, weighs 190 pounds. Calculate the residual (prediction error) for this individual.

5.11   Refer to the scatterplot for exercise 5.2 showing the relationship between the average math SAT score and the percent of high school graduates taking the test for the 50 states and District of Columbia. The regression line for these data is

$$\text{Average Math} = 575 - 1.11\ \text{Percent Took}$$

a. The slope of the equation is -1.11. Interpret this value in the context of how average math SAT changes when the percent of graduates taking the test changes.

b. In Missouri, only 8% of graduates take the SAT test. What is the predicted average math SAT score for Missouri?

c. In 1998, the average math SAT score for Missouri was 573. What is the residual (prediction error) for Missouri?

5.12   The equation for converting a temperature from x = degrees Celsius to y = degrees Fahrenheit is $y = 32 + 1.8x$. Does this equation describe a statistical relationship or a deterministic relationship? Briefly explain your answer.

5.13   The average August temperatures (y) and geographic lat-itudes (x) of 20 cities in the United States were given in the table for exercise 5.6. The regression equation for these data is

$$\hat{y} = 113.6 - 1.01x$$

a. What is the slope of the line? Interpret the slope in terms of how mean August temperature is affected by a change in latitude.

b. Estimate the mean August temperature for a city with latitude of 32.

c. San Francisco has a mean August temperature of 64 and its latitude is 38. Use the regression equation to es-timate the mean August temperature in San Francisco, and then calculate the prediction error (residual) for San Francisco.

d. Why should we not use this regression equation to es-timate the mean August temperature at the equator (latitude = 0)? If we did use the equation for this pur-pose, what would be the estimated August tempera-ture at the equator?

5.14   A regression equation for y = handspan (cm) and x = height (in.) was discussed in Section 5.2. If the roles of the variables are reversed and only women are considered, the regression equation is Average height = 51.1 + 0.7 Handspan.

a. Interpret the slope of 0.7 in terms of how height changes as handspan increases.

b. What is the estimated average height of women with a handspan of 20 centimeters?

c. Molly has a handspan of 20 centimeters and is 66.5 inches tall. What is the prediction error (residual) for Molly?

5.15   Imagine a regression line that relates y = average systolic blood pressure to x = age. The average blood pressure for people 30 years old is 120, while for those 50 years old the average is 130.

a. What is the slope of the regression line?

b. What is the estimated average systolic blood pressure for people who are 34 years old?

5.16   Iman (1994) reports that for professional golfers, a regression equation relating x = putting distance (in feet) and y = success rate (percent) based on observations of distances ranging from 5 feet to 15 feet is

$$\text{Success rate} = 76.5 - 3.95\ \text{Distance}$$

a. What percent success would you expect for these pro-fessional golfers if the putting distance was 10 feet?

b. Explain what the slope of -3.95 means in terms of how success changes with distance.

5.17   Refer to the previous exercise. The original data for the putting success of professional golfers included values beyond those used for that exercise (5 feet to 15 feet), in both directions. At a distance of 2 feet, 93.3% of the putts were successful, while at a distance of 20 feet, 15.8% of the putts were successful.

a. Use the equation in exercise 5.16 to predict success rates for those two distances (2 feet and 20 feet). Com-pare the predictions to the actual success rates.

b. Utilize your results from part (a) to explain why it is not a good idea to use a regression equation to predict information beyond the range of values from which the equation was determined.

c. Draw a picture of what you think the relationship between putting distance and success rate would look like for the entire range from 2 feet to 20 feet.

5.18 The *least squares* regression equation for the data immediately below is $\hat{y} = 5 + 2x$.

x	4	4	7	10	10
y	15	11	19	21	29

a. Calculate the value of $\hat{y}$ for each data point.

b. Calculate the sum of squared errors for this equation.

Section 5.3

5.19 Which of the numbers $0, 0.25, -1.7, -0.5, 2.5$ could not be values of a correlation coefficient? In each case, explain why.

5.20 For $n = 188$ students, the correlation between $y =$ fastest speed ever driven and $x =$ number "randomly" picked between 1 and 10 is about $r = 0$. Describe what this correlation indicates about the association between the maximum speed driven and picking a number between 1 and 10.

5.21 For 19 female bears, the correlation between $x =$ length of the bear (inches) and $y =$ chest girth (inches) is $r = +0.82$ (*Data source:* **bears-female** dataset on the CD for this text).

a. Describe how chest girth will change when length is increased.

b. Assuming there are no outliers and the relationship is linear, explain what the correlation indicates about the strength of the relationship.

c. If the measurements were made in centimeters rather than inches, what would be the value of the correlation coefficient?

5.22 Suppose a value of r^2 is 100% for the relationship between two variables.

a. What is indicated about the strength of the relationship?

b. What are the two possible values for the correlation coefficient for the two variables?

5.23 Which implies a stronger linear relationship, a correlation of $+0.4$ or a correlation of -0.6? Briefly explain.

5.24 In Figure 5.9, we observed that the correlation between the left and right handspans of college students was 0.95. The handspans were measured in centimeters. What would be the correlation if the handspans were converted to inches? Explain.

5.25 Explain how two variables can have a perfect curved relationship and yet have zero correlation. Draw a picture of a set of data meeting those criteria.

5.26 For Example 5.7 in this chapter, the correlation between verbal SAT and grade point average is $r = +0.485$. Calculate r^2 and write a sentence that interprets this value.

5.27 The following figure shows four graphs. Assume that all four graphs have the same numerical scales for the two axes.

a. Which graph shows the strongest relationship between the two variables? Which graph shows the weakest?

b. In scrambled order, correlation values for these four graphs are $-0.9, 0, +0.3, +0.6$. Match these correlation values to the graphs.

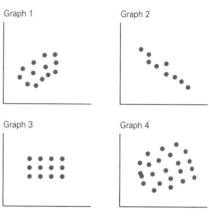

EXERCISE 5.27

5.28 Refer to exercise 5.8 and the table for exercise 5.8 in which heights and mid-parent heights are given for 18 college students (*Data source:* **UCDchap5** dataset on the CD for this text). Draw a scatterplot for the data, using different symbols for males and females as instructed in part (b) of that exercise. Based on the scatterplot, would you say that the correlation between height and mid-parent is higher for the females in the sample, or for the males? Or, are the correlation values about the same for males and females? Explain your reasoning.

5.29 In a regression analysis the total sum of squares (SSTO) is 800 and the error sum of squares (SSE) is 200. What is the value for r^2?

5.30 Suppose you know that the slope of a regression line is $b_1 = +3.5$. Based on this value, explain what you know and do not know about the strength and direction of the relationship between the two variables.

Sections 5.4 and 5.5

5.31 Explain why a strong correlation would be found between weekly sales of firewood and weekly sales of cough drops over a 1-year period.

5.32 An article in the *Sacramento Bee* (29 May, 1998, p. A17) noted "Americans are just too fat, researchers say, with 54 percent of all adults heavier than is healthy. If the trend continues, experts say that within a few generations virtually every U.S. adult will be overweight." This prediction is based on extrapolation, which assumes the current rate of increase will continue indefinitely. Is that a reasonable assumption? Do you agree with the prediction? Explain.

5.33 The **physical** dataset on the CD for this book gives heights (inches) and head circumferences (cm) for a sample of college students. For females only, the correlation between the two variables is 0.05, while for males only, the correlation is 0.19. For the combined sample of males and females, however, the correlation is 0.42. Explain why the correlation in the combined sample is higher than in the separate samples of males and females. Refer to Example 5.11 for guidance.

5.34 The **pennstate2** dataset on the book's CD includes heights and the total number of ear pierces for each person in a sample of college students. The correlation between the two variables is -0.495. What third variable may explain this observed correlation? Explain how that third variable could create the negative correlation.

5.35 Based on the data for the past 50 years in the United States, there is a strong correlation between yearly beer sales and yearly per capita income. Would you interpret this to mean that increasing a person's income will cause him or her to drink more beer? Explain.

5.36 Refer back to exercise 5.7 about stopping distance and vehicle speed. The least squares line for these data is

$$\text{Average distance} = -45 + 5.7 \text{ Speed}$$

a. Use this equation to estimate the average stopping distance when the speed is 80 miles per hour. Do you think this is an accurate estimate? Explain.

b. Draw a scatterplot of the data, as instructed in exercise 5.7(b). Use the scatterplot to estimate the average stopping distance for a speed of 80 mph.

c. Do you think the data on stopping distance and vehicle speed shown in exercise 5.7 describe the relationship between these two variables for all situations? What are some other variables that should be considered when the relationship between stopping distance and vehicle speed is analyzed?

5.37 When a correlation value is reported in research journals there often is not an accompanying scatterplot. Explain why reported correlation values should be supported with either a scatterplot or a description of the scatterplot.

5.38 A memorization test is given to ten women and ten men. The researchers find a negative correlation between scores on the test and height. Explain which of the reasons listed at the beginning of Section 5.4 for misleading correlations might explain this finding. Sketch a scatterplot for the relationship between the variables that is consistent with your explanation.

5.39 Sketch a scatterplot in which the presence of an outlier decreases the observed correlation between the response and explanatory variables.

5.40 Suppose a positive relationship had been found between each of the following sets of variables. For each set, discuss possible reasons why the connection may not be causal. Refer to the list of possible reasons for an observed association in Section 5.5.

a. Number of deaths from automobiles and soft drink sales for each year from 1950 to 2000.

b. Amount of daily walking and quality of health for men over 65 years old.

c. Number of ski accidents and average wait time for the ski lift for each day during one winter at a ski resort.

5.41 Suppose that in an observational study, it is observed

that the risk of heart disease increases as the amount of dietary fat consumed increases. Write a paragraph discussing why this result does not imply that diets high in fat cause heart disease.

Chapter Exercises

5.42 For each pair of variables, identify whether the pair is likely to have a positive correlation, a negative correlation, or no correlation. Briefly indicate your reasoning.

a. Hours of television watched per day and grade point average for college students.

b. Number of liquor stores and number of ministers in Pennsylvania cities.

c. Performance on a strength test and age for people between 40 and 80 years old.

d. Verbal skills and age for children under 12 years old.

e. Height of husband and height of wife.

5.43 The regression line relating verbal SAT scores and GPA for the data exhibited in Figure 5.10 is

$$\text{Average GPA} = 0.539 + 0.00362 \text{ Verbal SAT}$$

a. Estimate the average GPA for those with verbal SAT scores of 600.

b. Explain what the slope of 0.00362 represents in terms of the relationship between GPA and SAT.

c. For two students whose verbal SAT scores differ by 100 points, what is the estimated difference in college GPAs?

d. The lowest possible SAT score is 200. Does the intercept have any useful meaning for this example? Explain.

5.44 For Example 5.2 in this chapter, the correlation between sign legibility distance and age is $r = -0.8$. Calculate r^2 and write a sentence that interprets this value.

5.45 The heights (inches) and foot lengths (centimeters) of 33 college men are shown in the following table. (These data are in the dataset **heightfoot** on the CD for the text.)

Height (inches) and Foot Length (cm) for 33 College Students

Student	Height	Foot Length	Student	Height	Foot Length
1	66.5	27.0	18	68.0	25.0
2	73.5	29.0	19	72.5	28.0
3	70.0	25.5	20	78.0	31.5
4	71.0	27.9	21	79.0	30.0
5	73.0	27.0	22	71.0	28.0
6	71.0	26.0	23	74.0	29.0
7	71.0	29.0	24	66.0	25.5
8	69.5	27.0	25	71.0	26.7
9	73.0	29.0	26	71.0	29.0
10	71.0	27.0	27	71.0	28.0
11	69.0	29.0	28	84.0	27.0
12	69.0	27.2	29	77.0	29.0
13	73.0	29.0	30	72.0	28.0
14	75.0	29.0	31	70.0	26.0
15	73.0	27.2	32	76.0	30.0
16	72.0	27.5	33	68.0	27.0
17	69.0	25.0			

Data Source: William Harkness.

a. Draw a scatterplot with y = foot length (cm) and x = height (inches). Does the relationship appear to be linear? Are there any outliers? If so, do you think the outliers are legitimate data values?

b. Use statistical software or a calculator to calculate the correlation between height and foot length. If heights were converted to centimeters, what would be the correlation between height and foot length?

c. If there are any outliers, remove them and recalculate the correlation. Describe how the correlation changed from part (b).

5.46 Refer to the previous exercise about y = foot length and x = height (*Data source:* **heightfoot** dataset). If the person who reportedly is 84 inches tall is excluded, the regression equation for the remaining 32 men is $\hat{y} = 0.25 + 0.384x$.

a. How much does average foot length increase for each 1-inch increase in height?

b. Predict the difference in the foot lengths of men whose heights differ by 10 inches.

c. Suppose Max is 70 inches tall and has a foot length of 28.5 centimeters. Based on the regression equation, what is the predicted foot length for Max? What is the value of the prediction error (residual) for Max?

5.47 Refer to Case Study 5.1, in which regression equations are given for males and females relating ideal weight to actual weight. The equations are

$$\text{Women:} \quad \text{Ideal} = 44 + 0.6 \text{ Actual}$$
$$\text{Men:} \quad \text{Ideal} = 53 + 0.7 \text{ Actual}$$

a. Predict the ideal weight for a man who weighs 140 pounds and for a woman who weighs 140 pounds. Compare the results.

b. Do the intercepts have logical physical interpretations in the context of this example? Explain.

c. Do the slopes have logical interpretations in the context of this example? Explain.

5.48 Calculate r^2 for Example 5.9 in this chapter (about hours of sleep and hours of study) in which the correlation is -0.36. Write a sentence that interprets this value.

5.49 The winning time in the Olympic men's 500-meter speed skating race over the years 1924 to 1992 can be described by the regression equation

$$\text{Winning time} = 255 - 0.1094 \text{ Year}$$

a. Is the correlation between winning time and year positive or negative? Explain.

b. In 1994, the actual winning time for the gold medal was 36.33 seconds. Use the regression equation to predict the winning time for 1994, and compare the prediction to what actually happened.

c. Explain what the slope of -0.1094 indicates in terms of how winning times change from year to year.

d. Why should we not use this regression equation to predict the winning time in the 2050 Winter Olympics?

5.50 The following lists the number of pages and the price for 15 books. Eight of the books are hardcover and 7 are softcover. (These data are in the dataset **ProfBooks** on the data CD.)

Pages versus Price for Books,
Type H = hardcover, S = softcover

Pages	Price	Type	Pages	Price	Type
104	32.95	H	417	4.95	S
188	24.95	H	417	39.75	H
220	49.95	H	436	5.95	S
264	79.95	H	458	60.00	H
336	4.50	S	466	49.95	H
342	49.95	H	469	5.99	S
378	4.95	S	585	5.95	S
385	5.99	S			

a. Draw a scatterplot of y = price versus x = pages. Use different symbols for hardcover and softcover books.

b. For all 15 books, determine the correlation between price and pages.

c. Separate the books by type. Determine the correlation between price and pages for hardcover books only. Determine the correlation between price and pages for softcover books only.

d. Which of the reasons listed in Section 5.4 for misleading correlations is illustrated in this exercise?

5.51 Give an example of a prediction that is an extrapolation. Do not give an example that is already in this chapter.

5.52 Researchers have shown that there is a positive correlation between the average fat intake and the breast cancer rate across countries. In other words, countries with higher fat intake tend to have higher breast cancer rates. Does this correlation prove that dietary fat is a contributing cause of breast cancer? Explain.

5.53 United States Census Bureau estimates of the average number of persons per household in the United States for census years between 1850 and 2000 are shown in the following table. (On the data CD, these data are in the file **perhouse**.)

Persons per Household
in the United States

Year	Per House	Year	Per House
1850	5.55	1930	4.11
1860	5.28	1940	3.67
1870	5.09	1950	3.37
1880	5.04	1960	3.35
1890	4.93	1970	3.14
1900	4.76	1980	2.76
1910	4.54	1990	2.63
1920	4.34	2000	2.59

Data Source: *The World Almanac and Book of Facts,* 1999, p. 383, and Bureau of the Census.

a. Draw a scatterplot for the relationship between persons per household and year. Is the relationship linear or curvilinear? Is the association between persons per household and year positive or negative?

b. On your scatterplot, add a line that you believe fits the data pattern. Extend this line to the year 2010. Based on this line, estimate the persons per household in the United States in the year 2010.

5.54 Refer to the previous exercise about the trend in persons per household.

a. Using statistical software, determine the least squares line for these data. Use the equation of this line to estimate the persons per household in the year 2010 (*Data source:* **perhouse** dataset).

b. What is the slope of the line? Interpret the slope in the context of these variables.

c. Based on the regression line, what would be the predicted persons per household in the year 2200? What is the lowest possible value of the persons per household number? How does the estimate for 2200 compare to this value?

d. Part (c) illustrates that the observed pattern can't possibly continue in the same manner forever. Sketch the pattern for the trend in persons per household that you think might occur between now and the year 2200.

5.55 For a statistics class project at a large northeastern university, a student examined the relationship between

x = body weight (in pounds)
y = time to chug a 12-ounce beverage (in seconds)

We'll leave it to you to imagine the beverage. The student collected data from 13 individuals, and those data are in the following table. (On the data CD, this dataset is named **chugtime.**)

**Body Weight (pounds) and
Chug Time (seconds) for
13 College Students**

Person	Weight	Chug Time
1	153	5.6
2	169	6.1
3	178	3.3
4	198	3.4
5	128	8.2
6	183	3.5
7	177	6.1
8	210	3.1
9	243	4.0
10	208	3.2
11	157	6.3
12	163	6.9
13	158	6.7

Data Source: William Harkness.

a. Draw a scatterplot of the measurements. Characterize the relationship between chug time and body weight.

b. The heaviest person appears to be an outlier. Do you think that observation is a legitimate observation or do you think an error was made in recording or entering the data?

c. Outliers should not be thrown out unless there's a good reason, but there are several reasons why it may be legitimate to conduct an analysis without them (for instance, see part (e)). Delete the data point for the heaviest person, and determine a regression line for the remainder of the data.

d. Use the regression line from part (c) to estimate the chug time for an individual who weighs 250 pounds. Do you think this time could be achieved by anybody?

e. Sometimes the relationship between two variables is linear for a limited range of x values, then changes to a different line or curve. Using this idea, draw a sketch that illustrates what you think the actual relationship between weight and chug time might be for the range of weights from 100 to 300 pounds.

f. Discuss plausible reasons why the heaviest person appears to be an outlier with regard to his combination of weight and chug-time measurements.

5.56 The data for this exercise are

x	1	2	3	4
y	4	10	14	16

a. Determine the sum of squared errors (SSE) for each of the following two lines:

$$\text{Line 1: } \hat{y} = 3 + 3x$$
$$\text{Line 2: } \hat{y} = 1 + 4x$$

b. By the least squares criterion, which of the two lines is better for these data? Why is it better?

5.57 Measure the heights and weights of ten friends of the same sex.

a. Draw a scatterplot of the data, with weight on the vertical axis and height on the horizontal axis. Draw a line onto the scatterplot that you believe describes the average pattern. Based on two points on this line, estimate the slope of the relationship between weight and height.

b. Using statistical software, compute the least squares line, and compare the slope to your estimated slope from part (a).

5.58 The data in the following table come from the time when the United States still had a maximum speed limit of 55 miles per hour. An issue of some concern at that time was whether lower speed limits reduce the highway death rate. (On the CD, these data are called **speedlimit.**)

a. In the relationship between death rate and speed limit, which variable is the response variable and which is the explanatory variable?

b. Plot the data in the table, and discuss the result. Does there appear to be an association? Are there any outliers? If so, what is their influence on the correlation?

Highway Death Rates and Speed Limits

Country	Death Rate (per 100 million veh. miles)	Speed Limit (in miles per hour)
Norway	3.0	55
United States	3.3	55
Finland	3.4	55
Britain	3.5	70
Denmark	4.1	55
Canada	4.3	60
Japan	4.7	55
Australia	4.9	60
Netherlands	5.1	60
Italy	6.1	75

Source: "Fifty-five mph speed limit is no safety guarantee," D. J. Rivkin, *New York Times* (letters to the editor), Nov. 25, 1986, p. 26.

Dataset Exercises

5.59 Use the dataset **sats98** for this exercise. The variable **Verbal** contains the average scores on the verbal SAT in 1998 for the 50 states and the District of Columbia. **PctTook** is the percent of high school graduates, in each state, who took the SAT that year.

a. Make a scatterplot showing the connection between average *verbal* SAT and the percent of graduates who took the SAT in a state. Describe the relationship between these two variables.

b. Compute the least squares regression line for the relationship between these two variables. Write a sentence that interprets the slope of this equation in a way that could be understood by people who don't know very much about statistics.

c. Based on the appearance of the scatterplot, do you think that a straight line is an appropriate mathematical model for the connection between **Verbal** and **PctTook**? Why or why not?

d. Explain why the intercept of the equation computed in part (b) would not have a sensible interpretation for these two variables.

5.60 Use the **sats98** dataset.

a. Plot the relationship between average verbal (**Verbal**) and average math (**Math**) SAT scores in the 50 states. Describe the characteristics of the relationship.

b. What states are outliers? In what specific way are they outliers?

5.61 Use the dataset **idealwtmen**. It contains data for the men used for Case Study 5.1. The variable **diff** is the difference between actual and ideal weights and was computed as **diff = actual − ideal.**

a. Plot **diff** versus **actual** (actual weight). Does the relationship appear to be linear or is it curvilinear?

b. Compute the equation of the regression line for the relationship between **diff** and **actual**. Estimate the average difference for men who weigh 150 pounds. On average, do 150-pound men want to weigh more or less than they actually do?

c. Repeat part (b) for men who weigh 200 pounds.

d. What is the value of r^2 for the relationship between **diff** and **actual**?

5.62 Use the dataset **ceodata**. The ages of 60 CEOs for America's best small companies were given in exercise 2.32 of Chapter 2. The annual salaries (in thousands of dollars) for 59 of these CEOs are in the dataset along with the ages.

a. Plot **Salary** versus **Age**.

b. Compute the correlation coefficient and r^2.

c. Characterize the relationship between annual salary and age. What is the pattern of the relationship? How strong is the association?

5.63 Use the dataset **UCDwomht**. For a sample of college women, the variable **height** is student's height (in inches) and the variable **midparent** is the average height of the student's parents, in inches as reported by the student.

a. Compute the regression equation for predicting a student's height from the average of her parents' heights.

b. Use the regression equation to predict the height for a college woman with parents who have an average height of 68 inches.

c. Use the regression equation to predict the height of a college woman whose mother is 62 inches tall and whose father is 70 inches tall.

d. What other summaries of the data should be done in order to determine the strength of the relationship between **height** and **midparent** height?

5.64 Use the dataset **temperature.** A portion of this dataset was presented in exercise 5.6 in which the relationship between mean August temperature and geographic latitude was analyzed. For predicting mean April temperature (**AprTemp**), which of these two variables in the dataset is a stronger predictor: geographic latitude (**latitude**) or mean January temperature (**JanTemp**)? Support your answer with relevant statistics and plots.

Turn On Your Computer

*For these exercises, use the **Correlation** applet described in Section 5.3. It is on the CD for this text. In each exercise, you are asked to sketch a facsimile of a graph you create with the applet. Alternatively, you might use "Print Screen" on your keyboard to copy the screen image; then paste it to a word processing document.*

5.65 Using the applet, create a plot for the target correlation $r = +0.5$. Don't include any outliers. Sketch an approximate facsimile of your resulting graph.

5.66 Using the applet, create a plot for the target correlation $r = -0.8$. Don't include any outliers. Sketch an approximate facsimile of your resulting graph.

5.67 Using the applet, create a plot for the target correlation $r = 0$. Don't include any outliers. Sketch an approximate facsimile of your resulting graph.

5.68 Using the applet, create a plot for the target correlation $r = +0.5$ in which one point is an outlier that decreases the correlation. Make the plot such that if the outlier were removed, the correlation for the remaining points would be greater than $r = 0.7$. Sketch an approximate facsimile of your resulting graph.

5.69 Using the applet, create a plot for the target correlation $r = -0.8$ in which one point is an outlier that inflates the correlation. Make the plot such that if the outlier were removed, the correlation for the remaining points would be between -0.2 and $+0.2$. Sketch an approximate facsimile of your resulting graph. *Hint:* Start by putting points in the upper left corner of the plot.

5.70 Using the applet, with the target correlation $r = 0$, make a plot that has a curvilinear pattern for which the correlation is 0. Sketch an approximate facsimile of your resulting graph.

REFERENCES

Holbrook, M. B., and R. M. Schindler (1989). "Some exploratory findings on the development of musical tastes," *Journal of Consumer Research,* Vol. 16, pp. 119–124.

Iman, R. L. (1994). *A Data-Based Approach to Statistics,* Belmont, CA: Wadsworth.

Relationships Between Categorical Variables

CHAPTER 6

What else do they have on their bodies? See Example 6.2 and Exercise 6.45.

Whenever risk statistics are reported, there is a risk that they are misreported. Journalists often present risk data in a way that produces the best story rather than in a way that provides the best information. Very commonly, news reports either don't contain or don't emphasize the information you need to understand risk.

This chapter is about the analysis of the relationship between two categorical variables, so let's begin by recalling the meaning of the term *categorical variable.* The raw data from categorical variables consist of group or category names that don't necessarily have any ordering. Eye color and hair color, for instance, are categorical variables.

We can also use the methods of this chapter to examine *ordinal* variables. Ordinal variables can be thought of as categorical variables for which the categories have a natural ordering. For example, a researcher might define categories for quantitative variables, like age, income, or years of education.

Although there are many questions that we can and will ask about two categorical variables, in most cases the principal question that we ask is: Is there a relationship between the two variables, so that the category into which individuals fall for one variable seems to depend on the category they are in for the other variable? ◆

6.1

turn on your mind

Hair color and eye color are related characteristics. What exactly does it mean to say that these two variables are related? Suppose that you know the hair colors and the eye colors of 200 individuals. How would you assess whether the two variables are related for those individuals?*

6.1 | DISPLAYING RELATIONSHIPS BETWEEN CATEGORICAL VARIABLES

We have already encountered several examples of the type of problem we will study in this chapter. In Chapter 2, for instance, we described a study of 479 children that found that children who slept either with a nightlight or in a fully lit

*HINT: Read the sentence just before the Turn on Your Mind box.

room before the age of two had a higher incidence of myopia (nearsightedness) later in childhood. The following table was used to examine this relationship:

Slept with:	No Myopia	Myopia	High Myopia	Total
Darkness	155 (90%)	15 (9%)	2 (1%)	172
Nightlight	153 (66%)	72 (31%)	7 (3%)	232
Full Light	34 (45%)	36 (48%)	5 (7%)	75
Total	342 (71%)	123 (26%)	14 (3%)	479

The table displays the number of children in each combination of the categories of two categorical variables: the sleep-time lighting condition and the child's eyesight classification at the time of the study. The table also gives row percents, the percents of children within each row who fall into the different eyesight categories. For example, 90% (155/172) of those who slept in darkness had no myopia, but only 45% (34/75) of those who slept in full light had no myopia.

Notice that the row percents show us that the two variables are related. From these percents we learn that the incidence of myopia increases when the amount of sleep-time light increases. This result doesn't prove that sleeping with light *causes* myopia, but we can say that, for some reason, the characteristics of eyesight and sleep-time lighting are *associated* characteristics.

Displays like the table above are often called **contingency tables** because they cover all contingencies for the combinations of the two variables. Because the categories of two variables are used to create the table, a contingency table is also referred to as a **two-way table.** Each row and column combination of the table is called a **cell.**

The first step in analyzing the relationship between two categorical variables is to count how many observations fall into each cell of the contingency table. It's difficult, however, to look at a table of counts and make useful judgments about a relationship. Usually, we need to consider the **conditional percents** within either the rows or the columns of the table.

There are two types of conditional percents that we can compute for a contingency table. **Row percents,** like the percents in the myopia table, are the percents across a row of a contingency table. These percents are based on the total number of observations in the row. **Column percents** are the percents down a column of a contingency table. These percents are based on the total number of observations in the column.

In some cases, one variable can be designated as the *explanatory variable* and the other variable as the *response variable.* As we learned in Chapter 2, in these situations it is customary to define the rows using the categories of the explanatory variable and the columns using the categories of the response variable. When this is done, the row percents can be used to examine the relationship because they tell us what percent *responded* in each possible way. We can then see if individuals responded equivalently for each category of the explanatory variable. For instance, we can see if the myopia response was the same for different lighting conditions.

EXAMPLE 6.1
Smoking and the Risk of Divorce

Table 6.1 displays data on smoking habits and divorce history for the 1669 respondents who had ever been married in the 1991 and 1993 General Social Surveys done by the National Opinion Research Center at the University of Chicago.

TABLE 6.1 ■ **Smoking and Divorce, GSS Surveys 1991–1993**

Smoke?	Ever Divorced?		
	Yes	No	Total
Yes	238	247	485
No	374	810	1184
Total	612	1057	1669

Data Source: SDA archive at UC Berkeley website (www.csa.berkeley.edu:7502/).

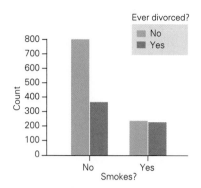

FIGURE 6.1   **Frequency counts for smoking habits and divorce**

The frequency counts in the cells of this table are visually summarized in Figure 6.1, using a bar graph as described in Chapter 2. In the bar graph, we see evidence of an association between smoking habits and the likelihood of divorce because the set of bars for smokers has a different pattern than the set of bars for nonsmokers. Among those who smoke, the number who have ever been divorced is about equal to the number who have not been divorced. Among those who do not smoke, the proportion ever divorced is clearly a minority. We also see that the majority of the sample does not smoke.

To describe the relationship, we should calculate row percents because we are interested in comparing divorce rates between smokers and nonsmokers. We are not interested in comparing smoking rates between those who have been divorced and those who have not. Among the respondents who smoke, 49% (238/485) have been divorced and 51% have not been divorced. Among those who don't smoke, only 32% (374/1184) have been divorced, while 68% have not been divorced. The difference between the two sets of row percents indicates a relationship. ◆

EXAMPLE 6.2
Tattoos and Ear Pierces

From 1996 through 1998, students in several statistics classes at Penn State provided responses to these two questions:

1. Do you have a tattoo?
2. How many total ear pierces do you have?

TABLE 6.2 ■ **Ear Pierces and Tattoos for Men (*n* = 565)**

Ear Pierces	No Tattoo	Tattoo	Total
0	381	43	424
1	54	16	70
2 or more	45	26	71
Total	480	85	565

The responses from 565 men are displayed in Table 6.2. In the table, the very few men who had more than two ear pierces are lumped with the men who had two ear pierces.

For this table, we could ask questions about the conditional percents in either direction because there is no clear explanatory and response variable distinction. For instance, does the likelihood that a man has a tattoo differ based on whether or not he has an ear pierce? Or, we could ask if men with tattoos are more likely to have an ear pierce. Let's consider the row percents.

◆ Among men with no ear pierces, 43/424 = 10% have a tattoo.

◆ Among men with one ear pierce, 16/70 = 23% have a tattoo.

◆ Among men with two or more ear pierces, 26/71 = 37% have a tattoo.

Clearly, the percent with a tattoo increases as the number of ear pierces increases, so the two characteristics are related. We also see the relationship when we consider the column percents. Those percents are shown in the bar chart of Figure 6.2. Men with tattoos are more likely to have an ear pierce.

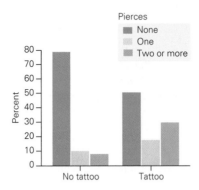

FIGURE 6.2   **Column percents for the ear pierce and tattoo data**

Percents based on the total sample can also be of interest. These percents don't provide information about the relationship, but they do provide useful descriptions of the overall group. For example, 381/565 = 67% of the men in the overall sample have neither a tattoo nor an ear pierce, while (16 + 26)/565 = 42/565 = 7.4% have both. The overall percent with an ear pierce is (70 + 71)/565 = 25%, and the overall percent with a tattoo is 85/565 = 15%. ◆

EXAMPLE 6.3
Gender and Reasons for Taking Care of Your Body

In a 1997 poll conducted by the *Los Angeles Times,* 1218 southern California residents were surveyed about their health and fitness habits. The respondents were selected using random-digit dialing methods. One of the questions was: "What is the most important reason why you try to take care of your body: Is it mostly because you want to be attractive to others, or mostly because you want to keep healthy, or mostly because it helps your self-confidence, or what?"

The percent distribution of the responses is shown for men and for women in Table 6.3. Notice that the pattern of responses is more or less the same for men and women. It seems reasonable to conclude that the response to the question was not related to the gender of the respondent. ◆

TABLE 6.3 ■ Reasons for Taking Care of Body and Gender

	Healthy	**Self-Confidence**	**Attractive**	**Don't Know**
Men	76%	16%	7%	1%
Women	74%	20%	4%	2%

Source: www.latimes.com, poll archives, study #401.

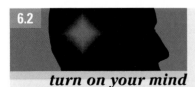

6.2

turn on your mind

In Example 6.1, the risk of divorce was associated with smoking habits. Were the data collected in an observational study or in an experiment? Do you think cigarette smoking may cause an increase in the divorce rate? Do you think getting divorced may cause a person to start smoking? How would you explain the association between smoking habits and the likelihood of divorce?*

HINT: Do you think participants were randomly assigned to smoke or not? Think about possible confounding variables.

6.2 | RISK, RELATIVE RISK, ODDS RATIO, AND INCREASED RISK

When a particular outcome is undesirable, researchers and journalists may describe the *risk* of that outcome. The **risk** that a randomly selected individual within a group falls into the undesirable category is simply the proportion in that category.

$$\text{Risk} = \frac{\text{Number in category}}{\text{Total number in group}}$$

It is commonplace to express risk as a percent rather than as a proportion. Suppose, for instance, that within a group of 200 individuals, asthma affects 24 people. In this group the *risk* of asthma is 24/200 = 0.12, or 12%.

Relative Risk

We often want to know how the risk of an outcome relates to an explanatory variable. One statistic used for this purpose is **relative risk,** which is the ratio of the risks in two different categories of an explanatory variable.

$$\text{Relative risk} = \frac{\text{Risk in category 1}}{\text{Risk in category 2}}$$

Relative risk describes the risk in one group as a multiple of the risk in another group. For example, suppose that a researcher states that, for those who drive while under the influence of alcohol, the relative risk of an automobile accident is 15. This means that the risk of an accident for those who drive under the influence is 15 times the risk for those who don't drive under the influence.

Some features of relative risk are

- When two risks are the same, the relative risk is 1.
- When two risks are different, the relative risk is different from 1; and when the category in the numerator has higher risk, the relative risk is greater than 1.
- The risk in the denominator (the bottom) of the ratio is often the **baseline risk,** which is the risk for the category in which no additional treatment or behavior is present.

6.3

turn on your mind

Based on the study in Section 6.1, the relative risk of developing any myopia later in childhood is 5.5 for babies sleeping in full light compared with babies sleeping in darkness. Restate this information in a sentence that the public would understand.*

EXAMPLE 6.1—
CONTINUED
Smoking and the
Relative Risk of Divorce

Refer to Table 6.1 on page 173. To compute the relative risk of divorce for smokers, we first find the risk of divorce in each smoking category:

- For those who smoke, the risk of divorce is 238/485 = 0.491, or about 49%.
- For nonsmokers, the risk of divorce is 374/1184 = 0.316, or about 32%. This can be considered to be the *baseline risk* of divorce.

*HINT: See the interpretation of relative risk in Example 6.1—continued.

The relative risk of divorce is the ratio of these two risks.

$$\text{Relative risk} = \frac{49\%}{32\%} = 1.53$$

In this sample, the risk of divorce for smokers is 1.53 times the risk of divorce for nonsmokers. ◆

Percent Increase or Decrease in Risk

Sometimes an increase or decrease in risk is presented as a percent change instead of a multiple. The **percent increase** (or decrease) **in risk** can be calculated as follows:

$$\text{Percent increase in risk} = \frac{\text{Difference in risks}}{\text{Baseline risk}} \times 100\%$$

Equivalently, when the relative risk has already been determined, the percent increase in risk can be calculated using the relationship

$$\text{Percent increase in risk} = (\text{relative risk} - 1) \times 100\%$$

When a risk is smaller than the baseline risk, the relative risk is less than 1 and the percent "increase" will actually be negative. In this situation, the term *percent decrease* should be used to describe the percent change in the risks.

EXAMPLE 6.1—CONTINUED
Percent Increase in the Risk of Divorce for Smokers

Because we've already calculated the relative risk, the easiest way to determine the percent increase in the risk of divorce for smokers is

$$\text{Percent increase in risk} = (1.53 - 1) \times 100\% = 53\%$$

We can get the same answer for the percent increase in risk by calculating

$$\text{Percent increase in risk} = \frac{\text{Difference in risks}}{\text{Baseline risk}} \times 100\%$$

$$= \frac{49 - 32}{32} \times 100\% = 53\%$$

So, the risk of divorce is 53% higher for smokers than it is for nonsmokers. ◆

Odds Ratio

Sometimes counts for the outcomes of a categorical variable are summarized by comparing the **odds** of one outcome to another, rather than by comparing one outcome to the total. For instance, the *odds* of getting a divorce to *not* getting a divorce for nonsmokers are 374 divorced to 810 not divorced, or equivalently about 0.46 divorced to 1 not divorced (divide both counts by 810 to get this). For smokers, on the other hand, the odds are 238 divorced to 247 not divorced, or 0.96 to 1, approximately even odds. Notice that odds are expressed using a phrase with the structure "a to b," so a ratio is implied but not actually computed.

The **odds ratio** is used to compare the odds of a certain behavior or event within two different groups. For example, we may want to compare the odds of success versus failure for two different treatments of clinical depression. The formula for the odds ratio will be given in a moment, but a numerical example should make it easy for you to see how it is computed. In Example 6.1, the odds ratio comparing the odds of divorce for smokers and nonsmokers is

$$\frac{\text{Odds of divorce for smokers}}{\text{Odds of divorce for nonsmokers}} = \frac{238/247}{374/810} = \frac{0.96}{0.46} = 2.1$$

This odds ratio tells us that the odds of ending up divorced instead of still married for smokers are about double the odds of ending up divorced for nonsmokers.

A useful characteristic of the odds ratio is that its value stays the same if the roles of the response and explanatory variables are reversed. In the previous paragraph, we compared the odds of divorce for smokers and nonsmokers. If we had, instead, compared the odds of being a smoker for those who have been divorced and those who have never divorced, the answer would still be 2.1. The calculation is

$$\frac{\text{Odds of smoking for divorced}}{\text{Odds of smoking for never divorced}} = \frac{238/374}{247/810} = \frac{0.636}{0.305} = 2.1$$

The interpretation here is that the odds of being a smoker for those who have been divorced are about double the odds of being a smoker for those who have never been divorced.

in summary — Statistics on Risk, Relative Risk, Odds, and Odds Ratios

Let's summarize the various ways in which risk, relative risk, odds, and odds ratios are constructed from a two-way contingency table. These measures are usually employed when there is a definable explanatory and response variable and when there is a baseline condition, so we present them using those distinctions. In situations where those distinctions cannot be made, simply be clear about which condition is in the numerator and which is in the denominator.

Explanatory Variable	Response Variable		
	Category 1	Category 2	Total
Category of Interest	A_1	A_2	T_A
Baseline Category	B_1	B_2	T_B

- Risk (of response 1) for category of interest = A_1/T_A
- Odds (of response 1 to response 2) for category of interest = A_1 to A_2
- Relative risk = $\dfrac{A_1/T_A}{B_1/T_B}$
- Odds ratio = $\dfrac{A_1/A_2}{B_1/B_2}$

6.3 | MISLEADING STATISTICS ABOUT RISK

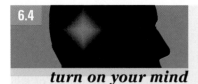

6.4

turn on your mind

Suppose a newspaper article claims that drinking coffee doubles your risk of developing a certain disease. Assume the statistic was based on legitimate, well-conducted research. What additional information would you want about the risk before deciding whether or not to quit drinking coffee?*

Whenever risk statistics are reported, there is a risk that they are misreported. Unfortunately, journalists often present risk data in a way that produces the best story rather than in a way that provides the best information. Very commonly, news reports either don't contain or don't emphasize the information you need to understand risk. You should always ask the following questions when you encounter statistics about risk:

1. What are the actual risks? What is the baseline risk?
2. What is the population for which the reported risk or relative risk applies?
3. What is the time period for this risk?

EXAMPLE 6.4
Disaster in the Skies?
Case Study 1.2
Revisited

Case Study 1.2 described a *USA Today* article that told us: "Errors by air traffic controllers climbed from 746 in fiscal 1997 to 878 in fiscal 1998, an 18% increase." To airplane travelers this may sound frightening, but a look at the risk of controller error per flight should ease their fear. In 1998, there were only 5.5 errors per million flights compared to 4.8 errors per million flights in 1997. The risk of controller error did increase, but the actual risk is extremely small. ◆

EXAMPLE 6.5
Dietary Fat and
Breast Cancer

"Italian scientists report that a diet rich in animal protein and fat—cheeseburgers, french fries, and ice cream, for example—increases a woman's risk of breast cancer threefold," according to *Prevention Magazine's Giant Book of Health Facts* (1991, p. 122). The statement attributed to the Italian scientists is nearly useless information for at least two reasons:

1. We don't how the data were collected, so we don't know what population these women represent.
2. We don't know the ages of the women studied, so we don't know the baseline rate of breast cancer for these women.

Age is a critical factor. A frequently stated statistic about breast cancer is that one in nine women will develop breast cancer, but this is actually an accumulated lifetime risk to the age of 85, so a woman who dies before she reaches age 85 does not have a one in nine risk. According to the *University of California at Berkeley Wellness Letter* (July 1992, p. 1) the accumulated lifetime risk of a woman developing breast cancer by certain ages is as follows:

by age 50: 1 in 50
by age 60: 1 in 23
by age 85: 1 in 9

Also, the annual risk of developing breast cancer is only about 1 in 3700 for women in their early 30s (Fletcher, Black, Harris, Rimer, and Shapiro, 1993, p. 1644). If the Italian study had been done on very young women, the threefold increase in risk represents a small increase. Unfortunately, *Prevention Magazine's Giant Book of Health Facts* did not even give enough information to lead us to the original research report, so it is impossible to intelligently evaluate the claim. ◆

EXAMPLE 6.6
Is Smoking More Dangerous for Women?

"Higher heart risk in women smokers" was the headline of an April 3, 1998, article at the YAHOO® Health news website. The article, due to the Reuters news agency, described Danish research that was interpreted as evidence that smoking affects the risk of a heart attack more for women than for men. Here is part of the article:

> Women who smoke have a greater than 50% higher risk of a heart attack than male smokers according to a study from Denmark. The researchers suggest that this difference may be related to the interaction of tobacco smoke and the female hormone, estrogen.
>
> "Women may be more sensitive than men to some of the harmful effects of smoking," the team writes in the April 4th edition of the *British Medical Journal*.
>
> Analyzing data from nearly 25,000 Danish men and women, Dr. Eva Prescott of the University of Copenhagen and colleagues report that women who smoke have a 2.24 relative risk of myocardial infarction, or heart attack, compared with nonsmokers. This is significantly higher than the relative risk of male smokers compared with nonsmokers.

Unfortunately, the only information we're given, for each sex, is the *relative risk* of a heart attack for smokers. What's missing from the article is any mention of the estimated risk of a heart attack for any group of interest. A look at those risks makes the research interpretation debatable.

The *British Medical Journal* article (Prescott, Hippe, Schnor, and Vestbo, 1998) provided more complete information about the risks. That information is displayed in Table 6.4. Based on the data in the table, do you agree with the conclusion stated in the Web article?

TABLE 6.4 ■ Smoking, Gender, and the Risk of a Heart Attack

	Sample Size	Heart Attacks	Risk of Heart Attack
Men			
Smokers	8490	902	10.62% (902/8490)
Nonsmokers	4701	349	7.42% (349/4701)
Women			
Smokers	6461	380	5.88% (380/6461)
Nonsmokers	5011	132	2.63% (132/5011)

We see from the table that the risk of a heart attack is generally greater for men than it is for women. In particular, for men who smoke the risk of a heart attack is 10.62%, which is clearly higher than the risk of 5.88% for women who smoke. These numbers contradict the Web article's headline and first sentence, which both imply that the risk of a heart attack is greater for women smokers than for men smokers.

The researchers focused on relative risk, which is the ratio of two risks. We can easily verify that the relative risks provided in the Web article are correct.

- For women who smoke, the relative risk of a heart attack is 5.88%/ 2.63% = 2.24.

- For men who smoke, the relative risk of a heart attack is 10.62%/ 7.42% = 1.43.

The first sentence of the article actually refers to a comparison of these two *relative* risks, which are not the *actual* risks for women and men smokers. In other words, the *relative risk* value for women, 2.24, is 50% higher than the *relative risk* for men, 1.43.

In a subsequent issue of the *British Medical Journal,* several letter writers argued that the correct interpretation of these data is

- For both smokers and nonsmokers, men have a higher risk of heart attack than women do.

- For both men and women, smoking *adds* about 3.2% to the risk experienced by nonsmokers. In other words, the effect of smoking is the same for the two sexes.

The researchers' response to these letters was that they believed that it was valid to consider the multiplicative effect of smoking on risk. Dr. Prescott and her colleagues believe it is important that, for women, smoking multiplies the risk of a heart attack by 2.24, but for men, smoking multiplies the risk by only 1.43. ◆

The moral of the story in Example 6.6 is that a *relative risk* is affected by the *baseline risk,* so it's important to know the baseline risk. If a risk increases to 3% from a baseline risk of 1%, the relative risk is 3. If a risk increases to 22% from a baseline risk of 20%, the relative risk is only 22/20 = 1.1, although the difference in risks is still 2%. Is an increase in risk from 1% to 3% more serious than an increase from 20% to 22%? That may depend on the situation.

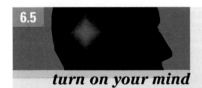

6.5

turn on your mind

If you were a frequent beer drinker and were worried about getting colon cancer, would it be more informative to you to know the *risk* of colon cancer for frequent beer drinkers or the *relative risk* of colon cancer for frequent beer drinkers compared to nondrinkers? Which of those statistics would likely be of more interest to the media? Explain your responses.*

*HINT: Suppose the relative risk is 3.0. Does that mean the same to you as an individual if the disease is rare as it does if it is common?

6.4 | THE EFFECT OF A THIRD VARIABLE AND SIMPSON'S PARADOX

In previous chapters, you have seen several examples in which a confounding or lurking variable may have affected the relationship between an explanatory variable and a response variable. In observational studies, a confounding variable might explain an apparent relationship between two variables or, in some instances, it can mask a relationship. Whenever the data are the product of an observational study, you should carefully consider the possibility that a third variable may affect the observed relationship.

EXAMPLE 6.7
Educational Status and Driving after Substance Use

In December of 1998, the U.S. government's Substance Abuse and Mental Health Services Administration released a report with information about drug use, alcohol use, and driving that was gathered in a 1996 nationwide survey of 11,847 individuals aged 16 and over (SAMSHA, 1998). A principal response variable throughout the report was a categorical variable that described whether a respondent had on any occasion in the previous year driven within two hours of using either alcohol or drugs. This response variable had three categories:

- ♦ never drove while impaired
- ♦ drove within two hours of alcohol use, but never after drug use
- ♦ drove within two hours of drug use and possibly after alcohol use

Figure 6.3 displays the association between this variable and educational status. The vertical axis of the bar chart is the proportion within an educational category that falls into a particular category of the response variable. The bar chart clearly shows that as the amount of education increases, the proportion who drove within two hours of alcohol use also increases. The data suggest a possible way to reduce the problem of drinking and driving: Let's not give out

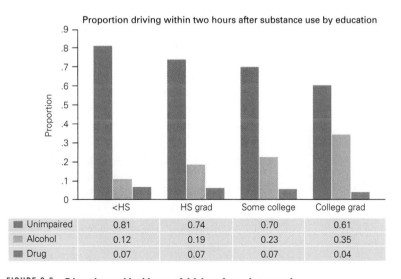

	<HS	HS grad	Some college	College grad
Unimpaired	0.81	0.74	0.70	0.61
Alcohol	0.12	0.19	0.23	0.35
Drug	0.07	0.07	0.07	0.04

FIGURE 6.3   **Education and incidence of driving after substance abuse**

any more college degrees. We hope, for many reasons, that you disagree with this solution.

What's going on here? This is an observational study, so we should consider the possibility of confounding. We should ask if the various educational groups differ in ways that may affect the response variable. One likely difference between the educational groups has to do with age. In the survey, about 13% of the respondents were under 21 years old. It is probable that almost none of these younger respondents were in the college degree group. It is also possible that many members of the "less than high school group" were under 21.

Age differences could at least partially explain the results. Another table in the report showed that about 80% of those under 21 fell into the "never while impaired category," but only 70% of the 21 or over age group fell into this category. Unfortunately, the report didn't contain the information necessary to sort out the relative effects of age and educational status, so we'll have to classify age as a lurking variable.

The amount of alcohol consumed within two hours of driving should also be considered. The report indicates that compared to younger drivers, older drivers were more likely to say that they had never had more than one drink before driving. ◆

Simpson's Paradox

Occasionally, the effect of a confounding factor is strong enough to produce a paradox known as **Simpson's Paradox.** The paradox is that the relationship appears to be in a different direction when the confounding variable is not considered than when the data are separated into the categories of the confounding variable.

EXAMPLE 6.8
Blood Pressure and Oral Contraceptive Use

We will illustrate Simpson's Paradox with a hypothetical example of an observational study done to examine the association between oral contraceptive use and blood pressure. Although the data are hypothetical, they are similar to data from several actual studies of the same problem. Suppose that 2400 women are categorized according to whether or not they use oral contraceptives and whether or not they have high blood pressure. The results shown in Table 6.5 indicate that the percent with high blood pressure is about the same among oral contraceptive users as it is for nonusers. In fact, a slightly higher percentage of the nonusers have high blood pressure.

We're certain that you can think of many factors that affect blood pressure. If the users and nonusers of oral contraceptives differ with respect to one of

TABLE 6.5 ■ Percent with High Blood Pressure for Users and Nonusers of Oral Contraceptives

	Sample Size	Number with High B.P.	% with High B.P.
Use Oral Contraceptives	800	64	64 of 800 = 8.0%
Don't Use Oral Contraceptives	1600	136	136 of 1600 = 8.5%

TABLE 6.6 ■ **Controlling for the Effect of Age**

	Age 18–34		Age 35–49	
	Sample Size	n and % with High B.P.	Sample Size	n and % with High B.P.
Use Oral Contraceptives	600	36 (6%)	200	28 (14%)
Don't Use Oral Contraceptives	400	16 (4%)	1200	120 (10%)

these factors, that factor confounds the results in Table 6.5. As in the previous example, age is a critical factor. The users of oral contraceptives tend to be younger than the nonusers are. This is important because blood pressure increases with age.

One way to control for the effect of age is to create separate contingency tables for women in different age groups. Table 6.6 divides the data from our hypothetical study into two age groups. In each age group, the percentage with high blood pressure is higher for the users than for the nonusers. This contradicts the direction of the relationship that we see when the age factor is not considered. In Table 6.5, not controlling for age differences masked the true nature of the relationship. Notice that the older women were less likely to be oral contraceptive users but more likely to have high blood pressure. ◆

6.5 | ASSESSING THE STATISTICAL SIGNIFICANCE OF A 2 × 2 TABLE

Most of the material you have seen thus far involves *descriptive statistics*, which are methods for describing the data in hand. This section introduces a special case of *hypothesis testing*, one of the two most common statistical procedures making up *inferential statistics*. The other common inference procedure is the use of a *confidence interval* to estimate a population value. Confidence intervals were introduced in Chapter 4. Hypothesis testing and confidence intervals will be covered in detail in Chapters 10 to 16, but this brief introduction will help you become familiar with some basic concepts and will help prepare you for those chapters.

Remember that observed data often represent a sample from a larger population. When this is the case, the purpose of collecting the data is usually to use the sample information to make generalizations about the population. Because this involves *inferring* something rather than being sure, the statistical methods used are called *inferential statistics*. When the data represent two categorical variables, the question of interest is whether a relationship observed in the sample data can also be inferred to hold in the population represented by the data.

When we see differences in the conditional distributions within the rows (or columns) of the contingency table, we say that the variables are related in the sample. A large difference between conditional distributions may convince us that the relationship is real. Small differences in conditional distributions, however, might not convince us. It is unlikely that any set of observed conditional distributions would ever be exactly the same for all categories of an explanatory variable. Small differences in observed conditional distributions could simply be the result of chance and may not represent a real relationship in the population.

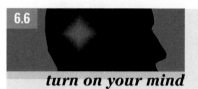

6.6

turn on your mind

A random sample includes 110 women and 90 men. Of the women, approximately 9% are left-handed, while approximately 11% of the men are left-handed. Based on this observed data, do you think there is a relationship between gender and handedness in the population represented by this sample? Why or why not?*

Case Study 1.6 described the Physician's Health Study in which about 22,000 physicians were randomized to take either aspirin or a placebo daily. Over a five-year period, the percentage of physicians experiencing heart attacks was lower in the aspirin group than in the placebo group. The difference was found to be **statistically significant,** and it was inferred that the observed difference reflected an actual difference in the population.

In the definitions following Case Study 1.6, the term *statistically significant* was defined as follows:

> A **statistically significant relationship** or difference is one that is large enough to be unlikely to have occurred in the observed sample if there is no relationship or difference in the population.

The practical upshot of calling an observed relationship statistically significant is that we are inferring that the relationship exists in the population. We make this inference by establishing that if there were really no relationship, it would be unlikely that we would have observed such a strong relationship in the sample.

The Five Steps to Determining Statistical Significance

There are five steps required in any hypothesis-testing situation. These steps lead to a decision about whether a statistically significant effect can be inferred about the population. The steps are

Step 1: Determine the null and alternative hypotheses.
Step 2: Verify necessary data conditions, and if met, summarize the data into an appropriate test statistic.
Step 3: Assuming the null hypothesis is true, find the *p*-value.
Step 4: Decide whether or not the result is statistically significant based on the *p*-value.
Step 5: Report the conclusion in the context of the situation.

In this section we describe how these steps are implemented for a 2×2 contingency table. In Chapter 15 you will learn how to implement them for any contingency table.

Step 1: Null and Alternative Hypotheses
The question we are considering is whether or not the sample data permit us to conclude that two variables are related. Another way to describe our objective is that we are deciding between two possible hypotheses about the population.

***HINT:** Consider *how many* men and women in the samples are left-handed. Now consider how the percents would change if one more or one fewer had been left-handed.

In statistical language, the possibilities are called the **null hypothesis** and the **alternative hypothesis,** and they can be stated as follows:

null hypothesis: The two variables are not related.
alternative hypothesis: The two variables are related.

The two hypotheses just given are specific to the context of two-way tables. In Chapters 11 and 13, you will learn how to write null and alternative hypotheses for a broad range of research questions.

Step 2: The Chi-square Statistic

The statistical significance of the association between two categorical variables can be examined using a value known as the **chi-square statistic.** This statistic measures the difference between the **observed counts** in the contingency table and the counts that would be **expected** if there were no relationship (the null hypothesis). A large difference between the observed and expected counts occurs when a relationship is present in the observed table. To simplify our discussion of the chi-square statistic, we will consider only 2×2 contingency tables so each variable will have just two categories. The same principles apply to larger tables, and we'll learn the details for larger tables in Chapter 15.

What follows is a brief discussion of the computation of the chi-square value. Don't worry too much about technical detail. Statistical software can be used to compute the chi-square statistic. You should, at this point, principally worry about the general ideas and how to use the computer output to decide whether you can declare statistical significance.

The first step in the calculation of a chi-square statistic is to determine a table of **expected counts.** These are the counts that would be expected to fall into the cells of the contingency table if there were no relationship between the two variables. They are found by making the row percents across columns equal for each row. To accomplish this, simply determine what percent of the total sample falls into each column, then split the total for each row into columns according to those same percents.

A table of expected counts has these two properties:

1. There is no association between the two variables. For instance, there are no differences between the conditional distributions in the rows (or columns) of a table of expected counts.
2. The expected counts in each row and column sum to the same totals as the observed numbers.

EXAMPLE 6.9
A Table of Expected Counts

Suppose that 90 men and 110 women are each asked about his or her handedness. Of the 200 total individuals, 20 are left-handed and 180 are right-handed. If gender and handedness are not related, what counts would we see in the cells of a contingency table? A strategy for determining these "expected counts" is to first calculate that $20/200 = 10\%$ of the overall sample is left-handed (falls into the first column), and to then make 10% of each gender be left-handed. The result is

	Left-handed	Right-handed	Total
Men	9 (10%)	81 (90%)	90
Women	11 (10%)	99 (90%)	110
Total	20 (10%)	180 (90%)	200

Keep in mind that these counts are what we would see if there were no relationship between gender and handedness. The row percents are the same for the two genders, and the total counts for each row and each column match the characteristics of the observed data. ◆

Calculating Expected Counts

formula   The expected count for each cell can be calculated as

$$\frac{\text{Row total} \times \text{Column total}}{\text{Total } n \text{ for table}}$$

For example, in Example 6.9, the expected count for the number of women who are left-handed could be calculated as

$$\frac{110 \times 20}{200} = 11$$

Verifying Conditions for the Test

tech note   The method described in this section won't work if the sample is too small. Proceed as long as at least three of the four expected counts are 5 or more and all of them are 1 or more.

Calculating the Chi-square Statistic

The specific way in which the chi-square statistic (χ^2) measures the difference between the observed and expected counts is relatively simple:

◆ First, for each cell in the table, we compute

$$\frac{(\text{Observed count} - \text{Expected count})^2}{\text{Expected count}}$$

◆ Then, we total these quantities over all cells of the table:

$$\chi^2 = \text{Sum of } \frac{(\text{Observed count} - \text{Expected count})^2}{\text{Expected count}}$$

Step 3: The p-value of the Chi-square Test

Now we're faced with a bit of a mystery. A large chi-square value indicates a relationship, but how large should the value be in order for us to declare statistical significance? In practice, this question is transformed into a different but equivalent question, which is: If there is actually no relationship in the population, what is the likelihood that the chi-square statistic could be as large as it is or larger?

definition   The *p-value* for a chi-square test is computed by assuming the null hypothesis is true and then determining the likelihood of observing data that would produce a chi-square statistic as large as the one observed, or larger. It answers the question, "How likely is it that a relationship of the magnitude observed, or one even stronger, would occur in the sample if there is no relationship in the population?"

The *p*-value is used to decide if the relationship observed in the sample is statistically significant and can be inferred to hold in the population. The *p*-value will be part of the computer output provided by almost any statistical software. If you are computing the chi-square statistic "by hand" instead of using statistical software, you can use Excel to find the *p*-value; see the Excel Tip on page 188. If that isn't possible, you can still determine whether the result is statistically significant using a standard rule, but the reason it works will remain a mystery until you encounter it again in Chapter 15. The decision process is as follows.

Steps 4 and 5: Making and Reporting a Decision

Generally, a large chi-square statistic and subsequently small *p*-value provide evidence that a real relationship exists in the population. To determine whether to decide in favor of the alternative hypothesis, that a real relationship exists, we'll use the rule that most researchers commonly use:

♦ When the *p*-value is less than or equal to 0.05 (5%) we will assume that the observed relationship did not occur by chance. In that case, we *can* say that the relationship is statistically significant, and we reject the null hypothesis in favor of the alternative hypothesis. For a chi-square statistic based on a 2 × 2 contingency table, this is equivalent to finding a chi-square statistic of at least 3.84.

♦ When the *p*-value is greater than 0.05 (5%), we will say that the observed relationship could have occurred just by chance. In that case, we *cannot* say that the relationship is statistically significant, and we cannot reject the null hypothesis. For a chi-square statistic based on a 2 × 2 contingency table, this is equivalent to finding a chi-square statistic of less than 3.84.

EXAMPLE 6.10
Randomly Pick S or Q

Example 2.2 of Chapter 2 described an experiment in which 92 college students were given a form that read, "Randomly choose one of the letters *S* or *Q*." Of these 92 students, 66% (61/92) picked *S*. Another 98 students were given a form that read, "Randomly choose one of the letters *Q* or *S*." Of these students only 46% (45/98) picked *S*. Is this sufficient evidence to generalize that the order of the letters on the form and the response are related?

Figure 6.4 displays computer output for testing the significance of the relationship for this experiment. The output was produced using the Minitab com-

Minitab v. 14 update on CD

```
Expected counts are printed below observed counts

                Letter Picked
    Form          S        Q       Total
    S first       61       31        92
                51.33    40.67

    Q first       45       53        98
                54.67    43.33

    Total        106       84       190

    Chi-Sq = 1.823 +  2.301 +
             1.712 +  2.160 = 7.995
    DF = 1, P-Value = 0.005
```

FIGURE 6.4 **Minitab output for the *S* or *Q* experiment (Example 6.10)** puter program. From the output, we learn that the *p*-value is 0.005. Because the *p*-value is less than 0.05, we can say that the relationship is statistically significant, and we reject the null hypothesis that there is no relationship. In other words, we infer that the observed result also holds in the population represented by these students.

Notice that the value of the chi-square statistic is 7.995. The *p*-value tells us that the chance is only 0.005 (which is 5 in 1000) that we would get a chi-square as large as 7.995 (or larger) if there really is no relationship between the order of the letters on the form and the letter that would be picked by people in this population. ◆

Minitab Tip

Minitab v. 14 update on CD

 tech note   **Computing a Chi-square Test for a Two-Way Table**

- ◆ If the raw data are stored in columns of the worksheet, use **Stat>Tables>Cross Tabulation.** In the dialog box, specify the two columns of interest and also click on "Chi-square analysis."

- ◆ If the data are already summarized into counts, enter the table of counts into columns of the worksheet and then use **Stat> Tables>Chi-Square Test.** In the dialog box, specify the columns containing the counts.

Excel Tip

tech note   The *p*-value can also be computed using Microsoft Excel. The function CHIDIST(x,df) provides the *p*-value, where x is the value of the chi-square statistic and df is a number called "degrees of freedom," which will be explained later in this book. The formula for df is (#rows − 1)(#columns − 1). For instance, corresponding to the information in Figure 6.4, $x = 7.995$, df $= (2 − 1)(2 − 1) = 1$ and the *p*-value is CHIDIST(7.995,1) = 0.00469, or about 0.005 as given by Minitab. Excel will also calculate the *p*-value directly from the data using the function CHITEST, but unlike Minitab, the user must provide the expected counts as well as the observed counts. Use the command *CHITEST(array1,array2)*, where the observed counts are in table format covered by the cells in *array1* and the expected counts are in *array2*. For example, if the observed counts are in a 2×2 table in the first four cells of a worksheet, *array1* would be given as A1:B2. Unlike Minitab and other statistical software, Excel gives only the *p*-value, and not the value of the chi-square statistic.

EXAMPLE 6.11
Breast Cancer Risk Stops Hormone Replacement Therapy Study

On July 17, 2002, the *Journal of the American Medical Association* published the results of a study that affected the lives of millions of women. The study was the first large randomized experiment to test the effects of combined estrogen and progestin, the most commonly prescribed hormone replacement therapy for postmenopausal women. The big news was that "On May 31, 2002, after a mean of 5.2 years of follow-up, the data and safety monitoring board recommended stopping the trial of estrogen plus progestin vs placebo because the test statistic for invasive breast cancer exceeded the stopping boundary for this adverse effect and the global index statistic supported risks exceeding benefits" (Writing Group for the Women's Health Initiative Investigators, 2002, p. 321).

TABLE 6.7 ■ Results of a Randomized Experiment Comparing Hormone Therapy and Placebo

	Invasive Breast Cancer?			Risk of Breast Cancer
	Yes	*No*	*Total*	
Hormones	166	8,340	8,506	.0195
Placebo	124	7,978	8,102	.0153
Total	290	16,318	16,608	.0175

The study measured several medical outcomes, but the two that received the most attention were breast cancer and coronary heart disease. Table 6.7 shows the results for breast cancer; results for coronary heart disease are given in the exercises. Is there a statistically significant effect of the hormone treatment on breast cancer?

Step 1: The Hypotheses

Null hypothesis: Occurrence of breast cancer is unrelated to hormone therapy.

Alternative hypothesis: Occurrence of breast cancer is related to hormone therapy.

Steps 2 and 3: The Conditions and Test Statistic

The expected count for the "Hormones, Yes" cell is $[(8,506)(290)]/16,608 = 148.53$. The other expected counts can be found similarly, or they can be found by subtraction because the row and column totals are the same as they are for the observed counts. For instance, the expected count for the "Hormones, No" cell is $290 - 148.53 = 141.47$. The sample size condition is met, since all expected counts exceed 5. The chi-square statistic is

$$\frac{(166-148.53)^2}{148.53} + \frac{(8340-8357.47)^2}{8357.47} + \frac{(124-141.47)^2}{141.47} + \frac{(7978-7960.53)^2}{7960.53} = 4.288$$

Steps 4 and 5: The p-value and Decision in Context

Using Excel, the *p*-value is found as CHIDIST(4.288,1) = 0.038382, or about 0.04. Using the standard convention, the result is statistically significant because 0.04 < 0.05. In the context of the problem, this means that there is a statistically significant relationship between hormone therapy and occurrence of breast cancer *in the population similar to the women in this experiment.* Because the data are from a randomized experiment, it can be concluded that hormone therapy *causes* a change in the incidence of breast cancer. It was for this reason that the study was terminated early. ◆

Factors That Affect Statistical Significance

In general, whether or not we can infer that an observed relationship represents a "real" relationship depends on two factors:

♦ the strength of the observed relationship

♦ how many people were studied

For 2×2 tables, the strength of the relationship can be measured by the difference between the two categories of the explanatory variable with respect to the percent falling into a particular category of the response variable. For instance, in Example 6.10, we saw that 66% of those with the "*S* or *Q*" order picked *S*, while only 46% of those with the "*Q* or *S*" form picked *S*. The difference between these two percents reflects the strength of the observed relationship between the order of the letters on the form and letter choice.

The sample size of a study also affects the significance of the results. Imagine, for example, that a psychiatric researcher compares the effectiveness of two treatments for depression by conducting a randomized experiment in which only ten patients receive each treatment. The results are that after three months, eight of the ten patients using treatment A show improvement, but among those who used treatment B, only five of ten patients show improvement. Although the difference between the percentages showing improvement is large (80% − 50%), the study is too small to safely infer that treatment A is the better treatment for the larger population of patients. The chi-square statistic and *p*-value would be 1.978 and 0.16. On the other hand, if 100 patients were in each treatment group and the same percents were observed (80% and 50%), the chi-square statistic would be 19.78 and the *p*-value would be 8.7×10^{-8}. We would almost certainly believe that treatment A was really better than treatment B in that case.

Practical Versus Statistical Significance

Statistical significance does not mean the two variables have a relationship that you would necessarily consider to be of *practical* importance. A table based on a very large number of observations will have little trouble achieving the status of statistical significance even if the relationship between the two variables in the population is only minor.

On the other hand, an interesting relationship in a population may fail to achieve statistical significance if there are only a few observations in the sample. Whenever a study "fails to find a relationship" between two variables, it does not necessarily mean that a relationship was not observed or does not exist in the population. It means that whatever relationship was observed in the sample did not achieve statistical significance. It is hard to rule out chance unless you have either a very strong relationship or a sufficiently large sample.

In the following example, you will see that with a large sample size, a relatively small arithmetic difference can be statistically significant. Remember that "statistical significance" means that we are convinced that a real relationship exists in the larger population represented by the sample. It makes sense that if we have a very large sample and it is exhibiting a relationship, we would be more convinced that it is real than we would be if a very small sample exhibited that same relationship.

EXAMPLE 6.12
Aspirin and Heart Attacks

In Case Study 1.6, the possible categories for the explanatory and response variables were

Explanatory variable = took placebo or took aspirin
Response variable = heart attack or no heart attack

Minitab v. 14 update on CD

Expected counts are printed below observed counts			

Heart Attack?

	Yes	No	Total
Aspirin	104	10933	11037
	146.52	10890.48	
Placebo	189	10845	11034
	146.48	10887.52	
Total	293	21778	22071

Chi-Sq = 12.339 + 0.166 +
$\qquad$ 12.343 + 0.166 = 25.014
DF = 1, P-Value = 0.000

FIGURE 6.5　**Minitab output for aspirin study (Example 6.12)**

Figure 6.5 shows the results of asking the Minitab program to compute the chi-square statistic. We see that the p-value is 0.000, so we can declare statistical significance. This p-value tells us that there is almost no chance that we could observe such a strong relationship if there is really no difference between the effects of aspirin and placebo.

Because the p-value is less than 0.05, we can reject the null hypothesis of no relationship. Among those in the placebo group, 189/11,034 = 1.71% had a heart attack during the study period, while among those in the aspirin group, 104/11,037 = 0.94% had a heart attack. The difference has practical importance because the percent decrease in risk for those taking aspirin is $(1.71 - 0.94)/1.71$ = 45%. However, the arithmetic difference in the percent experiencing a heart attack is 1.71% − 0.94% = 0.77%, or less than 1%. If a smaller study had been done, the researchers may not have been able to declare statistical significance, but since they used a sample with over 22,000 participants they were able to detect this important relationship. ◆

Interpreting a Nonsignificant Result

When we cannot claim statistical significance, we have to be careful about how we state the conclusion. The correct interpretation of a **nonsignificant result** is that the sample results are not strong enough to safely conclude that there is a relationship in the population. A p-value that's too large for us to declare significance simply means that the observed relationship could have resulted by chance, even if there is no relationship in the population. This is not the same as saying that we believe there is no relationship.

In the following case study, the difference between the relevant conditional percentages is larger than the difference we saw in the aspirin example. The result of the chi-square test, however, will be that we cannot declare statistical significance. You will see that the sample size for the case study is much smaller than the sample size for the aspirin study. Remember that the sample size affects our ability to declare significance.

CASE STUDY 6.1
Drinking, Driving, and the Supreme Court

In the early 1970s a young man challenged an Oklahoma State law that prohibited the sale of 3.2% beer to males under 21 but allowed its sale to females in the same age group. The case (*Craig* v. *Boren,* 429 U.S. 190, 1976) was ultimately heard by the U.S. Supreme Court.

Laws are allowed to use gender-based differences as long as they "serve important governmental objectives" and "are substantially related to the achievement of these objectives" (Gastwirth, 1988, p. 524). The defense argued that traffic safety was an important governmental objective and that data clearly show that young males are more likely to have alcohol-related accidents than young females.

The Supreme Court examined evidence from a "random roadside survey" that measured information on age, gender, and whether or not the driver had been drinking alcohol in the previous two hours. Although the survey was called a "random" survey of drivers, it probably was not. In roadside surveys, police tend to stop all drivers at certain locations at the time of the survey. This procedure does not really provide a random sampling of drivers in an area, but we'll treat it as though it does. Table 6.8 gives the results of the roadside survey for the drivers under 20 years of age.

TABLE 6.8 ■ Results of Roadside Survey for Young Drivers

| | **Drank Alcohol in Last Two Hours?** | | | |
	Yes	No	Total	Percent Who Drank
Males	77	404	481	16.0%
Females	16	122	138	11.6%
Total	93	526	619	15.0%

Source: Gastwirth, 1988, p. 526.

Notice that the percentage of young men who had been drinking alcohol is slightly higher than the percentage of young women. The difference is 16% − 11.6% = 4.4%. However, we cannot rule out chance as a reasonable explanation for this difference. In other words, if there really is no difference between the percents of young male and female drivers in the population who drink and drive, we could possibly see a difference as large as the one observed in a sample of this size.

In Figure 6.6, we present the results of asking the Minitab program to compute the chi-square statistic for this example. The chi-square summary statistic is 1.637, and the *p*-value for this statistic is 0.201. This *p*-value tells us that if there's really no association in the population (the null hypothesis), there's about a 20% chance that the sample would have a chi-square statistic as large as 1.637, or larger. In other words, the observed relationship could easily have occurred even if there is no relationship in the population represented by the sample.

The Supreme Court overturned the law, concluding that "the showing offered by the appellees does not satisfy us that sex represents a legitimate, accurate proxy for the regulation of drinking and driving" (Gastwirth, 1988, p. 527). Based on the chi-square analysis, you can see why the Supreme Court was reluctant to conclude that the difference in the sample represented sufficient evidence for a real difference in the population.

Minitab v. 14 update on CD

Expected counts are printed below observed counts

| | Drank in Last 2 Hours? | | |
	Yes	No	Total
Males	77	404	481
	72.27	408.73	
Females	16	122	138
	20.73	117.27	
Total	93	526	619

Chi-Sq = 0.310 + 0.055 +
            1.081 + 0.191 = 1.637
DF = 1, P-Value = 0.201

FIGURE 6.6   Minitab output for Case Study 6.1

KEY TERMS

Section 6.1
contingency table, 172
two-way table, 172
cell, 172
conditional percents, 172
row percents, 172
column percents, 172

Section 6.2
risk, 175
relative risk, 175

baseline risk, 175
percent increase in risk, 176
odds, 176
odds ratio, 176

Section 6.4
Simpson's Paradox, 182

Section 6.5
statistically significant relationship, 184

statistical significance, 184
null hypothesis, 185
alternative hypothesis, 185
chi-square statistic, 185
observed counts, 185
expected counts, 185, 186
p-value, 186
practical significance, 190
nonsignificant result, 191

EXERCISES   Blue = Basics   **Bold, Bold** = Answer in back

Section 6.1

6.1 Each fall, auditions for the band and orchestra are held at a large university. Last fall, the numbers of males and females in each class who auditioned were

Class	Female	Male	Total
Freshman	170	100	270
Sophomore	50	50	100
Junior	60	20	80
Senior	20	30	50
Total	300	200	500

a. Calculate the row percent for freshman females and explain what it means.
b. Calculate the column percent for freshman females and explain what it means.
c. Which class had the highest percentage of female applicants? Support your answer with numbers.
d. Which gender had a higher percentage of sophomore applicants? Support your answer with numbers.

6.2 Anton and Edward often play a game together, so they decide to see if who goes first affects who wins. They keep track of 50 games, with each going first 25 times. Of the 25 times Anton went first, he won 15 times. Of the 25 times Edward went first, he won 12 times. In constructing a contingency table for these results, the two variables are "Who went first" and "Did the person who went first win?"
a. What are the categories for each of the two variables?
b. If the explanatory variable is used to define the rows of the contingency table, which variable would be the row variable?
c. Construct a contingency table for the results.
d. Overall, what percent of the games were won by the person who went first?
e. Is there an advantage to going first? Explain.

6.3 For each pair of variables, indicate whether or not a two-way table would be appropriate for summarizing the relationship. In each case, briefly explain why or why not.

a. Political party (Republican, Democrat, etc.) and opinion about new gun control law.
b. Age group (under 20, 21–29, etc.) and rating of a song on 1 to 5 scale (1 = hate it, 5 = love it).
c. Weight (pounds) and height (inches).
d. Gender and opinion about capital punishment.
e. Head circumference (centimeters) and gender.

6.4 Suppose a study on the relationship between gender and political party included 200 men and 200 women and found 180 Democrats and 220 Republicans. Is that information sufficient for you to construct a contingency table for the study? If so, construct the table. If not, explain why not.

6.5 Students in a class were asked whether they preferred an in-class or a take-home final exam and were then categorized as to whether or not they had received an A on the in-class midterm. Of the 25 A students, 10 preferred a take-home exam, while of the 50 non-A students, 30 preferred a take-home exam.
a. Display the data in a contingency table.
b. In the relationship between grade on the midterm and opinion about type of final, which variable is the response variable and which is the explanatory variable?
c. Determine an appropriate set of conditional percents for determining if there is a relationship between grade on the midterm and opinion about the type of final. Based on these percents, does it appear that there is a relationship? Why or why not?

6.6 Do grumpy old men have a greater risk of having coronary heart disease than men who aren't so grumpy? Harvard Medical School researchers examined this question in a prospective observational study reported in the November 1996 issue of *Circulation* (Kawachi et al., 1994). For seven years, the researchers studied men between the ages of 46 and 90 years old. All study participants completed a survey of anger symptoms at the beginning of the study period. Among 199 men who had no anger symptoms, there were 8 cases of coronary heart disease. Among 559 men who had the most anger symptoms, there were 59 cases of coronary heart disease.
a. Construct a contingency table for the relationship

between degree of anger and the incidence of heart disease.

b. Among those with no anger symptoms, what percent had coronary heart disease?

c. Among those with the most anger symptoms, what percent had coronary heart disease?

d. Draw a bar graph of these data. Based on this graph, does there appear to be an association between anger and the risk of coronary heart disease? Explain.

6.7 In a class survey, Penn State statistics students were asked, "Regarding your weight, do you think you are: About right? Overweight? Underweight?" The following table displays the results by sex:

Gender and Perception of Weight

| Sex | Perception of Weight | | | |
	About Right	Overweight	Underweight	Total
Female	87	39	3	129
Male	64	3	16	83
Total	151	42	19	212

Source: The authors.

a. Write a sentence that explains what would be measured by the row percents for this table. Make your answer specific to this situation.

b. Determine the row percents.

c. Draw a bar graph of the row percents.

d. Briefly describe how males and females differ in their perceptions of weight.

e. An important objective of statistics is the use of sample information to make generalizations about a larger population. What population do you think is represented by this sample?

6.8 Refer back to Example 6.1 about smoking and divorce.

a. Figure 6.1 is a bar graph of the observed counts. Draw a bar graph of the row percents. Does this graph confirm the relationship? Explain.

b. Among those who were ever divorced, what percent smoke?

c. Among those who were never divorced, what percent smoke?

d. What percent of all respondents in the table have ever been divorced?

6.9 In a case-control study done in England, Voss and Mulligan (2000) collected data on height (short or not) and whether or not the student had ever been bullied in school for 209 secondary school students. The following table displays a contingency table of the data. A student was categorized as short if he or she was below the third percentile for height upon school entry, but because this was a case-control study, short people (cases) constitute almost half of the sample.

Height and Bullying in School

| Height | Ever Bullied | | |
	Yes	No	Total
Short	42	50	92
Not Short	30	87	117
Total	72	137	209

a. Among students in the "short" category, what percent have ever been bullied?

b. Among students in the "not short" category, what percent have ever been bullied?

c. Is there a relationship between height and the likelihood of having been bullied? Briefly justify your answer.

Sections 6.2 and 6.3

6.10 For each of the following measures, give the value that implies no difference between the two groups being compared.

a. Relative risk.

b. Odds ratio.

c. Percent increase in risk.

For Exercises 6.11 to 6.14: A study is done to compare side effects for those taking a drug versus those taking a placebo. One hundred people are given the drug and 100 are given the placebo. Results are as shown in the following table.

| | Headache? | | Nausea? | | Insomnia? | |
	Yes	No	Yes	No	Yes	No
Drug	10	90	15	85	6	94
Placebo	5	95	5	95	4	96

6.11 For each of the following side effects, compute the risk of the side effect when taking the drug.

a. Headache.

b. Nausea.

c. Insomnia.

6.12 For each of the following side effects, compute the relative risk of the side effect when taking the drug compared to when taking the placebo.

a. Headache.

b. Nausea.

c. Insomnia.

6.13 For each of the following side effects, compute the odds ratio for the side effect for those taking the drug compared to those taking the placebo.

a. Headache.

b. Nausea.

c. Insomnia.

6.14 For each of the following side effects, compute the percent increase (or decrease) in risk of side effect for those taking the drug compared to those taking the placebo.

a. Headache.

b. Nausea.

c. Insomnia.

6.15 If the baseline risk of a certain disease for nonsmokers is 1% and the relative risk of the disease is 5 for smokers compared to nonsmokers, what is the risk of the disease for smokers?

6.16 For a relative risk of 2.1, what is the percent increase in risk?

6.17 For an increase in risk of 40%, what is the relative risk?

6.18 The relative risk of contracting a certain coronary disease is 2.0 for male smokers compared to male nonsmokers and 3.0 for female smokers compared to female nonsmokers. Is this enough information to determine whether male smokers or female smokers are more likely to contract the disease? If so, make that determination. If not, explain what additional information would be needed to make that determination.

6.19 *Science News* (Feb. 25, 1995, p. 124) reported a study of 232 people, aged 55 or over, who had heart surgery. The patients were asked whether their religious beliefs give them feelings of strength and comfort and whether they regularly participate in social activities. Of those who said yes to both, about 1 in 50 died within six months after their operation. Of those who said no to both, about 1 in 5 died within six months after their operation. What is the relative risk of death (within six months) for the two groups? Write your answer in a sentence or two that would be understood by someone with no training in statistics.

6.20 Refer back to exercise 6.6 about anger symptoms and the incidence of heart disease.
  a. For those with the most anger symptoms, what is the relative risk of heart disease (compared to those with no anger symptoms)?
  b. For those with the most anger symptoms, what is the percent increase in the risk of heart disease?

6.21 Refer back to exercise 6.9 about height and the risk of being bullied.
  a. For each height category, calculate the risk of having been bullied.
  b. What is the relative risk for short students of having been bullied? Write a sentence that interprets this relative risk.
  c. What is the increased risk of having been bullied for short students? Write a sentence that interprets this increased risk.
  d. Calculate the odds ratio that compares the odds of having been bullied for the short students to the odds for students who are not short. Write a sentence that interprets this ratio.

6.22 Using the terminology of this chapter, what name applies to each of the boldface numbers in the following quotes (e.g., odds, risk, relative risk)?
  a. "Fontham found increased risks of lung cancer with increasing exposure to secondhand smoke, whether it took place at home, at work, or in a social setting. A spouse's smoking alone produced an overall **30** percent increase in lung-cancer risk" (*Consumer Reports*, Jan. 1995, p. 28).
  b. "What they found was that women who smoked had a risk [of getting lung cancer] **27.9** times as great as nonsmoking women; in contrast, the risk for men who smoked regularly was only **9.6** times greater than that for male nonsmokers" (Taubes, 1993, p. 1375).
  c. "**One student in five** reports abandoning safe-sex practices when drunk" (*Newsweek*, Dec. 19, 1994, p. 73).

6.23 The Roper Organization (1992) conducted a study as part of a larger survey to ascertain the number of American adults who had experienced phenomena such as seeing a ghost, "feeling as if you left your body," and seeing a UFO. A representative sample of adults (18 and over) in the continental United States were interviewed in their homes during July, August, and September 1991. The results when respondents were asked about seeing a ghost are shown in the following table.

Reportedly Has Seen a Ghost			
	Yes	No	Total
Aged 18 to 29	212	1313	1525
Aged 30 or over	465	3912	4377
Total	677	5225	5902

Data Source: The Roper Organization (1992), *Unusual Personal Experiences,* Las Vegas: Bigelow Holding Corp., p. 35.

  a. In each age group, find the percent who reported seeing a ghost.
  b. What is the *relative risk* of reportedly seeing a ghost for one group compared to the other? Write your answer in the form of a sentence that could be understood by someone who knows nothing about statistics.
  c. Repeat part (b) using *increased risk* instead of relative risk.
  d. What are the odds of reportedly seeing a ghost to not seeing one in the older group?

6.24 Suppose a newspaper article states that drinking three or more cups of coffee per day doubles the risk of gall bladder cancer. Before giving up coffee, what questions should be asked by a person who drinks this much coffee?

Section 6.4

6.25 Refer to Case Study 1.5, "Does Prayer Lower Blood Pressure?" One of the results quoted in that study was, "People who attended a religious service once a week and prayed or studied the Bible once a day were 40% less likely to have high blood pressure than those who don't go to church every week and prayed and studied the Bible less."
  a. What is the explanatory variable in this study?
  b. What is the response variable in this study?
  c. Give an example of a third variable that might at least partially account for the observed relationship.

6.26 Refer to the previous exercise, in which one or more third variables are at least partially likely to account for the observed relationship between religious activities and reduced incidence of high blood pressure. Is this an example of Simpson's Paradox? Explain.

6.27 This exercise presents a real example of Simpson's paradox (Wagner, 1982). The total income and total taxes paid in each of five income categories are given for two years, 1974 and 1978, in the following table.
  a. The "tax rate" for any income category is calculated as Tax divided by Income. Calculate the tax rates for each income bracket in each year.

	1974		1978	
Adjusted Gross Income	Income	Tax	Income	Tax
Under $5,000	41,651,643	2,244,467	19,879,622	689,318
$5,000 to $9,999	146,400,740	13,646,348	122,853,315	8,819,461
$10,000 to $14,999	192,688,922	21,449,597	171,858,024	17,155,758
$15,000 to $99,999	470,010,790	75,038,230	865,037,814	137,860,951
$100,000 or more	29,427,152	11,311,672	62,806,159	24,051,698
Total	880,179,247	123,690,314	1,242,434,934	188,577,186

b. Calculate the overall tax rate for each of the two years.

c. Did the tax rates within each income bracket increase or decrease from 1974 to 1978?

d. Did the overall tax rate increase or decrease from 1974 to 1978?

e. Compare your results in parts (c) and (d). Explain them in terms of Simpson's Paradox.

6.28 In a 1997 Marist College Institute for Public Opinion survey of 995 randomly selected Americans, 31% of the men and 12% of the women surveyed said they have dozed off while driving (*Source:* www.mipo.marist.edu). Think of a third variable that might at least partially explain the observed relationship between gender and dozing when driving. Briefly explain your answer.

6.29 Suppose two hospitals are willing to participate in an experiment to test a new treatment, and both hospitals agree to include 1100 patients in the study. Because the researchers conducting the experiment are on the staff of hospital A, they decide to perform the majority of cases with the new procedure. They randomly assign 1000 patients to the new treatment, with the remaining 100 receiving the standard treatment. Hospital B, which is a bit reluctant to try something new on too many patients, agrees to randomly assign 100 patients to the new treatment, leaving 1000 to receive the standard treatment. The following table displays the results.

Survival Rates for Standard and New Treatments at Two Hospitals

	Hospital A			Hospital B		
Treatment	*Survive*	*Die*	*Total*	*Survive*	*Die*	*Total*
Standard	5	95	100	500	500	1000
New	100	900	1000	95	5	100
Total	105	995	1100	595	505	1100

a. Which treatment was more successful in hospital A? Justify your answer with relevant percents.

b. Which treatment was more successful in hospital B? Justify your answer with relevant percents.

c. Combine the data from the two hospitals into a single contingency table that shows the relationship between treatment and outcome. Which treatment has the higher survival rate in this combined table? Justify your answer.

6.30 A researcher observes that, compared to students who do not procrastinate, students who admit to frequent procrastination are more likely to miss class due to illness. Does this mean that procrastinating increases illness? What is another explanation?

6.31 The success rates of two treatments (A and B) for clinical depression are being compared. The research team included five doctors, and the participants were 200 patients with depression. The doctors were supposed to randomly assign treatments to patients, but two doctors didn't do this. Instead, they often assigned more severely depressed patients to use treatment A and less severely depressed patients to use treatment B.

a. Three variables measured by the investigators were the outcome of treatment (successful or not), method

of treatment (A or B), and a rating of the initial severity of the depression (mild or severe). Which of these variables is the response variable in this investigation?

b. The doctors were surprised by the final result, which was that treatment B had a higher success rate than treatment A. Based on results attained by other investigators, they had expected the opposite to occur. Explain how the actions of the two doctors who did not always randomly assign treatments may have caused an unfair comparison of the two treatments.

c. Taking Example 6.8 into consideration, describe how the researchers might control for the effect of the initial severity of patient depression when they analyze the relationship between treatment method and outcome.

Section 6.5

6.32 The following 2×2 contingency table shows data for gender and opinion on the death penalty for respondents in the 1993 General Social Survey (**GSS-93** dataset on the CD for this text).

Opinion on Death Penalty

	Oppose	Favor	All
Male	113	535	648
Female	224	616	840
All	337	1151	1488

a. State null and alternative hypotheses about the relationship between the two variables.

b. Calculate the expected counts for a chi-square test.

6.33 The following 2×2 contingency table shows data for gender and opinion about the legalization of marijuana for respondents in the 1993 General Social Survey (**GSS-93** dataset on the CD for this text).

Opinion on Marijuana

	Don't Legalize	Legalize	All
Male	295	118	413
Female	475	116	591
All	770	234	1004

a. Verify that the value of the chi-square statistic is approximately 10.9.

b. The *p*-value is 0.001. Are the results statistically significant? Explain.

c. Write a conclusion in the context of the situation.

6.34 For each of the following results, explain what conclusion can be made about the null hypothesis that there is no relationship between two variables that form a 2×2 contingency table.

a. *p*-value = 0.001.

b. *p*-value = 0.101.

c. *p*-value = 0.900.

d. *p*-value = 0.049.

e. *p*-value = 0.755.

f. Chi-square statistic = 4.01.

g. Chi-square statistic = 2.98.

6.35 If a relationship is statistically significant, does that guarantee it also has practical significance? Explain.

6.36 If a relationship has practical significance, does it guarantee that statistical significance will be achieved in every study that examines it? Explain.

6.37 Explain whether each of the following is possible.

a. A relationship exists in the observed sample but not in the population from which the sample was drawn.

b. A relationship does not exist in the observed sample but does exist in the population from which the sample was drawn.

c. A relationship does not exist in the observed sample but an analysis of the sample shows that there is a statistically significant relationship, so it is inferred that there is a relationship in the population.

6.38 Refer back to exercise 6.23 in which data were presented for the relationship between age group and whether a person reports having ever seen a ghost (see table).

a. Write null and alternative hypotheses about the possible relationship between the two variables.

b. The Minitab output for a chi-square test of the relationship follows. Based on this output, what can be concluded about the relationship? What is the basis for this conclusion?

c. Notice in the Minitab output that the *p*-value is 0.001. Write a sentence that interprets this number.

Minitab v. 14 update on CD

```
Expected counts are printed below observed counts

               Yes          No         Total
18-29          212         1313         1525
               174.93      1350.07

30+            465         3912         4377
               502.07      3874.93

Total          677         5225         5902

Chi-Sq = 7.857 + 1.018 +
         2.737 + 0.355 = 11.967
DF = 1, P-Value = 0.001
```

6.39 Refer to the Minitab output for the previous exercise.

a. Demonstrate how the *expected count* was computed for the "age 18–29 yes" cell of the table.

b. Verify that the expected count for the "age 30+ yes" cell can be determined by subtracting the answer for part (a) from the total count for the "yes" column of the table.

c. In each row of the table, express the *expected count* for yes as a percent of the total count for that row. How do these two percents compare to each other?

6.40 Imagine that 50 men and 50 women are asked, "Do you favor or oppose capital punishment for those convicted of murder?" In the observed data, 38/50 = 76% of the men favor capital punishment compared to 32/50 = 64% of the women.

a. Write null and alternative hypotheses about the possible relationship between gender and opinion about capital punishment.

b. A chi-square test of the statistical significance of the relationship has a *p*-value of 0.19. Is this evidence that there is or is not a relationship? What is the justification for this conclusion?

6.41 Refer to the previous exercise about the relationship between gender and opinion about capital punishment.

a. Calculate the value of the chi-square statistic.

b. In the previous exercise, the *p*-value was given as 0.19. Write a sentence that interprets this number.

c. Suppose that 500 men and 500 women had been surveyed, rather than 50 of each sex. Further suppose that the proportions in favor of capital punishment remained the same, at 380/500 or 76% of the men and 320/500 or 64% of the women. Calculate the chi-square statistic in this case.

d. The *p*-value for the chi-square statistic in part (c) is 0.000035. Is this evidence that there is a relationship between gender and opinion about capital punishment?

e. Explain the reason for the discrepancy between the result in exercise 6.40b and the result in part (d) of this exercise .

6.42 Considering the effect of sample size on the chi-square test, explain why a finding that a relationship is "not statistically significant" should not be interpreted as absolute proof that there is no relationship in the population.

6.43 Refer to Example 6.11, in which a statistically significant relationship was found between hormone therapy and invasive breast cancer. In the same study, the following observed counts were found for death from coronary heart disease (CHD). Carry out the five steps leading to a conclusion about whether there is a statistically significant relationship between hormone therapy and death from CHD.

	Death from CHD?		
	Yes	*No*	*Total*
Hormones	33	8473	8506
Placebo	26	8076	8102
Total	59	16,549	16,608

Chapter Exercises

6.44 Wechsler and Kuo (2000) used data from the 1999 College Alcohol Study to examine the relationship between student alcohol use and student definitions of binge drinking. In the study, approximately 14,000 college students from 119 schools answered questions about drinking habits. Using the definition that binge drinking is five consecutive drinks for men and four consecutive drinks for women, the researchers categorized students into four "type of drinker" categories based on their answers to questions about personal alcohol use. Students in the survey were also asked how many drinks in a row they thought constituted binge drinking for men and for women. The following is a contingency table showing the relationship between the student definitions of binge drinking for a man and student alcohol use:

Type of Drinker and Personal Definition of Binge Drinking

Type of Drinker	Definition of Binge Drinking for Men (Drinks in a Row)				
	≤4	5	6	≥7	Total
Abstainer	814	659	484	702	2659
Nonbinge	714	1129	997	2223	5063
Occasional Binge	231	471	480	1780	2962
Frequent Binge	179	442	367	2147	3135
Total	1938	2701	2328	6852	13,819

a. Among all students, what percent defined binge drinking for men as being seven or more drinks in a row?

b. Write a sentence describing what would be measured by row percents in the table. Make your answer specific to these variables.

c. Write a sentence describing what would be measured by column percents in this table. Make your answer specific to these variables.

d. Calculate the row percents in the table. In a few sentences, describe the relationship between type of drinker and the personal definition of drinking.

6.45 Example 6.2 presented data on how many college men have ear pierces and tattoos. The following table contains data on ear pierces and tattoos for a sample of 678 college women. The ear pierce response is the total number of ear pierces for a woman, and this has been categorized.

Ear Pierces and Tattoos, 678 College Women

Pierces	No Tattoo	Have Tattoo
2 or less	245	19
3 or 4	210	26
5 or 6	91	32
7 or more	25	30

Data Source: One of the authors.

a. For each ear-pierce category, determine the percent with a tattoo. Based on these percents, would you say there is a relationship between these two variables? Explain.

b. Draw a bar graph of the column percents. What does the graph show us about the relationship between having a tattoo and the number of ear pierces?

c. Calculate the percent with at least five ear pierces for women who have a tattoo, and separately for women who don't have a tattoo.

d. What percent of the total sample has a tattoo?

e. What percent of the total has two or fewer ear pierces and also does not have a tattoo?

6.46 Refer back to Example 6.2.

a. Redraw the bar chart in Figure 6.2 using counts rather than percents. What features of the data does this plot reveal that are not revealed by Figure 6.2?

b. Briefly explain what would be measured by the row percents for the data in Table 6.2.

6.47 In a retrospective observational study, researchers asked women who were pregnant with planned pregnancies how long it took them to get pregnant (Baird and Wilcox,

1985; see also Weiden and Gladen, 1986). Length of time to pregnancy was measured according to the number of cycles between stopping birth control and getting pregnant. Women were also categorized on whether or not they smoked, with smoking defined as having at least one cigarette per day for at least the first cycle during which they were trying to get pregnant. The following table summarizes the observed counts.

Time to Pregnancy for Smokers and Nonsmokers

	Pregnancy Occurred After	
	First Cycle	Two or More Cycles
Smoker	29	71
Nonsmoker	198	288

a. Among smokers, what percent was pregnant after the first cycle?

b. Among nonsmokers, what percent was pregnant after the first cycle?

c. Draw a bar graph that can be used to examine the relationship between smoking and how long it took to get pregnant.

d. Among those who were pregnant after the first cycle, what percent smoked?

e. Among those who took two or more cycles to become pregnant, what percent smoked?

f. Do you think there is a relationship between smoking and how long it takes to get pregnant? Explain.

6.48 In the 1996 General Social Survey, religious preference and opinion about when premarital sex might be wrong were among the measured variables. The contingency table of counts for these variables is shown in the following table.

Religious Preference and Opinion about Premarital Sex

Religion	When Is Premarital Sex Wrong?				
	Always	Almost Always	Sometimes	Never	Total
Protestant	355	117	227	384	1083
Catholic	62	37	120	226	445
Jewish	0	3	14	34	51
None	20	13	45	147	225
Other	15	13	23	40	91
Total	452	183	429	831	1895

Source: SDA archive at csa.berkely.edu:7502.

a. For each religious preference category, determine the percent of respondents who think premarital sex is always wrong.

b. Do the percents computed in part (a) indicate that there is a relationship between the two variables? Briefly explain why or why not.

6.49 A case-control study in Berlin, reported by Kohlmeier, Arminger, Bartolomeycik, Bellach, Rehm, and Thamm (1992) and by Hand et al. (1994), asked 239 lung-cancer patients and 429 controls (matched to the cases by age and sex) whether or not they had kept a pet bird during adulthood. Of the 239 lung-cancer cases, 98 said yes. Of the 429 controls, 101 said yes.

   a. Construct a contingency table for the data.

   b. Compute the risk of lung cancer for bird owners and for those who had never kept a bird.

   c. Can the risks computed in part (b) be used as baseline risks for the populations of those who have and have not owned birds? Explain.

   d. What is the relative risk of lung cancer for bird owners?

   e. What additional information about the risk of lung cancer would you want before you made a decision about whether or not to own a pet bird?

   f. Can these data be used to establish a causal connection between owning a bird and the risk of lung cancer? Explain.

   g. For any study of lung cancer, the effect of smoking should be considered. How would you determine if smoking might be a confounding variable in this study?

6.50 Compute the chi-square statistic and assess the statistical significance for the relationship between bird ownership and lung cancer, based on the data in the previous exercise. State a conclusion about the relationship.

6.51 Pagano and Gauvreau (1993, p. 133) reported data for women participating in the first National Health and Nutrition Examination Survey (Carter, Jones, Schatzkin, and Brinton, 1989). The explanatory variable was whether or not the woman gave birth to her first child at the age of 25 or older, and the outcome variable was whether or not she developed breast cancer. Observed counts are in the following table.

Age at Birth of First Child and Breast Cancer

Age at First Child	Breast Cancer	No Breast Cancer	Total
25 or Older	31	1597	1628
Under 25	65	4475	4540
Total	96	6072	6168

Source: Pagano and Gauvreau (1993).

   a. Calculate the risk of breast cancer for women who were under 25 years old when they had their first child.

   b. Calculate the risk of breast cancer for women who were 25 or older when they had their first child.

   c. Calculate the relative risk of breast cancer for women who were 25 or older when they had their first child compared with women who were under 25. Write a sentence that interprets this relative risk in the context of this problem.

   d. What is the percent increase in the risk of breast cancer for women who were 25 or older when they had their first child?

   e. Can these data be used as evidence of a causal connection between age at first child and the risk of breast cancer? Explain.

6.52 Refer to Example 6.11, in which a statistically significant relationship was found between hormone therapy and invasive breast cancer. There were 166 cases of invasive breast cancer out of 8506 women taking hormones and 124 cases of invasive breast cancer out of 8102 women taking placebo.

   a. What is the relative risk of invasive breast cancer when taking hormones compared with taking placebo?

   b. What is the baseline risk of invasive breast cancer, assuming that the placebo group is representative of the relevant population of women?

   c. Explain why it is important to report the baseline risk found in part (b) when reporting the relative risk found in part (a).

6.53 A well-known example of Simpson's Paradox, published by Bickel, Hammel, and O'Connell (1975), examined admission rates for men and women who had applied to graduate programs at the University of California at Berkeley. The actual breakdown of data for specific programs is confidential, but the point can be made with similar, hypothetical numbers. For simplicity, we will assume there are only two graduate programs. The figures for acceptance to each program are shown in the following table.

Data for Exercise 6.31

	Program A		Program B	
	Admit	Deny	Admit	Deny
Men	400	250	50	300
Women	50	25	125	300

   a. Combine the data for the two programs into one table. What percent of all men who applied was admitted? What percent of all women who applied was admitted? Which sex was more successful in the admissions process?

   b. What percent of the men who applied did program A admit? What percent of the women who applied did program A admit? Repeat the question for program B. Which sex was more successful in getting admitted to program A? program B?

   c. Explain how this problem is an example of Simpson's Paradox. Provide a potential explanation for the observed figures by guessing what type of programs A and B might have been.

6.54 Compute the chi-square statistic and assess the statistical significance for the relationship between smoking and time to pregnancy in exercise 6.47. State a conclusion about the relationship.

6.55 Refer to Example 6.11, in which a statistically significant relationship was found between hormone therapy and invasive breast cancer. One of the results reported in the paper was "Absolute excess risks per 10,000 person-years attributable to estrogen plus progestin were . . . 8 more invasive breast cancers" (Writing Group for the Women's Health Initiative Investigators, p. 321). In other words, for every 10,000 "person-years" of women taking the hormones instead of a placebo, there would be 8 additional cases of invasive breast cancer. Discuss practical versus statistical significance in this situation.

6.56 Exercise 6.6 concerned the relationship between anger and the risk of heart disease. Some computer output for a chi-square test is shown below. What do the results indicate about the relationship? Explain.

Minitab v. 14 update on CD

```
Expected counts are printed below observed counts

                      Heart Disease
                  Yes           No
No Anger           8           191          199
                  17.59        181.41

Most Anger        59           500          559
                  49.41        509.57

Total             67           691          758

Chi-Sq = 5.228 +  0.507 +
         1.861 + 0.180 = 7.777
DF = 1, P-Value = 0.0005
```

6.57 Refer to the data in the previous exercise, showing the frequency of heart disease for men with no anger and men with the most anger. Show how the numbers in the following statement were calculated: The odds of remaining free of heart disease versus getting heart disease are about 24 to 1 for men with no anger, whereas those odds are only about 8.5 to 1 for men with the most anger.

6.58 Refer to the previous two exercises. Find the odds ratio for remaining free of heart disease for men with no anger compared with men with the most anger. Give the result in a sentence that someone with no training in statistics would understand.

6.59 "Saliva test predicts labor onset" was the headline of a Reuters Health story that appeared May 23, 2000 at the Yahoo Health News website, and the story described a medical test called SalEst. A positive SalEst test indicates an elevated estrogen level, and this knowledge may help predict how soon a pregnant woman will give birth because estrogen level increases in the 2 to 3 weeks before delivery. Researchers employed by the manufacturers of SalEst tested 642 women who had been pregnant for 39 weeks. Among the 615 women who delivered before the 42nd week of the pregnancy, about 59% had a positive SalEst test. Among the 27 women who delivered in the 42nd week or later, 33% had a positive SalEst test.

a. Construct a contingency table of counts for the relationship between the SalEst test result and the time of delivery.

b. Which of the two variables used to make the contingency table is the explanatory variable and which is the response variable?

c. Among all 642 women, what percent delivered before 42 weeks?

d. The article states that if a test result is positive, there is a 98% chance the woman will deliver before 42 weeks. Based on the contingency table, justify this claim.

e. Among those who had a negative test, what percent delivered before 42 weeks?

f. Would you advise a pregnant woman to spend $90 for the test? Explain why or why not.

6.60 According to a study on partner abuse reported by the *Sacramento Bee* (July 14, 2000, p. A6, Associated Press), 25% of women with male partners had been assaulted by their current or a former partner, whereas 11% of women with female partners had been assaulted. What is the rel-

ative risk of assault for women with male partners compared to women with female partners? Write your answer in a sentence that would be understood by someone with no training in statistics.

6.61 In exercise 6.53 data were given for admissions to two graduate programs for men and women. The data are given again here for use in this exercise, with the combined data presented as well.

	Program A		Program B		Combined	
	Admit	Deny	Admit	Deny	Admit	Deny
Men	400	250	50	300	450	550
Women	50	25	125	300	175	325

a. Give the odds of being admitted versus being denied to program A separately for men and women. Then give the odds ratio for women compared to men. Write a sentence explaining the odds ratio in words.

b. Give the odds of being admitted versus being denied for the combined programs separately for men and women. Then give the odds ratio for women compared to men. Write a sentence explaining the odds ratio in words.

c. Using the results from parts (a) and (b), write one sentence that sounds as if the university favors women applicants and another sentence that sounds as if the university favors male applicants.

Dataset Exercises

6.62 Use the dataset **UCDavis2** for this exercise. The variable **WtFeel** contains UC Davis student responses to the question, "Do you think you are: Underweight? About Right? Overweight?"

a. Create a contingency table for the relationship between **sex** (female or male) and perception of weight (**WtFeel**).

b. For each sex category, calculate the conditional percents for the perception of weight variable. Of the men, what percent thinks they are underweight? Of the women, what percent thinks they are underweight?

c. In Section 6.5, the chi-square test was only used for 2 × 2 tables, but it can be used for larger tables as well. Use software to do a chi-square test of statistical significance for the relationship between sex and perception of weight. What is the *p*-value of the test? Based on this *p*-value, what conclusion can be made about the relationship?

d. Penn State student responses to essentially the same question are summarized in exercise 6.7 (see table). Do the Penn State and UC Davis students differ in how they respond to the question about weight? If so, describe the differences.

6.63 Use the **UCDavis2** dataset for this exercise. The variable **Cheat** contains answers to a question about whether the respondent would tell the instructor if he or she saw another student cheating on an examination.

a. Create a two-way table for the relationship between **sex** (male or female) and the variable **Cheat.** Of all students, what percent said they would tell the instructor about witnessing cheating?

b. Of the females, what percent said they would tell the instructor about witnessing cheating? What percent of the males said they would do this?

c. Is the observed relationship between sex and response to the cheating question statistically significant? Justify your answer by reporting the results of a chi-square test.

d. For this question about cheating, what population do you think is represented by the students in the data set?

6.64 Use the **GSS-93** dataset for this exercise. The variable **gunlaw** has responses to a question about whether the respondent would favor a law requiring a police permit to be obtained before the purchase of any gun. The variable **owngun** indicates whether or not there are any guns in the respondent's home (including garages and sheds).

a. Write null and alternative hypotheses about the possible relationship between **owngun** and **gunlaw**.

b. Create a two-way table that summarizes the relationship between **owngun** and **gunlaw**. What percent of all respondents had a gun in their home?

c. Among those with a gun in their home, what percent favors a gun permit law? Among those with no gun in their home, what percent favors a gun permit law? Based on the difference between these two percents, would you say there is a relationship between the two variables? Why or why not?

d. What is the p-value of a chi-square test of statistical significance for the relationship between **owngun** and **gunlaw**? Is the observed relationship statistically significant?

e. Explain what it means to say that an observed relationship is statistically significant.

6.65 Use the dataset **GSS-93** for this exercise. Write a paragraph that describes the relationship between **race** (respondent's race) and **cappun** (opinion about capital punishment). Include a two-way table of counts and a relevant set of conditional percents.

REFERENCES

Baird, D. D., and A. J. Wilcox (1985). "Cigarette smoking associated with delayed conception," *Journal of the American Medical Association,* Vol. 253, pp. 2979–2983.

Bickel, P. J., E. A. Hammel, and J. W. O'Connell (1975). "Sex bias in graduate admissions: Data from Berkeley," *Science,* Vol. 187, pp. 298–304.

Carter, C. L., D. Y Jones, A. Schatzkin, and L. A. Brinton (1989). "A prospective study of reproductive, familial, and socioeconomic risk factors for breast cancer using NHANES I data," *Public Health Reports,* Vol. 104, January–February 1989, pp. 45–49.

Fletcher, S. W., B. Black, R. Harris, B. K. Rimer, and S. Shapiro (1993). "Report of the international workshop on screening for breast cancer," *Journal of the National Cancer Institute,* Vol. 85, No. 20, Oct. 20, 1993, pp. 1644–1656.

Gastwirth, J. L. (1988). *Statistical Reasoning in Law and Public Policy,* Academic Press: New York.

Hand, D. J., E. Daly, A. D. Lunn, K, J. McConway, and E. Ostrowski (1994). *A Handbook of Small Data Sets,* Chapman and Hall: London.

Kawachi, I., D. Sparrow, A. Spiro III, P. Vokonas, and S. T. Weiss (1994). "A prospective study of anger and coronary heart disease: The normative aging study," *Circulation,* Vol. 94, pp. 2090–2095.

Kohlmeier, L., G. Arminger, S. Bartolomeycik, B. Bellach, J. Rehm, and M. Thamm (1992). "Pet birds as an independent risk factor for lung cancer: Case-control study," *British Medical Journal,* Vol. 305, pp. 986–989.

Pagano, M., and K Gauvreau (1993). *Principles of Biostatistics,* Duxbury Press: Belmont, CA.

Prescott, E., M. Hippe, H. O. H. Schnor, and J. Vestbo (1998). "Smoking and risk of myocardial infarction in women and men: Longitudinal population study," *British Medical Journal,* Vol. 316, pp. 1043–1047.

Prevention Magazine's Giant Book of Health Facts (1991). John Feltman (ed.), New York: Wings Books.

SAMSHA (1998). "Driving After Drug and Alcohol Use: Findings from the 1996 National Household Survey on Drug Abuse." United States Department of Health and Human Services.

Taubes, G. (1993). "Claim of higher risk for women smokers attacked," *Science,* Vol. 262, Nov. 26, 1993, p. 1375.

The Roper Organization (1992). *Unusual Personal Experiences: An Analysis of the Data from Three National Surveys,* Las Vegas: Bigelow Holding Corp.

Voss, L. D., and J. Mulligan (2000). "Bullying in school: Are short pupils at risk? Questionnaire study in a cohort," *British Medical Journal,* Vol. 320, pp. 612–613.

Wagner, C. H. (1982). "Simpson's Paradox in real life," *The American Statistician,* Vol. 36, No. 1, pp. 46–48.

Wechsler, H., and M. Kuo (2000). "College students define binge drinking and estimate its prevalence: Results of a national survey," *Journal of American College Health,* Vol. 49, pp. 57–64.

Weiden, C. R., and B. C. Gladen (1986). "The beta-geometric distribution applied to comparative fecundability studies," *Biometrics,* Vol. 42, pp. 547–560.

Writing Group for the Women's Health Initiative Investigators (2002). "Risks and benefits of estrogen plus progestin in healthy postmenopausal women: Principal results from the Women's Health Initiative randomized controlled trial," *Journal of the American Medical Association,* Vol. 288, No. 3, pp. 321–333.

Probability

How is a teacher to choose? See Case Study 7.1.

We rarely stop to think about the precise meaning of the word *probability*. When we speak of the probability that we will win a lottery based on buying a single ticket, are we using the word in the same way that we do when we speak about the probability that we will buy a new house in the next five years?

Statistical methods are used to evaluate information in uncertain situations and probability plays a key role in that process. Remember our definition of statistics from Chapter 1: *Statistics is a collection of procedures and principles for gathering data and analyzing information in order to help people make decisions when faced with uncertainty.* Decisions like whether to buy a lottery ticket, whether to buy an extended warranty on a computer, or which of two courses to take are examples of decisions that you may have to make that involve uncertainty and the evaluation of probabilities.

Probability calculations also are a key element of *statistical inference*. In Chapter 6 we introduced p-values, which are probabilities used to determine if the results of a study are statistically significant. As a reminder of how p-values are used, consider Case Study 1.6 in which 22,071 physicians were randomly assigned to take either aspirin or a placebo. There were 189 heart attacks in the placebo group but only 104 in the aspirin group. Could this have happened just by the luck of how the physicians were randomized to the treatment groups?

Suppose that regardless of which group they were in, $104 + 189 = 293$ of the men would have had heart attacks anyway. What is the probability that, just by the luck of random assignment, the numbers of heart attacks in the two groups would have been so different? In other words, if aspirin and placebo are equally effective (or ineffective), what is the probability that we would see such a large discrepancy in the proportion of heart attacks in the two groups? The answer is the p-value, which is less than .00001. This is strong evidence that these results did not just occur by chance. From this, we conclude that aspirin really did reduce the number of heart attacks in the group that took it. ◆

7.1 | RANDOM CIRCUMSTANCES

The next case study is hypothetical but contains the kinds of situations that people encounter every day involving probability. We will use the elements of this story throughout the chapter to illustrate the concepts and calculations necessary to understand the role probability plays in our lives.

CASE STUDY 7.1
A Hypothetical Story—Alicia Has a Bad Day

Last week, Alicia went to her physician for a routine medical exam. This morning her physician phoned to tell her that one of her tests came back positive, indicating that she may have a disease we will simply call D. Thinking there must be some mistake, Alicia inquired about the accuracy of the test. The physician told her that the test is 95% accurate whether someone has disease D or not. In other words, when someone has D, the test detects it 95% of the time. When someone does not have D, the test is rightly negative 95% of the time.

Therefore, according to the physician, even though only 1 out of 1000 women of Alicia's age actually has D, the test is a pretty good indicator that Alicia may have the disease. Alicia doesn't know it yet, but her physician is wrong to imply that it's likely that Alicia has the disease. Actually, her chance of hav-

ing the disease is small, even given the positive test result. Later in this chapter you will discover why this is true.

Understandably, the physician's call upsets Alicia. She had planned to spend the morning studying for her statistics class, but instead she uses the time to search the Web for information about disease D. Her statistics professor randomly selects 3 separate students at the beginning of each class to answer questions about the material. Alicia reasons that with 50 students in the class she is unlikely to be one of the 3 selected. The students' names are placed on slips of paper in a bag, and one name is drawn (without replacement) for each question. Probability didn't go in Alicia's favor this day. After twice breathing a sigh of relief, she is picked to answer the third question.

Random Circumstances in Alicia's Day

A **random circumstance** is one in which the outcome is unpredictable. In many cases, *the outcome is not determined until we observe it*. It was not predetermined that Alicia would be selected to answer one of the questions in class. This happened when the professor drew her name out of the bag. In other cases, *the outcome is already determined, but our knowledge of it is uncertain*. Alicia either has disease D or not, but she and her physician don't know which possibility is true.

One lesson of this chapter is that the probabilities associated with random circumstances sometimes depend on other random circumstances. Alicia's test results were positive. The probability that Alicia would have a positive test depends on another random circumstance, which is whether or not she actually has disease D. In her statistics class, Alicia was not selected to answer questions 1 or 2. So, at the time of the drawing of a name for the third question, the probability that she would be selected was 1/48 because there were only 48 names left in the bag instead of the original 50. If she had been selected to answer either question 1 or question 2, her probability of being selected to answer question 3 would have been 0.

Here is a list of the random circumstances in Alicia's story and the possible outcomes for each of them:

Random Circumstance 1: Disease status

 ◆ Alicia has D.

 ◆ Alicia does not have D.

Random Circumstance 2: Test result

◆ Test is positive.

◆ Test is negative.

Random Circumstance 3: First student's name is drawn

◆ Alicia is selected.

◆ Alicia is not selected.

Random Circumstance 4: Second student's name is drawn

◆ Alicia is selected.

◆ Alicia is not selected.

Random Circumstance 5: Third student's name is drawn

◆ Alicia is selected.

◆ Alicia is not selected.

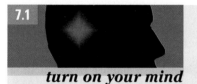

7.1

turn on your mind

At the beginning of Alicia's day, the outcomes of the five random circumstances listed were uncertain to her. Which of them were uncertain because *the outcome was not yet determined* and which were uncertain because of Alicia's *knowledge of the outcome?* *

Assigning Probabilities to the Outcomes of a Random Circumstance

We said the probability was 1/48 that Alicia would be picked for the third question given that she had not been picked for either of the first two. Note that this probability is expressed as a fraction. A probability is a value between 0 and 1 and is written either as a fraction or as a decimal fraction. From a purely mathematical point of view, a **probability** simply is a number between 0 and 1 that is assigned to a possible outcome of a random circumstance. Additionally, for the complete set of distinct possible outcomes of a random circumstance, the total of the assigned probabilities must equal 1.

In practice, of course, we should assign probabilities to outcomes in a meaningful way. A probability should provide information about how likely it is that a particular outcome will be the result of a random circumstance. In the next section, we discuss two different ways to assign and interpret probabilities. One of these ways, the relative frequency definition of probability, forms the foundation for the statistical inference methods that we will examine in later chapters.

7.2

turn on your mind

Based on your understanding of probability and random events, assign probabilities to the two possible outcomes for Random Circumstance 3.**

*HINT: Only two of them were already determined.

**HINT: See Example 7.3.

7.2 | INTERPRETATIONS OF PROBABILITY

The word *probability* is so common that in all probability you may encounter it today in everyday language. But, we rarely stop to think about the precise meaning of the word. When we speak of the probability that we will win a lottery based on buying a single ticket, are we using the word in the same way that we do when we speak about the probability that we will buy a new house in the next five years? We can quantify the chance of winning the lottery exactly, but our assessment of the chance that we will buy a new house is based on our personal and subjective belief about how life will evolve for us.

The conceptual difference illustrated by these two different situations creates two different interpretations of what is meant by the word *probability*. In most situations in statistics, a probability is assigned to a possible outcome based on what will or has happened over the long run of repeatedly observing a random circumstance. There are situations, however, where a probability may be assigned based on the expert assessment of an individual.

The Relative Frequency Interpretation of Probability

The relative frequency interpretation of probability applies to situations in which we can envision repeatedly observing the results of a random circumstance. For example, it is easy to imagine flipping a coin over and over again and counting the number of heads and tails. It makes sense to interpret the probability that the coin lands with heads up to be the relative frequency, over the long run, with which it lands with heads up. When we say that the probability of flipping heads is 1/2, we can interpret this to mean that about half of a large number of flips will result in heads.

Here are some other situations to which the relative frequency interpretation of probability can be applied:

- Buying lottery tickets regularly and observing how often you win.
- Drawing a student's name out of a hat and seeing how often a particular student is selected.
- Commuting to work daily and observing whether or not a certain traffic signal is red when we encounter it.
- Surveying many adults and determining what proportion smokes.
- Observing births and noting how often the baby is a female.

> *definition* In situations that we can imagine repeating many times, we define the **probability** of a specific outcome as *the proportion of times it would occur over the long run*. This also is called the **relative frequency** of that particular outcome.

Notice the emphasis on what happens *in the long run*. We cannot accurately assess the probability of a particular outcome by observing it only a few times. For example, consider a family with five children in which only one child is a boy. We should not take this as evidence that the probability of having a boy is only 1/5. To more accurately assess the probability that a baby is a boy, we have to observe many births.

TABLE 7.1 ■ Relative Frequency of Male Births over Time

Weeks of Watching	Total Births	Total Boys	Proportion of Boys
1	30	19	.633
4	116	68	.586
13	317	172	.543
26	623	383	.615
39	919	483	.526
52	1237	639	.517

EXAMPLE 7.1
Probability of Male Versus Female Births

According to the *Information Please Almanac* (1991, p. 815), the long-run relative frequency of males born in the United States is about .512. In other words, over the long run, there are 512 male babies and 488 female babies born per 1000 births.

Suppose we were to tally births and the relative frequency of male births in a certain city for the next year. Table 7.1, which was generated with a computer simulation, shows what we might observe. In the simulation, the chance was .512 that any individual birth was a boy, and the average number of births per week was 24. Notice how the *proportion* or *relative frequency* of male births is relatively far from .512 over the first few weeks, but then settles down to around .512 in the long run. After 52 weeks, the *relative frequency* of boys was 639/ 1237 = .517. If we used only the data from the first week to estimate the probability of a male birth, our estimate would have been far from the actual value. We need to look at a large number of observations to accurately estimate a probability. ◆

Determining the Relative Frequency Probability of an Outcome
There are two ways to determine a relative frequency probability. The first method involves *making an assumption about the physical world*. The second method involves *making a direct observation of how often something happens*.

Method 1: Make an Assumption about the Physical World Sometimes, it's reasonable to assign probabilities to possible outcomes based on what we think about physical realities. We generally assume, for example, that coins are manufactured in such a way that they are equally likely to land with heads or tails up when flipped. Therefore, we conclude that the probability of a flipped coin showing heads up is 1/2.

EXAMPLE 7.2
A Simple Lottery

A lottery game run by many states in the United States is one in which players choose a three-digit number between 000 and 999. A player wins if his or her three-digit number is chosen. If we assume that the physical mechanism used to draw the winning number gives each possibility an equal chance, we can determine the probability of winning. There are 1000 possible three-digit numbers (000, 001, 002, . . . , 999), so the probability that the state picks the player's number is 1/1000. In the long run, a player should win about 1 out of 1000 times. Note that this does not mean that a player will win *exactly* once in every thousand plays. ◆

EXAMPLE 7.3
The Probability That Alicia Has to Answer a Question

In Case Study 7.1, there were 50 student names in the bag when a student was selected to answer the first question. If we assume the names have been well mixed in the bag, then each student is equally likely to be selected. Therefore, the probability that Alicia would be selected to answer the first question was 1/50. ◆

Method 2: Observe the Relative Frequency   Another way to determine the probability of a particular outcome is by *observing the relative frequency of the outcome over many, many repetitions of the situation*. We illustrated this method with Example 7.1 when we observed the relative frequency of male births in a given city over a year. If we consider many births, we can get a very accurate figure for the probability that a birth will be a male. As mentioned, the relative frequency of male births in the United States has been consistently close to .512 (*Information Please Almanac*, 1991, p. 815). In 1987, for example, there were a total of 3,809,394 live births in the United States, of which 1,951,153 were males. Assuming that the probability does not change in the future, the probability that a live birth will result in a male is about 1,951,153/3,809,394 = .5122.

EXAMPLE 7.4
The Probability of Lost Luggage

According to Krantz (1992, p. 33; original source U.S. Department of Transportation), 1 in 176 passengers on U.S. airline carriers will temporarily lose their luggage. This number is based on data collected over the long run and is found by dividing the number of passengers who temporarily lost their luggage by the total number of passengers. Another way to state this fact is to say that the probability is 1/176, or about .006, that a randomly selected passenger on a U.S. carrier will temporarily lose luggage. ◆

Proportions and Percentages as Probabilities

Relative frequency probabilities are often derived from the proportion of individuals who have a certain characteristic, or a certain outcome in a random circumstance. The relative frequency probabilities are therefore numbers between 0 and 1. It is commonplace, however, to be somewhat loose in how relative frequencies are expressed. For instance, here are four ways to express the relative frequency of lost luggage:

- The proportion of passengers who lose their luggage is 1/176 or about .006.
- About 0.6% of passengers lose their luggage.
- The probability that a randomly selected passenger will lose his or her luggage is about .006.
- The probability that you will lose your luggage is about .006.

This last statement is not exactly correct, because *your* probability of lost luggage depends on a number of factors, including whether or not you check any luggage and how late you arrive at the airport for your flight. The probability of .006 really only applies to the long-run relative frequency across all passengers, or the probability that a *randomly selected* passenger is one of the unlucky ones.

Estimating Probabilities from Observed Categorical Data

When appropriate data are available, the long-run relative frequency interpretation of probability can be used to estimate probabilities of outcomes and combinations of outcomes for categorical variables. Assuming the data are representative, the probability of a particular outcome is estimated to be the relative frequency (proportion) with which that outcome was observed. As we learned in Case Study 1.3 and also in Chapter 4, such estimates are not precise, and an approximate margin of error for the estimated probability is calculated as $1/\sqrt{n}$. We'll learn a more precise way to calculate the margin of error in Chapter 10.

TABLE 7.2 ■ Nighttime Lighting in Infancy and Eyesight

Slept with:	No Myopia	Myopia	High Myopia	Total
Darkness	155 (90%)	15 (9%)	2 (1%)	172
Nightlight	153 (66%)	72 (31%)	7 (3%)	232
Full Light	34 (45%)	36 (48%)	5 (7%)	75
Total	342 (71%)	123 (26%)	14 (3%)	479

EXAMPLE 7.5
Nightlights and
Myopia Revisited

In an example discussed in Chapters 2 and 6, we investigated the relationship between nighttime lighting in early childhood and the incidence of myopia later in childhood. The results are shown again in Table 7.2.

Assuming these data are representative of a larger population, what is the approximate probability that someone from that population who sleeps with a nightlight in early childhood will develop some degree of myopia? There were 232 infants who slept with a nightlight. Of those, 72 developed myopia and 7 developed high myopia, for a total of 79 who developed some degree of myopia. Therefore, the relative frequency of some myopia among nightlight users is $(72 + 7)/232 = 79/232 = .34$. From this we can conclude that the approximate probability of developing some myopia, given that an infant slept with a nightlight, is .34. This estimate is based on a sample of 232 people, so there is a margin of error of about $1/\sqrt{232} = .066$. ◆

The Personal Probability Interpretation

The relative frequency interpretation of probability clearly is limited to repeatable conditions. Yet, uncertainty is a characteristic of most events, whether or not they are repeatable under similar conditions. We need an interpretation of probability that can be applied to situations even if they will never happen again.

If you decide to drive downtown this Saturday afternoon, will you be able to find a good parking space? Should a movie studio release its potential new hit movie before Christmas, when many others are released, or wait until January when it might have a better chance to be the top box-office attraction? If the United States enters into a trade alliance with a particular country, will that cause problems in relations with another country? Will you have a better time if you go to Cancun or to Florida for your next spring break?

These are unique situations, not likely to be repeated. They require people to make decisions based on an assessment of how the future will evolve. We could each assign a personal probability to these events, based on our own knowledge and experiences, and we could use that probability to help us with our decision. Different people may not agree on what the probability of an event happening is, but nobody would be considered to be wrong.

definition   We define the **personal probability** of an event to be the degree to which a given individual believes the event will happen. Sometimes the term **subjective probability** is used because the degree of belief may be different for each individual.

There are very few restrictions on personal probabilities. They must fall between 0 and 1 (or, if expressed as a percent, between 0 and 100%). They must also fit together in certain ways if they are to be **coherent.** By being coherent, we mean that your personal probability of one event doesn't contradict your personal probability of another. For example, if you thought that the probability of finding a parking space downtown Saturday afternoon was .20, then to be coherent you must also believe that the probability of not finding one is .80. We will explore some of these logical rules later in this chapter.

How We Use Personal Probabilities

People routinely base decisions on personal probabilities. This is why committee decisions are often so difficult. For example, suppose a committee is trying to decide which candidate to hire for a job. Each member of the committee has a different assessment of the candidates, and they may disagree on the probability that a particular candidate would best fit the job. We are all familiar with the problem juries sometimes have when trying to agree on someone's guilt or innocence. Each member of the jury has his or her own personal probability of guilt and innocence. One of the benefits of committee or jury deliberations is that they may help members reach some consensus in their personal probabilities.

Personal probabilities often take relative frequencies of similar events into account. For example, the late astronomer Carl Sagan believed that the probability of a major asteroid hitting the Earth soon is high enough to be of concern. "The probability that the Earth will be hit by a civilization-threatening small world in the next century is a little less than one in a thousand" (Jane Arraf, "Leave Earth or perish: Sagan," *China Post,* Taiwan, Dec. 14, 1994, p. 4). To arrive at that probability, Sagan obviously could not use the long-run frequency definition of probability. He would have to use his own knowledge of astronomy, combined with past asteroid behavior.

Notice that unlike relative frequency probabilities, personal probabilities assigned to unique events are not equivalent to proportions or percentages. For instance, it does not make sense to say that major asteroids will hit approximately 1 out of 1000 Earths in the next century.

Mysteries of the Past

There is some debate about how to represent probability when an outcome has been determined but is still unknown, such as if you have flipped a coin but not yet looked at it. Technically, any particular outcome has either happened or not. If it has happened, its probability of happening is 1, and if it hasn't, its probability of happening is 0.

For instance, in Case Study 7.1, Alicia either has the disease or not, but no one knows which is true. Because 1 out of 1000 women of her age have the disease, without any additional information, we might say that the probability that Alicia has it is 1/1000. But one could argue that this probability statement is meaningless, because she either has the disease or not.

As another example, suppose a woman is pregnant but no one knows the sex of the fetus. Is the probability that she will have a boy either 0 or 1, or is it .512? Suppose you flip a coin but cover it with your hand before anyone sees it. Is the probability of a head still 1/2, until it is revealed?

The answer to the dilemma in these kinds of situations is that technically, once an outcome has been determined, a relative frequency probability statement no longer makes sense. However, a personal probability statement about what outcome has occurred or may ultimately be revealed does make sense. This interpretation is often what is meant when we discuss such occurrences in everyday life, even in situations where relative frequency information is used to determine the probability value, such as in the three examples just discussed.

Another example of this type of situation is the construction of a 95% confidence interval, introduced in Chapter 4, which we will study in detail in Chapters 10 and 12. Before the sample is chosen, a probability statement makes sense. The probability is .95 that a sample will be selected for which the computed 95% confidence interval covers the truth. After the sample has been chosen, "the die is cast." Either the computed confidence interval will cover the truth, or it won't. That's why we say that we have *95% confidence* that the interval is correct, rather than saying that *the probability* that it is correct is .95.

7.3

turn on your mind

You are about to enroll in a course for which you know that 20% of the students will receive a grade of A. Do you think that the probability that *you* will receive an A in the class is .20? Do you think the probability that a randomly selected student in the class will receive an A is .20? Explain the difference in these two probabilities, using the distinction between relative frequency probability and personal probability in your explanation.*

---

in summary   Interpretations of Probability

The *relative frequency probability* of an outcome is the proportion of times the outcome would occur over the long run. Relative frequency probabilities can be determined by any of these methods:

- Making an assumption about the physical world and using it to find relative frequencies
- Observing outcomes over the long run
- Measuring a representative sample and observing relative frequencies of the sample that fall into various categories

The *personal probability* of an outcome is the degree to which a given individual believes it will happen. Sometimes data from similar events in the past and other knowledge are incorporated when determining personal probabilities.

7.3 | PROBABILITY DEFINITIONS AND RELATIONSHIPS

One of the key factors for being able to calculate and interpret probabilities is to be able to define specific outcomes or events of interest and how these events are related to each other. There are just a few definitions of these relationships. The first set of definitions has to do with the possible outcomes of a random circumstance. A specific possible outcome is called a **simple event,** and the

*HINT: Look at the "In Summary" box immediately following. Which of the four methods described apply in each case?

collection of all possible outcomes is called the **sample space** for the random circumstance. The term **event** is used to describe any collection of one or more possible outcomes.

> *definition* ◆ The **sample space** is the collection of unique, nonoverlapping possible outcomes of a random circumstance.
>
> ◆ A **simple event** is one outcome in the sample space. In other words, a simple event is a possible outcome of a random circumstance.
>
> ◆ An **event** is a collection of one or more simple events in the sample space. Events are often written using capital letters A, B, C, and so on.

EXAMPLE 7.6
***Days per Week
of Drinking Alcohol***

Suppose that we randomly sample college students and ask each student how many days he or she drinks alcohol in a typical week. The *simple events* in the *sample space* of possible outcomes are 0 days, 1 day, 2 days, and so on, up to 7 days. These simple events certainly are not equally likely. We might be interested to learn the probability of the *event* that a randomly selected student drinks alcohol on 4 or more days in a typical week. The *event* "4 or more" is comprised of the *simple events* {4 days, 5 days, 6 days, and 7 days}. ◆

Assigning Probabilities to Simple Events

A probability is assigned to each *simple event* (unique outcome) using relative frequency or personal probabilities. The relative frequency method is used most often. The notation $P(A)$ is used for the probability of an event A.

Conditions for Valid Probabilities

The assigned probabilities for the possible outcomes of a random circumstance must meet the following two conditions:

1. Each probability is between 0 and 1.
2. The sum of the probabilities over all possible simple events is 1. In other words, the total probability for all possible outcomes of a random circumstance is equal to 1.

Equally Likely Simple Events

If there are k simple events in the sample space and they are all equally likely, then the probability of the occurrence of each one is $1/k$. For instance, suppose a fair 6-sided die is rolled and the number that lands face up is observed. Then the sample space is {1,2,3,4,5,6} and each outcome has probability 1/6. Suppose two fair coins are tossed and the outcome of interest is the combination of results on the first toss then the second toss. There are four equally likely events, {HH, HT, TH, TT}, where H = heads and T = tails, so each outcome has probability 1/4.

EXAMPLE 7.2—*CONTINUED*
***Probabilities for
Some Lottery Events***

Random Circumstance: A three-digit winning lottery number is selected.
Sample Space: {000,001,002,003, . . . ,997,998,999}. There are 1000 simple events.
Probabilities for Simple Event: The probability that any specific three-digit number is a winner is 1/1000. We assume that all three-digit numbers are equally likely.

Probabilites for Some Events:

♦ Let event A = the last digit is a 9. Notice that event A contains the subset of all simple events that end with a 9. We can write this as A = {009,019, ...,999}. Notice that since one out of ten numbers will be included in this set, $P(A) = 1/10$.

♦ Let event B = the three digits are all the same. This event includes ten simple events, which are {000,111,222,333,444,555,666,777,888,999}. Because event B contains 10 of the 1000 equally likely simple events, $P(B) = 10/1000$, or 1/100. ♦

Complementary Events

definition One event is the **complement** of another event if the two events do not contain any of the same simple events *and* together they cover the entire sample space. For an event A, the notation A^C represents the complement of A.

An event either happens or it doesn't. A randomly selected person either has blue eyes or does not have blue eyes. Alicia either has a disease or does not. "Opposite" events are called **complementary events.** Because the total probability for a sample space must be equal to 1, the probabilities of complementary events must sum to 1. In symbols, $P(A) + P(A^C) = 1$.

EXAMPLE 7.2—CONTINUED
The Probability of
Not Winning the Lottery

If A is the event that a player buying a single lottery ticket wins, then A^C is the event that he or she does not win. As we have seen, $P(A) = 1/1000$, so it makes sense that $P(A^C) = 999/1000$. ♦

Mutually Exclusive Events

definition Two events are **mutually exclusive** if they do not contain any of the same simple events (outcomes). Equivalent terminology is that the two events are **disjoint.**

We sometimes are interested in two events that involve distinctly separate simple events, but the two events may not cover all possible outcomes in the sample space. For example, we may want to know the probability that a randomly selected man is unusually short or tall. We might want to know the probability that he is either at least 4 inches taller than the mean height, or at least 4 inches shorter than the mean height. The events "at least 4 inches taller than the mean" and "at least 4 inches shorter than the mean" are separate events that cannot occur simultaneously, and they are called *mutually exclusive* or *disjoint events.*

EXAMPLE 7.2—CONTINUED
Mutually Exclusive Events
for Lottery Numbers

Here are two mutually exclusive (disjoint) events when a winning lottery number is drawn from the set 000 to 999. Notice that they do not cover all possibilities, so they are not complements:

A = All three digits are the same.
B = The first and last digits are different. ◆

Independent and Dependent Events

> *definition* ◆ Two events are **independent** of each other if knowing that one will occur (or has occurred) does not change the probability that the other occurs.
>
> ◆ Two events are **dependent** if knowing that one will occur (or has occurred) changes the probability that the other occurs.

The definitions of independent and dependent events have to do with whether the fact that one event has taken place affects the probability that a second event could happen. The definitions can apply *either* to events *within the same random circumstance* or to events *from two separate random circumstances.* As an example of two dependent events within the same random circumstance, suppose a random digit from 0 to 9 is selected. If we know that the randomly selected digit is an *even number,* our assignment of a probability to the event that *the number is 4* changes from 1/10 to 1/5. Often, we examine questions about the dependence of the outcomes of two separate random circumstances. For instance, does the knowledge of whether a randomly selected person is male or female (random circumstance 1) affect the probability that he or she is opposed to abortion (random circumstance 2)? If the answer is yes, the outcomes of the two random circumstances are dependent events.

EXAMPLE 7.7
Winning a Free Lunch

Some restaurants display a glass bowl in which customers deposit their business cards, and a drawing is held once a week for a free lunch. Suppose you and your friend Vanessa each deposit a card in two consecutive weeks. Define:

Event A = You win in week 1.
Event B = Vanessa wins in week 1.
Event C = Vanessa wins in week 2.

Notice that events A and B refer to the same random circumstance, the drawing in week 1. They are *not* independent. If you win (event A), then Vanessa cannot win (event B). So if A occurs, then $P(B) = 0$. Knowing that A has or will occur changes $P(B)$.

Notice that events A and C refer to *different* random circumstances, the drawings in separate weeks. Those events *are* independent. Knowing that A occurred does not give any information about the probability that C will occur. ◆

EXAMPLE 7.3—*CONTINUED*
The Probability That Alicia Has to Answer a Question

It is not always the case that events related to two separate random circumstances are independent. Define:

Event A = Alicia is selected to answer question 1.
Event B = Alicia is selected to answer question 2.

In Example 7.3 we determined that $P(A) = 1/50$. Are A and B independent? Notice that they refer to different random circumstances, but $P(B)$ is affected by knowing whether or not A occurred. If Alicia answered question 1, her name is already out of the bag, so $P(B) = 0$. If event A does not occur, there are 49 names left in the bag including Alicia's, so $P(B) = 1/49$. Therefore, knowing whether or not A occurred changes $P(B)$. ◆

Conditional Probabilities

Suppose that two events are dependent. From the definition of dependent events, it follows that knowing that one event will occur (or has occurred) *does* change the probability that the other occurs. Therefore, we need to distinguish between two probabilities:

- ◆ $P(B)$, the unconditional probability that the event B occurs
- ◆ $P(B|A)$, read "probability of B given A," the conditional probability that the event B occurs *given that we know A has occurred or will occur*

> **definition**   The **conditional probability of the event B, given that the event A occurs,** is the long-run relative frequency with which event B occurs when circumstances are such that A also occurs. This probability is written as $P(B|A)$.

EXAMPLE 7.8
Probability That a Teenager Gambles Depends upon Gender

Based on a survey of nearly all of the 9th and 12th graders in Minnesota public schools in 1998, a total of 78,564 students, Stinchfield (2001) reported proportions admitting that they gambled at least once a week during the previous year. He found that these proportions differed considerably for males and females, as well as for 9th and 12th graders. One result he found was that 22.9% of the 9th-grade boys but only 4.5% of the 9th-grade girls admitted gambling at least weekly. Assuming these figures are representative of all Minnesota 9th graders, we can write these as conditional probabilities for 9th-grade Minnesota students:

P(student is weekly gambler|student is boy) = .229
P(student is weekly gambler|student is girl) = .045

Notice the dependence between the outcomes for the random circumstance "weekly gambling habit" and the outcomes for the random circumstance "gender." Knowledge of a student's gender changes the probability that he or she is a weekly gambler. ◆

7.4

turn on your mind

Remember that there were 50 students in Alicia's statistics class and that student names were not put back in the bag after being selected. Define the events A = Alicia is selected to answer question 1 and B = Alicia is selected to answer question 2. Describe each of these four conditional probabilities in words and determine a value for each.

$$P(B|A), \qquad P(B^C|A), \qquad P(B|A^C), \qquad P(B^C|A^C)$$

Now, based on your answers, can you formulate a rule about the value of $P(B|A) + P(B^C|A)$?*

*HINT: For $P(B|A)$ and $P(B|A^C)$, see Example 7.3—*CONTINUED*.

> *in summary*   **Key Probability Definitions**
>
> ◆ The **sample space** is the collection of unique, nonoverlapping possible outcomes of a random circumstance.
>
> ◆ A **simple event** is one outcome in the sample space. In other words, a simple event is a possible outcome of a random circumstance.
>
> ◆ An **event** is a collection of one or more simple events in the sample space. Events are often written using capital letters A, B, C, and so on.
>
> ◆ One event is the **complement** of another event if the two events do not contain any of the same simple events *and* together they cover the entire sample space. For an event A, the notation A^C represents the complement of A.
>
> ◆ Two events are **mutually exclusive** if they do not contain any of the same simple events (outcomes). Equivalent terminology is that the two events are **disjoint.**
>
> ◆ Two events are **independent** of each other if knowing that one will occur (or has occurred) does not change the probability that the other occurs.
>
> ◆ Two events are **dependent** if knowing that one will occur (or has occurred) changes the probability that the other occurs.
>
> ◆ The **conditional probability of the event B, given that the event A occurs,** is the long-run relative frequency with which event B occurs when circumstances are such that A also occurs. This probability is written as $P(B|A)$.

7.4 | BASIC RULES FOR FINDING PROBABILITIES

We already have been able to determine some probabilities based on our knowledge of physical situations and long-run relative frequencies. We also have stated two conditions that must apply to probabilities. Namely, that they are always between 0 and 1, and that the sum of the probabilities for all simple events must be 1.

In this section, we introduce four more rules that can be used to find probabilities for complicated events. When stating these rules, we use the letters A and B to represent events, and $P(A)$ and $P(B)$ for their respective probabilities. Each rule is followed by examples that translate the formulas into words. Common words that signal the use of the rule are shown in *italic* in the examples.

Probability an Event Does Not Occur

> *definition*   **Rule 1 (for "not the event"):** To find the probability of A^C, the complement of A, use $P(A^C) = 1 - P(A)$.

Rule 1 follows directly from the observation that $P(A) + P(A^C) = 1$, which was stated as an immediate consequence after the definition of A^C. Therefore, if $P(A)$ is known, it's easy to find the probability that A does not occur. Simply subtract $P(A)$ from 1. For instance, in Example 7.2—*CONTINUED,* it was easy to determine that the probability of *not* winning the lottery was .999, once we knew that the probability of winning was .001.

Rule 1 is simple but surprisingly useful. When you are confronted with finding the probability of an event, always ask yourself if it would be easier to find the probability of its complement, and then subtract that from 1.

EXAMPLE 7.9
Probability a Stranger Does Not Share Your Birth Date

Assuming all 365 birth dates are equally likely, the probability that the next stranger you meet will share your birthday is 1/365. Therefore, the probability that they will *not* share your birthday is $1 - (1/365) = (364/365) = .9973$. This does not hold if you were born on Feb. 29, but if that's you, you are used to ingenuity when it comes to birthdays! ◆

Probability That Either of Two Events Happens

> *definition* **Rule 2 (addition rule for "either/or"):** To find the probability that either A *or* B happens:
>
> **Rule 2a (general):** $P(A \text{ or } B) = P(A) + P(B) - P(A \text{ and } B)$
> **Rule 2b (for mutually exclusive events):** *If* A and B are *mutually exclusive* events, $P(A \text{ or } B) = P(A) + P(B)$
>
> Rule 2b is a special case of Rule 2a because when A and B are mutually exclusive events, $P(A \text{ and } B) = 0$.

To find the probability that either of two *mutually exclusive* events will be the outcome of a random circumstance, we add the probabilities for the two events. For instance, the probability that a randomly selected integer between 1 and 10 is either 3 or 7 is $.1 + .1 = .2$. When the two events are not mutually exclusive, we modify the sum of the probabilities in the manner described in Rule 2a and illustrated in Example 7.10.

EXAMPLE 7.10
Roommate Compatibility

Brett is off to college. There are 1000 additional new male students, and one of them will be randomly assigned to share Brett's dorm room. He is hoping it won't be someone who likes to party or who snores. The 2×2 table for the 1000 students is

	Snores	Doesn't Snore	Total
Likes to Party	150	100	250
Doesn't Like to Party	200	550	750
Total	350	650	1000

What is the probability that Brett will be disappointed and get a roommate who *either* likes to party or snores? Relevant events and their probabilities are

A = likes to party $\qquad\qquad P(A) = \dfrac{250}{1000} = .25.$

B = snores $\qquad\qquad\qquad P(B) = \dfrac{350}{1000} = .35$

A and B = likes to party *and* snores $\qquad P(A \text{ and } B) = \dfrac{150}{1000} = .15$

Using Rule 2a, $P(A \text{ or } B) = P(A) + P(B) - P(A \text{ and } B) = .25 + .35 - .15 = .45$. The probability is .45 that Brett will be assigned a roommate who either likes to party, or snores, or both. Notice that the probability is therefore $1 - .45 = .55$ that he will get a roommate who is acceptable, using Rule 1. We could also determine this probability from the table of counts because there are 550 suitable roommates out of 1000 possibilities. ◆

We can use Example 7.10 to illustrate why Rule 2a makes sense. In order to find $P(A \text{ or } B)$ we need to include all possible outcomes involved in either event. In Brett's case, 250 (.25) of the potential roommates like to party and 350 (.35) snore, for a total of 600 (.6). But notice that the 150 guys who like to party *and* who snore have been counted twice. So, we need to subtract them once, resulting in $250 + 350 - 150 = 450$ people out of the 1000 total who have one or both traits. So, Rule 2a tells you to add the probabilities for all of the outcomes that are part of event A, plus the probabilities for all of the outcomes that are part of event B, then to subtract the probabilities for all of the outcomes that have been counted twice, $P(A \text{ and } B)$.

EXAMPLE 7.11
*Probability of **Either** Two Boys or **Two** Girls in Two Births*

What is the probability that a woman who has two children has *either* two girls or two boys? The question asks about two *mutually exclusive* events because with only two children she can't have two girls and two boys. Rule 2b applies to this situation. We'll use the information in Example 7.1 that the probability of a boy is .512 and the probability of a girl is .488, and we'll also take a sneak peek at Rule 3 introduced next.

Event A = two girls. From Rule 3b (next), $P(A) = (.488)(.488) = .2381$.
Event B = two boys. From Rule 3b (next), $P(B) = (.512)(.512) = .2621$.

Therefore, the probability that the woman has *either* two girls *or* two boys is

$$P(A \text{ } or \text{ } B) = P(A) + P(B) = .2381 + .2621 = .5002 \quad ◆$$

Probability That Two or More Events Occur Together

Rule 2 allowed us to answer "either or both" questions about events. But what if we want to restrict consideration to the combination of both events only? In that case, we want to find $P(A \text{ and } B)$. For instance, if you flip a nickel and a dime, what is the probability that they *both* land heads up? You probably know intuitively that the answer is 1/4, or could be convinced of that by noting that the four possibilities, HH, HT, TH and TT, are equally likely. A more general way of arriving at the answer is to notice that 1/4 is the result of multiplying $1/2 \times 1/2$, the two individual probabilities of heads. This general method works for finding $P(A \text{ and } B)$ as long as A and B are independent events.

If A and B are dependent events, then finding the probability that they both occur requires that we know more than their individual probabilities. Sometimes we can find $P(A \text{ and } B)$ directly as a relative frequency. In Example 7.10, the probability that Brett's roommate will be someone who likes to party *and* who snores can be found directly as $150/1000 = .15$. If we can't find $P(A \text{ and } B)$ directly, then we need to know either $P(A|B)$ and $P(B)$, or $P(B|A)$ and $P(A)$.

---

definition **Rule 3 (multiplication rule for "and"):** To find the probability that two events both occur simultaneously or in a sequence:

Rule 3a (general): $P(A \text{ and } B) = P(A)P(B|A)$

Rule 3b (for independent events): *If* A and B are *independent* events, $P(A \text{ and } B) = P(A)P(B)$

Extension of Rule 3b to more than two independent events: For several independent events, $P(A_1 \text{ and } A_2 \ldots \text{ and } A_n) = P(A_1)P(A_2) \cdots P(A_n)$

Rule 3b is a special case of Rule 3a because when A and B are independent, $P(B|A)$ is equal to $P(B)$.

To find the probability that two or more independent events occur together, multiply the probabilities of the events. For instance, the probability is $(.5)(.5)(.5) = .125$ that the next three people of your sex you see all are taller than the median height for your sex. (Remember that the median is the value for which half of a group has a greater value and half has a smaller value, so the probability is .5 that a randomly selected person is taller than the median. We are assuming the probability of being exactly at the median is essentially 0.)

EXAMPLE 7.8—CONTINUED
Probability a Randomly Selected 9th Grader Is a Male and a Weekly Gambler

In a 1998 survey of most of the 9th-grade students in Minnesota, 22.9% of the boys and 4.5% of the girls admitted that they gambled at least once a week during the previous year (Stinchfield, 2001). The population consisted of 50.9% girls and 49.1% boys. Assuming these students represent all 9th-grade Minnesota teens, what is the probability that a randomly selected student will be a male *who also* gambles at least weekly?

Event A = A male is selected.
Event B = A weekly gambler is selected.

Events A and B are dependent, so we use Rule 3a (general multiplication rule) to determine the probability that A and B occur together. We know that $P(A) = .491$. We also know that the *conditional* probability that the selected teen is a weekly gambler, given that a male is selected, is $P(B|A) = .229$. The answer to our question is therefore:

$$P(\text{male } and \text{ weekly gambler}) = P(A \text{ and } B) = P(A)P(B|A) = (.491)(.229)$$
$$= .1124$$

About 11% of all 9th-grade teenagers are males *and* weekly gamblers. ◆

EXAMPLE 7.12
Probability Two Strangers Both Share Your Birth Month

Assuming birth months are equally likely, what is the probability that the next two unrelated strangers you meet *both* share your birth month? Because the strangers are unrelated, we assume independence between the two events that have to do with them sharing your birth month. Define:

Event A = first stranger shares your birth month; $P(A) = 1/12$.
Event B = second stranger shares your birth month; $P(B) = 1/12$.

The events are independent, so Rule 3b applies. The probability that *both* individuals share your birth month is the probability that A occurs *and* B occurs:

$$P(A \text{ and } B) = P(A)P(B) = \frac{1}{12} \times \frac{1}{12} = \frac{1}{144} = .007$$

The Extension of Rule 3b allows this result to be generalized. For instance, the probability that four unrelated strangers *all* share your birth month is $(1/12)^4$. ◆

Determining a Conditional Probability

> **definition Rule 4 (conditional probability):** This rule is simply an algebraic restatement of Rule 3a, but it is sometimes useful to use the following form of that rule:
>
> $$P(B|A) = \frac{P(A \text{ and } B)}{P(A)}$$
>
> The assignment of letters to events A and B is arbitrary, so it is also true that
>
> $$P(A|B) = \frac{P(A \text{ and } B)}{P(B)}$$

Often, a conditional probability is given to us directly. In Example 7.8, for instance, it was reported that the estimated conditional probability is .045 that a 9th grader is a weekly gambler if the student is a female. In other cases, the physical situation can make it easy to determine a conditional probability. For instance, in drawing 2 cards from an ordinary deck of 52 cards, the probability that the second card drawn is red, *given* that the first card drawn was red, is 25/51 because 25 of the remaining 51 cards are red. Occasionally, it is useful to use Rule 4, an algebraic restatement of the multiplication rule (Rule 3a), to calculate a conditional probability.

EXAMPLE 7.13
Probability Alicia Is Picked for the First Question Given That She's Picked to Answer a Question

Here is an example for which it is easy to see the logical answer. If *we know* that Alicia is picked to answer one of the questions, what is the probability that it was the first question? Once we know she had to answer one of the three questions, the conditional probability that it was the first one should be 1/3. All three questions would be equally likely to be the question for which Alicia was picked.

Let's use Rule 4 to verify this logic. Define:

A = Alicia is called on to answer the first question; $P(A) = 1/50$.
B = Alicia is called on to answer any one of the questions; $P(B) = 3/50$, since there are 50 students and 3 students are chosen to answer questions.

Notice that the event that A and B *both* occur is synonymous with the event A occurring, since A is a subset of B. If A has occurred, then B also has occurred. Therefore, $P(A \text{ and } B) = 1/50$. Applying Rule 4:

$$P(A|B) = \frac{P(A \text{ and } B)}{P(B)} = \frac{1/50}{3/50} = \frac{1}{3} \quad ◆$$

EXAMPLE 7.14
The Probability of Guilt and Innocence Given a DNA Match

This example is modified from one given by Paulos (1995, p. 72). Suppose there has been a crime and it is known that the criminal is a person within a population of 6,000,000. Further, suppose it is known that in this population only about one person in a million has a DNA type that matches the DNA found at the crime scene, so let's assume there are six people in the population with this DNA type. Someone in custody has this DNA type. *We know* the person's DNA matches, but what is the probability that he is actually innocent? In the absence of any other evidence beyond the DNA match, the probability of guilt is 1/6 because the criminal is one of the six individuals who have this DNA type in the population. Therefore, the probability of innocence is 5/6.

We now know the answer, but to illustrate Rule 4 we'll look at how it can be used to determine this probability. Imagine randomly selecting someone from the population. Define:

Event A = DNA of randomly chosen person matches DNA at the crime scene
Event B = person selected is innocent of the crime

We want to determine

$$P(B|A) = \frac{P(A \text{ and } B)}{P(A)}$$

The event "A and B" is the event that the selected person is innocent *and* the DNA matches. There are five such people in the population so $P(A \text{ and } B) = 5/6,000,000$. The probability of a DNA match is $P(A) = 6/6,000,000$, and

$$P(B|A) = \frac{P(A \text{ and } B)}{P(A)} = \frac{5/6,000,000}{6/6,000,000} = \frac{5}{6}$$

If you were on the jury, it would be important to realize that without additional evidence, the probability that this person is *innocent* is 5/6, even though the DNA matches. The prosecutor surely would emphasize a different conditional probability, specifically, the very small probability that an innocent person's DNA type would match the crime scene DNA (remember, there are 5,999,999 innocent people). Do not confuse these two different conditional probabilities. ◆

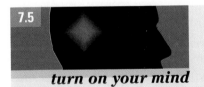

7.5

turn on your mind

Continuing the DNA example, verify that the conditional probability

$$P(\text{DNA match} \mid \text{innocent person}) = \frac{5}{5,999,999} = .00000083$$

Then provide an explanation that would be understood by a jury for the distinction between the two statements:

◆ The probability that a person who has a DNA match is innocent is 5/6.

◆ The probability that a person who is innocent has a DNA match is .00000083.*

---

*HINT: It might be helpful to construct a 2 × 2 table showing where all 6 million people fall, where row categories are "innocent or not," and column categories are "DNA match (yes, no)."

in summary — Independent and Mutually Exclusive Events and Probability Rules

Students sometimes confuse the definitions of independent and mutually exclusive events. Remember these two concepts:

◆ When two events are *mutually exclusive* and one happens, it turns the probability of the other one to 0.

◆ When two events are *independent* and one happens, it leaves the probability of the other one alone.

Probabilities under these two situations follow these rules:

When Events Are:	P(A or B) is:	P(A and B) Is:	P(A|B) Is:
Mutually Exclusive	P(A) + P(B)	0	0
Independent	P(A) + P(B) − P(A)P(B)	P(A)P(B)	P(A)
Any	P(A) + P(B) − P(A and B)	P(A)P(B|A)	$\dfrac{P(A \text{ and } B)}{P(B)}$

Sampling with and without Replacement

Suppose a classroom has 30 students in it and 3 of the students are left-handed. If the teacher randomly picks one student, the probability a left-handed student is selected is 3/30 or 1/10. The probability that a right-handed student is selected is 27/30 or 9/10. Now suppose the teacher randomly picks a student again. What is the probability that another left-handed student is chosen? The answer is that it depends on whether the first student selected is eligible to be selected again.

definition ◆ A **sample** is drawn **with replacement** if individuals are returned to the eligible pool for each selection.

◆ A **sample** is drawn **without replacement** if sampled individuals are not eligible for subsequent selection.

To determine the probability of a sequence of events, we use the multiplication rule. If sampling is done *with replacement,* the Extension of Rule 3b holds. If sampling is done *without replacement,* probability calculations are more complicated because the probabilities of the possible outcomes at any specific time in the sequence are *conditional* on previous outcomes.

Consider randomly drawing two left-handed students from a class of 30 where 3 students are left-handed. Here is how you would find the probability that a left-handed student is drawn both times if sampling is with replacement and if sampling is without replacement:

$$\textit{Sampled with replacement: } P(\text{Left and Left}) = \left(\frac{3}{30}\right)\left(\frac{3}{30}\right) = \frac{9}{900} = \frac{1}{100}$$

$$\textit{Sampled without replacement: } P(\text{Left and Left}) = \left(\frac{3}{30}\right)\left(\frac{2}{29}\right) = \frac{6}{870} = \frac{1}{145}$$

The probability that the second student is left-handed, given that the first student was left-handed, is a conditional probability. It is found by noticing that there are 29 students remaining, of whom 2 are left-handed.

If a sample is drawn from a very large population, the distinction between sampling with and without replacement becomes unimportant. For example, suppose a city has 300,000 people of whom 30,000 are left-handed. If two people are sampled, the probability that they are both left-handed is $1/100 = .01$ if drawn with replacement, and .0099997 if drawn without replacement. In most polls, individuals are drawn without replacement, but the analysis of the results is done as if they were drawn with replacement. The consequences of making this simplifying assumption are negligible.

7.5 | STRATEGIES FOR FINDING COMPLICATED PROBABILITIES

Much like we build complicated sentences using basic words and simple rules, we can compute complicated probabilities using the basic rules for probability relationships. Finding probabilities of complicated events is somewhat like trying to drive from one city to another. There often are many different routes that get you there. Some routes are faster than others, but all will get you to your destination. When you attempt to solve a probability problem, remember that you may take a different approach than this book or your professor or friend. The important thing is that you all get to the same answer eventually.

---

in summary **Probability Rules 1 Through 4**

Rule 1 (for "not the event"): To find the probability of A^C, the complement of A, use $P(A^C) = 1 - P(A)$

Rule 2 (addition rule for "either or"): To find the probability that either A *or* B happens:

> **Rule 2a (general):** $P(A \text{ or } B) = P(A) + P(B) - P(A \text{ and } B)$

> **Rule 2b (for mutually exclusive events):** *If* A and B are *mutually exclusive* events: $P(A \text{ or } B) = P(A) + P(B)$

Rule 3 (multiplication rule for "and"): To find the probability that two events both occur simultaneously or in a sequence:

> **Rule 3a (general):** $P(A \text{ and } B) = P(A)P(B|A)$

> **Rule 3b (for independent events):** *If* A and B are *independent* events, $P(A \text{ and } B) = P(A)P(B)$

> **Extension of Rule 3b to more than two independent events:** For several independent events, $P(A_1 \text{ and } A_2 \dots \text{ and } A_n) = P(A_1)P(A_2) \cdots P(A_n)$

Rule 4 (conditional probability): To find the probability that B occurs given that A has occurred:

$$P(B|A) = \frac{P(A \text{ and } B)}{P(A)}$$

The assignment of letters to events A and B is arbitrary, so it is also true that

$$P(A|B) = \frac{P(A \text{ and } B)}{P(B)}$$

EXAMPLE 7.2—CONTINUED
Winning the Lottery

Suppose you have purchased a lottery ticket with the number 956. What is the probability that you will win? Let the event A = winning number is 956. Here are two methods for finding $P(A)$:

> *Method 1:* With the physical assumption that all 1000 possibilities are equally likely, it is clear that $P(A) = 1/1000$.

> *Method 2:* Define three events,

>> B_1 = first digit drawn is 9
>> B_2 = second digit drawn is 5
>> B_3 = third digit drawn is 6

Event A occurs if and only if all three of these events occur. With the physical assumption that each of the ten digits $(0, 1, \ldots, 9)$ is equally likely on each draw, $P(B_1) = P(B_2) = P(B_3) = 1/10$. Since these events are all independent, we can apply the Extension of Rule 3b to find $P(A) = (1/10)^3 = 1/1000$.

Both of these methods are effective in determining the correct answer, but the first method is faster. ◆

Hints and Advice for Finding Probabilities

- To find $P(A \text{ and } B)$, define the combined event in physical terms and see if you know its probability. If this fails, try the multiplication rule (Rule 3).
- To find the probability that a series of independent events all happen, simply multiply all individual probabilities (Extension of Rule 3b).
- To find the probability that one of a collection of mutually exclusive events happens, simply add all of the individual probabilities (Rule 2b extended).
- It is sometimes easier to find the probability of the *complement* of the event of interest and then subtract it from 1 (applying Rule 1).
- To find the probability that *none* of a collection of mutually exclusive events happens, find the probability that one of them happens, and then subtract that from 1.
- To find a conditional probability, define the event in physical terms and see if you know its probability. If this fails, try Rule 4. See the next bullet as well.
- If you know $P(B|A)$ but are trying to find $P(A|B)$, you can use Rule 3a to find the two pieces in $P(B) = P(A \text{ and } B) + P(A^C \text{ and } B)$, then use Rule 4 to find $P(A|B)$.

Bayes Rule

formula   The last tip is an informal statement of **Bayes Rule,** which states:

$$P(A|B) = \frac{P(A \text{ and } B)}{P(B|A)P(A) + P(B|A^C)P(A^C)}$$

This rule looks complicated, but note that the denominator is simply $P(B)$ broken into two parts, $P(A \text{ and } B) + P(A^C \text{ and } B)$, each written using Rule 3a.

A Specific Set of Steps for Finding Probabilities

Here are a series of steps that usually work for solving probability problems. You may prefer to take a different route to solve problems, which is fine as long as you reach the same final destination.

Step 1: List each separate random circumstance involved in the problem.

Step 2: List the possible outcomes for each random circumstance.

List mutually exclusive events covering all possibilities. Determine the simplest events for which you want to specify probabilities. For instance, when we determined whether or not Alicia was called upon to answer question 1, we did not need to list each of the possible outcomes, the 50 students, separately. We simply listed whether or not Alicia was selected.

Step 3: Assign whatever probabilities you can with the knowledge you have.

Sometimes this will require making physical assumptions, sometimes it will be based on long-run frequencies, and sometimes it will be based on personal probabilities. Sometimes the probabilities you know how to assign are conditional probabilities.

Step 4: Specify the event for which you want to determine the probability.

Do this in terms of the possible outcomes listed in step 2.

Step 5: Determine which of the probabilities from step 3 and which probability rules can be combined to find the probability of interest.

Remember that there may be several ways to do this.

EXAMPLE 7.15
Family Composition

Assume the probability that each birth in a family is a boy is .512 and that the outcomes of successive births are independent. If the family has three children, what is the probability that they have two boys and one girl (in any order)? Here are two equally correct solutions to this problem:

Example 7.15, Solution 1

Define each birth as a separate "random circumstance."

Step 1: List all of the separate random circumstances involved in the problem.

Step 2: List the possible outcomes for each random circumstance.

Step 3: Assign whatever probabilities you can with the knowledge you have.

For these three steps, define three random circumstances to be birth of first child, birth of second child, and birth of third child. For each of these circumstances, there are two possible outcomes, boy and girl. Let B_1, B_2, and B_3 be the event that birth 1, birth 2, and birth 3, respectively, result in a boy. Then $P(B_1) = P(B_2) = P(B_3) = .512$. Similarly, let G_1 be the event that the first birth results in a girl. (Notice that we could write this as B_1^C, but the notation is awkward.) Similarly define G_2 and G_3 as girls for births 2 and 3. Then $P(G_1) = P(G_2) = P(G_3) = 1 - .512 = .488$.

Step 4: Specify the event for which you want to determine the probability.

The event "two boys and one girl" can happen in any of the following three ways:

$(B_1$ and B_2 and $G_3)$ *or* $(B_1$ and G_2 and $B_3)$ *or* $(G_1$ and B_2 and $B_3)$.

Step 5: Determine which of the probabilities from step 3 and which probability rules can be combined to find the probability of interest.

Since birth outcomes are independent, and since the events in step 4 involve "and" statements, use the Extension of Rule 3b to find:

$$P(B_1 \text{ and } B_2 \text{ and } G_3) = (.512)(.512)(.488) = .1279$$
$$P(B_1 \text{ and } G_2 \text{ and } B_3) = (.512)(.488)(.512) = .1279$$
$$P(G_1 \text{ and } B_2 \text{ and } B_3) = (.488)(.512)(.512) = .1279$$

Finally, because these are disjoint, $P[(B_1 \text{ and } B_2 \text{ and } G_3)$ *or* $(B_1 \text{ and } G_2 \text{ and } B_3)$ *or* $(G_1 \text{ and } B_2 \text{ and } B_3)] = .1279 + .1279 + .1279 = .3837$. There is about a 38% chance that the family will have two boys and one girl.

Example 7.15, Solution 2

Define the collection of births in the family as one "random circumstance." This requires listing all possible sets of three births. We use the notation BGB to indicate that the first birth is a boy, the second is a girl, and the third is a boy, and similarly for all of the possible outcomes.

Step 1: List all of the separate random circumstances involved in the problem.

Step 2: List the possible outcomes for each random circumstance.

For steps 1 and 2, the random circumstance is the three births, and the possible outcomes are {GGG,GGB,GBG,BGG,BBG,BGB,GBB,BBB}.

Step 3: Assign whatever probabilities you can with the knowledge you have.

Since the births are independent, the probability of each outcome is found by multiplying .512 for each B and .488 for each G. For instance, $P(\text{BGG}) = (.512)(.488)(.488) = .1219$.

Step 4: Specify the event for which you want to determine the probability.

The event "two boys and one girl" consists of the following subset of the sample space: {BBG,BGB,GBB}.

Step 5: Determine which of the probabilities from step 3 and which probability rules can be combined to find the probability of interest.

Using Rule 2b, add the probabilities of the three disjoint outcomes listed in step 4. Each one has probability $(.512)(.512)(.488) = .1279$, so the combined probability of two boys and one girl is $.1279 + .1279 + .1279 = .3837$. ◆

EXAMPLE 7.16
Prizes in Cereal Boxes

A particular brand of cereal contains a prize in each box. There are four possible prizes, and any box is equally likely to contain each of the four prizes. What is the probability that you will receive two different prizes if you purchase two boxes? Here are the known facts:

◆ The probability of receiving each prize is 1/4 for each box.

◆ The prizes you receive in two boxes are independent of each other.

There are a number of equally correct ways to solve this problem. The simplest method is to recognize that no matter what you receive in the first box, the probability of receiving a different prize in the second box is 3/4. Therefore, the probability of receiving two different prizes is 3/4. Here is a more detailed method.

Label the four prizes as a,b,c,d and define an "outcome" as the pair of prizes from the first and second box. Since outcomes are independent, the probability of any pair is found from Rule 3b to be $(1/4)(1/4) = 1/16$. The sample space is

{aa,ab,ac,ad,ba,bb,bc,bd,ca,cb,cc,cd,da,db,dc,dd}

There are 12 disjoint ways to get two different prizes, so reapplying Rule 2b, the probability of getting two different prizes is $12(1/16) = 12/16 = 3/4$. ◆

VIDEO

EXAMPLE 7.17
Optimism for Alicia —
She Is Probably Healthy

If you were in Alicia's position, you would be most interested in finding the probability of having the disease, *given* that the test was positive. Let's use these steps to find that probability.

Steps 1 to 3: Random circumstances, possible outcomes and known probabilities.

Random circumstance 1:   Alicia's disease status

Possible Outcomes: A = disease

A^C = no disease

Probabilities: $P(A) = 1/1000 = .001$ (from her physician's knowledge)

$P(A^C) = 999/1000 = .999$

Random circumstance 2:   Alicia's test results

Possible Outcomes: B = test is positive

B^C = test is negative

Probabilities: We only know conditional probabilities, from her physician:

$P(B|A) = .95$ (positive test given disease)

$P(B^C|A) = .05$ (negative test given disease)

$P(B|A^C) = .05$ (positive test given no disease)

$P(B^C|A^C) = .95$ (negative test given no disease)

Step 4: Specify the event for which you want to determine the probability.

We would like to know $P(\text{disease}|\text{positive test}) = P(A|B)$.

Step 5: Determine which of the probabilities from step 3 and which probability rules can be combined to find the probability of interest.

The first thing to notice is that this is a conditional probability, so we first try to use either the definition of conditional probability or Rule 4. Neither of these works immediately, but notice that we have conditional probabilities in one direction and want them in the other direction. Using the last of the above tips, we follow these steps:

$P(B) = P(A \text{ and } B) + P(A^C \text{ and } B)$, since these are the disjoint and complete ways B can happen.

From Rule 3a, find $P(A \text{ and } B) = P(B|A)P(A)$, similarly find $P(A^C \text{ and } B)$. Add them to get $P(B)$.

$$P(A \text{ and } B) = P(B|A)P(A) = (.95)(.001) = .00095$$
$$P(A^C \text{ and } B) = P(B|A^C)P(A^C) = (.05)(.999) = .04995$$
$$P(B) = .0509$$

From Rule 4 and the pieces you just found, you can now find $P(A|B)$:

$$P(A|B) = \frac{P(A \text{ and } B)}{P(B)} = \frac{.00095}{.0509} = .019$$

Notice what this tells Alicia! Even though her test was positive, the probability that she *actually has disease D is only .019.* There is less than a 2% chance that Alicia has the disease, even though her test was positive. The fact that so many more people are disease-free than diseased means that the 5% of disease-free people who falsely test positive far outweigh the 95% of those with the disease who legitimately test positive. ◆

Two-Way Tables and Tree Diagrams—Two Useful Tools

There are two useful tools for illustrating combinations of events and answering questions about them. The two tools, constructing a hypothetical two-way table and constructing a tree diagram, are particularly useful when conditional probabilities are already known in one direction, such as $P(A|B)$. The tools can then be used to find $P(A$ and $B)$, $P(B|A)$, and other probabilities of interest.

Constructing a Two-Way Table:
The "Hypothetical Hundred Thousand"

When conditional or joint probabilities are known for two events, it is sometimes useful to construct a hypothetical two-way table of outcomes for a round number of individuals, and 100,000 is a good number because usually it results in whole numbers of people in each cell. Therefore, we will call this method the *hypothetical hundred thousand.* Let's illustrate the idea with an example.

EXAMPLE 7.8—CONTINUED
Teens and Gambling

In Example 7.8, a study was reported in which 9th-grade Minnesota teens were asked whether they had gambled at least once a week during the past year. The sample consisted of 49.1% boys and 50.9% girls. The proportion of boys who had gambled weekly was .229, while the proportion of girls who had done so was only .045. If we were to construct a table for a hypothetical group of 100,000 teens reflecting these proportions, there would be $(.491)(100,000) = 49,100$ boys, and 50,900 girls. Of the 49,100 boys, $(.229 \times 49,100)$ or 11,244 would be weekly gamblers, and of the 50,900 girls, .045, or 2,291 girls would be weekly gamblers. The "hypothetical hundred thousand" table can be constructed by starting with these numbers and filling in the rest by subtraction and addition. Here is the resulting table:

	Weekly Gambler	Not Weekly Gambler	Total
Boy	11,244	37,856	49,100
Girl	2,291	48,609	50,900
Total	13,535	86,465	100,000

It's easy to find various probabilities from this table. Here are some examples:

◆ P(boy and weekly gambler) $= 11,244/100,000 = .11244$, so about 11% of 9th graders are boys who gamble weekly.

◆ P(boy|weekly gambler) $= 11,244/13,535 = .8307$, so about 83% of weekly gamblers are boys.

◆ P(weekly gambler) $= 13,535/100,000 = .13535$, so about 13.5% of 9th graders are weekly gamblers. ◆

Tree Diagrams

Another useful tool for combinations of events is a tree diagram. A **tree diagram** is a schematic representation of the sequence of events and their probabilities, including conditional probabilities based on previous events for events that happen sequentially. It is easiest to illustrate by example.

EXAMPLE 7.18
Alicia's Possible Fates

Let's consider what could have happened with Alicia's medical test, and the probabilities associated with those options. There are two random circumstances involved. The first circumstance is whether or not she has the disease (*D* or *not D*), and the second circumstance is whether the test result is positive or negative (*positive test* or *negative test*). We have already specified the probabilities of each of the pieces, and using Rule 3a we can find the probabilities of the various combinations, like the probability that Alicia has the disease but the test result is negative.

Figure 7.1 illustrates these probabilities in a tree diagram. The initial set of branches shows the two possibilities and associated probabilities for her disease status. The next set of branches shows the two possibilities for the outcome of the test and associated *conditional probabilities* depending on whether or not she has the disease. The "final probability" shown for each combination of branches is found by multiplying the probabilities on the branches, applying Rule 3a. Thus, for instance, the probability that Alicia has D *and* has a positive test is .00095. ◆

Here are the general steps for creating a tree diagram:

Step 1: Determine the first random circumstance in the sequence, and create the first set of branches to illustrate possible outcomes for it. Create one branch for each outcome, and write the associated probability on the branch.

Step 2: Determine the next random circumstance, and append branches for the possible outcomes to each of the branches in step 1. Write the associated *conditional probabilities* on the branches, where the outcome for step 2 is *conditional* on the branch taken in step 1.

Step 3: Continue this process for as many steps (and sets of branches) as necessary.

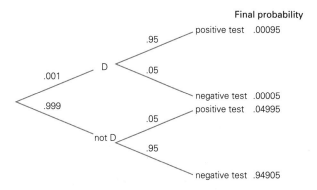

FIGURE 7.1   Tree diagram for Alicia's medical test

Step 4: To determine the probability of following any particular sequence of branches, multiply the probabilities on those branches. This is an application of Rule 3a.

Step 5: To determine the probability of any collection of sequences of branches, add the individual probabilities for those sequences, as found in step 4. This is an application of Rule 2b.

EXAMPLE 7.18—*CONTINUED*
The Probability That
Alicia Has a Positive Test

Notice that Alicia's test is positive if she follows branch D and then branch "positive test" *or* if she follows branch "not D" and then branch "positive test." The overall probability that the test is positive can be found by adding the probabilities from these two sets of branches. Hence,

$$P(\text{positive test}) = (.001)(.95) + (.999)(.05) = .00095 + .04995 = .0509$$

In other words, for all women like Alicia who are tested, including the .001 of them with the disease and the .999 of them without the disease, a positive test will occur with probability .0509. This is equivalent to saying that .0509 or 5.09% of these women will have a positive test. ◆

It is always the case that the first set of branches in a tree diagram displays probabilities for the outcomes of one random circumstance, then the second set of branches displays *conditional* probabilities for the outcomes of another random circumstance, *given* what happened on the first one. In general, the first set of branches corresponds to the circumstance that occurs first chronologically, if the events have a time order. If they don't have a time order, then think about what probabilities are known. You must know the unconditional probabilities of outcomes for the events on the first set of branches.

EXAMPLE 7.8—*CONTINUED*
Teens and Gambling

Let's construct a tree diagram using the information provided in Example 7.8 about the gambling behavior of 9th-grade Minnesota teenagers. For a randomly selected 9th grader, the two random circumstances are gender (boy or girl) and weekly gambler (yes or no). We know the unconditional probabilities for gender: $P(\text{boy}) = .491$ and $P(\text{girl}) = .509$. We know the *conditional* probabilities of weekly gambling, *given* gender: $P(\text{weekly gambler}|\text{boy}) = .229$ and $P(\text{weekly gambler}|\text{girl}) = .045$. Therefore, the first set of branches represents gender and the second set represents weekly gambler or not. Figure 7.2 illustrates the completed tree diagram.

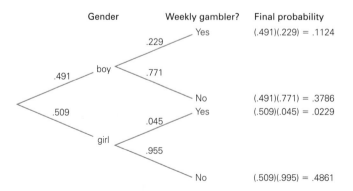

FIGURE 7.2   **Tree diagram for 9th-grade teens and weekly gambling**

Once the tree diagram is completed, we can easily find probabilities for any combination of gender *and* weekly gambling by multiplying the probabilities across branches. For instance,

$$P(\text{boy and weekly gambler}) = (.491)(.229) = .1124$$

$$P(\text{girl and not weekly gambler}) = (.509)(.955) = .4861$$

We can find other probabilities by adding outcomes from two sets of branches. For instance,

$$P(\text{weekly gambler}) = P(\text{boy and weekly gambler}) + P(\text{girl and weekly}$$
$$\text{gambler}) = (.1124) + (.0229) = .1353$$

Finally, we can find conditional probabilities in the opposite direction to those known by finding probabilities such as those we just found and then applying Rule 4. For instance,

$$P(\text{boy}|\text{weekly gambler}) = \frac{P(\text{boy and weekly gambler})}{P(\text{weekly gambler})} = \frac{.1124}{.1353} = .8307. \; \blacklozenge$$

Probabilities, Proportions, and Percents

tech note   In many of the examples in this chapter we have discussed probabilities associated with a "randomly selected individual" from a population. These are equivalent to percents in the population with specific characteristics, but multiplying percents doesn't work. For instance, in Example 7.8, 49.1% or .491, of the teens were boys and of those, 22.9%, or .229, were weekly gamblers. To find what *percent* of the teens were boys *and* weekly gamblers, it would not work to multiply (49.1%)(22.9%) = 1124.39%. Instead, convert everything to proportions, multiply, then convert back to a percent by multiplying by 100%: (.491)(.229) = .1124, or .1124 × 100% = 11.24% of the teens are boys who gamble weekly. Proportions can always be interpreted as percents in this manner. For instance, in Example 7.18— *CONTINUED*, we found that for all women like Alicia who are tested for this disease, .0509 × 100% = 5.09% will have a positive test, including 4.995% who do not have the disease and test positive and only .095% who have the disease and test positive.

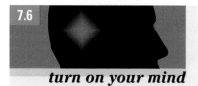

7.6

turn on your mind

Explain why the tree diagram in Figure 7.1 displayed disease status first and test results second rather than the other way around.*

7.6 | USING SIMULATION TO ESTIMATE PROBABILITIES

Some probabilities are so difficult or time-consuming to calculate that it is easier to simulate the situation repeatedly using a computer or calculator and observe the relative frequency of the event of interest. If you simulate the random circumstance *n* times and the outcome of interest occurs in *x* out of those *n*

*HINT: See the paragraph just before the previous Example, 7.8—CONTINUED.

times, then the estimated probability for the outcome of interest is x/n. This is an estimate of the long-run relative frequency with which the outcome would occur in real life.

EXAMPLE 7.19
Getting All the Prizes

Refer to the situation in Example 7.16. Suppose cereal boxes each contain one of four prizes, and any box is equally likely to contain any of the four prizes. You would like to collect all four prizes. If you buy six boxes of cereal, how likely is it that you will get all four prizes?

It would be quite time-consuming to solve this problem using the rules presented in the previous sections. It is much simpler to simulate the situation using a computer or calculator. Table 7.3 shows 50 simulations using the Minitab computer software. In each case, six digits were drawn, each equally likely to be 1, 2, 3 or 4. The outcomes that contain all four digits are highlighted in bold.

Counting the outcomes in bold reveals that 19 out of 50, or .38, of these simulations satisfy the condition that all four prizes are collected. In fact, this simulation had particularly good results. There are $4^6 = 4096$ equally likely possible outcomes for listing six digits randomly chosen from 1, 2, 3, 4, and 1560 of those outcomes include all four digits. So the actual probability of getting all four prizes is $1560/4096 = .3809$. ◆

In general, determining the accuracy of simulation for estimating a probability is similar to determining the margin of error when estimating a proportion from a sample survey. The probability estimated by simulation will be no farther off than $1/\sqrt{n}$ most of the time, where n is the number of simulated repetitions of the situation.

This information about the accuracy of the simulation approach for estimating probabilities should make it clear that you need to repeat the simulated situation a large number of times to obtain accurate results. Therefore, it is most useful if you can use a computer to generate the simulated outcomes and also to count how many of them satisfy the event of interest.

EXAMPLE 7.20
Finding Gifted ESP Participants

An ESP test is conducted by randomly selecting one of five video clips and playing it in one building, while a participant in another building tries to describe what is playing. Later, the participant is shown the five video clips and is asked to determine which one best matches the description he or she had given. By chance, the participant would get this correct with probability 1/5. Individual participants are each tested eight times, with five new video clips each time. They are identified as "gifted" if they guess correctly at least five times out of the

TABLE 7.3 ■ **50 Simulations of Prizes from Cereal Boxes**

112142	443222	323324	321223	**332314**
224123	342324	**232413**	121123	422244
244412	**342431**	121333	**244132**	434224
234323	**144332**	121142	**432121**	144313
443313	211141	**421342**	441211	111134
113432	433424	**312314**	241114	313411
214443	222422	144441	**213141**	312232
312341	411124	111422	312213	**323314**
143124	111131	**441233**	121223	424433
213341	114333	**243311**	442244	**214133**

eight tries. Suppose people actually do have some ESP and can guess correctly with probability .30 (instead of the .20 expected by chance). What is the probability that a participant will be identified as "gifted"?

In Chapter 8 you will learn how to solve this kind of problem, but we can simulate the answer using a random number table or generator that produces the digits 0, 1, 2, . . . , 9 with equal likelihood. Table 4.1 (page 95) is an example of a random number table. Many calculators and computers will simulate these digits. Here are the steps needed for one "repetition":

- ◆ Each "guess" is simulated with a digit, equally likely to be 0 to 9.
- ◆ For each participant, we simulate eight "guesses" resulting in a string of eight digits.
- ◆ If a digit is 7, 8, or 9, we count that guess as "correct" so P(correct) = 3/10 = .3, as required in the problem. If the digit is 0 to 6, the guess is "incorrect." (There is nothing special about 7, 8, 9; we could have used any three digits.)
- ◆ If there are five or more "correct" guesses (digits 7, 8, 9), we count that as "gifted."

The entire process is repeated many times, and the proportion of times the result is a "gifted" participant is an estimate of the desired probability.

Process and Results Using Minitab

1. Generate 1000 rows of 8 columns, using random integers from 0 to 9. Each row represents the 8 guesses for one participant.
2. Use the "code" feature to recode all of the values, (7,8,9)→1, (0,1,2,3,4,5,6)→0.
3. Sum the 8 columns; for each row (each participant) the sum is the number of "correct" guesses. This is equivalent to the number of 1s in the row after recoding.
4. The sums are now in a column with 1000 rows. Tally this column to find out how many of the participants had five or more "correct guesses."
5. The results of doing this were as follows:

Number Correct	0	1	2	3	4	5	6	7	8
Frequency	64	187	299	271	119	44	14	2	0

Notice that there were 44 + 14 + 2 + 0 = 60 individuals who got five or more correct, so the probability of finding a "gifted" participant each time is about 60/1000 = .06. In other words, if everyone is equally talented, and each guess is correct with probability .3, then there will be five or more correct guesses out of eight tries with probability .06, or about 6% of the time. ◆

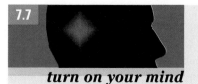

7.7

turn on your mind

Does using simulation to estimate probabilities rely on the relative frequency or the personal probability interpretation of probability? Explain. Whichever one you chose, could simulation be used to find probabilities for the other interpretation? Explain.*

7.7 | COINCIDENCES AND INTUITIVE JUDGMENTS ABOUT PROBABILITY

People do not have good intuition about probability assessments. In Case Study 7.1, Alicia's physician made a common mistake by informing her that the medical test was a good indicator of her disease status since it was 95% accurate whether or not she had the disease. As we learned in Example 7.17, the probability that she actually has the disease is small, about 2%. In this section, we explore this phenomenon, called "Confusion of the Inverse," and some other ways in which intuitive probability assessments can be seriously flawed.

Confusion of the Inverse

Eddy (1982) posed the following scenario to 100 physicians:

> One of your patients has a lump in her breast. You are almost certain that it is benign, in fact you would say there is only a 1% chance that it is malignant. But just to be sure, you have the patient undergo a mammogram, a breast X-ray designed to detect cancer.
>
> You know from the medical literature that mammograms are 80% accurate for malignant lumps and 90% accurate for benign lumps. In other words, if the lump is truly malignant, the test results will say that it is malignant 80% of the time and will falsely say it is benign 20% of the time. If the lump is truly benign, the test results will say so 90% of the time and will falsely declare that it is malignant only 10% of the time.
>
> Sadly, the mammogram for your patient is returned with the news that the lump is malignant. What are the chances that it is truly malignant?

Most of the physicians to whom Eddy posed this question thought the probability that the lump was truly malignant was about 75% or .75. In truth, given the probabilities described in the scenario, *the probability is only .075.* The physicians' estimates were ten times too high! When Eddy asked the physicians how they arrived at their answers, he realized they were confusing the answer to the actual question with the answer to a different question. "When asked about this, the erring physicians usually report that they assumed that the probability of cancer given that the patient has a positive X-ray was approximately equal to the probability of a positive X-ray in a patient with cancer" (1982, p. 254).

Robyn Dawes has called this phenomenon **confusion of the inverse** (Plous, 1993, p. 132). The physicians were confusing the conditional probability of cancer *given a positive X-ray* with the inverse, the conditional probability of a positive X-ray *given that the patient has cancer!*

It is not difficult to see that the correct answer is indeed .075. Let's construct a hypothetical table of 100,000 women who fit this scenario. In other words, these are women who would present themselves to the physician with a lump for which the probability that it was malignant seemed to be about 1%. Thus, of the 100,000 women, about 1%, or 1000 of them, would have a malignant lump. The remaining 99%, or 99,000, would have a benign lump.

Further, given that the test was 80% accurate for malignant lumps, it would show a malignancy for 800 of the 1000 women who actually had one. Given that it was 90% accurate for the 99,000 women with benign lumps, it would show benign

TABLE 7.4 ■ Breakdown of Actual Status Versus Test Status for a Rare Disease

	Test is Malignant	Test is Benign	Total
Actually Malignant	800	200	1000
Actually Benign	9900	89,100	99,000
Total	10,700	89,300	100,000

for 90%, or 89,100 of them, and malignant for the remaining 10%, or 9900 of them. Table 7.4 shows how the 100,000 women would fall into these possible categories.

Let's return to the question of interest. Our patient has just received a positive test for malignancy. Given that her test showed malignancy, what is the actual probability that her lump is malignant? Of the 100,000 women, 10,700 of them would have an X-ray show malignancy. But of those 10,700 women, only 800 of them actually have a malignant lump! Thus, given that the test showed a malignancy, the probability of malignancy is just 800/10,700 = 8/107 = .075.

Sadly, many physicians are guilty of confusion of the inverse. Remember, in a situation where the *base rate* for a disease is very low and the test for the disease is less than perfect, there will be a relatively high probability that a positive test result is a false positive. If you ever find yourself in a situation similar to the one just described, you may wish to construct a table like the one above.

To determine the probability of a positive test result being accurate, you only need three pieces of information.

1. What the base rate or probability that you are likely to have the disease is, without any knowledge of your test results.
2. What the **sensitivity** of the test is, which is the proportion of people who correctly test positive when they actually have the disease.
3. What the **specificity** of the test is, which is the proportion of people who correctly test negative when they don't have the disease.

Notice that items 2 and 3 are measures of the accuracy of the test. They do not measure the probability that people have the disease when they test positive or the probability that they do not have the disease when they test negative. Those probabilities, which are obviously the ones of interest to the patient, can be computed by constructing a table similar to Table 7.4 or by the other methods shown earlier in this chapter.

Specific People Versus Random Individuals

According to the Federal Aviation Administration, between 1987 and 1996 there were 39 fatal airline accidents for regularly scheduled flights on U.S. carriers. During that same time period, there were 76,557,000 flight departures. That means that the relative frequency of fatal accidents was about one accident per 2 million departures. Based on these kinds of statistics you will sometimes hear statements like "The probability that you will be in a fatal plane crash is 1 in 2 million" or "The chance that your marriage will end in divorce is 50%."

Do these probability statements really apply to you personally? In an attempt to personalize the information, reporters express probability statements in terms of individuals when they actually apply to the aggregate. Obviously, if you never fly, the probability that you will be in a fatal plane crash is 0. Here are two equivalent, correct ways to restate the aggregate statistics about fatal plane crashes:

◆ In the long run, about 1 out of every 2 million flight departures end in a fatal crash.

◆ The probability that a randomly selected flight departure ends in a fatal crash is about 1/2,000,000.

If you have had a terrific marriage for 30 years, it is not likely to end in divorce now. Your probability of a divorce surely is less than the 50% figure often reported in the media. Here are two correct ways to express aggregate statistics about divorce:

◆ In the long run, about 50% of marriages end in divorce.

◆ At the beginning of a randomly selected marriage, the probability that it will end in divorce is about .50.

Notice the emphasis on "random selection" in the second version of each statement. Sometimes the phrase "randomly selected" is omitted, but it should always be understood that randomness is part of the communication.

Coincidences

When one of the authors of this book was in college in upstate New York, she visited Disney World in Florida during summer break. While there, she ran into three people she knew from her college, none of whom were there together. A few years later, she visited the top of the Empire State Building in New York City and ran into two friends (who were there together) and two additional unrelated acquaintances. Years later, when traveling from London to Stockholm she ran into a friend at the airport in London while waiting for the flight. Not only did the friend turn out to be taking the same flight but had been assigned an adjacent seat.

These events are all examples of what would commonly be called coincidences. They are certainly surprising, but are they improbable? Most people think that coincidences have low probabilities of occurring, but we shall see that our intuition can be quite misleading regarding such phenomena. We will adopt the definition of **coincidence** proposed by Persi Diaconis and Fred Mosteller:

> *definition* A **coincidence** is a surprising concurrence of events, perceived as meaningfully related, with no apparent causal connection. (1989, p. 853)

The mathematically sophisticated reader may wish to consult the article by Diaconis and Mosteller, in which they provide some instructions on how to compute probabilities for coincidences. For our purposes, we will need nothing more sophisticated than the simple probability rules we encountered earlier in this chapter.

Following are some examples of coincidences that at first glance seem highly improbable.

EXAMPLE 7.21
Two George D. Brysons   "My next-door neighbor, Mr. George D. Bryson, was making a business trip some years ago from St. Louis to New York. Since this involved weekend travel and he was in no hurry, . . . and since his train went through Louisville, he asked the conductor, after he had boarded the train, whether he might have a stopover in Louisville.

"This was possible, and on arrival at Louisville he inquired at the station for the leading hotel. He accordingly went to the Brown Hotel and registered. And then, just as a lark, he stepped up to the mail desk and asked if there was any mail for him. The girl calmly handed him a letter addressed to Mr. George D. Bryson, Room 307, that being the number of the room to which he had just been assigned. It turned out that the preceding resident of Room 307 was another George D. Bryson" (Weaver, 1963, pp. 282–283). ◆

EXAMPLE 7.22
Identical Cars and Matching Keys

Plous (1993, p. 154) reprinted an Associated Press news story describing a coincidence in which a man named Richard Baker and his wife were shopping on April Fool's Day at a Wisconsin shopping center. Mr. Baker went out to get their car, a 1978 maroon Concord, and drove it around to pick up his wife. After driving for a short while, they noticed items in the car that did not look familiar. They checked the license plate, and sure enough, they had someone else's car. When they drove back to the shopping center (to find the police waiting for them), they discovered that the owner of the car they were driving was a Mr. Thomas Baker, no relation to Richard Baker. Thus, both Mr. Bakers were at the same shopping center at the same time, with identical cars and with matching keys. The police estimated the odds as "a million to one." ◆

EXAMPLE 7.23
Winning the Lottery Twice

Moore (1997, p. 330) reported on a *New York Times* story of February 14, 1986, about Evelyn Marie Adams who won the New Jersey lottery twice in a short time period. Her winnings were $3.9 million the first time and $1.5 million the second time. When Ms. Adams won for the second time, the *New York Times* claimed that the odds of one person winning the top prize twice were about 1 in 17 trillion. Then, in May 1988, Robert Humphries won a second Pennsylvania lottery, bringing his total winnings to $6.8 million. ◆

Most people think that the events just described are exceedingly improbable, and they are. *What is not improbable is that someone, somewhere, someday will experience those events or something similar.*

When we examine the probability of what appears to be a startling coincidence, we ask the wrong question. For example, the figure quoted by the *New York Times* of 1 in 17 trillion is the probability that a *specific* individual who plays the New Jersey State Lottery exactly twice will win both times (Diaconis and Mosteller, 1989, p. 859.) However, millions of people play the lottery every day, and it is not surprising that someone, somewhere, someday would win twice.

In fact, Purdue professors Stephen Samuels and George McCabe (cited in Diaconis and Mosteller, 1989, p. 859) calculated those odds to be practically a sure thing. They calculated that there was at least a 1 in 30 chance of a double winner in a four-month period and better than even odds that there would be a double winner in a seven-year period somewhere in the United States. And they used conservative assumptions about how many tickets past winners had purchased.

When you experience a coincidence, remember that there are over 6 billion people in the world and over 280 million in the United States. If something has a 1 in a million probability of occurring to each individual on a given day, it will occur to an average of over 280 people in the United States *each day* and over 6000 people in the world *each day*. Of course probabilities of specific events depend on individual circumstances, but we hope you can see that it is not unlikely that something surprising will happen quite often.

EXAMPLE 7.24
Sharing the Same Birthday

Here is a famous example that you can use to test your intuition about surprising events. How many people would need to be gathered together to be at least 50% sure that two of them share the same birthday (month and day, not necessarily year)? Most people provide answers that are much higher than the correct one, which is only 23 people.

There are several reasons why people have trouble with this problem. If your answer was somewhere close to 183, or half the number of birthdays, then you may have confused the question with another one, such as the probability that someone in the group has *your* birthday or that two people have a specific date as their birthday.

It is not difficult to see how to calculate the appropriate probability, using our probability rules. Notice that the only way to avoid two people having the same birthday is if all 23 people have different birthdays. To find that probability, we simply use the rule that applies to the word *and* (Rule 3), thus multiplying probabilities. The probability that the first three people have different birthdays is the probability that the second person does not share a birthday with the first, which is 364/365 (ignoring February 29), and the third person does not share a birthday with either of the first two, which is 363/365. (Two dates were already taken.)

Continuing this line of reasoning, the probability that none of the 23 people share a birthday is

$$\frac{(364)(363)}{(365)(365)} \cdots \frac{(343)}{(365)} = .493$$

The probability that at least two people share a birthday is the probability of the complement, or $1 - .493 = .507$.

If you find it difficult to imagine this could be correct, picture it this way. Imagine each of the 23 people shaking hands with the other 22 people and asking them about their birthday. There would be 253 handshakes and birthday conversations. Surely there is a relatively high probability that at least one of those pairs would discover a common birthday.

By the way, the probability of a shared birthday in a group of ten people is already better than one in nine, at .117. (There would be 45 handshakes.) With only 50 people it is almost certain, with a probability of .97. (There would be 1225 handshakes.) ◆

The technique used in Example 7.24 can be used to illustrate why some coincidences are not at all unlikely. Consider the situation described in Example 7.21 in which two George D. Brysons held the same hotel room in succession. It is not at all unlikely that occasionally two successive occupants of a hotel room somewhere, sometime will have the same name. To see that this is so, consider the probability of the complement—that no matter who occupies a hotel room, the next occupant will have a different name. Even ignoring common names, adding these probabilities across many thousands of hotel rooms and many days, weeks, and years, it would be extremely unlikely that successive occupants would always have different names.

EXAMPLE 7.25
Unusual Hands in Card Games

As a final example of unlikely coincidences, consider a card game such as bridge, in which a standard 52-card deck is dealt to four players, so they each receive 13 cards. Any specific set of 13 cards is equally likely, each with a probability of about 1 in 635 billion. You would probably not be surprised to get a mixed hand, say the 4, 7, and 10 of hearts; 3, 8, 9, and jack of spades; 2 and queen of diamonds; and 6, 10, jack, and ace of clubs. Yet, that specific hand is just as unlikely as getting all 13 hearts! The point is that any very specific event, surprising or not, has extremely low probability, but there are many, many surprising events and their combined probability is quite high.

Magicians sometimes exploit the fact that many small probabilities add up to one large probability by doing a trick in which they don't tell you what to expect in advance. They set it up so that *something* surprising is almost sure to happen. When it does, you are likely to focus on the probability of *that particular outcome,* rather than realizing that a multitude of other outcomes would have also been surprising and that one of them was likely to happen.

To summarize, most coincidences only seem improbable if we ask for the probability of that specific event occurring at that time, to us. If, instead, we ask the probability of it occurring some time, to someone, the probability can become quite large. Further, because of the multitude of experiences we each have every day, it is not surprising that some of them may appear to be improbable. That specific event is undoubtedly improbable. What is not improbable is that something "unusual" will happen to each of us once in awhile. ◆

The Gambler's Fallacy

Another common misperception about random events is that they should be self-correcting. Another way to state this is that people think the long-run frequency of an event should apply even in the short run. This misperception has classically been called the **gambler's fallacy.** Researchers William Gehring and Adrian Willoughby have discovered that this fallacy and related decision making while gambling are reflected in activity in the brain, in "a medial-frontal region in or near the anterior cingulate cortex" (p. 2279).

The gambler's fallacy can lead to poor decision making, especially if applied to gambling. For example, people tend to believe that a string of good luck will follow a string of bad luck in a casino. Unfortunately, independent chance events have no such memory. Making ten bad gambles in a row doesn't change the probability that the next gamble will also be bad.

Notice that the gambler's fallacy primarily applies to independent events. The gambler's fallacy may not apply to situations where knowledge of one outcome affects probabilities of the next. For instance, in card games using a single deck, knowledge of what cards have already been played provides information about what cards are likely to be played next, and gamblers routinely make use of that information. If you normally receive lots of mail but have received none for two days, you would probably (correctly) assess that you are likely to receive more than usual the next day.

A somewhat different but related misconception is what Tversky and Kahneman (1982) call the belief in the **law of small numbers,** "according to which [people believe that] even small samples are highly representative of the populations from which they are drawn" (p. 7). They report that "in considering tosses of a coin for heads and tails, for example, people regard the sequence HTHTTH to be more likely than the sequence HHHTTT, which does not appear random, and also more likely than the sequence HHHHTH, which does not represent the fairness of the coin" (p. 7). Remember that any specific sequence of heads and tails is just as likely as any other sequence if the coin is fair, so the idea that the first sequence is more likely is a misperception.

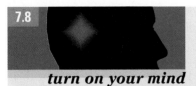

7.8

turn on your mind

If you wanted to pretend to be psychic you could do a "cold reading" on some-one you do not know. Suppose you are doing this for a 25-year-old woman. You make statements such as the following:

♦ I see that you are thinking of two men, one with dark hair, and the other one with slightly lighter hair or complexion. Do you know who I mean?

♦ I see a friend who is important to you but who has disappointed you recently.

♦ I see that there is some distance between you and your mother that bothers you.

Using the material in this section, explain why this would often work to con-vince people you are psychic.*

KEY TERMS

Section 7.1
random circumstances, 204
probability, 205, 206

Section 7.2
relative frequency, 206
personal probability, 209
subjective probability, 209
coherent probabilities, 210

Section 7.3
simple event, 211–212
sample space, 212
event, 212
complement, 213
complementary events, 213

mutually exclusive events, 213
disjoint events, 213
independent events, 214
dependent events, 214
conditional probability, 215

Section 7.4
probability rule 1: complements, 216
probability rule 2: addition, 217
addition rule, 217
probability rule 3: multiplication, 219
multiplication rule, 219
probability rule 4: conditional
   probability, 220
sample with replacement, 222
sample without replacement, 222

Section 7.5
Bayes Rule, 224
tree diagram, 229

Section 7.6
simulation, 231–232

Section 7.7
confusion of the inverse, 234
sensitivity, 235
specificity, 235
coincidence, 236
gambler's fallacy, 239
law of small numbers, 239

EXERCISES   Blue = Basics   **Bold, Bold** = Answer in back

Sections 7.1 and 7.2

7.1 According to a U.S. Department of Transportation web-site (http://www.dot.gov/airconsumer), 76.1% of domes-tic flights flown by the top ten U.S. airlines from June 1998 to May 1999 arrived on time. Represent this in terms of a random circumstance and an associated probability.

7.2 Suppose you live in a city that has 125,000 households with telephones and a polling organization randomly se-lects 1000 of them to phone for a survey. What is the prob-ability that your household will be selected?

7.3 Jan is a member of a class with 20 students. Each day for a week (Monday to Friday) a student in Jan's class is ran-domly selected to explain how to solve a homework prob-

lem. Once a student has been selected, he or she is not se-lected again that week. If Jan was not one of the four stu-dents selected earlier in the week, what is the probability that she will be picked on Friday? Explain how you found your answer.

7.4 Is each of the following values a legitimate probability value? Explain any "no" answers.
   a. .50
   b. .00
   c. 1.00
   d. 1.25
   e. −.25

7.5 Which interpretation of probability applies to each of the following situations? If it's the relative frequency interpretation, explain which of the three methods listed in the "In Summary" box at the end of Section 7.2 applies.

a. If a spoon is tossed 10,000 times and lands with the rounded head face up 3,000 of those times, we would say that the probability of the rounded head landing face up for that spoon is about .30.

b. In a debate with you, a friend says that she thinks there is a 50–50 chance that God exists.

c. Based on data from the 1991 to 1993 General Social Survey, the probability in those years was about 612/1669 = .367 that a randomly selected adult who had ever been married had divorced.

7.6 A car dealer has noticed that one out of 25 new-car buyers returns the car for warranty work within the first month.

a. Write a sentence expressing this fact as a proportion.

b. Write a sentence expressing this fact as a percent.

c. Write a sentence expressing this fact as a probability.

7.7 Explain which interpretation of probability (relative frequency or personal) applies to each of these statements and how you think the probability was determined.

a. According to Krantz (1992, p. 161), the probability that a randomly selected American will be injured by lightning in a given year is 1/685,000.

b. According to my neighbor, the probability that the tomato plants she planted last month will actually survive to produce fruit is only 1/2.

c. According to the nursery where my neighbor purchased her tomato plants, if a plant is properly cared for, the probability it will produce tomatoes is .99.

d. The probability that a husband will outlive his wife (for U.S. couples) is 3/10 (Krantz, 1992, p. 163).

7.8 Identify three "random circumstances" in the following story and give the possible outcomes for each of them:

> It was Robin's birthday and she knew she was going to have a good day. She was driving to work and when she turned on the radio her favorite song was playing. Besides, the traffic light at the main intersection she crossed to get to work was green when she arrived, something that seemed to happen less than once a week. When she arrived at work, rather than having to search as she usually did, she found an empty parking space right in front of the building.

7.9 Refer to the previous exercise. Suppose Robin wants to find the probability associated with the outcomes in the random circumstances contained in the story. Identify one of the circumstances and explain how she could determine the probabilities associated with its outcomes.

7.10 Casino games often use a "fair die" that has six sides with 1 to 6 dots on them. When the die is tossed or rolled, each of the six sides is equally likely to come out on top. Using the physical assumption that the die is fair, determine the probability of each of the following outcomes for the number of dots showing on top after a single roll of a die:

a. Six dots.

b. One or two dots.

c. An even number of dots.

7.11 A computer solitaire game uses a standard 52-card deck and randomly shuffles the cards for play. Theoretically, it should be possible to find optimal strategies for playing and then to compute the probability of winning based on the best strategy. Not only would this be an extremely complicated problem but also the probability of winning for any particular person will depend on that person's

skill. Explain how an individual could determine his or her probability of winning this game. (*Hint:* It might take a while to determine the probability!)

7.12 Find information on a random circumstance in the news. Identify the circumstance and possible outcomes, and assign probabilities to the outcomes. Explain how you determined the probabilities.

7.13 Give an example of a situation for which a probability statement makes sense, but for which the relative frequency interpretation could not apply, such as the probability given by Carl Sagan for an asteroid hitting the Earth.

7.14 Alicia's statistics class meets 50 times during the semester, and each time it meets the probability that she will be called on to answer the first question is 1/50. Does this mean that Alicia will be called on to answer the first question exactly once during the semester? Explain.

7.15 Every day, John buys a lottery ticket with the number 777 for the simple lottery described in Example 7.2. He has now played 999 times and has not won a single time. He reasons that since tomorrow will be his 1000th time, and since the probability of winning is 1/1000, he will have to win tomorrow. Explain whether or not John's reasoning is correct.

7.16 Refer to Example 7.5. What is the probability that a randomly selected child who slept in darkness would develop some degree of myopia?

7.17 When 190 students were asked to pick a number from 1 to 10, the number of students selecting each number were as follows:

Number	1	2	3	4	5	6	7	8	9	10	Total
Frequency	2	9	22	21	18	23	56	19	14	6	190

a. What is the approximate probability that someone asked to pick a number from 1 to 10 will pick the number 3?

b. What is the approximate probability that someone asked to pick a number from 1 to 10 will pick one of the two extremes, 1 or 10?

c. What is the approximate probability that someone asked to pick a number from 1 to 10 will pick an odd number?

Section 7.3

7.18 A student wants to send a bouquet of roses to her mother for Mother's Day. She can afford to buy only two types of roses, and decides to randomly pick two from the following four varieties: Blue Bell, Yellow Success, Sahara, and Aphrodite. Label the varieties B, Y, S, A.

a. Make a list of the six simple events in the sample space.

b. Assuming all outcomes are equally likely, what is the probability that she will pick Sahara and Aphrodite?

7.19 Remember that the event A^C is the complement of the event A.

a. Are A and A^C mutually exclusive? Explain.

b. Are A and A^C independent? Explain.

7.20 A penny and a nickel are tossed. Explain whether the outcomes for the two coins are:

a. Independent events.

b. Complementary events.

c. Mutually exclusive events.

7.21 Suppose events A and B are mutually exclusive with $P(A) = 1/2$ and $P(B) = 1/3$.
a. Are A and B independent events? Explain how you know.
b. Are A and B complementary events? Explain how you know.

7.22 Suppose A, B, and C are all disjoint possible outcomes for the same random circumstance. Explain whether each of the following sets of probabilities is possible.
a. $P(A) = 1/3$, $P(B) = 1/3$, $P(C) = 1/3$.
b. $P(A) = 1/2$, $P(B) = 1/2$, $P(C) = 1/4$.
c. $P(A) = 1/4$, $P(B) = 1/4$, $P(C) = 1/4$.

7.23 Jill and Laura have lunch together. They flip a coin to decide who pays for lunch, and then flip a coin again to decide who pays the tip. Define a "possible outcome" to be who pays for lunch and who pays the tip, in order, for example "Jill,Jill."
a. List the simple events in the sample space.
b. Are the simple events equally likely? If not, why not?
c. What is the probability that Laura will have to pay for lunch and leave the tip?

7.24 Refer to Example 7.7 in which you and your friend Vanessa each enter a drawing for a free lunch in week 1 and again in week 2. Events defined were: A = you win in week 1, B = Vanessa wins in week 1, C = Vanessa wins in week 2.
a. Are events B and C independent? Explain.
b. Suppose that after week 1, the cards not drawn as the winning card are retained in the bowl for the drawing in week 2. Are events B and C independent? Explain.

7.25 When a fair die is tossed, each of the six sides (numbers 1 to 6) is equally likely to land face up. Two fair dice, one red and one green, are tossed. Explain whether the following pairs of events are disjoint:
a. A = red die is a 3; B = red die is a 6.
b. A = red die is a 3; B = green die is a 6.
c. A = red die and green die sum to 4; B = red die is a 3.
d. A = red die and green die sum to 4; B = red die is a 4.

7.26 Refer to the previous exercise. For each part, explain whether the two events are independent.

7.27 According to Krantz (1992, p. 102), "Women between the ages of 20 and 24 have a 90 percent chance of being fertile while women between 40 and 44 have only a 37 percent chance of bearing children [i.e., being fertile]." Define appropriate events, and write these statements as conditional probabilities, using appropriate notation.

7.28 Refer to the previous exercise. Which method of finding probabilities do you think was used to find the "90 percent chance" and "37 percent chance"? Explain.

7.29 Refer to the previous two exercises. Suppose an American woman is randomly selected. Are her age and her fertility status independent? Explain.

7.30 Refer to Case Study 7.1. Define C_1, C_2, and C_3 to be the events that Alicia is called upon to answer questions 1, 2, and 3, respectively.
a. Based on the physical situation used to select students, what is the (unconditional) probability of each of these events? Explain.
b. What is the *conditional* probability of C_3, *given* that C_1 occurred?
c. Are C_1 and C_3 independent events? Explain.

7.31 Use the information given in Case Study 7.1 and the "physical assumption" method of assigning probabilities to argue that on any given day, the probability that Alicia has to answer one of the three questions is 3/50.

Section 7.4

7.32 A fair coin is tossed three times. The event A = getting all heads, has probability = 1/8.
a. Describe in words what the event A^C is.
b. What is the probability of A^C?

7.33 Two fair dice are rolled. The event A = "getting the same number on both dice" has probability = 1/6.
a. Describe in words what the event A^C is.
b. What is the probability of A^C?

7.34 Two fair coins are tossed. Define:

A = Getting a head on the first toss.
B = Getting a head on the second toss.
A and B = Getting a head on both the first and second tosses.
A or B = Getting a head on the first toss, or the second toss, or both tosses.

a. Find $P(A)$ = the probability of A.
b. Find $P(B)$ = the probability of B.
c. Using the multiplication rule (Rule 3b), find $P(A \text{ and } B)$.
d. Using the addition rule (Rule 2a), find $P(A \text{ or } B)$.

7.35 Julie is taking English and history. Suppose at the outset of the term, her probabilities for getting A's are:

P(grade of A in English class) = .70.
P(grade of A in history class) = .60.
P(grade of A in both English and history classes) = .50.

a. Are the events "grade of A in English class" and "grade of A in history class" independent? Explain how you know.
b. Use the addition rule for two events (Rule 2) to find the probability that Julie will get at least one A between her English and history classes.

7.36 In a recent election, 55% of the voters were Republicans and 45% were not. Of the Republicans, 80% voted for Candidate X and of the non-Republicans, 10% voted for Candidate X. Consider a randomly selected voter. Define events:

A = Voter is Republican.
B = Voted for Candidate X.

a. Write values for $P(A)$, $P(A^C)$, $P(B|A)$, and $P(B|A^C)$.
b. Find $P(A \text{ and } B)$ and write in words what outcome it represents.
c. Find $P(A^C \text{ and } B)$ and write in words what outcome it represents.
d. Using the results in parts (b) and (c), find $P(B)$. (*Hint:* The events in parts (b) and (c) cover all of the ways B can happen.)
e. Use the result in part (d) to state what percent of the vote Candidate X received.

7.37 In each of the following situations, explain whether the selection is made with replacement or without replacement.
a. The three digits in the simple lottery in Example 7.2.
b. The three students selected to answer questions in Case Study 7.1.
c. The five people selected for extra security screening while boarding a particular flight.
d. The two football teams selected to play in the Rose Bowl in a given year.
e. The cars stopped by the police for speeding in five consecutive mornings on the same stretch of highway.

7.38 In Example 7.12 we found the probability that both of two unrelated strangers share your birth month. In this exercise, we find the probability that at least one of the two strangers shares your birth month. Assume all 12 months are equally likely.

a. What is the probability that the first stranger does *not* share your birth month?

b. What is the probability that the second stranger does *not* share your birth month?

c. What is the probability that *neither* of them shares your birth month?

d. Use the result from part (c) to find the probability that at least one of them shares your birth month.

7.39 Refer to the previous exercise. In this exercise another method is used for finding the probability that at least one of two unrelated strangers shares your birth month.

a. What is the probability that the first stranger shares your birth month?

b. What is the probability that the second stranger shares your birth month?

c. What is the probability that both of them share your birth month?

d. Use Rule 2a (for "either A or B") to find the probability that at least one of the strangers shares your birth month.

7.40 In Example 7.11 we found the probability that a woman with two children (not twins) has two girls or two boys is .5002. What is the probability that she has one child of each sex?

7.41 Harold and Maude plan to take a cruise together, but they live in separate cities. The cruise departs from Miami, and they each book a flight to arrive in Miami an hour before they need to be on the ship. Their travel planner explains that Harold's flight has an 80% chance of making it on time for him to get to the ship, and Maude's flight has a 90% chance of making it on time.

a. How do you think the travel planner determined these probabilities?

b. Assuming the probabilities quoted are long-run relative frequencies, do you think whether Harold's plane is on time is independent of whether Maude's plane is on time on the particular day they travel? Explain. (*Hint:* Consider reasons why planes are delayed.)

c. Whether realistic or not, assuming the probabilities are independent, what is the probability that Harold and Maude will both arrive on time?

d. What is the probability that one of the two will be cruising alone?

7.42 A robbery has been committed in an isolated town. Witnesses all agree that the criminal was driving a red pickup truck and had blond hair. From evidence at the scene, the criminal also smoked cigarettes. Police determine that 1/50 of the vehicles in town are red pickup trucks, 30% of the residents smoke, and 20% of the residents have blond hair. The next day they notice a red pickup truck whose driver has blond hair and is smoking. They arrest the driver for the robbery.

a. Discuss whether or not you think it's reasonable to assume that whether someone drives a red pickup truck, smokes cigarettes, and has blond hair are all independent traits.

b. Assuming the traits listed are all independent, what proportion of the vehicle owners in town are red pickup truck owners who smoke and have blond hair?

c. Continuing to assume the traits listed are independent, if there are 10,000 vehicle owners in town, how many of them fit the description of the criminal?

d. Assuming there is no other evidence, what is the probability that the driver arrested by the police is innocent of the robbery?

e. The prosecuting attorney argues that the probability that someone would own a red pickup truck and smoke cigarettes and have blond hair is very small, so the person arrested by the police must be guilty. Do you agree with this reasoning? Explain.

7.43 A raffle is held in a club in which 10 of the 40 members are good friends with the president. The president draws two winners.

a. If the two winners are drawn *with replacement,* what is the probability that a friend of the president wins each time?

b. If the two winners are drawn *without replacement,* what is the probability that a friend of the president wins each time?

c. If the two winners are drawn *with replacement,* what is the probability that neither winner is a friend of the president?

d. If the two winners are drawn *without replacement,* what is the probability that neither winner is a friend of the president?

Section 7.5

7.44 A public library carries 50 magazines that focus on either news or sports. Thirty of them focus on news and the remaining 20 focus on sports. Among the 30 news magazines, 20 include international news and 10 include national, state, or local news only. Among the 20 sports magazines, 5 focus on international sports and the remaining 15 focus on national, state, or local sports.

a. Create a table illustrating these numbers, with type of magazine (news, sports) as the row variable.

b. A customer randomly picks a magazine from the shelf and notices that it is a news magazine. Given that it is a news magazine, what is the probability that it includes international news?

c. What proportion of the magazines include international news or sports?

7.45 In a large general education class, 60% (.6) are science majors and 40% (.4) are liberal arts majors. Twenty percent (.2) of the science majors are seniors, while 30% (.3) of the liberal arts majors are seniors.

a. If there are 100 students in the class, how many of them are science majors?

b. If there are 100 students in the class, how many of them are science majors and seniors?

c. Create a hypothetical table of 100 students with major (science, liberal arts) as the row variable and class (senior, nonsenior) as the column variable, illustrating the proportions given in the exercise.

d. Use the table created in part (c) to determine what percent of the class are seniors.

7.46 Refer to the previous exercise.

a. Create a tree diagram for this situation.

b. Use the tree diagram in part (a) to determine what percent of the class are seniors.

7.47 In a computer store, 30% (.3) of the computers in stock are laptops and 70% (.7) are desktops. Five percent (.05) of the laptops are on sale, while 10% (.1) of the desktops are on sale. Use a tree diagram to determine what percent of the computers in the store are on sale.

7.48 In an Italian restaurant, a waitress has observed that 80% of her customers order coffee, and 25% of her customers order both biscotti and coffee. Define events:

A = a randomly selected customer orders coffee.
B = a randomly selected customer orders biscotti.

a. Express the waitress' observations as probability statements involving events A and B.
b. What is the conditional probability that a randomly selected customer would order a biscotti, given that he or she orders coffee?
c. What is the conditional probability that a customer would not order a biscotti, given that he or she orders coffee?

7.49 Two students each use a random number generator to pick a number between 1 and 7. What is the probability that they pick the same number?

7.50 A standard poker deck of cards contains 52 cards, of which four are aces. Suppose two cards are drawn sequentially, so that one random circumstance is the result of the first card drawn and the second random circumstance is the result of the second card drawn. Go through the five steps listed under "Specific Set of Steps for Finding Probabilities" to find the probability that the first card is an ace and the second card is not an ace.

7.51 Refer to the example toward the end of Section 7.4, of drawing two students without replacement from a class of 30 students in which 3 are left-handed. Draw a tree diagram to illustrate that the probability of selecting two left-handed students is 1/145.

7.52 Suppose that a magnet high school includes grades 11 and 12, with half of the students in each grade. Half of the senior class and 30% of the junior class are taking calculus. Create a hypothetical table of 100 students for this situation. Use grade (junior, senior) as the row variable.

7.53 Refer to the previous exercise. Suppose a calculus student is randomly selected to accompany the math teachers to a conference. What is the probability that the student is a junior?

7.54 An airline has noticed that 40% of its customers who buy tickets don't take advantage of "advance-purchase" fares, and the remaining 60% do. The "no-show" rate for those who don't have advance-purchase fares is 30%, while for those who do have them it is only 5%.
a. Create a tree diagram for this situation.
b. Create a "hypothetical hundred thousand" table for this situation.
c. Using the tree diagram from part (a) or the table from part (b), find the percent of customers who are "no-shows."
d. Given that a customer is a no-show, what is the probability that he or she had an advance-purchase fare?
e. Refer to the answer you found in part (d). Write a sentence presenting this result that someone with no training in statistics would understand.

7.55 In a test for ESP, a picture is randomly selected from four possible choices and hidden away. Participants are asked to describe the hidden picture, which is unknown to anyone in the environment. They are later shown the four possible choices and asked to identify the one they thought was the hidden target picture. Because the correct answer had been randomly selected from the four choices before the experiment began, the probability of guessing correctly by chance alone is 1/4 or .25. Each time the experiment is repeated, four new pictures are used.
a. Researchers are screening people for potentially gifted ESP participants for a later experiment. They will only accept people who get three correct answers in three repetitions of the screening experiment. What is the probability that someone who is just guessing will be accepted for the later experiment?
b. Refer to part (a). Suppose that an individual has some ESP ability and can correctly guess with probability .40 each time. What is the probability that such a person will be accepted for the later experiment?
c. Refer to parts (a) and (b). Suppose that half of the people in the population being tested are just guessing and have no ESP ability, while the other half can actually guess correctly with probability .40 each time. Given that someone is accepted for the upcoming experiment, what is the probability that they actually have ESP ability?

Section 7.6

7.56 Five fair dice were tossed and the sum of the resulting tosses was recorded. This process was repeated 10,000 times using a computer simulation. The number of times the sum of the five tosses equaled 27 was 45. What is the estimated probability that the sum of the five dice will be 27?

7.57 The observed risk of an accident per month at a busy intersection without any stoplights was 1%. The potential benefit of adding a stoplight was studied using a computer simulation modeling the typical traffic flow for a month at that intersection. Based on 10,000 repetitions, a total of 50 simulated accidents occurred.
a. What is the estimated probability of an accident in a month at the intersection with a stoplight added?
b. Would adding the stoplight be a good idea? Explain.

7.58 Refer to Example 7.19 and use the results given in Table 7.3 to estimate probabilities of the following outcomes:
a. Prize 4 is received at least three times.
b. Prize 4 is received at least three times *given* that the full collection of all four prizes was not received.
c. Prize 4 is received at least three times *and* the full collection of all four prizes was received.

7.59 Refer to the simulation given for Example 7.20. What is the estimated probability that a participant would guess four or more correctly?

7.60 Refer to Example 7.20. Explain how you would change the simulation procedure if the assumption was that everyone was randomly guessing, so that the probability of a correct guess was .20 instead of .30 each time.

7.61 Refer to Example 7.20. Suppose the probability of a correct guess each time is .40.
a. Explain how you would simulate this situation.
b. Carry out the simulation and estimate the probability that a participant will be identified as gifted.

7.62 Suppose that in a lottery game players choose three digits, each from the set 0 to 9. A winning sum is then selected from the possibilities 0 (0 + 0 + 0) to 27 (9 + 9 + 9) by randomly choosing three digits and using their sum. For instance, if the three digits 483 are selected, then a winning ticket is any ticket in which the 3 digits sum to $4 + 8 + 3 = 15$. Suppose the winning sum is 15.

a. Using the physical assumption that all 1000 choices $(000, 001, \ldots, 999)$ are equally likely, find the probability that three digits chosen at random will sum to 15.

b. Simulate playing this game 1000 times, randomly selecting the three digits each time. Based on the simulation, find the approximate probability of winning when the winning sum is 15.

c. Based on information about how people respond when asked to pick random digits, do you think the proportion of people who would actually pick three digits summing to 15 is close to the probabilities you found in parts (a) and (b)? Explain.

7.63 Janice has noticed that on her drive to work there are several things that can slow her down. First, she hits a red light with probability .3. If she hits the red light, she also has to stop for the commuter train with probability .4, but if she doesn't hit the red light she only has to stop with probability .2.

a. Find the probability that Janice has to stop for both the red light and the train.

b. Find the probability that Janice has to stop for the red light *given* that she has to stop for the commuter train.

c. Explain how you would conduct a simulation to estimate the probabilities in parts (a) and (b).

d. Conduct the simulation described in part (c). Use at least 50 repetitions.

e. Compare the answers you computed in parts (a) and (b) to the estimated probabilities from your simulation in part (d).

Section 7.7

7.64 The PAP smear is a screening test to detect cervical cancer. Estimate the sensitivity and specificity of the test if a study of 200 women with cervical cancer resulted in 160 testing positive, and in another 200 women without cervical cancer, 4 tested positive.

7.65 In tossing a fair coin ten times, if the first nine tosses resulted in all tails, will the chance be greater than .5 that the tenth toss will turn up heads? Explain.

7.66 Which of the following sequences resulting from tossing a fair coin five times is most likely: HHHHH, HTHHT, or HHHTT? Explain your answer.

7.67 Using material from Section 7.7, explain what is wrong with the following statement: "The probability that you will win the million dollar lottery is about the same as the probability that you will give birth to quintuplets."

7.68 The *U.C. Berkeley Wellness Encyclopedia* contains the following statement in its discussion of HIV testing (1991, p. 360): "In a high-risk population, virtually all people who test positive will truly be infected, but among people at low risk the false positives will outnumber the true positives. Thus, for every infected person correctly identified in a low-risk population, an estimated ten noncarriers [of HIV] will test positive." Suppose you have a friend who is part of this low-risk population, but who has just tested positive.

a. Using the numbers in the statement above, what is the probability that your friend actually carries the virus?

b. Understandably, your friend is upset and doesn't believe that the probability of being infected with HIV isn't really near 1. After all, the test is accurate and it came out positive! Explain to your friend how the quoted statement can be true, even though the test is very accurate both for people with HIV and for people who don't carry it. If it's easier, you can make up numbers to put in a table to support your argument.

7.69 A rare disease occurs in about 1 out of 1000 people who are similar to you. A test for the disease has sensitivity of 95% and specificity of 90%.

a. Create a "hypothetical one hundred thousand" table illustrating this situation, where the row categories are disease (yes, no) and the column categories are test results (positive, negative).

b. Draw a tree diagram for this situation.

c. Using the table from part (a) or the tree diagram from part (b), find the proportion of people who will test positive for the disease.

d. Find the probability that you actually have the disease given that your test results are positive.

7.70 Suppose there are 30 people in your statistics class and you are divided into 15 teams of 2 students each. You happen to mention that your birthday was last week, upon which you discover that your teammate's mother has the same birthday you have (month and day, not necessarily year). Assume the probability is 1/365 for any given day.

a. What is the probability that your teammate's mother would have the same birthday as yours?

b. Suppose your teammate has two siblings and two parents, for a family of size 5. What is the probability that at least one of your teammate's family has the same birthday as yours, assuming their birth dates are all independent?

c. Is the fact that your teammate's mother has the same birthday as yours a surprising coincidence? Explain.

7.71 Give an example of a situation where the gambler's fallacy would not apply because the events are not independent.

7.72 A friend has three boys and would like to have a girl. She explains to you that the probability that her next baby will be a girl is very high, because the law of averages says she should have half of each and she already has three boys. Is she correct? Explain.

7.73 A friend, quite upset, calls you because she had a dream that a building had been bombed and she was helping to search for survivors. The next day, a terrorist bombs an embassy building in another country. Your friend is convinced that her dream was a warning and that she should have told someone in authority. Comment on this situation.

7.74 Tomorrow morning when you first arise, pick a three-digit number (anything from 000 to 999). You can choose randomly or simply decide what number you want to use. As you go through the day, notice whether or not you encounter that number. It could be in a book, in a newspaper or magazine, on a digital clock when you happen to glance at it, part of the serial number on a bill, and so on. Write down all of the instances in which you encounter your chosen three-digit number. Are you surprised at how many or how few times you encountered the number? Explain.

7.75 Give an example of a coincidence that has occurred in your life. Using the material from this chapter, try to approximate the probability of exactly that event happening. Discuss whether or not the answer convinces you that something very odd happened to you.

7.76 Suppose you are seated next to a stranger on an airplane and you start discussing various topics like where you were born (what state or country), what your favorite movie of all time is, your spouse's occupation, and so on. For simplicity, assume the probability that your details "match" for any given topic is 1/50 and is independent from one topic to the next. If you discuss ten topics, how surprising would it be to find that you "match" on at least one of them?

7.77 Do you think that all coincidences can be explained by random events? Explain why or why not, using probability as the focus of your explanation. (There is no correct answer; your reasoning is what counts.)

Chapter Exercises

7.78 New spark plugs have just been installed in a small airplane with a four-cylinder engine. For each spark plug, the probability that it is defective and will fail during its first 20 minutes of flight is 1/10,000, independent of the other spark plugs.
  a. For any given spark plug, what is the probability that it will *not* fail during the first 20 minutes of flight?
  b. What is the probability that none of the four spark plugs will fail during the first 20 minutes of flight?
  c. What is the probability that at least one of the spark plugs will fail?

7.79 A psychological test identifies people as being one of eight types. For instance, Type 1 is "Rationalist" and applies to 15% of men and 8% of women. Type 2 is "Teacher" and applies to 12% of men and 14% of women. Each person fits one and only one type.
  a. What is the probability that a randomly selected male is either "Rationalist" or "Teacher"?
  b. What is the probability that a randomly selected female is not a "Teacher"?
  c. Suppose college roommates have a particularly hard time getting along with each other if they are both "Rationalists." A college randomly assigns roommates of the same sex. What proportion of male roommate pairs will have this problem?
  d. Refer to part (c). What proportion of female roommate pairs will have this problem?
  e. Using your answers to parts (c) and (d) and assuming that half of college roommate pairs are male and half are female, what proportion of all roommate pairs will both be "Rationalists"?

7.80 Refer to the previous exercise. A psychologist has noticed that "Teachers" and "Rationalists" get along particularly well with each other, and she thinks they tend to marry each other. One of her colleagues disagrees and thinks that the "types" of spouses are independent of each other.
  a. If the "types" are independent, what is the probability that a randomly selected married couple would consist of one "Rationalist" and one "Teacher"?
  b. In surveys of thousands of randomly selected married couples, she has found that about 5% of them have one "Rationalist" and one "Teacher." Does this contradict her colleague's theory that the types of spouses are independent of each other? Explain.

7.81 In Chapter 6 we learned that if there is no relationship be-

tween two categorical variables in a population, a "statistically significant" relationship will appear in samples from that population 5% of the time. Suppose two researchers independently conduct studies to see if there is a relationship between drinking coffee (regularly, sometimes, never) and having migraine headaches (frequently, occasionally, never).
  a. If there really is no relationship in the population, what is the probability that the first researcher finds a statistically significant relationship?
  b. If there really is no relationship in the population, what is the probability that both researchers find a statistically significant relationship?

7.82 A professor has noticed that, even though attendance is not a component of the grade for his class, students who attend regularly obtain better grades. In fact, 40% of those who attend regularly receive A's in the class, while only 10% of those who do not attend regularly receive A's. About 70% of students attend class regularly. Find the following percents:
  a. The percent who receive A's *given* that they attend class regularly.
  b. The percent who receive A's *given* that they do not attend class regularly.
  c. The overall percent who receive A's.

7.83 Refer to the previous exercise. Draw a tree diagram and use it to find the overall percent who receive As.

7.84 Refer to the previous two exercises. Construct a "hypothetical hundred thousand" table for this situation.

7.85 Refer to the previous three exercises. Given that a randomly chosen student receives an A grade, what is the probability that he or she attended class regularly?

7.86 Recall from Chapter 2 that the median of a dataset is the value with half of the observations at or above it and half of the observations at or below it. Suppose four individuals are randomly selected *with replacement* from a large class and asked how many hours they studied last week. Assume that there are an even number of students in the class and that nobody has a response exactly equal to the median.
  a. What is the probability that a particular individual will give a response that is above the median for the class?
  b. Use the Extension of Rule 3b to find the probability that all four individuals will give a response that is above the median for the class.
  c. If the individuals had been selected without replacement, would the Extension of Rule 3b still be applicable? Explain.

7.87 About 1/3 of all adults in the United States have Type O+ blood. If three randomly selected adults donate blood, find the probability of each of these events:
  a. All three are Type O+.
  b. None of them is Type O+.
  c. Two out of the three are Type O+.

7.88 In October 1999, experts at the U.S. Geological Survey's San Francisco Bay Area Earthquake Hazards Project announced that "the likelihood of a powerful earthquake occurring in the next 30 years in the Bay Area is 70 percent" (*Sacramento Bee*, Eddie Lau, Oct. 15, 1999, p. A1). Do you think this figure is based on personal probability, long-run relative frequency, physical assumptions about the world, or some combination? Explain.

7.89 According to the U.S. Census Bureau "only 35 percent of the foreign-born people in the United States in 1997 were naturalized citizens, compared with 64 percent in 1970" (*Sacramento Bee,* Oct. 15, 1999, p. A1).

   a. What is the probability that two randomly selected foreign-born people in the United States in 1997 were both naturalized citizens?

   b. Refer to part (a) of this exercise. Suppose a married couple was randomly selected instead of two separate individuals. Do you think the probability that they were both naturalized citizens would be the same as the probability in part (a)? Explain.

   c. Do you think the probabilities reported by the Census Bureau are relative frequency probabilities or personal probabilities? Explain.

   d. A student writing a report about these statistics wrote, "If you had lived in the United States in 1970 and were foreign-born, the probability that you would be a naturalized citizen would have been .64. But by 1997 if you still lived in the United States, the probability that you would be a naturalized citizen would be only .35." Rewrite this sentence in a way that conveys the information correctly.

7.90 Recall that the Empirical Rule in Chapter 2 stated that for bell-shaped distributions about 68% of the values fall within 1 standard deviation of the mean. The heights of women at a large university are approximately bell-shaped, with a mean of 65 inches and standard deviation of 2.5 inches. Use this information to answer the following questions:

   a. What is the probability that a randomly selected woman from this university is 67.5 inches or taller?

   b. What is the probability that two randomly selected women from this university are both 62.5 inches or shorter?

   c. What is the probability that of two randomly selected women, one is 62.5 inches or shorter and the other is 62.5 inches or taller?

   d. What is the probability that two randomly selected women are both 65 inches or taller?

The following scenario applies to the remaining exercises. In Chapter 6 we discussed a random sample of people who had ever been married and demonstrated the proportions who smoked and who had ever been divorced. The numbers are shown again here.

Smoking and divorce, GSS surveys 1991–1993

Smoke?	Ever Divorced?		
	Yes	*No*	*Total*
Yes	238	247	485
No	374	810	1184
Total	612	1057	1669

Data Source: SDA archive at UC Berkeley website (http://csa.berkeley.edu:7502/).

Because this survey was based on a random sample in the United States in the early 1990s, the data should be representative of the adult population who had ever been married at that time. Answer these questions for a randomly selected member of that population.

7.91 Draw a tree diagram illustrating this situation, where the first set of branches represents smoking status and the second set represents "ever divorced."

7.92 What is the approximate probability that the person smoked?

7.93 What is the approximate probability that the person had ever been divorced?

7.94 Given that the person smoked, what is the probability that he or she had been divorced?

7.95 Given that the person had been divorced, what is the probability that he or she smoked?

7.96 Suppose two people were randomly selected from that population, without replacement. What is the probability that they had both been divorced?

7.97 Suppose two people were randomly selected from that population, without replacement. What is the probability that one of them smoked but the other one did not (in either order)?

7.98 Suppose two people were randomly selected from that population. Given that one of them had been divorced, what is the probability that the other one had also been divorced?

REFERENCES

Diaconis, P., and F. Mosteller (1989). "Methods for studying coincidences," *Journal of the American Statistical Association,* Vol. 84, pp. 853–861.

Eddy, D. M. (1982). "Probabilistic reasoning in clinical medicine: Problems and opportunities," Chapter 18 in D. Kahneman, P. Slovic, and A. Tversky (eds.), *Judgment under Uncertainty: Heuristics and Biases,* Cambridge, U.K.: Cambridge University Press.

Gehring, W. J., and A. R. Willoughby (2000). "The medial frontal cortex and the rapid processing of monetary gains and losses," *Science,* Vol. 295, pp. 2279–2284.

Information Please Almanac (1991). Otto Johnson (ed.), Boston: Houghton Mifflin.

Krantz, Les (1992). *What the Odds Are,* New York: Harper Perennial.

Moore, D. S. (1997). *Statistics: Concepts and Controversies,* 4th ed., New York: W. H. Freeman.

Paulos, John Allen (1995). *A Mathematician Reads the Newspaper,* New York: Basic Books.

Plous, S. (1993). *The Psychology of Judgment and Decision Making,* New York: McGraw-Hill.

Stinchfield, R. (2001). "A comparison of gambling by Minnesota public school students in 1992, 1995 and 1998," *Journal of Gambling Studies,* Vol. 17, No. 4, pp. 273–296.

Tversky, A., and D. Kahneman (1982). "Judgment under uncertainty—Heuristics and biases," Chapter 1 in D. Kahneman, P. Slovic, and A. Tversky (eds.), *Judgment under Uncertainty: Heuristics and Biases,* Cambridge, U.K.: Cambridge University Press.

University of California, Berkeley (1991). *The Wellness Encyclopedia,* Boston: Houghton Mifflin.

Weaver, W. (1963). *Lady Luck: The Theory of Probability,* Garden City, NY: Doubleday.

CHAPTER 8

Random Variables

Joseph Sohm; ChromoSohm Inc./CORBIS

What's random about this event? See Example 8.1.

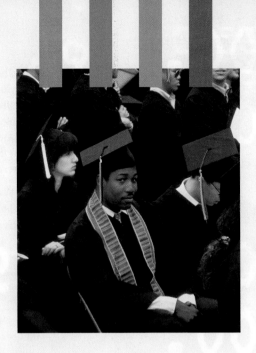

If a veterinarian had to learn a different pattern for treating every different breed, it might be impossible for any individual to learn enough to be able to treat dogs in general. Similarly, situations involving uncertainty and probability fall into certain broad classes, and we can use the same set of rules and principles for all situations within a class.

The numerical outcome of a random circumstance is called a *random variable.* In this chapter, we'll learn how to characterize the pattern of the distribution of the values that a random variable may have, and we'll learn how to use the pattern to find probabilities.

Patterns make life easier to understand and decisions easier to make. For instance, dogs come in a variety of breeds, sizes, and temperaments, but all dogs fit certain patterns that veterinarians can rely upon when treating nearly any type of dog. If a veterinarian had to learn a different pattern for treating every different breed, it might be nearly impossible for any individual to learn enough to be able to treat dogs in general.

Similarly, situations involving uncertainty and probability fall into certain broad classes, and we can use the same set of rules and principles for all situations within a class. We encountered this idea in Chapter 2 when we learned the Empirical Rule for bell-shaped distributions. All variables that follow a bell-shaped distribution have the same pattern, which is that about 68% of the values fall within 1 standard deviation of the mean, about 95% of the values fall within 2 standard deviations of the mean, and so on. This pattern holds whether the variable is heights of adult males, handspans of college-age females, or SAT scores of high school seniors. ◆

8.1 | WHAT IS A RANDOM VARIABLE?

definition A **random variable** assigns a number to each outcome of a random circumstance, or, equivalently, a random variable assigns a number to each unit in a population.

It is easier to create rules for broad classes of situations and then identify how a specific example fits into a class than it is to create rules for each specific example. We can employ this strategy quite effectively for working with a wide variety of situations involving probability and random outcomes. We categorize these situations by defining a generic numerical outcome, or "random variable," for similar random circumstances. Identifying the type of random variable appropriate in a given situation makes it easy to find probabilities and other information that would be difficult to derive from first principles.

There are two different broad classes of random variables:

1. A **continuous random variable** can take any value in an interval or collection of intervals.
2. A **discrete random variable** can take one of a countable list of distinct values.

Discrete or Continuous?

> *tech note*   Sometimes a random variable fits the technical definition of a discrete random variable but it is more convenient to treat it as a continuous random variable. Examples include incomes, prices, and exam scores. Sometimes continuous random variables are rounded off to whole units, giving the appearance of a discrete random variable, such as age in years or pulse rate to the nearest beat. In most of these situations, the number of possible values is large and we are more interested in probabilities concerning intervals than specific values, so the methods for continuous random variables will be used.

EXAMPLE 8.1
Random Variables at an Outdoor Graduation or Wedding

Suppose you are participating in a major outdoor event, like a graduation or wedding ceremony. There are a number of random factors that will determine how enjoyable the event will be, such as the temperature and the number of airplanes that fly overhead during the important speeches. If you are responsible for setting the date for the event, you might try to get some information on likely values for these random outcomes on different dates, so you can optimize the likelihood of an enjoyable event. In this context, *temperature* is a *continuous random variable* because it can take on any value in an interval. We often round off continuous random variables to the nearest whole number, like temperature in degrees, but conceptually the value can be anything in an interval. In contrast, the *number of airplanes* that fly overhead during the event is a *discrete random variable.* The value can be 0, 1, 2, 3, and so on but cannot be anything in between the integers, like 2.5 airplanes. ◆

EXAMPLE 8.2
Probability an Event Occurs Three Times in Three Tries

As an example of a class of random variables, think about what the following three questions have in common:

◆ What is the probability that three tosses of a fair coin will result in three heads?

◆ Assuming boys and girls are equally likely each time (which we learned in Chapter 7 is not quite true), what is the probability that three births will result in three girls?

◆ Assuming the probability is 1/2 that a randomly selected individual will be taller than the median height of a population, what is the probability that three randomly selected individuals will all be taller than the median?

These three questions all have the same answer, which is 1/8, found using the multiplication rule in Chapter 7. Furthermore, if each question had asked about the probability of getting one outcome of one type and two of the other type, such as one head and two tails or one boy and two girls, the answers to all three questions would also have been the same. It would have been 3/8.

Probabilities for the number of heads in three tosses, girls in three births, and individuals falling above the median in three random selections are all found using the exact same process. In each case, a random variable can be defined as X = number of times the "outcome of interest" occurs in three independent tries. It doesn't matter if the "outcome of interest" is a head on a coin toss, a girl in a birth, or someone taller than the median in a randomly selected male. The pattern for finding probabilities will be the same. ◆

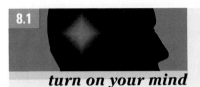

8.1

turn on your mind

If you know that the number of possible values a random variable can have is finite, do you know whether the random variable is discrete or continuous? Answer the same question for a random variable that can have an infinite number of possible values.*

8.2 | DISCRETE RANDOM VARIABLES

We consider discrete and continuous random variables separately because probabilities are computed and used differently for them. For discrete random variables, we are interested in probabilities of exact outcomes or a series of them. For continuous random variables, we are interested in the probability of the outcome falling into a specific interval. For instance, suppose you are waiting to fly on standby and you have first priority. You would want to know the probability that the number of standby passengers who will be able to board the plane (a discrete random variable) is one or more. You might also want to know the probability that the flying time (a continuous random variable) will be no more than the time specified in the flight schedule. The probability that the flying time is *exactly* that amount would be essentially zero, and it doesn't make sense to talk about probabilities of exact values for continuous random variables.

Probability Notation for Discrete Random Variables

> **formula** X = the random variable.
>
> k = a number the discrete random variable could assume.
>
> $P(X = k)$ is the probability that X equals k.

Remember that a *discrete random variable* is one that can only result in a countable set of possibilities. Although a discrete random variable often has a finite number of outcomes, that is not always the case. Example 8.3 illustrates a discrete random variable with an infinite number of possible values.

*HINT: An interval has an infinite number of values in it. Also, see Example 8.3.

EXAMPLE 8.3
It's Possible to Toss Forever

Consider repeatedly tossing a fair coin, and define the random variable:

X = number of tosses until the first head occurs

Notice that, theoretically, X could equal any value from 1 to infinity because the coin could continue to come up tails indefinitely. The number of tosses required to get the first head is not likely to be large, but any number of flips is a possible outcome.

To find the probability that k tosses are needed to get the first head, we can use the multiplication rule in Chapter 7 (extension of Rule 3b):

$$P(X = k) = (1/2)^k$$

For instance, $P(X = 1) = 1/2$, $P(X = 2) = (1/2)(1/2) = 1/4$, and so on. The same pattern of probabilities applies to the number of births until the first girl occurs (assuming boys and girls are equally likely), the number of people randomly selected until the first person taller than the population median is found, and so on. ◆

The Probability Distribution of a Discrete Random Variable

The purpose of defining a random variable usually is to find probabilities associated with its possible values. In Chapter 7 we learned that the sample space for a random circumstance consists of a list of the unique possible outcomes, called simple events. When a random variable X is defined, a value for X is given to each simple event in the sample space. The probability for any particular value of X can then be found by adding the probabilities for all of the simple events that have that value of X.

Using the Sample Space to Find Probabilities for Discrete Random Variables

Step 1: List all simple events in the sample space.

Step 2: Find the probability for each simple event (often they are equally likely).

Step 3: List the possible values for the random variable X and identify the value for each simple event.

Step 4: Find all simple events for which $X = k$, for each possible value k.

Step 5: $P(X = k)$ is the sum of the probabilities for all simple events for which $X = k$.

EXAMPLE 8.4
How Many Girls Are Likely?

If a family has three children and the probability of a girl is 1/2 for each birth,[1] what are the probabilities of having 0, 1, 2, or 3 girls? We could use the rules in Chapter 7 to solve this problem, but to organize the solution we list the sample space, assign probabilities to each simple event, and then, for each simple event, determine a value of the random variable "number of girls":

Sample Space   For each birth, write either B for boy or G for girl. There are eight possible arrangements of B and G for three births. These are the *simple events*.

Sample Space and Probabilities   The eight simple events are equally likely.

---

[1] In Chapter 7 we learned that P(girl) is about .488, but for simplicity here we use .5.

Random Variable X = number of girls in three births. For each simple event, the value of the random variable X is the number of Gs listed. The following table displays the value of X for each simple event.

Simple Event	BBB	BBG	BGB	GBB	BGG	GBG	GGB	GGG
Probability	1/8	1/8	1/8	1/8	1/8	1/8	1/8	1/8
X = # of Girls	0	1	1	1	2	2	2	3

We can find the probability for each value of X by adding the probabilities of the simple events that have that value. The probabilities for the possible values of X are

$P(X = 0) = 1/8$

$P(X = 1) = 3/8$ (added probabilities for the simple events with $X = 1$)

$P(X = 2) = 3/8$ (added probabilities for the simple events with $X = 2$)

$P(X = 3) = 1/8$

Notice that these probabilities sum to 1 over all possible values of X. ◆

---

definition The **probability distribution function (pdf)** for a discrete random variable X is a table or rule that assigns probabilities to the possible values of the random variable X.

It is often useful to represent a probability distribution function wih a picture similar to a histogram or bar graph. The possible values are placed on the horizontal axis and their probabilities on the vertical axis. A bar is drawn centered on each possible value, with the height of the bar equal to the probability for that value. Sometimes the bars are scaled so that their total area is 1. Then the probability for any combination of values can be found as the area of the bars (rectangles) for those values. This method will be displayed in the next example.

EXAMPLE 8.4—CONTINUED
The Probability Distribution for Number of Girls

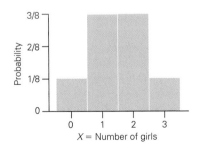

FIGURE 8.1   **The pdf for X = Number of Girls in Three Births**

The probabilities for all possible values of X = number of girls in three births make up the probability distribution of X. These probabilities can be written in the table:

k	0	1	2	3
$P(X = k)$	1/8	3/8	3/8	1/8

Figure 8.1 displays the pdf. Notice that the bar over each value forms a rectangle that has a width of 1 and a height equal to the probability for that value. Therefore, the area of each bar is equivalent to the probability for that value. For instance, $P(X = 2) = 3/8$ coincides with the area of the rectangle centered on $X = 2$, a rectangle that has height = 3/8 and width = 1. Later in this chapter, you will learn that equating probability to an area is a crucial and necessary idea when working with continuous variables. For discrete variables, however, we

generally can use more direct ways to find the probability that X takes a particular value, so, for discrete variables, we usually don't have to find the area of a rectangle to find a probability. ◆

Conditions for Probabilities for Discrete Random Variables

Two conditions must always apply to the probabilities for discrete random variables:

> *Condition 1*   The sum of the probabilities over all possible values of a discrete random variable must equal 1. Stated mathematically
>
> $$\sum_k P(X = k) = 1$$
>
> *Condition 2*   The probability of any specific outcome for a discrete random variable must be between 0 and 1. Stated mathematically
>
> $$0 \le P(X = k) \le 1 \qquad \text{for any value } k$$

The Cumulative Distribution Function of a Discrete Random Variable

Sometimes the question of interest involves the probability that a random variable is less than or equal to a specific value. For instance, imagine that you take a quiz that has ten multiple-choice questions. You are not prepared so you have to guess on every question. Suppose that if you get six or fewer correct, you fail the test. Define the random variable X = number correct. The probability of interest to you is $P(X \le 6)$.

> **definition**   The **cumulative distribution function (cdf)** for a random variable X is a rule or table that provides the probabilities $P(X \le k)$ for any real number k. Generally, the term **cumulative probability** refers to the probability that X is less than or equal to a particular value.

EXAMPLE 8.4—CONTINUED
Cumulative Distribution for the Number of Girls

The cumulative distribution function for the number of girls in a family with three children is

k	0	1	2	3
$P(X \le k)$	1/8	4/8	7/8	1

For example, the probability of two or fewer girls among three children is $P(X \le 2) = 7/8$. Notice that this is simply $P(X = 0) + P(X = 1) + P(X = 2) = 1/8 + 3/8 + 3/8$. ◆

Finding Probabilities for Complex Events Based on Random Variables

The probability rules in Chapter 7 all apply to events defined for random variables and can be used to find probabilities of complicated events.

EXAMPLE 8.4—CONTINUED
A Mixture of Children

What is the probability that a family with three children will have at least one child of each sex? In Example 8.4 we defined the random variable X = number of girls in three children. If there is at least one child of each sex, then either the family has one girl and two boys ($X = 1$) or two girls and one boy ($X = 2$). The probability can be found as

$$P(X = 1 \text{ or } X = 2) = P(X = 1) + P(X = 2) = 6/8 = 3/4$$

Notice that we can simply add these probabilities because the events $X = 1$ and $X = 2$ are mutually exclusive events. A family with three children cannot have both exactly one girl ($X = 1$) and exactly two girls ($X = 2$). ◆

8.2

turn on your mind

In the preceding example, we added the probabilities of $X = 1$ and $X = 2$ because they were mutually exclusive events. For any discrete random variable X, is it always true that $X = k$ and $X = m$ are mutually exclusive events, where k and m represent two values that X can have? Explain.*

EXAMPLE 8.5
Two Dice

Many gambling games use dice in which the number of dots rolled is equally likely to be 1, 2, 3, 4, 5, or 6. For instance, craps is a game in which two dice are rolled and the dots on the two dice determines whether the gambler wins, loses, or continues rolling the dice. Let X = sum of two fair dice. Here is the sample space for the two dots and the corresponding values of the random variable X:

Dots	X	Dots	X	Dots	X	Dots	X	Dots	X	Dots	X
1,1	2	2,1	3	3,1	4	4,1	5	5,1	6	6,1	7
1,2	3	2,2	4	3,2	5	4,2	6	5,2	7	6,2	8
1,3	4	2,3	5	3,3	6	4,3	7	5,3	8	6,3	9
1,4	5	2,4	6	3,4	7	4,4	8	5,4	9	6,4	10
1,5	6	2,5	7	3,5	8	4,5	9	5,5	10	6,5	11
1,6	7	2,6	8	3,6	9	4,6	10	5,6	11	6,6	12

There are 36 equally likely simple events in the sample space, so each simple event has probability 1/36. To construct the probability distribution (pdf) for X, count the number of simple events resulting in each value of X and divide by 36.

The probability distribution for the sum of two dice is

k	2	3	4	5	6	7	8	9	10	11	12
$P(X = k)$	1/36	2/36	3/36	4/36	5/36	6/36	5/36	4/36	3/36	2/36	1/36

Figure 8.2 displays the pdf with the rectangles for $X = 4, 5$, and 6 shaded. We can find $P(a \le X \le b)$ by adding the areas of the rectangles over those values, or by

*HINT: See step 4 on page 252. Are there simple events for which $X = k$ and $x = m$ at the same time?

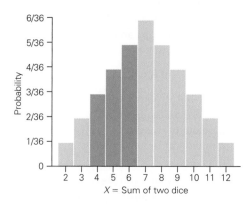

FIGURE 8.2   **Probability Distribution Function for Sum of Two Dice Showing $P(4 \leq X \leq 6)$**

adding the probabilities for all individual values in that range. For instance, $P(4 \leq X \leq 6) = 3/36 + 4/36 + 5/36 = 12/36 = 1/3.$ ◆

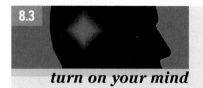

8.3

Refer to the probability distribution for the sum of two dice. What is the value of $P(X \leq 4)$? What does this probability measure? Explain what is measured by the value of $1 - P(X \leq 4)$.*

turn on your mind

8.3 | EXPECTATIONS FOR RANDOM VARIABLES

Generally, for any random variable in which the probabilities of the outcomes are known, the average value over the long run can be predicted. Suppose you play a lottery every week. Although the specific outcome for a given week cannot be predicted in advance, the average amount you can expect to win or lose per play over the long run can be predicted accurately. This long-run prediction is provided by the *expected value* or, synonymously, the **mean value of the random variable** that describes the possible outcomes on each play.

> *definition*   The **expected value** of a random variable is the mean value of the variable X in the sample space, or population, of possible outcomes. *Expected value* can also be interpreted as the mean value that would be obtained from an infinite number of observations of the random variable.

For a discrete random variable, calculating the expected value, or mean value, is simple and the logic of the calculation is somewhat intuitive. Imagine, for instance, a gambling game in which on any play, the probabilities are .3 that a player wins $2 and .7 that a player loses $1. In ten plays of the game, the relative frequencies of wins and losses will not necessarily correspond exactly with the theoretical probabilities, but to develop an intuitive feel for expected value let's suppose that ten games go exactly according to theory.

Because the probabilities for winning and losing are .3 and .7, respectively, theory suggests that ten games would result in three wins and seven losses. With three wins of $2 and seven losses of $1, the total money won and lost is $2 \times 3 + (-\$1) \times 7$, and the average result per play is

$$\frac{\$2 \times 3 + (-\$1) \times 7}{10} = \frac{-\$1}{10} = -\$0.10$$

This is the *expected value* of X = amount won or lost per play, and in this instance, the outlook isn't good for the player. If the game goes according to theory, which it will in the long run, a player can expect to lose an average of 10 cents per play.

To do our calculation, we assumed ten plays would follow theory exactly, but in fact, the calculation of the expected value of a random variable is not tied to any particular number of plays. We could use the strategy described in the previous paragraph for any arbitrary number of plays, and the answer would still be $-\$0.10$ per play. The probability distribution of a discrete random variable determines the expected value of the variable. In our imagined gambling game, the expected value can be computed as $\$2 \times .3 + (-\$1) \times .7 = -\$0.10$, which is the result of summing up "value $\times$ probability" over all possible values of the random variable.

Calculating Expected Value for a Discrete Random Variable

> ***formula*** If X is a random variable with possible values $x_1, x_2, x_3, \ldots$, occurring with probabilities $p_1, p_2, p_3, \ldots$, then the **expected value** of X is calculated as
>
> $$E(X) = \sum x_i p_i$$
>
> ◆ The formula says to calculate "value times probability" separately for each possible value of X, and then calculate the sum of these "value times probability" quantities.
>
> ◆ The Greek letter "mu" (μ) can be used to represent the mean or expected value of X. In other words, $\mu = E(X)$.

EXAMPLE 8.6
California Decco Lottery Game

The California lottery has offered a number of games over the years. One such game is Decco, in which a player chooses one card from each of the four suits in a regular deck of playing cards. For example, the player might choose "4 of hearts," "3 of clubs," "10 of diamonds," and "jack of spades." A winning card is then drawn from each suit. If one or more of the choices matches the winning cards drawn, a prize is awarded. It costs $1.00 for each play, so the net gain for any prize is $1.00 less than the prize. Define the random variable to be X = net gain for any single ticket. Table 8.1 shows the prizes, net gain, and probability of winning each prize, taken from the back of a Decco game card. Notice that we count the free ticket as an even trade because it is worth $1.00, the same amount it cost to play the game, but see the technical note following this example for further explanation.

TABLE 8.1 ■ Probabilities for the Decco Game

Number of Matches	Prize	X = Net Gain	Probability
4	$5,000	$4,999	.000035
3	$50	$49	.00168
2	$5	$4	.0303
1	Free ticket	0	.242
0	None	−$1	.726

How much would you win or lose per ticket in this game, over the long run? The expected value of X, denoted as $E(X)$, is the answer, and the calculation is

$$E(X) = (\$4{,}999 \times .000035) + (\$49 \times .00168) + (\$4 \times .0303)$$
$$+ (\$0 \times .242) + (-\$1 \times .726)$$
$$= -\$0.35$$

This tells us that over many repetitions of the game, players will *lose* an *average* of 35 cents per play. From the perspective of the Lottery Commission, this means that they pay out about 65 cents in prizes for each $1 ticket they sell for this game. ◆

Free Tickets Are Not Worth the Price

tech note The situation is actually worse than presented in Example 8.6. Many lottery games give a free ticket as a prize, and you may think the value of that ticket is equal to its cost, usually $1.00. It's not, unless you can sell the ticket to someone else for that price. The real value of the ticket is the expected win for one play of the game. In the computation in Example 8.6, for simplicity we used a "net gain" of 0 when the prize was a free ticket, but because it is actually worth less than that, the average loss per play is even more than $.35. We will continue to use an expected loss of $.35 for simplicity in discussions of the Decco game.

Notice that the expected value in Example 8.6 is not one of the possible outcomes, so the phrase "expected value" is a bit misleading. In other words, the expected value may not be a value that is ever expected on a single random outcome. Instead, it is the average over the long run.

8.4

turn on your mind

Suppose the probability of winning in a gambling game is .001, and when a player wins, his or her net gain is $999. When a player loses, the net amount lost is $1 (the cost to play). Is this game fair? Why or why not? How would you define a fair game? Does the number of times the game is played affect your view of whether the game is fair? Explain.*

Standard Deviation for a Discrete Random Variable

We first encountered the idea of standard deviation in Chapter 2, where we learned how to compute the standard deviation as a measure of spread for a quantitative variable and then used it in the Empirical Rule. Similarly, the **standard deviation of a discrete random variable** can be useful to quantify how spread out the possible values of a discrete random variable might be, weighted by how likely each value is to occur. As with the standard deviation of a set of measurements, the standard deviation of a random variable is essentially the average distance the random variable falls from its mean, or expected value, over the long run.

Variance and Standard Deviation of a Discrete Random Variable

> *formula*   If X is a random variable with possible values of $x_1, x_2, x_3, \ldots$, occurring with probabilities $p_1, p_2, p_3, \ldots$, and with expected value $E(X) = \mu$, then,
>
> **Variance of X** $= V(X) = \sigma^2 = \Sigma(x_i - \mu)^2 p_i$
>
> **Standard deviation of X** $=$ square root of $V(X) = \sigma = \sqrt{\Sigma(x_i - \mu)^2 p_i}$
>
> The sum is taken over all possible values of the random variable X.

EXAMPLE 8.7
Stability or Excitement— Same Mean, Different Standard Deviations

Suppose you decide to invest $100 in a scheme you hope will make some money. You have two choices of investment plans, and you must decide which one to choose. For each $100 invested under the two plans, the possible net gains one year later and their probabilities are as shown:

Plan 1		Plan 2	
$X =$ Net Gain	Probability	$Y =$ Net Gain	Probability
$5,000	.001	$20	.3
$1,000	.005	$10	.2
$0	.994	$4	.5

Examine the possible gains and associated probabilities. Which plan would you choose? Let's calculate the expected value for each plan. The expected value will tell us what the average long-run gain (loss?) per investment is for a particular plan.

For Plan 1,

$$E(X) = \$5,000 \times (.001) + \$1,000 \times (.005) + \$0 \times (.994) = \$10.00$$

For Plan 2,

$$E(Y) = \$20 \times (.3) + \$10 \times (.2) + \$4 \times (.5) = \$10.00$$

In other words, the expected net gain is the same for the two plans.

What's different about the two plans is the variability in the possible net gains. Under the first plan, you could make a lot of money, but with a low probability. Most likely, you will make nothing at all. Under the second plan, you will be sure to make a small amount, but there is no possibility of a big gain. This difference in the variability in possible amounts under the two plans can be quantified by finding the standard deviations.

TABLE 8.2 ■ Standard Deviation Calculations for Example 8.7

Plan 1			Plan 2		
$(X - \mu)^2$	p	$(X - \mu)^2 p$	$(Y - \mu)^2$	p	$(Y - \mu)^2 p$
($5,000 - $10)^2 = $24,900,100	.001	$24,900.1	($20 - $10)^2 = $100	.3	$30
($1,000 - $10)^2 = $980,100	.005	$4,900.5	($10 - $10)^2 = 0	.2	0
($0 - $10)^2 = 100	.994	$99.4	($4 - $10)^2 = $36	.5	$18

The calculations of the standard deviations are displayed in Table 8.2. Summing the third column under each plan provides the variances. Taking the square root of a variance provides the standard deviation of the net gain for a plan.

For Plan 1,

$$V(X) = \$29,900.00 \qquad \text{and} \qquad \sigma = \sqrt{29,900.00} = \$172.92$$

For Plan 2,

$$V(Y) = \$48.00 \qquad \text{and} \qquad \sigma = \sqrt{48} = \$6.93$$

These values demonstrate that the possible outcomes for Plan 1 are much more variable than the possible outcomes for Plan 2. If you wanted to invest cautiously, you would choose Plan 2, but if you wanted to have the chance to gain a large amount of money, you would choose Plan 1. ◆

Expected Value (Mean) and Standard Deviation for a Population

When we have a population of measurements, we can think of each individual measurement as the value of a random variable, and we can calculate the mean and standard deviation in either of two ways. We can create a probability distribution for the measurements and use the definitions of expected value and standard deviation in this section. Or, we can find the mean and standard deviation by using the individual measurements directly in the formulas of Chapter 2. The two methods will give the same answers.

TABLE 8.3 ■ Hours of Study Yesterday for a Population of College Students

X = Hours of Study	Number of Students	Probability
0	16	16/200 = .08
1	28	28/200 = .14
2	50	50/200 = .25
3	36	36/200 = .18
4	26	26/200 = .13
5	22	22/200 = .11
6	14	14/200 = .07
7	6	6/200 = .03
8	2	2/200 = .01
Total	200	1.00

EXAMPLE 8.8
Mean Hours of Study for the Class Yesterday

A class of $N = 200$ students is asked, "To the nearest hour, how many hours did you study yesterday?" The instructor is interested only in the responses of the students in his class, so the 200 students in the class constitute a population. The probability distribution of the responses is shown in Table 8.3.

What is the mean number of hours studied the previous day in this population? The answer can be found using the formula $E(X) = \Sigma x_i p_i$, and the calculation is

$$E(X) = (0 \times .08) + (1 \times .14) + (2 \times .25) + (3 \times .18) + (4 \times .13)$$
$$+ (5 \times .11) + (6 \times .07) + (7 \times .03) + (8 \times .01)$$
$$= 2.96 \text{ hours}$$

This mean can also be calculated in the usual way that we average 200 individual values, which is that we can total the 200 individual values and divide that total by $N = 200$. ◆

Mean and Standard Deviation for a Population

formula Suppose a population has N individuals and a measurement X is of interest. Define:

k_i = value of X for individual i.

$x_1, x_2, x_3, \ldots$ as the distinct possible values for the measurement X.

$p_1, p_2, p_3, \ldots$ as the proportions of the population with the values $x_1, x_2, x_3, \ldots$.

Then, the population mean and standard deviation are

$$E(X) = \mu = \frac{1}{N}\sum k_i = \sum x_i p_i$$

$$\text{Standard deviation of } X = \sigma = \sqrt{\frac{\sum(k_i - \mu)^2}{N}} = \sqrt{\sum(x_i - \mu)^2 p_i}$$

8.4 | BINOMIAL RANDOM VARIABLES

We learned in Section 8.2 that a random variable is either *discrete* or *continuous*. Within each of these broad types, certain classes of random variables are so common that they also have a name. There are formulas for the probability distribution functions of these classes, and many computer programs and calculators provide probabilities for them.

In this section, we consider an important class of *discrete* random variables. A variable in this class is called a *binomial random variable*. Certain conditions must be met for a variable to fall into this class, but the basic idea is that a binomial random variable is a count of how many times an event occurs (or does not occur) in a particular number of observations or trials that make up a random circumstance. Other types of discrete random variables are discussed in Chapter 17.

Binomial Experiments and Binomial Random Variables

The number of heads in three tosses of a fair coin, the number of girls in six independent births, and the number of men who are six feet tall or taller in a random sample of ten adult men from a large population are all examples of binomial random variables. A binomial random variable is the result of a binomial experiment.

definition A **binomial experiment** is defined by the following conditions:

1. There are n "trials" where n is determined in advance and is not a random value.
2. There are two possible outcomes on each trial, called "success" and "failure" and denoted S and F.
3. The outcomes are independent from one trial to the next.
4. The probability of a "success" remains the same from one trial to the next, and this probability is denoted by p. The probability of a "failure" is $1 - p$ for every trial.

A **binomial random variable** is defined as X = number of successes in the n trials of a binomial experiment. Each row in Table 8.4 contains an example of a

TABLE 8.4 ■ Examples of Binomial Experiments and Binomial Random Variables

Random Circumstance	Random Variable	Success	Failure	n	p
(1) Toss three fair coins	X = number of heads	Head	Tail	3	1/2
(2) Roll a die eight times	X = number of 4s and 6s	4, 6	1, 2, 3, 5	8	2/6 = 1/3
(3) Randomly sample 1000 U.S. adults	X = number who have seen a UFO	Seen UFO	Have not seen UFO	1000	Proportion of all adults who have seen a UFO
(4) Roll two dice once	X = number of times sum is 7	Sum is 7	Sum not 7	1	6/36 = 1/6

binomial experiment and a binomial random variable. For each example, go through the conditions for a binomial experiment and make sure you understand why those conditions apply to that example.

The following four examples illustrate a few subtle features of binomial random variables that could initially be confusing. Keep these points in mind when trying to determine whether a random variable fits the binomial description:

- *There may be more than two possible outcomes for each trial,* but the random variable counts how many times a particular subset of the possibilities occurs. Anything in that subset is a "success" and anything not in the subset is a "failure." For example, a single die can display either 1, 2, 3, 4, 5, or 6 dots, but a success can be defined by a particular set of them, while a failure is anything else. Look at the second example in Table 8.4.

- *Sample surveys can produce a binomial random variable* when we count how many individuals in the sample have a particular opinion or trait (see the third example in Table 8.4). A "trial" is one sampled individual. A "success" is that the individual has the opinion or trait. The probability of "success" is the proportion in the population who have the opinion or trait. If a random sample is taken without replacement from a large population, the conditions of a binomial experiment are considered to be met, although the probability of a "success" actually changes very slightly from one trial to the next as each sampled individual is removed from eligibility.

- *Sampling without replacement from a small population does not produce a binomial random variable.* Suppose a class consists of ten boys and ten girls. Five children are randomly selected to be in a play. X = the number of girls selected. Notice that X is *not* a binomial random variable because the probability that a girl is selected each time depends on who is already in the sample and who is left, violating the condition that the probability of "success" must remain the same on each trial. X is an example of a *hypergeometric random variable,* discussed in Chapter 17.

- *Any individual random circumstance can be treated as a binomial experiment* with n = 1 and p = probability of a particular outcome. In this case, the random variable X is either 0 or 1, and the random variable may also be called a **Bernoulli random variable.** The fourth example in Table 8.4 illustrates such a variable.

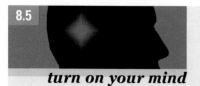

8.5

turn on your mind

The word *binomial* is from the Latin *bi* = "two," and *nomen* = "name." Explain why the word *binomial* is appropriate for a binomial random variable.*

Finding Probabilities for Binomial Random Variables

For a binomial random variable, the probabilities for the possible values of X are given by the formula

$$P(X = k) = \frac{n!}{k!(n - k)!} p^k(1 - p)^{n-k} \qquad \text{for } k = 0, 1, 2, \ldots, n$$

The formula for $P(X = k)$ is made up of two parts.

1. The first part, $n!/k!(n - k)!$, gives the number of simple events in the sample space (consisting of all possible listings of successes and failures in n trials) that have exactly k successes. The notation $n!$ is read "n-factorial," and it is the product of the integers from 1 to n. For instance, $3! = 1 \times 2 \times 3 = 6$. By convention, $0! = 1$.

2. The second part, $p^k(1 - p)^{n-k}$, gives the probability for each of the simple events for which $X = k$. It follows from the extension of Rule 3b in Chapter 7. If $X = k$, there are k successes and $(n - k)$ failures. Multiply p for each success and $(1 - p)$ for each failure to get $p^k(1 - p)^{n-k}$.

Remember that a *probability distribution function* for a discrete random variable is a table or rule that assigns probabilities to the possible values of the random variable X. It is synonymous to say that a random variable X is a *binomial random variable* and to say that X has a **binomial distribution.** In both cases, the *probability distribution function* for X is the formula just given for $P(X = k)$.

EXAMPLE 8.9
Probability of Two Wins in Three Plays

Suppose that the probability that you win a game is .2 for each play and plays of the game are independent of one another. Let X = number of wins in three plays. What is $P(X = 2)$, the probability that you win exactly twice in three plays? For this problem, X is a binomial random variable with $n = 3$ and success probability $p = .2$. Also, $1 - p = .8$ and $k = 2$ in the formula for $P(X = k)$.

◆ There are

$$\frac{3!}{2!(3 - 2)!} = \frac{6}{2(1)} = 3$$

simple events that produce $X = 2$. The three sequences with two wins are WWL, WLW, and LWW.

◆ For each of these simple events, the probability is $p^k(1 - p)^{n-k} = (.2)^2(.8)^1 = .032$.

◆ So the probability of exactly two wins is $P(X = 2) = 3(.032) = .096$.

The complete probability distribution function (pdf) for X = number of wins in three plays is graphed in Figure 8.3. See if you can verify that $P(X = 1) = .384$. ◆

FIGURE 8.3 **Probability distribution function for binomial random variable with $n = 3$ and $p = .2$**

Using Excel to Find Binomial Probabilities

 tech note   Fortunately, binomial probabilities are available in many computer software programs and calculators. To find binomial probabilities using Excel,

- $P(X = k)$, is calculated with the command BINOMDIST(k,n,p,false). The "false" indicates that we want the probability for exactly k successes.

- $P(X \leq k)$ is calculated with the command BINOMDIST(k,n,p,true). The "true" indicates that the desired probability is "cumulative" and that we want the probability of k or fewer successes.

EXAMPLE 8.10
Excel Calculations for Number of Girls in Ten Births

Let X = number of girls in ten births, and assume that $p = .488$ is the probability that any birth is a girl. As noted in Chapter 7, this value of p is based on birth records in the United States. We can use Excel to find that the probability of exactly seven girls in ten births is = BINOMDIST(7,10,.488,false) = .106. The probability of seven or fewer girls is = BINOMDIST(7,10,.488,true) = .953. ◆

Using Minitab to Find Binomial Probabilities

 tech note   ◆ Use **Calc**>**Probability Distributions**>**Binomial**.

- In the dialog box, select "Probability" or "Cumulative Probability" depending on whether you want $P(X = k)$ or $P(X \leq k)$. Specify the number of trials and the probability of success.

- Click on "Input constant" and fill in the corresponding box with the value of k.

Note: To find probabilities for several values of k at once, first store the values in a column of the worksheet, and then specify that column as "Input Column."

EXAMPLE 8.11
Guessing Your Way to a Passing Score

You've been busy lately, so busy you're surprised to learn when you arrive at today's statistics class that a 15-question True-False quiz is on the agenda. The quiz is about readings that haven't been discussed during class and you haven't done the readings, so you're forced to guess at every question. You'll pass the quiz if you get ten or more correct answers, so you wonder about $P(X \geq 10)$, where X = number of correct answers. X is a binomial random variable with $n = 15$ trials, and $p = .5$ is the success probability for any question. To find the probability of ten or more correct answers, note that you'll either get ten or more questions right, or you'll get nine or fewer right. So, the answer for $P(X \geq 10)$ can be found by determining the cumulative probability $P(X \leq 9)$, and then subtracting that value from 1. Here's Minitab output with the value of $P(X \leq 9)$:

```
Cumulative Distribution Function
Binomial with n = 15 and p = 0.500000
       x        P(X <= x)
      9.0         0.8491
```

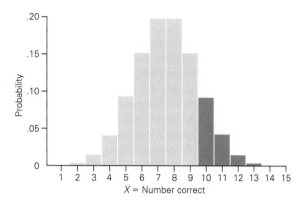

FIGURE 8.4   **Probabilities for guessing on a 15-question true-false quiz**

The probability of a passing score is $1 - .8491 = .1509$. You've got a chance to pass, although the probability you don't pass is .8491! Figure 8.4 illustrates this probability. ◆

Expected Value (Mean) and Standard Deviation for a Binomial Random Variable

To find the mean value of a binomial random variable, we don't have to use the formula from Section 8.3 for expected value. There is a much simpler formula that applies to all binomial random variables, and this formula is easy to understand. If you flip a fair coin 100 times, for instance, how many heads would you expect to result, on average? The answer is $100 \times 1/2 = 50$ heads, which is the number of flips times the probability of heads for any single flip. Or, if you guess the answer for every question on a multiple-choice test with 25 questions, each with 5 choices, you could expect to get about $25 \times 1/5 = 5$ right answers. For both of these examples, we determined the mean, μ, by calculating np, the product of the number of trials and the probability of success.

There also is a simplified formula for the standard deviation of a binomial random variable. The formulas for the mean (expected value) and standard deviation of a binomial random variable were derived using algebra and the formulas introduced earlier in this chapter for expected value, standard deviation, and $P(X = k)$ when X is a binomial random variable. You won't have to know how to do these derivations, but the results are useful for later applications in this textbook, so remember where to find them.

Mean, Standard Deviation, and Variance of a Binomial Random Variable

formula   For a *binomial random variable X* based on *n* trials with success probability *p*,

The **mean** is $\mu = E(X) = np$.

The **standard deviation** is $\sigma = \sqrt{np(1 - p)}$.

Note: For any random variable, the standard deviation is the square root of the **variance.** For a binomial random variable, the variance is $V(X) = np(1 - p)$, which leads to the formula for standard deviation.

EXAMPLE 8.12
Is There Extraterrestrial Life?

Suppose that 50% of a large population would say "yes" if they were asked, "Do you believe there is extraterrestrial life?" A sample of $n = 100$ is taken from this population, and each person in the sample is asked the question about belief in extraterrestrial life. The random variable X = number in the sample who say "yes" is approximately a binomial random variable. For this variable, $n = 100$ and $p = .5$, so the mean of X is $\mu = (100)(.5) = 50$, and the standard deviation is $\sigma = \sqrt{100(.5)(1 - .5)} = 5$. In other words, in repeated samples of size 100 from this population, on average 50 people would say "yes" to the question. The amount by which that number would differ from sample to sample is represented by the standard deviation of 5. We will discuss this kind of "sample variability" in much more detail in Chapter 9. ◆

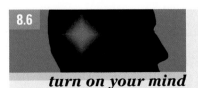

8.6

turn on your mind

In Example 8.11, we determined the probability that you could guess your way to a passing score on a quiz with 15 True-False questions. If you did guess at each of 15 True-False questions, what is the expected value of X = number of correct answers? Is the expected value a possible score on the quiz? What exactly does the expected value tell us?*

8.5 | CONTINUOUS RANDOM VARIABLES

> *definition* A **continuous random variable** is one for which the outcome can be any value in an interval or collection of intervals.

In practice, all measurements are rounded to a specified number of decimal places, so we may not be able to accurately observe all possible outcomes of a continuous variable. For example, the limitations of weighing scales keep us from observing that a weight may actually be 128.3671345993 pounds. With most scales, we would be forced to round such a weight to 128 pounds, or, with a digital scale, perhaps to 128.4 pounds. Generally, however, we call a random variable a continuous random variable if there are a large number of observable outcomes covering an interval or set of intervals. For instance, measurements of height, time, weight, and monetary values are all treated as continuous random variables.

The way that we determine probabilities for continuous random variables differs in one important respect from how we determine probabilities for discrete random variables. For a discrete random variable, we can find the probability that the variable X exactly equals a specified value. We can't do this for a continuous random variable. For a continuous random variable, we are only able to find the probability that X falls between two values. We do this by determining the area between the two values under a curve called the **probability density function** of the random variable. In other words, unlike discrete random variables, continuous random variables do not have probability distribution functions specifying the exact probabilities of specified values. Instead, they have probability density functions, which are used to find probabilities that the random variable falls into a specified interval of values.

> *definition*   The **probability density function** for a continuous random variable *X* is a curve such that the area under the curve over an interval equals the probability that *X* is in that interval. In other words, the probability $P(a \leq X \leq b)$ is the area under the density curve over the interval between the values *a* and *b*.

8.7

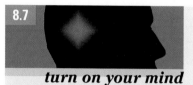

turn on your mind

The total area under the probability density function over the entire range of values the random variable *X* can possibly have is the same for all continuous random variables. What is that total area? What probability does it represent?

Suppose the random variable is the amount of rainfall in Davis, California, for a randomly selected year and can range from 0 inches to 50 inches. What is the area under the appropriate density curve over the range from 0 to 50 inches? What probability does it represent?*

EXAMPLE 8.13
***Time Spent
Waiting for the Bus***

A bus arrives at a bus stop every 10 minutes. If a person arrives at the bus stop at a random time, how long will he or she have to wait for the next bus? Define the random variable *X* = waiting time until the next bus arrives. The value of X could be any value between 0 and 10 minutes, and *X* is a continuous random variable. (In practice, the limitations of watches would force us to round off the exact time.) Figure 8.5 shows the probability density function for the waiting time. Possible waiting times are along the horizontal axis, and the vertical axis is a density scale. The height of the "curve" is .1 for all *X* between 0 and 1, so the total area between 0 and 10 minutes is (10)(.1) = 1. ◆

The density function shown in Figure 8.5 is a flat line that covers the interval of possible values for *X*. There is a "uniformity" to this density curve in that every interval with the same width has the same probability. A random variable with this property is called a **uniform random variable** and is the simplest example of a continuous random variable.

EXAMPLE 8.13—CONTINUED

Suppose we want to find the probability that the waiting time *X* was in the interval from 5 to 7 minutes. The general principle for any continuous random vari-

FIGURE 8.5   **A uniform probability density function for Example 8.13**

*HINT: For X = Davis rainfall as described, what is $P(0 \leq X \leq 50)$?

FIGURE 8.6   **The probability that waiting time is between 5 and 7 minutes**

able is that the probability $P(a \leq X \leq b)$ is the "area under the curve" over the interval from a to b. In this example, the "area under the curve" is the area of a rectangle that has width $= 7 - 5 = 2$ minutes and height $= .1$. This area is $(2)(.1) = .2$, which is the probability that the waiting time is between 5 and 7 minutes. In Figure 8.6, the shaded area represents the desired probability. ◆

Theoretically, the use of calculus is needed to find an area under a density curve (unless it's a simple rectangle or other simple shape, as in Example 8.13). In practice, however, tables of appropriate probabilities usually are available. Computer software and graphing calculators also can be used to "automate" the calculations for the most commonly encountered models.

8.6 | NORMAL RANDOM VARIABLES

We first encountered normal random variables in Chapter 2, as a special case of bell-shaped distributions. If a population of measurements follows a normal curve, and if X is the measurement for a randomly selected individual from that population, then X is said to be a **normal random variable.** X is also said to have a **normal distribution.** Any normal random variable can be completely characterized by specifying its mean, μ, and standard deviation, σ.

Normal random variables are the most common type of *continuous random variables.* As with any continuous random variable, the probability that X falls into some specified interval is equivalent to the area under a probability density curve, which in this case is the bell-shaped **normal curve.** Nature provides numerous examples of measurements that follow a normal curve. The fact that so many different kinds of measurements follow a normal curve is not surprising. On many attributes, the majority of people are somewhat close to average, and as you move further from the average, either above or below, there are fewer people with such values.

Here are some features shared by all normal curves and normal random variables (X):

1. As with any continuous random variable, $P(X = k) = 0$ for any value k.
2. The curve is symmetric on either side of the mean, μ, so $P(X \leq \mu) = P(X \geq \mu) = .5$. For instance, half of all women's heights are above the mean and half are below the mean.
3. $P(X \leq \mu - d) = P(X \geq \mu + d)$ for any number d. For instance, the probability that a randomly selected woman is at least two inches shorter than

the mean is the same as the probability that she is at least two inches taller than the mean. In this case, X = woman's height and $d = 2$.

4. The Empirical Rule holds:

- $P(\mu - \sigma \leq X \leq \mu + \sigma) \approx .68$
- $P(\mu - 2\sigma \leq X \leq \mu + 2\sigma) \approx .95$
- $P(\mu - 3\sigma \leq X \leq \mu + 3\sigma) \approx .997$

EXAMPLE 8.14

College Women's Heights

Data collected in several classes by the authors of this text and some of their colleagues suggest that the distribution of the heights of college women can be described reasonably well by a normal curve with mean $\mu = 65$ inches (5 foot 5 inches) and standard deviation $\sigma = 2.7$ inches. A normal curve with these characteristics is shown in Figure 8.7. Heights are shown along the horizontal axis, with tick marks given at the mean and at 1, 2, and 3 standard deviations above and below the mean. Notice that half (.5) of the area is above the mean of 65.0 and half is below it.

Most of the normal curve is over the interval of height from 56.9 inches to 73.1 inches. This is consistent with the Empirical Rule that we encountered in Chapter 2. One part of the Empirical Rule is that for a bell-shaped curve, about 99.7% of the values fall within 3 standard deviations of the mean in either direction. Here, the interval of heights within 3 standard deviations of the mean is $65 \pm 3 \times 2.7$, which is 65 ± 8.1 inches, or 56.9 and 73.1 inches. The Empirical Rule is generally used as an approximate rule for describing sample data with a bell-shaped curve, but the features of that rule are exact characteristics of a normal curve model. ◆

Standardized Scores

Once the mean, μ, and the standard deviation, σ, have been specified, finding probabilities is a simple process for a normal random variable. We simply have to convert the endpoints of an interval of interest to **standardized scores** and then use a table, calculator, or computer software to find probabilities associated with these standardized scores. A standardized score is also called a ***z-score***. We first encountered z-scores in Chapter 2; the definition is repeated here, using notation for random variables.

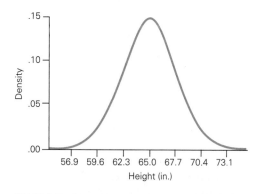

FIGURE 8.7   **Normal curve for heights of college women**

Calculating a *z*-Score

> **formula** The formula for converting any value *x* to a **z-score** is
>
> $$z = \frac{\text{Value} - \text{Mean}}{\text{Standard deviation}} = \frac{x - \mu}{\sigma}$$
>
> A *z*-score measures the number of standard deviations that a value falls from the mean.

EXAMPLE 8.14 — CONTINUED

The *z*-Score for a Height of 62 Inches

For a population of college women, the *z*-score corresponding to a height of 62 inches is

$$z = \frac{\text{Value} - \text{Mean}}{\text{Standard deviation}} = \frac{62 - 65}{2.7} = \frac{-3}{2.7} = -1.11$$

This *z*-score tells us that 62 inches is 1.11 standard deviations below the mean height for this population. If you are a college woman who is 62 inches tall, your height is 1.11 standard deviations below the mean height for other college women. ◆

Finding Probabilities for *z*-Scores

Converting any normal random variable to *z*-scores is equivalent to converting the random variable of interest to a normal random variable that has mean $\mu = 0$ and standard deviation $\sigma = 1$. A normal random variable with these characteristics is said to be a **standard normal random variable** and to have a **standard normal distribution.** Tables and software packages are designed to give probabilities for the standard normal distribution. The Appendix of this book contains a table of standard normal probabilities. To read Table A.1, Standard Normal Probabilities,

- ◆ The body of the table contains $P(Z \leq z^*)$, which is the probability that the value of a standard normal variable *Z* takes a value that is less than the specific value z^*.
- ◆ The left-most column of the table shows the algebraic sign, the digit before the decimal place, and the first decimal place for z^*.
- ◆ The second decimal place of z^* is in the column heading across the top of the table.

Technical Detail: $P(Z = z^*) = 0$, so it is equivalent to write $P(Z \leq z^*)$ or $P(Z < z^*)$.

As an example, Table A.1 shows that $P(Z \leq 1.82) = .9656$. To see this, look in the portion of the table of positive values of z. In the first column (labeled z), locate the row for 1.8. In that row, move over to the ".02" column to find the cumulative probability for z = 1.82. The general idea is

z	.00	.01	.02	.03	.04	.05	.06	.07	.08	.09
1.8	$\longrightarrow$		.9656							

Here are some other examples that you should verify to be sure you understand how to read the table:

$$P(Z \leq -2.59) = .0048$$
$$P(Z \leq 1.31) \quad = .9049$$
$$P(Z \leq 2.00) \quad = .9772$$

Notice at the bottom of both pages of the table, there is a section titled "In the Extreme" where cumulative probabilities are given for selected "extreme" z-scores. For instance, from that section of the table you can learn that $P(Z \leq -4.75) = .000001$ (one in a million!).

To find probabilities for other types of intervals, simply add and subtract appropriate probabilities of the form $P(Z \leq z^*)$. When using a table like ours, it is usually best to draw a picture of the problem, highlight the desired area, and then figure out how the area relates to probabilities of the form $P(Z \leq z^*)$.

There are some simple probability relationships that are useful for solving normal curve problems. We state them for a standard normal random variable Z here, and for any normal random variable X later. In the Exercises, you will be asked to show how they follow from the probability rules in Chapter 7.

1. $P(Z > a) = 1 - P(Z \leq a)$
2. $P(Z > a) = P(Z < -a)$
3. $P(a \leq Z \leq b) = P(Z \leq b) - P(Z \leq a)$

Examples 8.15 and 8.16 demonstrate these relationships.

EXAMPLE 8.15
Probability That Z Is Greater Than 1.31

There are two equivalent ways to find the probability that Z is greater than 1.31:

$$P(Z > 1.31) = 1 - P(Z \leq 1.31)$$
$$= 1 - .9049 = .0951$$

This is an application of the general relationship $P(Z > a) = 1 - P(Z \leq a)$. Figure 8.8 illustrates the area of interest and the solution. Notice that the desired probability equals the area to the right of 1.31. We could also use the symmetry relationship,

$$P(Z > a) = P(Z < -a),$$

so

$$P(Z > 1.31) = P(Z < -1.31) = .0951 \quad \blacklozenge$$

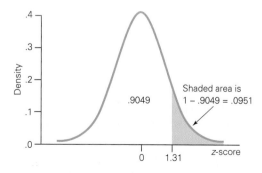

FIGURE 8.8   **Area to the right of $Z = 1.31$**

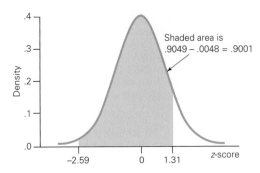

FIGURE 8.9   **Area between** $Z = -2.59$ **and** $Z = 1.31$

EXAMPLE 8.16
Probability That Z Is
Between −2.59 and 1.31

The probability that a standard normal variable Z is between -2.59 and 1.31 is found by taking the difference between the probabilities to the left of these two values.

$$P(-2.59 \le Z \le 1.31) = P(Z \le 1.31) - P(Z \le -2.59)$$
$$= .9049 - .0048 = .9001$$

Figure 8.9 shows the area under the normal curve for this example. This is an example of using the relationship:

$$P(a \le Z \le b) = P(Z \le b) - P(Z \le a) \quad \blacklozenge$$

How to Solve General Normal Curve Problems

Every probability problem about a normal curve can be solved using z-scores. The following example illustrates the process.

EXAMPLE 8.14—*CONTINUED*
Probability That Height Is
Less Than 62 Inches

If we assume that college women's heights follow a normal curve with $\mu = 65$ inches and $\sigma = 2.7$ inches, we can find probabilities associated with any possible range of heights. For example, what is the probability that a randomly selected college woman is 62 inches or shorter? Equivalently, what proportion of college women are 62 inches or shorter?

$$P(X \le 62) = P\left(Z \le \frac{62 - 65}{2.7}\right)$$
$$= P(Z \le -1.11) = .1335$$

In other words, about 13% of college women are 62 inches or shorter. Figure 8.10 illustrates the area of interest under the normal curve for college women's heights. ◆

The area displayed in Figure 8.10, the probability that a randomly selected college woman has a height less than 62 inches, is called the cumulative probability for 62 inches. A cumulative probability is a "less than" type of probability. We encountered this idea for discrete random variables when we defined the *cumulative distribution function* as $P(X \le k)$. The same definition holds for continuous random variables, and the probability is simply the area under the density curve to the left of the value k.

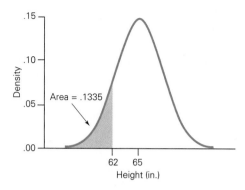

FIGURE 8.10 **College women's heights and** *P*(height ≤ 62 inches)

To find a cumulative probability for any random variable with a normal distribution, we simply need to know its mean and standard deviation, find the *z*-score corresponding to *k*, and find the area under the standard normal curve to the left of that *z*-score.

Finding a Cumulative Probability *P*(*X* ≤ *k*) for any Normal Random Variable

If *X* is a normal random variable with known mean μ and standard deviation σ, to find $P(X \leq k)$, the probability that *X* is less than or equal to *k*:

Step 1: Calculate a *z*-score for the value *k*; call it *z**.

Step 2: Use a table, calculator, or computer to find $P(Z \leq z^*)$.

EXAMPLE 8.14 —
CONTINUED
Proportion of Women
Who Are Taller
Than 68 Inches

Suppose we want to find the proportion of college women who are taller than 68 inches. The simplest solution is to first find the probability that a woman is shorter than 68 inches and then subtract that answer from 1. The relationship is $P(X > 68) = 1 - P(X \leq 68)$. To find $P(X \leq 68)$, we convert it to an equivalent question about the *z*-score for 68 inches. The *z*-score for 68 inches is $(68 - 65)/2.7 = 1.11$, and

$$P(Z > 1.11) = 1 - P(Z \leq 1.11)$$
$$= 1 - .8665 = .1335$$

About 13% of the women in this population are at least 68 inches tall. Figure 8.11 illustrates the problem and the solution. ◆

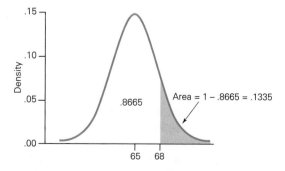

FIGURE 8.11 **Probability that college woman is taller than 68 inches**

You may have noticed that the proportion taller than 68 inches equals the proportion shorter than 62 inches. The equivalence is not a coincidence. Both problems have to do with being more than 3 inches from the mean. By the symmetry of the normal curve, the probability that a woman is at least 3 inches shorter than the mean equals the probability that a woman is at least 3 inches taller than the mean.

Useful Probability Relationships for Normal Random Variables

The following relationships are useful for working with probabilities for a normal random variable X.

- To find the probability that X **is greater than b,** subtract the probability that X is less than b from 1. In mathematical terms, $P(X > b) = 1 - P(X \le b)$.

- The probability that X is *more* than d units *above* the mean is the same as the probability that X is *less* than d units *below* the mean. In mathematical terms, $P(X > \mu + d) = P(X < \mu - d)$.

- To find the probability that X **is between a and b,** take the difference between the probability that X is less than b and the probability that X is less than a. In mathematical terms, $P(a \le X \le b) = P(X \le b) - P(X \le a)$.

EXAMPLE 8.14—CONTINUED
The Proportion of Women Between 62 and 68 Inches Tall

To find the probability that a college woman's height is between 62 and 68 inches, use the relationship $P(62 \le X \le 68) = P(X \le 68) - P(X \le 62)$. In other words, take the difference between the cumulative probabilities for 62 inches and 68 inches. The probabilities that we need were determined in previous parts of this example, and the answer is $.8665 - .1335 = .733$. About 73% of college women are between 62 and 68 inches tall. ◆

Finding Percentiles

There probably have been several instances in your life where information about your percentile ranking on a particular variable has been provided to you. For example, you may have been told that your Verbal SAT score was at the 80th percentile of all Verbal SAT scores, or that your height was at the 50th percentile for individuals of your sex and age. Remember from Chapter 2 that the kth percentile for a dataset is a number that has $k\%$ of the data values at or below it. The same definition holds for random variables. If your Verbal SAT score was at the 80th percentile, then your score was higher than the scores of 80% of the other test takers (and lower than 20%).

In some problems, we may want to know what value of a variable defines a specified percentile ranking. We may, for example, want to know what IQ score is the 98th percentile, or what pulse rate is the 25th percentile of pulse rates for your gender. You should notice that the word **percentile** refers to the *value* of a variable, while **percentile ranking** refers to the *proportion* below that value. For instance, if the 25th percentile of pulse rates is 64 beats per minute, then 25% of pulse rates are below 64 and 75% are above 64. The *percentile* is 64 beats per minute, and the *percentile ranking* is 25% or .25.

In probability terms, the percentile rank for the value of a variable corresponds to the *cumulative probability* for that value. For example, the 25th percentile of pulse rates is the pulse rate for which .25 is the cumulative probability (area to the left under the density curve). Figure 8.12 illustrates this concept.

To determine a percentile for a specified percentile ranking (or cumulative probability) for a normal random variable, there are two steps:

FIGURE 8.12   The 25th percentile for a random variable with a normal distribution

Step 1: Find the z-score that has the specified cumulative probability. This step does not involve calculation; it involves a table search (or a computerized version of a table search). For instance, using Table A.1 in the Appendix, find the percentile rank in the body of the table and read the z-score from the row and column headings.

Step 2: Calculate the value of the variable that has the z-score found in step 1. This can be done using the relationship $x = z\sigma + \mu$.

EXAMPLE 8.17
The 75th Percentile of Systolic Blood Pressures

Suppose that the blood pressures of men aged 18 to 29 years old can be described with a normal curve having mean $\mu = 120$ and standard deviation $\sigma = 10$. What is the 75th percentile? In other words, what is the blood pressure value x such that P(Blood pressure $\leq x$) = .75?

Using the table of standard normal probabilities, we learn that $z = .67$ is approximately the z-score for which the cumulative probability is the desired .75. (Search for .75 within the table, then look at the row and column labels to read the value of z.) The blood pressure with this z-score is determined from $x = z\sigma + \mu$, and the answer is .67(10) + 120 = 6.7 + 120 $\approx$ 127. ◆

Finding a Percentile or a Value with Specified Cumulative Probability

If X is a normal random variable with mean μ and standard deviation σ, and p is a specified probability, then to find the value x for which $P(X \leq x) = p$:

Step 1: Find the value z^* for which $P(Z \leq z^*) = p$ using a table, computer, or calculator.

Step 2: Compute $x = z^*\sigma + \mu$.

8.8

Which of the following measurements do you think are likely to have a normal distribution: heights of college men, incomes of 40-year-old women, pulse rates of college athletes? Explain your reasoning for each variable. For those variables that are likely to be normally distributed, give approximate values for the mean and standard deviation.*

turn on your mind

Minitab Tip

 tech note **Calculating Normal Curve Probabilities and Percentiles**

- ◆ Use **Calc>Probability Distributions>Normal.**
- ◆ Specify the mean and standard deviation.
- ◆ To find $P(X \leq k)$, select "Cumulative Probability" and also specify the value of k in the box labeled "Input Constant."
- ◆ To find a percentile, select "Inverse Cumulative" and also specify the cumulative probability for the percentile in the box labeled "Input Constant."

Note: It's not necessary to compute z-scores when using Minitab for determining normal curve probabilities.

8.7 | APPROXIMATING BINOMIAL DISTRIBUTION PROBABILITIES

One useful application of the normal distribution is that it can be used to approximate probabilities for some other types of random variables. In this section, we learn how to use the normal distribution to approximate cumulative probabilities for binomial random variables. When X has a binomial distribution with a large number of trials, the binomial probability formula is difficult to use because the factorial expressions in the formula become very large. The work required to find binomial probabilities when n is large is enormous, and even computer programs and calculators may not be able to do the task. Fortunately, the normal distribution can be used to approximate probabilities for a binomial random variable when this situation occurs.

The **normal approximation to the binomial distribution** is based on the following result, derived from mathematics. If X is a binomial random variable based on n trials with success probability p, and n is large, then the random variable X is also approximately a normal random variable. For this normal random variable,

Mean $= \mu = np$

Standard deviation $= \sigma = \sqrt{np(1 - p)}$

Conditions: The approximation works well when both np and $n(1 - p)$ are at least 10.

EXAMPLE 8.18
The Number of Heads in 30 Flips of a Coin

Figure 8.13 displays the probability distribution function of a binomial random variable based on $n = 30$ trials and a success probability $p = .5$. This distribution could, for example, describe $X =$ number of heads when a coin is flipped $n = 30$ times, or the number correct if you are just guessing on a 30-question True-False quiz. The bell-shaped pattern of the distribution is clear, and a normal curve could be used to approximate this distribution because both np and $n(1 - p)$ are greater than 10. Recall that 99.7%, or nearly all, of a normal curve falls within 3 standard deviations of the mean. For the binomial random variable displayed in Figure 8.13, $\mu = 15$, the standard deviation is $\sigma = \sqrt{30 \times .5 \times (1 - .5)} = 2.74$, and the interval $\mu \pm 3\sigma$ is approximately 15 ± 8, or 7 to 23. The probability is small that X would take a value outside this interval. ◆

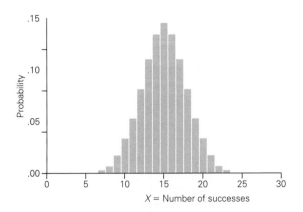

FIGURE 8.13   Binomial distribution for $n = 30$ and $p = .5$

Approximating Cumulative Probabilities for a Binomial Random Variable

The normal curve approximation can be used to find cumulative probabilities for binomial random variables, as follows:

$$P(X \leq k) \approx P(Z \leq z^*)$$

where

$$z^* = \frac{k - np}{\sqrt{np(1 - p)}}$$

In other words, to find a cumulative probability, first calculate a z-score for the value of interest, k, using the formulas for the mean and the standard deviation of a binomial variable. Then, use the standard normal curve to determine the cumulative probability for this z-score.

More complex probabilities can be found as well, using the same relationships as for any normal random variable X. For example, find the probability that X is between two numbers a and b as follows:

$$P(a \leq X \leq b) = P(X \leq b) - P(X \leq a) \approx P\left(Z \leq \frac{b - np}{\sqrt{np(1 - p)}}\right)$$
$$- P\left(Z \leq \frac{a - np}{\sqrt{np(1 - p)}}\right)$$

It is important to note that you should not use the normal curve approximation if your statistical software or calculator is able to determine a binomial probability exactly. In this regard, the capabilities of different software programs and calculators vary greatly. For instance, versions 13 and 14 of Minitab can provide an exact answer, but version 12 of Minitab cannot.

EXAMPLE 8.19
Political Woes

A politician is convinced that at least 50% of her constituents favor a woman's right to choose abortion but is concerned because the latest poll of 500 adults in her district found that slightly less than half, only 240, or 48% of the sample, supported this position. If indeed 50% of adults in the district support abortion, what is the probability that a random sample of 500 adults would find 240 or fewer of them holding this position?

To solve this problem, notice that if X = number in sample who support abortion, and if 50% of all adults in the district actually support it, then X is a binomial random variable with $n = 500$ and $p = .5$. The desired probability is $P(X \leq 240)$. It would be quite tedious to find this probability by adding the individual probabilities from 0 to 240! It is much easier to use the normal approximation, with

$$\mu = np = 250$$
$$\sigma = \sqrt{np(1 - p)} = 11.2$$

The solution is $P(X \leq 240) \approx \left(P(Z \leq \frac{240 - 250}{11.2}\right) = P(Z \leq -.89) = .1867$.

This means that it is not unlikely that she could observe a minority of 48% or less of the sample favoring abortion, even if 50% of the district favors it. ◆

EXAMPLE 8.18 — *CONTINUED*
Guessing and Passing
a True-False Test

In Example 8.18 and Figure 8.13 we displayed probabilities for a binomial random variable with $n = 30$ and $p = .5$. Suppose you need 70%, or 21 questions, correct to pass a True-False test on which you are just guessing. Find $P(X \geq 21)$. Notice that the complement of this event is *not* $P(X \leq 21)$; it's $P(X \leq 20)$. Using the normal approximation to the binomial with $n = 30$ and $p = .5$:

$$P(X \geq 21) = 1 - P(X \leq 20) \approx 1 - P\left(Z \leq \frac{20 - np}{\sqrt{np(1 - p)}}\right) = 1 - P(Z \leq 1.83)$$
$$= 1 - .9664 = .0336.$$

Using Minitab, the exact probability that $X \geq 21$ is $1 - .9786 = .0214$.

Figure 8.13 illustrates the exact binomial pdf for this situation. Notice that $P(X \geq 21)$ corresponds to the areas of the rectangles above 21, 22, . . . , 30, of which only the ones above 21, 22, and 23 are visible. After that, the probabilities, and thus heights of the rectangles, are essentially 0. For instance, $P(X = 24) = .0006$. ◆

Continuity Correction

tech note   Notice in Figure 8.13 that the rectangle centered on $X = 21$ actually begins at 20.5, not at 21.0. But in the normal approximation to the binomial calculation we found the area under a normal curve starting at 21, not 20.5. In doing so, we omitted half of the rectangle centered at 21. To make the approximation more accurate, sometimes a **continuity correction** is used by adding or subtracting .5, based on which rectangles are desired. For Example 8.18,

$$P(X \geq 21) \approx P\left(Z \leq \frac{20.5 - (30)(.5)}{\sqrt{(30)(.5)(1 - .5)}}\right) = P(Z \geq 2.01) = P(Z \leq -2.01)$$
$$= .0222$$

which is much closer to the exact probability of .0214 than was the approximation of .0336 found without the continuity correction.

8.8 | SUMS, DIFFERENCES, AND COMBINATIONS OF RANDOM VARIABLES

There are many instances where we want information about combinations of random variables. Suppose, for example, that your final score in a class is found by adding together 25% of each of two midterm exam scores (Exam1 and Exam2) and 50% of your final exam score (Exam3). Expressed as an equation, the final score is calculated as

Final score = $.25 \times$ Exam1 + $.25 \times$ Exam2 + $.50 \times$ Exam3

If you know the mean for each exam, it is easy to calculate the mean final score. It probably makes sense to you that the mean final score can be found by combining the mean exam scores in the same way that an individual's scores are combined. If the means on the three exams are μ_1, μ_2, and μ_3, the formula for the mean final score is

Mean of final scores = $.25 \times \mu_1 + .25 \times \mu_2 + .50 \times \mu_3$

For instance, if the exam means are 72, 76, and 80, respectively, then the mean of the final scores is

$$\text{Mean of final scores} = (.25 \times 72) + (.25 \times 76) + (.50 \times 80)$$
$$= 18 + 19 + 40 = 77$$

We are assuming the class constitutes the entire population of interest so the means are represented as population means (μ) instead of sample means ($\bar{x}$). The formula for calculating the mean final score, however, would hold for sample means as well.

The combination of exam scores that we just considered is an example of a "linear combination" of random variables. In a linear combination, we add (or subtract) variables, but some of the combined variables may be multiplied by a numerical value, as occurred in calculating the final score. For two random variables X and Y, the most commonly encountered linear combinations are the sum $X + Y$ and the difference $X - Y$.

> *definition* A **linear combination** of random variables, X, Y, ... is a combination of the form
>
> $$L = aX + bY + \cdots$$
>
> where a, b, etc. are numbers, which could be positive or negative. The most commonly encountered linear combinations of two variables are
>
> Sum $= X + Y$
>
> Difference $= X - Y$

A number of rules tell us about the statistical properties of a linear combination of random variables. The most basic rule is for the mean. This rule applies to any set of random variables, whether they are discrete or continuous, independent of each other or not, and regardless of their distributions. The only assumption is that the variables combined all have finite means.

Mean of a Linear Combination of Random Variables (Including Sums and Differences)

> *formula* If X, Y, ... are random variables, a, b, ... are numbers, either positive or negative, and
>
> $$L = aX + bY + \cdots$$
>
> The mean of L is
>
> $$\text{Mean}(L) = a\,\text{Mean}(X) + b\,\text{Mean}(Y) + \cdots$$
>
> In particular,
>
> $$\text{Mean}(X + Y) = \text{Mean}(X) + \text{Mean}(Y)$$
>
> $$\text{Mean}(X - Y) = \text{Mean}(X) - \text{Mean}(Y)$$

Rules for the variance and standard deviation of a linear combination are complicated, and we will only consider a specific situation, which is that the random variables are independent. Two random variables are statistically **independent** if the probability for any event associated with one random variable is not altered by whether or not any event for the other random variable has happened. In a more practical sense, it is usually safe to say that *two random variables are independent if there is no physical connection between the two variables, or if there is no apparent reason why the value of one variable should influence the value of the other.*

Variance and Standard Deviation of a Linear Combination of Independent Random Variables

formula   If X, Y, etc. are independent random variables, a, b, etc. are numbers, and

$$L = aX + bY + \cdots$$

then, the *variance* and *standard deviation* of L are

$$\text{Variance}(L) = a^2\,\text{Variance}(X) + b^2\,\text{Variance}(Y) + \cdots$$

$$\text{Standard Deviation of } L = \sqrt{\text{Variance}(L)}$$

In particular,

$$\text{Variance}(X + Y) = \text{Variance}(X) + \text{Variance}(Y)$$

$$\text{Variance}(X - Y) = \text{Variance}(X) + \text{Variance}(Y)$$

Keep in mind that the standard deviation is the square root of the variance. Notice also that the variance of the *difference* is the *sum* of the variances because $b = -1$ and $b^2 = +1$.

Simply knowing the mean and standard deviation of a combination of random variables is of little practical use in most cases. To find probabilities associated with various outcomes, we also need to know the distribution.

Combining Independent Normal or Binomial Random Variables

Some classes of random variables keep the same shape when linear combinations of variables within the class are formed. This occurs for normal curve distributions. Any linear combination of normally distributed variables also has a normal distribution, but we only consider independent variables and focus on rules for the **sum** and for the **difference** of two such variables.

Linear Combinations of Independent Normal Random Variables

formula   If X, Y, etc. are independent, normally distributed random variables, and a, b, etc. are numbers, either positive or negative, then the random variable $L = aX + bY + \cdots$ is normally distributed. In particular,

$X + Y$ is normally distributed with mean $\mu_X + \mu_Y$ and standard deviation $\sqrt{\sigma_X^2 + \sigma_Y^2}$.

$X - Y$ is normally distributed with mean $\mu_X - \mu_Y$ and standard deviation $\sqrt{\sigma_X^2 + \sigma_Y^2}$.

EXAMPLE 8.20
Will Meg Miss Her Flight?   Meg travels often and lately has become lax about giving herself enough time to get to the airport. She leaves home 45 minutes before the last call for her flight

will occur. Her travel time from her front door to the airport parking lot is normally distributed with mean of 25 minutes and standard deviation of 3 minutes. From the parking lot, she must then take a shuttle bus to the terminal and go through security screening. The mean time for doing this is 15 minutes and the standard deviation is 2 minutes, and this time also is normally distributed. The driving time and the airport time are independent of each other. What is the probability that Meg will miss her flight because her total time for getting to the plane is more than 45 minutes?

Let's define some random variables:

X = driving time; normally distributed with μ_X = 25 min. and σ_X = 3 min.
Y = airport time; normally distributed with μ_Y = 15 min. and σ_Y = 2 min.
$T = X + Y$ = total time.

The random variable T has a normal distribution because it is a sum of two independent, normally distributed random variables. For T, the mean is the sum of the means for X and Y, which is $\mu = 25 + 15 = 40$ min., and the standard deviation is $\sigma = \sqrt{\sigma_X^2 + \sigma_Y^2} = \sqrt{3^2 + 2^2} = \sqrt{9 + 4} = \sqrt{13} = 3.6$. Using Table A.1 in the Appendix (or statistical software or the right calculator), we find that the probability the total time T exceeds 45 minutes is

$$P(T > 45) = P\left(Z > \frac{45 - 40}{3.6}\right)$$
$$= P(Z > 1.39) = 1 - P(Z < 1.39)$$
$$= 1 - .9177 = .0823$$

Meg should leave sooner. If she continues this behavior, she will miss her flight about 8% of the time. ◆

EXAMPLE 8.21
Can Alison Ever Win?

Alison and her sister Julie each swim a mile every day. Alison's times are normally distributed with mean = 37 minutes and standard deviation = 1 minute. Julie is faster but less consistent than Alison, and her times are normally distributed with mean = 33 minutes and standard deviation = 2 minutes. On any day, their times are independent of each other. Can Alison ever beat Julie?

Let's define three random variables for a given day:

X = Alison's time; normally distributed with μ_X = 37 min. and σ_X = 1 min.
Y = Julie's time; normally distributed with μ_Y = 33 min. and σ_Y = 2 min.
$D = X - Y$ = Difference in their times.

The random variable D is normally distributed because it is the difference of two normally distributed variables. For D, the difference between the sisters' times, the mean is $\mu = 37 - 33 = 4$ minutes and the standard deviation is $\sigma = \sqrt{\sigma_X^2 + \sigma_Y^2} = \sqrt{1^2 + 2^2} = \sqrt{5} = 2.236$. Alison beats Julie when $X < Y$ or, equivalently, when $D = X - Y < 0$. The probability that $D < 0$ is

$$P(D < 0) = P\left(Z < \frac{0 - 4}{2.236}\right) = P(Z < -1.79) = .0367$$

Alison swims a faster time about 3.7% of the time, which is about one day per month. ◆

8.9

turn on your mind

In Example 8.21, the time it takes Alison to swim a mile on any given day was assumed to be *independent* of the time it takes Julie to swim a mile on that same day. Refer to the explanation of when two random variables are independent, given earlier in this section. Explain whether it is reasonable to assume Alison's and Julie's times on the same day are independent in each of the following scenarios:

1. They swim at different times, and the one who swims first does not report her time to the other until after they are both done.
2. They swim side-by-side in two lanes of a pool.
3. They swim at different times, but the one who swims second knows the time it took the other to swim the mile that day.*

Combining Independent Binomial Random Variables

A linear combination of binomial random variables is generally not a binomial variable, but there is one situation where it is. The sum of independent binomial variables, each with the same success probability, has a binomial distribution. Put another way, independent binomial experiments with the same success probability can be combined and analyzed as a single binomial experiment. Imagine that you flip a coin five times while one of your friends flips a coin ten times. Probability questions about the total number of heads flipped by you and your friend can be answered using the binomial distribution for $n = 5 + 10 = 15$ trials and $p = .5$ (assuming you both flipped fairly!).

Adding Binomial Random Variables with the Same Success Probability

If X, Y, etc. are independent binomial random variables with n_X, n_Y, etc. trials and all have the same success probability p, then the sum $X + Y + \cdots$ is a binomial random variable with $n = n_X + n_Y + \cdots$, and success probability p.

Note: If the success probabilities differ, the sum is not a binomial random variable.

EXAMPLE 8.22
Donations Add Up

A volunteer organization is having a fund drive. Three volunteers agree to call potential donors. In past drives, about 20% of those called agreed to make a donation, independent of who called them. If the three volunteers make 10, 12, and 18 calls, respectively, what is the probability that they get at least ten donors?
    Let's define a random variable for each volunteer:

X = number of donors secured by volunteer 1; binomial $n = 10$, $p = .2$
Y = number of donors secured by volunteer 2; binomial $n = 12$, $p = .2$
W = number of donors secured by volunteer 3; binomial $n = 18$, $p = .2$

The total number of donors, $T = X + Y + W$, is a binomial random variable with $n = 10 + 12 + 18 = 40$ trials and $p = .2$. Notice that either there will be ten or more donors or there will be nine or fewer, so $P(T \geq 10) = 1 - P(T \leq 9) = 1 -$

.7318 (from Minitab) = .2682. So, in about 27% of the cases in this type of fund drive, there will be at least ten donors when 40 people are called. ◆

EXAMPLE 8.23
Strategies for Studying When
You Are Out of Time

You have a two-part multiple-choice test tomorrow, with 10 questions on one part and 10 on the other; all questions have four choices. You don't have time to study all the material well. From past experience you know that if you study all of the material in the time you have, you can narrow all 20 questions down to two choices and get them correct with probability .5, so your grade would be a binomial random variable with $n = 20$ and $p = .5$. If you study only the first set of material, you will get the questions right for that set with probability .8 but have to guess completely on the other set, which means $p = .25$ for those 10 questions. You need to get 13 or more questions right to pass the test. What should you do?

Strategy 1: Study all material. The score for the test has a binomial distribution with $n = 20$ and $p = .5$. The probability you pass is $P(\text{Score} \geq 13) = 1 - P(\text{Score} \leq 12)$, which is $1 - .8684$ (from Minitab) $= .1316$. The mean and standard deviation of the score will be useful for comparing the two strategies:

$$\text{Mean(Score)} = np = (20)(.5) = 10 \text{ questions}$$

$$\text{Variance(Score)} = np(1 - p) = (20)(.5)(1 - .5) = 5$$

$$\text{Standard deviation} = 2.24 \text{ questions}$$

Strategy 2: Study only set one. Score $= X + Y$, where $X =$ number correct on first part and $Y =$ number correct on second part. First, determine the mean and variance for each of X and Y:

X is binomial with $n = 10$, $p = .8$	Y is a binomial with $n = 10$, $p = .25$
$\text{Mean}(X) = np = 10(.8) = 8$	$\text{Mean}(Y) = np = 10(.25) = 2.5$
$\text{Variance}(X) = np(1 - p)$ $= 10(.8)(.2) = 1.6$	$\text{Variance}(Y) = np(1 - p)$ $= 10(.25)(.75) = 1.875$

With these values, we can determine the mean and standard deviation of Score $= X + Y$.

$$\text{Mean(Score)} = \text{Mean}(X) + \text{Mean}(Y) = 8 + 2.5 = 10.5 \text{ questions}$$

$$\text{Variance(Score)} = \text{Variance}(X) + \text{Variance}(Y) = 1.6 + 1.875 = 3.475$$

$$\text{Standard deviation} = \sqrt{3.47} = 1.86 \text{ questions}$$

Strategy 1 has a lower mean (10) than strategy 2 (10.5), so in the long run, strategy 2 will be slightly better. But, you are only taking a single exam, so an effective long-run strategy is not your concern. Strategy 1 has a higher standard deviation (2.24) than strategy 2 (1.86), so perhaps the chance of a score well above the mean is greater for strategy 1. This turns out not to be the case. For strategy 1, the passing score of 13 is $(13 - 10)/2.24 = 1.34$ standard deviations above the mean. For strategy 2, a score of 13 is $(13 - 10.5)/1.86 = 1.34$ standard deviations above the mean. It does not appear to matter which strategy you use.

Unfortunately, when the success probabilities are unequal (as in strategy 2), it can be tedious to find probabilities for a sum of binomial random variables with a small number of trials. For strategy 2, calculating the probability of passing involves finding probabilities for every possible combination of X and Y for which the sum is at least 13. This is a case where simulation may be an effective way to determine the probability. Using the Minitab package, strategy 2 was simulated by randomly generating values for X and Y and adding them together. This was done 1000 times, and a passing grade of 13 or more was achieved in 130 of the 1000 repetitions. The estimated probability that you pass the test using strategy 2 is about $130/1000 = .130$, almost identical to the probability of .1316 for strategy 1. Neither strategy offers you much hope! ◆

CASE STUDY 8.1
Does Caffeine Enhance the Taste of Cola?

Soft drink manufacturers claim that caffeine enhances the flavor of cola drinks, but in a study reported in the *Archives of Family Medicine* (Griffiths, 2000) researchers from Johns Hopkins University dispute this claim. In a taste test done by the researchers, 25 participants each tried 20 times to identify which of two cola drinks was "sample A" and which was "sample B." Unknown to the participants, one of the cola samples (A or B) contained caffeine, while the other cola sample did not. The particular sample (A or B) containing caffeine was the same for all 20 trials done by an individual. On a single trial, an individual tasted drinks from two unlabeled cups, and then guessed which cup contained sample A and which contained sample B. After each trial, the participant was told whether he or she was correct. Prior to the 20 trials that counted, each participant performed 5 practice trials.

If a participant could make 15 or more correct identifications during the 20 trials that counted, his or her flavor detection performance was called "significant" by the researchers. Only 2 of the 25 participants achieved this "significant" level. Further, over the 500 total trials done by the 25 participants, correct identifications were made only 53% of the time, a performance level that, according to the investigators, was not significantly better than chance performance (50%).

The research team used the binomial distribution to develop the criterion for individual "significance," which was 15 or more correct identifications in 20 tries. Recall that after each trial a participant learned whether he or she was correct. If this feedback were effective, the probability of success would increase during the 20 trials, so this study would not be a binomial experiment. For a binomial experiment, we need the same probability of success for each trial. If we assume, however, that a participant cannot detect the difference and makes a random guess every trial, X = the number of correct guesses for an individual is a binomial random variable with $n = 20$ and $p = .5$. For this binomial distribution, $P(X \leq 14) \cong .98$, and $P(X \geq 15) = 1 - .98 = .02$. The probability is only .02 that somebody who randomly guesses every time could get 15 or more correct, so doing this well would be "significant."

If all 25 participants randomly guessed for all trials, the "expected" number of "significant" participants would be $25 \times .02 = .5$, a value that could loosely be interpreted as about 0 or 1 participants. Only 2 participants achieved "significance," which is just slightly above the number we would expect if all participants randomly guess every time. A final piece of evidence is that correct identifications occurred only in 265 (53%) of all 500 trials performed by the 25 participants. This result would not be unusual if a random guess were made on every trial. For a binomial distribution with $n = 500$ and $p = .5$, $P(X \geq 265) \cong .10$.

Soft drink manufacturers were critical of this study. Some of the criticisms had to do with the sample sizes, both for participants and trials. The most pointed criticism, though, concerned the preparation of the two drink samples. The investigators created the two drink samples by either adding or not adding caffeine to caffeine-free Coca Cola. According to soft drink manufacturers, other ingredients must also be added along with caffeine in order to gain the flavor enhancement effects of caffeine. In their report, the investigators admitted that this was a weakness of the study, and suggested that additional studies might be needed to determine if caffeine does affect flavor.

KEY TERMS

Section 8.1
random variable, 249
continuous random variable, 250, 266
discrete random variable, 250

Section 8.2
probability distribution function (pdf), 253
cumulative distribution function (cdf), 254
cumulative probability, 254

Section 8.3
mean value of a discrete random variable, 256
expected value, 256

EXERCISES   Blue = Basics   Bold, **Bold** = Answer in back

Sections 8.1 and 8.2

8.1 A book is randomly chosen from a library shelf. For each of
the following characteristics of the book, decide whether
the characteristic is a continuous or a discrete random
variable:
a. Weight of the book (e.g., 2.3 pounds).
b. Number of chapters in the book (e.g., 10 chapters).
c. Width of the book (e.g., 8 inches).
d. Type of book (0 = hardback or 1 = paperback).
e. Number of typographical errors in the book (e.g.,
4 errors).

8.2 Decide whether each of the following characteristics of a
television news broadcast is a continuous or a discrete
random variable:
a. Number of commercials shown (e.g., five commercials).
b. Length of the first commercial shown (e.g., 15 seconds).
c. Whether there were any fatal car accidents reported
(0 = there were no fatal accidents reported, 1 = there
was at least one reported).
d. Whether or not rain was forecast for the next day
(0 = no rain forecast, 1 = rain forecast).

8.3 What is the missing value, represented by "?" in each of the
following probability distribution functions?
a.

k	1	4	5	7
P(X = k)	1/6	1/6	1/6	?

b.

k	100	200	300	400	500
P(X = k)	.1	.2	.3	.3	?

8.4 A professor gives a weekly quiz with varying numbers of
questions and uses a randomization device to decide how
many questions to include. Let the random variable X = the
number of questions on an upcoming quiz. The probability
distribution for X is given in the following table, but one
probability is missing. What is the probability that $X = 30$?

No. of Questions, X	10	15	20	25	30
Probability	.05	.20	.50	.15	?

8.5 Let the random variable X = number of phone calls you
will get in the next 24 hours. Suppose the possible values
for X are 0, 1, 2, or 3 and their probabilities are .1, .1, .3, and
.5, respectively. For instance, the probability that you will
receive no calls is .1.
a. Verify that the "Conditions for Probabilities for Discrete
Random Variables" given in Section 8.2 are met.
b. Draw a picture of the probability distribution function.
c. Find the cumulative distribution function.

8.6 For a fair coin tossed three times, the eight possible simple
events are: HHH, HHT, HTH, THH, HTT, THT, TTH, TTT. Let
X = number of heads. Find the probability distribution for
X by writing a table showing each possible value of X along
with the probability that value occurs.

8.7 Suppose the probability distribution for X = number of
jobs held during the past year for students at a school is as
follows:

No. of Jobs, X	0	1	2	3	4
Probability	.14	.37	.29	.15	.05

a. Find $P(X \leq 2)$, the probability that a randomly selected
student held two or fewer jobs during the past year.
b. Find the probability that a randomly selected student
held either one or two jobs during the past year.
c. Find $P(X > 0)$, the probability that a randomly selected
student held at least one job during the past year.

8.8 A kindergarten class has three left-handed children and
seven right-handed children. Two children are selected
without replacement for a shoe-tying lesson. Let X = the
number who are left-handed.
a. Write the simple events in the sample space. For instance, one simple event is RL, indicating that the first
child is right-handed and the second one is left-handed.
b. Find the probability for each of the simple events in the
sample space. (*Hint:* A tree diagram may help you solve
this.)

c. Find the probability distribution function for X.

d. Draw a picture of the probability distribution function for X.

8.9 Refer to Example 8.5, in which the probability distribution is given for the sum of two fair dice. Use the distribution to find the probability that the sum is even.

8.10 Consider three tosses of a fair coin. Write the sample space, and then find the probability distribution function for each of the following random variables:

a. X = number of tails.

b. Y = the difference between the number of heads and the number of tails.

c. Z = the sum of the number of heads and the number of tails.

8.11 A woman decides to have children until she has her first girl, or until she has four children, whichever comes first. Let X = number of children she has. For simplicity, assume that the probability of a girl is .5 for each birth.

a. Write the simple events in the sample space. Use B for boy and G for girl. For instance, one simple event is BG, because the woman quits once she has a girl.

b. Find the probability for each of the simple events in the sample space.

c. Find the probability distribution function for X.

d. Draw a picture of the probability distribution function for X.

8.12 Refer to Example 8.1, the scheduling of an outdoor event. Give an example of another continuous random variable (in addition to temperature) and another discrete random variable (in addition to number of planes flying overhead) that would influence the enjoyment of the event. Give the sample space for those random variables.

8.13 Refer to Example 8.5. Find the cumulative distribution function (cdf) for the sum of two fair dice.

8.14 Explain which would be of more interest in each of the following situations, the pdf or the cdf for X. If you think both would be of interest, explain why.

a. A politician wants to know how her constituents feel about a proposed new law. She hires a survey firm to take a random sample of voters in her district, and X = number in the sample who oppose the new law.

b. Each time someone is exposed to a virus he or she could become infected with it. Let X = number of exposures a person can sustain before becoming infected. For instance, if $X = 0$, they become infected on the first exposure; if $X = 1$, they become infected on the second exposure, and so on.

c. A pharmaceutical company wants to show that its new drug is effective for lowering blood pressure. The drug is administered to a sample of 800 patients with high blood pressure, and X = the number for whom the drug reduces blood pressure.

d. A caterer who serves dinner at wedding receptions is told in advance the number of people who have said they will attend; X is the number of people who actually attend the dinner.

Section 8.3

8.15 Burt pays $30 a year for "towing" insurance. He thinks that the probability that he will need to have his car

towed once in the next year is 1/10 (.1) and the probability that he will have to have it towed more than once is zero. It will cost $100 if his car is towed if he doesn't have insurance, but will cost nothing if he does have insurance. Let X = Burt's cost next year for towing and/or insurance.

a. If Burt buys the insurance, X has only one possible value. What is it?

b. If Burt doesn't buy the insurance, X has two possible values. List the two values and their probabilities.

c. Refer to parts (a) and (b). In each case, find $E(X)$. Using these two values for $E(X)$, explain whether Burt should buy the insurance.

8.16 The Brann family wants to have children. They are working with a financial advisor who informs them that based on families with similar characteristics, the probability distribution for the random variable X = number of children they might have is

No. of Children	0	1	2	3
Probability	.05	.60	.30	.05

a. What is the expected value of X?

b. Explain what "expected value" means in this example.

c. Is $E(X)$ a possible outcome for the number of children the Brann family will have?

8.17 The probability that Mary will win a game is .01, so the probability she will not win is .99. If Mary wins, she will be given $100, while if she loses, she must pay $5. If X = amount of money Mary wins (or loses), what is the expected value of X?

8.18 A random variable X has the following probability distribution:

Values of X	−1	0	1
Probability	.25	.50	.25

a. Calculate the mean of X.

b. Calculate the variance of X.

c. Calculate the standard deviation of X.

8.19 A random variable X has the following probability distribution:

Values of X	−2	0	2
Probability	.25	.50	.25

a. Calculate the mean of X.

b. Calculate the variance of X.

c. Calculate the standard deviation of X.

8.20 Find the expected value for the number of girls in a family with three children, assuming boys and girls are equally likely. (*Hint:* The probability distribution function was found in Example 8.4.)

8.21 Find the expected value for the sum of two fair dice. (*Hint:* The probability distribution function was found in Example 8.5.)

8.22 Find the standard deviation for the sum of two fair dice.

8.23 Find the standard deviation for the net gain in the California Decco lottery game in Example 8.6.

8.24 An insurance company expects 10% of its policyholders to collect claims of $500 this year, and the remaining 90%

to collect no claims. What is the expected value for the amount they will pay out in claims per person?

8.25 Refer to the previous exercise. If the insurance company wants to make a net profit of $10 per policyholder for the year, how much should they charge each person for insurance? Ignore administrative and other costs.

8.26 In 1991, 72% of children in the United States were living with both parents, 22% were living with mother only, 3% were living with father only, and 3% were not living with either parent (*World Almanac*, 1993, p. 945). What is the expected value for the *number* of parents a randomly selected child was living with in 1991? Does the concept of expected value have a meaningful interpretation for this example? Explain.

8.27 Suppose the probability that you get an A in any class you take is .3 and the probability that you get a B is .7. To construct a grade point average, an A is worth 4.0 and a B is worth 3.0. What is the expected value for your grade point average? Would you expect to have this grade point average for each quarter or semester of your school career? Explain.

8.28 Suppose you have to cross a train track on your commute. The probability that you will have to wait for a train is 1/5, or .20. If you don't have to wait, the commute takes 15 minutes, but if you have to wait it takes 20 minutes.
  a. What is the expected value of the time it takes you to commute?
  b. Is the expected value ever equal to the actual commute time? Explain.

Section 8.4

8.29 A fair coin is flipped 200 times. The random variable X = number of heads out of the 200 tosses.
  a. Is X a binomial random variable? If so, specify n and p. If not, explain why not.
  b. What is the expected value of X?

8.30 For each of the examples below, decide if X is a binomial random variable. If so, specify n and p. If not, explain why not.
  a. X = number of heads from flipping the same coin ten times, where the probability of a head = 1/2.
  b. X = number of heads from flipping two coins five times each, where the probability of a head for one coin = 1/2 and the probability of a head for the other coin = 1/4.
  c. X = number of cities in which it will rain tomorrow among five neighboring cities located within ten miles of each other.
  d. X = number of children who will get the flu this winter in a kindergarten class with 20 children.

8.31 For each of the following binomial random variables, specify n and p:
  a. A fair die is rolled 30 times. X = number of times a 6 is rolled.
  b. A company puts a game card in each box of cereal and 1/100 of them are winners. You buy ten boxes of cereal and X = number of times you win.
  c. Jack likes to play computer solitaire and wins about 30% of the time. X = number of games he wins out of his next 20 games.

8.32 Refer to the previous exercise. Find μ in each case.

8.33 Assuming X is a binomial random variable with $n = 10$ and $p = .20$, find the probability for each of the following values of X:
  a. $X = 5$.
  b. $X = 2$.
  c. $X = 1$.
  d. $X = 9$.

8.34 Find the mean and standard deviation for a binomial random variable X with:
  a. $n = 10, p = .50$.
  b. $n = 1, p = .40$.
  c. $n = 100, p = .90$.
  d. $n = 30, p = .01$.

8.35 Explain which of the conditions for a binomial experiment is *not* met for each of the following random variables:
  a. A football team plays 12 games in its regular season. X = number of games the team wins.
  b. A woman buys a lottery ticket every week for which the probability of winning anything at all is 1/10. She continues to buy them until she has won three times. X = the number of tickets she buys.
  c. A poker hand consists of 5 cards drawn from a standard deck of 52 cards. X = the number of aces in the hand.

8.36 A computer chess game and a human chess champion are evenly matched. They play ten games. Find probabilities for the following events.
  a. They each win five games.
  b. The computer wins seven games.
  c. The human chess champion wins at least seven games.

8.37 In an ESP test, a "participant" tries to draw a hidden "target" photograph that is unknown to anyone in the room. After the drawing attempt, the participant is shown four choices and asked to determine which one had been the real target. The real target is randomly selected from the four choices in advance, so the probability of a correct match by chance is 1/4. The test is repeated ten times using four new photographs each time.
  a. Go through the conditions for a binomial experiment, and explain how this situation fits each one of them, assuming the participant is just guessing each time.
  b. Let X = number of correct choices in the ten tests. If the participant is just guessing, is X a binomial random variable? If not, explain why not. If so, specify n and p.
  c. If the participant is just guessing, find $P(X \geq 6)$.
  d. Suppose the participant actually has some psychic ability and can get each answer correct with probability .5 instead of .25. Find $P(X \geq 6)$.
  e. Compare the answers in parts (c) and (d). If the participant actually selects six of the ten answers correctly, would you believe he or she was just guessing, or that he or she was using some psychic ability? Explain your answer. (Note that there is no correct answer here; your reasoning is what counts.)

8.38 For each of the following situations, assume X is a binomial random variable with the specified n and p, and find the requested probability.
  a. $n = 10, p = .5, P(X = 4)$.
  b. $n = 10, p = .3, P(X \geq 4)$.
  c. $n = 10, p = .3, P(X \leq 3)$.
  d. $n = 5, p = .1, P(X = 0)$.
  e. $n = 5, p = .1, P(X \geq 1)$.

8.39 For each of the following scenarios, write the desired probability in a format such as $P(X \le 10)$ and specify n and p. Do not actually compute the desired probability. If you cannot specify a numerical value for p, write in words what it represents.

    a. A random sample of 1000 adults is drawn from the United States. $X =$ the number in the sample who are living with a partner but are not married. The desired probability is the probability that at least 1/4 of the sample fits this description.

    b. A pharmaceutical company claims that 20% of those taking its new allergy medication will experience drowsiness. To test this claim, it randomly assigns 500 people to take the new medication and measures $X =$ the number who experience drowsiness. The desired probability is the probability that 110 or more people in the sample experience drowsiness if the claim is true.

    c. A student has not studied the material for a 20-question True-False test and simply guesses on each question. A passing grade is 70%; the desired probability is the probability that the student passes the test.

8.40 Find the expected value and standard deviation for a binomial random variable with each of the following values of n and p:

    a. $n = 10, p = 1/2$.
    b. $n = 100, p = 1/4$.
    c. $n = 2500, p = 1/5$.
    d. $n = 1, p = 1/10$.
    e. $n = 30, p = .4$.

8.41 Use the direct formulas for expected value and standard deviation to verify that the mean and standard deviation for a binomial random variable with $n = 2$ and $p = .5$ are $\mu = 1$ and $\sigma = 0.7071$, respectively. (*Hint:* The only possible values of X are 0, 1, 2. Find their probabilities and use the formulas at the end of Section 8.3.)

Sections 8.5 and 8.6

8.42 Suppose that the time students wait for a bus can be described by a uniform random variable X, where X is between 0 minutes and 60 minutes.

    a. What is the probability that a student will wait between 0 and 30 minutes for the next bus?

    b. What is the probability that a student will have to wait at least 30 minutes for the next bus?

8.43 In each situation below, calculate the standardized score (or z-score) for the value x:

    a. Mean $\mu = 0$, standard deviation $\sigma = 1$, value $x = 1.5$.
    b. Mean $\mu = 10$, standard deviation $\sigma = 6$, value $x = 4$.
    c. Mean $\mu = 10$, standard deviation $\sigma = 5$, value $x = 0$.
    d. Mean $\mu = -10$, standard deviation $\sigma = 15$, value $x = -25$.

8.44 For each value of z^*, find the cumulative probability $P(Z \le z^*)$:

    a. $z^* = 0$.
    b. $z^* = -0.35$.
    c. $z^* = 0.35$.
    d. $z^* = 1.96$.
    e. $z^* = -2.33$.
    f. $z^* = 2.58$.
    g. $z^* = 1.65$.

8.45 Weights (X) of men in a certain age group have a normal distribution with mean $\mu = 180$ pounds and standard deviation $\sigma = 20$ pounds. Find each of the following probabilities:

    a. $P(X \le 200) =$ probability the weight of a randomly selected man is less than or equal to 200 pounds.

    b. $P(X \le 165) =$ probability the weight of a randomly selected man is less than or equal to 165 pounds.

    c. $P(X > 165) =$ probability the weight of a randomly selected man is more than 165 pounds.

8.46 A game is played with a spinner on a circle, like the minute hand on a clock. The circle is marked evenly from 0 to 100, so, for example, the 3:00 position corresponds to 25, the 6:00 position to 50, and so on. The player spins the spinner and the resulting number is the number of seconds he or she is given to solve a word puzzle. Let $X =$ amount of time the player is given to solve the puzzle.

    a. Is X a uniform random variable, a binomial random variable, a normal random variable, or none of these? Explain how you know.

    b. Write down the density function for X. Be sure to specify the range of values for which it holds.

    c. Find $P(X \le 15$ seconds).

    d. Find $P(X \ge 40$ seconds).

    e. Draw a picture of the density function of X, and use it to illustrate the probabilities you found in parts (c) and (d).

    f. What is the expected value of X? This was not covered in the text, so explain your reasoning.

8.47 Give an example of a uniform random variable that might occur in your daily life.

8.48 Draw the "curve" corresponding to each of the following random variables, then shade the area corresponding to the desired probability. You *do not* need to compute the probability.

    a. X is a uniform random variable from 10 to 20, $P(10 \le X \le 13)$.

    b. X is a normal random variable with $\mu = 75$ and $\sigma = 5$, $P(70 \le X \le 85)$.

    c. X is a normal random variable with $\mu = 15$ and $\sigma = 10$, $P(-5 \le X \le 20)$.

8.49 Find the following probabilities for a standard normal random variable Z:

    a. $P(Z \le -1.4)$.
    b. $P(Z \le 1.4)$.
    c. $P(-1.4 \le Z \le 1.4)$.
    d. $P(Z \ge 1.4)$.

8.50 Find the following probabilities for a standard normal random variable Z:

    a. $P(Z \le -3.72)$.
    b. $P(-3.72 \le Z \le 3.72)$.
    c. $P(Z \ge 15)$.

8.51 Find the value z^* that satisfies each of the following probabilities for a standard normal random variable Z:

    a. $P(Z \le z^*) = .025$.
    b. $P(Z \le z^*) = .975$.
    c. $P(-z^* \le Z \le z^*) = .95$.

8.52 For each of the following formulas given in this chapter, use the rule referenced from Chapter 7 to verify that the relationship is correct for a standard normal random variable Z:

    a. Use Rule 1 to verify that $P(Z > a) = 1 - P(Z \le a)$.

b. Use Rule 2b to verify that $P(Z \le b) = P(Z \le a) + P(a \le Z \le b)$. Use that result to show that $P(a \le Z \le b) = P(Z \le b) - P(Z \le a)$.

8.53 Find the following probabilities for Verbal SAT test scores X, for which the mean is 500 and the standard deviation is 100. Assume that SAT scores are described by a normal curve.
   a. $P(X \le 500)$.
   b. $P(X \le 650)$.
   c. $P(X \ge 700)$.
   d. $P(500 \le X \le 700)$.

8.54 Refer to Example 8.14. Find the proportion of college women whose heights fall into the following ranges:
   a. Between 62 inches and 65 inches.
   b. Between 60 inches and 70 inches.
   c. Less than 70 inches.
   d. Greater than 60 inches.
   e. Either less than 60 inches *or* greater than 70 inches.

8.55 Refer to Example 8.14. Find the height such that about 25% of college women are shorter than that height. What is the percentile ranking for that height?

8.56 Refer to Example 8.14. Find the height such that about 10% of college women are taller than that height. What is the percentile ranking for that height?

8.57 Based on data from Penn State students, assume that right handspan measurements for college women are approximately normally distributed, with mean 20 cm and standard deviation 1.8 cm, and for men they are normally distributed, with mean 22.5 cm and standard deviation 1.5 cm. Measure your own right handspan.
   a. Assuming you fit with the population of college students of your sex, what is the *z*-score corresponding to your right handspan measurement? (Be sure to give your sex and right handspan measurement with your answer.)
   b. What proportion of students of your sex have right handspan measurements smaller than yours?

8.58 Suppose the yearly rainfall totals for a city in northern California follow a normal distribution, with mean of 18 inches and standard deviation of 6 inches. For a randomly selected year, what is the probability that total rainfall will be in each of the following intervals?
   a. Less than 10 inches.
   b. Greater than 30 inches.
   c. Between 15 and 21 inches.
   d. Greater than 35 inches.

8.59 Refer to the previous exercise. Suppose that in a given year the total rainfall is only 6 inches. You work for the local newspaper, and your editor has asked you to write a story about the terrible drought the town is suffering and how abnormal the situation is. Write a few sentences that you could use to explain the statistical facts to your readers (and your editor). Be sure to comment on whether or not you agree that the situation is terribly abnormal.

Section 8.7

8.60 Suppose that a fair coin is flipped $n = 100$ times.
   a. Calculate the mean and standard deviation of $X =$ number of heads. (*Hint:* The variable X has a binomial distribution.)
   b. Use the normal approximation to the binomial distri-

bution to estimate the probability that the number of heads is greater than or equal to 60.
   c. Repeat part (b) using the continuity correction.

8.61 Suppose $p = .512$ is the proportion of one-child families in which the child is a boy.
   a. For a random sample of $n = 50$ one-child families, estimate the probability that there will be 20 or fewer boys. Use the normal approximation to the binomial distribution.
   b. Repeat part (a) using the continuity correction.

8.62 Use the normal approximation to the binominal distribution to approximate the stated probability in each of the following scenarios:
   a. A random sample of 1000 eligible voters is drawn and $X =$ number who actually voted in the last election. It is known that 60% of all eligible voters did vote. Find (approximately) $P(X \le 620)$.
   b. To be eligible for a certain job, women must be at least 62 inches tall, and 87% of women meet this criterion. In a random sample of 2000 women, $X =$ number who qualify for the job (based on height). Find (approximately) $P(X \le 1700)$.

8.63 In a test for extrasensory perception, the "participant" repeatedly guesses the suit of a card randomly sampled with replacement from an ordinary deck of cards. There are four suits, all equally likely. The participant guesses 100 times, and $X =$ number of correct guesses.
   a. Explain why X is a binomial random variable, and specify n and p, assuming the participant is just guessing.
   b. Find the mean and standard deviation for X if the participant is just guessing.
   c. Suppose the participant guesses correctly 33 times. Find the approximate probability of guessing this well or better by chance.
   d. Suppose the participant guesses correctly 50 times. Find the approximate probability of guessing this well or better by chance. In that circumstance, would you be convinced that the participant was doing something other than just guessing? Explain.

Section 8.8

8.64 Suppose the heights of adult males in a population have a normal distribution with mean $\mu = 70$ inches and standard deviation $\sigma = 2.8$ inches. Two unrelated men will be randomly sampled. Let $X =$ height of the first man and $Y =$ height of the second man.
   a. Consider $D = X - Y$, the difference between the heights of the two men. What type of distribution will the variable D have?
   b. What is the mean value for the distribution of D?
   c. Assuming independence between the two men, find the standard deviation of D.
   d. Determine the probability that the first man is more than 3 inches taller than the second man. That is, find $P(D > 3)$.
   e. Find the probability that one of the men is at least 3 inches taller than the other. That is, find the probability that either $(D > 3)$ or $(D < -3)$.

8.65 Suppose the length of time a person takes to use an ATM machine is normally distributed with mean $\mu = 100$ seconds and standard deviation $\sigma = 10$ seconds. There are

$n = 4$ people ahead of Jackson in a line of people waiting to use the machine. He is concerned about T = total time the four people will take to use the machine.

a. What is the mean value of T = total time for the four people ahead of Jackson?

b. Assuming the times for the four people are independent of each other, determine the standard deviation of T.

c. Jackson hopes the total time T is less than or equal to 360 seconds (6 minutes). Find $P(T < 360)$, the probability that the total waiting time is less than 360 seconds.

8.66 Ed and Taylor work together on a group quiz that has 15 multiple-choice questions, each with four choices for the possible answer. Unfortunately, neither had time to study so they decide to randomly guess at all answers. Ed guesses answers for the first 7 questions, while Taylor guesses for the other 8 questions.

a. For any single question on the quiz, what is p = chance of a correct guess?

b. Let X = number of correct guesses that Ed makes. What are the values of n and p for the binomial distribution that describes X?

c. Let Y = number of correct guesses that Taylor makes. What are the values of n and p for the binomial distribution that describes Y?

d. Consider $X + Y$ = total correct guesses that Ed and Taylor make. What are the values of n and p for the binomial distribution that describes $X + Y$?

8.67 The variable X has a normal distribution with mean $\mu = 75$ and standard deviation $\sigma = 6$. The variable Y has a normal distribution with mean $\mu = 70$ and standard deviation $\sigma = 8$. Variables X and Y are independent variables.

a. Find the mean and standard deviation of the sum $X + Y$.

b. Find the mean and standard deviation of the difference $X - Y$.

8.68 Give the mean, variance, and standard deviation of the sum $X + Y$ in each of the following cases. If possible, name what distribution the sum has as well.

a. X is a binomial random variable with $n = 10$, $p = .5$. Y is a binomial random variable with $n = 20$, $p = .4$. X and Y are independent.

b. X is a normal random variable with $\mu = 100$ and $\sigma = 15$. Y is a normal random variable with $\mu = 50$ and $\sigma = 10$. X and Y are independent.

c. X is a normal random variable with $\mu = 100$ and $\sigma = 15$. Y is a binomial random variable with $n = 200$, $p = .25$. X and Y are independent.

8.69 Ethan has a midterm in his statistics class, which starts 10 minutes after the scheduled end of his biology class. The biology teacher rarely ends class on time though, and the amount of time he is overtime is approximately normally distributed, with mean of 2 minutes and standard deviation of 1/2 minute. The time it takes Ethan to get from one class to another is also normal, with mean 6 minutes and standard deviation 1 minute.

a. Is it reasonable to assume that the two times are independent? Explain.

b. Assuming the two times are independent, what is the probability that Ethan will be late for his exam?

8.70 Joe performs remarkably well on "remote viewing" ESP tests, which require the "viewer" to draw a picture, then determine which of four possible photos was the intended "target." Because the target is randomly selected from among the four choices, the probability of a correct match by chance is 1/4, or .25. Joe participates in three experiments with $n = 10$, 20, and 50 trials, respectively. In the three experiments, his numbers correct are 4, 8, and 20, respectively, for a total of 32 correct out of 80 trials.

a. For each experiment, what is the expected number correct if Joe is just guessing?

b. Over all 80 trials, what is the expected number correct? Explain how you found your answer.

c. For each experiment separately, find the approximate probability that someone would get as many correct as Joe did, or more, if they were just guessing.

d. Out of the 80 trials overall, find the approximate probability that someone would get as many correct as Joe did (32), or more, if they were just guessing.

e. Compare your answers in parts (c) and (d). If Joe were trying to convince someone of his ability, would it be better to show the three experiments separately, or would it be better to show the combined data? Explain.

8.71 Annmarie is trying to decide whether to take the train or the bus into the city. The train takes longer but is more predictable than the bus because there are no traffic delays. Train times are approximately normally distributed, with mean 60 minutes and standard deviation 2 minutes, while bus times are approximately normally distributed, with mean 50 minutes and standard deviation 8 minutes. The bus and train times are independent of each other. Find the probability that the train is faster on any given day.

8.72 Charles, Julia, and Alex are in grades 4, 3, and 2, respectively, and are representing their school at a spelling bee. The school's team score is the sum of the number of words the individual students spell correctly out of 50 words each. Different words are given for each grade level. From practicing at school, it is known that the probability of spelling each word correctly is .9 for Charles and .8 for the younger two, Julia and Alex.

a. Find the mean and standard deviation for the number of correct words for each child.

b. Find the mean and standard deviation for the team score.

c. Assume the individual scores are independent. Does the team score have a binomial distribution? Explain.

d. Although not obvious from the material in this chapter, the team score would be approximately normal with the mean and standard deviation you were asked to find in part (b). If last year's team score was 131, what is the approximate probability that this year's team scores as well or better?

Chapter Exercises

8.73 Find the following probabilities for X = pulse rates of women, for which the mean is 75 and the standard deviation is 8. Assume a normal distribution.

a. $P(X \leq 71)$.

b. $P(X \geq 85)$.

c. $P(59 \leq X \leq 95)$.

8.74 Refer to the previous exercise about the pulse rates of women. What is the 10th percentile of women's pulse rates?

8.75 The vehicle speeds at a particular interstate location can be described by a normal curve. The mean speed is 67 mph and the standard deviation is 6 mph. What proportion of vehicle speeds at this location are faster than 75 mph?

8.76 Do you think that a normal curve would be a good approximation to the distribution of the ages of all individuals in the world? Briefly explain your answer.

8.77 A histogram is drawn of the weights of all students in a class of 100 men and 100 women. Would this histogram have a bell shape? Briefly explain your answer.

8.78 Suppose that 10% of a population is left-handed. What is the probability that in a sample of $n = 10$ individuals, 3 or more individuals are left-handed?

8.79 New spark plugs have just been installed in a small airplane with a four-cylinder engine. There is one spark plug per cylinder, so four spark plugs have been installed. For each spark plug, the probability that it is defective and will fail during its first 20 minutes of flight is 1/10,000, independent of the other spark plugs.
   a. For each spark plug, what is the probability that it will not fail?
   b. If one spark plug fails, the plane will shake and not climb higher, but it can be landed safely. What is the probability that this happens?
   c. If two or more spark plugs fail, the plane will crash. What is the probability that this happens?

8.80 In the casino game of roulette, a gambler can bet on which of 38 numbers will be the result when the roulette wheel is spun. On a $2 bet, a gambler gains $70 if he or she picks the right number, but loses the $2 otherwise.
   a. Let X = amount gained or lost on a $2 bet on a roulette number. Write out the probability distribution of X.
   b. Calculate $E(X)$, the mean value of X. What does this value indicate about the advantage that a casino has over roulette players?

8.81 Suppose that a college determines the following distribution for X = number of courses taken by full-time students this semester:

k	3	4	5	6	7
$P(X = k)$	.07	.14	.52	.25	.02

   a. Write out the cumulative distribution function of X.
   b. What is the probability that a randomly selected full-time student is taking five or fewer courses this semester?
   c. What is the probability that a randomly selected full-time student is taking more than five courses?

8.82 Refer to exercise 8.81. What is the mean number of courses taken by full-time students?

8.83 Explain which of the conditions for a binomial experiment is *not* met for each of the following random variables:
   a. A ten-question quiz has five True-False questions and five multiple-choice questions, each with four possible choices for the answer. A student randomly picks an answer for every question. X = number of answers that are correct.
   b. Four students are randomly picked without replacement from a class of ten women and ten men. X = number of women among the four selected students.

8.84 The standard medical treatment for a certain disease is successful in 60% of all cases.
   a. The treatment is given to $n = 200$ patients. What is the probability that the treatment is successful for 70% or more of these 200 patients? (*Tip:* 70% of $n = 200$ is 140 patients.)
   b. The treatment is given to only $n = 20$ patients. What is the probability that the treatment is successful for 70% or more of these 20 patients?

8.85 Kim and her sister Karen each plan to have four children. Assume the probability of a girl is .50, independent across births.
   a. If X = number of girls Kim will have, what is the distribution of X?
   b. If Y = number of girls Karen will have, what is the distribution of Y?
   c. What is the distribution for T, the total number of girls Kim and Karen have?

8.86 Shaun (3 years old) and Patrick (4 years old) are each allowed to pick one book for bedtime stories, and the parent on duty reads the two books sequentially. The lengths of time it takes to read the books from which they select are approximately normally distributed, with mean 5 minutes and standard deviation 2 minutes.
   a. Does it make sense to assume the times for the two books are independent? Explain.
   b. On about what proportion of nights will storytime exceed 15 minutes, assuming the times for the two books are independent?

8.87 You will need a computer program that simulates a large number of binomial random variables at once for this exercise. Refer to Example 8.23, "Strategies for Studying When You Are Out of Time." Simulate what would happen if you used strategy 2, in which the total score was the sum of two binomial random variables, one with $n = 10$ and $p = .8$, the other with $n = 10$ and $p = .2$. Run 1000 simulations, and determine the proportion of them for which you would have received a passing grade (13 or more). Explain what you did.

REFERENCES

Griffiths, R. R., and E. M. Vernotica (2000). "Is caffeine a flavoring agent in cola soft drinks?" *Arch. Fam. Med.,* Vol. 9, pp. 727–734.

World Almanac and Book of Facts (1993). Mark S. Hoffman (ed.), New York: Pharos Books.

CHAPTER 9

Means and Proportions As Random Variables

Video games again? Has their favorite TV show been cancelled? See Example 9.7.

This chapter introduces the reasoning that allows researchers to make conclusions about entire populations using relatively small samples of individuals. The secret to understanding how things work is to understand what kind of dissimilarity we should expect to see among different samples from the same population.

In Chapter 8 we learned how to work with random variables, including how to make probability statements about ranges of possible values, what to expect as the long-run average, and how to identify special classes of random variables. This chapter considers what happens if we regard a summary statistic for an entire random sample as a random variable. For instance, if we take a random sample of adults and find the proportion who have a certain trait or opinion, that proportion is the numerical outcome of a random circumstance. In other words, the sample proportion is itself a random variable.

This chapter serves as an introduction to the reasoning that allows researchers to make conclusions about entire populations on the basis of a relatively small sample of individuals. By understanding how a sample mean or proportion behaves as a random variable, we can begin to understand the population from which it was found.

The basic idea is that we must work backwards, from a sample to a population. We start with a question about a population like: How many teenagers are infected with HIV? At what average age do left-handed people die? What is the average income of all students at a large university? We collect a sample from the population about which we have the question and measure the variable of interest. We can then answer the question of interest for the sample. Finally, based on statistical theory, we will be able to determine how close our sample answer is to what we really want to know, the true answer for the population. ◆

9.1 | UNDERSTANDING DISSIMILARITY AMONG SAMPLES

The secret to understanding how things work is to understand what kind of dissimilarity we should expect to see in various samples from the same population. For example, suppose we knew that most samples were likely to provide an answer that is within 10% of the population answer. Then we would also know the population answer should be within 10% of whatever our specific sample gave. Armed only with our sample value, we could make a good guess about the population value. You have already seen this idea at work in Chapter 4, when we used the margin of error for a sample survey to estimate results for the entire population. Statisticians have worked out similar techniques for a variety of sample measurements.

Statistics and Parameters

We have already encountered the idea of a single random variable derived from an entire sample in Chapter 8. Remember that if we take a random sample from a population and define X = number in the sample with a certain opinion, then X is a binomial random variable. If we repeated the entire sampling procedure over and over again, we would get different values for X each time.

A **statistic** is a numerical value computed from a sample. Examples of statistics when the sample consists of a quantitative variable are the mean, the median, and the standard deviation. For example, for the sample of 103 Penn State women whose right handspan measurements were discussed in Chapter 2, the mean was about 20 cm and the standard deviation was about 1.8 cm. These values are examples of *statistics* associated with that sample of measurements. In contrast, a **parameter** is a number associated with a population, and it is generally assumed that the values of parameters are fixed, unchanging numbers. The average right handspan for the population of all college women is an example of a parameter.

For categorical variables, statistics associated with a sample include the number or the proportion of the sample who fall into various categories. For example, when Penn State students were asked to choose a number from 1 to 10, 19 out of 190, or exactly 10%, chose the number 8. Both the frequency of 19 and the percentage of 10% are statistics associated with that sample.

The distinction between a sample statistic and a population parameter is crucial. To be clear which is being referred to in formulas, different symbols are used to represent a statistic and the corresponding population parameter. You might recall that in Chapter 2 we used the symbols $\bar{x}$ and s to represent the mean and standard deviation of a sample. In Chapter 8, however, we used the symbols μ and σ for the mean and standard deviation of a population. For instance, if the mean pulse rate is 75.3 for a sample of 16 women, we should express this mean as $\bar{x}$ = 75.3. If we used this sample mean to estimate the mean pulse rate of all women, we would write that $\bar{x}$ = 75.3 estimates the unknown value of the population mean μ.

For a categorical variable, the symbol p is used to represent the proportion of a population that has a particular characteristic or trait. The symbol $\hat{p}$ (read as "p hat") is used to represent the proportion of a sample that has the characteristic or trait of interest. For example, if we sample 100 students at our school and there are 53 females in the sample, we would write that $\hat{p}$ = .53 is the proportion of females in the sample. On the other hand, if we learn from the school registrar that 51.2% of all enrolled students are female, we would write that p = .512 is the proportion of all students who are female.

Sampling Distributions

Every time a new sample is taken, the sample statistics will change. For instance, if we were to sample a different collection of 103 Penn State women, we would get a different mean right handspan value. *The distribution of possible values of a*

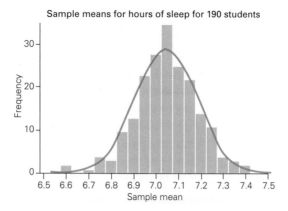

FIGURE 9.1   **The sampling distribution for mean hours of sleep for 200 different samples, each with $n = 190$ students**

statistic for repeated samples of the same size from a population is called the **sampling distribution** *of the statistic.* Think of the statistic as a random variable, and the sampling distribution as the distribution for that random variable. As we will see in this and subsequent chapters, many statistics of interest have sampling distributions that are approximately normal distributions.

EXAMPLE 9.1
Mean Hours of Sleep for College Students

The sample of Penn State students first discussed in Chapter 2 included the question "How many hours of sleep did you get last night?" The mean of the 190 responses was 7.1 hours. Suppose we were to ask another 190 students that same question. It is unlikely that the mean response would be exactly 7.1 hours again.

Figure 9.1 shows a histogram of possible sample means we might get, found from simulating this experiment 200 times. A normal curve is superimposed on top of the histogram to demonstrate that the sampling distribution of possible sample means is approximately a normal distribution. We will return to this idea later in this chapter.

To understand how this picture was developed, think about a university that has 50,000 students. You and 199 friends (you are very popular) each go out and take a random sample of 190 of those students. You ask the students in your sample how many hours they slept last night and then calculate the mean of the 190 answers. Finally, you and your friends (200 of you) get together and draw a histogram of your 200 different sample means. You would get a picture similar to Figure 9.1. ◆

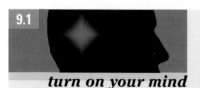

9.1

turn on your mind

Suppose you and your friends each asked 1900 people how many hours they slept last night, instead of asking 190 people each, so that you each compute a sample mean based on 1900 answers. When the 200 of you created the histogram of your results, how do you think it would compare to Figure 9.1? Would it be less spread out? More spread out? In other words, would possible sample means based on 1900 people be more variable or less variable than possible sample means based on 190 people, with the samples taken from the same population?*

> ### *in summary* Statistic, Parameter, and Sampling Distribution
>
> ◆ A **statistic** is a numerical summary of a sample. Its value may differ for different samples.
>
> ◆ A **parameter** is a numerical summary of a population. It is considered to be fixed and unchanging.
>
> ◆ The **sampling distribution** for a statistic is the distribution of possible values of the statistic for repeated random samples of the same size taken from a population.

9.2 | SAMPLING DISTRIBUTIONS FOR SAMPLE PROPORTIONS

Suppose we conduct a binomial experiment with n trials and get successes on x of the trials. Or, suppose we measure a categorical variable for a representative sample of n individuals, and x of them have responses in a certain category. In each case, we can compute the statistic $\hat{p}$ = the **sample proportion** = x/n, the proportion of trials resulting in success, or the proportion in the sample with responses in the specified category. If we repeated the binomial experiment or collected a new sample, we would probably get a different value for the sample proportion.

For instance, suppose we would like to know what proportion of a large population carries the gene for a certain disease. We sample 25 people and use the sample proportion from that sample to estimate the true answer. Suppose that in truth 40% of the population carries the gene, although this fact is unknown to us.

What will happen if we randomly sample 25 people from this population? Will we always find 10 people (40%) with the gene and 15 people (60%) without? Or, will the number and proportion with the gene differ for different samples of $n = 25$? You may recognize that this situation is a binomial experiment and that if X = number (out of 25) who carry the gene, then X is a *binomial random variable* with $n = 25$ and $p = .4$.

Many Possible Samples

Consider four different random samples of 25 people taken from this population. Remember that we are trying to estimate the proportion of the population with the gene, based on the sample *statistic,* which is the *sample proportion.* We do not know that in truth the population proportion (the parameter) is actually 40%. Here is what we would have concluded about the proportion of people who carry the gene, given four possible samples with X as specified:

Sample 1: $X = 12$, proportion with gene = $12/25 = .48 = 48\%$
Sample 2: $X = 9$, proportion with gene = $9/25 = .36 = 36\%$
Sample 3: $X = 10$, proportion with gene = $10/25 = .40 = 40\%$
Sample 4: $X = 7$, proportion with gene = $7/25 = .28 = 28\%$

Notice that each one of these samples would have given a different answer, and the sample answer may or may not have matched the truth about the population, which is that 40% carry the gene.

In practice, when a researcher conducts a study similar to this one or a polling organization randomly samples a group of people to measure public opinion, only one sample is collected. There is no way to determine whether or not the sample is an accurate reflection of the population. However, statisticians have calculated what to expect for the vast majority of possible samples.

The Normal Curve Approximation Rule for Sample Proportions

In Chapter 8 we learned if X is a binomial random variable with large n, then X is also approximately a normal random variable. This result can also be applied to the proportion X/n, since dividing every value by n does not change the shape of the distribution of possible values. In other words, the *sampling distribution for a sample proportion is approximately normal.*

Let's give this result a name. We will call it the **Normal Curve Approximation Rule for Sample Proportions.** Sometimes we will refer to it simply as the **Rule for Sample Proportions.** This rule can be applied in two different situations:

Situation 1: A random sample is taken from an actual population.
Situation 2: A binomial experiment is repeated numerous times.

Conditions for Which the Normal Curve Approximation Rule for Sample Proportions Applies

In each situation, three conditions must all be met for the Normal Curve Approximation Rule for Sample Proportions to apply:

Condition 1: The Physical Situation   Situation 1: There is an actual population with a fixed proportion who have a certain trait, opinion, disease and so on.

or

Situation 2: There is a repeatable situation for which a certain outcome is likely to occur with a fixed relative frequency probability.

Condition 2: Data Collection   In Situation 1, a random sample is selected from the population, thus ensuring that the probability of observing the characteristic is the same for each sample unit.

or

In Situation 2, the situation is repeated numerous times, with the outcome each time independent of all other times.

Condition 3: The Size of the Sample or Number of Trials   The size of the sample or the number of repetitions is relatively large. The necessary size depends on the proportion or probability under investigation. It must be large enough so that we are likely to see at least five of each of the two possible responses or outcomes, in other words, np and $n(1 - p)$ must be at least 10.

Examples of Scenarios for Which the Rule for Sample Proportions Applies

From your study of binomial random variables in Chapter 8, you should recognize that these are simply the conditions for a binomial random variable with large n. Here are some examples of scenarios that meet these conditions.

Scenario 1: Election Polls   A pollster wants to estimate the proportion of voters who favor a certain candidate. The voters are the population units and favoring the candidate is the opinion of interest. A large sample is taken.

Scenario 2: Television Ratings   A television rating firm wants to estimate the proportion of households with television sets that are tuned to a certain television program. The collection of all households with television sets makes up the population and being tuned to that particular program is the trait of interest.

Scenario 3: Consumer Preferences   A manufacturer of soft drinks wants to know what proportion of consumers prefer a new mixture of ingredients compared with the old recipe. The population consists of all consumers and the response of interest is preference for the new formula over the old.

Scenario 4: Testing ESP   A researcher studying extrasensory perception wants to know the probability that people can successfully guess which of five symbols is on a hidden card. Each card is equally likely to contain each of the five symbols. There is no physical population. The repeatable situation of interest is a guess and the response of interest is a successful guess. The researcher wants to see if the probability of a correct guess is higher than .20, the chance for a successful guess if there were no such thing as extrasensory perception. The number of guesses made by each person should be at least 50 for Condition 3 (number of trials) to hold. With random guessing, $p = 1/5 = .2$, so $n = 50$ trials would lead to $np = 10$ and $n(1 - p) = 40$.

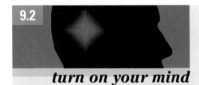

9.2

turn on your mind

Each of scenarios 1 to 4 can be thought of as a binomial experiment. Specify what constitutes a "success" in each case. Then construct your own example of a scenario for which the Rule for Sample Proportions would hold, and specify what constitutes a "success" for your example.*

definition   The **Normal Curve Approximation Rule for Sample Proportions** can be defined as follows:

Let p = population proportion of interest or binomial probability of success.

Let $\hat{p}$ = corresponding sample proportion or proportion of successes.

If numerous samples or repetitions of the same size n are taken, the distribution of possible values of $\hat{p}$ is approximately a normal curve distribution with

Mean = p

Standard deviation = s.d.$(\hat{p})$ = $\sqrt{\dfrac{p(1 - p)}{n}}$

This approximate normal distribution is called the **sampling distribution of $\hat{p}$.**

Technical Note: The n individuals in the sample or the repetitions must be independent, equivalent to Condition 3 in a binomial experiment.

EXAMPLE 9.2
Proportions for a Candidate

Suppose that of all voters in the United States, 40% are in favor of Candidate X for president. Pollsters take a sample of 2400 voters. What proportion of the sample would be expected to favor Candidate X? The rule tells us that the pro-

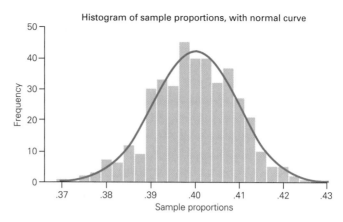

FIGURE 9.2   **Histogram of 400 sample proportions based on $n = 2400$, $p = .4$**

portion of the sample who favor Candidate X is a random variable that has a normal distribution. The mean and standard deviation for the distribution are

$$\text{Mean} = p = .40 \ (40\%)$$

$$\text{s.d.}(\hat{p}) = \sqrt{\frac{p(1-p)}{n}} = \sqrt{\frac{(.4)(1-.4)}{2400}} = \sqrt{\frac{.24}{2400}} = \sqrt{.0001} = .01$$

Figure 9.2 shows a histogram of the sample proportions resulting from simulating this situation 400 times, with the appropriate normal curve superimposed on the histogram. Notice that as expected from the Empirical Rule in Chapter 2, the possible values cover the range $\mu \pm 3(\text{s.d.})$, in this case, $0.40 \pm 3(.01)$. ◆

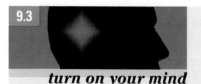

9.3

turn on your mind

For Example 9.2, into what range of possible values should the sample proportion fall 95% of the time, based on the Empirical Rule? If the polling organization used a sample of only 600 voters instead, would this range of possible sample proportions be wider, more narrow, or the same? Explain your answer, and explain why it makes intuitive sense.*

EXAMPLE 9.3
Caffeinated or Not?

Case Study 8.1 described a study in which 25 people each tried 20 times to guess whether a cola they taste is caffeinated or not caffeinated. For some of the 20 trials, the cola being tasted contained caffeine while in the others it did not. Suppose we take the view that people cannot judge merely by taste whether caffeine is present. If this were so, then participants would actually be randomly guessing on each trial, so $p = .5$, where p is the probability of correctly guessing whether caffeine is present or not on any trial. If this is the case for all individuals, then the overall experiment can be seen as a binomial experiment with $n = 500$ trials (25 people $\times$ 20 trials each) and $p = .5$.

We can use the Rule for Sample Proportions to approximate the sampling distribution of potential values of $\hat{p}$, the proportion of correct guesses in the sample of 500 trials. The sampling distribution will be approximately a normal curve with the following mean and standard deviation:

FIGURE 9.3 **Approximate sampling distribution of possible sample proportions when** $n = 500$, $p = .5$

$$\text{Mean} = p = .5$$

$$\text{Standard deviation} = \text{s.d.} (\hat{p}) = \sqrt{\frac{.5 \, (1 - .5)}{500}} = .0224$$

Figure 9.3 shows this normal curve. Based on the Empirical Rule, we know values fall in the interval $\mu \pm (2 \times \text{s.d.})$ about 95% of the time. So, in this situation there's about a 95% chance that the proportion of correct guesses in a sample of 500 trials will be in the interval $.5 \pm (2 \times .0224)$, or between about .455 and .545. In the actual experiment, there were 265 correct guesses made, so the sample proportion was $\hat{p} = 265/500 = .53$. This is well within the interval of values that would occur for about 95% of samples of $n = 500$ trials where a random guess is made on every trial. In other words, $\hat{p} = .53$ is not strong evidence against the hypothesis that in general people cannot taste whether a cola contains caffeine or not.

Estimating the Population Proportion from a Single Sample Proportion

In practice, when we actually take a random sample of 2400 voters in a political poll, we have only one sample proportion and we don't know the true population proportion. However, we do know how far apart the sample proportion and the true population proportion are likely to be. That information is contained in the **standard deviation of $\hat{p}$,**

$$\text{s.d.}(\hat{p}) = \sqrt{\frac{p(1 - p)}{n}}$$

which can be estimated using the sample proportion $\hat{p}$ in the formula. The estimated version is called the **standard error of $\hat{p}$,**

$$\text{s.e.}(\hat{p}) = \sqrt{\frac{\hat{p}(1 - \hat{p})}{n}}$$

For instance, if $\hat{p} = .39$ and $n = 2400$, the standard error is $\sqrt{.39 \, (1 - .39)/2400} = .01$. This value estimates the theoretical standard deviation of the sampling distribution for sample proportions, based on a single sample. Because we know that the mean, which is the true proportion p, is almost surely within 3 standard deviations of the observed value $\hat{p}$, we know that p is almost surely within the range $\hat{p} \pm 3(\text{s.e.}) = .39 \pm 3(.01) = .39 \pm .03$. So we know the true proportion who support the candidate is almost surely between .36 and .42. Notice that the only numerical value we needed to determine this was the sample proportion $\hat{p}$, and of course the sample size n. We will explore this idea in much more detail in Chapter 10, "Estimating Proportions with Confidence."

9.3 | WHAT TO EXPECT OF SAMPLE MEANS

In the previous section, the value of interest was the proportion falling into one category of a categorical variable or the probability of success in a binomial experiment. We saw that if we knew the size of the sample and the magnitude of the true proportion, we could determine an interval of values that was likely to cover the sample proportion.

We now turn to the case where the information of interest involves the mean or means of quantitative variables. For example, researchers might want to find the mean difference in age at death for left- and right-handed siblings. A company that sells oat products might want to know the mean cholesterol level people would have if everyone had a certain amount of oat bran in their diet. To help determine financial aid levels, a large university might want to know the mean income of all students on campus who work.

Many Possible Samples

Suppose a population consists of thousands or millions of individuals, and we are interested in estimating the mean of a quantitative variable. If we sample 25 people and compute the mean of the variable for that sample, how close will that *sample mean* be to the *population mean* we are trying to estimate? Each time we take a sample we will get a different sample mean. Can we say anything about what we expect those means to be?

For example, suppose we are interested in estimating the average weight loss for everyone who attends a national weight-loss clinic for ten weeks. Suppose, unknown to us, the distribution of weight losses for everyone in this population is approximately normal with a mean of 8 pounds and a standard deviation of 5 pounds. Figure 9.4 shows a normal curve with these characteristics. If the weight losses are approximately normal, we know from the Empirical Rule in Chapter 2 that 95% of the individuals will fall within 2 standard deviations, or 10 pounds, of the mean of 8. In other words, 95% of the individual weight losses will fall between -2 (a gain of two pounds) and 18 pounds lost.

Table 9.1 lists some possible samples that could result from randomly sampling 25 people from this population; these were indeed the first four samples produced by a computer simulation of this situation. The weight losses have been put into increasing order for ease of reading. A negative value indicates a weight gain. (As an aside, notice that, as expected, there are $5/100 = 5\%$ of the values outside of the range -2 and 18.) We are interested in the behavior of the possible sample means for samples such as these.

Here are the sample means and standard deviations for each of these four samples. You can see that the sample means, although all different, are relatively

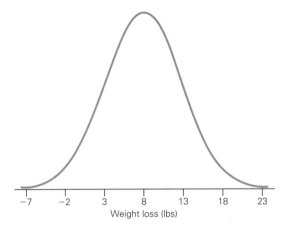

FIGURE 9.4   **Normal curve for individual weight losses with mean $\mu = 8$ pounds and standard deviation $\sigma = 5$ pounds**

TABLE 9.1   ■   **Four Potential Samples from a Population with $\mu = 8$, $\sigma = 5$**

Sample 1:	1, 1, 2, 3, 4, 4, 4, 5, 6, 7, 7, 7, 8, 8, 9, 11, 11, 13, 13, 14, 14, 15, 16, 16
Sample 2:	−2, −2, 0, 0, 3, 4, 4, 4, 5, 5, 6, 8, 8, 9, 9, 9, 9, 9, 10, 11, 12, 13, 13, 16
Sample 3:	−4, −4, 2, 3, 4, 5, 7, 8, 8, 9, 9, 9, 9, 9, 10, 10, 11, 11, 11, 12, 12, 13, 14, 16, 18
Sample 4:	−3, −3, −2, 0, 1, 2, 2, 4, 4, 5, 7, 7, 9, 9, 10, 10, 10, 11, 11, 12, 12, 14, 14, 14, 19

close to the population mean of 8. You can also see that the sample standard deviations are relatively close to the population standard deviation of 5. This latter fact is important, because although we are interested in using the sample mean to estimate the population mean, we will need to use the sample standard deviation to help find a margin of error for this estimate.

Sample 1: Mean = 8.32 pounds, standard deviation = 4.74 pounds
Sample 2: Mean = 6.76 pounds, standard deviation = 4.73 pounds
Sample 3: Mean = 8.48 pounds, standard deviation = 5.27 pounds
Sample 4: Mean = 7.16 pounds, standard deviation = 5.93 pounds

Conditions for Which the Normal Curve Approximation Rule for Sample Means Applies

The mean of a sample is another example of a statistic. Because it is a numerical outcome of a random circumstance, it is also a random variable. As with sample proportions, statisticians have developed a rule to tell us what to expect for the possible distribution of sample means in repeated sampling from the same population. We call this rule the **Normal Curve Approximation Rule for Sample Means,** or simply the **Rule for Sample Means.** The Rule for Sample Means applies in both of the following types of situations:

Situation 1   The population of the measurements of interest is bell-shaped and a random sample of any size is measured.

Situation 2   The population of measurements of interest is not bell-shaped, but a *large* random sample is measured. Thirty is usually used as an arbitrary demarcation of "large," but if there are extreme outliers it is better to have a larger sample.

There are actually only a limited number of situations for which the Rule for Sample Means does *not* apply. It does not apply at all if the sample is not random, and it does not apply for small random samples unless the original population is bell-shaped. In practice, it is often difficult to get a random sample. Researchers are usually willing to use the Rule for Sample Means as long as they can get a representative sample with no obvious sources of confounding or bias.

Examples of Scenarios for Which the Rule for Sample Means Applies
Here are some examples of scenarios that meet the conditions.

> *Scenario 1: Average Weight Loss*   A weight-loss clinic is interested in measuring the average weight loss for participants in its program. It makes the assumption that the weight losses will be bell-shaped, so the Rule for Sample Means will apply for any sample size. The population of interest is all current and potential clients and the measurement of interest is weight loss.
>
> *Scenario 2: Average Age At Death*   A researcher is interested in estimating the average age at which left-handed adults die, assuming they have lived to be at least 50. Because ages at death are not bell-shaped, the researcher should measure at least 30 such ages at death. The population of interest is all left-handed people who live to be at least 50 years old. The measurement of interest is age at death.
>
> *Scenario 3: Average Student Income*   A large university wants to know the mean monthly income of students who work. The population consists of all students at the university who work. The measurement of interest is monthly income. Because incomes are not bell-shaped and there are likely to be outliers (a few people with high incomes), the university should use a large random sample of students. They should take particular care to reach the people who are actually selected to be in the sample. There could be a large bias if, for example, they were willing to replace the desired respondent with a roommate who happened to be home when the researcher called. The students working the longest hours, and thus making the most money, would probably be hardest to reach by phone and the least likely to respond to a mail questionnaire.

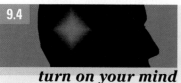

9.4

turn on your mind

Construct an example of interest to you personally for which the Rule for Sample Means applies and for which a study could be done to estimate a population mean.*

> *definition* The **Normal Curve Approximation Rule for Sample Means** can be defined as follows:
>
> Let μ = mean for the population of interest.
> Let σ = standard deviation for the population of interest.
> Let $\bar{x}$ = mean for the sample = sample mean.
>
> If numerous random samples of the same size n are taken, the distribution of possible values of $\bar{x}$ is approximately normal, with
>
> Mean = μ
>
> Standard deviation = s.d.$(\bar{x})$ = $\dfrac{\sigma}{\sqrt{n}}$
>
> This approximate normal distribution is called the **sampling distribution of** $\bar{x}$ or the **sampling distribution of the mean.**
>
> *Technical Note:* The n observations in each sample must all be independent, which they will be if random samples are used.

Be careful that you do not confuse the standard deviation for the original population of measurements, σ, with the **standard deviation of the sample means,** $\sigma/\sqrt{n}$. The parameter σ is a measure of the variability among individual measurements within the population. The parameter $\sigma/\sqrt{n}$ is a measure of the variability among sample means for the many different random samples of size n that can be taken from the population.

EXAMPLE 9.4
Hypothetical Mean Weight Loss

For our hypothetical weight-loss example, the population mean and standard deviation were μ = 8 pounds and σ = 5 pounds, respectively, and we were taking random samples of size 25. The rule tells us that potential sample means $\bar{x}$ could be anything from a normal curve with a mean μ of 8 pounds and standard deviation s.d. $(\bar{x})$ of $\sigma/\sqrt{n}$ = 5/5 = 1.0. (We divide the population standard deviation of 5 by the square root of n = 25, which also happens to be 5, to find the standard deviation of potential sample means.)

From the Empirical Rule in Chapter 2, we know the following facts about possible sample means in this situation, based on intervals extending 1, 2, and 3 standard deviations from the mean of 8:

- There is a *68%* chance that the sample mean will be between *7 and 9.*
- There is a *95%* chance that the sample mean will be between *6 and 10.*
- It is *almost certain* that the sample mean will be between *5 and 11.*

If you look at the four hypothetical samples we chose (Table 9.1), you will see that the sample means range from 6.76 to 8.48, well within the range we expect to see using these criteria. Figure 9.5 displays a histogram and superimposed normal curve for 400 simulated sample means, each based on a sample of 25 weight losses from a population with a mean of 8 pounds and a standard deviation of 5 pounds. ◆

This discussion presumes that we know the population mean and the standard deviation. Obviously, that's not much use to us in real situations when the population mean is what we are trying to determine. In Chapter 12 we will see

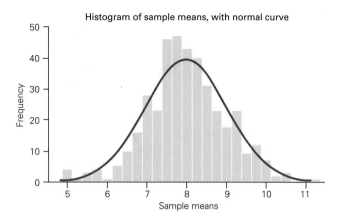

FIGURE 9.5   **Histogram of 400 sample means for samples of $n = 25$, $\mu = 8$, $\sigma = 5$**

how to modify the Rule for Sample Means to accurately estimate the population mean when all we have available is a single sample for which we can compute the sample mean and the sample standard deviation.

Standard Error of the Mean

In practice, the population standard deviation σ is rarely known, so the sample standard deviation s is used in its place when determining the standard deviation for the sampling distribution of sample means. Consistent with what we did for sample proportions in the previous section, when making this substitution we call the result the **standard error of the mean.** The terminology makes sense because the standard error measures roughly how much, on average, the sample mean $\bar{x}$ is in error as an estimate of the population mean μ.

We denote the standard error of the mean as s.e.$(\bar{x})$, and the formula is

$$\text{s.e.}(\bar{x}) = \frac{s}{\sqrt{n}}$$

For the $n = 25$ weight losses in Sample 1 shown in Table 9.1, the sample standard deviation is $s = 4.74$ pounds. Based on this sample, the standard error of the mean is $s/\sqrt{n} = 4.74/\sqrt{25} = 0.948$. This value estimates $\sigma/\sqrt{n}$, the theoretical standard deviation of the sampling distribution for sample means. If we were to use a different sample of $n = 25$ observations from Table 9.1, the standard error of the mean would have a different value.

Increasing the Size of the Sample

Suppose we had considered samples of 100 people in the weight-loss example rather than samples of 25. Notice that the mean of the possible sample means would still be 8 pounds, but the standard deviation, s.d.$(\bar{x})$, would decrease. It would now be $\sigma/\sqrt{n} = 5/\sqrt{100} = 0.5$, instead of $5/\sqrt{25} = 1.0$. Therefore, for samples of size $n = 100$, the standard deviation of potential sample means is only half of what it is for samples of size $n = 25$. In Example 9.4 we applied the Empirical Rule to see that for samples of $n = 25$, the sample means are likely to range between 8 ± 3 pounds, or between about 5 pounds and 11 pounds. In contrast, for samples of $n = 100$, the sample means are likely to range only between about 8 ± 1.5 pounds, or between 6.5 pounds and 9.5 pounds. Figure 9.6 compares the sam-

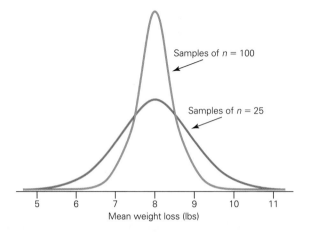

FIGURE 9.6 **Sampling distributions of potential sample means for samples of $n = 25$ and $n = 100$ taken from population that has mean $\mu = 8$ and standard deviation $\sigma = 5$**

pling distributions for samples of $n = 25$ and $n = 100$ and shows clearly that there is less variability among potential sample means for the larger sample size.

In general, a fourfold increase in sample size cuts the standard deviation of the distribution of possible means, and thus the range of likely means, in half. A ninefold increase in sample size cuts the standard deviation of possible means to a third of what it was, and so on. It's obvious that the Rule for Sample Means tells us the same thing our common sense tells us: Larger samples tend to result in more accurate estimates of population values than smaller samples.

9.5

turn on your mind

From the weight-loss example discussed in this section, we learned that increasing the sample size fourfold would about halve the range of possible sample means. Would the range of *individual* weight losses in the sample be likely to increase, decrease, or remain about the same if the sample size were increased fourfold? Explain.*

Sampling for a Long, Long Time: The Law of Large Numbers

The Rule for Sample Means tells us how much variability in possible sample means to expect for different sample sizes. There is a much simpler technical result in statistics called the **Law of Large Numbers,** which guarantees that the sample mean $\bar{x}$ will eventually get "close" to the population mean μ *no matter how small a difference* you use to define "close." The result requires that the population mean be finite (not infinite) and that the observations in the sample be independent of each other. "Eventually" could be a very long time, depending on how much variability there is in the population of measurements. In practice, the Law of Large Numbers says that for any specific population, the larger the sample size, the more you can count on $\bar{x}$ to be an accurate representation of μ.

The Law of Large Numbers is also used by insurance companies, casinos, and so on to give them peace of mind about what will happen to their average

*HINT: Which sample size might be more likely to include some of the more extreme weight losses for individuals in the population, leading to a wider range?

profit (or loss) "in the long run." They know that eventually, after enough gamblers have bet or customers have bought insurance and claims have been paid, they can count on the mean net profit to be close to the theoretical expected value or mean. This peace of mind comes at a price; they must make sure they have enough cash on hand to pay the claimant or gambler through the natural fluctuations in the process.

EXAMPLE 9.5
The Long Run for the Decco Lottery Game

In Chapter 8 we computed the expected value of the California Decco lottery game and found that the state gains an average of $0.35 for every dollar played. In other words, over all tickets sold, it pays out an average of $0.65 and keeps $0.35. Obviously, this average is not achieved in the short run. The state knows that to achieve this average it must sell hundreds of thousands of tickets. For instance, the first ten tickets sold might all be losers, or there might be a $5000 prize winner. The state must be prepared to pay winnings in the short run. States and casinos that operate gambling games have large funds they can use to pay winning players. For this reason, even when a game is an even bet, a gambler with limited funds will still eventually lose while the "house" weathers the lows (and highs). ◆

turn on your computer

Finding the Pattern in Sample Means

The main idea for any sampling distribution is that it gives the pattern for how the potential value of a statistic may vary from sample to sample. The Rule for Sample Means tells us that in two common situations, a normal curve approximates the sampling distribution of the sample mean. The **SampleMeans** applet lets us see the pattern that emerges when we look at the means of many different random samples from the same population. Figure 9.7 illustrates the appearance of the applet when it is started. The histogram shown is a population

FIGURE 9.7 **The SampleMeans applet starting point**

FIGURE 9.8   **Example of the SampleMeans applet after two samples of *n* = 25 have been selected**

FIGURE 9.9   **Example of the SampleMeans applet after 500 samples of *n* = 25 have been selected**

of individual measurements with $\mu = 8$ and $\sigma = 5$, as for the weight-loss example in this section. Notice that the distribution of individual measurements is bell-shaped, so the Rule for Sample Means should hold for any sample size. At the start, the number of observations per sample is set at $n = 25$.

What Happens

The applet simulates choosing simple random samples from the population. For each sample selected, the mean is calculated, displayed in the box under "Sample Means," and graphed in a histogram. The individual measurements in the most recently selected sample are shown at the bottom of the display. The buttons under "# Samples" can be used to select either 1, 10, 100, or 500 different samples at a time. Figure 9.8 illustrates an example in which two different samples of $n = 25$ per sample have been selected, while Figure 9.9 shows an example in which 500 different samples of $n = 25$ were selected. Notice that in Figure 9.8, the two distinct sample means are evident in the bottom histogram, while in Figure 9.9 the histogram of the 500 sample means (the bottom histogram) looks bell-shaped.

You can control the speed of the applet with the menu under the "# Samples" button. With slow and fast sampling you see the histogram change as samples are added. With "batch" sampling, the histogram of sample means is not updated until all requested samples have been selected. The "# Observations per sample" box is used to change the sample size per sample. The "Clear" button clears all results and should be used whenever changing the sample size ("# Observations per sample").

What to Do

With the number of observations per sample set at 25, generate many different samples. Take note of the resulting shape of the histogram of sample means. Take note of the center and range of the distribution of sample means. Use the

Clear button to clear the histogram, and then enter 100 as "# Observations per sample." Generate many different samples for samples of this size. Again, take note of the shape, center, and spread of the histogram of sample means. Verify that the range of the histogram of sample means is approximately $\mu \pm 3 \times (\sigma/\sqrt{n})$. Do the same things for "# Observations per sample" = 200. Experiment with the applet to formulate answers to the following questions. What feature of the distribution of sample means is affected by changing the sample size? What features are not affected by changing the sample size? How does changing the sample size affect the chance that a sample mean will fall in the interval 7.5 to 8.5 (within 0.5 pound of the true mean)?

Lessons Learned

This applet lets you see that a normal curve approximates the distribution of sample means for different random samples of the same size from a population. This is evident from the shape of the histogram for sample means, assuming you select many different samples of the same size. You can also see that increasing the number of observations in a sample decreases the variability among sample means. Notice that regardless of sample size, the histogram of sample means is approximately a normal curve centered at 8 pounds, the population mean. But, increasing the sample size increases the likelihood that a sample mean will be close to 8 pounds, the population mean.

9.4 | WHAT TO EXPECT IN OTHER SITUATIONS: CENTRAL LIMIT THEOREM

We have discussed two common situations that arise in assessing public opinion, conducting medical research, and many other applications. The first situation arises when we want to know what proportion of a population fall into one category of a categorical variable or want to estimate the probability of success in a binomial experiment. The second situation occurs when we want to know the mean of a population for a measurement variable.

There are numerous other situations for which researchers would like to use results from a sample to say something about a population or to compare two or more populations. Statisticians have determined rules similar to those in this chapter for most of the situations researchers are likely to encounter. Most of these rules are derived from the same basic statistical result, called the **Central Limit Theorem**.

definition The **Central Limit Theorem** states that if n is sufficiently large, the sample means of random samples from a population with mean μ and finite standard deviation σ are approximately normally distributed with mean μ and standard deviation $\sigma/\sqrt{n}$.

Technical Note: The mean and standard deviation given in the Central Limit Theorem hold for any sample size; it is only the "approximately normal" shape that requires n to be sufficiently large.

You may notice that the Central Limit Theorem is nothing more than a restatement of the Rule for Sample Means, except that in the latter we specified the conditions needed to satisfy "if n is sufficiently large." The Central Limit Theorem is often used in situations where it isn't obvious that the statistic of interest is actually a sample mean. In fact, both of the rules in this chapter are special cases of the Central Limit Theorem.

The Rule for Sample Proportions follows from the Central Limit Theorem by defining each observation in the sample to be either 1 or 0. The observation is 1 if the individual sampled has the desired trait or opinion, and the observation is 0 otherwise. The sample mean is simply the average of the 1s and 0s in the sample, which is the sample proportion $\hat{p}$. The population mean $= p$, the proportion of 1s in the population. The population standard deviation is the standard deviation of the 1s and 0s in the population, which can be shown to be $\sigma = \sqrt{p(1 - p)}$, so $\sigma/\sqrt{n} = \sqrt{p(1 - p)/n}$.

9.6

turn on your mind

Verify that if the raw data for each individual in a sample is 1 when the individual has a certain trait, and 0 otherwise, then the sample mean is equivalent to sample proportion with the trait. You can do this by using a formula, explaining it in words, or constructing a numerical example.*

EXAMPLE 9.5—CONTINUED
California Decco Winnings

In Chapter 8 and in Example 9.5 of this chapter we discussed the California Decco lottery game, for which the mean amount lost per ticket over millions of tickets sold is $\mu = \$0.35$. The standard deviation is $\sigma = \$29.67$, reflecting the large variability in possible amounts won or lost, ranging from a net win of $4999 to a net loss of $1. Suppose a store sells 100,000 tickets in a year. What are the possible values for the average loss per ticket? The Central Limit Theorem tells us the answer: Assuming $n = 100,000$ is sufficiently large, the distribution of possible sample means is approximately normal, with a mean (loss) of $0.35 and a standard deviation of $\sigma/\sqrt{n} = \$29.67/\sqrt{100,000} = \$0.09$. Using the Empirical Rule, the mean loss is almost surely in the range $0.35 \pm 3(\$0.09)$, or between 8 cents and 62 cents. Remember that this is the average loss per ticket for a collection of 100,000 tickets. Notice that the *total* loss for the 100,000 is therefore likely to be between ($0.08 \times 100,000$) and ($0.62 \times 100,000$), or between $8000 and $62,000! Certainly you can think of better ways to invest $100,000. ◆

Mean or Median as "Central"?

The Central Limit Theorem provides information about what to expect of the mean in large samples, but the median remains the statistic of choice for representing the "central" value in an extremely skewed set of numbers. For instance, consider two random samples of net winnings for ten Decco tickets:

Net Amount Won ($n = 10$)	Mean	Median
−1,−1,−1,−1,−1,−1,−1, 0, 0	−$0.80 (80 cent loss)	−$1.0 ($1.00 loss)
−1,−1,−1,−1,−1,−1,−1, 0, 0, 4999	$499.20 (won)	−$1.0 ($1.00 loss)

*HINT: Write out a short list of 0s and 1s; then find the mean for the list and compare that to the proportion of 1s in the list.

Clearly, the median is a better measure of the "typical" net win in ten plays of this game. The problem is that for the occasional sample with a large outlier, the mean will be inflated, but this is not representative of the entire population. Remember that in this case, the population mean is 35 cents lost, and the population median is −$1.00. In fact, 72.6% of all tickets (as shown in Chapter 8) result in $1.00 lost or a −$1.00 net win.

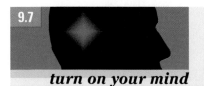

9.7

turn on your mind

The Central Limit Theorem does not specify what is meant by "a sufficiently large sample." What factor(s) about the population of values do you think determine how large is large enough for the approximate normal shape to hold? Consider the California Decco example. Do you think $n = 30$ would be large enough for the distribution of possible values for the average loss to be approximately normal? Why or why not?

Now consider the handspan measurements of females. Do you think $n = 30$ would be large enough for the approximate normal shape to hold? What is different about these two examples?*

9.5 | SAMPLING DISTRIBUTION FOR ANY STATISTIC

In Section 9.1 we defined the sampling distribution of a statistic as the distribution of possible values of a statistic for repeated samples of the same size from a population. Two special cases discussed in this chapter were the sampling distribution of a sample proportion $\hat{p}$ and the sampling distribution of the sample mean. In both cases, under specified conditions the sampling distribution was approximately normal.

Every statistic has a sampling distribution, but the appropriate distribution may not always be normal, or even be approximately bell-shaped. Statisticians have developed a theory that helps determine the appropriate sampling distribution for many common statistics. It is also possible to construct an approximate sampling distribution for a statistic by actually taking repeated samples of the same size from a population and constructing a relative frequency histogram or table for the values of the statistic over the many samples. The next example illustrates this method.

EXAMPLE 9.6
Winning the Lottery by Betting on Birthdays

In the Pennsylvania Cash 5 lottery game, players select five numbers from the integers 1 to 39, and the grand prize is won by anyone who selects the same five numbers as the ones randomly drawn by the lottery officials. Suppose that in such a game someone bets numbers corresponding to the days of the month on which five family members were born, which is anecdotally a common strategy. Obviously, the highest integer someone using that strategy could select is 31, so there is no chance of winning the grand prize at all if the highest of the winning draws is bigger than 31—that is, if it is 32 to 39. What is the probability of that happening?

The statistic in question is the *highest* of five integers randomly drawn without replacement from the integers 1 to 39; let's call it *H*. For instance, if the num-

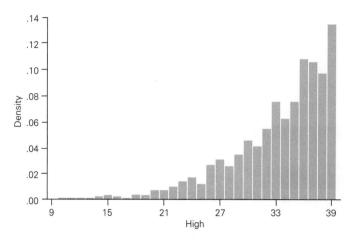

FIGURE 9.10   **Approximate sampling distribution for highest Cash 5 number**

bers selected are 3, 12, 22, 36, and 37, then $H = 37$. Determining the actual sampling distribution of H is a complicated exercise in probability. However, we can approximate the sampling distribution of H using lottery data supplied by the Pennsylvania Lottery Commission for repeated plays of the Cash 5 game.

The relative frequency histogram in Figure 9.10 shows the values of H for the 1560 games between April 23, 1992 and May 18, 2000, obtained from the website www.palottery.com. (The game was played weekly at first, then daily.) Table 9.2 displays some of the possible values of H and the frequency with which they occurred in the 1560 games. For instance, for one game, the highest of the five numbers was only 10, while for 70 out of the 1560 games (about 4.5% of the time) the high was 30. (Examples noted in this paragraph are in boldface in the table to

TABLE 9.2 ▦ **Partial Sampling Distribution for Highest Cash 5 Number (examples from text are in bold)**

H = high	Count	Cumulative	Percent	Cum. Percent
10	**1**	1	0.06	0.06
11	1	2	0.06	0.13
12	1	3	0.06	0.19
13	1	4	0.06	0.26
14	2	6	0.13	0.38
15	3	9	0.19	0.58
16	2	11	0.13	0.71
⋮	⋮	⋮	⋮	⋮
29	55	309	3.53	19.81
30	**70**	379	**4.49**	24.29
31	65	**444**	4.17	**28.46**
32	85	529	5.45	33.91
33	118	647	7.56	41.47
34	98	745	6.28	47.76
35	118	863	7.56	55.32
36	169	1032	10.83	66.15
37	165	1197	10.58	76.73
38	152	1349	9.74	86.47
39	**211**	1560	**13.53**	100.00

help you find them.) Notice that the high was 31 or less for 444 of the games, which represents 28.46% of the games. In other words, all of the winning numbers are contained in the integers 1 to 31 in just over 28% of the games. For the remaining 72% of the games, the highest number is over 31. In fact, the most common outcome is $H = 39$, the highest possible number, which occurred in 211/1560, or about 13.5% of all games played.

Notice that the shape of the probability distribution for H is not symmetric and is skewed to the left. This example illustrates that not all sample statistics have bell-shaped sampling distributions. ◆

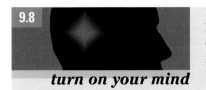

9.8

turn on your mind

Example 9.6 described a sample statistic, H = highest number drawn, for the Cash 5 lottery game. Give another example of a sample statistic for the Cash 5 game, and describe what you think the shape of its sampling distribution would be.*

9.6 | STANDARDIZED STATISTICS

In Chapter 2 we learned that sometimes it is useful to transform a raw score into a standardized score, or z-score, measuring how many standard deviations the raw score falls above or below the mean. The z-scores for raw scores from a normal population with mean μ and standard deviation σ are determined by the following formula:

$$z = \frac{x - \mu}{\sigma}$$

The resulting z-scores form a *standard normal population,* with $\mu = 0$, $\sigma = 1$.

Similarly, sometimes it is useful to transform a raw sample statistic into its standardized version. Doing this transformation requires knowledge of the mean and standard deviation of the sampling distribution for the sample statistic. If the sampling distribution is approximately normal, then the standardized statistic has, approximately, a standard normal distribution.

As we learned earlier in this chapter, many sample statistics have approximately normal sampling distributions. Researchers often conduct studies in which they observe one value of such a statistic (such as a sample proportion or a sample mean), and they want to know how far that value falls from a hypothetical center of the sampling distribution. Sometimes it is useful to obtain that information in standardized form. In other words, it is useful to find the standardized version of the observed sample value. Let's look at how to find the standardized version of sample means and proportions.

Standardized z-Statistics for Sample Means and Proportions

Assuming appropriate conditions are met for the Rule for Sample Proportions and the Rule for Sample Means to hold, as given in Sections 9.2 and 9.3, the following **standardized statistics** have, approximately, a standard normal distribution:

*HINT: Some possibilities are lowest value, median value, and the mean value.

$$\text{For sample means: } z = \frac{\bar{x} - \mu}{\text{s.d.}(\bar{x})} = \frac{\bar{x} - \mu}{\sigma/\sqrt{n}} = \frac{\sqrt{n}(\bar{x} - \mu)}{\sigma}$$

$$\text{For sample proportions: } z = \frac{\hat{p} - p}{\text{s.d.}(\hat{p})} = \frac{\hat{p} - p}{\sqrt{\dfrac{p(1-p)}{n}}}$$

EXAMPLE 9.7

Unpopular TV Shows

It is common for television networks to cancel shows with consistently low ratings. Ratings are based on a random sample of households, and the sample proportion of households tuned to the show ($\hat{p}$) is an estimate of p, the population proportion watching the show. Suppose a network wants p to be at least .20 or the show will be cancelled. In a random sample of 1600 households, 288, or a proportion of .18, are watching. Is this close enough to the desired population proportion of .20 to be attributable to random fluctuations in possible sample proportions, even if p really is .20 or higher?

The sampling distribution of $\hat{p}$ is approximately normal, with p = actual population proportion watching the show. If $p = .20$, then the mean and standard deviation of the sampling distribution are $p = .20$ and s.d. $(\hat{p}) = .01$. Therefore, the standardized version of $\hat{p} = .18$ is

$$z = \frac{\hat{p} - p}{\text{s.d.}(\hat{p})} = \frac{\hat{p} - p}{\sqrt{\dfrac{p(1-p)}{n}}} = \frac{.18 - .20}{\sqrt{\dfrac{(.2)(.8)}{1600}}} = \frac{.18 - .20}{.01} = -2.00$$

The sample proportion of .18 is about 2 standard deviations below the mean of .20. That probably isn't far enough below .20 to rule out sampling variability as the reason for such a low proportion, but it probably would be if the ratings were consistently that low. ◆

9.7 | STUDENT'S *t*-DISTRIBUTION: REPLACING σ WITH s

We have a dilemma. Computing a standardized score for $\bar{x}$ requires knowledge of the *population* standard deviation σ, which is rarely available. The best we can do is to approximate σ with the *sample* standard deviation s, and thus approximate s.d.$(\bar{x}) = \sigma/\sqrt{n}$ with s.e.$(\bar{x}) = s/\sqrt{n}$. Unfortunately, for small sample sizes this approximation is often off the mark, so the standardized statistic doesn't exactly conform to the standard normal distribution. Instead, under certain conditions (see In Summary box on the next page), it has a probability distribution called **Student's *t*-distribution,** or just ***t*-distribution** for short.

Theoretically, the *t*-distribution arises when a standardized score or statistic is calculated for a normally distributed score or statistic using the *sample* standard deviation in place of the *population* standard deviation. The Student's *t*-distribution was named for William Gosset, who published a paper about it in 1908 using the pseudonym "A Student" so as not to reveal that his place of employment, the Guinness Brewery, was using statistical methods. A *t*-distribution has a bell shape, it's centered at 0, and it is more spread out than the standard normal curve in that there's more probability in the extreme areas than there is for the standard normal curve.

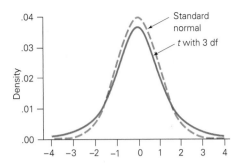

FIGURE 9.11 **A t-distribution with df $= 3$ and a standard normal distribution**

A parameter called **degrees of freedom,** abbreviated as **df,** is associated with any t-distribution. In most applications, this parameter is a function of the sample size for the problem, but the specific formula for the degrees of freedom depends on the type of problem. For the standardized version of a sample mean $\bar{x}$, df $= n - 1$, where n is the sample size. A property of the t-distribution is that as the degrees of freedom value increases, the distribution gets closer to the standard normal curve. The t-distribution with df $= \infty$ (infinity) is identical to the standard normal curve. In practice, if degrees of freedom are very large, the t-distribution is so close to the standard normal distribution that they are used interchangeably.

Figure 9.11 shows a t-distribution with df $= 3$ along with the standard normal curve. As can be seen in this figure, the t-curve is more spread out than the standard normal. For the t-distribution, there's slightly more probability in the extreme tails of the distribution than there is for the standard normal. This difference diminishes as the degrees of freedom increase.

---

in summary   Standardized z- and t-Statistics

If a small random sample is taken from a normal population or a large random sample is taken from any population (in which case these results are approximate), then the standardized statistic

$$z = \frac{\bar{x} - \mu}{\text{s.d.}(\bar{x})} = \frac{\bar{x} - \mu}{\sigma/\sqrt{n}} = \frac{\sqrt{n}(\bar{x} - \mu)}{\sigma}$$ has a standard normal distribution, and

$$t = \frac{\bar{x} - \mu}{\text{s.e.}(\bar{x})} = \frac{\bar{x} - \mu}{s/\sqrt{n}} = \frac{\sqrt{n}(\bar{x} - \mu)}{s}$$ has a t-distribution with df $= n - 1$.

VIDEO

EXAMPLE 9.8
Standardized Mean Weights

In Section 9.3 we considered four hypothetical samples of $n = 25$ people who were trying to lose weight at a clinic. We played the role of the all-knowing sage and assumed we knew that $\mu = 8$ and $\sigma = 5$. If the value for μ is correct, then the standardized statistic

$$t = \frac{\bar{x} - \mu}{s/\sqrt{n}} = \frac{\bar{x} - 8}{s/\sqrt{25}}$$

has a *t*-distribution with df = 25 − 1 = 24. If we were to generate thousands of random samples of size 25 and draw a histogram of the resulting standardized *t*-statistics, they would adhere to this *t*-distribution.

In practice, we do not draw thousands of samples and we do not know μ. Suppose we speculated that μ = 8 pounds and drew one random sample, the first one given in Table 9.1, for which $\bar{x}$ = 8.32 pounds and *s* = 4.74 pounds. Are the sample results consistent with the speculation that μ = 8 pounds? In other words, is a sample mean of 8.32 pounds reasonable to expect if μ = 8 pounds? The standardized statistic is

$$t = \frac{\bar{x} - \mu}{s/\sqrt{n}} = \frac{\bar{x} - 8}{s/\sqrt{25}} = \frac{8.32 - 8}{4.74/5} = 0.34$$

This statistic tells us that the sample mean of 8.32 is only about one-third of a standard error above 8, which is certainly consistent with a population mean weight loss of 8 pounds. ◆

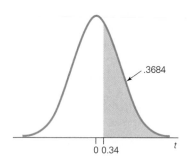

FIGURE 9.12   **A *t*-distribution with df = 24, illustrating location of *t* = 0.34**

Areas and Probabilities for Student's *t*-Distribution

Because Student's *t*-distribution differs for each possible df value, we can't summarize the probability areas in one table like we could for the standard normal distribution. We would need a separate table for each possible df value. Instead, tables for the *t*-distribution are tailored to specific uses, and we will explore two types of *t*-distribution tables in Chapters 12 and 13.

Many calculators and computer software programs provide probabilities (areas) for specified *t*-values and *t*-values for specified areas. For example, the Excel command TDIST(k,d,1) provides the area above the value *k* for a *t*-distribution with degrees of freedom *d*. (The reason for including the third value, 1, will become clear in Chapter 13.) In other words, it provides $P(t > k)$. For the situation in Example 9.8, TDIST(0.34,24,1) = 0.3684. Figure 9.12 illustrates this area.

Excel Tip

> ***tech note***   To find a *t*-value corresponding to a specified area, the Excel command is a bit tricky. The command TINV(p,d) provides the value *k* from the *t*-distribution with df = *d* for which *p* is the *combined* area in the region above *k* and below −*k*. For instance, TINV(0.05,24) = 2.06. This indicates that the area below −2.06 plus the area above 2.06 is 0.05, for a *t*-distribution with df = 24. Notice that this also implies that 95% of the area is between −2.06 and +2.06, in contrast to the standard normal curve for which 95% of the area is between about −2 and +2.

9.8 | STATISTICAL INFERENCE

The results in this chapter will be used for the remainder of the book to make conclusions about populations on the basis of samples. The procedures we will use for making these conclusions are a subset of **statistical inference** procedures. The two most common procedures are to find *confidence intervals* and conduct *hypothesis tests*. Here, we give just a brief introduction to these procedures, which will be developed throughout the rest of the book.

Confidence Intervals

In the next several chapters, we will explore the two basic techniques researchers use to summarize their statistical results. The first technique is to create a **confidence interval,** which is an interval of values that the researcher is fairly sure covers the true value for the population parameter.

We encountered confidence intervals in Chapter 4 when we learned about the margin of error. Adding and subtracting the margin of error to the reported sample proportion creates an interval that we are 95% "confident" covers the truth. That interval is the confidence interval. We will explore confidence intervals further in Chapter 10 and again in Chapters 12 and 14.

Hypothesis Testing

The second statistical technique is called **hypothesis testing** or **significance testing.** Hypothesis testing uses sample data to attempt to reject the hypothesis that nothing interesting is happening—that is, to reject the notion that chance alone can explain the sample results.

We encountered this idea in Chapter 6 when we learned how to determine whether or not the relationship between two categorical variables is "statistically significant." The hypothesis researchers set about to reject in that setting is that two categorical variables are unrelated to each other. In most research settings, the *desired* conclusion is that the variables under scrutiny are related. Achieving **statistical significance** is equivalent to rejecting the idea that chance alone can explain the observed results. For instance, in Example 9.8, in which we obtained a random sample of weight-loss values for a clinic, we might want to see if we can reject the idea that there is *no* weight loss on average—that is, that $\mu = 0$ pounds. For our sample mean and standard deviation of 8.32 and 4.74 pounds, if $\mu = 0$ the standardized sample mean would be $t = \sqrt{25}(8.32 - 0)/4.74 = 8.78$. Clearly, a sample mean that is 8.78 standard errors above the population mean is virtually impossible, so we would *reject* the notion that $\mu = 0$. We will explore hypothesis testing further in Chapters 11 and 13.

CASE STUDY 9.1
Do Americans Really Vote When They Say They Do?

On November 8, 1994, a historic election took place, in which the Republican Party won control of both houses of Congress for the first time since 1952. But how many people actually voted? On November 28, 1994, *Time Magazine* (p. 20) reported that in a telephone poll of 800 adults taken during the two days following the election, 56% reported that they had voted. Considering that only about 68% of adults are registered to vote, that isn't a bad turnout.

But *Time* reported another disturbing fact along with the 56% reported vote. *Time* reported that, in fact, only 39% of American adults had voted, based on information from the Committee for the Study of the American Electorate.

Could it be the case that the results of the poll simply reflected a sample that, by chance, voted with greater frequency than the general population? The Rule for Sample Proportions can answer that question. Let's suppose that the truth about the population is, as reported by *Time,* that only 39% of American adults voted. Then the Rule for Sample Proportions tells us what kind of sample proportions we can expect in samples of 800 adults, the size used by the *Time Magazine* poll. The mean of the possibilities is .39, or 39%. The standard deviation is the square root of $(.39)(.61)/800 = .017$, or 1.7%.

Therefore, we are almost certain that the sample percent based on a sample of 800 adults should fall within $3 \times 1.7\% = 5.1\%$ of the truth of 39%. In other words, if respondents were telling the truth, the sample percent should be no higher than 44.1%—nowhere near the reported percent of 56%.

In fact, if we combine the Rule for Sample Proportions with what we learned about normal curves in Chapter 8, we can say even more about how unlikely this sample result would be. If in truth only 39% (0.39) of the population voted, the standardized score for the reported value of 56% (0.56) is $z = (.56 - .39)/.017 = 10.0$. We know from Chapter 8 that it is virtually impossible to obtain a standardized score of 10.

Another example of the fact that *reported* voting tends to exceed *actual* voting occurred in the 1992 U.S. presidential election. According to the *World Almanac* (1995, p. 631), 61.3% of American adults reported voting in the 1992 election. In a footnote, the Almanac explains:

> Total reporting voting compares with 55.9 percent of population actually voting for president, as reported by Voter News Service. Differences between data may be the result of a variety of factors, including sample size, differences in the respondents' interpretation of the questions, and the respondents' inability or unwillingness to provide correct information or recall correct information.

Unfortunately, because figures are not provided for the size of the sample, we cannot assess whether or not the difference between the actual percent of 55.9 and the reported percent of 61.3 can be explained by the natural variability among possible sample proportions.

KEY TERMS

Section 9.1
statistic, 294, 296
parameter, 294, 296
sampling distribution, 294–295, 296

Section 9.2
sample proportion, 296
normal curve approximation rule for sample proportions, 297, 298
sampling distribution of $\hat{p}$, 298
standard deviation of $\hat{p}$, 300
standard error of $\hat{p}$, 300

Section 9.3
normal curve approximation rule for sample means, 302–303, 304

sampling distribution of $\bar{x}$, 304
sampling distribution of the mean, 304
standard deviation of the sample mean, 304
standard error of the mean, 305
law of large numbers, 306

Section 9.4
central limit theorem, 309

Section 9.6
standardized statistic, 313–314
standardized z-statistic, 313–314, 315

Section 9.7
Student's t-distribution, 314
t-distribution, 314
degrees of freedom, (df), 315
standardized t-statistic, 315

Section 9.8
statistical inference, 316
confidence interval, 317
hypothesis testing, 317
significance testing, 317
statistical significance, 317

EXERCISES   Blue = Basics   Bold, **Bold** = Answer in back

Section 9.1

9.1 In each situation, explain whether the value given in bold print is a statistic or a parameter:
   a. A polling organization samples 1000 adults nationwide and finds that **72%** of those sampled favor tougher penalties for persons convicted of drunk driving.
   b. In their year 2000 census, the United States Census Bureau found that the median age of all American citizens was about **35** years.

   c. For a sample of 20 men and 25 women, there is a **14** centimeter difference in the mean heights of the men and women
   d. A writer wants to know how many typing mistakes there are in his manuscript, so he hires a proofreader who reads the entire manuscript and finds **15** errors.

9.2 A polling organization plans to sample 1000 adult Americans to estimate the proportion of Americans who think crime is a serious problem in this country. In the context of this poll, explain the difference between a statistic and a parameter. What is the parameter of interest to the polling organization? What will be the statistic?

9.3 Explain whether $\hat{p}$ or p is the correct statistical notation for each proportion described:

a. The proportion that smokes in a randomly selected sample of $n = 300$ students in the 11th and 12th grades.

b. The proportion that smokes among all students in the 11th or 12th grade in the United States.

c. The proportion that is left-handed in a sample of $n = 250$ individuals.

9.4 Suppose that a simple random sample of $n = 2$ numbers will be selected from the list of values 1, 3, 5, 7, 9.

a. There are ten possible equally likely samples of $n = 2$ numbers that could be selected (1 and 3, 1 and 5, etc.). List the ten possible samples, and calculate the sample mean $\bar{x}$ for each sample.

b. Summarize the results of part (a) into a table showing the sampling distribution of possible sample means. To do this, list each possible value for the sample mean along with the probability the value would occur.

9.5 Explain whether $\bar{x}$ or μ is the correct statistical notation for each mean described.

a. The mean hours of study per week was 15 hours for a sample of $n = 30$ students at a college.

b. A university administrator determines that the mean age of all students at a college is 20.2 years.

9.6 Some stockholders want to know if the mean salary for male employees in a large company is higher than the mean salary for female employees. The company allows them access to salary information for a random sample of 100 male and 100 female employees, and the mean salaries are $41,000 for the males and $39,500 for the females.

a. Are the mean salaries in this example statistics or parameters? Explain.

b. Based on these means, can the shareholders determine that the mean salary for males in the company is higher than the mean salary for females? Explain.

c. If they selected new random samples of 100 males and 100 females, would the average salaries of those samples be $41,000 and $39,500? Explain, using the concept of sampling distribution in your explanation.

9.7 Refer to the previous exercise. Suppose the 100 men and 100 women in the sample were all of the company's employees. Now answer questions a and b, repeated here:

a. Are the mean salaries in this example statistics or parameters? Explain.

b. Based on these means, can the shareholders determine that the mean salary for males in the company is higher than the mean salary for females? Explain.

9.8 A soft drink called Crash is so popular that vending machines allocate the top two selection buttons to it. Over the long run, of those who buy Crash 60% push the top button and 40% push the lower button, and this behavior seems to be independent from one purchase to the next. Suppose a sample of $n = 10$ people buy Crash at a vending machine one morning and $X =$ the number who push the top button.

a. Is X a statistic? Explain.

b. Describe the sampling distribution of X.

9.9 In a large retirement community, 60% of households have no dogs, 30% have one dog, and 10% have two dogs. Following a news story about how the elderly have trouble walking their dogs in the winter weather, a magnanimous group of teenagers decides that it will randomly select two households each and call and offer to walk their dog(s). For each teen who participates, T is the *total* number of dogs in the two households called.

a. List all possible values for T.

b. Find the sampling distribution for T. (*Hint:* This is simply a list of possible values for T and their probabilities.)

c. If there are no dogs in the two households ($T = 0$), the teen starts over with a new sample of two households. If 1000 teens participate, for about how many of them will $T = 0$ for their first sample?

9.10 The *I Ching* is an ancient Chinese system of asking for advice. In one version, three pennies are flipped and if the number of heads is odd (1 or 3), a "broken line" is recorded, while if it is even (0 or 2), a "solid line" is recorded. This process is repeated six times, then the results are used to identify one of the $2^6 = 64$ possible patterns of solid and broken lines and corresponding advice. Assume the coins are fair and the flips are independent. Define B to be the number of broken lines out of the six.

a. For each of the six lines, what is the probability that it is a "broken line"?

b. What are the possible values for B?

c. What is the probability that $B = 0$, so that all of the lines are solid?

d. Find the sampling distribution for B.

9.11 Are there situations for which the sampling distribution of a statistic consists of a single possible value, which has probability 1 of occurring each time? If not, explain why not. If so, explain how that could happen and give an example.

Section 9.2

9.12 Assuming the Rule for Sample Proportions applies, calculate the mean and the standard deviation of the sampling distribution of possible sample proportions for each combination of sample size (n) and population proportion (p).

a. $n = 400, p = .5$.

b. $n = 1600, p = .5$.

c. $n = 64, p = .8$.

d. $n = 256, p = .8$.

9.13 Refer to the previous exercise, and notice that for parts (a) and (b), the values of n are different while p is the same. This pattern also occurs in parts (c) and (d). Use the answers to the previous exercise to describe how changing the sample size affects the mean and standard deviation of the distribution of possible sample proportions.

9.14 A polling organization polls $n = 100$ randomly selected registered voters in order to estimate the proportion of a large population that intends to vote for Candidate Y in an upcoming election. Although it is not known by the polling organization, $p = .55$ is the actual proportion of the population that prefers Candidate Y.

a. Give the numerical value of the mean of the sampling distribution of $\hat{p}$.

b. Calculate the standard deviation of the sampling distribution of $\hat{p}$.

c. Use the Empirical Rule to find values that fill in the blanks in the following sentence. In about 99.7% of all randomly selected samples of $n = 100$ from this population, the sample proportion preferring Candidate Y will be between ____ and ____.

9.15 In a random sample of $n = 200$ drivers, 50 individuals say they never wear a seatbelt when driving.

a. Give a numerical value of $\hat{p}$ = sample proportion that never wears a seatbelt when driving.

b. Calculate the standard error of $\hat{p}$.

c. In a different survey of $n = 400$ drivers, 100 people say they never wear a seatbelt when driving. Give numerical values for $\hat{p}$ and the standard error of $\hat{p}$.

9.16 In a random sample of $n = 500$ adults, 300 individuals say they believe in love at first sight.

a. Calculate the value of $\hat{p}$ = sample proportion that believes in love at first sight.

b. Calculate the standard error of $\hat{p}$.

9.17 Suppose the probability is $p = .2$ that a person purchasing an instant lottery ticket wins money, and this probability holds for every ticket purchased. Consider different random samples of $n = 64$ purchased tickets. Let $\hat{p}$ = sample proportion of winning tickets in a sample of 64 tickets.

a. Give the numerical value of the mean of the sampling distribution of $\hat{p}$.

b. Calculate the standard deviation of the sampling distribution of $\hat{p}$.

c. Use the Empirical Rule to find values that fill in the blanks in the following sentence. In about 68% of all randomly selected samples of $n = 64$ instant lottery tickets, the proportion of tickets that will be money winners will be between ____ and ____.

d. Use the Empirical Rule to find values that fill in the blanks in the following sentence. In about 95% of all randomly selected samples of $n = 64$ instant lottery tickets, the proportion of tickets that will be money winners will be between ____ and ____.

9.18 Suppose medical researchers think that $p = .70$ is the proportion of all teenagers with high blood pressure whose blood pressure would decrease if they took calcium supplements. To test this theory, the researchers plan a clinical trial (experiment) in which $n = 200$ teenagers with high blood pressure will take regular calcium supplements.

a. Assume $p = .70$ actually is the population proportion that would experience a decrease in blood pressure. What are the numerical values of the mean and standard deviation of the sampling distribution of $\hat{p}$, the sample proportion, for samples of $n = 200$ teenagers?

b. Use the results of part (a) to calculate an interval that will contain the sample proportion $\hat{p}$ for about 99.7% of all samples of $n = 200$ teenagers.

c. In the clinical trial, 120 of the 200 teenagers taking calcium supplements experienced a decrease in blood pressure. What is the value of $\hat{p}$ for this sample? Is this value a parameter or a statistic?

d. Given the answers for parts (b) and (c), explain why the observed value of $\hat{p}$ could be used as evidence that the researchers might be wrong to think that $p = .70$.

9.19 An automobile club comes to the aid of stranded motorists who are members. Over the long run, about 5% of the members utilize this service in any 12-month period. A small town has 400 members of the club. Consider the 400 members to be representative of the larger club membership. Let $\hat{p}$ be the proportion of them who will utilize the service in the coming 12-month period.

a. Explain why conditions 1 to 3 for the Normal Curve Approximation Rule for Sample Proportions are met in this scenario. State any assumptions you need to make to do so.

b. What are the values of n and p in this situation?

c. Use the Rule for Sample Proportions to describe the sampling distribution of $\hat{p}$, including values for the mean and standard deviation.

d. The towing service in town contracts with the club to come to the aid of up to 28 members in the next 12-month period. What proportion is that of the 400 members in town? What is the approximate probability that the actual proportion requiring aid will exceed that value?

9.20 Refer to the previous exercise where $\hat{p}$ is the sample proportion of 400 randomly selected club members who require aid in a 12-month period. Based on the Empirical Rule, into what interval should $\hat{p}$ fall about 95% of the time?

9.21 Explain the difference between the *standard error of $\hat{p}$*, and the *standard deviation of the sampling distribution of $\hat{p}$*. Write down the formula for each one. Which one is more likely to be used in practice? Why?

9.22 For each of the following, *if* the scenario fits the Rule for Sample Proportions, describe the sampling distribution of $\hat{p}$ and draw a picture of it. If the scenario doesn't fit, explain why not (in which case you do not have the tools to describe the sampling distribution of $\hat{p}$).

a. An auto insurance company has 1500 customers in a city, and over the long run 6% of them will file a claim in any given year. Define $\hat{p}$ to be the proportion who file a claim in the next year.

b. A dean has found that over the years, about 15% of students who begin their college work in her program do not finish. This year, 90 students will begin her program; define $\hat{p}$ to be the proportion who will finish.

c. An essay contest has three winners, but the number of entrants varies each time it is offered, so the proportion of entrants who are winners also varies. Define $\hat{p}$ to be the proportion of entrants who will be winners the next time the contest is offered.

9.23 Recent studies have shown that about 20% of American adults fit the medical definition of being obese. A large medical clinic would like to estimate what percent of their patients are obese, so they take a random sample of 100 patients and find that 18 are obese. Suppose in truth, the same percent holds for the patients of the medical clinic as for the general population, 20%. Give a numerical value for each of the following:

a. The population proportion of obese patients in the medical clinic.

b. The proportion of obese patients for the sample of 100 patients.

c. The standard error of $\hat{p}$.

d. The mean of the sampling distribution of $\hat{p}$.

e. The standard deviation of the sampling distribution of $\hat{p}$.

9.24 Refer to the previous exercise, in which 20% of the patients in a medical clinic are obese. If the clinic took repeated random samples of 100 observations and found the sample proportion who were obese, into what interval should those sample proportions fall about 95% of the time?

9.25 A student commented, "Once a poll has been taken, $\hat{p}$ (the sample proportion with a certain opinion) is a known value, so it cannot have a distribution of possible values. Therefore, I don't understand what you mean by the sampling distribution of $\hat{p}$." Write an explanation for this student.

Sections 9.3 and 9.4

9.26 Explain whether the Rule for Sample Means can be used to describe the possible values of the sample mean in each scenario.

a. Mean normal body temperature will be determined for a randomly selected sample of 18 individuals. In the population of all humans, normal body temperature has approximately a normal distribution with mean $\mu = 98.2$ degrees Fahrenheit and standard deviation $\sigma = 0.5$.

b. Mean number of music CDs owned will be determined for a randomly selected sample of 4 college students. In the population of all college students, the distribution of number of CDs owned is skewed to the right.

c. Refer to the previous part. The mean number of music CDs owned will be determined for a randomly selected sample of 900 college students.

9.27 The weights of men in a particular age group have mean $\mu = 170$ pounds and standard deviation $\sigma = 24$ pounds.

a. For randomly selected samples of $n = 16$ men, what is the standard deviation of the sampling distribution of possible sample means?

b. For randomly selected sample of $n = 64$ men, what is the standard deviation of the sampling distribution of possible sample means?

c. In general, how does increasing the sample size affect the standard deviation of the sampling distribution of possible sample means? The previous two parts provide a hint.

9.28 Example 9.1 concerned selecting random samples of $n = 190$ from a population of student responses to the question "How many hours of sleep did you get last night?" Suppose that in a large population of students the mean amount of sleep the previous night was $\mu = 7.05$ hours and the standard deviation $\sigma = 1.75$ hours. Consider randomly selected samples of $n = 190$ students.

a. What is the value of the mean of the sampling distribution of possible sample means?

b. Calculate the standard deviation of the sampling distribution of possible sample means.

c. Use the Empirical Rule to find values that fill in the blanks at the end of the following sentence. For 68% of all randomly selected samples of $n = 190$ students, the mean amount of sleep the previous night will be between ____ and ____ hours.

d. Use the Empirical Rule to fill in the blanks at the end of the following sentence. For 95% of all randomly selected samples of $n = 190$ students, the mean amount of sleep will be between ____ and ____ hours.

9.29 The **cholest** dataset on the CD accompanying this book includes cholesterol levels measured for 28 heart attack patients two days after their attacks. The sample mean for this sample is 253.9 and the standard deviation is 47.7.

a. Assuming this was a sample from a larger population of heart attack patients, what symbols should be used to represent the values given for the mean and standard deviation?

b. Calculate the standard error of the sample mean.

9.30 A randomly selected sample of $n = 60$ individuals over 65 years old takes a test of memorization skills. The sample mean is $\bar{x} = 53$ and the standard deviation is $s = 7.2$. Give the numerical value of the standard error of the mean.

9.31 Vehicle speeds at a certain highway location are believed to have approximately a normal distribution with mean $\mu = 60$ mph and standard deviation $\sigma = 6$ mph. The speeds for a randomly selected sample of $n = 36$ vehicles will be recorded.

a. Give numerical values for the mean and standard deviation of the sampling distribution of possible sample means for randomly selected samples of $n = 36$ from the population of vehicle speeds.

b. Use the Empirical Rule to find values that fill in the blanks in the following sentence. For a random sample of $n = 36$ vehicles, there is about a 95% chance that mean vehicle speed in the sample will be between ____ and ____ mph.

c. Sample speeds for a random sample of 36 vehicles are measured at this location, and the sample mean is 66 mph. Given the answer to part (b), explain whether this result is consistent with the belief that the mean speed at this location is $n = 60$ mph.

9.32 Small planes cannot fly well if the cargo (people and luggage) weigh too much. Suppose an airline runs a commuter flight that holds 40 people. They know that the weights of passenger + luggage for typical customers on this flight is approximately normal with a mean of 210 pounds and a standard deviation of 25 pounds.

a. Draw a picture of this distribution.

b. Using the Rule for Sample Means, describe the sampling distribution of the mean weight of passenger + luggage for a random sample of 40 customers.

c. Draw a picture superimposing the distributions from parts (a) and (b). (Label it clearly, and remember that the total area under each curve must equal 1.)

d. Assume customers on any particular flight are similar to a random sample. If the *total* weight of passengers and their luggage should not exceed 8800 pounds, what is the probability that a sold-out flight (40 passengers and their luggage) will exceed the weight limit? (*Hint:* Rewrite the desired limit as an average per passenger.)

9.33 Explain the difference between the standard deviation of the sampling distribution of the mean and the standard error of the mean. Explain which one is more likely to be used in practice and why.

9.34 Describe the sampling distribution for the statistic of interest in each of the following stories, including numerical values for the mean and standard deviation of the sampling distribution.

a. A fisherman takes tourists out fishing and has noticed that over the long run the weights of a certain type of fish typically caught is approximately normal with a mean of 10 pounds and a standard deviation of 2 pounds. During a good day, the boatload of eight people catches the limit of 2 fish each, or a total of 16 fish. The statistic of interest is the mean weight of the 16 fish caught on a good day.

b. A fisherman hangs out at the local bar in the evening and hears tourists talking about the weights of the fish they caught that day. He notices that the weights mentioned seem to have a mean of about 20 pounds and a standard deviation of about 4 pounds. He records what he hears as the weights for the next 50 "fish stories" at the bar. The statistic of interest is the mean weight for the 50 stories. Assume they represent a random sample of all such stories.

9.35 Consider the population of net gains for tickets in the California Decco lottery game described in Example 9.5. The possible values are $4999, $49, $4, 0 and −$1; the mean is −$0.35 and the standard deviation is $29.67. Suppose someone buys ten tickets each week.

a. Explain why the Rule for Sample Means (or the Central Limit Theorem) would *not* hold for the distribution of possible mean net gains for the weekly purchases of ten tickets.

b. Although the possible means for samples of ten tickets are *not* approximately normal, you can still specify numerical values for the mean and standard deviation of the sampling distribution of the mean in this situation. What are they?

9.36 A rental car company has noticed that the distribution of the number of miles customers put on rental cars per day is somewhat skewed to the right, with an occasional high outlier. The distribution has a mean of 80 miles and a standard deviation of 50 miles.

a. Draw a picture of a normal curve with this mean and standard deviation (80 miles and 50 miles), and use the information in the picture to verify that the distribution of number of miles must be skewed to the right and/or have occasional high outliers rather than be normally distributed.

b. What is the approximate distribution for the *mean* number of miles per day put on a typical one of the company's rental cars in a year (365 days)?

c. What is the approximate distribution for the *total* number of miles put on a typical one of the rental cars in a year (365 days)?

d. Do you think the necessary conditions for using the Rule of Sample Means to find the answer in part (b) are met in this situation? Explain.

Section 9.5

Suppose that a simple random sample of $n = 2$ numbers will be selected from the list of values 1, 2, 3, 4, 5.

a. There are ten possible equally likely samples of $n = 2$ numbers that could be selected (1 and 2, 1 and 3, etc.).

List the ten possible samples, and give H = highest number for each sample.

b. Summarize the results of part (a) into a table showing the sampling distribution of H. To do this, list each possible value for H along with the probability the value would occur.

c. Which value is most likely to be the highest number in the sample?

9.38 Refer to the previous exercise.

a. For each of the ten possible samples listed in part (a) of the previous exercise, give the value of the range $R = H - L$, where H is the highest number and L is the lowest number.

b. Use the results of part (a) to find the probability that $R = 4$.

c. Summarize the results of part (a) into a table showing the sampling distribution of R. To do this, list each possible value for R along with the probability the value would occur.

d. Which value of R is most likely to occur? Which value is least likely to occur?

9.39 Refer to the discussion of the Cash 5 lottery game in Example 9.6 and the accompanying Table 9.2.

a. In what proportion of the games is the highest number 35 or more?

b. What is the median value of H for the 1560 Cash 5 games represented in Example 9.6?

c. In what proportion of the games is the highest number 15 or less?

9.40 Define L to be the *lowest* of the five numbers drawn in the Cash 5 lottery game. Describe the shape of the sampling distribution of L, and explain how you know.

9.41 Use a computer to simulate the numbers drawn for the Cash 5 lottery game 1000 times, and draw a histogram for the values of H = highest number drawn. Compare your histogram to the one shown in Figure 9.10.

9.42 Suppose two fair dice are rolled and H = higher of the two faces. If both dice show the same value, H is that value. Construct the sampling distribution of H, listing all possible values and their probabilities.

9.43 Suppose someone who takes a ten-question True-False test doesn't know any of the answers and randomly selects True or False for each question. Define the statistic X = number correct.

a. Describe the sampling distribution of X. (*Hint:* This distribution was introduced in Chapter 8.)

b. What is the mean of this sampling distribution?

c. What is the standard deviation of this sampling distribution?

9.44 Refer to the previous exercise. Now define the statistic of interest to be the *proportion* correct, $X/10$. What is the *mean* of the sampling distribution of this statistic? Explain how you found your answer.

Section 9.6

In each part, give the value of the standardized statistic (z-score) for the sample proportion.

a. $\hat{p} = .60, p = .50$, $n = 100$.

b. $\hat{p} = .60$, $p = .50$, $n = 200$.

c. $\hat{p} = .78$, $p = .80$, $n = 400$.

d. $\hat{p} = .82$, $p = .80$, $n = 400$.

e. $\hat{p} = .25$, $p = .40$, $n = 900$.

9.46 Based on past history, a car manufacturer knows that 10% ($p = .10$) of all newly made cars have an initial defect. In a random sample of $n = 100$ recently made cars, 13% ($\hat{p} = .13$) have defects. Find the value of the standardized statistic (z-score) for this sample proportion.

9.47 In each part, give the value of the standardized statistic (z-score) for the sample mean:

a. $\bar{x} = 74$, $\mu = 72$, $\sigma = 10$, $n = 25$.

b. $\bar{x} = 70$, $\mu = 72$, $\sigma = 10$, $n = 25$.

c. $\bar{x} = 92$, $\mu = 100$, $\sigma = 16$, $n = 4$.

d. $\bar{x} = 92$, $\mu = 100$, $\sigma = 16$, $n = 64$.

e. $\bar{x} = 10.15$, $\mu = 10.0$, $\sigma = 5$, $n = 10,000$.

9.48 Home prices in a city have mean $\mu = \$200,000$ and standard deviation $\sigma = \$25,000$. For $n = 25$ randomly selected homes, the mean price is $\bar{x} = \$208,000$. Find the value of the standardized statistic (z-score) for this sample mean.

9.49 In addition to the sample size n, what population value(s) must be known in order to find a standardized score for a sample proportion $\hat{p}$?

9.50 In addition to the sample size n, what population value(s) must be known in order to find a standardized score for a sample mean $\bar{x}$?

9.51 A blind taste test is done to compare Cola 1 and Cola 2. Among $n = 75$ participants, $\hat{p} = .64$ is the proportion preferring Cola 1. If, actually, the two colas are equally preferred in a larger population of tasters, $p = .5$ is the corresponding population proportion. Assuming $p = .5$, find the value of the standardized statistic (z-score) for the sample proportion observed in the taste test.

9.52 Suppose a random sample of 36 IQ scores is drawn from a population of IQ scores with mean = 100 and standard deviation = 15.

a. Find the standardized statistic for $\bar{x} = 97$.

b. Find the standardized statistic for $\bar{x} = 105$.

c. Give numerical values for the mean and standard deviation of the sampling distribution of $\bar{x}$.

d. Draw a picture of the sampling distribution of $\bar{x}$, and identify where $\bar{x} = 97$ and $\bar{x} = 105$ fall on the picture.

e. Draw a picture of the standard normal curve, and identify where the standardized statistics found in parts (a) and (b) fall on the picture.

f. What is the relationship between the pictures in parts (d) and (e)?

9.53 Suppose an ESP test consists of $n = 75$ independent tries, for which there are four possible choices on each try. Suppose someone is just guessing, so the probability of a correct guess on each try is $p = .25$. Consider two possible sample proportions, $\hat{p} = .20$ (15 correct) and $\hat{p} = .33$ (25 correct).

a. Find the standardized z-statistics for $\hat{p} = .20$ and $\hat{p} = .33$.

b. Draw a picture of the sampling distribution of $\hat{p}$, and identify where $\hat{p} = .20$ and $\hat{p} = .33$ fall on the picture.

c. Draw a picture of the standard normal curve, and identify where the standardized statistics found in part (a) fall on the picture.

d. What is the relationship between the pictures in parts (b) and (c)?

Sections 9.7 and 9.8

9.54 Explain whether or not each of the following would ever differ for two random samples of the same size from the same population.

a. The sample mean $\bar{x}$.

b. The standard deviation of the sampling distribution of $\bar{x}$.

c. The standard error of $\bar{x}$.

d. The standardized z-score for $\bar{x}$.

e. The population mean μ.

9.55 In each situation, find the value of the t-statistic for the sample mean $\bar{x}$ and give the value of degrees of freedom (df).

a. $\bar{x} = 5$, $\mu = 10$, $s = 20$, $n = 16$.

b. $\bar{x} = 15$, $\mu = 10$, $s = 20$, $n = 16$.

c. $\bar{x} = 56$, $\mu = 50$, $s = 15$, $n = 25$.

d. $\bar{x} = 50$, $\mu = 50$, $s = 20$, $n = 9$.

9.56 Explain the difference between what is represented by s and σ, the two different symbols for standard deviation.

9.57 In each situation, find the value of the standardized statistic for the sample mean, and indicate whether the standardized statistic is a t-statistic or a z-statistic.

a. $\bar{x} = 175$, $\mu = 170$, $s = 24$, $n = 4$.

b. $\bar{x} = 175$, $\mu = 170$, $\sigma = 20$, $n = 4$.

c. $\bar{x} = 161$, $\mu = 170$, $s = 18$, $n = 36$.

9.58 Draw a picture of each of the following distributions. Shade the area above the value given, and label the proportion of the distribution that falls in that region, as in Figure 9.12. You do not have to label the vertical axis. *Note:* For parts (b) and (c), the use of either statistical software or Excel is necessary.

a. Standard normal distribution; $z = .70$.

b. t-distribution with df = 10, $t = .70$.

c. t-distribution with df = 10, $t = -.70$.

9.59 Are there any (finite) values for which the area above that value is the same for all t-distributions? If so, specify the value(s) and the area(s). If not, explain why not.

9.60 In each of the following situations, which of the two distributions would be more spread out?

a. Standard normal distribution or t-distribution with df = 10.

b. t-distribution with df = 5 or t-distribution with df = 25.

c. t-distribution with df = 100 or normal distribution with standard deviation = 100.

9.61 Compute the standardized statistic for the sample mean in each of these situations.

a. A random sample of 25 IQ scores from a population with mean 100; $\bar{x} = 105$, $s = 10$.

b. A random sample of 100 salaries of women who work in a traditional male job for which the population mean salary is $\$80,000$; $\bar{x} = \$78,000$, $s = \$4000$.

9.62 Refer to the previous exercise. Draw a picture of the appropriate distribution for the standardized statistic in each case, and find the area below the value of the standardized statistic. (*Note:* The use of either statistical software or Excel is necessary.)

9.63 Refer to the parts of the previous two exercises that concern the 100 salaries of women. Suppose allegations were made that the population mean salary of women who work in that job is lower than the known population mean salary of $80,000 for men.
   a. Does the sample information provided for the women seem to support that allegation? Explain.
   b. The question in part (a) is an example of hypothesis testing, the goal of which was described in Section 9.8 as "to reject the notion that chance alone can explain the sample results." Explain how the question in part (a) fits that description of hypothesis testing.

9.64 Draw a picture illustrating each of the following distributions, showing the *t*-values marking the specified areas and the proportion of the *t*-distribution that falls in that area. *Note:* The use of either statistical software or Excel is necessary.
   a. *t*-distribution with df = 10, middle 95%.
   b. *t*-distribution with df = 20, area above 1.725.
   c. *t*-distribution with df = 20, area between 1.725 and −1.725.
   d. *t*-distribution with df = 10, upper 5%.
   e. *t*-distribution with df = 10, area below −*t* and above +*t* (same *t*) totals 5%.

Chapter Exercises

9.65 Suppose you want to estimate the proportion of students at your college who are left-handed. You decide to collect a random sample of 200 students and ask them which hand is dominant. Go through the Conditions for Which the Rule for Sample Proportions Applies and explain why the rule would apply to this situation.

9.66 Refer to the previous exercise. Suppose the truth is that .12 or 12% of the students are left-handed and you take a random sample of 200 students. Use the Rule for Sample Proportions to draw a picture similar to Figure 9.3, showing the possible sample proportions for this situation.

9.67 According to the *Sacramento Bee* (2 April 1998, p. F5), "A 1997–98 survey of 1027 Americans conducted by the National Sleep Foundation found that 23% of adults say they have fallen asleep at the wheel in the last year."
   a. Conditions 2 and 3 needed to apply the Rule for Sample Proportions are met because this result is based on a large random sample of adults. Explain how condition 1 is also met.
   b. The article also said that (based on the same survey) "37 percent of adults report being so sleepy during the day that it interferes with their daytime activities." If in truth 40% of all adults have this problem, find the interval in which about 95% of all sample proportions should fall, based on samples of size 1027. Does the result of this survey fall into that interval?
   c. Suppose a survey based on a random sample of 1027 college students was conducted and 25% reported being so sleepy during the day that it interferes with their daytime activities. Would it be reasonable to conclude that the population proportion of college students who have this problem differs from the proportion of all adults who have the problem? Explain.

9.68 A recent Gallup Poll found that of 800 randomly selected drivers surveyed, 70% thought that they were better-than-average drivers. In truth, in the population, at most only 50% of all drivers can be "better than average."
   a. Draw a picture of the possible sample proportions that would result from samples of 800 people from a population with a true proportion of .50.
   b. Would you be unlikely to see a sample proportion of .70, based on a sample of 800 people, from a population with a proportion of .50? Explain, using your picture from part (a).

9.69 Suppose you are interested in estimating the average number of miles per gallon of gasoline your car can get. You calculate the miles per gallon for each of the next nine times you fill the tank. Suppose in truth the values for your car are bell-shaped, with a mean of 25 miles per gallon and a standard deviation of 1. Draw a picture of the possible sample means you are likely to get based on your sample of nine observations. Include the intervals into which 68%, 95%, and almost all of the potential sample means will fall, using the Empirical Rule from Chapter 2.

9.70 Refer to the previous exercise. Redraw the picture under the assumption that you will collect 100 measurements instead of only 9. Discuss how the picture differs from the one in the previous exercise.

9.71 Give an example of a scenario of interest to you for which the Rule for Sample Proportions would apply. Explain why the conditions allowing the rule to be applied are satisfied for your example.

9.72 Suppose the population of IQ scores in the town or city where you live is bell-shaped, with a mean of 105 and a standard deviation of 15. Describe the distribution of possible sample means that would result from random samples of 100 IQ scores.

9.73 Suppose that 35% of the students at a university favor the semester system, 60% favor the quarter system, and 5% have no preference. Would a random sample of 100 students be large enough to provide convincing evidence that the quarter system is favored? Explain.

9.74 According to *USA Today* (20 April 1998 Snapshot), a poll of 8709 adults taken in 1976 found that 9% believed in reincarnation, while a poll of 1000 adults taken in 1997 found that 25% held that belief.
   a. Assuming a proper random sample was used, verify that the sample proportion for the 1976 poll represents the population proportion to within about 1%.
   b. Based on these results, would you conclude that the proportion of all adults who believe in reincarnation was higher in 1997 than it was in 1976? Explain.

9.75 Suppose 20% of all television viewers in the country watch a particular program.
   a. For a random sample of 2500 households measured by a rating agency, describe the distribution of the possible sample proportions who watch the program.
   b. The program will be cancelled if the ratings show less than 17% watching in a random sample of households. Given that 2500 households are used for the ratings, is the program in danger of getting cancelled? Explain.

9.76 Use the Rule for Sample Means to explain why it is desirable to take as large a sample as possible when trying to estimate a population value.

9.77 According to the *Sacramento Bee* (2 April 1998, p. F5), Americans get an average of 6 hours and 57 minutes of sleep per night. A survey of a class of 190 statistics students at a large university found that they averaged 7.1 hours of sleep the previous night, with a standard deviation of 1.95 hours.

a. Assume that the population average for adults is 6 hours and 57 minutes, or 6.95 hours of sleep per night, with a standard deviation of 2 hours. Draw a picture illustrating how the Rule for Sample Means would apply to sample means for random samples of 190 adults.

b. Would the mean of 7.1 hours of sleep obtained from the statistics students be a reasonable value to expect for the sample mean of a random sample of 190 adults? Explain.

c. Can the sample taken in the statistics class be considered a representative sample of all adults? Explain.

9.78 Explain whether or not each of the following scenarios meets the conditions for which the Rule for Sample Proportions applies. If not, explain which condition is violated.

a. Unknown to the government, 10% of all cars in a certain city do not meet appropriate emissions standards. The government wants to estimate that percentage, so they take a random sample of 30 cars and compute the sample proportion that do not meet the standards.

b. The Census Bureau would like to estimate what proportion of households have someone at home between 7 P.M. and 7:30 P.M. on weeknights to determine whether or not that would be an efficient time to collect census data. They survey a random sample of 2000 households and visit them during that time to see whether or not someone is at home.

c. You want to know what proportion of days in typical years have rain or snow in the area where you live. For the months of January and February, you record whether or not there is rain or snow each day, and then you calculate the proportion.

d. A large company wants to determine what proportion of its employees are interested in on-site day care. For a random sample of 100 employees, the company calculates the sample proportion who are interested.

9.79 Explain whether or not you think the Rule for Sample Means applies to each of the following scenarios. If it does apply, specify the population of interest and the measurement of interest. If it does not apply, explain why not.

a. A researcher wants to know what the average cholesterol level would be if people restricted their fat intake to 30% of calories. He gets a group of patients who had heart attacks to volunteer to participate, puts them on a restricted diet for a few months, then measures their cholesterol and calculates the mean value.

b. A large corporation would like to know the average income of the spouses of its workers. Rather than go to the trouble to collect a random sample, they post someone at the exit of the building at 5 P.M. Everyone who leaves between 5 P.M. and 5:30 P.M. is asked to complete a short questionnaire on the issue; there are 70 responses.

c. A university wants to know the average income of its alumni. They select a random sample of 200 alumni and mail them a questionnaire. They follow up with a phone call to those who do not respond within 30 days.

d. An automobile manufacturer wants to know the average price for which used cars of a particular model and year are selling in a certain state. It is able to obtain a list of buyers from the state motor vehicle division, from which a random sample of 20 buyers is selected. The manufacturer makes every effort to find out what those people paid for the cars and is successful in doing so.

9.80 In Case Study 9.1, we learned that about 56% of American adults actually voted in the presidential election of 1992, whereas about 61% of a random sample claimed that they had voted. The size of the sample was not specified, but suppose it was based on 1600 American adults, a common size for such studies.

a. Into what interval of values should the sample proportion fall 68%, 95%, and almost all of the time?

b. Do you think the observed value of 61% is reasonable, based on your answer to part (a)?

c. Now suppose the sample had been of only 400 people. Compute a standardized score to correspond to the reported percentage of 61%. Comment on whether or not you believe people in the sample could all have been telling the truth, based on your result.

9.81 Suppose the population of grade point averages (GPAs) for students at the end of their first year at a large university has a mean of 3.1 and a standard deviation of 0.5.

a. Draw a picture of the distribution of the sample mean GPA for a random sample of 100 students.

b. Find the probability that the sample mean will be 3.15 or greater.

9.82 The administration of a large university wants to take a random sample to measure student opinion of a new food service on campus. It plans to use a continuous scale from 1 to 100, where 1 is complete dissatisfaction and 100 is complete satisfaction. The administration knows from past experience with such questions that the standard deviation for the responses is going to be about 5, but it does not know what to expect for the mean. It wants to be almost sure that the sample mean is within plus or minus 1 point of the true population mean value. How large will the random sample have to be?

Dataset Exercises

9.83 The data for this exercise are in the **GSS-93** dataset. The variable ***gunlaw*** is whether a respondent favors or opposes stronger gun control laws. Of the 1055 respondents, 870 are in favor. *For this exercise, assume these respondents represent a population.*

a. What is the population proportion p who favor stronger gun control laws?

b. Suppose random samples of $n = 60$ are drawn from this population. Describe and draw the approximate sampling distribution for the proportion $\hat{p}$ who would favor stronger gun control laws.

c. Simulate 200 samples of size 60 from this population, and draw a histogram of the sample proportions who favor stronger gun control laws.

d. Compare your simulated distribution in part (c) with the sampling distribution you described in part (b).

9.84 Use the **pennstate1** dataset for this exercise. The data for the variable **HrsSleep** are responses by $n = 190$ students to the question, "How many hours did you sleep last night?" For this exercise, suppose they represent a population.
 a. Simulate 200 samples of $n = 10$ responses to **HrsSleep**. Show the commands you used or explain how you did the simulation.
 b. Draw a histogram of the 200 sample means.
 c. Describe the simulated sampling distribution you found in part (b).
 d. Draw a histogram of the 200 sample medians for the samples you found in part (a).
 e. Compare the sampling distributions for the sample mean and the sample median.

9.85 Use the data in the **GSS-93** dataset for this exercise. One of the questions asked was age in years. A five-number summary for age is 18, 32, 43, 58, and 89.
 a. Draw a boxplot of the ages for the entire dataset.
 b. Let H = highest age in the sample for samples of size $n = 10$. Do any of the rules given in this chapter provide information about the sampling distribution of H? If so, explain.
 c. Simulate 200 samples of size 10 and find H for each one. Create a histogram for the 200 values of H.
 d. Using the histogram in part (c) and any other numerical or graphical information you wish to provide, describe the sampling distribution of H.

9.86 Use the **Fantasy5** dataset for this exercise, with data from 2318 games of California's version of the same scheme as the Pennsylvania Cash 5 game.
 a. Draw a picture of the sampling distribution of H, the highest number each day, and compare it to the equivalent picture for the Cash 5 game, shown in Figure 9.10. Do you expect them to be similar? Are they?
 b. Consider each game to be a random sample of size 5, drawn without replacement from the integers 1 to 39. Which column in the dataset represents the median for each of these random samples?
 c. Draw a picture of the sampling distribution of the median.
 d. Find the mean of each of the 2318 samples of size 5. Draw a picture of the sampling distribution of the mean.
 e. Compare the sampling distributions of the median and the mean. Which one is more spread out? Why do you think that is the case?

9.87 Use the **Cash5** dataset for this exercise, with data from 1560 games of Pennsylvania's Cash 5 lottery game described in Section 9.5. The variables **High** and **Low** give the highest and lowest numbers picked in each game.
 a. For each game, find $R =$ **High** $-$ **Low**, the range between the highest and lowest numbers. Draw a histogram of R, and describe the shape of this histogram.
 b. Find the mean for each of the variables R, **High**, and **Low**. Verify that the mean for R equals the difference between the means for **High** and **Low**.
 c. What is the maximum possible value of the range R in the Cash 5 game? What values of **High** and **Low** would lead to this value?
 d. Tally how often each value of R occurred in the 1560 games given in the dataset. How often did the maximum possible value of R occur?

Turn On Your Computer

For Exercises 9.88 to 9.90, use the **SampleMeans** *applet described in Section 9.3. It is on the CD for this text. The applet selects random samples from a bell-shaped population in which the mean $\mu = 8$ and the standard deviation $\sigma = 5$.*

9.88 This exercise concerns possible sample means for random samples of $n = 36$.
 a. Use the applet to generate 500 different random samples of $n = 36$ observations. What is the approximate shape of the resulting histogram of sample means?
 b. Approximately, what were the smallest and largest values of the sample mean among the 500 samples you generated?
 c. For random samples of $n = 36$, use the Empirical Rule to find the interval of values in which about 99.7% of all sample means will fall. Compare that interval to your answer for part (b).

9.89 This exercise concerns possible sample means for random samples of $n = 75$.
 a. Use the applet to generate 500 different random samples of $n = 75$ observations. What is the approximate shape of the resulting histogram of sample means?
 b. Approximately, what were the smallest and largest values among the sample mean in the 500 samples you generated?
 c. For random samples of $n = 75$, use the Empirical Rule to find the interval of values into which the sample mean will fall about 99.7% of the time. Compare the interval to your answer for part (b).
 d. Describe any difference between the histograms created for part (a) of this exercise and part (a) of the previous exercise. (If you have not already generated the histogram for part (a) of the previous exercise, do so now.)

9.90 This exercise compares sample means for random samples of $n = 10$ and random samples of $n = 100$.
 a. Use the applet to generate ten different random samples of $n = 10$. List the ten sample means, and identify the lowest and highest values in the list.
 b. Use the applet to generate ten different random samples of $n = 100$. List the ten sample means, and identify the lowest and highest values in the list.
 c. Compare the results of parts (a) and (b). What is indicated about the effect of sample size on the variation among sample means for different samples?

9.91 Use the **TVMeans** applet, which selects random samples from a population modeled using responses given by college students to a question asking how many hours they watch television in a typical week. The mean for the population is $\mu = 8.352$ hours and the standard deviation is $\sigma = 7.723$ hours.
 a. Generate 500 different random samples of $n = 5$ observations. Draw a sketch showing the shape of the histogram of sample means for random samples of $n = 5$.
 b. Generate 500 different random samples of $n = 49$ observations. Draw a sketch showing the shape of the histogram of sample means for random samples of $n = 49$.
 c. Refer to the two situations, given in Section 9.3, for which the Normal Curve Approximation Rule for Sample Means holds. Explain which situation is pres-

ent in the **TVMeans** applet. Also, explain how parts (a) and (b) illustrate that situation.

9.92 Refer to the previous exercise about the **TVMeans** applet. Assume the Normal Curve Approximation Rule for Sample Means holds for random samples of $n = 49$ observations.

a. Use the Empirical Rule to find the interval of values into which the sample mean will fall about 95% of the time for random samples of $n = 49$.

b. Use the Empirical Rule to find the interval of values into which the sample mean will fall about 68% of the time for random samples of $n = 49$.

REFERENCE

World Almanac and Book of Facts (1995). Robert Famighetti. Mahwah, NJ: Funk and Wagnalls.

CHAPTER 10

Estimating Proportions With Confidence

Ron Chapple/GettyImages

How many teens have dated someone of another race or ethnic group?
See Example 10.1.

One of the most common types of inference procedures is to construct a *confidence interval* estimate of the unknown value of a population parameter. Here is a typical confidence interval statement: Based on this sample, we have 95% confidence that somewhere between 33% and 39% of all Americans suffer from allergies.

In this chapter, we will learn how to use a sample of data to estimate the proportion of a population that falls into a particular category of a categorical variable. Suppose, for instance, that we have been asked to estimate the proportion of students at your school who think that marijuana should be legalized. How can we go about developing our estimate? One possibility that may occur to you is that we could use information from a random sample of students to estimate the proportion for the whole population.

Let's imagine that we survey 150 randomly selected students and that about 41% of these students think that marijuana should be legalized. Figure 10.1 illustrates the situation.

Because we only have information from a sample of 150 students, we should be concerned about whether the sample result (41%) is an accurate estimate of the population value. Some questions about accuracy that typify those that we will examine in this chapter are

FIGURE 10.1 **Population and sample results**

- ◆ How much difference might there be between the sample estimate, 41%, and the true population value?
- ◆ How does the sample size affect the accuracy of our sample estimate?
- ◆ If we report that between 33% and 49% of all students at the college think that marijuana should be legalized, how confident can we be that we are correct?

The last of these questions is connected to the idea of a **confidence interval,** an interval of estimates that is likely to capture the population value. The primary objective of this chapter is to describe how to calculate and interpret a confidence interval estimate of a population proportion. The general concepts that we learn here will also apply to many other estimation problems that we will encounter later in the text.

Remember from Chapters 3 and 9 that the term *statistical inference* is used to describe the use of sample data to make judgments or decisions about populations. The method of using a confidence interval for estimating a population value is one of the two principal techniques of statistical inference. In this chapter, we introduce confidence intervals by illustrating their use for proportions. In Chapter 12, we broaden the discussion and show how to find confidence intervals for a variety of other situations.

The second method of statistical inference is hypothesis testing, which is a method for using sample data to determine whether a specific statement about a population may be true. In Chapter 6, we used a chi-square test to determine if there was a statistically significant relationship between two categorical variables. That procedure was an example of a hypothesis test. We will illustrate the concepts of hypothesis testing by demonstrating their use for proportions in the next chapter (11), then elaborate on the use of hypothesis tests for some other situations in Chapter 13. ◆

10.1 | THE LANGUAGE AND NOTATION OF ESTIMATION

To make our discussion of estimation clear, let's summarize and review some of the language and notation associated with the estimation problem.

- ◆ A **unit** is an individual person or object to be measured.
- ◆ The **population** (or **universe**) is the entire collection of units about which we would like information or the entire collection of measurements we would have if we could measure the whole population. The population might be imagined rather than actual, such as the population of outcomes if 100,000 coin flips were to be conducted.
- ◆ The **sample** is the collection of units we will actually measure or the collection of measurements we will actually obtain.
- ◆ The **sample size,** denoted by the letter n, is the number of units or measurements in the sample.
- ◆ The symbol p denotes the **population proportion,** the fraction of the population that has a certain trait or characteristic or the probability of success in a binomial experiment. The value of the *parameter p* is not known; the sample proportion estimates it.
- ◆ The symbol $\hat{p}$ denotes the **sample proportion,** the fraction of the sample that has a certain trait or characteristic. The *statistic $\hat{p}$* is an estimate of *p.*

Finally, remember that we would ideally like to have a *randomly selected sample* from the population, but in Chapters 3 and 4 we learned that is not always possible and that sometimes we can settle for less. Let's review the crucial criterion for statistical inference to be valid.

> *definition* The **Fundamental Rule for Using Data for Inference** is that available data can be used to make inferences about a much larger group *if the data can be considered to be representative with regard to the question(s) of interest.*

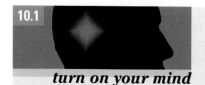

10.1

turn on your mind

Each day, Maria gets dozens of e-mail messages. She keeps track of what proportion of the messages are "junk" and what proportion are interesting each day. If the collection of messages on a single day is considered to be a sample of all e-mail messages she ever receives, explain the meaning of each of the definitions in this section in the context of this example. Discuss whether you think the Fundamental Rule for Using Data for Inference would allow Maria to draw conclusions about the population proportion based on a sample proportion.*

10.2 | MARGIN OF ERROR

In our imaginary survey of student opinion about the legalization of marijuana, we raised questions about the accuracy of a sample estimate. One commonly reported measure of accuracy is the margin of error, a number that provides a likely upper limit for the difference between $\hat{p}$, the sample proportion, and p, the unknown population proportion that we're trying to estimate.

The Meaning of the Margin of Error Given with Survey Results

The **margin of error** provided in media descriptions of survey results has these characteristics:

- The difference between the sample proportion and the population proportion is *less* than the margin of error *about 95% of the time,* or for about 19 of every 20 sample estimates.

- The difference between the sample proportion and the population proportion is *more* than the margin of error *about 5% of the time,* or for about 1 of every 20 sample estimates.

In other words, for most sample estimates, about 95% of them, the actual error is quite likely to be smaller than the margin of error. Occasionally, about 5% of the time, the error might be larger than the margin of error. Unfortunately, we never know the actual amount of error in a particular estimate. We only know that most of the time the actual error is less than the reported margin of error.

Some polling organizations use the phrase "95% confidence" in conjunction with the margin of error. As we'll see later, researchers occasionally determine a value for the margin of error associated with a "confidence level" other than 95%. For instance, a margin of error associated with 90% confidence has the characteristic that the actual sampling error will be less than this number for about 90% of all sample estimates.

EXAMPLE 10.1
Teens and Interracial Dating:
Case Study 1.3 Revisited

In Case Study 1.3, we described a 1997 *USA Today*/Gallup Poll of teenagers in which one finding was that 57% of the 496 teens in the sample who date said they had been out on a date with someone of another race or ethnic group. The margin of error for this estimate was about 4.5%.

*HINT: A "unit" is a single e-mail. Are some days of the week more likely than others to bring junk e-mail messages?

What does the margin of error indicate about the difference between *the sample estimate* of 57% and *the true percent* of all dating American teens who have been out on a date with someone of another race or ethnic group? The answer is that in surveys of this size, the difference between the sample and population percents is likely to be less than 4.5% one way or the other. There is, however, a small chance that the sample estimate might be off by more than 4.5%. ◆

EXAMPLE 10.2
If I Won the Lottery I'd Consider Quitting

What would you do for the rest of your life if you won a lot of money in a lottery? In a 1997 poll conducted by the Gallup Organization, one of the questions was "If you won 10 million dollars in the lottery would you continue to work, or would you stop working?" The results were reported at the Gallup Organization's website (www.gallup.com).

Surprisingly, 59% of the 616 employed respondents answered that they would continue working, 40% said they would stop working, and 1% had no opinion. The website article also gave this information about the poll:

> The current results are based on telephone interviews with a randomly selected sample of 1,014 adults, conducted August 22–25, 1997. Among this group, 616 are employed full-time or part-time. For results based on this sample of "workers," one can say with 95% confidence that the error attributable to sampling could be plus or minus 4 percentage points.

Gallup describes the margin of error with the phrase "could be plus or minus 4 percentage points," but the phrase "could be *as large as* plus or minus 4 percentage points" would have been more informative. The margin of error is a likely upper limit on the sampling error. Here, there is a 95% chance that the sampling error is actually smaller than 4 percentage points. ◆

10.3 | CONFIDENCE INTERVALS

One of the most common types of inference procedures is to construct a *confidence interval* estimate of the unknown value of a population parameter.

> *definition* A **confidence interval** is an interval of values computed from sample data that is likely to include the true population value.

Here is a typical confidence interval statement: "Based on this sample, we have 95% confidence that somewhere between 33% and 39% of all Americans suffer from allergies." Notice that this statement conveys a range of estimates of a population value as well as the likelihood that the interval does indeed capture the population value.

The phrase **confidence level** is used to describe the chance that an interval actually contains the true population value in the following sense. Most of the time (quantified by the confidence level) intervals computed in this way will capture the truth about the population, but occasionally they will not. In any given instance, the interval either captures the truth or it does not, but we will never know which is the case. Therefore, our confidence is in the *procedure*—it works most of the time—and the "confidence level" or "level of confidence" is the percent of the time we expect it to work. For instance, a 95% confidence level means that confidence intervals computed using the same procedure will include the true population value for 95% of all possible random samples from the population. In practice, we can't know for sure whether our confidence interval actually does capture the population value. We can feel fairly certain that it does because the procedure that we use will be successful most of the time.

---

in summary   Interpreting the Confidence Level

♦ For a confidence interval, the **confidence level** is the probability that the procedure used to determine the interval will provide an interval that includes the population parameter.

♦ The confidence level has a relative frequency probability interpretation: If we consider all possible randomly selected samples of the same size from a population, the *confidence level* is the fraction or percent of those samples for which the confidence interval includes the population parameter.

Note: It is commonplace to express the confidence level as a percent.

The most common confidence level that researchers use is 95%. In other words, researchers define "likely to include the true population value " to mean that they are 95% certain the procedure will work. So, they are willing to take a 5% risk that the interval does not actually include the true value. Sometimes researchers employ only 90% confidence, and occasionally they use higher confidence levels like 98% or 99%.

It would be wonderful to be 100% confident that a confidence interval contains the population value. Unfortunately, 100% confidence is impossible unless either we sample the entire population or we provide an absurdly wide interval of estimates. For instance, if we said that the percent of American college students who think marijuana should be legalized is somewhere between 0% and 100%, we would have 100% confidence that we're right. The cost of a high confidence level might be that the interval is so wide that it's not informative.

Constructing a 95% Confidence Interval for a Population Proportion

formula   An approximate 95% confidence interval for a population proportion is easy to calculate once the appropriate margin of error is known. The interval is

Sample estimate ± Margin of error

In the long run, about 95% of all confidence intervals computed in this way will capture the population value of the proportion, and about 5% of them will miss the population value.

Be careful when giving information about a specific confidence interval computed from an observed sample. The confidence level only expresses how often the confidence interval procedure works in the long run. It does not tell us the probability that a specific interval includes the population value. For example, suppose that a 95% confidence interval estimate of the percent of all Americans suffering from allergies is 33% to 39%. This interval either does or does not include the true unknown value of the population percent, but we can never know which is the case because we can't know the population value.

EXAMPLE 10.1—*CONTINUED*
Teens and Interracial Dating

The *USA Today*/Gallup Poll finding was that 57% of the dating teens in the sample had been out with somebody of another race or ethnic group. The margin of error for this estimate was reported as 4.5%. Consequently, a 95% confidence interval is 57% ± 4.5%, which is 52.5% to 61.5%. We have 95% confidence that somewhere between 52.5% and 61.5% of all American teens who date have gone out with somebody of another race or ethnic group. ◆

EXAMPLE 10.2—*CONTINUED*
Winning the
Lottery and Work

What percent of the entire American employed adult population would say they would quit working if they won the lottery? In the Gallup Poll described in Example 10.2, the sample estimate of this percent was 40% and the margin of error was 4%. A 95% confidence interval estimate is

Sample percentage ± Margin of error

40% ± 4%

36% to 44%

With 95% confidence, we can say that somewhere between 36% and 44% of working Americans would say they would quit working if they won $10 million in the lottery.

Notice that the interval does not cover 50%; it resides completely below 50%. Therefore, it would be fair to conclude, with high confidence, that fewer than half of all working Americans think they would quit if they won the lottery (and more than half think they would continue working). ◆

10.2

turn on your mind

Explain in your own words what it means to say that we have 95% confidence in the interval estimate. Then give an example of something you do in your life that illustrates the same concept—you follow the same procedure each time, and it either works (most of the time) or does not work to produce the desired result. What confidence level would you assign to the procedure in your example; i.e., what percent of the time do you think it produces your desired result?*

10.4 | CALCULATING A MARGIN OF ERROR FOR 95% CONFIDENCE

The margin of error of $1/\sqrt{n}$ given in Chapter 4 is actually a conservative approximation that works best when p is close to .5. The formula given next describes a more accurate margin of error for the *proportion* of the sample that has a particular trait. A proportion is a fraction or decimal fraction that has a value between 0 and 1. To determine the margin of error for a percent, first calculate the margin of error for the proportion and then multiply by 100.

*HINT: What is a task you frequently perform that you can't always do successfully?

Calculating the 95% Margin of Error for a Proportion

formula For a 95% confidence level, a formula for the approximate **margin of error for a sample proportion is**

$$\text{Margin of error} \approx 2\sqrt{\frac{\hat{p}(1-\hat{p})}{n}}$$

Recall that in Section 9.2, we learned that the **standard error of a sample proportion** is

$$\text{s.e.}(\hat{p}) = \sqrt{\frac{\hat{p}(1-\hat{p})}{n}}$$

Thus, the "95% margin of error" is simply two standard errors, or 2 s.e.$(\hat{p})$.

Notice from the formula that three factors determine the margin of error.

1. *The sample size, n.* When the sample size is increased, the margin of error decreases. The effect of sample size on margin of error is intuitive. It makes sense that the accuracy of an estimate will improve when the sample size is increased.
2. *The sample proportion, $\hat{p}$.* Within this context, $\hat{p}$ is a measure of the natural variability among the units. If the proportion is close to either 1 or 0 most individuals have the same trait or opinion, so there is little natural variability and the margin of error is smaller than if the proportion is near .5.
3. *The "multiplier" 2.* This multiplier is connected to the "95% of the time" aspect of the margin of error. Later in the chapter, you'll learn how to change the multiplier in order to change the confidence level. The general theory for the multiplier comes from the normal curve approximation rule for sample proportions described in Section 9.2.

Technical Detail: For reasons explained in Section 10.5, the precise multiplier is 1.96, so the 95% margin of error would be more precisely computed as 1.96 s.e.$(\hat{p})$. Using the value 2 as the multiplier provides a 95.44% confidence level, which is of course approximately 95%. Using the value 2 rather than 1.96 gives a more direct connection to the Empirical Rule, and we generally will compute the 95% margin of error as 2 s.e.$(\hat{p})$.

10.3

turn on your mind

As noted toward the end of Section 9.2, s.e.$(\hat{p})$ is an estimate of s.d.$(\hat{p})$, and it is generally very close. Use the Empirical Rule from Chapters 2 and 8, and the Normal Curve Approximation Rule for Sample Proportions from Section 9.2 to draw a picture of possible sample proportions, using s.e.$(\hat{p})$ in place of s.d.$(\hat{p})$. Show what percent of sample proportions you expect to fall into the ranges:

Population proportion $\pm$ s.e.$(\hat{p})$
Population proportion $\pm$ 2 s.e.$(\hat{p})$
Population proportion $\pm$ 3 s.e.$(\hat{p})$

Then identify what connection margin of error has with the value "95%" in this picture.*

EXAMPLE 10.3
***The Pollen Count
Must Be High Today***

In April of 1998, the Marist College Institute for Public Opinion surveyed a random sample of 883 American adults about allergies. According to a report posted at the Institute's website (www.mipo.marist.edu), 36% of the sample answered "yes" to the question "Are you allergic to anything?" The sample proportion is $\hat{p} = .36$ and $n = 883$, so the margin of error is

$$95\% \text{ margin or error} \approx 2 \times \sqrt{\frac{\hat{p}(1-\hat{p})}{n}}$$

$$= 2 \times \sqrt{\frac{.36(1-.36)}{883}} = .032$$

A 95% confidence interval is $.36 \pm .032$, which is roughly .33 to .39. We can be 95% confident that somewhere between 33% and 39% of all adult Americans have allergies. Remember that the confidence level describes our confidence in the procedure used to determine the interval. In the long run, the procedure will be correct about 95% of the time. ◆

EXAMPLE 10.4
College Men and Ear Pierces

During the 1990s, it became more common for American men, particularly young men, to wear an earring. In class surveys done between 1996 and 1998 in statistics classes at Penn State, 141 of 565 college men indicated that they had at least one ear pierce. The sample proportion with an ear pierce was $\hat{p} = 141/565 = .25$. The approximate margin of error for a 95% confidence level is

$$2\sqrt{\frac{.25(1-.25)}{565}} = .036$$

A 95% confidence interval for the population proportion is $.25 \pm .036$. This is .214 to .286 which, in percentage terms, is 21.4% to 28.6%.

There is an important question that should be asked about this sample, which is: What population does it represent? Because all 565 men were from the same school, the sample may not represent all college men in the United States. It may not even represent all Penn State men because the surveyed classes were taken only by students in the liberal arts and social sciences. Perhaps the population represented is Penn State men in the liberal arts and social sciences.

Another difficulty here is that the sample was a convenience sample that simply included the men who happened to be in certain classes. In the theoretical development of a 95% confidence interval, the assumption is that we have a random sample. Rather than attempting to use the sample to estimate a population proportion, maybe we only should report that 25% of these 565 men had at least one ear pierce. It can be argued, however, that even though the sample was not randomly selected, it is likely to be representative of a larger group and therefore can be used to make inferences about that group. ◆

The Conservative Estimate of Margin of Error

The formula $1/\sqrt{n}$ provides a **conservative estimate of the margin of error** in that it usually overestimates the actual size of the margin of error. This conservative number is used by polling organizations so they can report a single margin of error for all questions on a survey. The more precise formula involves the sample proportion, so the exact margin of error might change from one question to the next within the same survey. The formula $1/\sqrt{n}$ provides a margin of error that works (conservatively) for all questions based on the same sample size, even if the sample proportions differ from one question to the next.

How do we reconcile the two different formulas for the margin of error? The two formulas for the margin of error are equivalent when the proportion used in the more complicated formula is .5. When we use .5 for $\hat{p}$, the margin of error is

$$2 \times \sqrt{\frac{.5(1 - .5)}{n}} = \frac{1}{\sqrt{n}}$$

An important part of this story is that when you use any proportion other than .5 in the more exact formula, the answer is smaller than what you get when you use .5. This means that the conservative formula will almost always overestimate the size of the margin of error.

EXAMPLE 10.5
The Gallup Poll Margin of Error for n = 1000

The Gallup Organization often uses random samples of about 1000 randomly selected Americans. For $n = 1000$ the conservative estimate of margin of error is $1/\sqrt{1000} = .032$, or about 3.2%. ◆

EXAMPLE 10.3 —
CONTINUED
Allergies and Really Bad Allergies

In the Marist Institute allergy survey, about 3% said they experienced "severe" allergy symptoms. The sample size is $n = 883$, so the conservative margin of error is $1/\sqrt{883} = .034$, or 3.4%. Using this conservative margin of error, a 95% confidence interval estimate of the percent of all Americans who have severe allergies is 3% ± 3.4%, or −0.4% to 6.4%. Notice that the lower end of this interval is negative, an impossible value for a population percentage.

Generally, when the sample proportion $\hat{p}$ is far from .5, as it is for the severe allergy symptoms group, the conservative margin of error is too conservative. Using $\hat{p} = .03$ in the more complicated exact formula produces .011 or 1.1% as the margin of error value. The corresponding 95% confidence interval estimate of the population percent with severe allergy symptoms is 3% ± 1.1%, which is 1.9% to 4.1%. ◆

10.4

turn on your mind

Suppose the legislature in a particular state wanted to know what proportion of students graduating from the state university last year were permanent residents of the state. The university had information for all students, and it showed that 3900 of the 5000 graduates were state residents. Is a confidence interval appropriate for this situation? If so, compute the appropriate interval. If not, explain why not.*

10.5 | GENERAL THEORY OF CONFIDENCE INTERVALS FOR A PROPORTION

First, we'll consider the theory behind the development of a 95% confidence interval and then we'll learn how to compute intervals with other confidence levels.

Developing the 95% Confidence Interval

The key to developing a 95% confidence interval for a proportion is the Normal Curve Approximation Rule for Sample Proportions described in Chapter 9. That rule describes the distribution of the sample proportions that would occur if we took many different random samples of the same size from a population. By this rule, if the sample size is sufficiently large, the distribution of possible values of $\hat{p}$ is approximately a normal curve. For this normal distribution,

- The mean is p, the population proportion.
- The standard deviation is

$$\sqrt{\frac{p(1-p)}{n}}$$

Because the distribution of possible sample proportions has a normal curve, we can use the Empirical Rule to make the following statement: *In about 95% of all samples, the sample proportion $\hat{p}$ will fall within 2 standard deviations of p, the true population proportion, so for about 95% of all samples,*

$$-2 \text{ standard deviations} < \hat{p} - p < 2 \text{ standard deviations}$$

This means that in about 95% of all samples, the difference between the sample proportion and the true population proportion is less than 2 standard deviations. The previous sentence basically defines the 95% margin of error.

We have one difficulty. The standard deviation formula involves p, the true population proportion, a value that we don't know. The obvious solution is to put $\hat{p}$, the sample proportion, in the place of p, the standard deviation formula. As discussed in Chapter 9, when this is done, many researchers use the phrase **standard error** to describe the result. Let's review that terminology.

---

definition The **standard error of a sample proportion,** denoted by s.e.$(\hat{p})$, is

$$\text{s.e.}(\hat{p}) = \sqrt{\frac{\hat{p}(1-\hat{p})}{n}}$$

If we assume that the standard error of $\hat{p}$ is a reasonably close approximation to the standard deviation of $\hat{p}$, we can use the standard error in place of the standard deviation. So, for approximately 95% of all random samples from a population, the difference between the sample proportion and the population proportion is described by the mathematical inequality

$$-2 \text{ standard errors} < \hat{p} - p < 2 \text{ standard errors}$$

This mathematical inequality says that in about 95% of all samples, the difference between the sample proportion and the population proportion is less than 2 standard errors. With two steps of algebra, this inequality can be changed to

$$\hat{p} - 2 \text{ standard errors} < p < \hat{p} + 2 \text{ standard errors}$$

This holds for about 95% of all samples and thus provides the formula for the 95% confidence interval.

in summary ## General Description of the Approximate 95% Confidence Interval for a Proportion

An approximate 95% confidence interval for a population proportion is

$\hat{p} \pm 2$ standard errors

The standard error is

$$\text{s.e.}(\hat{p}) = \sqrt{\frac{\hat{p}(1 - \hat{p})}{n}}$$

Interpretation: For about 95% of all randomly selected samples from the population, the confidence interval computed in this manner captures the population proportion.

Necessary Conditions: The confidence interval derived from the normal curve can safely be used when $n\hat{p}$ and $n(1 - \hat{p})$ are both at least 10. It is also assumed that the sample is randomly selected.

Reminders

1. The confidence level describes the chance that the procedure we use to determine the interval will work. For a specific sample, we won't know whether or not the confidence interval actually contains the population proportion. With a 95% confidence level, we know there's a good chance that the interval contains the population value, because it will for approximately 95% of all possible random samples from the population.
2. The exact multiplier for a 95% confidence interval is 1.96, but it is common practice to round this multiplier to 2.

Other Levels of Confidence and Other Types of Confidence Intervals

Researchers don't always use the 95% confidence level. Occasionally, they use a lower level like 90%, and sometimes they use a higher level like 98%. Additionally, researchers estimate many different types of parameters. In Chapter 12, we will learn procedures for finding confidence intervals for means, differences in two means, and differences in two proportions. All of these situations can be accommodated by one general formula. The details for the formula change based on the situation, but the format of the calculation remains the same.

General Format for Confidence Intervals for Proportions and Means

formula   For any confidence level, whether it's 95% or some other value, a confidence interval for either a population proportion or a population mean can be expressed as

Sample estimate $\pm$ Multiplier $\times$ Standard error

The multiplier is affected by the choice of confidence level.

TABLE 10.1 ■ **Confidence Intervals for a Population Proportion**

Confidence Level	Multiplier	Confidence Interval
90	1.645 or 1.65	$\hat{p} \pm 1.65$ standard errors
95	1.96, often rounded to 2	$\hat{p} \pm 2$ standard errors
98	2.33	$\hat{p} \pm 2.33$ standard errors
99	2.58	$\hat{p} \pm 2.58$ standard errors

For a 95% confidence level, the appropriate **multiplier** is close to 2 and is often rounded off to 2. To change the *confidence level,* we simply change the *multiplier.* The larger the multiplier, the greater the confidence. A multiplier larger than 2 provides a confidence level higher than 95% because the interval will be wider than the 95% confidence interval. If the multiplier is smaller than 2, the confidence level will be lower than 95%.

For confidence intervals for proportions, the normal curve is used to determine the multiplier for a specific confidence level. For example, for any normal curve, about 98% of the values are within plus or minus 2.33 standard deviations of the mean. Using the Normal Curve Approximation Rule for Sample Proportions in the same way we did for a 95% confidence interval, we can determine that the correct multiplier for a 98% confidence interval is 2.33 and the associated confidence interval is $\hat{p} \pm 2.33$ standard errors.

Table 10.1 shows the multipliers and the corresponding confidence intervals for four frequently used confidence levels. Notice that when the confidence level increases, the value of the multiplier also increases. With a greater multiplier, the width of the interval will increase. With a wider interval, we'll be more confident that we've captured the population proportion.

EXAMPLE 10.6
Is There Intelligent Life on Other Planets?

In a 1997 Marist Institute survey of 935 randomly selected Americans, 60% of the sample answered "yes" to the question "Do you think there is intelligent life on other planets?" (*Source:* www.mipo.marist.edu). Let's use this sample estimate to calculate a 90% confidence interval for the proportion of all Americans who believe there is intelligent life on other planets. The sample proportion is $\hat{p} = .60$ and $n = 935$. The standard error of the sample proportion is

$$\text{s.e.}(\hat{p}) = \sqrt{\frac{.6(1 - .6)}{935}} = .016$$

For a 90% confidence level, the multiplier in Table 10.1 is 1.65, and a 90% confidence interval for the population proportion is

$.60 \pm 1.65 \times .016$

$.60 \pm .026$

With 90% confidence, we can say that in 1997 the proportion of Americans who believed there is intelligent life on other planets was in the range $.60 \pm .026$ (or, $60\% \pm 2.6\%$).

For 98% confidence, the interval is *sample proportion $\pm$ 2.33 standard errors.* The calculation is $.60 \pm 2.33 \times .016$, which is $.60 \pm .037$. When converted to percents, the 98% confidence interval is $60\% \pm 3.7\%$. We can be fairly sure that in

1997 roughly between 56% and 64% of Americans believed there is life on other planets. Because the entire interval is above 50%, we can also state with high confidence that in 1997 a majority of Americans believed there is intelligent life on other planets. ◆

---

in summary Formula for a Confidence Interval for a Population Proportion p

A formula for the confidence interval for a population proportion is

$$\hat{p} \pm z^* \sqrt{\frac{\hat{p}(1 - \hat{p})}{n}}$$

where

- ◆ $\hat{p}$ is the sample proportion.
- ◆ z^* denotes the multiplier.
- ◆ $\sqrt{\frac{\hat{p}(1 - \hat{p})}{n}}$ is the standard error of the sample proportion.

Table 10.1 has z^* multipliers for the most commonly used confidence levels. The multiplier z^* is such that the area under the standard normal curve between $-z^*$ and z^* corresponds to the confidence level.

Conditions for Using the Formula

Whenever we use a sample to infer something about a population, we have to be sure that the sample satisfies certain conditions. In order to use the procedure described here, we should verify that data meet two conditions:

1. The sample is a randomly selected sample from the population.
2. The normal curve approximation to the distribution of possible sample proportions assumes a "large" sample size. Both $n\hat{p}$ and $n(1 - \hat{p})$ should be at least 10 (although some authors say these quantities need only to be at least 5).

The first condition is an idealistic one that is sometimes difficult to achieve. A more lenient condition is given by the Fundamental Rule for Using Data for Inference, which we repeat here: Available data can be used to make inferences about a much larger group *if the data can be considered to be representative with regard to the question(s) of interest.* In the context of this chapter, a confidence interval based on a sample proportion can be used to estimate the population proportion if the individuals in the sample can be considered to be representative of the individuals in the population for the trait, opinion, etc. of interest.

More About the Multiplier z^*

In the formula for a confidence interval for a proportion, the multiplier z^* is the standardized score such that the area between $-z^*$ and $+z^*$ under the standard normal curve corresponds to the desired confidence level. Figure 10.2 illustrates this relationship. In practice, it would be unusual for anybody to use a confidence

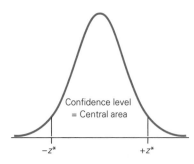

FIGURE 10.2   **The relationship between the confidence level and the multiplier *z**. The confidence level is the probability between −*z** and *z** in the standard normal curve.**

FIGURE 10.3   **Finding *z** for the 50% confidence level**

level other than the four shown in Table 10.1. Theoretically, however, any confidence level could be specified, and the standard normal curve would be used to determine the associated z^* multiplier.

EXAMPLE 10.6—
CONTINUED
50% Confidence
Interval for Proportion
Believing Intelligent
Life Exists Elsewhere

Suppose we want a 50% confidence interval for the 1997 proportion that believed intelligent life exists elsewhere. Figure 10.3 illustrates the problem that we have to solve in order to find the z^* value for a 50% confidence level. To use the Standard Normal Curve table in Appendix A.1, we observe from Figure 10.3 that if the area between $-z^*$ and $+z^*$ is .50, the area to the right of z^* is .25 and so the area to the left of z^* is .75. From Appendix A.1, we learn that the z-value for which the cumulative probability is .75 is about $z^* = .67$. This means that a 50% confidence interval for a population proportion is *sample proportion* $\pm$ *.67 standard errors,* which in this example is $.60 \pm .67 \times .016$, or $.60 \pm .011$. Notice that this interval is more narrow than the 90% and 98% confidence intervals computed previously. Using a lower confidence level produces a more narrow interval. ◆

10.5

turn on your mind

Suppose that we collect data on "hours of sleep last night" from 64 students who are present at an 11:00 A.M. statistics class. The sample mean is 6.8 hours, the standard deviation of the data is 1.6 hours, and the standard error of the mean is 0.2 hours. What population, if any, is represented by this sample? Assuming there is a population represented by this sample, explain why we can not use the formula

$$\hat{p} \pm z^* \sqrt{\frac{\hat{p}(1 - \hat{p})}{n}}$$

to compute a 95% confidence interval for the population mean. What general formula in this section could be used to calculate the interval? Calculate the confidence interval.*

*HINT: Is the given formula appropriate for a mean? Find the definition box with a title referring to a general format for proportions and means. Table 10.1 can be used for the multiplier.

Minitab Tip

> *tech note* **Calculating a Confidence Interval for a Proportion**
>
> ◆ To compute a confidence interval for a proportion, use **Stat>Basic Statistics>1 Proportion.** (This procedure is not in versions earlier than Version 12.)
>
> ◆ If the raw data are in a column of the worksheet, specify that column. If the data have already been summarized, click on "Summarized Data," and then specify the sample size and the count of how many observations have the characteristic of interest.
>
> *Note:* To calculate intervals in the manner described in this chapter, use the **Options** button, and click on "Use test and interval based on normal distribution." Note also that the confidence level can be changed by using the ***Options*** button.

10.6 | CHOOSING A SAMPLE SIZE FOR A SURVEY

When surveys are planned, the choice of a sample size is an important issue. One commonly used strategy is to use a **sample size** that provides a **desired margin of error** for a 95% confidence interval. Table 10.2 displays the margin of error for several different sample sizes. The margin of error calculations were done using the "conservative" formula $1/\sqrt{n}$. This is commonly done by polling organizations because, before the sample is observed, there is no way to know what sample proportion to use in the more exact margin of error formula. With a table like Table 10.2, researchers can pick a sample size that provides suitable accuracy for any sample proportion, within the constraints of the time and money available for the survey.

Two important features of Table 10.2 are

1. When the sample size is *increased,* the margin of error *decreases.*
2. When a large sample size is made even larger, the improvement in accuracy is relatively small. For example, when the sample size is increased from 2500 to 10,000, the margin of error only decreases from 2% to 1%. In general, cutting the margin of error in half requires a fourfold increase in sample size.

Polling organizations determine a sample size that is accurate enough for their purposes and is also economical. Many national surveys use a sample size of about 1000 which, as you can see from Table 10.2, makes the margin of error roughly 3%. This is a reasonable degree of accuracy for most questions asked in these surveys.

Some federal government surveys utilize much larger sample sizes, sometimes as large as $n = 120,000$, to make accurate estimates of quantities like the unemployment rate. Also, when researchers want to make accurate estimates for subgroups within the population, they have to use a very large overall sample size. For instance, to get information from approximately 1000 African-American women in the 18- to 29-year-old age group, a random sample of 120,000 Americans might be necessary.

TABLE 10.2 ■ **Relationship Between Sample Size and Margin of Error for 95% Confidence**

Sample Size n	Margin of Error $= 1/\sqrt{n}$
100	.10 (10%)
400	.05 (5%)
625	.04 (4%)
1000	.032 (3.2%)
1600	.025 (2.5%)
2500	.02 (2%)
10,000	.01 (1%)

The Effect of Population Size

You might wonder how the number of people in the population affects the accuracy of a survey. The surprising answer is that for most sample surveys, the number of people in the population has almost no influence on the accuracy of sample estimates. The margin of error for a sample size of 1000 is about 3% whether the number of people in the population is 30,000 or 200 million.

The formulas for margin of error in this chapter were derived assuming that the number of units in the population is essentially infinite. In practice, as long as the population is at least 10 times as large as the sample, we consider only how sample size affects accuracy and we ignore the specific size of the population. For small populations a "finite population correction" is used. We will not discuss it in this book, but you can consult any book on survey sampling for details.

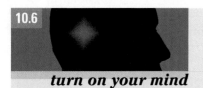

10.6

turn on your mind

Suppose that a survey of 400 students at your school is conducted to assess student opinion about a new academic honesty policy. Based on Table 10.2, about what will be the margin of error for the poll? How many students attend your school? Given this figure, do you think the values in Table 10.2 should be used to estimate the margin of error for a survey of students at your school? Explain.*

10.7 | USING CONFIDENCE INTERVALS TO GUIDE DECISIONS

Two research questions that investigators often consider about population proportions are

- ◆ Is it reasonable to reject a specific hypothesized value of a population proportion?
- ◆ Is it reasonable to conclude that two proportions are different from each other?

In later chapters, we will learn more efficient and powerful procedures that we can use to study these questions. Confidence intervals, however, can be used as a somewhat informal method for evaluating these issues.

Principles for Using Confidence Intervals to Guide Decision Making

Two important principles are useful for examining these questions. Both principles can be applied to any situation in which confidence intervals are calculated. We state the principles only in the context of proportions, but they can be used to answer questions about population means or any other population parameter.

> **Principle 1.** A value not in a confidence interval can be rejected as a possible value of the population proportion. A value in a confidence interval is an "acceptable" possibility for the value of a population proportion.

Principle 2. When the confidence intervals for proportions in two different populations do not overlap, it is reasonable to conclude that the two population proportions are different.

In Chapter 12 we will learn how to find a confidence interval directly for the difference between two population proportions, and we will also learn how to find a confidence interval estimate for the difference between two means. A direct confidence interval for the difference is more efficient than the use of Principle 2 for comparing two population proportions (or means), but if a difference is found using Principle 2, it provides clear evidence of a real difference in the population proportions.

To better understand Principles 1 and 2, let's look at some examples.

EXAMPLE 10.7
Which Drink
Tastes Better?

Imagine a taste test done to compare two different soft drinks. In the test, 60 people taste both drinks and 55% of these 60 participants say they like the taste of drink A better than the taste of drink B. The makers of drink A will be excited that their drink was picked by more than half of the sample and might even create commercials to brag about the results. The makers of drink B, however, can use a 95% confidence interval to show that there is not enough evidence to claim that drink A would be preferred by a majority of the population represented by the sample.

An approximate 95% confidence interval for *p*, the population proportion that prefers drink A, is .42 to .68. The calculation is sample proportion $\pm$ margin of error:

$$.55 \pm 2\sqrt{\frac{.55(1 - .55)}{60}}$$

$$.55 \pm .13$$

Here, the confidence interval leaves the issue unsettled. Most of the interval is above .50 so the makers of drink A may argue that it's reasonable to claim that more than half of the population prefers their drink. On the other hand, the makers of drink B can point out that it is possible that a majority of the population actually prefers their drink because the confidence interval also includes some estimates that are less than .50 (50%)—in fact as low as only 42%.

The difficulty with this hypothetical experiment is that it's too small to be definitive. In actual practice, a much larger sample size should be used to generate a more precise estimate of the proportion that prefers drink A. ◆

CASE STUDY 10.1
Extrasensory Perception Works with Movies

Extrasensory perception (ESP) is the apparent ability to obtain information in ways that exclude ordinary sensory channels. Early ESP research focused on having people try to guess at simple targets, like symbols on cards, to see if they could guess at a better rate than would be expected by chance. In recent years, experimenters have used more interesting targets, like photographs, outdoor scenes, or short movie segments.

In a study of ESP reported by Bem and Honorton (1994), subjects called "receivers" described what another person, the "sender," was seeing on a television screen in another room. The receivers were shown four pictures and asked to pick which one they thought the sender had actually seen. The actual image shown to the sender had been randomly picked by the investigators from among these four choices.

Sometimes the sender was looking at a single, "static" image on the television screen and sometimes at a "dynamic" short video clip, played repeatedly. The additional three choices shown to the receiver for judging were always of the same type (static or dynamic) as the actual target to eliminate biases due to a preference for one type of picture.

Results from the experiment are shown in Table 10.3. Because the right answer is one of four possibilities, receivers who are randomly guessing will be right about 25% of the time. For the static pictures, the 27% success rate is about what would be expected from random guessing. The 41% success rate for the dynamic pictures, however, is noticeably higher than the 25% that would result from random guessing.

A skeptic might raise the possibility that the results for the dynamic pictures were the product of good luck. Is there enough evidence to safely say that the percent of correct guesses for the dynamic pictures is significantly above 25%, the percent expected by chance? A confidence interval helps us answer this question.

For the dynamic pictures, the 95% margin of error is

TABLE 10.3 ■ **Results of the Bem and Honorton ESP Study**

	Successful Guess?			
	Yes	No	Total	Percent Success
Static Picture	45	119	164	27.4%
Dynamic Picture	77	113	190	40.5%

Source: Bem and Honorton, 1994.

$$\text{Margin of error} = 2 \times \sqrt{\frac{.405 \times (1 - .405)}{190}} = .072$$

So, an approximate 95% confidence interval for the proportion is .405 ± .072, which is .333 to .477. In percentage terms, the interval ranges from 33.3% to 47.7%, an interval that does not include 25%. This means that it's reasonable to reject the possibility that the true percent of correct guesses is 25%. The researchers can claim that the true percent of correct guesses is, in fact, significantly better than what would occur from random guessing.

CASE STUDY 10.2
Nicotine Patches Versus Zyban®

Some of you may know from personal experience that quitting smoking is difficult. Many people trying to quit use nicotine replacement methods like nicotine patches or nicotine gum to ease nicotine withdrawal symptoms. Recently, medical researchers have begun investigating whether the use of an antidepressant medication might be a more effective aid to those attempting to give up cigarettes. In a study reported in the March 4, 1999 *New England Journal of Medicine,* Dr. Douglas Jorenby and colleagues compared the effectiveness of nicotine patches to the effectiveness of the antidepressant buproprion, which is marketed with the brand name Zyban.

The 893 participants were randomly allocated to four treatment groups: placebo, nicotine patch only, Zyban only, and Zyban plus nicotine patch. To keep participants blind as to their treatments, they all used a patch (nicotine or placebo) and took a pill (Zyban or placebo). For instance, in the placebo-only group, participants used both a placebo patch and took placebo pills. The Zyban group also used placebo patches, and so on. The treatments were used for nine weeks.

TABLE 10.4 ■ **95% Confidence Intervals for Proportion Not Smoking after Six Months**

Treatment	Subjects	Proportion Not Smoking	Approx. 95% CI
Placebo only	160	.188	.13 to .25
Nicotine patch	244	.213	.16 to .26
Zyban	244	.348	.29 to .41
Zyban and nicotine patch	245	.388	.33 to .44

Table 10.4 displays, for each treatment group, an approximate 95% confidence interval for the proportion not smoking six months after the start of the experiment. Each interval is a range of estimates of the proportion that would not be smoking after six months in a population of individuals using that particular method. A more direct way to compare the four success rates and determine if there are differences will be given in Chapter 15. At this point, we can use the intervals shown in the table to gauge how the four treatments might differ.

FIGURE 10.4 **95% Confidence intervals for proportion not smoking after six months**

The display in Figure 10.4 compares the 95% confidence intervals graphically, clearly showing an appropriate conclusion for this experiment.

Based on the graph, we can make the following generalizations:

♦ Zyban is effective. The Zyban groups had higher success rates and the confidence intervals for the two groups that used Zyban do not overlap the intervals for the two groups not using Zyban.

♦ The nicotine patch is not particularly effective. There is substantial overlap in the range of estimates in the intervals for the nicotine patch and placebo groups, so we can't conclude that the patch is better than the placebo. Similarly, there is substantial overlap between the intervals for the Zyban-only group and the Zyban-plus-patch group, so these two treatments do not significantly differ.

CASE STUDY 10.3
What a Great Personality

Students in a statistics class at Penn State were asked, "Would you date someone with a great personality even though you did not find them attractive?" By gender, the results were

♦ 61.1% of 131 women answered "yes."

♦ 42.6% of 61 men answered "yes."

There clearly is a difference between the men and the women in these samples. Can this difference be generalized to the populations represented by the samples? This question brings up another question. What are the populations that we're comparing? These men and women weren't randomly picked from any particular population, but for this example we'll assume they are like a random sample from the populations of all American college men and women.

A comparison of the 95% confidence intervals for the population percent will help us make a generalization.

♦ For men, the approximate 95% confidence interval is 30.2% to 55%.

♦ For women, the approximate 95% confidence interval is 52.7% to 69.4%.

The two intervals are compared graphically in Figure 10.5. There's a slight overlap in the range of estimates contained in the two intervals. For the most part, however, the interval

FIGURE 10.5 **95% Confidence intervals for proportions that would date someone with a great personality who wasn't attractive**

for the men is below the interval for the women. It's probably safe to generalize that, compared to college men, a higher proportion of college women would say, "Yes, I would date someone with a great personality even if that someone wasn't attractive." In Chapter 12, we will learn how to find a confidence interval for the *difference* in the proportions of men and women who would answer yes to the question.

We can also see the effect of sample size on interval width in Figure 10.5. The confidence interval for women is not as wide as the interval for men. There were more than twice as many women as there were men in the class, and the larger sample size for women leads to a more precise estimate of the population proportion.

10.7

turn on your mind

A randomly selected sample of 400 students is surveyed about whether additional coed dorms should be created at their school. Of those surveyed, 57% say there should be more coed dorms. The 95% margin of error for the survey is 5%. Compute a 95% confidence interval for the population percent in favor of more coed dorms. Based on this confidence interval, can we conclude that more than 50% of all students favor more coed dorms? Explain. Can we reject the possibility that the population proportion is .60?*

in summary   Finding a Confidence Interval for a Population Proportion

In this chapter, each method covered for finding a confidence interval estimate of a population proportion (p) is an instance of the same general format, which is

Sample p ± Multiplier × Standard error

The methods covered were:

Confidence interval for p with exact multiplier for any confidence level:

$$\hat{p} \pm z^* \sqrt{\frac{\hat{p}(1 - \hat{p})}{n}}$$

where the symbol $\hat{p}$ represents the sample proportion, and the multiplier z^* is such that the area under the standard normal curve between $-z^*$ and z^* is the confidence level. For 95% confidence, the exact multiplier is $z^* = 1.96$.

Approximate 95% confidence interval for p:

$$\hat{p} \pm 2 \sqrt{\frac{\hat{p}(1 - \hat{p})}{n}}$$

This formula results from rounding the multiplier $z^* = 1.96$ to the value 2. The corresponding confidence level is 95.44%, or approximately 95%.

Conservative 95% confidence interval for p:

$$\hat{p} \pm \frac{1}{\sqrt{n}}$$

This method, which is the simplest, was covered in Chapter 4 as well. It is used by many polling organizations and results from substituting $\hat{p} = .5$ in the formula for the standard error when determining the approximate 95% confidence interval. The confidence level for this procedure generally is greater than 95%.

KEY TERMS

*HINT: See Example 10.7 in Section 10.7 and Example 10.2—CONTINUED at the end of Section 10.3.

EXERCISES   Blue = Basics   Bold, **Bold** = Answer in back

Sections 10.1, 10.2, and 10.3

10.1 In each part of this question, explain whether the proportion described is a sample proportion or a population proportion. Also in each part, indicate which of the two notations for a proportion, p or $\hat{p}$, should be used to denote the given value.

a. In the 1990 United States Census, it was found that about 1 in 9 Americans were at least 65 years old at that time.

b. A randomly selected group of 500 registered voters in your state is asked whom they intend to vote for in an upcoming election for governor. Based on the proportion choosing each candidate, the polling organization predicts the election winner.

c. In a clinical trial done to assess the effectiveness of a new medication for asthma, a satisfactory relief from symptoms was experienced by 55% of $n = 80$ participants.

10.2 In a 1997 survey done by the Marist College Institute for Public Opinion, 36% of a randomly selected sample of $n = 935$ American adults said they do not get enough sleep each night (*source:* www.mipo.marist.edu). The margin of error was reported to be 3.5%.

a. Use the survey information to create a 95% confidence interval for the percent that feels they don't get enough sleep each night. Write a sentence that interprets this interval. Specify the population.

b. Do you think this sample could be used to estimate the percent of college students who think they don't get enough sleep each night? Why or why not?

10.3 Suppose that a polling organization reports that the margin of error is 4% for a sample survey. Explain what this indicates about the possible difference between a percent determined from the survey data and the population value of the percent.

10.4 A CNN/Time Poll conducted in the United States October 23–24, 2002 asked, "Do you favor or oppose the legalization of marijuana?" In the nationwide poll of $n = 1007$ adults, 34% said they favored legalization. The margin of error was given as 3.1% (*Source:* www.pollingreport.com).

a. Find a 95% confidence interval estimate of the percent of American adults who favored the legalization of marijuana at the time of the poll.

b. Write a sentence that interprets the interval computed in part (a).

10.5. A Gallup Organization poll of $n = 1004$ randomly selected American adults in July 2002 found that 55% of those surveyed felt their weight was about right. The margin of error for the survey was given as 3% (*Source:* www.gallup .com).

a. Find a 95% confidence interval estimate of the percent of American adults who think their weight is about right.

b. Based on the interval computed in part (a), explain whether it is reasonable to say that more than 50% of American adults think their weight is about right.

10.6 In each situation, explain why you think the sample proportion should or should not be used to estimate the population proportion.

a. An Internet news organization asks visitors to its website to respond to the question "Are you satisfied with the president's job performance?" Of $n = 3500$ respondents, 61% say they are not satisfied with the president's performance. Based on this survey, the organization writes an article in which they say that a majority of Americans are not satisfied with the president.

b. A convenience sample of $n = 400$ college students in two classes at the same university is used to estimate the proportion that is left-handed in the nationwide population of college students.

10.7 A federal law that lowered the limit for a legal blood alcohol level for automobile drivers was signed by President Clinton in October 2000. At their website, the Gallup Organization reported the following survey results in an article dated October 26, 2000:

> 72% of Americans support lowering the drunk driving limit to 0.08% BAC. . . . A total 1,002 telephone interviews were conducted between July 20–Aug. 3, 2000, with a representative sample of the U.S. public age 16 and older. The findings are based on 930 respondents who identified themselves as licensed drivers. The margin of error equals plus or minus three percentage points.

a. The first sentence of the information from the Gallup Organization states, "72% of Americans support lowering the drunk driving limit to 0.08% BAC." Considering the difference between a sample and a population, explain why this sentence is not necessarily correct. Rewrite the sentence so that it is correct.

b. The article provides the information that "the margin of error equals plus or minus three percentage points." Write one or two sentences interpreting this value that could be understood by someone who does not know anything about statistics.

c. Using the information provided by the Gallup Organization, calculate a 95% confidence interval for the proportion that supports the lower legal blood alcohol limit. Write a sentence that interprets this interval. Be sure to specify the appropriate population.

d. Based on the interval computed in part (c), is it reasonable to conclude that more than half of American licensed drivers support the lower blood alcohol limit?

10.8 Suppose 200 different researchers all randomly select samples of $n = 400$ individuals from a population. Each researcher uses his or her sample to compute a 95% confidence interval for the proportion that has blue eyes in the population. About how many of the confidence intervals will capture the population proportion? About how many of the intervals will not capture the population proportion? Briefly explain how you determined your answers.

10.9 Suppose that a margin of error for a sample percent is reported to be 3%. What is the probability that the difference between the sample percent and the population percent will be more than 3%? Interpret the meaning of this probability.

10.10 Taking into account the purpose of a confidence interval described in Section 10.3, explain what is wrong with the following statement: "Based on the survey data, a 95% confidence interval estimate of the sample proportion is .095 to .117."

Section 10.4

10.11 For each combination of sample size and sample proportion, find the approximate margin of error for the 95% confidence level.
   a. $n = 100, \hat{p} = .56$.
   b. $n = 400, \hat{p} = .56$.
   c. $n = 400, \hat{p} = .20$.
   d. $n = 400, \hat{p} = .80$.
   e. $n = 1000, \hat{p} = .50$.

10.12 Suppose a new treatment for a certain disease is given to a sample of $n = 200$ patients. The treatment was successful for 166 of the patients. Assume that these patients are representative of the population of individuals who have this disease.
   a. Calculate the sample proportion successfully treated.
   b. Determine a 95% confidence interval for the proportion successfully treated. Write a sentence that interprets this interval.

10.13 In a survey of 190 college students, 134 students said they believe there is extraterrestrial life (*Data source:* **pennstate3** dataset on the CD for this book).
   a. Find $\hat{p} =$ sample proportion that believes there is extraterrestrial life.
   b. Calculate the approximate margin of error for the 95% confidence level.
   c. Using the margin of error found in part (b), find a 95% confidence interval estimate of the proportion of all college students who believe there is extraterrestrial life.

10.14 Suppose a polling organization is conducting a survey to estimate the proportion of Americans who regularly attend religious services. They plan to gather data from a randomly selected sample of 550 individuals. Based on the "conservative" formula, what will be the margin of error for the survey?

10.15 In a poll done by Princeton Survey Research Associates in October of 1998 for *Newsweek Magazine,* 56% of 753 randomly sampled American adults answered "yes" to the question "Do you personally believe that abortion is wrong?" (*Source:* www.pollingreport.com/abortion.htm.)
   a. Calculate the margin of error necessary for a 95% confidence interval.
   b. Calculate a 95% confidence interval for the percent of all Americans who believe that abortion is wrong.
   c. Based on this confidence interval, is it reasonable to conclude that more than half of all Americans think abortion is wrong? Explain.

10.16 In a Gallup Youth Survey done in 2000, $n = 501$ randomly selected American teenagers were asked about how well they get along with their parents.
   a. According to the Gallup Organization, the margin of error for the poll was 5%. Verify that this figure is approximately correct.
   b. A survey result was that 54% of the sample said they get along "very well" with their parents. Using the reported margin of error, calculate a 95% confidence interval for the population proportion that gets along "very well" with their parents. Write a sentence that interprets this interval.
   c. Using the more exact formula for margin of error in Section 10.4, calculate a 95% confidence interval. Compare the answer to the answer in part (b).
   d. Another result in the same survey was that only 5% said that cheating on exams was "not at all serious." Using the reported margin of error, compute a 95% confidence interval for the population proportion that would say this. Also, compute the 95% confidence interval using the more exact formula for margin of error. Which interval is narrower?
   e. You should have found that the intervals in parts (b) and (c) were quite similar to each other, but the two intervals computed in part (d) were not. Explain why the two intervals agreed more closely in parts (b) and (c) than they did in part (d).

10.17 In Example 10.3—*CONTINUED* the information given was that 3% of a nationwide randomly selected sample of $n = 883$ suffered from severe allergies.
   a. Determine the standard error of the sample proportion in this problem.
   b. Verify that the margin of error is about 1.1% for a 95% confidence interval for the percent that suffers from severe allergies in the population represented by this sample.

Section 10.5

10.18 For each confidence interval procedure, provide the confidence level.
   a. Sample proportion $\pm 1.645 \times$ standard error.
   b. Sample proportion $\pm 2 \times$ standard error.
   c. Sample proportion $\pm 2.33 \times$ standard error.
   d. Sample proportion $\pm 2.58 \times$ standard error.

10.19 Explain whether the width of a confidence interval would increase, decrease, or remain the same as a result of each of the following changes:
   a. Increase the confidence level from 95% to 99%.

b. Decrease the confidence level from 95% to 90%.

10.20 In a randomly selected sample of $n = 400$ registered voters in a community, 220 individuals say they plan to vote for Candidate Y in the upcoming election.
   a. Find $\hat{p}$ = sample proportion planning to vote for candidate Y.
   b. Calculate the standard error of the sample proportion.
   c. Find a 95% confidence interval for the proportion of the registered voter population who plan to vote for Candidate Y.
   d. Find a 98% confidence interval for the proportion of the registered voter population who plan to vote for Candidate Y.

10.21 In an ABC News.com nationwide poll done in 2001, the proportion of respondents who thought it should be illegal to use a handheld cellular telephone while driving a car was $\hat{p} = .69$ (69%). The poll's sample size was $n = 1027$.
   a. Find the value of the standard error of the sample proportion.
   b. Find a 95% confidence interval estimate of the population proportion that thinks it should be illegal to use a cellular telephone while driving.
   c. Find a 90% confidence interval estimate of the population proportion that thinks it should be illegal to use a cellular telephone while driving.

10.22 Refer to Example 10.6 of this chapter where for a sample of $n = 935$ American adults, $\hat{p} = .60$ was the proportion that thinks intelligent life exists elsewhere. Compute a 99% confidence interval for the population proportion.

10.23 Refer to exercise 10.12. Calculate a 98% confidence interval for the proportion successfully treated. Is this interval wider or narrower than the interval computed in part (b) of exercise 10.12?

10.24 A randomly selected sample of $n = 15$ individuals is asked whether they are right-handed or left-handed. In the sample, only one person is left-handed. Explain why a confidence interval for the population proportion should not be computed using the methods described in this chapter.

10.25 Determine the value of the z^* multiplier that would be used to compute an 80% confidence interval for a population proportion.

Sections 10.6 and 10.7

10.26 Suppose that a researcher is designing a survey to estimate the proportion of adults in your state who oppose a proposed law that requires all automobile passengers to wear a seat belt.
   a. What would be the approximate margin of error if the researcher randomly sampled 400 adults?
   b. What sample size would be needed to provide a margin of error of about 2%?

10.27 Explain whether the width of a confidence interval would increase, decrease, or remain the same as a result of each of the following changes.
   a. Increase the sample size from 1000 to 2000.
   b. Decrease the sample size from 1000 to 500.
   c. Increase the population size from 10 million to 20 million.

10.28 Suppose a health expert has claimed that 28% of college students smoke cigarettes. To investigate this claim, researchers survey a random sample of college students. Using this sample, a 95% confidence interval for the percent of college students who smoke is found to be 19% to 24%. Based on this interval, explain whether the claim that 28% of all college students smoke is reasonable or not.

10.29 A university is contemplating switching from the quarter system to a semester system. The administration conducts a survey of a random sample of 400 students and finds that 240 of them prefer to remain on the quarter system.
   a. Construct a 95% confidence interval for the true proportion of all students who would prefer to remain on the quarter system.
   b. Does the interval you computed in part (a) provide convincing evidence that more than half of all students prefer to remain on the quarter system? Explain.

10.30 In which of the following three samples will the margin of error be the smallest? Explain. Assume that each sample is a random sample.
   Sample A: sample of $n = 1000$ from a population of 10 million.
   Sample B: sample of $n = 2500$ from a population of 200 million.
   Sample C: sample of $n = 400$ from a population of 50,000.

10.31 Suppose that a 95% confidence interval for the proportion of men who experience sleep apnea (irregular breathing during sleep) is .11 to .17 and a 95% confidence interval for the proportion of women who experience sleep apnea is .04 to .08. Based on these intervals, is it reasonable to conclude that the population proportions experiencing sleep apnea differ for men and women? Explain.

10.32 A college dean plans a student survey to estimate the percent of presently enrolled students who plan to take classes during the next summer session. She wants the 95% margin of error for the estimate to be at most 5%. What is the necessary sample size?

10.33 An epidemiologist plans a survey to estimate the prevalence of Alzheimer's disease in the population of adults aged 65 years or older. It's desired that the 95% margin of error for the sample estimate be no more than 3%. What sample size is needed?

10.34 Exercise 10.5 describes a 2002 Gallup Organization poll in which the finding was that 55% of adult Americans surveyed said their weight was about right. The margin of error for the survey was 3%. The website article about this survey also contained the information that in a 1990 survey done by Gallup, 46% of the respondents thought their weight was about right. Assuming the margin of error for the 1990 survey was also 3%, discuss whether it is reasonable to conclude that the population percent thinking their weight is about right is higher in 2002 than in 1990.

Chapter Exercises

10.35 In Case Study 10.1, we saw the results of an ESP experiment in which there was a 1/4 chance of correct response if the "receiver" was merely guessing at what image a "sender" in another room was viewing. For static images, the "receivers" responded correctly in 45 of 164 trials.

a. Calculate the sample proportion of correct guesses at the static images.

b. Calculate the standard error for this sample proportion.

c. Calculate a 95% confidence interval for the true proportion of correct guesses at static images.

d. Based on the confidence interval in part (c), do you think that these data provide evidence that there is extrasensory perception? Explain.

10.36 Suppose that in a random sample $n = 300$ of employed Americans, there are 57 individuals who say they would fire their boss if they could. Calculate a 95% confidence interval for the population proportion. Write a sentence or two that interprets this interval.

10.37 Refer to the previous exercise. Calculate a 90% confidence interval for the population proportion.

10.38 *U.S. News and World Report* (Dec. 19, 1994, pp. 62–71) reported on a survey of 1000 American adults, conducted by telephone on December 2–4, 1994, designed to measure beliefs about apocalyptic predictions. One of the results reported was that 59% of the sample said they believe the world will come to an end.

a. Calculate the standard error of the sample proportion who believe the world will come to an end.

b. Calculate a 95% confidence interval for the population proportion. Interpret the interval in a way that could be understood by a statistically naïve reader.

10.39 In Chapter 2, we saw data from a statistics class activity in which 190 students were asked to randomly pick one of the numbers 1, 2, 3, 4, 5, 6, 7, 8, 9, 10. The number 7 was picked by 56 students.

a. For the sample, calculate the proportion of students who picked 7.

b. Calculate the standard error for this sample proportion.

c. Calculate a 90% confidence interval for the population proportion.

d. Calculate a 95% confidence interval for the population proportion.

e. Calculate a 98% confidence interval for the population proportion.

f. What do the results of parts (c–e) indicate about the effect of confidence level on the width of a confidence interval?

g. Based on these confidence intervals, do you think that students choose numbers "randomly"? Explain.

10.40 A study reported in the *Journal of Occupational and Environmental Medicine* (Franke et al., 1998, 40:441– 444) found that retired male Iowa policemen are more likely to have heart disease compared to other men with similar ages but different occupations. Data from the study were that about 32% of a sample of 232 retired policemen had heart disease, but only 18% of a sample of 817 men with other occupations had heart disease.

a. In Chapter 3, we learned about the various types of studies. Is this an observational study or a randomized experiment? Explain.

b. Do you think that we can attribute the observed difference in heart disease percentages to job-related factors? Explain.

c. Graphically compare the policemen and the other men using a bar chart.

d. Calculate a 95% confidence interval for the proportion of retired Iowa policemen.

e. Calculate a 95% confidence interval for the proportion of other retired men who have heart disease.

f. Based on these two confidence intervals, do you think it is safe to conclude that police officers are more likely to experience heart disease than other men? Explain.

g. Why are the widths of the two confidence intervals different from each other?

10.41 Use an example from this chapter to explain the difference between a sample proportion and a population proportion.

10.42 A professor wants to know if her class of 60 students would prefer the final exam to be given as a take-home exam or an in-class exam. She surveys the class and learns that 45 of the 60 students would prefer a take-home. Explain why she should not use these data to compute a 95% confidence interval for the proportion preferring a take-home exam.

10.43 Suppose that a margin of error for a survey is reported to be 4%. Write a brief explanation of what this margin of error tells us about the accuracy of sample percentages from this survey.

10.44 The Gallup Organization used two different wordings of a question in a February 1999 poll about the death penalty. The two different questions were asked of two different random samples of Americans. Results reported at the Gallup website were

Question 1: Are you in favor of the death penalty for a person convicted of murder? (Based on 543 adults: margin of error plus or minus 5 percentage points)

	For	Against	No Opinion
99 Feb 8–9	71%	22%	7%

Question 2: What do you think should be the penalty for murder—the death penalty or life imprisonment with absolutely no possibility of parole?

(Based on 511 adults: margin of error plus or minus 5 percentage points)

	Death Penalty	Life Imprisonment	No Opinion
99 Feb 8–9	56%	38%	6%

a. Compare the two questions using a bar chart.

b. The margin of error for each question was reported as 5%. Verify that this is approximately true.

c. For the first question, calculate a 95% confidence interval for the percent of all Americans who are "For" the death penalty. Use the margin of error reported by the Gallup Organization. Write a sentence that interprets this confidence interval.

d. For the second question, calculate a 95% confidence interval for the percent of all Americans who would answer "Death Penalty" when asked whether the penalty for murder should be the death penalty or life imprisonment. Use the margin of error reported by the Gallup Organization. Write a sentence that interprets this interval.

10.45 In a survey reported in a special edition of *Newsweek* (Spring/Summer 1999, *Health for Life*), only 3% of a sample of 757 American women responded "Not at all satisfied" to the question "How satisfied are you with your overall physical appearance?"

a. The margin of error for the poll was reported as 3.5%. Use the conservative formula for margin of error to verify that this is approximately correct.

b. Based on the reported margin of error, calculate a 95% confidence interval for the percent of all American women who are "not at all satisfied" with their physical appearance. Are all values in this interval valid estimates of the population percent? Explain.

c. Use the "exact" formula to calculate the margin of error.

d. Based on the margin of error in part (c), calculate a 95% confidence interval for the population percent. Write a sentence that interprets this confidence interval for somebody who doesn't know very much about statistics.

10.46 What is the probability that a 95% confidence interval will not cover the true population value? Explain.

10.47 In 1998, two statistics classes for students in the liberal arts at a large northeastern university were asked, "Do you have a tattoo?" The responses were

	Yes	No
Women	46	227
Men	32	175

a. What proportion of the women in this sample has a tattoo?

b. What proportion of the men in this sample has a tattoo?

c. Compare the men and women using a bar graph.

d. What population do you think is represented by this sample? In other words, to whom do you think the results can be generalized?

e. For the college women, calculate a 95% confidence interval for the proportion of the population that has a tattoo. Write a sentence that interprets this interval.

f. For the college men, calculate a 95% confidence interval for the proportion of the population that has a tattoo. Write a sentence that interprets this interval.

g. Use the results of parts (e) and (f) to compare college men and women with regard to the proportion that has a tattoo.

10.48 Find the reported results of a poll in which a margin of error is also reported. Look in a weekly news magazine such as *Newsweek* or *Time,* in a newspaper such as the *New York Times,* or on the Internet.

a. Explain what question was asked in the poll and what margin of error was reported. Verify the margin of error with a calculation.

b. Present a 95% confidence interval for the results.

Explain in words what the interval means for your example.

10.49 For Case Study 10.2 (nicotine patches and Zyban), verify the confidence intervals reported in Table 10.4.

10.50 *Parade Magazine* reported that "nearly 3200 readers dialed a 900 number to respond to a survey in our Jan. 8 cover story on America's young people and violence" (Feb. 19, 1995, p. 20). Of those responding, "63.3% say they have been victims or personally know a victim of violent crime." Can the methods in this chapter legitimately be used to compute a 95% confidence interval for the proportion of Americans who fit that description? Explain why or why not.

Dataset Exercises

10.51 Use the **GSS-93** dataset for this exercise. The dataset contains data from the 1993 General Social Survey, and the variable *polparty* indicates the political party preference of the respondent. Calculate a 95% confidence interval for the proportion of Americans who indicated that they prefer the Democratic party.

10.52 Use the **GSS-93** dataset for this exercise. The dataset contains data from the 1993 General Social Survey, and the variable *marijuan* indicates the respondent's opinion about the legalization of marijuana. Calculate a 95% confidence interval for the proportion of Americans who are opposed to the legalization of marijuana.

10.53 Use the **pennstate3** dataset for this exercise. The data are from a student survey in a statistics class at Penn State. The variable *atfirst* gives data for whether or not a student believes in love at first sight.

a. What is the sample proportion that believes in love at first sight?

b. Find a 95% confidence interval for the population proportion that believes in love at first sight.

c. Write a sentence that interprets the interval found in part (b).

10.54 Refer to the previous exercise about the variable *atfirst* in the **pennstate3** dataset.

a. Create a two-way table that shows the relationship between the variables *atfirst* and *Sex.* (See Sections 2.3 and 6.1 to review two-way tables.)

b. What proportion of the females in the sample believes in love at first sight? What proportion of the males believes in love at first sight?

c. For each sex separately, find a 95% confidence interval for the population proportion that believes in love at first sight.

d. Compare the two confidence intervals found in part (c). What does the comparison indicate about the difference between the proportions of males and females who believe in love at first sight?

| REFERENCE

Bem, D., and C. Honorton (1994). "Does psi exist? Replicable evidence for an anomalous process of information transfer," *Psychological Bulletin,* Vol. 115, No. 1, pp. 4–18.

CHAPTER 11

Testing Hypotheses About Proportions

How do pharmaceutical companies test medication side effects?
See Examples 11.1 and 11.1—CONTINUED.

The logic of statistical hypothesis testing is similar to the "presumed innocent until proven guilty" logic of the American judicial system. In hypothesis testing, we assume that the null hypothesis is a possible truth until the sample data conclusively demonstrate otherwise.

Suppose that your state legislature is considering a proposal to lower the legal limit for the blood alcohol level that constitutes drunk driving. A legislator wants to determine if a majority of adults in her district favor this proposal. To gather information, she surveys 200 randomly selected individuals from her district and learns that 59% of these people favor the proposal.

Can the legislator conclude that a majority of all adults in her district favor this proposal? Because the result is based on a sample, there is the possibility that the observed majority might have occurred just by the "luck of the draw." If a majority of the whole population actually opposes the proposal, how likely is it that 59% of a random sample would favor the proposal?

In this chapter, we will learn how to use the method of *statistical hypothesis testing* to analyze this type of issue. The hypothesis testing method uses data from a sample to judge whether or not a statement about a population may be true. For example, is it true that a majority of all adults in the legislator's district favor the proposal for the tougher blood alcohol standard?

Hypothesis testing and confidence interval estimation are related methods, and often both methods can be used to analyze the same situation. As we learned in Chapter 10, a confidence interval is a numerical answer to the question "What is the population value?" A hypothesis test is used to answer questions about particular values for a population parameter, or particular relationships in a population, based on information in the sample data.

In this chapter, we focus on testing hypotheses about a population proportion, but the basic steps are the same for any hypothesis test. Additional situations will be considered in Chapters 13 to 16. The five steps for any hypothesis test, first introduced in Chapter 6, are as follows.

Steps in Any Hypothesis Test

Step 1: *Determine the* null *and* alternative *hypotheses.*

Step 2: *Verify necessary data conditions, and if met, summarize the data into an appropriate* test statistic.

Step 3: *Assuming the null hypothesis is true, find the p-value.*

Step 4: *Decide whether or not the result is* statistically significant *based on the p-value.*

Step 5: *Report the conclusion in the context of the situation.* ◆

11.1 | FORMULATING HYPOTHESIS STATEMENTS

Many of the questions that researchers ask can also be expressed as questions about which of two statements might be true for a population. Consider, for instance, the nature of these research questions:

- ◆ Does a majority of the population favor a new legal standard for the blood alcohol level that constitutes drunk driving?
- ◆ Do female students study, on average, more than male students do?
- ◆ Will side effects be experienced by fewer than 20% of people who take this new medication?

All of these questions can be answered either with a "no" or a "yes," and each possible answer makes a specific statement about the situation. For instance, we can view the research question about the proportion favoring a new legal standard for drunk driving as a choice between two competing *hypothesis* statements:

Hypothesis 1: The proportion favoring the new standard *is not a majority.*
Hypothesis 2: The proportion favoring the new standard *is a majority.*

Or, for the question about the comparative study habits of women and men, we can view the research question in terms of a choice between these two statements:

Hypothesis 1: On average, women *do not* study more than men do.
Hypothesis 2: On average, women *do* study more than men do.

Terminology for the Two Choices

In the language of statistics, the two possible answers to questions like the ones we just encountered are called the **null hypothesis** and the **alternative hypothesis.**

definition The **null hypothesis,** represented by the symbol H_0, is a statement that there is nothing happening. The specific null hypothesis varies from problem to problem, but generally it can be thought of as the status quo, or no relationship, or no difference. In most situations, the researcher hopes to disprove or reject the null hypothesis.

The **alternative hypothesis,** represented by the symbol **H_a,** is a statement that something is happening. In most situations, this hypothesis is what the researcher hopes to prove. It may be a statement that the assumed status quo is false, or that there is a relationship, or that there is a difference.

Each of the following statements is an example of a null hypothesis:

- There is no extrasensory perception.
- There is no difference between the mean pulse rates of men and women.
- There is no relationship between exercise intensity and the resulting aerobic benefit.

Some examples of alternative hypotheses are

- There is extrasensory perception.
- Men have lower mean pulse rates than women do.
- Increasing exercise intensity increases the resulting aerobic benefit.

EXAMPLE 11.1
Are Side Effects Experienced by Fewer Than 20% of Patients?

Suppose that a pharmaceutical company wants to be able to claim that for its newest medication the proportion of patients who experience side effects is less than 20%. The issue can be addressed with these null and alternative hypotheses:

Null: 20% (or more) of users will experience side effects.
Alternative: Fewer than 20% of users will experience side effects.

Notice that the claim that the company hopes to prove is used as the alternative hypothesis. ◆

EXAMPLE 11.2
Does a Majority Favor the Proposed Blood Alcohol Limit?

For the blood alcohol question, suppose that the legislator's plan is to vote for the proposal if there is conclusive evidence that a majority of her constituents favor the proposal. In statistical terms, the null and alternative hypotheses for the legislator are

$H_0: p \le .5$ (not a majority)
$H_a: p > .5$ (a majority)

As in Chapters 9 and 10, p represents a population proportion. In this case, p is the proportion of her constituents that favors the proposal. The alternative hypothesis is a "something happening" statement in the sense that its truth will cause the legislator to vote for the new limit. ◆

In Example 11.1, the alternative hypothesis is that the percent experiencing side effects is *less than* 20%. In Example 11.2, the alternative hypothesis is that the proportion of the population in favor of the legislator's plan is *more than* .5. Both of these alternative hypotheses describe values that differ from the null hypothesis in a particular numerical direction. When the alternative hypothesis specifies a single direction, the test is called a **one-sided** or **one-tailed hypothesis test.** In practice, most hypothesis tests are one-sided tests because investigators usually have a particular direction in mind when they consider a question.

When the alternative hypothesis includes values in either direction from a specific standard, the test is called a **two-sided** or **two-tailed hypothesis test.** Suppose, for example, that an alternative hypothesis is that men and women have different verbal abilities. This statement does not specify a particular direction for the difference, so the hypothesis test is a two-sided test.

11.1

turn on your mind

Confidence intervals and hypothesis testing are the two major categories of "statistical inference." Based on this information, do you think null and alternative hypotheses are generally statements about populations, samples, or both? Explain.*

11.2 | THE LOGIC OF HYPOTHESIS TESTING: WHAT IF THE NULL IS TRUE?

The logic of statistical hypothesis testing is similar to the "presumed innocent until proven guilty" logic of the American judicial system. In hypothesis testing, we assume that the null hypothesis is a possible truth until the sample data conclusively demonstrate otherwise. The "something is happening" hypothesis is chosen only when the data show us that we can reject the "nothing is happening" hypothesis.

Unfortunately, the hypothesis test method is a somewhat indirect strategy for making a decision. We will not be able to determine the chance that a hypothesis statement is either true or false. We can only assess whether or not the observed data are consistent with an assumption that the null hypothesis is true about the population, within the reasonable bounds of sampling variability discussed in Chapter 9. If the data would be unlikely when the null hypothesis is true, we *reject* the statement made in the null hypothesis.

The Probability Question on Which Hypothesis Testing Is Based

If the null hypothesis is true about the population, what is the probability of observing sample data like that observed? Let's look at two examples that illustrate how the answer to this question is useful for deciding whether or not to reject the null hypothesis.

EXAMPLE 11.3
Psychic Powers

A popular cartoon once featured two characters playing a coin-flipping game. After one had correctly guessed the outcome 100 times, supposedly using psychic abilities, the other still insisted that it must have been a coincidence. When the first character successfully guessed that the next flip would result in seven rotations followed by hovering in the air, the skeptic resorted to chanting the word "luck" repeatedly, still refusing to believe in psychic abilities.

This cartoon nicely captures the thought process of hypothesis testing. We all realize that correctly guessing 100 times in a row is unlikely to be a coincidence. It seems nearly impossible that anybody could be this lucky. Suppose, however, that we accept the skeptic's explanation and assume that the cartoon character is indeed only guessing. If this is true, how likely is it that he could correctly guess 100 consecutive times in a fair coin toss?

We can use the methods of Chapter 7 to calculate that the probability of 100 correct guesses in a row is $(1/2)^{100}$. As you might expect, the answer is extraordinarily small. After the decimal point, there are 30 zeroes before the first nonzero digit appears. Basically, there is no chance that anybody could be this lucky. It is reasonable to conclude that the character must have been using psychic powers, or some method other than just guessing. We would reject the null hypothesis that he was just guessing because the sample results are extremely inconsistent with that hypothesis. ◆

EXAMPLE 11.4
Stop the Pain
Before It Starts

A headline at the CNN website on October 21, 1996 read, "Study: Painkillers More Effective Before Surgery Than Afterward." The article described a University of Pennsylvania research study done to test the theory that the most effective pain relief strategy is to stop the pain before it starts. In the study, men undergoing surgery for prostate cancer were randomly assigned either to an "experimental" group that began taking painkillers before the operation or to a "control" group that followed the standard practice of waiting until after the operation to begin taking painkillers. The author of the website article presented the key result in terms of the likelihood that the observed data could have occurred just by chance:

> But 9 1/2 weeks later . . . only 12 members of the 60 men in the experimental group were still feeling pain. Among the 30 control group members, 18 were still feeling pain. Gottschalk said the likelihood of this difference being due to chance was only 1 in 500.

The null hypothesis, the statement that nothing is happening, is that the effectiveness of the painkillers would be the same whether they were started before or after the operation. According to the investigator, if we assume that this null hypothesis is true, the probability is only 1 in 500 that the observed difference could have been as large as it was or larger. Based on this evidence, it seems reasonable to reject the null hypothesis that the two timings are equally effective. ◆

11.2

turn on your mind

For each of the two examples in this section, explain what the null hypothesis and the alternative hypothesis would be. Then write a sentence answering "the probability question at the center of a hypothesis test" for each example. The sentence should be of the form: "If [fill in null hypothesis] is true, then the likelihood that [fill in event] would have happened is [fill in likelihood]."*

11.3 | REACHING A CONCLUSION ABOUT THE TWO HYPOTHESES

When we do a hypothesis test, the objective is to decide between the null and alternative hypotheses.

- ◆ The data summary that we use to evaluate the two hypotheses is called the **test statistic.**
- ◆ The likelihood that we would have observed a *test statistic* as extreme as what we did, or something even more extreme, if the null hypothesis is true is called the ***p*-value.**
- ◆ The decision is made to accept the alternative hypothesis if the *p*-value is smaller than a designated **level of significance,** denoted by the Greek letter **α** (alpha), and usually set by researchers at 0.05, less commonly at 0.10 or 0.01.

Computing the *p*-Value of a Hypothesis Test

> *definition* The ***p*-value** is computed by assuming the null hypothesis is true and then determining the probability of a result as extreme (or more extreme) as the observed test statistic in the direction of the alternative hypothesis.

In the language of Chapter 7, the *p*-value is the *conditional probability of observing a test statistic as extreme as that observed or more so, given that the null hypothesis is true.* "Extreme" is understood to mean in the direction supporting the alternative hypothesis. We will learn details of how to compute a "test statistic" for specific situations in this and subsequent chapters. In general, a test statistic is simply a summary that compares the sample data to the null hypothesis. The chi-square statistic introduced in Chapter 6 is an example of a test statistic.

Notice that because the *p*-value is a probability, it must be between 0 and 1. It indicates the strength of the evidence against the null hypothesis. The smaller the *p*-value is, the more conclusive the evidence is against the null hypothesis. For example, the evidence against the null hypothesis is more conclusive when the *p*-value is 0.001 than when it is 0.25. As we learned earlier in this chapter, the logic of hypothesis testing is indirect. A *p*-value *does not* tell us the probability that the null hypothesis is true. Instead, it only tells us the probability that our test statistic could have been as extreme as it is, if we assume the null hypothesis is true.

An Advanced Technical Note — Bayesian Statistics

 tech note The *p*-value represents the *conditional probability* $P(A|B)$, where

A = observing a test statistic value as extreme as that observed or more so, and

B = the null hypothesis is true.

We would really prefer to know $P(B|A)$, the probability that the null hypothesis is actually true, *given* the observed data. Suppose we try using Rule 3 or 4 from Chapter 7 to find this conditional probability. We would need to find $P(A$ and $B)$, and to find that, we would need to know $P(B)$. But that's the probability that the null hypothesis is true. If we knew that, we wouldn't have to conduct the hypothesis test! That's why we must rely on the indirect information provided by the *p*-value to make our conclusion. In fact, the null hypothesis is either true or not, so even considering $P(B)$ only makes sense if we think of it as a personal probability representing our belief that the null hypothesis is true. There is a branch of statistics, called **Bayesian statistics,** that utilizes the approach of assessing $P(B)$ and then combining it with the data to find an updated $P(B)$.

Rejecting the Null Hypothesis

The phrase **statistically significant** is used to describe the results when the researcher has decided that the *p*-value is small enough to decide in favor of the alternative hypothesis. The **level of significance,** also called the **α level** ("alpha" level), is the borderline for deciding that the *p*-value is small enough to justify choosing the alternative hypothesis. A result is statistically significant when the *p*-value is less than the chosen level of significance.

Most researchers use the convention that we can reject the null hypothesis when the *p*-value is less than 0.05 (or 5%). In other words, the *level of significance* is usually set at $\alpha = 0.05$, and the result is *statistically significant* when the *p*-value is less than 0.05. In some research disciplines, the phrase *highly significant* is used to describe the result when the *p*-value is less than 0.01 (1%), and *marginally significant* is used when the *p*-value is between 0.05 and 0.10. *p*-Values greater than 0.10 are rarely considered sufficient evidence to reject the null hypothesis. Researchers frequently criticize these somewhat arbitrary borderlines, but they are used by many scientific journals.

> *in summary*   Interpreting a *p*-Value
>
> In any statistical hypothesis test, the smaller the *p*-value is, the stronger the evidence is against the null hypothesis. It is common research practice to call a result *statistically significant* when the *p*-value is less than 0.05.

Stating the Two Possible Conclusions

The two possible conclusions of a hypothesis test are

- When the *p*-value is small, we *reject the null hypothesis* or, equivalently, we *accept the alternative hypothesis.* "Small" is defined as a *p*-value $\leq \alpha$, where α = level of significance (usually 0.05).

- When the *p*-value is not small, we conclude that we *cannot reject the null hypothesis* or, equivalently, *there is not enough evidence to reject the null hypothesis.* "Not small" is defined as a *p*-value $> \alpha$, where α = level of significance (usually 0.05).

The conclusion that we "cannot reject the null hypothesis" may seem wordy to you, and you may be tempted to say that you accept the null hypothesis. Resist that temptation because *it is almost never correct to say,* "I accept the null hypothesis." The problem with "accepting the null" is that this wording makes it seem that we are convinced that the null hypothesis is true. The strategy of hypothesis testing is to reject the null hypothesis when the evidence is convincingly against it. If we only collect a small sample amount of data, we may not see convincing evidence for the alternative hypothesis because the sampling error is so large. That is certainly *not* equivalent to accepting that the null hypothesis is true. For instance, if we observe three births and two are boys, we would certainly not be willing to *accept* the hypothesis "2/3 of all births are boys" even though the observed data don't provide evidence that it is false.

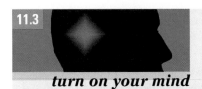

11.3

turn on your mind

Suppose an ESP test is conducted by having someone guess whether each of n coin flips will result in heads or tails. The null hypothesis is that $p = .5$, and the alternative is that $p > .5$, where p = probability of guessing correctly. Suppose one participant guesses $n = 10$ times and gets 6 right, while another guesses $n = 100$ times and gets 60 right. In each case, the percent correct was 60%. Do you think the p-value would be lower in one case than in the other, or would it be the same? Explain.*

EXAMPLE 11.5
A Jury Trial

If you are on a jury in the American judicial system, you must presume that the defendant is innocent unless there is enough evidence to conclude that he or she is guilty. Therefore the two hypotheses are

Null hypothesis: The defendant is innocent.
Alternative hypothesis: The defendant is guilty.

The prosecution collects evidence, much like researchers collect data, in the hope that the jurors will be convinced that such evidence would be extremely unlikely if the assumption of innocence were true. Consistent with our thinking in hypothesis testing, in many cases we would not *accept* the hypothesis that the defendant is innocent. We would simply conclude that the evidence was not strong enough to rule out the possibility of innocence. In fact, in the United States the two conclusions juries are instructed to choose from are "guilty" and "not guilty." A jury would never conclude, "the defendant is innocent." ◆

There are two different errors that can be made in hypothesis testing, just as in the jury system. The first type of error is equivalent to a "guilty" verdict when in fact the defendant is innocent. This might be called a false positive. In hypothesis testing, this is equivalent to rejecting the null hypothesis when it is true. The second type of error is equivalent to a "not guilty" verdict when the defendant actually committed the crime. This might be called a false negative. In hypothesis testing, this is equivalent to not rejecting the null hypothesis when in fact the alternative hypothesis is true.

***HINT:** The p-value is the conditional probability of observing 60% or more correct, given that the participant is just guessing. Would it be harder to guess at least 6 correct in 10 tries, or at least 60 correct in 100 tries?

In the judicial system in the United States, the first type of error is seen as more serious. This is what leads to the reasoning that a defendant should be considered "innocent until proven guilty." In any hypothesis testing situation, the two types of errors should be considered to determine which is more serious, and the level of significance should be adjusted accordingly. These errors and how they affect the decision will be covered in more detail in Section 11.7 and in Chapter 13, Section 13.6.

11.4 | TESTING HYPOTHESES ABOUT A PROPORTION

In this section, we illustrate the five steps to hypothesis testing in the context of testing a specified value for a population proportion.

Steps in any Hypothesis Test

Step 1: *Determine the* null *and* alternative *hypotheses.*

Step 2: *Verify necessary data conditions, and if met, summarize the data into an appropriate* test statistic.

Step 3: *Assuming the null hypothesis is true, find the p-value.*

Step 4: *Decide whether or not the result is* statistically significant *based on the p-value.*

Step 5: *Report the conclusion in the context of the situation.*

Examples 11.1 and 11.2 were illustrations of how to write the null and alternative hypotheses in words and in formulas when the parameter of interest is a population proportion. For instance, in Example 11.2 about the proportion favoring a lower legal blood alcohol limit, the null hypothesis was H_0: $p \leq .5$ and the alternative was H_a: $p > .5$. The specific value specified in the hypotheses for the population proportion of interest is called the **null value** and is denoted by the symbol p_0. In Example 11.2 just described, the value of p_0 is .5.

The possible null and alternative hypotheses are one of these three choices, depending on the research question:

1. H_0: $p = p_0$ versus H_a: $p \neq p_0$ (two-sided)
2. H_0: $p \geq p_0$ versus H_a: $p < p_0$ (one-sided)
3. H_0: $p \leq p_0$ versus H_a: $p > p_0$ (one-sided)

Often the null hypothesis for a one-sided test is written as H_0: $p = p_0$ instead. We will write it both ways in this chapter, to give you practice with the two ways you may encounter the null hypothesis in journal articles. Remember that a *p*-value is computed by assuming the null hypothesis is true, and the specific null value p_0 is what is assumed to be the truth about the population for that computation.

The *z*-Test for a Proportion

When we have a sufficiently large random sample, we can use a **z-test** to examine hypotheses about a population proportion. The test is called a "*z*-test" because the **test statistic** is a standardized score (*z*-score) for measuring the difference between the sample proportion and the null hypothesis value of the population proportion. The logic of the test is as follows:

◆ Determine the sampling distribution of possible sample proportions when the true population proportion is p_0 (called the *null value*), the value specified in H_0.

♦ Using properties of this sampling distribution, calculate a standardized score (z-score) for the observed sample proportion $\hat{p}$.

♦ If the standardized score has a large magnitude, conclude that the sample proportion $\hat{p}$ would be unlikely if the null value p_0 is true, and reject the null hypothesis.

The principal idea is that we determine how likely it would have been for the sample proportion to have occurred if the null hypothesis is true. For example, if the null value is $p_0 = .5$ and the observed sample proportion is $\hat{p} = .65$, we would determine how likely it is that we could observe a sample proportion as large as .65 (or larger) if the population proportion was .5.

Conditions for Conducting the z-Test

Whenever we conduct a hypothesis test, we must make sure the data meet certain conditions. These conditions are the assumptions that were made when the theory was derived for the test statistic that we are using to test our hypotheses. For testing proportions with the procedure described in this section, there are two conditions that should be true for the sample:

1. The sample should be a random sample from the population.
2. The quantities np_0 and $n(1 - p_0)$ should both be at least 10.

The first condition is not always practical, and most researchers are willing to use the test procedure as long as the sample is considered to be representative of the population for the question of interest. For instance, a college statistics class may be representative of all college students on questions like which of two soft drinks is preferred or whether their right foot is longer than their left foot.

The second condition is a sample size requirement. To test the null hypothesis that $p = .5$, for example, we would need a sample of at least $n = 20$ to meet this condition. If the null hypothesis is that $p = .1$, the sample size would have to be at least $n = 100$. Our standard may be somewhat conservative; some authors suggest that np_0 and $n(1 - p_0)$ only need to be larger than 5 (e.g., Ott, 4th ed., p. 370).

Let's go through the five steps of hypothesis testing for an example, and then formalize the details.

EXAMPLE 11.6
The Importance of Order

This example was first introduced in Chapter 2. In a student survey, a statistics teacher asked his students to "randomly pick a letter from these two choices—S or Q." For about half of the students, the order of the letters S and Q was reversed so that the end of the instruction read "Q or S." The purpose of the activity was to determine if there might be a preference for choosing the first letter. A tendency to pick the first choice offered has been noted in several settings, including in elections.

Step 1: *Determine the null and alternative hypotheses.*
Use the letter p to represent the proportion of the population that would pick the first letter. The null hypothesis is a statement of "nothing happening." If there is no general preference for either the first or second letter, $p = .5$ (because there are two choices). So the "null value" for p is $p_0 = .5$.

The alternative hypothesis usually states the researcher's belief or speculation. The purpose of the activity was to see if there is a general preference for picking the first choice, and a preference for the first letter would mean that p is greater than .5. The null and alternative hypotheses can be summarized as

$H_0: p = .5$

$H_a: p > .5$

◆ These hypotheses are statements about the larger population that this sample represents. Hypothesis tests are always used to say something about the population from which the sample has been taken.

◆ A more literal statement of the null hypothesis would be $H_0: p \leq .5$ because it is possible that less than half of the population would choose the first letter. Even if the null hypothesis is modified in this way, the value .5 is used as the assumed null value in the calculations.

Step 2: *Verify necessary data conditions, and if met, summarize the data into an appropriate test statistic.*

Before computing the test statistic, we verify that our sample meets the necessary conditions for using the z-statistic. With $n = 190$ and $p_0 = .5$, both np_0 and $n(1 - p_0)$ equal 95, a quantity larger than 10, so the sample size condition is met. The sample, however, is not really a random sample—it's a convenience sample of students who were enrolled in this class. It does not seem that this will bias the results for this question, so we will behave as though the sample was a random sample.

Among 92 students who saw the order "*S* or *Q*," 61 picked *S*, the first choice. Among the 98 students who saw the order "*Q* or *S*," 53 picked *Q*, the first choice. In all, 114 of 190 students picked the first choice of letter. Expressed as a proportion, this is $114/190 = .60$.

Data analysts nearly always use statistical software to do the calculations for a hypothesis test. Figure 11.1 displays output for this problem from the Minitab statistical program. We first work through the remainder of the hypothesis test utilizing this output. Following that, we will show the detailed calculations for a hypothesis test for a population proportion.

Test of p = 0.5 vs p > 0.5					
X	N	Sample p	95.0 % CI	Z-Value	P-Value
114	190	0.60	(0.53, 0.67)	2.76	0.003

FIGURE 11.1 **Minitab (Version 12) output for the *S* or *Q* choice**

The output items labeled "X," "N," and "Sample p" summarize the calculation of the sample proportion. The sample proportion, $\hat{p} = .60$, is the primary information for the test statistic, but we need to determine the corresponding z-statistic to find the p-value. From the output, we see under the label "Z-Value" that the value of the z-statistic is 2.76. The z-value is a standardized score for measuring the difference between the sample proportion, $\hat{p} = .60$, and the null hypothesis value, $p_0 = .50$.

Step 3: *Assuming the null hypothesis is true, find the p-value.*

The *p*-value for this hypothesis test is the answer to this question: *If the true p is .5, what is the probability that, for a sample of 190 people, the sample proportion could be as large as .60 (or larger)?*

The answer to this question is also the answer to a corresponding question about the *z*-statistic: *If the null hypothesis is true, what is the probability that the z-statistic could be as large as 2.76 (or larger)?*

We can either use computer output to find this value, or we can find it from Table A.1, the Standard Normal Table in the Appendix. Details of using the table will be given later in this chapter. In the Minitab output, we can locate the answer, "0.003," at the far right under the heading "P-Value." If the population proportion is really .5, the probability is only 0.003, about 3 in 1000, that we could observe a sample proportion as large as .60 or larger.

> *Note:* In the specific context of hypothesis tests for proportions, it is unfortunate that the letter "*p*" could have two different meanings. Do not confuse the population proportion "*p*" with the *p*-value for the test, which is sometimes abbreviated to just "*p*" in research articles and computer output. You should be able to sort out which is which based on the context.

Step 4: *Decide whether or not the result is statistically significant based on the p-value.*

The convention used by most researchers is to declare statistical significance when the *p*-value is smaller than 0.05. The *p*-value in this example is 0.003, so we can reject the null hypothesis.

Step 5: *Report the conclusion in the context of the situation.*

The statistical conclusion was to reject the null hypothesis that $p = .50$. In this situation, the conclusion is that *there is statistically significant evidence that the first letter presented is preferred.* This conclusion is a generalization that applies to the population represented by this sample. In other words, we conclude that for the entire population of individuals represented by these students, there would be a preference for choosing the first letter presented.

You may have noticed that the output in Figure 11.1 also provided a 95% confidence interval for the population proportion. With 95% confidence we can say that the population proportion who would choose the first letter presented is somewhere between .53 and .67. With only 190 responses, the margin of error is somewhat large (.07 or 7%) so the confidence interval is not particularly precise. The true proportion that picks the first letter may be only a slight majority (.53) or it could be as much as a two-thirds majority (.67).

Conclusions are strengthened when another study replicates the results. The instructor repeated this activity in another class of 327 students. In that class, 58% of the students picked the first letter, which is very close to the 60% found in the original study. ◆

Details for Calculating the *z*-Statistic

Our fundamental concern is the difference between the sample proportion and the null hypothesis value for the population proportion. If the sample proportion is sufficiently far from the null value, we can reject the null hypothesis. For example, if the null hypothesis is that $p = .5$, a sample proportion of .9 based on a relatively large random sample would be strong evidence against the null hypothesis. A sample proportion equal to .52 would be far less convincing evidence against the statement that the true $p = .5$.

We can measure the difference between the sample proportion and the null hypothesis value with a standardized score. From the Rule for Sample Proportions in Chapter 9, we know that the possible values of the sample proportion $\hat{p}$, based on a sample of size n, are approximately normal with

- mean $= p =$ true population proportion

- standard deviation $= \sqrt{\dfrac{p(1-p)}{n}}$

When we assume the null hypothesis is true, so that $p = p_0$, we assign special names to the mean and standard deviation. In particular, the term *null value* is used for p_0, and the term *null standard error* is used for the standard deviation when p_0 is substituted for p. Under the null hypothesis, the sampling distribution of $\hat{p}$ is approximately normal with

- mean $=$ null value $= p_0$

- standard deviation $=$ null standard error $= \sqrt{\dfrac{p_0(1-p_0)}{n}}$

The z-statistic for the significance test is

$$z = \frac{\text{Sample estimate} - \text{Null value}}{\text{Null standard error}} = \frac{\hat{p} - p_0}{\text{Null standard error}} = \frac{\hat{p} - p_0}{\sqrt{\dfrac{p_0(1-p_0)}{n}}}$$

- $\hat{p}$ represents the sample estimate of the proportion

- p_0 represents the specific value in the null hypothesis

- n is the sample size

EXAMPLE 11.6—*CONTINUED*
Calculating
the z-Statistic for
the S or Q Problem

For the "random" choice between S and Q, the null hypothesis was $p = .5$, and the proportion of the sample of $n = 190$ students that picked the first letter was $\hat{p} = .6$. In Figure 11.1, we saw that the z-statistic for the hypothesis test is 2.76. The calculation is

$$z = \frac{\text{Sample estimate} - \text{Null value}}{\text{Null standard error}} = \frac{.6 - .5}{\sqrt{\dfrac{.5(1 - .5)}{190}}} = \frac{.10}{.0362} = 2.76 \; \blacklozenge$$

EXAMPLE 11.1—
CONTINUED
Do Fewer Than
20% Experience
Medication Side Effects?

Suppose that a pharmaceutical company wants to claim that side effects will be experienced by fewer than 20% of the patients who use a particular medication. In a clinical trial with $n = 400$ patients, they find that 68 patients experienced side effects.

Step 1: *The null and alternative hypotheses are*

$H_0: p \geq .20$ (company's claim is not true)

$H_a: p < .20$ (company's claim is true)

Notice that the pharmaceutical company's claim is used as the alternative hypothesis. Also notice that the null hypothesis is a region of values. To calculate the z-statistic, it is standard practice to use the value that separates the null and alternative hypothesis regions, in this case $p_0 = .20$.

Step 2: *Verify necessary data conditions, and if met, summarize the data into an appropriate test statistic.*

There were 400 patients, so both np_0 and $n(1 - p_0)$ are large enough to proceed with a test based on the z-statistic. Although most clinical trials use volunteer patients, presumably the company would choose volunteers who are representative of the larger population of potential users of the medication, so for practical purposes we will accept that the necessary conditions are met. Out of 400 patients, 68 experienced side effects. The sample proportion that experienced side effects is $\hat{p} = 68/400 = .17$. The z-statistic for the hypothesis test is

$$z = \frac{\text{Sample estimate} - \text{Null value}}{\text{Null standard error}} = \frac{.17 - .20}{\sqrt{\dfrac{.20(1 - .20)}{400}}} = \frac{-.03}{.02} = -1.5$$

Notice that the z-statistic is negative. This occurs because the sample proportion is less than the null value, the result that the pharmaceutical company wanted. We will complete the remaining three steps for this example after we present the general format for finding a p-value. ◆

Computing the p-Value for the z-Test

Recall that a p-value is computed by assuming the null hypothesis is true and then determining the probability of a result as extreme, or more extreme, as the observed test statistic in the direction of the alternative hypothesis. For the z-test of hypotheses about a population proportion, the p-value probability is found using the normal curve. This is a consequence of the Normal Curve Approximation Rule for Sample Proportions introduced in Chapter 9. The details of "determining the probability of a result as extreme (or more extreme) as the observed test statistic" are consistent with the direction specified in the alternative hypothesis:

◆ For a *greater than* alternative hypothesis, find the probability that the test statistic z could have been *equal to or greater than* what it is.

◆ For a *less than* alternative, find the probability that the test statistic z could have been *equal to or less than* what it is.

◆ For a *two-sided alternative hypothesis,* the p-value includes the probability areas in both extremes of the distribution of the test statistic z.

The correspondence between the type of alternative hypothesis and the p-value area is summarized in Table 11.1.

TABLE 11.1 ■ **Alternative Hypothesis Regions and p-Value Areas**

Statement of H_a		p-Value Area	Normal Curve Region
$p < p_0$	(less than)	Area to the left of z (even if $z > 0$)	
$p > p_0$	(greater than)	Area to the right of z (even if $z < 0$)	
$p \neq p_0$	(not equal)	$2 \times$ area to the right of $|z|$	

The reasoning behind including both extremes for the two-sided alternative hypothesis is as follows. Suppose we want to test whether the true percentage of heads is 50% when we spin a coin on its edge and watch how it lands. We spin the coin on its edge 100 times, and it lands "heads-up" 60 times. Remember that "*the p-value is computed by assuming the null hypothesis is true and then determining the probability of a result as extreme (or more extreme) as the observed test statistic in the direction of the alternative hypothesis.*" In this case, "in the direction of the alternative hypothesis" includes both above and below 50%. So 40% heads is as extremely far into the alternative hypothesis region (as far from the null value of 50%) as is 60% heads. Both extremes provide equivalent support for the alternative hypothesis.

If we use Minitab to do the hypothesis test, the program will automatically compute and report the *p*-value. To calculate the *p*-value "by hand," we can use Table A.1, the Standard Normal Table, or we could use a spreadsheet program like Excel (the command NORMSDIST(z) finds the area to *left* of z).

EXAMPLE 11.1—
CONTINUED
The Medication
Side Effects Problem

We already have done steps 1 and 2 for this example. The null and alternative hypotheses are $H_0: p \geq .20$ and $H_a: p < .20$, and the *z*-statistic is $z = -1.5$. We now complete the final three steps.

Step 3: *Assuming the null hypothesis is true, find the p-value.*

The alternative hypothesis is that the *proportion is less than* a specified value, so the *p*-value is the *area to the left* of the observed *z*-statistic. We can use Table A.1, the Standard Normal Table, to determine this probability, or we can use the Excel instruction NORMSDIST(-1.5). For $z = -1.5$, the area to the left is about 0.067, and this is the *p*-value. It is the answer to this question: *If the true p is .2, what is the probability that, for a sample of 400 people, the sample proportion could be as small as .17 (or smaller)?*

The relationship between the *p*-value and the *z*-statistic for this problem is illustrated in Figure 11.2.

Step 4: *Decide whether or not the result is statistically significant based on the p-value.*

If the usual 0.05 standard for statistical significance is used, the *p*-value of 0.067 is not quite small enough to reject the null hypothesis.

Step 5: *Report the conclusion in the context of the situation.*

The sample evidence was in the desired direction for the pharmaceutical company, but it was not strong enough to conclusively reject the null hypothe-

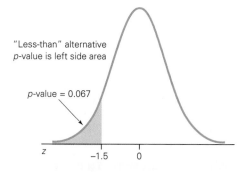

FIGURE 11.2   *p*-**Value for medication side effects example**

sis. The company cannot reject the idea that the proportion of the population who would experience side effects is .20 (or more). If this is an important issue for the company, they should consider gathering additional data. As we will learn in the next section, sample size affects statistical significance. This example also illustrates why it is not a good idea to "accept the null hypothesis." We would not be convinced that 20% or more of the population would experience side effects when only 17% of the sample did so. ◆

EXAMPLE 11.6—CONTINUED
The p-Value for the "Greater Than" Alternative in the S or Q Example

For the "choose S or Q" problem, the alternative hypothesis is $H_a: p > .5$. When an alternative hypothesis is that a *proportion is greater than* a specified value, the p-value is the *area to the right* of the observed z-statistic. We determined before that $z = 2.76$ and the corresponding p-value is 0.003. The p-value, which is the area to the right of 2.76, is illustrated in Figure 11.3.

To determine this p-value "by hand," we can use Table A.1 in the Appendix to learn that for $z = 2.76$, the area to the left is roughly 0.997. This is the probability that we could get a z-score smaller than 2.76, so the probability we could get a z-score greater than 2.76 is $1 - 0.997 = 0.003$. To calculate this p-value with Excel, enter $=1-@NORMSDIST(2.76)$ into a cell of the spreadsheet. (The @ symbol tells Excel to invoke that function.) ◆

EXAMPLE 11.7
A Two-Sided Test: Are Left and Right Foot Lengths Equal?

Students in a statistics class measured the lengths of their right and left feet to the nearest millimeter. The right and left foot measurements were equal for 103 of the 215 students, but the two foot lengths were different for 112 students.

Let's consider the 112 individuals with unequal measurements, and use them to represent the population of adults with unequal foot lengths. Among those with unequal lengths, we can use the letter p to represent the proportion that has a longer right foot. If we have no advance belief that either foot will tend to be longer, the null and alternative hypotheses for p are

$H_0: p = .5$
$H_a: p \neq .5$

These two hypotheses are statements about the population represented by this sample of students. The null hypothesis states that among the population of people who have different foot lengths, exactly half will have a longer right foot

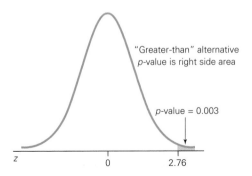

FIGURE 11.3   *p-Value for the "S or Q" example*

Test of p = 0.5 vs p not = 0.5					
X	N	Sample p	95.0 % CI	Z-Value	P-Value
63	112	0.5625	(0.471, 0.654)	1.32	0.186

FIGURE 11.4   **Minitab output for foot length example**

and exactly half will have a longer left foot. The alternative states that this is not the case, but does not specify a direction.

Both np_0 and $n(1 - p_0)$ are greater than 10, so we can proceed with the test, assuming that we think the students in the sample are representative of the larger population. In the sample, a longer right foot was reported by 63 of the 112 students who reported different foot lengths. The corresponding proportion with a longer right foot is $\hat{p} = 63/112 = .56$. Minitab output for the hypothesis test is shown in Figure 11.4. We see that the *p*-value is 0.186, so we cannot reject the null hypothesis. Although there was a tendency toward a longer right foot in the sample, there is insufficient evidence to conclude that the proportion in the population with a longer right foot is different from the proportion with a longer left foot.

The calculation of the *z*-value (the answer is shown in the output as Z-value = 1.32) is

$$z = \frac{\text{Sample estimate} - \text{Null value}}{\text{Null standard error}}$$

$$= \frac{.5625 - .5}{\sqrt{\dfrac{.5(1 - .5)}{112}}} = \frac{0.0625}{0.0472} = 1.32$$

The two-tailed *p*-value is the total of the area to the right of $z = 1.32$ and the area to the left of $z = -1.32$, an area illustrated in Figure 11.5. ◆

The Rejection Region Approach to Hypothesis Testing

Before computers and calculators capable of finding *p*-values were widely available, hypothesis testing was routinely conducted by using what is called the

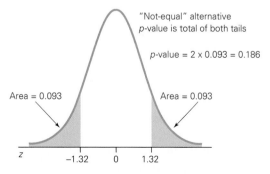

FIGURE 11.5   **Two-tailed *p*-value for foot length example**

"rejection region" approach. In this approach, instead of finding a p-value and comparing it to the desired level of significance, the computed test statistic is compared to a "rejection region." If the test statistic falls in the rejection region, the null hypothesis is rejected. If not, the null hypothesis cannot be rejected.

The **rejection region** in a hypothesis test is the region of possible values for the test statistic that would lead to rejection of the null hypothesis. *If* the null hypothesis is true, the probability that the computed test statistic will fall in the rejection region is α, the desired level of significance.

Like p-values, rejection regions depend on whether the alternative hypothesis is one-sided or two-sided. Here are the rejection regions and rules for a z-test with levels of significance $\alpha = 0.05$ and 0.01:

	Rejection Region Rule	
Alternative Hypothesis	$\alpha = 0.05$	$\alpha = 0.01$
H_a: $p < p_0$	Reject H_0 if $z < -1.645$	Reject H_0 if $z < -2.33$
H_a: $p > p_0$	Reject H_0 if $z > +1.645$	Reject H_0 if $z > +2.33$
H_a: $p \neq p_0$	Reject H_0 if $|z| > 1.96$	Reject H_0 if $|z| > 2.58$

Notice that the form of the rejection region (upper tail, lower tail, or both tails) is the same as the form of the p-value areas given in Table 11.1. In both cases, the decision to reject the null hypothesis is only made if the test statistic falls into a tail that supports the alternative hypothesis. To understand how the rejection region boundaries are found, examine the table of Standard Normal Probabilities in the Appendix. In each case, the rejection region is the part of the standard normal curve that

◆ includes values of the test statistic z that support the alternative hypothesis

◆ has total area of α in that region

In other words, the probability that a standard normal z-score falls into the rejection region is the specified α. When H_0 is true, the z-statistic is simply a standard normal z-score, so using this method ensures that the probability that the null hypothesis will be rejected if it is actually true is α. To find rejection regions for other values of α, find the region of the standard normal curve that supports the alternative hypothesis and has total area of α.

EXAMPLE 11.6—*CONTINUED*
Rejecting the
Hypothesis of Equal Choices

For the "choose S or Q" problem, the test statistic is $z = 2.76$ and the alternative hypothesis is that $p > .5$. Based on the rejection region rule for $\alpha = 0.05$, which is to reject H_0 if $z > +1.645$, we conclude that the null hypothesis can be rejected at $\alpha = 0.05$. Using a more stringent $\alpha = 0.01$, we can also reject the null hypothesis because the test statistic $z = 2.76$ is greater than 2.33. The conclusion would be stated, "The null hypothesis can be rejected at $\alpha = 0.01$." In the context of the example, the result might be stated, "There is statistically significant evidence that the first letter presented is preferred using $\alpha = 0.01$." ◆

11.4

turn on your mind

Here are two questions about *p*-values and one-sided versus two-sided tests:

1. Under what conditions would the *p*-value for a one-sided *z*-test be greater than 0.5?
2. When the data are consistent with the direction of the alternative hypothesis for a one-sided test, the *p*-value for the corresponding two-sided test is double what it would be for the one-sided test. Use this information to explain why it would be "cheating" to look at the data before deciding whether to do a one- or two-sided test.*

Minitab Tip

 tech note   **Testing Hypotheses about a Proportion**

◆ To test hypotheses about a proportion, use **Stat**>**Basic Statistics**>**1 Proportion.** (This procedure is not in versions earlier than Version 12.)

◆ If the raw data are in a column of the worksheet, specify that column. If the data have already been summarized, click on "Summarized Data," then specify the sample size and the count of how many observations have the characteristic of interest.

◆ Use the *Options* button, specify the value of p_0, select the type of alternative hypothesis, and click on "Use test and interval based on normal distribution."

◆ If the "Use . . . normal distribution" option is not selected, then the reported *p*-value will be exact, based on adding the relevant binomial probabilities. The confidence interval will also be exact. See the Minitab Help feature for details on how it is computed.

◆ In Version 13, if the alternative hypothesis selected is one-sided, the reported confidence interval will also be one-sided. For instance, if the alternative hypothesis is "greater than p_0" a lower bound will be given for the confidence interval. The interval extends from the lower bound to 1.0. To produce a two-sided confidence interval, go back to the default option of "not equal" as the alternative hypothesis.

11.5 | THE ROLE OF SAMPLE SIZE IN STATISTICAL SIGNIFICANCE

There is a big difference between the following two beginnings to an advertisement for a toothpaste:

◆ 2 out of 3 dentists recommend . . .

◆ 2/3 of the 500 dentists surveyed recommend . . .

*HINT: Q1: Study Table 11.1. Q2: The *p*-value is supposed to reflect how unlikely such extreme data would be by chance. Would all possibilities be covered if the alternative hypothesis was altered based on looking at the data?

We should be suspicious that the first claim may be based literally on the recommendations of only three dentists. If this is the case, we know that the claim is meaningless. A sample of just three individuals is not large enough to be conclusive. On the other hand, the second claim seems stronger. A sample of 500 dentists seems like a pretty big sample, although we should ask how the 500 dentists were selected for whatever survey was done.

The size of the sample affects our ability to make firm conclusions based on that sample. With a small sample, we may not be able to conclude anything. With larger samples, there is more chance that we can make firm conclusions in hypothesis testing situations, and in fact, with large samples we may find statistical significance even though the effect is minor and unimportant. The results of hypothesis testing should always be interpreted with the size of the sample in mind. We will learn more about this idea in the next section and in Chapter 13.

Cautions About Sample Size and Statistical Significance

- ◆ When there is a small to moderate effect in the population, a small sample has little chance of providing statistically significant support for the alternative hypothesis.
- ◆ With a large sample, even a small and unimportant effect in the population may lead to a conclusion of statistical significance.

EXAMPLE 11.8
How the Same Sample Proportion Can Produce Different Conclusions

In Example 10.7 of Chapter 10, we used a confidence interval to analyze a taste test in which 55% of the 60 participants liked the taste of drink A better than the taste of drink B. Based on the 95% confidence interval for the "true" proportion that prefers drink A, we were unable to conclude whether a majority of the population prefers the taste of drink A.

We can also examine these data with a hypothesis test in which the null hypothesis is that there is no general preference for either drink. The null and alternative hypotheses are

$H_0: p = .5$ (no preference)
$H_a: p \neq .5$ (preference for one or the other)

where the letter p represents the proportion in the population that would prefer drink A.

Now, suppose that a much larger sample size of 960 was used and that the result was still that 55% of the sample prefers drink A. What effect will the larger sample size have on the statistical significance of the data?

Figure 11.6 displays hypothesis test and confidence interval results (from using Minitab) for both sample sizes, assuming the sample proportion remains at 55%. When the sample size is 60, the p-value is 0.439, so we cannot reject the null hypothesis as a possible truth. When the sample size is 960, the p-value is 0.002, so the result is *statistically significant*. In that case, we can reject the null hypothesis and conclude that a majority of the population would prefer drink A.

The sample size also influences the width of the 95% confidence intervals for the true p. When the sample size is only 60, the 95% confidence interval for the true p is .42 to .68, an imprecise, wide range of estimates. When the sample size is 960, the 95% confidence interval is .52 to .58, a relatively precise range of estimates and an interval range that is entirely in the region defined by the alternative hypothesis. ◆

```
Test of p = 0.5 vs p ≠ 0.5

Results when n = 60 and sample p = 0.55

X         N        Sample p      95.0 % CI       Z-Value     P-Value
33        60         0.55       (0.42, 0.68)       0.77        0.439

Results when n = 960 and sample p = 0.55

X         N        Sample p      95.0 % CI       Z-Value     P-Value
528       960        0.55       (0.52, 0.58)       3.10        0.002
```

FIGURE 11.6   **Taste test results for $n = 60$ and $n = 960$**

This example shows us again why it's not advisable to use the phrase "accept the null hypothesis." The sample of only 60 participants was too small to be conclusive about whether or not we should reject the statement that the true p is .5. The larger sample resulted in a sample proportion that would be very unlikely if in fact the true population proportion is .5 and thus allowed a stronger conclusion.

Why is the same sample proportion more significant for a relatively larger sample size? The conceptual, or intuitive, answer to this question is that, with more data, we have more accurate sample estimates and we reduce our uncertainty about what population values are likely. The technical answer is that increasing the sample size decreases the standard error of the sample proportion.

The z-value for the hypothesis test is

$$z = \frac{\text{Sample estimate} - \text{Null value}}{\text{Null standard error}}$$

For both of our hypothesis tests, the sample proportion is .55 and the null hypothesis value is .5, so the top portion of the calculation is the same for both cases. The z-value changes because the sample size affects the standard error.

- When the sample size is $n = 60$, the null standard error $= \sqrt{\dfrac{.5(1 - .5)}{60}} \approx$ .065.

- When the sample size is $n = 960$, the null standard error $= \sqrt{\dfrac{.5(1 - .5)}{960}} \approx$ .016.

Notice that increasing the sample size decreases the null standard error. Generally, a standard error roughly measures the "typical difference" between a statistic and a population parameter. If the null hypothesis that $p = .5$ were actually true, the consequence of taking a larger sample is that the sample proportion would be likely to fall closer to the null value (.5) than it would for a smaller sample. As a result, the difference between the sample proportion of .55 and the null value of .5 is relatively greater when the sample size is large than when it is small.

One more comment is in order about the example just presented. If in fact the null hypothesis were true, and $p = .50$, then a sample proportion as large as .55 would probably not have occurred in the larger sample. It is quite feasible for a sample of $n = 60$ to produce a sample proportion that differs from the truth by as much as .55 differs from .50. But it is quite unlikely for a sample of $n = 960$ to produce a sample proportion this far from the true proportion.

In summary, it is the size of the standard error that changes based on the size of the sample, and increasing the sample size gives a smaller standard error. As a result, a particular absolute difference between the sample proportion and null value is more significant with a large sample than with a small sample. No matter how large the sample, the sample proportion is likely to fall within two or so standard errors of the true population proportion. If the difference between the sample proportion and the null value is more than this, it is significant.

11.5

turn on your mind

If the null hypothesis is true, the correct conclusion is "cannot reject the null hypothesis." If the alternative hypothesis is true, the correct conclusion is "reject the null hypothesis." Increasing the sample size increases the probability of making the correct conclusion in one of these cases, but not in the other. Your job is to figure out which is which. Here is a hint: The conclusion "cannot reject the null hypothesis" is made as long as the sample proportion falls within a reasonable number of standard errors of the null value. When the null hypothesis is true, does the likelihood of this happening change as the sample size increases? Explain, and use this explanation to determine which conclusion has increased probability of being made correctly with increasing sample size.*

11.6 | REAL IMPORTANCE VERSUS STATISTICAL SIGNIFICANCE

The phrase *statistically significant* only means that the data are strong enough to reject a null hypothesis. A result that is significant in the statistical meaning of the word is not necessarily significant in the more common meaning of the word. One reason for this, of course, is that the research study might not address an important real-world issue. Even when the research issue is important, however, statistical significance does not guarantee practical significance.

A *p*-value provides information about the conclusiveness of the evidence against the null hypothesis, but it does *not* provide information about the *magnitude* of the effect. In some instances, the magnitude of a statistically significant effect can be so small that the practical effect is not important. If the sample size is large enough, almost any null hypothesis can be rejected be-

****HINT:** Think about the probability that the sample proportion falls within 2 standard errors of the null value p_0 when it is the truth. Does that probability depend on the sample size?

cause there is almost always at least a slight relationship between two variables, or a slight difference between two groups, or a slight deviation from the status quo.

EXAMPLE 11.9
Birth Month and Height

The headline of a Reuters news article posted at the Yahoo Health News website February 18, 1998 was "Spring Birthday Confers Height Advantage." The article describes an Austrian study of the heights of 507,125 military recruits. In an article in the journal *Nature,* the researchers reported their finding that men born in the spring were, on average, about 0.6 centimeters taller than men born in the fall (Weber et al., *Nature,* 1998, 391:754–755). This is a small difference; 0.6 centimeters is only about 1/4 of an inch. The sample size for the study is so large that even a very small difference will earn the title *statistically significant.* Do you think the practical import of this difference warranted the headline? ◆

CASE STUDY 11.1
The Internet and Loneliness: Case Study 1.7 Revisited

We encountered the concept that a statistically significant result may not have practical importance when we read Case Study 1.7. That case study examined research done at Carnegie Mellon University in which the principal finding was that Internet usage may lead to feelings of loneliness and depression. This finding received a lot of media attention, but a close look at the study shows that the actual effects were quite small. For example, according to the *New York Times,* "one hour a week on the Internet was associated, on average, with an increase of 0.03, or 1 percent on the depression scale." (Harman, 30 August 1998, p. A3. Copyright © 1998 by The New York Times Co. Reprinted by permission.)

11.6

turn on your mind

In this section, we noted that a research finding of "statistical significance" does not necessarily indicate that the finding is of practical importance. In situations where the hypothesis test is about a specified value for a population parameter, explain why it would be helpful to present a confidence interval for the parameter along with the finding of statistical significance.*

11.7 | WHAT CAN GO WRONG: THE TWO TYPES OF ERRORS

Whenever we use a sample of data to make a decision about a larger population, we may make a mistake. In testing statistical hypotheses, there are two potential decisions and each decision brings the possibility of an error. Decision makers and researchers should consider the consequences of these possible errors when they create and apply their decision-making rules.

*HINT: Think about what information a confidence interval for a proportion provides.

EXAMPLE 11.5—*CONTINUED*
The Courtroom Analogy

Let's use the courtroom analogy as an illustration. Viewed from within the hypothesis testing framework, the two hypothesis are

> *Null hypothesis:* The accused is *innocent.*
> *Alternative hypothesis:* The accused is *guilty.*

and the two possible decisions for a judge or jury are

> *Do not reject the null hypothesis:* The accused is *not guilty.*
> *Reject the null hypothesis:* The accused is *guilty.*

For trials in general, here are the possible errors and the consequences that accompany those errors:

- *Possible error 1:* A "guilty" verdict for a person who is really innocent.
  Consequence: An innocent person is falsely convicted. The guilty party remains free.

- *Possible error 2:* A "not guilty" verdict for a person who committed the crime.
  Consequence: A criminal is not punished.

In the American court system, a false conviction (possible error 1) is generally viewed as the more serious error. Not only is an innocent person punished but also a guilty one remains free. Courtroom rules and rules affecting pretrial investigations tend to reflect society's concern about incorrectly punishing an innocent person. ◆

EXAMPLE 11.10
A Medical Analogy

Imagine that you are tested to determine if you have a disease. The lab technician or physician who evaluates your results must make a choice between two hypotheses:

> *Null hypothesis:* You do not have the disease.
> *Alternative hypothesis:* You have the disease.

Unfortunately, many laboratory tests for diseases are not 100% accurate. There is a chance the result is wrong. Consider the two possible errors and their consequences:

- *Possible error 1:* You are told you have the disease, but you actually don't. The test result was a **false positive.**
  Consequence: You will be unnecessarily concerned about your health and you may receive unnecessary treatment.

- *Possible error 2:* You are told that you do not have the disease, but you actually do. The test result was a **false negative.**
  Consequence: You do not receive treatment for a disease that you have. If this is a contagious disease, you may infect others.

Which error is more serious? In most medical situations, the second possible error, a false negative, is more serious but this could depend on the disease and the follow-up actions that are taken. For instance, in a screening test for cancer, a false negative could lead to a fatal delay in treatment. Initial test results that are "positive" for cancer are usually followed up with a retest so a false positive may be discovered quickly. ◆

Type 1 and Type 2 Errors in Statistical Hypothesis Testing

The situation encountered in statistical hypothesis testing is about the same as in the courtroom and medical analogies. One notable difference is that in statistical hypothesis testing, random sampling error causes the errors defined in this section. When a sample is used, the "luck of the draw" could be such that the sample statistic does not properly represent the population value. This is particularly likely when the sample size is small and the corresponding margin of error is large.

In the terminology of statistical hypothesis testing, the two possible errors are called the **type 1** and **type 2** errors.

> *definition* A **type 1 error** can only occur when the null hypothesis is actually true. The error occurs by concluding that the alternative hypothesis is true.
>
> A **type 2 error** can only occur when the alternative hypothesis is actually true. The error occurs by concluding that the null hypothesis cannot be rejected.

The numbering of the error types indicates which of the two hypotheses, (1) null or (2) alternative, is actually true. For example, a type 1 error is the error that occurs when the first hypothesis, the null, is really true but we decide in favor of the alternative.

EXAMPLE 11.11
Calcium and the Relief of Premenstrual Symptoms

Suppose researchers are planning a randomized "blind" trial in which 400 women who suffer from premenstrual symptoms are randomly assigned either to take calcium supplements or a placebo. The research group believes that the use of calcium supplements will reduce the severity of the symptoms. For three months, the researchers will track the participants' premenstrual symptoms and will use the sample data to evaluate these hypotheses:

H_0: For reducing symptoms, calcium is not better than placebo.
H_a: For reducing symptoms, calcium is better than placebo.

A type 1 error would occur if the researchers decide that calcium is better than placebo when, in fact, it is not (the null hypothesis is true). A type 2 error would occur if the researchers decide that calcium is not better than placebo when, in fact, it is (the alternative hypothesis is true).

In 1998, a research article describing a study like our hypothetical example was published in the *American Journal of Obstetrics and Gynecology* (Thys-Jacobs et al., 1998). The authors concluded that there was a statistically significant difference between the effectiveness of calcium supplements and placebo. Ideally, this conclusion is correct and no error was made. If future research and experience proves that calcium supplements are actually not effective, these researchers will have made a type 1 error by choosing the alternative when the null hypothesis is true. If a type 1 error was made in this situation, women may be taking calcium to reduce their symptoms when it does not actually work. On the other hand, if the researchers had not found a difference but calcium really is effective (a type 2 error) then women would not know about a treatment that is actually effective. ◆

EXAMPLE 11.10—*CONTINUED*
The Medical Analogy

In Example 11.10, we saw the two possible errors that might occur when you are tested for the presence of disease. Let's rephrase the possible errors in statistical terminology.

♦ A type 1 error occurs when the diagnosis is that you have the disease when actually you do not. In other words, *a type 1 error is a false positive.*

♦ A type 2 error occurs when the diagnosis is that you do not have the disease when actually you do. In other words, *a type 2 error is a false negative.* ◆

The Trade-Off Between the Risks of the Two Possible Errors

There always is a trade-off between the chances of making the two types of errors. For instance, in the American court system, the procedures designed to minimize the probability of a false conviction carry with them an increased risk of erroneous not guilty verdicts. But, if court rules were made more lenient to decrease the risk of incorrect acquittals, the probability of false convictions would increase.

Determining the safer direction in which to err depends upon the situation as well as the consequences of each type of potential error. HIV screening tests tend to err on the side of false positives because the consequence of failing to detect HIV could be disastrous. Because HIV screening does produce many false positives, people who test positive with an inexpensive screening test for HIV generally are tested again with an extremely accurate but more expensive test before they are given the results.

In cases when the null hypothesis is true, *the probability of making a type 1 error is the level of significance* α. Therefore, the researcher actually determines the probability that a type 1 error will occur by deciding what to use as the significance level, α. A type 1 error occurs when the p-value $< \alpha$, leading to the conclusion that the null hypothesis is false. But when the null hypothesis is true, the probability of choosing a sample that results in a p-value less than α is simply α. For instance, with a z-statistic and a two-sided test, the p-value is less than 0.05 if $|z| > 1.96$, which happens with probability 0.05.

Determining the probability of a type 2 error is much more difficult. That probability depends on the sample size, the form of the test statistic, the true parameter value, and the level of significance. We will discuss probabilities associated with the two types of errors in more detail in Chapter 13.

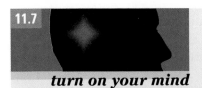

11.7

turn on your mind

Draw a tree diagram (see Chapter 7) where "H₀ true" and "Hₐ true" are the two choices for the first set of branches, and the two possible decisions are the choices for the second set of branches. For each of the four resulting pathways, identify whether the pathway results in a correct decision, a type 1 error, or a type 2 error.*

CASE STUDY 11.2
An Interpretation of a p-Value Not Fit to Print

One of the most common mistakes that researchers and journalists make has to do with the interpretation of the *p*-value. We learned that a *p*-value does not tell us the probability that a hypothesis is true. When a *p*-value is erroneously interpreted in this manner, the resulting statements can be completely misleading.

In an article entitled, "Probability Experts May Decide Pennsylvania Vote," the *New York Times* (April 11, 1994, p. A15) reported on the use of statistics to try to decide whether there was fraud in a special election held in Philadelphia. When a state senator from Pennsylvania's Second Senate District died in 1993, a special election was held to fill the seat. The Democratic candidate, William Stinson, defeated the Republican candidate, Bruce Marks, by the narrow margin of 20,518 votes to 20,057 votes, a difference of only 461 votes.

The Republicans were disturbed when they saw the comparison of voting booth results and absentee ballot results, shown in Table 11.2. The Republican Marks beat Stinson in the voting booths, but Stinson, the election winner, had a huge advantage in the absentee ballots. The Republicans charged that the election was fraudulent and asked the courts to disallow the absentee ballot votes on the basis of suspicion of fraud. In February 1994, three months after the election, Philadelphia Federal District Court Judge Clarence Newcomer disqualified all absentee ballots and ruled that the Republican, Marks, should be seated. The Democrats appealed this ruling, and both sides hired statisticians to help sort out what might have happened.

One statistical expert, Orley Ashenfelter, examined 22 previous senatorial elections in Philadelphia to determine the relationship between votes cast in the voting booth and those cast by absentee ballot. Using the previous data, he calculated a regression equation to predict the difference in absentee ballot votes for the two parties based on a given amount of difference between the parties in the voting booth.

In the voting booth, there was a difference of 564 votes in favor of the Republicans. Using his equation, Ashenfelter estimated that when there was this much difference in favor of the Republicans in the voting booths, there would be a difference of 133 votes in favor of the Republicans in the absentee ballots. Instead, the difference for absentee ballots in the disputed election was 1025 votes in favor of the *Democrats*.

Of course, everyone knows that chance events play a role in determining what happens in any given election. A hypothesis test can be used to evaluate the possible effects of randomness. Ashenfelter considered the null hypothesis that there was no fraud and calculated that the *p*-value was 6%. In other words, he determined that *if there was no fraud,* the probability was 6% that the Democratic advantage in absentee votes would be as large as 1025 votes when the Republican advantage in the voting booths was 565 votes.

Unfortunately, the author of the *New York Times* article misinterpreted this probability in a way that leads readers to believe that the election was probably fraudulent. When you read the following quote from the article, see if you can detect the mistake in interpretation:

> More to the point, there is some larger probability that chance alone would lead to a sufficiently large Democratic edge on the absentee ballots to overcome the Republican margin on the machine balloting. And the probability of such a swing . . . Professor Ashenfelter calculates, was about 6 percent. Putting it another way, if past elections are a reliable guide to current voting behavior, *there is a 94 percent*

TABLE 11.2 ■ **Election Results by Type of Vote**

	Stinson	Marks	Difference
Voting Booth	19,127	19,691	−564
Absentee	1391	366	+1025
Total	20,518	20,057	+461

chance that irregularities in the absentee ballots, not chance alone, swung the election to the Democrat, Professor Ashenfelter concludes. (Passell, 11 April 1994, p. A15. Copyright © 1994 by The New York Times Co. Reprinted by permission. Italics added).

The author's statement that there is a 94% chance that irregularities swung the election is wrong. He mistakenly interpreted the *p*-value to be the probability that the null hypothesis is true so he reported what he thought to be the probability that the election was fraudulent. In this case, the *p*-value only tells us that, if we assume there was no fraud, there is a 6% chance that we would see results like these. The *p*-value cannot be used to find the probability that the election was fraudulent.

The case went through two appeals, but the original decision made by Judge Newcomer to disallow the absentee ballots was upheld each time. The statistical evidence was not a major factor in these decisions. The appeal judges found that there was enough evidence of irregular absentee voting procedures to justify Judge Newcomer's decision.

KEY TERMS

Section 11.1
null hypothesis, 356
alternative hypothesis, 357
one-sided hypothesis test, 357
one-tailed hypothesis test, 357
two-sided hypothesis test, 358
two-tailed hypothesis test, 358

Sections 11.2 and 11.3
test statistic, 360
p-value, 360
level of significance, 360, 361
Bayesian statistics, 360–361

statistically significant, 361
α level, 361

Section 11.4
null value, 363
z-test for a proportion, 363
z-test statistic for proportion, 363, 367
p-value, test for proportion, 368
rejection region, 372

Section 11.5
sample size and statistical significance, 374

Section 11.6
statistical significance vs. importance, 376

Section 11.7
false positive, 378
false negative, 378
type 1 error, 379
type 2 error, 379

EXERCISES   Blue = Basics   **Bold, Bold** = Answer in back

Section 11.1

11.1 Determine whether each of these statements is an example of a null hypothesis or an alternative hypothesis.
   a. The average weight of Canadian geese is the same as the average weight of Canadian warblers.
   b. The proportion of books in the local public library that are novels is higher than the proportion of books in the university library that are novels.
   c. The average price of wool jackets in New York City is lower in the summer than in the winter.
   d. The proportion of students who receive A grades from Professor Harrington is the same as or higher than the proportion of students who receive A grades from Professor Cantor.

11.2 For each of the following situations, write the alternative hypothesis.
   a. The null hypothesis is H_0: $p = .30$ and it's a two-sided hypothesis test.
   b. The null hypothesis is H_0: $p \leq .45$.
   c. The null hypothesis is H_0: $p \geq .60$.

11.3 Is each of the following statements a valid null hypothesis? If not, explain why not.
   a. In a sample of students, the mean pulse rate for the men is equal to the mean pulse rate for the women.
   b. The average weight of newborn boys is the same as the average weight of newborn girls.
   c. The proportion of cars in California that are white is higher than the proportion of cars in Maine that are white.

11.4 State the null and alternative hypotheses for each of the following potential research questions. In each situation, also indicate whether the hypothesis test will be one-sided or two-sided.
 a. Do female college students study more, on average, than male college students do?
 b. Compared to men who are not bald, are bald men more likely to have heart disease?
 c. Is there a correlation between height and head circumference?
 d. Will an increase in the speed limits on interstate highways lead to an increase in the highway fatality rate?
 e. When a coin is spun on its edge, is the probability that it will land "heads-up" equal to .5?

11.5 Suppose that a statistics teacher asks his students to each randomly pick one of the numbers 0, 1, 2, 3, 4. His general theory is that the proportion who pick the number 0 will be less than what would be expected from random selection. Use p to represent the population proportion that would pick the number 0.
 a. What is the specific value of p that corresponds to true random selection?
 b. In terms of p, write null and alternative hypotheses for this situation.

11.6 Suppose the present success rate in the treatment of a particular psychiatric disorder is .65 (65%). A research group hopes to demonstrate that the success rate of a new treatment will be better than this standard. Use the letter p to represent the success rate of the new treatment. Write null and alternative hypotheses for p.

11.7 Do you think researchers should determine whether to use a one-sided or two-sided hypothesis test before they look at the sample data or after they look at it? Explain.

11.8 For each of the following situations, write the null and alternative hypotheses in words (as in Example 11.1) and in symbols (as in Example 11.2).
 a. In the last census, taken five years ago, it was determined that 6% of school-aged children in a certain state lived with their grandparent(s). To support a bill on tax breaks for seniors, a congressional member plans to take a random sample of school-aged children to determine if that percent has increased.
 b. An anthropologist is trying to determine if the people in a certain region descend from the same ancestors as another region she has studied. She knows that in the region she has studied, 15% of the people have a certain unique genetic trait. She plans to take a random sample of people in the new region and test them for that trait.
 c. Paul likes a certain candy that comes in a bag with mixed colors, each with their own flavor. The candy company's website claims that 30% of these candies are red, which is Paul's favorite. He doubts their claim and thinks less than 30% are red. He is willing to assume that the candy is randomly placed into bags, and he plans to buy several bags of candy and count the proportion of red pieces.

11.9 Refer to the previous exercise. In each situation, the hypotheses are about a population proportion p. Explain what the population is and what the proportion of interest is in each case.

Sections 11.2 and 11.3

11.10 Consider this quote: "In a recent survey, 61 out of 100 consumers reported that they preferred plastic bags instead of paper bags for their groceries. If there is no difference in preference in the population, the chance of such extreme results in a sample of this size is about 0.03. Because 0.03 is less than 0.05, we can conclude that there is a statistically significant difference in preference." Give a numerical value for each of the following.
 a. The p-value.
 b. The level of significance, α.
 c. The sample proportion.
 d. The sample size.

11.11 State the conclusion that would be made in each of the following situations.
 a. Level of significance = 0.05, p-value = 0.10.
 b. Level of significance = 0.05, p-value = 0.01.
 c. Level of significance = 0.01, p-value = 0.99.
 d. Level of significance = 0.01, p-value = 0.002.

11.12 Suppose that a woman thinks she might be pregnant so she takes a pregnancy test. Considering this situation to be analogous to hypothesis testing, write each of the following in words.
 a. The null hypothesis. (*Hint:* Remember that the null hypothesis generally states that there is nothing going on.)
 b. The alternative hypothesis.
 c. The conclusion if there is not enough evidence to reject the null hypothesis.
 d. The conclusion if the null hypothesis is rejected.

11.13 Consider testing the null hypothesis that there is no relationship between smoking and getting a certain disease versus the alternative hypothesis that smokers are more likely than nonsmokers to get the disease. Explain in words what would be concluded about the relationship in the population if
 a. The p-value is 0.33 and the level of significance is 0.05.
 b. The p-value is 0.03 and the level of significance is 0.05.

11.14 Refer to part (a) of the previous exercise. Use the scenario to explain why it is not appropriate to *accept* the null hypothesis.

11.15 About 10% of the human population is left-handed. Suppose that a researcher speculates that artists are more likely to be left-handed than are other people in the general population. The researcher surveys 150 artists and finds that 18 of them are left-handed.
 a. State the researcher's null and alternative hypotheses.
 b. What proportion of the sample of artists is left-handed?
 c. To calculate a p-value for the hypothesis test, what probability should the researcher calculate? Make your answer specific to this situation.

11.16 An article in *USA Today* (Elias, June 3, 1999, p. D1) describes a study done by Georgia State University psychologist James Dabbs in which he found that women who are trial lawyers (litigators) are more likely to have male children than women lawyers who are not trial lawyers. According to the article, "58% of litigators' kids were boys vs. 44% [boys] for the others. The odds of this happening by chance are less than 5%, he says."

a. Consider the second sentence in the quote. What statistical term describes this result?

b. The quote from the article tells us that a p-value is less than 5%. Write one sentence that describes exactly what the p-value measures in this setting.

c. Dabbs believes the result can be attributed to the women's testosterone levels because, in a previous study, he found that, on average, women litigators have higher testosterone levels than women lawyers who aren't litigators. Do you think this conclusion is justified? Explain why or why not.

11.17 Given the convention of declaring that a result is "statistically significant" if the p-value is 0.05 or less, what decision would be made concerning the null and alternative hypotheses in each of the following cases? Be explicit about the wording of the decision.

a. p-value = 0.35.

b. p-value = 0.001.

c. p-value = 0.04.

11.18 Two researchers are testing the null hypothesis that a population proportion p is .25 and the alternative hypothesis that $p > .25$. Both take a sample of 100 observations. Researcher A finds a sample proportion of .29 and Researcher B finds a sample proportion of .33. For which researcher will the p-value of the test be smaller? Explain without actually doing any computations.

11.19 A physician claims that as soon as his patients have a positive pregnancy test he is generally able to predict the sex of the baby and that his probability of being right is greater than the one-half that would be expected if he was just guessing. A skeptic challenges his claim, and the physician decides to collect data for the next ten pregnancies to try to support his claim.

a. From the perspective of the physician, what are the null and alternative hypotheses for this situation?

b. Out of the ten pregnancies, he is correct six times. Write in words what probability needs to be computed to find the p-value.

c. The p-value for this test is 0.377. Does that prove that the skeptic is correct? Explain.

d. State the appropriate conclusion, using $\alpha = 0.05$.

e. Assuming the physician really does have some ability to make these predictions, what would you recommend that he do to increase the chance of proving his claim?

Section 11.4

11.20 Find the p-value for each of these situations, taking into account whether the test is one-sided or two-sided.

a. z-statistic = 2.10, H_0: $p = .10$, H_a: $p \neq .10$.

b. z-statistic = -2.00, H_0: $p = .6$, H_a: $p < .6$.

c. z-statistic = -1.09, H_0: $p = .5$, H_a: $p < .5$.

d. z-statistic = 4.25, H_0: $p = .25$, H_a: $p > .25$.

11.21 Refer to the previous exercise. For each of parts (a) to (d), specify the rejection region for $\alpha = 0.05$, then make a conclusion for the test using the rejection region rule.

11.22 In each of the following, determine whether the conditions for conducting a z-test for a proportion are met. If not, explain why not.

a. Twenty students are randomly selected from the list of all sorority and fraternity members at a university, to determine if a majority of sorority and fraternity students favor a new policy on alcohol on campus. The hypotheses are

H_0: $p = .50$.

H_a: $p > .50$.

b. Twenty employees of a large company are randomly selected to determine if the proportion of company employees who are left-handed exceeds the national proportion of 10% who are left-handed. The hypotheses are

H_0: $p = .10$.

H_a: $p > .10$.

c. A company employs 500 stockbrokers. They are all surveyed to find out if a majority believes the market will go up in the next year.

H_0: $p = .50$.

H_a: $p > .50$.

d. A market research firm wants to know if more than 30% of the people who visit a mall actually buy something. A researcher stands by the exit door starting at noon and asks 50 people as they are leaving whether they bought anything.

H_0: $p = .30$.

H_a: $p > .30$.

11.23 Refer to the previous exercise. In each of the parts define in words the population parameter p.

11.24 For each of the following calculate the z-statistic:

a. Sample size $n = 30$; sample proportion $\hat{p} = .60$.

H_0: $p = .50$.

H_a: $p \neq .50$.

b. Sample size $n = 60$; sample proportion $\hat{p} = .10$.

H_0: $p = .25$.

H_a: $p < .25$.

c. Sample size $n = 500$; sample proportion $\hat{p} = .30$.

H_0: $p = .20$.

H_a: $p > .20$.

d. Sample size $n = 200$; sample proportion $\hat{p} = .50$.

H_0: $p = .80$.

H_a: $p < .80$.

11.25 Refer to the previous exercise. In each case, calculate the p-value for the test.

11.26 Explain the difference between the *null standard error* used in the denominator of a z-statistic for a test for a proportion and the *standard error* used in finding a confidence interval for a proportion.

11.27 *Time Magazine* reported that in a 1994 survey of 507 randomly selected adult American Catholics, 59% answered yes to the question "Do you favor allowing women to be priests?" (*Time*, 26 December–2 January 1995, pp. 74–76).

a. Set up the null and alternative hypotheses for deciding whether more than half of American Catholics favor allowing women to be priests.

b. What conditions are necessary for using a z-statistic to test the hypotheses in part (a)? Are those conditions met here? Explain.

c. Compute the test statistic for this situation.

d. Calculate the p-value for the test.

e. Based on the p-value, make a conclusion for this situation. Write it in both statistical language and in words that someone with no training in statistics would understand.

f. Calculate a 95% confidence interval for the proportion of American Catholics who favor allowing women to be priests. Does this confidence interval confirm the conclusion that a majority of American Catholics favor allowing women priests? Explain.

11.28 In Exercise 11.15, we described a survey done to determine whether artists are more likely to be left-handed than others in the general population. If we use p to represent the proportion of all artists who are left-handed, the hypotheses are $H_0: p = .10$ and $H_a: p > .10$. The sample result was that 18 artists of 150 surveyed (or 12%) are left-handed.

a. From the given information, do we know if the conditions necessary to use a z-statistic for this test are met? Explain.

b. Conduct a hypothesis test. Clearly give the details of the five steps. Be sure to write a conclusion.

11.29 Suppose a one-sided test for a proportion resulted in a p-value of 0.04. What would the p-value be if the test was two-sided instead?

11.30 Suppose a two-sided test for a proportion resulted in a p-value of 0.07.

a. Using the usual criterion for hypothesis testing, would we conclude that the population proportion was different from the null hypothesis value? Explain.

b. Suppose the test had been constructed as a one-sided test instead and that the sample proportion was in the direction to support the alternative hypothesis. Using the usual criterion for hypothesis testing, would we be able to decide in favor of the alternative hypothesis? Explain.

11.31 In Example 11.7, results were given for a survey of students in a statistics class who measured the lengths of their right and left feet to the nearest millimeter. The right and left foot measurements were equal for 103 of the 215 students. Assuming this class is representative of all college students, is there evidence that the proportion of college students whose feet are the *same* length differs from one-half? Go through the five steps of hypothesis testing for this situation.

11.32 A Gallup Poll released on October 13, 2000 (Chambers, 2000) found that 47% of the 1052 U.S. adults surveyed classified themselves as "very happy" when given the choices of "very happy," "fairly happy," or "not too happy." Suppose that a journalist who is a pessimist took advantage of this poll to write a headline titled "Poll finds that U.S. adults who are very happy are in the minority." If p = proportion of all U.S. adults who are very

happy, go through the five steps of hypothesis testing and determine whether the headline is justified. Be sure to comment on the headline in your conclusion in step 5.

Sections 11.5 and 11.6

11.33 Suppose the null and alternative hypotheses in a test are
$$H_0: p = .70.$$
$$H_a: p \neq .70.$$
For each of the following sample sizes calculate the null standard error.
a. $n = 40$.
b. $n = 100$.
c. $n = 500$.
d. $n = 1000$.

11.34 Refer to the previous exercise. In each case, suppose the sample proportion is $\hat{p} = .75$ and compute the test statistic z and the p-value. Then make a conclusion using a level of significance of 0.05. Comment on the relationship between the sample size and the conclusions.

11.35 Sometimes a result that is statistically significant does not have practical significance. Is this more likely to happen when the sample size is very large or when it is very small? Explain.

11.36 One possible problem in hypothesis testing is that researchers fail to find a statistically significant result even though the null hypothesis is false. Is this problem more likely to occur with a very small sample or with a very large sample? Explain.

11.37 Researchers know that 20% of a certain ethnic group has a distinctive trait. They want to know if the proportion with the trait is higher in a second ethnic group. They take a random sample of n members of the second group and find the sample proportion $\hat{p}$ with the trait. The parameter of interest is p, the population proportion with the trait for the second group. In each of the following scenarios, determine which situation is more likely to lead to a statistically significant finding, assuming all other conditions remain the same. Explain your answer.
a. $n = 100$ or $n = 1000$.
b. $p = .25$ or $p = .45$.
c. $\hat{p} = .25$ or $\hat{p} = .45$.

11.38 When reporting the results of a study, explain why a distinction should be made between "statistical" significance and significance as the term is used in ordinary language.

11.39 Which do you think is more informative when you are given the results of a hypothesis test, the p-value or the decision about whether or not to reject the null hypothesis? Explain.

11.40 Which do you think is more informative when you are given the results of a study, a confidence interval or a p-value? Explain.

11.41 Suppose you were to read in your local newspaper that a new study based on 100 men had found that there was *no difference* in heart attack rates for men who exercised regularly and men who did not. Do you think the study found *exactly* the same rate of heart attacks for the two groups of men? In the context of the material in this chapter, what would you suspect was the reason for that reported finding?

11.42 Suppose that a study is designed to choose between the hypotheses

Null hypothesis: Population proportion is .20.

Alternative hypothesis: Population proportion is higher than .20.

On the basis of a random sample of size 400, the sample proportion is .25.

a. Compute the *z*-score corresponding to the sample proportion of .25, assuming the null hypothesis is true.

b. What is the *p*-value for the standardized score computed in part (a)?

c. Based on the result from part (b), make a conclusion. Be explicit about the wording of your conclusion and justify your answer.

d. Suppose that the sample size had only been 100, rather than 400, but the sample proportion was again .25. Using this smaller sample size, repeat parts (a) to (c).

e. Explain what this problem illustrates about the effect of sample size on statistical significance.

11.43 An advertisement for Claritin, a drug for seasonal nasal allergies, made this claim: "Clear relief without drowsiness. In studies, the incidence of drowsiness was similar to placebo" (*Time,* Feb. 6, 1995, p. 43). The advertisement also reported that 8% of the 1926 Claritin takers and 6% of the 2545 placebo takers reported drowsiness as a side effect. A one-sided test of whether or not a higher proportion of Claritin takers than placebo takers would experience drowsiness in the population results in a *p*-value of about 0.005.

a. From this information, would you conclude that the incidence of drowsiness for the Claritin takers is statistically significantly higher than for the placebo takers?

b. Does the answer to part (a) contradict the statement in the advertisement that the "incidence of drowsiness was similar to placebo"? Explain.

c. Use this example to discuss the importance of making the distinction between the common use and the statistical use of the word *significant.*

Section 11.7

11.44 Explain whether each of the following statements is true or false.

a. The *p*-value is the probability that the null hypothesis is true.

b. If the null hypothesis is true, then the level of significance is the probability of making a type 1 error.

c. A type 2 error can only occur when the null hypothesis is true.

d. The probability of making a type 1 error plus the probability of making a type 2 error is 1.

11.45 Can the two types of error both be made in the same hypothesis test? Explain.

11.46 Define events as follows: A = Null hypothesis is true. A^C = Alternative hypothesis is true. B = Null hypothesis

is not rejected. B^C = Null hypothesis is rejected. For each of the following outcomes explain whether a type 1 error, a type 2 error, or a correct decision has been made.

a. A and B.

b. A^C and B.

c. A and B^C.

d. A^C and B^C.

11.47 A medical insurance company wants to know if the proportion of its customers requiring a hospital stay during a year will decrease if it provides coverage for certain types of alternative medicine. The company conducts a one-year study in which it gives insurance coverage for alternative medicine to 5000 randomly selected customers. Based on this study, the company tests the following hypotheses about the effect of offering the alternative medicine coverage to its customers:

H_0: proportion requiring a hospital stay will not decrease

H_a: proportion requiring a hospital stay will decrease

If the null hypothesis is rejected, the company will offer this coverage to all of its customers in the future.

a. Explain what a type 1 error would be in this situation.

b. Explain what a type 2 error would be in this situation.

c. Explain which type of error would be more serious for the insurance company.

d. Explain which type of error would be more serious for the customers.

11.48 Medical researchers now believe there may be a link between baldness and heart attacks in men.

a. State the null hypothesis and the alternative hypothesis for a study used to investigate whether or not there is such a relationship.

b. Discuss what would constitute a type 1 error in this study.

c. Discuss what would constitute a type 2 error in this study.

11.49 Consider medical tests, in which the null hypothesis is that the patient does not have the disease and the alternative hypothesis is that they do.

a. Give an example of a medical situation in which a type 1 error would be more serious.

b. Give an example of a medical situation in which a type 2 error would be more serious.

11.50 Explain which type of error, type 1 or type 2, could be made in each of the following cases.

a. The null hypothesis is true.

b. The alternative hypothesis is true.

c. The null hypothesis is not rejected.

d. The null hypothesis is rejected.

11.51 Refer to Example 11.6 in which convincing evidence was found for the importance of order in a "random" selection between two letters. Suppose a student in your class ran for president of the campus student association and lost. After reading Example 11.6, she realized that because her name is at the end of the alphabet, she was listed after all of her opponents on the ballot and the election was unfair to her. She demands a reelection with names listed in random order.

a. Which type of error, type 1 or type 2, could have been made in Example 11.6? Explain.

b. If the type of error you specified in part (a) was made, is the reelection justified? Explain.

c. Now suppose the other type of error had been made, and the student read the results of the study in that case instead. Explain what the consequences would be for the student and the student government.

d. Which type of error do you think is more serious for the student who ran for president? Explain.

11.52 A politician is trying to decide whether to vote for a new tax bill that calls for substantial reforms. A random sample of voters in his district led him to believe the alternative hypothesis, $H_a: p > .5$, where p is the proportion of all voters in his district who support the bill. As a consequence, he decides to vote for the bill.

a. What would a type 1 error be in this situation, and what would be the consequences for the politician?

b. What would a type 2 error be in this situation, and what would be the consequences for the politician?

c. Explain which error would be more serious, if either, for the politician in this example.

d. Given the situation described, could the politician have made a type 1 error or a type 2 error?

Chapter Exercises

11.53 Consider the two conditions necessary for conducting a z-test for a proportion. One is a sample size condition, and the other is a condition about how the sample was collected. For each of the two conditions, explain why it is required in order for the test and the results to be valid.

11.54 Refer to the five steps of any hypothesis test, given in the introduction to this chapter and again in Section 11.4. Which step(s) can be completed before collecting any data?

11.55 A student has been accused of cheating on an examination by copying another student's paper, and you have been asked to serve on the panel that must decide the student's fate. If the student is found guilty, he will fail the course and will have to write an essay on honesty for the school paper.

a. What are the null and alternative hypotheses for this situation?

b. What is a type 1 error in this situation, and what are the consequences?

c. What is a type 2 error in this situation, and what are the consequences?

d. As a member of the panel making the decision, you must evaluate the seriousness of the two types of error. Explain which type you think is more serious.

11.56 In a survey of 240 students in an elementary statistics class at the University of California, Davis (UCD), 20 said they were left-handed and 220 said they were right-handed. Assume the students are representative of all students at the school. Does this provide evidence that the proportion of UCD students who are left-handed differs from the national proportion of .10? Carry out the five steps of the hypothesis test for this situation.

11.57 Max likes to keep track of birthdays of people he meets. He has 170 birthdays listed on his birthday calendar. One cold January night, he comes up with the theory that people are more likely to be born in October than they would be if all 365 days were equally likely. He consults his birthday calendar and finds that 22 of the 170 birthdays are in October.

a. Write down the null and alternative hypotheses for Max's test. Be sure to carefully specify what p represents.

b. Carry out the remaining four steps of the hypothesis test for this situation.

11.58 Specify the null and alternative hypotheses in words for each of the following research questions.

a. Does listening to Mozart increase performance on an intelligence test?

b. Does talking to plants result in better growth for the plants?

c. Does drinking fluoridated water lead to increased bone fractures in elderly people?

11.59 For the following two situations, define the population parameter in words. Then specify the null and alternative hypotheses in symbols.

a. For people suffering from a certain type of chronic pain, 70% experience temporary relief when they take a standard medication. If a new medication appears to be more effective, it will replace the standard. Does the new medication provide pain relief to a larger proportion than the standard medication?

b. A student at a large university is thinking about starting a new service in which students can order their textbooks from his website and he will buy them at the bookstore and deliver them for a small service charge. He plans to survey a random sample of students and ask them if they would be willing to pay for such a service. If he is convinced that at least 5% of students would use his service, he will go forward with it.

11.60 Refer to the previous exercise. For each of the two situations described, explain the consequences of a type 1 error and the consequences of a type 2 error. For part (a), explain which type would be more serious for the patients, and for part (b) explain which type would be more serious for the student who is planning the new service.

11.61 Find the p-value for each of these situations, taking into account whether the test is one-sided or two-sided.

a. z-statistic = 1.45, $H_0: p = .50$, $H_a: p \neq .50$.

b. z-statistic = −1.78, $H_0: p = .3$, $H_a: p < .3$.

c. z-statistic = 1.78, $H_0: p = .3$, $H_a: p < .3$.

d. z-statistic = 2.25, $H_0: p = .75$, $H_a: p > .75$.

11.62 Refer to the previous exercise. For each of parts (a) to (d), specify the rejection region for $\alpha = 0.05$, then make a conclusion for the test using the rejection region rule.

11.63 For a test of $H_0: p = .25$ versus $H_a: p > .25$, for what range of values of the sample proportion $\hat{p}$ will the p-value for the test be *greater* than 0.5? Explain your answer.

11.64 For a one-sided hypothesis test for a proportion in which the alternative hypothesis is $H_a: p < p_0$, for what values of the test statistic z will the p-value be *greater* than 0.5? Explain your answer.

11.65 In Chapter 2, we saw data from a statistics class activity in which students were asked to "randomly" pick a number (integer) from 1 to 10. Of the 190 students, 56 picked the number 7. Carry out the five steps of hypothesis testing to determine if people similar to these students are more likely to pick the number 7 than would be expected by chance if all numbers were equally likely.

11.66 A professor planned to give an examination in a large class on the Monday before Thanksgiving vacation. Some students asked if he could change the date because so many of their classmates had at least one other exam on that date. They speculated that at least 40% of the class had this problem. The professor agreed to poll the class, and if there was convincing evidence that the proportion with at least one other exam on that date was greater than .40, he would change the date. Of the 250 students in the class, 109 reported that they had another exam on that date.

a. What proportion of the class reported that they had another exam on that date?

b. Is the proportion you found in part (a) a sample proportion or a population proportion?

c. The professor conducted a z-test of H_0: $p = .40$ versus H_a: $p > .40$ and found $z = 1.16$ and p-value = 0.125. He said he would not move the exam because the null hypothesis cannot be rejected, and there is not convincing evidence that the population proportion is greater than .40. What is wrong with his reasoning? [*Hint:* Refer to part (b).]

11.67 According to *USA Today* (Snapshot, April 20, 1998, also referenced in exercise 9.74), a random sample of 8709 adults taken in 1976 found that 9% believed in reincarnation. A poll of 1000 adults in 1997 found that 25% believed in reincarnation. Because the margin of error for the 1976 poll is so small, we will treat the sample proportion of .09 as if it were the population proportion with this belief in 1976. Test the hypothesis that the proportion of adults believing in reincarnation in 1997 was higher than the proportion in 1976.

11.68 In a survey of students at the University of California, Davis, students were asked which of two popular soft drinks they preferred—let's call them brand C and brand P. Of the 159 respondents, 80 preferred brand P

and 79 preferred brand C. Assuming these students represent all students at UCD, test whether there is a preference for one drink over the other for students at UCD.

11.69 Refer to the previous exercise. Some students were asked if they preferred C or P, while other students were asked if they preferred P or C. Of the 159 respondents, 86 responded with the first drink presented. Test the hypothesis that the first drink presented is more likely to be selected than the second drink presented.

Dataset Exercises

11.70 Use the dataset **UCDavis2**. The variable ***Friends*** is a response to the question "Who do you find it easiest to make friends with? People of the (circle one) Same Sex or Opposite Sex?" The responses are S = same or O = opposite.

a. Is there sufficient evidence that students find it easier to make friends with the opposite sex than with the same sex? What assumption(s) did you have to make about the sample in order for the testing procedure to be valid?

b. Conduct the analysis separately for men and for women, using the variable "Sex" to identify males (M) and females (F).

c. Refer to part (b). Compare the results for males and females by writing a summary of the results in words that would be understood by someone who has not studied statistics.

d. Compare your results in parts (a) and (b). Explain how they are similar and how they are different, if at all. Explain any discrepancies you found.

11.71 Use the dataset **UCDavis2.** The variable ***Seat*** is a response to the question "Where do you typically sit in a classroom?" Possible responses were F = Front, M = Middle, B = Back.

a. Test the hypothesis that a majority of students prefer to sit in the middle of the room.

b. Specify the population of interest, and comment on whether this sample is an appropriate representation of it.

REFERENCES

Chambers, Chris (2000). "Americans are overwhelmingly happy and optimistic about the future of the United States," Gallup Poll Press Release, October 13, 2000, www.gallup.com/poll/releases/pr001013.asp.

Ott, R. L. (1998). *An Introduction to Statistical Methods and Data Analysis,* 4th ed., Belmont, CA: Duxbury Press.

Passell, Peter (1994). "Probability experts may decide Pennsylvania vote," *New York Times,* April 11, 1994, p. A15.

Thys-Jacobs, S., P. Starkey, D. Bernstein, J. Tian, and the Premenstrual Syndrome Study Group (1998). "Calcium carbonate and the premenstrual syndrome: Effects on premenstrual and menstrual symptoms," *American Journal of Obstetrics and Gynecology,* 179, 2, pp. 444–452.

CHAPTER 12

More About Confidence Intervals

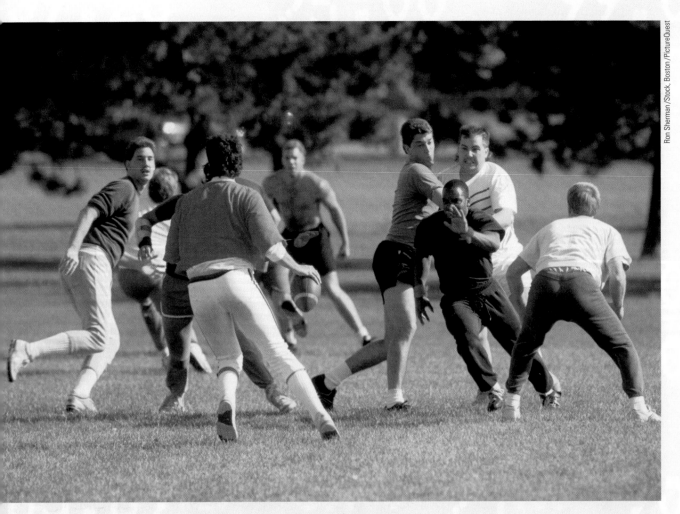

Ron Sherman /Stock, Boston /PictureQuest

Do men lose more weight from dieting or from exercising? See Example 12.4 on pages 398 and 402.

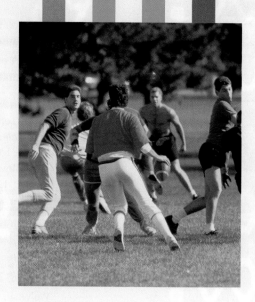

When complicated procedures are needed to compute confidence intervals, journal articles provide completed intervals. Your job as a reader is to interpret those intervals. Principles in this chapter for understanding confidence intervals for means and proportions can be used to understand any confidence interval.

In Chapter 10 we learned how to form and interpret a confidence interval for a population proportion. Remember that a *confidence interval* is an interval of values computed from sample data that is likely to include the true population value. For example, based on a sample survey of 1000 Americans, we may be able to say with 95% confidence that the interval from .23 to .29 (or 23% to 29%) will cover the true proportion of all Americans who favor the legalization of marijuana. The *confidence level* for an interval describes our confidence in the procedure we used to determine the interval. We are confident that most of the confidence intervals we compute using the procedure will contain the true population value.

In this chapter, we expand the use of confidence intervals to other characteristics of populations, like the mean of a quantitative variable, or the difference in means for two populations, or the difference between proportions in two populations. We may, for example, wish to know what is the mean number of hours per day that college students watch television. Or, we may want to know how much difference there is between the mean pulse rates of women and men. A psychiatric researcher may want to know how much difference there is between the proportions of successful outcomes for two different treatments of clinical depression. In each of these types of situations, the objective is the same. We use sample information to form a confidence interval estimate of a population value. ◆

12.1 | EXAMPLES OF DIFFERENT ESTIMATION SITUATIONS

To begin, let's review the following two definitions, which are relevant to our discussion of the estimation of population values:

◆ A **parameter** is a population characteristic. The numerical value of a parameter usually cannot be determined because we cannot measure all units in the population. We have to estimate the parameter using sample information.

♦ A **statistic,** or **estimate,** is a characteristic of a sample. A statistic estimates a parameter.

Figure 12.1 illustrates the statistical estimation problem. We wish to estimate a characteristic of a population. In more technical terms, we want to estimate a *parameter.* To do this, we take a random sample from the population and use the sample to calculate a sample *statistic* that estimates the parameter.

Following are descriptions of four common research situations that can be analyzed using a confidence interval estimate of the value of a population parameter. The parameters in these four different situations are

1. p = a population proportion
2. μ = a population mean
3. $p_1 - p_2$ = the difference between two population proportions
4. $\mu_1 - \mu_2$ = the difference between two population means

FIGURE 12.1 **Sample information is used to estimate a population parameter**

We've already covered the estimation of a proportion in Chapter 10, but we include that situation here to emphasize that the confidence intervals presented in this chapter all have the same general structure that a confidence interval for a proportion does. For each situation, we provide two examples, as well as the notation and description for the parameter and the sample estimate.

Situation 1. Estimating the Proportion Falling into a Category of a Categorical Variable

♦ *Example research questions:*

What proportion of American adults believe there is extraterrestrial life? In what proportion of British marriages is the wife taller than her husband?

♦ *Population parameter:* p = proportion in the population falling into that category

♦ *Sample estimate:* $\hat{p}$ = proportion in the sample falling into that category

Situation 2. Estimating the Mean of a Quantitative Variable

♦ *Example research questions:*

What is the mean time that college students watch TV per day? What is the mean pulse rate of women?

♦ *Population parameter:* μ (spelled "mu" and pronounced "mew") = population mean for the variable

♦ *Sample estimate:* $\bar{x}$ = the sample mean for the variable

Situation 3. Estimating the Difference Between Two Populations with Regard to the Proportion Falling into a Category of a Categorical Variable

♦ *Example research questions:*

How much difference is there between the proportions that would quit smoking if taking the antidepressant buproprion (Zyban) versus if wearing a nicotine patch?

How much difference is there between men who snore and men who don't snore with regard to the proportion who have heart disease?

 ◆ *Population parameter:* $p_1 - p_2$, where p_1 and p_2 represent the proportions in populations 1 and 2, respectively

 ◆ *Sample estimate:* $\hat{p}_1 - \hat{p}_2$, the difference between the two sample proportions

Situation 4. Estimating the Difference Between Two Populations with Regard to the Mean of a Quantitative Variable

 ◆ *Example research questions:*

 How much difference is there in average weight loss for those who diet compared to those who exercise to lose weight?

 How much difference is there between the mean foot lengths of men and women?

 ◆ *Population parameter:* $\mu_1 - \mu_2$, where μ_1 and μ_2 represent the means in populations 1 and 2, respectively

 ◆ *Sample estimate:* $\bar{x}_1 - \bar{x}_2$, the difference between the two sample means

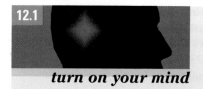

12.1

turn on your mind

Pick one of the example research questions provided in one of the four situations just described. Explain an appropriate sampling procedure, and then describe what the population parameter and the sample estimate would be for your example.*

Independent Samples

Situations 3 and 4 both involve finding the difference in parameters (means or proportions) for two independent samples. Two samples are called **independent samples** when the measurements in one sample are not related to the measurements in the other sample. For example, a comparison of the mean pulse rates of men and women is a comparison of two *independent samples* as long as the individuals sampled are not related pairs, like siblings or spouses.

Independent samples are generated in a variety of ways. Here are the most common methods:

 ◆ Random samples are taken separately from two populations and the same response variable is recorded for each individual.

 ◆ One random sample is taken and a variable recorded for each individual, but then units are categorized as belonging to one population or another, such as male and female.

 ◆ Participants are randomly assigned to one of two treatment conditions, such as diet or exercise, and the same response variable, such as weight loss, is recorded for each individual unit.

 ◆ Two random samples are taken from a population and a separate variable is measured in each sample. For example, one sample may be asked about hours of Internet use, whereas a separate sample is asked about hours of television viewing. Then the results are compared. This method for collecting data is not used very often.

*HINT: Sampling methods are covered in Chapter 4, Sections 4.2 and 4.3.

Depending upon whether the response variable is quantitative or categorical, a researcher might compare two independent groups by looking at the difference between two means or by looking at the difference between two proportions.

Paired Data: A Special Case of One Mean

An important special case of a single mean of a population occurs when two quantitative variables are collected in pairs and we desire information about the difference between the two variables. For example, we might measure the IQ score of each person in a sample after he or she listens to a relaxation tape and again after he or she listens to a Mozart tape. The difference between the two IQ measurements can be computed for each person, and we could use the sample differences to estimate the mean population difference. This would provide information about whether listening to Mozart produces higher IQ scores, on average, than listening to a relaxation tape does.

The term *paired data* (or *paired samples*) is used when pairs of variables are collected and used in this way. Generally, in such situations we only are interested in the population (and sample) of *differences,* and not in the original data. Therefore, for paired quantitative variables, it is appropriate to use methods designed for inferences about a single population mean. The mean in these situations is the mean for the population of differences that would result if we measured both variables on all units in the population.

> *definition*   The term **paired data** means that the data have been observed in natural pairs. Some ways paired data can occur are
>
> ◆ Each person is measured twice. The two measurements of the same characteristic or trait are made under different conditions.
>
> ◆ Similar individuals are paired prior to an experiment. During the experiment, each member of a pair receives a different treatment. The same response variable is measured for all individuals.
>
> ◆ Two different variables are measured for each individual. There is interest in the amount of difference between the two variables.

12.2 | STANDARD ERRORS

Whenever we use a sample to estimate a population parameter, there is the possibility of an "error" due to the act of sampling. It is unlikely, for example, that the mean pulse rate of a sample of 25 women would exactly equal the mean pulse rate in the population of women represented by the sample. For any particular observed sample estimate, we can't know the exact amount of difference between the sample estimate and the population parameter because we don't know the value of the parameter. However, we are able to determine approximately the "average" sampling error that would occur over all possible random samples that can be taken from a population.

We encountered the phrase *standard error* when we calculated confidence intervals for the population proportion in Chapter 10. Recall from Section 10.5 that the *margin of error* for an approximate 95% confidence interval for a proportion was expressed as $2 \times$ standard error. From this connection between margin of error and standard error, you should realize that a standard error somehow measures the accuracy of a sample statistic. The phrase *standard error* can be used in conjunction with any type of statistic, such as a mean or a difference between two means. In nontechnical terms, the standard error is a rough guide to the potential size of sampling error.

We learned in Chapter 9 that a sample statistic is a random variable because the value of a statistic varies over the different samples that could be taken from the same population. The mean pulse rate of a sample of 25 women, for instance, may be different for each possible random sample that could be taken from the population of all women. The **sampling distribution** of a statistic like a sample mean or a sample proportion is the distribution of possible values of the statistic for all possible random samples of a specified size from the population. In technical terms, the **standard error** of a statistic is the estimated standard deviation of the sampling distribution of the statistic.

definition *Rough Definition:* The **standard error** of a sample statistic measures, roughly, the average difference between the statistic and the population parameter. This "average difference" is over all possible random samples of a given size that can be taken from the population.

Technical Definition: The **standard error** of a sample statistic is the estimated standard deviation of the sampling distribution for the statistic.

Calculating the Standard Error

The concept of standard error is always the same, but the formula for determining the standard error depends upon the type of sample statistic. For instance, the formula for the standard error of a sample mean is different from the formula for the standard error of a sample proportion. The formula for the standard error of a sample proportion and the formula for the standard error of a sample mean are shown again here. These two formulas were first introduced in Chapter 9, and the formula for the standard error of a sample proportion was emphasized in Chapter 10. The notation s.e.(statistic) represents a standard error. As it has throughout, n represents the sample size.

Standard Errors for Sample Proportion and Sample Mean

formula   **Standard Error of a Sample Proportion**

$$\text{s.e.}(\hat{p}) = \sqrt{\frac{\hat{p}(1 - \hat{p})}{n}}, \qquad \hat{p} = \text{sample proportion}$$

Standard Error of a Sample Mean

$$\text{s.e.}(\bar{x}) = \frac{s}{\sqrt{n}}, \qquad s = \text{sample standard deviation}$$

EXAMPLE 12.1
Revisiting Intelligent
Life on Other Planets

In Chapter 10 we used data from a 1997 Marist Institute survey of 935 randomly selected Americans to estimate the proportion of Americans who would answer "yes" to the question, "Do you think there is intelligent life on other planets?" The sample proportion of "yes" answers was $\hat{p} = .60$. Based on the sample size of $n = 935$, the standard error is

$$\text{s.e.}(\hat{p}) = \sqrt{\frac{\hat{p}(1 - \hat{p})}{n}} = \sqrt{\frac{(.6)(.4)}{935}} = .016$$

The $\hat{p}$ observed in this sample was one of many possibilities for the sample proportion in the sense that a different sample of $n = 935$ might provide a different value of $\hat{p}$. The standard error of .016 is roughly the average difference between the statistic, $\hat{p}$, and the population parameter, p, for all possible random samples of $n = 935$ from this population. Theoretically, the standard error is the estimated standard deviation for the sampling distribution of possible sample proportions that would result from asking this question of different random samples of 935 individuals. The distribution of these sample proportions is centered on the unknown value of the parameter p, and the standard error, .016, estimates the standard deviation of this distribution. ◆

EXAMPLE 12.2
Mean Hours per
Day That Penn
State Students Watch TV

In a class survey at Penn State, a question was "In a typical day, about how much time do you spend watching television?" Here is a numerical summary, generated by Minitab, of the responses (in hours):

Variable	N	Mean	Median	TrMean	StDev	SE Mean
TV	175	2.09	2.000	1.950	1.644	0.124

Based on this information, what is our estimate of the mean hours of television per day in the *population* represented by this sample? The sample mean, $\bar{x} = 2.09$ hours per day, is the statistic that estimates μ, the population mean. From the Minitab output, we see that the standard error of the sample mean is $\text{s.e.}(\bar{x}) = 0.124$ (look under "SE Mean"). The calculation of this value is

$$\text{s.e.}(\bar{x}) = \frac{s}{\sqrt{n}} = \frac{1.644}{\sqrt{175}} = 0.124 \ ◆$$

The standard error in Example 12.2 measures, roughly, the average difference between the unknown value of the population mean μ and the sample mean for samples of $n = 175$ from this population. Technically, the standard error estimates the standard deviation of the sampling distribution of possible sample means. The sample mean $\bar{x} = 2.09$ hours observed in this sample is just one of a plethora of possibilities for the sample mean. The distribution of these possibilities for $\bar{x}$ is centered on the unknown value μ hours, and the estimated standard deviation of the distribution of sample means is 0.124 hours.

The Standard Error of a Difference for Independent Samples

Remember that two samples are called *independent samples* when the measurements in one are not related to the measurements in the other; situations in which such samples arise were discussed in Section 12.1. Depending upon whether the response variable is quantitative or categorical, a researcher might compare two independent groups either by looking at the difference between two means or by looking at the difference between two proportions. In either case, the **standard error of the difference** between statistics from independent samples is determined as

$$\text{s.e.(difference)} = \sqrt{[\text{s.e.(statistic 1)}]^2 + [\text{s.e.(statistic 2)}]^2}$$

This formula leads to the following formulas for the standard error of a difference between two sample proportions and the difference between two sample means. In these formulas, a subscript of 1 or 2 indicates that the measure is for the sample from population 1 or 2, respectively.

Standard Errors for Difference Between Two Proportions and Difference Between Two Means

formula **Standard Error of the Difference Between Two Sample Proportions**

$$\text{s.e.}(\hat{p}_1 - \hat{p}_2) = \sqrt{\frac{\hat{p}_1(1 - \hat{p}_1)}{n_1} + \frac{\hat{p}_2(1 - \hat{p}_2)}{n_2}}$$

Standard Error of the Difference Between Two Sample Means

$$\text{s.e.}(\bar{x}_1 - \bar{x}_2) = \sqrt{\frac{s_1^2}{n_1} + \frac{s_2^2}{n_2}}$$

EXAMPLE 12.3
Nicotine Patches or Antidepressants If You Want to Quit Smoking?

Case Study 10.2 presented the results of a study done to determine whether the antidepressant buproprion (also called Zyban) is more effective than the nicotine patch in helping people to quit smoking (Jorenby et al., 1999). Volunteers were randomly allocated to four possible treatments. For this example, we compare two of the treatments, nicotine patch (administered along with a placebo pill) and Zyban (administered along with a placebo patch). There were $n = 244$ participants randomly assigned to each of these two treatments.

After six months, 85 of the 244 Zyban users had quit smoking, while only 52 of the 244 nicotine patch users had quit. The corresponding sample proportions who quit and the difference between these proportions are

$$\hat{p}_1 = \frac{85}{244} = .348, \text{ for Zyban}$$

$$\hat{p}_2 = \frac{52}{244} = .213, \text{ for nicotine patch}$$

$$\hat{p}_1 - \hat{p}_2 = .348 - .213 = .135$$

We see that with Zyban the success rate is .135 (or 13.5%) higher than the success rate is with the nicotine patch. The sample difference $\hat{p}_1 - \hat{p}_2 = .135$ estimates the population parameter $p_1 - p_2$, the difference in the proportions who would quit under each method for the population of all smokers like those in this experiment. The standard error of the statistic $\hat{p}_1 - \hat{p}_2$ is

$$\text{s.e.}(\hat{p}_1 - \hat{p}_2) = \sqrt{\frac{\hat{p}_1(1 - \hat{p}_1)}{n_1} + \frac{\hat{p}_2(1 - \hat{p}_2)}{n_2}}$$

$$= \sqrt{\frac{.348(1 - .348)}{244} + \frac{.213(1 - .213)}{244}} = .040$$

In this example, the standard error roughly measures the average difference between the statistic $\hat{p}_1 - \hat{p}_2$ and the population parameter $p_1 - p_2$. Theoretically, the standard error estimates the standard deviation of the sampling distribution of possible sample differences in proportions, $\hat{p}_1 - \hat{p}_2$. Other samples of 244 in each treatment condition could produce other estimates of the difference, so the observed difference is just one of a plethora of possibilities for the sample difference. The distribution of all possibilities for the sample difference is centered on the unknown value of the parameter $p_1 - p_2$ and has an estimated standard deviation of .040. ◆

EXAMPLE 12.4
Do Men Lose More Weight by Diet or by Exercise?

In a study done by Wood et al. (1988), 42 sedentary men were placed on a diet while 47 previously sedentary men were put on an exercise routine (study also reported by Iman, 1994, p. 258). The group on a diet lost an average of 7.2 kg, with a standard deviation of 3.7 kg. The men who exercised lost an average of 4.0 kg, with a standard deviation of 3.9 kg.

The difference between the sample means is an estimate of $\mu_1 - \mu_2$, the difference in the means of two hypothetical populations defined by the two weight-loss strategies in the study. The first hypothetical population is the population of weight losses for all sedentary men like the ones in this study if they were to be placed on this diet for a year. The second hypothetical population is the population of weight losses for those same men if they were to follow the exercise routine for a year.

The difference between the two means is $\bar{x}_1 - \bar{x}_2 = 7.2 - 4.0 = 3.2$ kg (about 7 pounds), with the dieters losing more weight than the exercisers. The standard error of the statistic $\bar{x}_1 - \bar{x}_2$ is

$$\text{s.e.}(\bar{x}_1 - \bar{x}_2) = \sqrt{\frac{s_1^2}{n_1} + \frac{s_2^2}{n_2}}$$

$$= \sqrt{\frac{3.7^2}{42} + \frac{3.9^2}{47}} = 0.81$$

In this example, the standard error measures, roughly, the average difference between the statistic $\bar{x}_1 - \bar{x}_2$ and the population parameter $\mu_1 - \mu_2$. Technically, the standard error estimates the standard deviation of the sampling distribution of possible sample differences in means, $\bar{x}_1 - \bar{x}_2$, for samples of sizes 42 and 47 from the populations. The observed sample difference of 3.2 kg is just one of many possibilities for the sample difference. For the distribution of the many possible values of the difference in sample means, the estimated standard deviation is $\text{s.e.}(\bar{x}_1 - \bar{x}_2) = 0.81$ kg. ◆

12.2

turn on your mind

Notice that all of the standard error formulas in this section have the sample size(s) in the denominator. This tells us that if the sample size is increased, the standard error will decrease (assuming the sample statistics remain about the same). Refer to the "rough definition" of standard error and explain why this relationship between sample size and standard error makes sense, based on that definition.*

12.3 | APPROXIMATE 95% CONFIDENCE INTERVALS

The general format for determining a confidence interval is the same for each situation that we consider in this chapter. This format, which we first encountered in Chapter 10, can be written as

Sample estimate ± Multiplier × Standard error

In other words, we form a confidence interval by adding and subtracting an appropriate number of standard errors to (and from) the sample estimate. The **multiplier** depends upon the confidence level, the sample size, and the type of parameter under study.

In Chapter 10, we illustrated how to find approximate 95% confidence intervals for a population proportion. For an approximate 95% confidence interval for a proportion, the appropriate number of standard errors is 1.96, which is often rounded to 2. This same technique can be used to determine approximate 95% confidence intervals for the three new situations that we've introduced in this chapter. The Normal Curve Approximation Rules for means and proportions in Chapter 9 provide the theoretical foundation. With a sufficient sample size, for about 95% of all random samples from a population, the difference between the sample statistic and the population parameter is less than 2 standard errors. This leads to a simple and conceptually convenient way to determine an **approximate 95% confidence interval** for a parameter.

---

definition   For sufficiently large samples, the interval

Sample estimate ± 2 × Standard error

is an **approximate 95% confidence interval** for a population parameter. This is so for a proportion, a mean, the difference between two proportions, and the difference between two means.

Note: The 95% confidence level describes how often the procedure provides an interval that includes the population value. For about 95% of all random samples of a specific size from a population, the confidence interval captures the population parameter.

Necessary Conditions for Computing Approximate 95% Confidence Intervals

The approximate 95% confidence intervals described in this section are only valid if the sample sizes are large enough for the Normal Approximation Rules in Chapter 9 to hold. Conditions for which confidence intervals for one mean or a difference in two means can be computed with small sample sizes are presented in Sections 12.4 and 12.5. As guidelines for the situations we consider, the following sample size requirements should be met to use the method for approximate 95% confidence intervals in this section:

For one proportion: Both $n\hat{p}$ and $n(1 - \hat{p})$ are at least 5, preferably at least 10. These quantities are the number of observations in the category of interest and not in that category, respectively.

For one mean: n is greater than 30.

For two proportions: $n\hat{p}$ and $n(1 - \hat{p})$ are at least 5 (preferably 10) for each sample.

For two means: n_1 and n_2 are each greater than 30.

In addition to the sample size requirements, another requirement is that the samples are randomly selected. In practice, it is sufficient to assume that samples are representative of the population for the question of interest. For problems that concern the difference between two means or two proportions, there is one more condition:

Independence of samples: For the confidence intervals for the difference between two proportions or two means, the two samples must be *independent* of each other.

EXAMPLE 12.1—
CONTINUED
Confidence Interval
for Proportion of
Americans Who Believe
There Is Intelligent
Life on Other Planets

The sample proportion that answered "yes" to the question about extraterrestrial life was $\hat{p} = .60$. The sample size was $n = 935$. We determined that the standard error is s.e.$(\hat{p}) = .016$. An approximate 95% confidence interval for the proportion of Americans who think there is intelligent life on other planets is

Sample estimate $\pm 2 \times$ Standard error

$$.60 \pm 2 \times .016$$

$$.60 \pm .032$$

$$.568 \text{ to } .632$$

In percent terms, the interval ranges from 56.8% to 63.2%. We are approximately 95% confident that this interval is correct, meaning that it captures the population percent that believes there is intelligent life on other planets. The confidence level is a confidence in the procedure used to determine the interval. For about 95% of all random samples from the population, the corresponding confidence interval captures the population parameter. We don't know if this particular interval does or does not capture the population value. ◆

EXAMPLE 12.2—
CONTINUED
Confidence Interval
for Mean Hours
per Day That Penn
State Students Watch TV

The sample mean, based on a sample $n = 175$ students, was $\bar{x} = 2.09$ hours per day. The standard error is s.e.$(\bar{x}) = 0.124$, and an approximate 95% confidence interval estimate of μ, the population mean, is

Sample estimate $\pm 2 \times$ Standard error

$$\bar{x} \pm 2 \times \text{s.e.}(\bar{x})$$

$$2.09 \pm 2 \times 0.124$$

$$2.09 \pm 0.248$$

$$1.842 \text{ to } 2.338 \text{ hours}$$

In a research article, this confidence interval might be described as follows: *We are 95% confident that the mean time that Penn State students spend watching television per day is somewhere between 1.842 and 2.338 hours.* Actually the report would probably round off the numbers and declare that, with 95% confidence, the mean is between 1.8 and 2.3 hours. Remember what is meant technically by that statement: We are 95% confident that the interval from 1.842 to 2.338 hours is correct, in that it captures the mean television viewing hours for Penn State students. ◆

12.3

turn on your mind

◆ What population do you think is represented by the sample of 175 students in Example 12.2? Do you think the Fundamental Rule for Using Data for Inference holds in this case?

◆ Does the confidence interval in Example 12.2 tell us that *95% of the students* watch television between 1.842 and 2.338 hours per day? If not, what exactly does the interval tell us?*

The first of the next two examples involves finding an approximate 95% confidence interval for the difference between two population proportions, while the second involves finding an approximate 95% confidence interval for the difference between two population means. When considering a confidence interval for a difference between two parameters, it is informative to see whether the interval covers the value 0 or not. If a confidence interval for a difference between two parameters covers (includes 0), it is possible the two population values are equal because 0 is a plausible value for the difference between them. When a confidence interval for a difference between two parameters does not cover 0, we have evidence of a difference between the population parameters because 0 (no difference) is an implausible estimate of the difference between them.

VIDEO

**EXAMPLE 12.3 —
CONTINUED
Estimating the
Difference Between Nicotine
Patches and Antidepressants**

The sample difference $\hat{p}_1 - \hat{p}_2 = .135$ estimates the population parameter $p_1 - p_2$, the difference in the proportions who would quit smoking under each method for the population of all smokers like those in this experiment. The standard error of the statistic $\hat{p}_1 - \hat{p}_2$ is s.e.$(\hat{p}_1 - \hat{p}_2) = .040$. An approximate 95% confidence interval for the difference between population proportions $p_1 - p_2$ is

Sample difference $\pm$ 2 $\times$ Standard error

.135 $\pm$ 2 $\times$ .040

.135 $\pm$ .080

.055 to .215

With 95% confidence, we can say that this interval covers the difference between the proportions who would quit smoking in a population taking Zyban and a population using the nicotine patch. Expressed as percents, this represents a difference of somewhere between about 5.5% and 21.5%, with Zyban having a higher success rate. We also note that the interval does not include the value 0, so it is reasonable to rule out the possibility that there actually may be no difference between the success rates of the two methods. ◆

EXAMPLE 12.4 —
CONTINUED
Confidence Interval
for Difference
Between Mean Weight
Losses by Diet or Exercise

Group 1 is the "diet" group and group 2 is the "exercise" group. The sample difference between the mean weight losses for the two methods is $\bar{x}_1 - \bar{x}_2 = 7.2 - 4.0 = 3.2$ kg (about 7 pounds). We found that the standard error of the statistic $\bar{x}_1 - \bar{x}_2$ is s.e.$(\bar{x}_1 - \bar{x}_2) = 0.81$. An approximate 95% confidence interval for the difference between population means is

$$\text{Sample difference} \pm 2 \times \text{Standard error}$$
$$3.2 \pm 2 \times 0.81$$
$$3.2 \pm 1.62$$
$$1.58 \text{ to } 4.82 \text{ kg}$$

We are 95% confident that the interval from 1.58 to 4.82 kg covers the increased mean weight loss for dieters compared to those who exercised. Because the interval does not cover 0, we can be fairly certain that this difference was not just for this particular sample of men but also holds for the population as well. ◆

The Confidence Level in Action

In Chapter 10, we learned that a confidence level has a relative frequency probability interpretation. If we consider all possible randomly selected samples of the same size from a population, the confidence level is the fraction or percent of these samples for which the confidence interval includes the population parameter. The **ConfidenceLevel** applet on the CD for the text lets you see this concept in action.

What Happens

The applet simulates choosing random samples of $n = 16$ from a normally distributed population of measurements with mean $\mu = 170$ and standard deviation $\sigma = 20$, roughly the characteristics of weights of men aged 18 to 24 years old. For each sample, a confidence interval for the population mean is found and graphed. Figure 12.2 is an example of how the display looks when the applet is

FIGURE 12.2   **Example of the initial ConfidenceLevel applet display**

first started. Notice that the initial confidence level is 68% for the procedure being used to estimate the population mean. The red vertical line in the display is at the value of the population mean, $\mu = 170$. In Figure 12.2, the confidence interval shown, 167.7 to 177.7, covers the population mean.

Use the **sample!** button to generate a new sample. Figure 12.3 is an example of how the display might appear after this button is clicked. Notice that the confidence interval found from this new sample also covers the population mean (located at the red line), and at the top of the display the information "1 out of 1 (100%) captures" is given. For the first new sample generated, the confidence interval captured the population mean, so the procedure has been successful 100% of the time thus far in the simulation.

FIGURE 12.3   The ConfidenceLevel applet display after one new sample has been generated

The applet continues to tally how often the confidence interval captures the population value (170) as new samples are generated. The **sample!** button is used to generate one new sample, whereas using the **animate!** button causes the applet to repeatedly generate samples until **stop!** is clicked. (The **stop!** button appears after **animate!** is clicked.)

Figure 12.4 shows how the display looked after 103 different random samples had been generated. At the top of the display, we see that 70.87% of the 103 samples gave confidence intervals that captured the population parameter. This is close to what we would expect for the 68% confidence level. Over many different samples, the confidence interval procedure captures the population mean

FIGURE 12.4   The ConfidenceLevel applet display after 103 new samples have been generated

about 68% of the time. Notice, by the way, that the interval shown in Figure 12.4 did not cover the red line at 170, so that interval does not capture the population mean.

What to Do

Carry out the process just described. Keep the confidence level set at 68% and generate several new samples. Take note of how often the confidence intervals determined from these samples cover the population mean. Eventually, after you generate many new samples, you will see that about 68% of the samples you generate will capture the population mean (the red line). Change the confidence level by using the scrollbar at the bottom of the display. When this is done, the applet will automatically reset the tally of how often the procedure has captured the parameter. Using this new confidence level, generate many samples and take note of how often the samples give intervals that capture 170, the population mean. Over many samples, the percent shown at the top of the display will be close to the confidence level for the procedure. Examine what happens when the confidence level is set at 99%. You'll see that the confidence interval will capture the population mean (170) for nearly every sample generated.

It's informative to watch what happens to the displayed confidence intervals when the scrollbar is used to change the confidence level. You'll see an interval become wider as the confidence level is increased, and become narrower as the confidence level is decreased. Figure 12.5 shows the 68% and the 99% confidence intervals for the same random sample. The interval for the 99% level is much wider than the interval for the 68% level. Notice that the 99% confidence interval has captured the parameter, but the 68% confidence interval has not. Generally, the wider interval is more likely to capture the parameter value.

FIGURE 12.5   **The 68% and 99% confidence intervals for the population mean found using the same random sample**

Lessons Learned

The applet illustrates the idea that if we consider many different randomly selected samples of the same size from a population, the confidence level is approximately the fraction or percent of these samples for which the confidence interval includes the population parameter. In other words, the confidence level expresses our confidence that the procedure works in the long run. Also, you will see that for the same random sample, the higher the confidence level, the wider the interval. These principles apply to all confidence interval procedures, regardless of the parameter being estimated.

12.4 | GENERAL CONFIDENCE INTERVALS FOR ONE MEAN OR PAIRED DATA

In this section, we describe how to determine a confidence interval for a population mean using a sample of any size, large or small, and with any confidence level. We may, for instance, want to use a small sample of only $n = 9$ forearm lengths to create a 95% confidence interval for the mean forearm length in a population. Or, we may want to use a 98% confidence level rather than a 95% level.

A **confidence interval estimate of the mean of a population** is sometimes called a **t-interval** because the calculations involve the probability distribution called the *Student's t-distribution,* which was covered in Chapter 9. As a starting point for developing the t-interval, we can use the general format of a confidence interval:

$$\text{Sample estimate} \pm \text{Multiplier} \times \text{Standard error}$$

When the parameter is the mean of a population, the *sample estimate* is $\bar{x}$ = the sample mean for the variable and the *standard error* is s.e.$(\bar{x}) = s/\sqrt{n}$. The **multiplier** is denoted as t^*, and it is determined using the Student's t-distribution. After introducing an example, we describe how to find the value of t^*.

EXAMPLE 12.5
The Mean Forearm Length of Men

Suppose that the forearm lengths (cm) for a randomly selected sample of $n = 9$ men are

25.5, 24.0, 26.5, 25.5, 28.0, 27.0, 23.0, 25.0, 25.0

The sample mean is $\bar{x}$ = 25.5 cm and the sample standard deviation is s = 1.52 cm. The standard error of the mean is s.e.$(\bar{x}) = s/\sqrt{n} = 1.52/\sqrt{9} = 0.507$. A 95% confidence interval for the mean in the population is sample mean $\pm$ multiplier $\times$ standard error, which is $25.5 \pm t^* \times 0.507$. ◆

Determining the t^* Multiplier

Remember that theoretically the t-distribution arises when a standardized statistic is calculated using the sample standard deviation in place of the population standard deviation. This almost always is what happens when estimating means, because the population standard deviation σ is rarely known.

A *t*-distribution has a bell shape, it's centered at 0, and it is more spread out than the standard normal curve in that there's more probability in the extreme areas than there is for the standard normal curve. This additional variability results from using *s,* which differs for each sample from the same population, instead of *σ,* which is a fixed constant for a given population of measurements.

A parameter called **degrees of freedom,** abbreviated as **df,** is associated with any *t*-distribution. For problems involving a *single mean,* df = $n - 1$, where *n* is the sample size. As df gets large, the *t*-distribution gets closer to the standard normal curve because the sample standard deviation *s* becomes a more accurate estimate of the population standard deviation *σ*. The *t*-distribution with df = infinity is identical to the standard normal curve.

> *definition* For a confidence interval for a population mean, the ***t** multiplier** is the value in a *t*-distribution with df = $n - 1$ such that the area between $-t^*$ and $+t^*$ equals the desired confidence level.

Most statistical software packages have a facility for determining a confidence interval for a mean. If you are forced to do confidence interval calculations "by hand," it's relatively simple to complete the calculations. To find the multiplier for the desired confidence level, all you need is a table of the *t*-distribution (or the right calculator). Table A.2 in the Appendix shows t^* multipliers for four different confidence levels and varying degrees of freedom. To use this table, compute the degrees of freedom, choose a confidence level, and then look in the row and column for these values in the table.

EXAMPLE 12.5—*CONTINUED*
Calculating the 95%
Confidence Interval
for Forearm Length

We determined that a confidence interval for mean forearm length (cm) is $25.5 \pm t^* \times 0.507$. For the multiplier t^*, the degrees of freedom are df = $n - 1 = 9 - 1 = 8$. In Table A.2, we find that the appropriate multiplier is $t^* = 2.31$, so the 95% confidence interval is $25.5 \pm 2.31 \times 0.507$, which is 25.5 ± 1.17, or 24.33 cm to 26.67 cm. We can say with 95% confidence that in the population represented by the sample, the mean forearm length is between 24.33 and 26.67 cm. ◆

Conditions Required for Using the *t* Confidence Interval

The methods presented in this section are derived mathematically by assuming that the sample has been randomly selected from a population in which the response variable has a normal distribution. The *t*-interval procedure, however, is a *robust* procedure because it works well over a wide range of situations. Put another way, the stated confidence level for a *t*-interval is approximately correct in many situations where the assumption about a normally distributed response variable is not correct.

For large sample sizes, a *t*-interval is a valid estimate of the population mean, even in the presence of skewness. For small sample sizes, a *t*-interval can be used if the data are not skewed and contain no outliers.

Two Situations for Which a
t Confidence Interval for One Mean Is Valid

Situation 1: The population of the measurements is bell-shaped and a random sample of any size is measured. In practice, for small samples, the

FIGURE 12.6   **Dotplot of the forearm lengths for Example 12.5**

data should show no extreme skewness and should not contain any outliers.

Situation 2: The population of measurements is not bell-shaped, but a *large* random sample is measured. A somewhat arbitrary definition of a "large" sample is $n \geq 30$, but if there are extreme outliers, it is better to have a larger sample.

Before calculating a confidence interval for a mean, first check that one of the situations just described holds. To look for outliers or skewness, plot the data using either a histogram, boxplot, dotplot, or stemplot. The sample mean and median can also be compared to each other. Differences between the mean and the median usually occur if the data are skewed—that is, contain outliers in one direction. If $n \geq 30$, the procedure in this section usually produces a valid confidence interval estimate of the population mean, but if the data are obviously skewed, you should question whether the median might be a better measure of location than the mean.

EXAMPLE 12.5—CONTINUED
Checking the Necessary Conditions for the Forearm Length Confidence Interval

Figure 12.6 is a dotplot of the nine forearm lengths used to compute the confidence interval for the population mean. There is no obvious skewness and there are no outliers. Assuming the sample can be thought of as a random sample, Situation 1 holds and the necessary conditions for computing a confidence interval for the mean are present. ◆

in summary   Calculating a Confidence Interval for a Population Mean

A confidence interval for a population mean μ is

$$\bar{x} \pm t^* \times \text{s.e.}(\bar{x}), \quad \text{which is } \bar{x} \pm t^* \frac{s}{\sqrt{n}}$$

To determine this interval (also called a ***t*-interval**),

1. Make sure the appropriate conditions apply.
2. Choose a confidence level.
3. Determine the sample mean and standard deviation ($\bar{x}$ and s).
4. Calculate the standard error of the mean. $\text{s.e.}(\bar{x}) = \dfrac{s}{\sqrt{n}}$.
5. Calculate df $= n - 1$.
6. Use Table A.2 (or statistical software) to find the multiplier t^*.

EXAMPLE 12.6
What Type of
Students Sleep More?

On the second day of classes in Spring 2000, a Monday, students in two statistics classes at the University of California at Davis were given a survey. One of the questions asked was, "How many hours of sleep did you get last night, to the nearest half hour?" One class was Statistics 10, a statistical literacy course for liberal arts majors ($n = 25$), while the other was Statistics 13, a large introductory class in statistical methods for more technical majors ($n = 148$). Here is Minitab output that summarizes the information on this question for the two classes:

class	N	Mean	StDev	SE Mean
Stat 10	25	7.66	1.34	0.27
Stat 13	148	6.81	1.73	0.14

Notice that Minitab has computed the standard errors. For instance, for the Statistics 10 class, s.e.$(\bar{x}) = s/\sqrt{n} = 1.34/\sqrt{25} = 0.27$. The standard deviation for the Statistics 13 class is actually larger than for the Statistics 10 class, but the standard error is much smaller because the sample size n is so much larger. The larger the sample, the more accurate the sample mean is as an estimate of the population mean.

Let's find 95% confidence intervals for the mean hours of sleep for the *populations* represented by these two classes. Notice that there are some differences between the hypothetical populations. For the Statistics 10 class, students are liberal arts majors and the course is a general education elective. The Statistics 13 students represent a wide variety of majors, and the course is required for most of them. For both classes, the night in question was a Sunday and many of the students had just returned from spring break that day (skipping the first day of classes, a Friday).

Checking the Conditions

For each group, we must make the assumption that these students are equivalent to a random sample from the population of interest. Although not strictly true, the students in these two classes are probably representative of all students who take these and similar classes over the years. Second, for the Statistics 10 class the sample size is only 25, so we must assume that the population of sleep times is approximately bell-shaped. A histogram of the sample values (not shown) indicated that this assumption is reasonable.

Calculating the Confidence Intervals

Class	n	df = n − 1	Mean	t*	s.e.(Mean)	95% Confidence Interval
Stat 10	25	24	7.66	2.06	0.27	7.10 to 8.22 hours
Stat 13	148	147	6.81	1.98	0.14	6.53 to 7.09 hours

Interpreting the Confidence Intervals

Each interval indicates the range of values that we think covers the true *average* number of hours students in each population sleep. For instance, we can say with 95% confidence that the average hours of sleep for all students similar to those in Statistics 10 is covered by the interval from 7.10 to 8.22 hours. Notice that the interval is much wider for the Statistics 10 situation, because the sample size was much smaller. Notice also that the two intervals do not overlap.

Therefore, it appears that the liberal arts students' average hours of sleep really is higher than the general students' average hours of sleep. ◆

Paired Data: Confidence Interval for the Mean Difference in Paired Variables

Remember that an important special case of a single mean of a population occurs when two quantitative variables are collected in pairs and we desire information about the *difference* between the two variables. For example, we might want to know the average difference between Verbal SAT scores for individuals before and after taking a training course. Or we may want to know the average difference in manual dexterity for the dominant and nondominant hands. In both of these examples, we would collect two measurements from each individual in the sample.

To analyze paired data, we begin by calculating the difference in the two measurements for each pair in the sample. As notation for an analysis of the differences, we could simply use the notation already developed for a single mean, but it is preferable to use notation emphasizing that the data consist of differences. Here is notation that can be used for paired data:

Notation for Paired Differences

Data: two variables for each of n individuals or pairs; use the difference $d = x_1 - x_2$.

Population parameter: μ_d = mean of differences for the population = $\mu_1 - \mu_2$.

Sample estimate: $\bar{d}$ = sample mean of the differences = $\bar{x}_1 - \bar{x}_2$.

Standard deviation and standard error: s_d = standard deviation of the sample of differences; s.e.$(\bar{d}) = s_d/\sqrt{n}$.

Confidence interval for μ_d: $\bar{d} \pm t^* \times$ s.e.$(\bar{d})$, where df = $n - 1$ for the multiplier t^*.

EXAMPLE 12.7
Screen Time— Computer versus TV

The 25 students in a liberal arts course in statistical literacy were given a survey that included questions on how many hours per week they watched television and how many hours a week they used a computer. The responses are shown in Table 12.1, along with the difference d for each student. From these data, let's

TABLE 12.1 ■ **Data on Weekly Hours of Computer Use and TV Viewing**

Student	Computer	TV	Difference	Student	Computer	TV	Difference
1	30	2.0	28.0	14	5	6.0	−1.0
2	20	1.5	18.5	15	8	20.0	−12.0
3	10	14.0	−4.0	16	30	20.0	10.0
4	10	2.0	8.0	17	40	35.0	5.0
5	10	6.0	4.0	18	15	15.0	0.0
6	0	20.0	−20.0	19	40	5.0	35.0
7	35	14.0	21.0	20	3	13.5	−10.5
8	20	1.0	19.0	21	21	35.0	−14.0
9	2	14.0	−12.0	22	2	1.0	1.0
10	5	10.0	−5.0	23	9	4.0	5.0
11	10	15.0	−5.0	24	14	0.0	14.0
12	4	2.0	2.0	25	21	14.0	7.0
13	50	10.0	40.0				

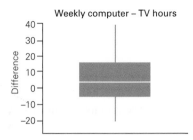

FIGURE 12.7 **Boxplot for difference between computer and TV hours per week, $n = 25$**

construct a 90% confidence interval for μ_d, the average *difference* in hours spent using the computer versus watching television (Computer use − TV) for the population of students represented by this sample.

First, check that the conditions for a confidence interval for one mean hold for the differences. The sample size is under 30, so it is important to make sure there are no major outliers or extreme skewness. The sample mean and median are 5.36 and 4.0 hours, respectively. In data ranging from −20 to +40, these statistics are close enough to rule out extreme skewness. The boxplot of the differences in Figure 12.7 provides further evidence that the appropriate conditions are satisfied.

The standard deviation of the differences is 15.24 hours, and with $n = 25$, the standard error is $15.24/\sqrt{25} = 3.05$. The df $= 25 - 1 = 24$, so, from Table A.2, for a 90% confidence interval $t^* = 1.71$. The computation for a 90% confidence interval is

$$\bar{d} \pm t^* \times \text{s.e.}(\bar{d})$$
$$5.36 \pm (1.71)(3.05)$$
$$5.36 \pm 5.22 \text{ or } 0.14 \text{ to } 10.58 \text{ hours}$$

Interpretation: We are 90% confident that the average difference between computer usage and television viewing for students represented by this sample is covered by the interval from 0.14 to 10.58 hours per week, with more hours spent on computer usage than on television viewing. ◆

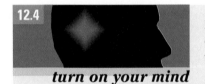

12.4

turn on your mind

For a fixed sample, explain why it is logical that a 95% confidence interval covers a wider range of values than a 90% confidence interval. Explain this in terms of our confidence that the procedure works in any given case.*

Minitab Tip

Minitab v. 14 update on CD

tech note **Computing a Confidence Interval for a Mean**

◆ To compute a confidence interval for a mean, use **Stat**>**Basic Statistics**>**1-sample t.** Specify the column that contains the raw data. To change the confidence level, use the ***Options*** button. The default confidence level is 95%.

◆ For *paired data,* first calculate a column of differences using **Calc**>**Calculator,** then use the *1-sample t* procedure.

◆ Alternatively, for *paired data,* use **Stat**>**Basic Stats**>**Paired t.** Specify the two columns that contain the raw data for the pair of measurements.

Note: Use the ***Graphs*** button to create visual displays of the data.

12.5 | GENERAL CONFIDENCE INTERVAL FOR THE DIFFERENCE BETWEEN TWO MEANS (INDEPENDENT SAMPLES)

As is the case in confidence intervals for a single mean, the t-distribution is used to determine the multiplier in the confidence interval for the difference in means for independent samples. The general format of a confidence interval for the difference in two means is

Difference in sample means $\pm\ t^* \times$ Standard error

The *standard error* of the difference in sample means is

$$\text{s.e.}(\bar{x}_1 - \bar{x}_2) = \sqrt{\frac{s_1^2}{n_1} + \frac{s_2^2}{n_2}}$$

Unfortunately, there's a muddy mathematical story underneath the calculation of a confidence interval for the difference between two population means. On the surface, however, the story appears to be easy and can be summarized as follows.

Confidence Interval for the Difference Between Two Means (Independent Samples)

formula   An approximate **confidence interval for $\mu_1 - \mu_2$** is

$$\bar{x}_1 - \bar{x}_2 \pm t^* \sqrt{\frac{s_1^2}{n_1} + \frac{s_2^2}{n_2}}$$

The multiplier t^* is a t-value such that the area between $-t^*$ and $+t^*$ in the appropriate t-distribution equals the desired confidence level. Appropriate degrees of freedom depend on whether we assume the two population variances are similar.

The difficulty is that for this confidence interval procedure, it's not exactly mathematically correct to use a t-distribution to determine the multiplier. It is approximately correct to do so, but the approximation involves a complicated formula for the degrees of freedom. That approximation formula (often called **Welch's approximation**) is

$$\text{df} \approx \frac{\left(\frac{s_1^2}{n_1} + \frac{s_2^2}{n_2}\right)^2}{\frac{1}{n_1 - 1}\left(\frac{s_1^2}{n_1}\right)^2 + \frac{1}{n_2 - 1}\left(\frac{s_2^2}{n_2}\right)^2}$$

Statistical software, such as Minitab, will determine this quantity and will also automate the process of finding the multiplier t^*. If software is not available, a conservative "by-hand" approach is to use the lesser of $n_1 - 1$ and $n_2 - 1$ for the degrees of freedom.

Equal Variances and Pooled Standard Error

tech note   If we are willing to make the assumption that the two populations have the same (or similar) variance, then there is a procedure for which the t^* multiplier *is* mathematically correct. It involves "pooling" both samples to estimate the common variance. The procedure is discussed at the end of this section.

Two Situations for Which a
t Confidence Interval for Two Means Is Valid

An important condition required for the formulas in this section to be valid is that the samples must be *independent* of each other. In addition, one of two different situations must hold:

> *Situation 1:* The populations of measurements are both bell-shaped, and random samples of any size are measured. In practice, for small samples the observed data should not show extreme skewness and there should not be any outliers.
>
> *Situation 2: Large* random samples are measured. A somewhat arbitrary definition of a "large" sample is $n \geq 30$, but if there are extreme outliers or extreme skewness in the measurements, it is better to have an even larger sample.

Note: For a confidence interval for the difference in two means, one of these situations must hold for both groups.

EXAMPLE 12.8
The Effect of a Stare on Driving Behavior

Social psychologists at the University of California at Berkeley wanted to study the effect that staring at drivers would have on driver behavior (Ellsworth, Carlsmith, and Henson, 1972). In a randomized experiment, the researchers either stared or did not stare at the drivers of automobiles stopped at a campus stop sign. The researchers timed how long it took each driver to proceed from the stop sign to a mark on the other side of the intersection. Suppose that the crossing times, in seconds, were

> No Stare Group ($n = 14$): 8.3, 5.5, 6.0, 8.1, 8.8, 7.5, 7.8, 7.1, 5.7, 6.5, 4.7, 6.9, 5.2, 4.7
> Stare Group ($n = 13$): 5.6, 5.0, 5.7, 6.3, 6.5, 5.8, 4.5, 6.1, 4.8, 4.9, 4.5, 7.2, 5.8

Based on these data, let's determine a 95% confidence interval for $\mu_1 - \mu_2$, the difference between the mean crossing times in the populations represented by these two independent samples.

Checking the Conditions and Preliminary Graphical Analysis

The two samples are *independent* because different drivers were measured in the two different experimental conditions and measurements are not paired in any way. Boxplots of the data (see Figure 12.8) indicate that there are no outliers

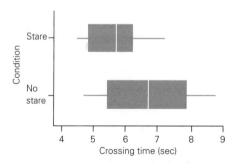

FIGURE 12.8   **Boxplots of the intersection crossing time data for Example 12.8**

Two-sample T for CrossTime

Group	N	Mean	StDev	SE Mean
NoStare	14	6.63	1.36	0.36
Stare	13	5.59	0.822	0.23

95% CI for mu (NoStare) − mu (Stare): (0.14, 1.93)
T-Test mu (NoStare) = mu (Stare) (vs not =): T = 2.41   P = 0.025   DF = 21

FIGURE 12.9   **Minitab output for Example 12.8**

and that the shape of the data is more or less symmetric within each group. The boxplots also illustrate that the crossing times in the stare sample were generally faster than crossing times in the no stare sample. It also looks like the data within the stare group are less variable than the data within the no stare group, so it is probably not reasonable to assume that the populations have equal variances.

Calculating the Confidence Interval

Figure 12.9 displays Minitab output that includes a 95% confidence interval for the difference between population means. This output was produced by the "two-sample t" procedure, and it includes descriptive statistics for each group, a 95% confidence interval for the difference between means, and a significance test concerning the population means. We'll examine this significance test in Chapter 13.

For the no stare data, the sample mean is $\bar{x}_1 = 6.63$ seconds; for the stare data, $\bar{x}_2 = 5.59$ seconds; and the difference between the sample means is $6.63 - 5.59 = 1.04$ seconds. The 95% confidence interval for the difference between population means, $\mu_1 - \mu_2$, is 0.14 to 1.93 seconds.

We see from the output that df = 21 for this problem. Minitab calculated this quantity using the approximation formula for degrees of freedom. Table A.2 can be used to determine that when df = 21, $t^* = 2.08$ at the 95% level. Minitab calculated the confidence interval reported on the output as

Sample estimate $\pm$ Multiplier $\times$ Standard error

$$\bar{x}_1 - \bar{x}_2 \pm 2.08 \times \sqrt{\frac{s_1^2}{n_1} + \frac{s_2^2}{n_2}}$$

$$(6.63 - 5.59) \pm 2.08 \times \sqrt{\frac{1.36^2}{14} + \frac{0.822^2}{13}}$$

0.14 to 1.93 seconds

(Actually, the numbers given produce a lower endpoint of 0.15 instead of 0.14. Minitab obtained 0.14 by using more decimal places in its calculations.) If the approximate df provided by Minitab had not been available, we could have used the "conservative" degrees of freedom, the minimum of the two values $n_1 - 1$ and $n_2 - 1$, which is $13 - 1 = 12$. With df = 12, $t^* = 2.18$ (see Table A.2) and the resulting interval is 0.11 to 1.97 seconds, an interval only slightly wider than the interval provided by Minitab. ◆

12.5

turn on your mind

In Section 12.4 we learned how to find a confidence interval for the mean of paired differences, which we used in Example 12.7 to estimate the mean difference in weekly computer and TV hours for a population of liberal arts students.

◆ Explain why it would not have been appropriate to use the methods in this section, for two independent samples, to estimate that mean difference, even though in either case the sample estimate is the difference in sample means, 5.36 hours.

◆ If the methods in this section had been erroneously used, by treating computer usage and TV viewing hours as independent samples, do you think the standard error of the sample estimate would have been larger or smaller than it was in Example 12.7? Explain your answer using common sense, not formulas. Think about how much natural variability there would be in the data for two independent samples, compared with measuring both sets of hours on the same individuals.*

The Equal Variance Assumption and the Pooled Standard Error

When estimating the difference between two population means, it may sometimes be reasonable to assume that the populations have equal standard deviations. The term *variance* describes the squared standard deviation, so the assumption of equal standard deviations for the two populations is the same as assuming the variances are equal. Using statistical notation, we can express the assumption of equal population variances as $\sigma_1^2 = \sigma_2^2 = \sigma^2$, where σ^2 represents the common value of the variance. With this assumption, information from both groups is combined in order to estimate the value of σ^2. The estimate of variance based on the combined or "pooled" data is called the **pooled variance.** The square root of the pooled variance is called the **pooled standard deviation,** and it is computed as

$$\text{Pooled standard deviation} = s_p = \sqrt{\frac{(n_1 - 1)s_1^2 + (n_2 - 1)s_2^2}{n_1 + n_2 - 2}}$$

> **tech note**   Notice that the pooled variance is a "weighted" average of the individual sample variances. The weights are $(n_1 - 1)/(n_1 + n_2 - 2)$ and $(n_2 - 1)/(n_1 + n_2 - 2)$, giving the variance of the larger sample more weight. If $n_1 = n_2$, then the weights are equal, and $s_p^2 = (s_1^2 + s_2^2)/2$.

*HINT: Remember that the standard error measures how much the sample estimate is likely to vary from sample to sample. Do you think the paired sample estimate $\bar{d}$ would be likely to vary more than or less than the difference in the two separate means?

Replacing the individual standard deviations s_1 and s_2 with the pooled version s_p in the formula for standard error leads to the **pooled standard error for the difference between two means:**

$$\text{Pooled s.e.}(\bar{x}_1 - \bar{x}_2) = \sqrt{\frac{s_p^2}{n_1} + \frac{s_p^2}{n_2}} = \sqrt{s_p^2\left(\frac{1}{n_1} + \frac{1}{n_2}\right)} = s_p\sqrt{\frac{1}{n_1} + \frac{1}{n_2}}$$

This all may seem quite complicated, but if the assumption of equal population variances is correct, this complication actually provides a cleaner mathematical solution for the determination of the multiplier t^*. In this case, the degrees of freedom are simply df $= n_1 + n_2 - 2$.

Pooled Confidence Interval for the Difference Between Two Independent Means

formula   If we assume the population variances are the same, then the confidence interval for $\mu_1 - \mu_2$, the difference between the population means, is

$$\bar{x}_1 - \bar{x}_2 \pm t^* \sqrt{s_p^2\left(\frac{1}{n_1} + \frac{1}{n_2}\right)}$$

t^* is found using a t-distribution with df $= n_1 + n_2 - 2$, and s_p is the pooled standard deviation.

Minitab Tip

tech note   In Minitab, the default option for the two-sample t procedure is the "unpooled" version in which the population variances are not assumed to be equal. To get the pooled version, check the dialog box item that says "Assume Equal Variances."

EXAMPLE 12.9
Pooled t-Interval for Difference Between Mean Female and Male Sleep Times

The introductory statistical methods class (Statistics 13) discussed in Example 12.6 consisted of 83 females and 65 males. Let's assume these students are equivalent to a random sample of all students who take introductory statistics. How much difference is there between how long female and male students represented by this sample slept the previous night? To answer that question, we'll find a 95% confidence interval for $\mu_1 - \mu_2 =$ difference in mean sleep hours for females versus males. Here is the Minitab output for the pooled procedure in which we assume equal population variances for males and females

Two-sample T for sleep [with "Assume Equal Variance" option]

Sex	N	Mean	StDev	SE Mean
Female	83	7.02	1.75	0.19
Male	65	6.55	1.68	0.21

Difference = mu (Female) − mu (Male   )
Estimate for difference: 0.461
95% CI for difference: (−0.103, 1.025)
T-Test of difference = 0 (vs not =): T-Value = 1.62 P-Value = 0.108 DF = 146
Both use Pooled StDev = 1.72

Notice that the sample standard deviations are very similar (1.75 for females and 1.68 for males), so it may be reasonable to assume that the population variances are similar as well. The sample mean given for females is $\bar{x}_1 = 7.02$ hours, the sample mean given for males is 6.55 hours, and the difference between these two sample means is reported on the output as "Estimate for difference: 0.461 hours." Notice that the sample means were rounded to two decimal places in the output, and that causes an inconsistency because the difference between 7.02 and 6.55 is 0.47. The value 0.461 is, however, the correct difference between the sample means.

The 95% confidence interval for the difference in mean hours of sleep for the populations of female and male students is −0.103 to 1.025 hours (in bold and underlined in the output display). Because the interval covers 0, we can't rule out the possibility that the population means are equal for men and women, although for this sample, women slept an average of about half an hour more than the men did.

The pooled standard deviation given at the bottom of the output is $s_p = 1.72$. Although not given in the output, the pooled standard error is 0.284. These two values can be found as follows:

$$s_p = \sqrt{\frac{(83-1)(1.75)^2 + (65-1)(1.68)^2}{83+65-2}} = \sqrt{2.957} = 1.72$$

$$\text{Pooled s.e.}(\bar{x}_1 - \bar{x}_2) = \sqrt{\frac{(1.72)^2}{84} + \frac{(1.72)^2}{65}} = 1.72\sqrt{\frac{1}{84} + \frac{1}{65}} = 0.284$$

In the pooled procedure, the degrees of freedom are found as df $= n_1 + n_2 - 2$, which in this example equals $83 + 65 - 2 = 146$. If Table A.2 is used, a conservative estimate of the multiplier t^* is 1.98 (for df $= 100$ and 0.95 confidence). The confidence limits found by Minitab can be verified by calculating $0.461 \pm (1.98)(0.284)$. ◆

Pooled or Unpooled?

In Example 12.9, the sample standard deviations for females and males had about the same values, so it was reasonable to use the assumption that the population standard deviations were equal. The confidence interval for the difference in means, however, would have been about the same even if the assumption of equal standard deviations had not been made. With the unpooled procedure, the 95% confidence interval for the difference in population means is −0.10 to 1.03 hours, quite close to the pooled version. One advantage for the pooled version is that finding the value of the degrees of freedom is simpler.

Sample standard deviations in two independent samples will almost never be identical in practice. So, how do we know when it might be reasonable to use the pooled version of a confidence interval for a difference between two population means? And, what is the risk of using the pooled procedure when, in truth, the population standard deviations differ? We will explore that question in detail when we consider hypothesis testing for the difference between two means in Chapter 13, but here we give some preliminary guidance.

If the larger value of the two sample standard deviations is from the group with the larger sample size, the pooled procedure will tend to give a wider confidence interval than the unpooled version and so would be a conservative estimate of the true difference, as the next example illustrates. Similar to when we used the conservative margin of error in a confidence interval for one proportion, it is acceptable to use the more conservative pooled procedure. But, it is not good practice to knowingly create an interval that is wider than necessary. On the other hand, if the smaller of the two sample standard deviations is from the group with the larger sample size, the pooled version of the procedure may produce a misleading narrow interval. Generally, it's best to use the unpooled procedure unless the sample standard deviations are quite similar.

EXAMPLE 12.6 —
CONTINUED
Sleep Time with and
without the Equal
Variance Assumption

In Example 12.6 we compared confidence intervals for a single mean for the average amount of sleep for two types of statistics students. We can also find a confidence interval for the difference in mean hours of sleep for the two types of students. Here is the Minitab output for doing so, with and without the "Assume Equal Variance" option:

Two-Sample T-Test and Confidence Interval

Two-sample T for sleep [without "Assume Equal Variance" option]

class	N	Mean	StDev	SE Mean
10	25	7.66	1.34	0.27
13	148	6.81	1.73	0.14

95% CI for mu (10) − mu (13): (0.23, 1.46)
T-Test mu (10) = mu (13) (vs not =): T = 2.79   P = 0.0083   DF = 38

Two-sample T for sleep [with "Assume Equal Variance" option]

class	N	Mean	StDev	SE Mean
10	25	7.66	1.34	0.27
13	148	6.81	1.73	0.14

95% CI for mu (10) − mu (13): (0.13, 1.57)
T-Test mu (10) = mu (13) (vs not =): T = 2.33   P = 0.021   DF = 171
Both used Pooled StDev = 1.68

For this example, it is not reasonable to assume equal variances. Notice that doing so increases the width of the confidence interval. Without the assumption the interval is from 0.23 to 1.46, but with the assumption it is 0.13 to 1.57. A comparison of the standard error of the difference for the two situations illustrates why the width increased. Using the pooled estimate increases the standard error of the mean for the sample with $n = 25$.

$$\text{Unpooled s.e.}(\bar{x}_1 - \bar{x}_2) = \sqrt{\frac{(1.34)^2}{25} + \frac{(1.73)^2}{148}} = \sqrt{0.07 + 0.02} = 0.30$$

$$\text{Pooled s.e.}(\bar{x}_1 - \bar{x}_2) = \sqrt{\frac{(1.68)^2}{25} + \frac{(1.68)^2}{148}} = \sqrt{0.11 + 0.02} = 0.36$$

This example illustrates that it is conservative to use the pooled procedure if the two sample sizes are decidedly different and the larger sample standard deviation accompanies the larger sample size. ◆

Minitab Tip

Minitab v. 14 update on CD

tech note **Computing a Confidence Interval for the Difference Between Two Independent Means**

- To compute a confidence interval for the difference in two means, use **Stat**>**Basic Statistics**>**2-sample t.** Specify the location of the data. The raw data for the response may be in one column (*Samples*), and the raw data for group categories (*Subscripts*) may be in a second column. Or, the raw data for the two independent groups may be in two separate columns.

- To change the confidence level, use the **Options** button.

- To use the *pooled* standard error, click on **Assume Equal Variances.**

Note: Use the **Graphs** button to create a comparative dotplot or a comparative boxplot.

12.6 | THE DIFFERENCE BETWEEN TWO PROPORTIONS (INDEPENDENT SAMPLES)

The objective in this section is to form a confidence interval for the difference between two population proportions. Again, we start with the same general format that we've used throughout this chapter, which is

$$\text{Sample estimate} \pm \text{Multiplier} \times \text{Standard error}$$

The *sample estimate* is $\hat{p}_1 - \hat{p}_2$, the difference between the two sample proportions. The *standard error* is

$$\text{s.e.}(\hat{p}_1 - \hat{p}_2) = \sqrt{\frac{\hat{p}_1(1 - \hat{p}_1)}{n_1} + \frac{\hat{p}_2(1 - \hat{p}_2)}{n_2}}$$

The multiplier is denoted as z^*, and it is determined using the standard normal distribution.

Confidence Interval for the Difference Between Two Population Proportions

formula A **confidence interval for $p_1 - p_2$** is

$$\hat{p}_1 - \hat{p}_2 \pm z^* \sqrt{\frac{\hat{p}_1(1 - \hat{p}_1)}{n_1} + \frac{\hat{p}_2(1 - \hat{p}_2)}{n_2}}$$

z^* is the value of the standard normal variable with the desired confidence level as the area between $-z^*$ and z^*.

Note: Values of z^* for common confidence levels can be found in the last row of Table A.2, because the *t*-distribution with infinite degrees of freedom is equivalent to the standard normal distribution. These simply are more precise versions of the z^* multipliers provided in Table 10.1, page 340.

Conditions for a Confidence Interval for the Difference in Two Proportions

As is the case for all statistical inference procedures, there are conditions that should be present in order to use the confidence interval just described. For a confidence interval for the difference between two proportions, the principal conditions have to do with the sample sizes for the observed samples. Both conditions must hold.

> *Condition 1:* Sample proportions are available based on independent, randomly selected samples from the two populations.
>
> *Condition 2:* All of the quantities $n_1\hat{p}_1$, $n_1(1 - \hat{p}_1)$, $n_2\hat{p}_2$, and $n_2(1 - \hat{p}_2)$ are at least 5 and preferably at least 10. These quantities represent the counts observed in the category of interest and not in that category, respectively, for the two samples.

EXAMPLE 12.10
Snoring and Heart Attacks

P. G. Norton and E. V. Dunn conducted a study to determine whether there is a relationship between snoring and risk of heart disease (Norton and Dunn, 1985). They found that of the 1105 snorers they sampled, 86 had heart disease, while only 24 of the 1379 nonsnorers had heart disease. We will define population 1 to be snorers and population 2 to be nonsnorers. In each case, p is the (unknown) population proportion with heart disease.

How much difference is there between the proportions with heart disease for the two populations? We can answer this question with a confidence interval. First recognize that the conditions for calculating this confidence interval are satisfied. Condition 1 is satisfied because the researchers collected independent samples from each population, and Condition 2 is satisfied because the relevant observed counts all are greater than 5. In the group that snores, the numbers with and without heart disease, respectively, are 86 and 1019, while in the group that doesn't snore, the numbers with and without heart disease are 24 and 1355, respectively.

The sample proportions with heart disease in the two groups and the standard error of the difference between the two sample proportions are

$$\hat{p}_1 = \frac{86}{1105} = .0778, \quad \hat{p}_2 = \frac{24}{1379} = .0174, \quad \text{and} \quad \hat{p}_1 - \hat{p}_2 = .0604$$

$$\text{s.e.}(\hat{p}_1 - \hat{p}_2) = \sqrt{\frac{\hat{p}_1(1 - \hat{p}_1)}{n_1} + \frac{\hat{p}_2(1 - \hat{p}_2)}{n_2}}$$

$$= \sqrt{\frac{.0778(.9222)}{1105} + \frac{.0174(.9826)}{1379}} = .0088$$

In general, the confidence interval estimate of the difference $p_1 - p_2$ is

$$\hat{p}_1 - \hat{p}_2 \pm z^* \text{ s.e.}(\hat{p}_1 - \hat{p}_2)$$

$$.0604 \pm z^*(.0088)$$

Based on this general statement, the following table includes confidence intervals for three different confidence levels. For each level, we show the appropriate z^* multiplier and the completed answer to $.0604 \pm z^*(.0088)$. The df $= \infty$ row of Table A.2 was used to determine the multipliers.

Confidence Level	z^*	Confidence Interval
90%	1.645	.046 to .075
95%	1.960	.043 to .078
99%	2.576	.038 to .083

As is always the case, the higher the level of confidence, the wider the interval. From these intervals, it appears that the proportion of snorers with heart disease in the population is about 4% to 8% higher than the proportion of non-snorers with heart disease.

As a final note about this example, we point out that it is also informative to compute the relative risk of heart disease for snorers compared to nonsnorers. This relative risk is the ratio of the sample proportions, which is $\hat{p}_1/\hat{p}_2 = .078/.0174 = 4.5$. In other words, we estimate that the risk of heart disease for snorers is about 4.5 times what the risk is for nonsnorers. Unfortunately, it is not straightforward to compute a confidence interval for relative risk in the population. Also, remember that on the basis of an observational study such as this, we can only conclude that there is a relationship between snoring and heart attacks, not a cause-and-effect influence. ◆

EXAMPLE 12.11
Do You Always Buckle Up When Driving?

How often do you wear a seatbelt when driving a car? This was asked in the 2001 Youth Risk Behavior Surveillance System, a nationwide biennial survey of 9th through 12th graders in the United States. Possible responses to the question about seatbelt use were as follows: never, rarely, sometimes, most times, and always. Exercise 2.15 presented data on the connection between seatbelt use and course grades usually attained in school for 12th graders who said they drive. In this example, we compare female and male 12th-grade drivers. To do so, we find a 95% confidence interval for $p_1 - p_2 =$ difference between the population proportions of 12th-grade female and male drivers who would say they always wear a seatbelt when driving. Following is the Minitab output for this confidence interval. (The data are in the **YouthRisk** dataset on the text CD.)

Test and CI for Two Proportions

Success = always

Sex	X	N	Sample p
Female	747	1443	0.517672
Male	677	1572	0.430662

Estimate for p(Female) − p(Male): 0.0870099
95% CI for p(Female) − p(Male): (0.0514590, 0.122561)
Test for p(Female) − p(Male) = 0 (vs not = 0): Z = 4.78 P-Value = 0.000

The sample proportions of females and males who said they always wear a seatbelt are given toward the top of the output under "Sample p." The values under "X" and "N" are, respectively, the number who said they always wear a seatbelt and the sample size for each sex. Rounded to three decimal places, the sample proportion for females is $\hat{p}_1 = 747/1443 = .518$ and the sample proportion for males is $\hat{p}_2 = 677/1572 = .431$. We see that females are more likely to say they always wear a seatbelt, and the difference between the sample proportions is $\hat{p}_1 - \hat{p}_2 = .518 - .431 = .087$ (8.7%). This value is given in the line of output that begins "Estimate for . . ."

The 95% confidence interval for $p_1 - p_2$ rounded to three decimal places is .051 to .123. This information is given in the next-to-last line of output. The confidence interval estimates the difference in population proportions of female and male 12th-grade drivers who would say they always wear a seatbelt when driving. As always, the 95% confidence level expresses our confidence in the procedure used to find the interval. Because the interval does not include 0, it's reasonable to conclude that there is a difference between the population proportions. Apparently, 12th-grade female drivers are more likely than 12th-grade male drivers to say they always buckle up when driving.

The last line of output gives results for a hypothesis test in which the null hypothesis is that there is no difference between the female and male population proportions. The details for that type of hypothesis test will be covered in Chapter 13. Based on your knowledge of the principles of hypothesis testing covered in Chapter 11, however, you may know how to use the results in the last line of output to make a conclusion in this situation. The *p*-value for the test, given as 0.000 at the end of the output, is small, so the null hypothesis of no difference can be rejected. This is consistent with the finding that the 95% confidence interval did not include 0, the value that would indicate no difference. ◆

Minitab Tip

tech note   **Determining the Interval for the Difference in Proportions**

◆ To compute a confidence interval for the difference in two proportions, use **Stat**>**Basic Statistics**>**2 Proportions.** See the following note for information about inputting data.

◆ To change the confidence level, use the **Options** button. The default confidence level is 95%.

Note: The raw data for the response may be in one column, and the raw data for group categories (Subscripts) may be in a second column. Or the raw data for the two independent groups may be in two separate columns. Or the data may already be summarized. If so, click **Summarized Data,** then specify the sample size and the *number* of successes for each group.

12.6

turn on your mind

An environmental group is suing a manufacturer because chemicals dumped into a nearby river may be harming fish. A sample of fish from upstream (no chemicals) is compared with a sample from downstream (chemicals), and a 95% confidence interval for the difference in proportions of healthy fish is .01 to .11 (with a higher proportion of healthy fish upstream). First, interpret this interval. The statistician for the manufacturer produces a 99% confidence interval ranging from −.01 to +.13, and tells the judge that because it includes 0, and because it has higher accompanying confidence than the other interval, we can't conclude that there is a problem. Comment.*

*HINT: Does knowing the 99% interval change the interpretation of the 95% interval? You could also turn the argument around by noting the upper end of the 99% interval.

12.7 | UNDERSTANDING ANY CONFIDENCE INTERVAL

The statistics and parameters that researchers examine are not limited to proportions, means, differences in means, or differences in proportions. Throughout this book, we've seen several other statistics, such as the median, the lower and upper quartiles, the correlation coefficient, and relative risk. For each of these statistics, there are formulas and procedures that allow us to compute confidence intervals for the corresponding population parameter. Some of those formulas and procedures, however, are quite complex.

In cases where a complicated procedure is needed to compute a confidence interval, authors of journal articles usually provide the completed intervals. Your job as a reader is to be able to interpret those intervals. The principles you have learned for understanding confidence intervals for means and proportions are directly applicable to understanding any confidence interval. As an example, let's consider the confidence intervals reported in one of our earlier case studies.

CASE STUDY 12.1
Confidence Interval for Relative Risk: Case Study 3.4 Revisited

In Case Study 3.4, we discussed a study relating heart disease to baldness in men. A parameter of interest in that study was the relative risk of heart disease based on the degree of baldness. The investigators focused on the relative risk (RR) of myocardial infarction (a heart attack) for men with baldness compared to men without any baldness. Here is how they reported some of the results:

> For mild or moderate vertex baldness, the age-adjusted RR estimates were approximately 1.3, while for extreme baldness the estimate was 3.4 (95% CI, 1.7 to 7.0). . . . For any vertex baldness (i.e., mild, moderate, and severe combined), the age-adjusted RR was 1.4 (95% CI, 1.2 to 1.9). (Lesko et al., 1993, p. 1000)

The 95% confidence intervals for age-adjusted relative risk are not simple to compute, and they don't take the form "sample estimate $\pm 2 \times$ standard error." This is evidenced by the fact that the intervals are not symmetric about the sample relative risks given. However, these intervals have the same interpretation as any other confidence interval. For instance, with 95% confidence, we can say that the ratio of risks of a heart attack in the populations of men with extreme baldness and men with no baldness is somewhere between 1.7 and 7.0. In other words, men with extreme baldness are probably somewhere between 1.7 to 7 times as likely to experience a heart attack as men of the same age who have no baldness. Of course, these results assume that the men in the study are representative of the larger population.

12.7

turn on your mind

Part of the quote in Case Study 12.1 said: "For any vertex baldness (i.e., mild, moderate, and severe combined), the age-adjusted RR was 1.4 (95% CI, 1.2 to 1.9)." Explain what is wrong with the following interpretation of this result, and write a correct interpretation:

Incorrect Interpretation: There is a .95 probability that the age-adjusted relative risk of a heart attack (for men with any vertex baldness compared to men without any) is between 1.2 and 1.9.*

CASE STUDY 12.2
Premenstrual Syndrome? Try Calcium

The front-page headline in the *Sacramento Bee* was "Study says calcium can help ease PMS" (Maugh, 1998). The article described a randomized, double-blind experiment in which women who suffered from premenstrual syndrome (PMS) were randomly assigned to take daily either a placebo or 1200 mg of calcium in the form of four Tums E-X tablets (Thys-Jacobs et al., 1998). The participants were 466 women with a history of PMS. In the experiment, 231 of these women were randomly assigned to the calcium treatment group and 235 women were in the placebo group.

The primary measure of interest was a composite score based on 17 PMS symptoms (insomnia, six mood-related symptoms, five water-retention symptoms, two involving food cravings, and three related to pain). Participants were asked to rate each of the 17 symptoms daily on a scale from 0 (absent) to 3 (severe). The actual "symptom complex score" was the mean rating for the 17 symptoms. Thus, a score of 0 would imply that all symptoms were absent, and a score of 3 would indicate that all symptoms were severe. The original article (Thys-Jacobs et al., 1998) presents results individually for each of the 17 symptoms in addition to the composite score.

The treatments were continued for three menstrual cycles. We report the symptom scores for the baseline premenstrual time (7 days) before treatments began and for the premenstrual time before the third cycle. Table 12.2 shows some results given in the journal article, including the mean symptom complex scores $\pm$ 1 standard deviation and the sample sizes. We see that the mean symptom scores for the two treatment groups were about equal at the baseline time but differed more noticeably before the third cycle. Notice also that the sample sizes for the third cycle are slightly reduced because patients dropped out of the study.

One interesting outcome is that the mean symptom score was substantially reduced for both the placebo and the calcium-treated groups. For the placebo group, the mean symptom score dropped by about a third; and for the calcium-treated group, the mean was more than cut in half. The purpose of the experiment is to see if taking calcium diminishes symptom severity. But we see that placebo alone can be responsible for reducing symptoms, so the appropriate comparison is between the placebo and calcium-treated groups. We should compare those two groups to each other rather than simply examine the reduction in scores for only the calcium-treated group.

We can use the results in Table 12.2 to compute a confidence interval for $\mu_1 - \mu_2$, the difference between the placebo and calcium mean symptom scores before the third cycle for the entire population of PMS sufferers. The difference in sample means (placebo − calcium) is $\bar{x}_1 - \bar{x}_2 = 0.60 - 0.43 = 0.17$. The standard error is

$$\text{s.e.}(\bar{x}_1 - \bar{x}_2) = \sqrt{\frac{0.52^2}{228} + \frac{0.40^2}{212}} = 0.044$$

An approximate 95% confidence interval for the difference in population means is

$$\bar{x}_1 - \bar{x}_2 \pm 2 \times \text{s.e.}(\bar{x}_1 - \bar{x}_2)$$

$$0.17 \pm 2(0.044)$$

$$0.08 \text{ to } 0.26$$

To put this in perspective, remember that the scores are averages over the 17 symptoms. Therefore, a reduction from a mean of 0.60 to a mean of 0.43 would, for instance, correspond to a reduction from $(0.6)(17) = 10.2$ or about 10 mild symptoms (rating of 1) to $(0.43)(17) = 7.31$, or just over 7 mild symptoms. In fact, examination of the full results of the study shows that for each one of the 17 symptoms, the reduction in mean severity score was greater in the calcium-treated group than in the placebo group. Because this is a randomized experiment and not an observational study, we can conclude that the calcium treatment actually *caused* the greater reduction in PMS symptoms.

TABLE 12.2   Results for Case Study 12.2

| | Symptom Complex Score: Mean $\pm$ s.d. | |
	Placebo Group	*Calcium-Treated Group*
Baseline	0.92 ± 0.55 ($n = 235$)	0.90 ± 0.52 ($n = 231$)
Third Cycle	0.60 ± 0.52 ($n = 228$)	0.43 ± 0.40 ($n = 212$)

SUMMARY OF CONFIDENCE INTERVAL PROCEDURES

The basic structure of intervals for the parameters in the following table is

Sample statistic $\pm$ Multiplier $\times$ Standard error

The table describes the specific details for each type of parameter.

	Parameter	Statistic	Standard Error	Multiplier	Example
One Mean	μ	$\bar{x}$	$\dfrac{s}{\sqrt{n}}$	t^* (see note 1)	12.5
Mean Difference in Paired Variables	μ_d	$\bar{d}$	$\dfrac{s_d}{\sqrt{n}}$	t^* (see note 1)	12.7
Difference Between Means (unpooled)	$\mu_1 - \mu_2$	$\bar{x}_1 - \bar{x}_2$	$\sqrt{\dfrac{s_1^2}{n_1} + \dfrac{s_2^2}{n_2}}$	t^* (see note 2)	12.8
Difference Between Means (pooled)	$\mu_1 - \mu_2$	$\bar{x}_1 - \bar{x}_2$	$s_p\sqrt{\dfrac{1}{n_1} + \dfrac{1}{n_2}}$	t^* (see note 3)	12.9
One Proportion	p	$\hat{p}$	$\sqrt{\dfrac{\hat{p}(1-\hat{p})}{n}}$	z^* (see note 4)	10.6
Difference Between Proportions	$p_1 - p_2$	$\hat{p}_1 - \hat{p}_2$	$\sqrt{\dfrac{\hat{p}_1(1-\hat{p}_1)}{n_1} + \dfrac{\hat{p}_2(1-\hat{p}_2)}{n_2}}$	z^* (see note 4)	12.10

[1] Use Table A.2, df $= n - 1$.
[2] Use Table A.2,

$$df \approx \frac{\left(\dfrac{s_1^2}{n_1} + \dfrac{s_2^2}{n_2}\right)^2}{\dfrac{1}{n_1 - 1}\left(\dfrac{s_1^2}{n_1}\right)^2 + \dfrac{1}{n_2 - 1}\left(\dfrac{s_2^2}{n_2}\right)^2}$$

or use df $= \min(n_1 - 1, n_2 - 1)$.
[3] Use Table A.2, df $= n_1 + n_2 - 2$; see Section 12.5 for the formula for s_p.
[4] Use last row of Table A.2, or the z^* multipliers:

Conf. Level	z^*
0.90	1.645 or 1.65
0.95	1.960 or 2.00
0.98	2.326 or 2.33
0.99	2.576 or 2.58

KEY TERMS

EXERCISES   Blue = Basics   **Bold, Bold** = Answer in back

Section 12.1

12.1 In a random sample of 300 parents in a school district, 104 supported a controversial new education program. The purpose of the survey was to estimate the proportion of all parents in the school district who support the new program.

a. What is the research question of interest for this survey?

b. What is the population parameter in this study? What is the appropriate statistical notation (symbol) for this parameter?

c. What is the value of the sample estimate (statistic) in this study? What is the appropriate statistical notation (symbol) for this estimate?

12.2 A large Internet provider conducted a survey of its customers. One question it asked was how many e-mail messages the respondent had received the previous day. The mean number was 13.2.

a. What is the research question of interest for this survey?

b. What is the population parameter in this study? What is the appropriate statistical notation (symbol) for this parameter?

c. What is the value of the sample estimate (statistic) in this study? What is the appropriate statistical notation (symbol) for this estimate?

12.3 A medical researcher wants to estimate the difference in the proportions of women with high blood pressure for women who use oral contraceptives versus women who do not use oral contraceptives. In an observational study involving a sample of 900 women, the researcher finds that .15 (15%) of the 500 women who used oral contraceptives had high blood pressure, whereas only .10 (10%) of the 400 women who did not use oral contraceptives had high blood pressure.

a. What is the research question of interest for this study?

b. What is the population parameter in this study? What is the appropriate statistical notation for this parameter?

c. What is the value of the sample estimate (statistic) in this study? What is the appropriate statistical notation for this estimate?

12.4 A student doing a project for a statistics class wants to estimate the difference between the mean number of music CDs owned by male and female students. He surveys 50 male students and 45 female students about the number of music CDs they own. The mean number for the males is 110 and the mean number for females is 90.

a. What is the research question of interest for this survey?

b. What is the population parameter in this study? What is the appropriate statistical notation for this parameter?

c. What is the value of the sample estimate (statistic) in this study? What is the appropriate statistical notation for this estimate?

12.5 In each of the following situations, explain whether the comparison involves paired data or data from two independent samples:

a. Twenty sophomores are randomly selected from campus dormitories at a college, and twenty other sophomores are randomly selected from students who live off-campus. The mean grade point averages of the two groups are compared.

b. Thirteen individuals each take a strength test both before and again after they participate in a six-week weight-training program. The investigator wants to estimate the average gain in strength due to the training program.

c. A polling organization compares the proportions of men and women who favor a particular candidate in an upcoming election.

12.6 Refer to exercise 12.2. The study was a retrospective study because the Internet customers were asked how many e-mail messages they received the previous day. Give one disadvantage for that method and one advantage compared to asking the customers in the sample to keep track for the next day.

12.7 A study was done by randomly assigning 200 volunteers with sore throats to either drink a cup of herbal tea or use a throat lozenge to ease their pain. The percents reporting relief for the two methods were compared. What are the research question, the population parameter, and the sample estimate for this study?

12.8   A sample of 100 students at a university was asked how many hours a week they spent studying and how many they spent socializing. The difference was computed for each student. What are the research question, the population parameter, and the sample estimate for this study?

12.9   Refer to the previous exercise. Does the situation described represent independent samples or paired data? Explain.

12.10 For each of the following research questions, would it make more sense to collect independent samples or paired data? Explain.
   a. In the United States, on average what is the age difference between husbands and wives?
   b. What is the difference in average ages at which teachers and plumbers retire?
   c. What is the difference in average salaries for high school graduates and college graduates?

12.11 Give an example of a research question for which the population parameter of interest would be $p_1 - p_2$.

12.12 Give an example of a research question for which the population parameter of interest would be $\mu_1 - \mu_2$.

Sections 12.2 and 12.3

12.13 In each part, use the information given to calculate the standard error of the mean, and find an approximate 95% confidence interval for the population mean.
   a. Mean height for a sample of $n = 81$ women is $\bar{x} = 64.2$ inches, and the standard deviation is $s = 2.7$ inches.
   b. Mean systolic blood pressure for a sample of $n = 100$ men is $\bar{x} = 123.5$, and the standard deviation is $s = 9$.
   c. Mean systolic blood pressure for a sample of $n = 324$ men is $\bar{x} = 123.5$, and the standard deviation is $s = 9$.

12.14 For a sample of 36 men, the mean head circumference is 57.5 cm with a standard deviation equal to 2.4 cm. For a sample of 36 women, the mean head circumference is 55.3 cm with a standard deviation equal to 1.8 cm.
   a. Find the value of the difference between the sample means for men and women.
   b. Find the standard error of the difference between the sample means in this context.
   c. Find an approximate 95% confidence interval for the difference between population mean head circumferences for men versus women.

12.15 Suppose a randomly selected sample of $n = 64$ men has a mean foot length of $\bar{x} = 27.5$ cm, and the standard deviation of the sample is 2 cm.
   a. Calculate the standard error of the sample mean. Write one or two sentences that interpret this value.
   b. Calculate an approximate 95% confidence interval for the mean foot length of men. Write a sentence that interprets this interval.

12.16 Refer to the previous exercise about the foot lengths of men. A randomly selected sample of $n = 100$ women has a mean foot length of $\bar{x} = 24.0$ cm, and the standard deviation of the sample is 2 cm.
   a. Calculate the standard error of the mean. Explain why the value is smaller than the standard error for the men in the previous exercise.
   b. Now consider the difference between the mean foot lengths of men and women. Compute the standard error for the difference between sample means for the men and women.
   c. Compute an approximate 95% confidence interval for the difference between the mean foot lengths of men and women. Write a sentence that interprets this interval.

12.17 In a survey of college students, 70% (.70) of the 100 women surveyed said they believe in love at first sight, whereas only 40% (.40) of the 80 men surveyed said they believe in love at first sight.
   a. Find the value of the difference between the sample proportions for men and women.
   b. Find the standard error of the difference between the sample proportions.
   c. Find an approximate 95% confidence interval for the difference between population proportions believing in love at first sight for men versus women.

12.18 The pulse rates of $n = 50$ people are measured. These people then march in place for one minute, and the pulse rates are measured again. We are interested in estimating the mean difference between the two pulse rates for the population represented by this sample. Explain how you would compute the standard error for the sample mean difference. Be specific about what formula you would use.

12.19 Example 12.4 of Sections 12.2 and 12.3 described a study that compared mean weight loss for a diet plan to the mean weight loss for an exercise plan. In the study, $n = 42$ men were put on a diet. The men who dieted lost an average of 7.2 kg, with a standard deviation of 3.7 kg.
   a. Compute the standard error of the mean for the men who dieted. Write a sentence or two to explain what is measured by this value.
   b. Compute an approximate 95% confidence interval for the mean weight loss for the men who dieted. Write a sentence that interprets this interval.

12.20 Suppose that a teacher asks her class how many hours they studied for the midterm exam. For the 40 students enrolled in the class, the mean was 5.5 hours and the standard deviation was 2 hours. Explain why the teacher should not calculate a standard error for the mean and should not calculate a confidence interval for the mean.

12.21 Example 12.2 in Sections 12.2 and 12.3 used a sample of $n = 175$ students to estimate the mean daily hours that Penn State students watch television. The following summary shows results for daily hours of television by gender:

Sex	N	Mean	St Dev	SE Mean
Female	116	1.95	1.51	0.14
Male	59	2.37	1.87	0.24

   a. Calculate the difference between the mean values for the females and males. Is this a statistic or a parameter? Explain.

b. For each gender, verify the standard error of the mean by substituting appropriate values into the correct formula.

c. An approximate 95% confidence interval for the difference between means is -0.14 to $+0.97$ hours. Based on this interval, what can we conclude, if anything, about the difference in the mean daily television watching times of men and women? Justify your answer.

d. What formula was used to compute the interval given in part (c)? Substitute appropriate values into this formula.

12.22 On January 30, 1995, *Time* magazine reported the results of a poll of adult Americans in which it asked, "Have you ever driven a car when you probably had too much alcohol to drive safely?" The exact results were not given, but from the information given we can guess at what they were. Of the 300 men who answered, 189 (63%) said yes and 108 (36%) said no. The remaining 3 men weren't sure. Of the 300 women, 87 (29%) said yes and 210 (70%) said no; the remaining 3 women weren't sure.

a. Compute an approximate 95% confidence interval for the difference in proportions of men and women who would say yes to this question.

b. Write a sentence or two interpreting this interval.

12.23 Refer to the previous exercise. Calculate an approximate 95% confidence interval for the proportion of men who would say they have ever driven a car when they probably have had too much to drink, and write a sentence or two that interprets the interval. Be specific about the population that the interval describes.

Section 12.4

12.24 In each case, use the information given to compute a confidence interval for the population mean μ. Assume all necessary conditions are present for using the method described in Section 12.4.

a. $\bar{x} = 76$, $s = 6$, $n = 9$, 95% confidence level.
b. $\bar{x} = 76$, $s = 6$, $n = 9$, 90% confidence level.
c. $\bar{x} = 76$, $s = 6$, $n = 16$, 90% confidence level.
d. $\bar{x} = 76$, $s = 8$, $n = 16$, 95% confidence level.
e. $\bar{x} = 100$, $s = 8$, $n = 16$, 95% confidence level.

12.25 What three factors affect the width of a confidence interval for a population mean? For each factor, indicate how an increase in the numerical value of the factor affects the interval width.

12.26 Use Table A.2 to find the multiplier t^* for calculating a confidence interval for a single population mean in each of the following situations:

a. $n = 22$; confidence level = 0.98.
b. $n = 5$; confidence level = 0.90.
c. $n = 5$; confidence level = 0.99.
d. $n = 25$; confidence level = 0.95.
e. $n = 81$; confidence level = 0.95.

12.27 A sample of $n = 9$ men between 30 and 39 years old is asked to do as many situps as they can in one minute. The mean number is $\bar{x} = 26.2$ and the standard deviation is $s = 6$.

a. Find the value of the standard error of the sample mean. Write a sentence that interprets this value (refer back to Section 12.2 for guidance).

b. Find a 95% confidence interval for the population mean.

c. Write a sentence that interprets the confidence interval in the context of this situation.

12.28 Volunteers who had developed a cold within the previous 24 hours were randomized to take either zinc or placebo lozenges every 2 to 3 hours until their cold symptoms were gone (Prasad et al., 2000). Twenty-five participants took zinc lozenges, and 23 participants took placebo lozenges. The mean overall duration of symptoms for the zinc lozenge group was 4.5 days, and the standard deviation of overall duration of symptoms was 1.6 days. For the placebo group, the mean overall duration of symptoms was 8.1 days, and the standard deviation of symptoms was 1.8 days.

a. Calculate a 95% confidence interval for the mean overall duration of symptoms in a population of individuals like those who used the zinc lozenges.

b. Calculate a 95% confidence interval for the mean overall duration of symptoms in a population of individuals like those who used the placebo lozenges.

c. Based on the intervals computed in parts (a) and (b), is it reasonable to conclude generally that taking zinc lozenges reduces the overall duration of cold symptoms? Explain why you think this is or is not an appropriate conclusion.

d. In their paper, the researchers say they tested the null hypothesis that the data were sampled from a normal curve population and this hypothesis was not rejected. How is this relevant to the calculations done in parts (a) and (b)?

12.29 Refer to the previous exercise. Compute a 99% confidence interval for the mean duration of symptoms for individuals using the placebo lozenges. Write one or two sentences that interpret this interval.

12.30 A randomly selected sample of $n = 12$ students at a university is asked, "How much did you spend for textbooks this semester?" The responses, in dollars, are

200 175 450 300 350 250 150 200 320 370 404 250

Computer output with a 95% confidence interval for the population mean is

Variable	N	Mean	StDev	SE Mean	95.0% CI
spent	12	284.9	96.1	27.7	(223.9, 346.0)

a. Draw either a dotplot or a boxplot of the data. Briefly discuss whether the necessary conditions for doing a confidence interval for the population mean are present or not.

b. The standard error of the mean given in the output is 27.7 (under "SE Mean"). Verify this value by giving the correct formula and substituting appropriate values into this formula.

c. What are the degrees of freedom for the t^* multiplier? What is the value of the t^* multiplier for the given confidence interval?

d. Write a sentence that interprets the confidence interval. Be specific about the population that is being described.

e. Calculate a 90% confidence interval for the population mean.

12.31 Refer to the previous exercise. Use the Empirical Rule to create an interval that estimates the textbook expenses of 95% of the individual students at the university.

12.32 Suppose that a highway safety researcher is studying the design of a highway sign and is interested in the mean maximum distance at which drivers are able to read the sign. The maximum distances (in feet) at which $n = 16$ drivers can read the sign are

$$\begin{array}{cccccccc}
440 & 490 & 600 & 540 & 540 & 600 & 240 & 440 \\
360 & 600 & 490 & 400 & 490 & 540 & 440 & 490
\end{array}$$

a. Verify the necessary conditions for computing a confidence interval for the population mean distance. Assume the sample can be considered to be a random sample from the population of interest.

b. Estimate the mean maximum distance at which drivers can read the sign. Use a 95% confidence interval, and write a sentence that interprets this interval.

12.33 For a sample of $n = 20$ women aged 18 to 29, responses to the question, "How tall would you like to be?" are recorded along with actual heights. In the sample, the mean desired height is 66.7 inches, the mean actual height is 64.9 inches, and the mean difference is $\bar{d} = 1.8$ inches. The standard deviation of the differences is $s_d = 2.1$ inches.

a. Compute a 95% confidence interval for the mean difference between desired and actual height, and write a sentence that interprets this interval. Be specific about what population is described by the interval.

b. What are the necessary conditions for the validity of the confidence interval computed in part (a)?

c. Explain why the following statement is not correct: "Based on the confidence interval computed in part (a), we can conclude that all women aged 18 to 29 would like to be taller."

Section 12.5

12.34 For each of the following situations identify whether it would be appropriate to use a paired procedure (described in Section 12.4) or a two-sample procedure (described in Section 12.5) to estimate the difference of interest:

a. A researcher wants to estimate the difference between the mean scores on a memorization test for people older than 60 compared to people under 40.

b. Women give their actual and desired weights in a survey. An estimate of the mean difference is desired.

c. Two treatments for reducing cholesterol are compared. Forty people with high cholesterol use the first treatment and 40 other people with high cholesterol use the second treatment. An estimate of the difference between the mean decrease in cholesterol for treatment 1 and the mean decrease in cholesterol for treatment 2 is desired.

12.35 Suppose you were given a 95% confidence interval for the difference in two population means. What could you conclude about the population means if

a. The confidence interval did *not* cover zero?

b. The confidence interval *did* cover zero?

12.36 Data on the testosterone levels (ng/dl in saliva) of men in different professions were given in a paper published in the *Journal of Personality and Social Psychology* (Dabbs et al., 1990). The researchers suggest that there are occupational differences in mean testosterone level. Medical doctors and university professors were two of the occupational groups for which means and standard deviations were given along with a dotplot of the raw data. Computer output for a 95% confidence interval for the difference in mean testosterone levels of medical doctors and university professors is

---

Two-sample T for MD vs Profs

	N	Mean	StDev	SE Mean
MD	16	11.60	3.39	0.85
Prof	10	10.70	2.59	0.82

Difference = mu MD − mu Profs
Estimate for difference: 0.90
95% CI for difference: (−1.54, 3.34)
DF = 22

---

a. The 95% confidence interval is given as −1.54 to 3.34. The unpooled procedure was used to compute this interval. Show how the interval was computed by giving the formula and identifying numerical values for all elements of the formula. Note the degrees of freedom are given as "DF = 22."

b. Interpret the confidence interval for the difference in means. Be specific about the populations that are described. What assumptions are you making in this interpretation?

c. Based on the 95% confidence interval given, what can be concluded generally about the difference in mean testosterone levels of medical doctors and professors?

12.37 What assumption is necessary in order to use the pooled procedure for finding a confidence interval for the difference between two population means?

12.38 Each of 63 students in a statistics class used their nondominant hand to print as many letters of the alphabet,

---

Two-sample T for Exercise 12.38

Sex	N	Mean	StDev	SE Mean
Female	29	12.55	4.01	0.74
Male	34	13.65	4.46	0.77

Difference = mu (Female) − mu (Male)
Estimate for difference: −1.10
95% CI for difference: (−3.23, 1.04)
T-Test of difference = 0 (vs not =):
T-Value = −1.03 P-Value = 0.309 DF = 60

---

in order, as they could in 15 seconds. The output for this exercise gives results for a 95% confidence interval for the difference in population means for females and males. The "unpooled" procedure was used (*Data source:* **letters** dataset on the CD for the text).

a. What was the mean number of letters printed by females in the sample? What was the mean number printed by males? What is the difference between the sample means for men and women?

b. What is the 95% confidence interval given in the output for the difference between population means? Write a sentence that interprets this interval.

c. Refer to the previous part. Based on the 95% confidence interval, what conclusion can be made about whether the population means for men and women differ or not? Explain.

d. Find the value of the (unpooled) standard error of the difference between sample means.

e. Note that the value for degrees of freedom for this problem are given at the end of the output as "DF=60." Use Table A.2 to find the value of t^* that was used in finding the interval.

12.39 Refer to the output for the previous exercise about letters printed with the nondominant hand.

a. Find the value of s_p = pooled standard deviation for these data.

b. For the pooled two-sample procedure, find the (pooled) standard error of the difference in sample means.

c. What would be the degrees of freedom for the pooled procedure?

12.40 Refer to exercise 12.28 about the effect of zinc lozenges on the duration of cold symptoms. For $n = 25$ in the zinc lozenge group, the mean overall duration of symptoms was 4.5 days and the standard deviation was 1.6 days. For $n = 23$ in the placebo group, the mean overall duration of symptoms was 8.1 days and the standard deviation was 1.8 days.

a. Calculate $\bar{x}_1 - \bar{x}_2$ = difference in sample means and also compute the unpooled s.e.$(\bar{x}_1 - \bar{x}_2)$ = standard error of the difference in means.

b. Compute a 95% confidence interval for the difference in mean days of overall symptoms for the placebo and zinc lozenge treatments, and write a sentence interpreting the interval. Use the unpooled standard error and use the smaller of $n_1 - 1$ and $n_2 - 1$ as a conservative estimate of degrees of freedom.

c. Is the interval computed in part (b) evidence that the population means are different? Explain.

12.41 Refer to the previous exercise. Use the pooled procedure to compute a 95% confidence interval for the difference in mean days of symptoms for the zinc and placebo treatments.

12.42 Following are data for $n = 20$ individuals on resting pulse rate and whether the individual regularly exercises or not. Calculate a 95% confidence interval for the difference in means for the two groups defined by whether the individual regularly exercises or not. Be sure to verify necessary conditions and state any assumptions.

Person	Pulse	Regularly Exercises
1	72	No
2	62	Yes
3	72	Yes
4	84	No
5	60	Yes
6	63	Yes
7	66	No
8	72	No
9	75	Yes
10	64	Yes
11	62	No
12	84	No
13	76	No
14	60	Yes
15	52	Yes
16	60	No
17	64	Yes
18	80	Yes
19	68	Yes
20	64	Yes

Section 12.6

12.43 In each situation explain whether the method covered in Section 12.6 for finding a confidence interval for the difference in two proportions should be used or not:

a. A survey is done to estimate the difference between the proportions of college students and high school students who smoke cigarettes. Data on cigarette smoking habits are gathered from randomly selected samples of 500 college and 500 high school students.

b. Two new treatments for a serious disease are compared. Each treatment is used for five patients. The first treatment is successful for three patients of the five patients who used it. The second treatment is successful for only one of the five patients who used that treatment.

c. An economist wants to estimate the difference between the mean annual income of college graduates and the mean annual income of high school graduates who did not go to college. Income data are collected from samples of 200 individuals in each educational degree group.

12.44 In the General Social Survey, an ongoing nationwide survey done by the National Opinion Research Center at the University of Chicago, a question asked is whether a respondent favors or opposes capital punishment (the death penalty) for persons convicted of murder. The output for this exercise compares the proportions who said they were opposed to the death penalty in the year 2000 and the year 1993 (*Sources:* **GSS-93** dataset for this text and http://csa.berkeley.edu:7502/archive.htm).

Sample	X	N	Sample p
2000	801	2565	0.312
1993	337	1488	0.226

Estimate for $p(1) - p(2)$: 0.086
95% CI for $p(1) - p(2)$: (0.058, 0.114)

a. What proportion of the year 2000 sample was opposed to the death penalty? What proportion of the year 1993 sample was opposed?

b. What is the estimated difference between the proportions opposed to the death penalty in the two years?

c. Write the 95% confidence interval given in the output. Then, interpret this interval in the context of this situation.

d. Provide the formula used to calculate the interval, and substitute appropriate numerical values into the formula.

12.45 Refer to the output given for the previous exercise.

a. Find the value of s.e.$(\hat{p}_1 - \hat{p}_2)$ = standard error of the difference between the two sample proportions.

b. Find a 90% confidence interval for the difference between the population proportions opposed to the death penalty in the years 2000 and 1993. Use Example 12.10 for guidance.

12.46 In a study done in Maryland, investigators surveyed individuals by telephone about how often they get tension headaches (Schwartz et al., 1998). One response variable measured was whether or not the respondent had experienced an episodic tension-type headache (ETTH) in the prior year. A headache pattern was called "episodic" if the headaches occurred less often than 15 times a month; otherwise, the headaches were called "chronic." Of the 1600 women in the survey aged 18 to 29, 653 said they had experienced episodic tension-type headaches in the last year. Of the 2122 women in the 30–39-year-old age group, the number having experienced episodic headaches was 995.

a. Estimate the proportion in each group that experienced an episodic headache in the prior year, and compute the difference in these two proportions.

b. Compute a 95% confidence interval for the difference between the proportions for these age groups in the population. Write a sentence that interprets this confidence interval.

12.47 Refer to the previous exercise. In the sample, there were 4594 individuals with at least a college degree and $n = 7076$ individuals with at least a high school diploma (but not a college degree). Following is Minitab output with a 95% confidence interval for the difference in proportions experiencing episodic tension-type headaches in the two educational groups:

Sample	X	N	Sample p
1	2140	4594	0.466
2	2690	7076	0.380

Estimate for p(1) − p(2): 0.086
95% CI for p(1) − p(2): (0.068, 0.104)

a. Write two or three sentences that interpret the results.

b. Provide the formula used to calculate the interval, and substitute appropriate numerical values into the formula.

12.48 Suppose a randomly selected sample of $n = 900$ registered voters is surveyed in order to estimate the proportions that will vote for the two candidates in an upcoming election. Fifty-five percent of those sampled say they will vote for one candidate, while 45% indicate a preference for the other candidate. Explain why the method in Section 12.6 for computing a confidence interval for a difference in proportions should not be used to estimate the difference in the proportions planning to vote for the two candidates.

12.49 In Case Study 10.3, students in a statistics class at Penn State were asked, "Would you date someone with a great personality even though you did not find them attractive?" By gender, the results were that 61.1% of 131 women answered "yes" while 42.6% of 61 men answered "yes" (*Source:* **pennstate** dataset).

a. Calculate a 95% confidence interval for the difference in proportions of men and women who would say "yes" to the question if asked. Write a sentence that interprets the interval.

b. What populations do you think are represented by these samples of men and women?

c. Does the interval computed in part (a) include the value 0? What does this tell us about whether there is a difference between college men and women for this question?

12.50 Refer to the previous exercise.

a. Compute a 99% confidence interval for the difference in proportions.

b. How does the width of the 99% confidence interval computed in part (a) of this problem compare to the width of the 95% confidence interval computed in part (a) of the previous exercise?

c. Would a 90% confidence interval for the difference in proportions be wider or narrower than the 99% confidence interval computed in part (a)? Explain.

Section 12.7 and Case Study 12.2

12.51 Responses from $n = 204$ college students to the question "About how many CDs do you own?" are used to determine 95% confidence intervals for the median and the mean. (*Note:* The data are in the dataset **pennstate2**.)

a. A 95% confidence interval for the population median is 50 to 60. Write two or three sentences interpreting this result. Be specific about the population that the interval describes, and also describe any implied assumptions about the sample.

b. A 95% confidence interval for the mean μ is 62.7 to 82.9. Compare this result to the result in part (a), and explain why this comparison suggests that the data for music CDs owned may be skewed.

c. Sample statistics summarizing the data are $\bar{x} = 72.8$, $s = 72.2$, $n = 204$. Use this information to verify the 95% confidence interval for the mean given in part (b).

12.52 Suppose you were given a 95% confidence interval for the relative risk of disease under two different conditions. What could you conclude about the risk of disease under the two conditions if

a. The confidence interval did not cover 1.0?

b. The confidence interval did cover 1.0?

12.53 In a twelve-year study done at ten medical centers, Cole et al. (2000) investigated whether the decrease in heart rate over the first two minutes after stopping treadmill exercise was a useful predictor of subsequent death during the study period. Based on observations of $n = 5234$ participants, the investigators concluded: "Abnormal heart rate recovery predicted death (relative risk 2.58 [CI, 2.06 to 3.20]). After adjustment for standard risk factors, fitness, and resting and exercise heart rates, abnormal heart rate recovery remained predictive (adjusted relative risk, 1.55 [CI, 1.22 to 1.98])." Explain why the given confidence intervals are evidence that abnormal heart rate recovery after treadmill exercise is a predictor of subsequent death during the study period.

12.54 Refer to Case Study 12.2. Use the baseline data for the placebo group to find an approximate 95% confidence interval for the mean PMS complex score for the population of all women like the ones in this study. Write a sentence interpreting the interval.

12.55 Refer to Case Study 12.2.
a. Use the baseline data to find an approximate 95% confidence interval for the difference in means for the populations represented by the placebo and calcium groups at the start of the study.
b. Explain why the researchers would want the interval computed in part (a) to cover zero.

Chapter Exercises

Be sure to include a check of the appropriate conditions as part of your work when you compute a confidence interval.

12.56 Would it be appropriate to use the ages of death of First Ladies given in Chapter 2 (Table 2.5) to find a 95% confidence interval for the mean? If your answer is yes, what is the parameter of interest? If your answer is no, explain why not.

12.57 A sample of college students was asked if they would return the money if they found a wallet on the street. Of the 93 women, 84 said "yes," and of the 75 men, 53 said "yes." Assume these students represent all college students (*Source:* **UCDavis2** dataset).
a. Find separate approximate 95% confidence intervals for the proportions of college women and college men who would say "yes" to this question.
b. Find an approximate 95% confidence interval for the difference in the proportions of college men and women who would say "yes" to this question.
c. Write a few sentences interpreting the intervals in parts (a) and (b).

12.58 Find an exact 95% confidence interval (use t^*) for the mean weight loss after a year of exercise for the population of men represented by those in the study in Example 12.4. There were 47 men in the exercise group in the sample, with mean weight loss of 4.0 kg and standard deviation of 3.9 kg.

12.59 In computing a confidence interval for a population mean μ, explain whether the interval would be wider, more narrow, or neither as a result of each of the following changes. Assume features not mentioned (confidence level, mean, standard deviation, sample size) remain the same.

a. The level of confidence is changed from 90% to 95%.
b. The sample size is doubled.
c. A new sample of the same size is taken, and $\bar{x}$ increases by 10.

12.60 Example 2.11 reported that a random sample of 199 married British women had a mean height of 1602 mm, with standard deviation of 62.4 mm.
a. Find a 99% confidence interval for the mean height of all women represented by this sample. Write a sentence interpreting the interval.
b. Give a 99% confidence interval for the mean height *in inches* of all women represented by those in this sample (1 mm = 0.03937 in.).

12.61 Refer to the previous exercise. The sample included the heights of the women's husbands as well. For these men, the mean height was 1732.5 mm, with standard deviation 68.8 mm.
a. Find a 90% confidence interval for the mean height of the population of men represented by this sample.
b. Given the information provided in this and the previous exercise, is it possible to compute an approximate 95% confidence interval for the average difference in height between the husband and wife in British couples? If so, compute the interval. If not, explain what additional numerical information you would need.

12.62 Example 2.7 gave the weights in pounds of the 18 men on the crew teams at Oxford and Cambridge universities in 1991–92. The data were

Cambridge: 188.5, 183.0, 194.5, 185.0, 214.0, 203.5, 186.0, 178.5, 109.0
Oxford: 186.0, 184.5, 204.0, 184.5, 195.5, 202.5, 174.0, 183.0, 109.5

a. Explain why it is not appropriate to use the methods in this chapter to compute a confidence interval for the mean weight of college crew team members, even if we assume these 18 men represent a random sample of all college crew team members.
b. The first 8 values listed for Cambridge represent the weights of the 8 rowers on the team. Assuming these 8 men represent a random sample of all Cambridge crew team rowers over the years, compute a 90% confidence interval for the mean of that population.
c. Make the same assumption as in part (b) about the weights of the first 8 men in the Oxford sample. Compute a 90% confidence interval for the difference in mean weights of Cambridge and Oxford rowers. Explain whether you used the pooled or unpooled standard error, and why.
d. Write a sentence interpreting the interval in part (c).

12.63 In Stroop's Word Color Test, 100 words that are color names are shown in colors different from the word. For example, the word *red* might be displayed in blue. The task is to correctly identify the display color of each word; in the example just given the correct response would be blue. Gustafson and Kallmen (1990) recorded the time needed to complete this test for $n = 16$ individuals after they had consumed alcohol and for $n = 16$ other individuals after they had consumed a placebo drink flavored to taste as if it contained alcohol. Each group included 8 men and 8 women. In the alcohol group, the mean completion time was 113.75 seconds and the standard deviation was 22.64 seconds. In the placebo group, the mean completion time was 99.87 seconds and the standard deviation was 12.04 seconds.

a. Calculate the unpooled standard error for the difference in the two sample means.

b. Calculate a 95% confidence interval for the difference in population means. For the unpooled procedure, the approximate degrees of freedom are df = 22.

c. Based on the confidence interval computed in part (b), can we conclude that the population means for the two groups are different? Why or why not?

d. Can we verify the necessary conditions for computing the confidence interval in part (b)? If so, verify the conditions. If not, explain why it is not possible.

12.64 Example 6.2 presented data collected between 1996 and 1998 on ear pierces and tattoos for male Penn State students. Assume these men represent a random sample of male Penn State students. The results were

	Tattoo?		
Ear Pierce?	*No*	*Yes*	*Total*
No	381	43	424
Yes	99	42	141
Total	480	85	565

a. For the population of men with no ear pierce, find a 90% confidence interval for the proportion with a tattoo.

b. For the population of men with an ear pierce, find a 90% confidence interval for the proportion with a tattoo.

c. Find a 99% confidence interval for the difference in proportions of men with a tattoo for the populations with and without an ear pierce. Write a sentence interpreting the interval.

12.65 An experiment was done by 15 students in a statistics class at the University of California at Davis to see if manual dexterity was better for the dominant hand compared to the nondominant hand (left or right). Each student measured the number of beans they could place into a cup in 15 seconds, once with the dominant hand and once with the nondominant hand. The order in which the two hands were measured was randomized for each student.

a. Explain whether a comparison of beans placed using the dominant and nondominant hands is a comparison of paired data or two independent samples.

b. Referring to material earlier in the book, explain why the order of the two hands was randomized rather than, for instance, having each student test the dominant hand first.

c. The data are presented in the table below and are given also in the **beans** dataset on the CD for this text. Assuming these students represent a random sample from the population, compute a 90% confidence interval for the mean difference in the number of beans that can be placed into a cup in 15 seconds by the dominant and nondominant hands.

Student	1	2	3	4	5	6	7	8	9	10	11	12	13	14	15
Dominant	22	19	18	17	15	16	16	20	17	15	17	17	14	20	26
Nondom.	18	15	13	16	17	16	14	16	20	15	17	17	16	18	25
Difference	4	4	5	1	−2	0	2	4	−3	0	0	0	−2	2	1

d. Write a sentence or two using the interval in part (c) to address the question of whether manual dexterity is better, on average, for the dominant hand.

12.66 In a random sample of 199 married British couples (described in Chapter 2), there were 170 for which both ages were reported. The average difference in husband's age and wife's age was 2.24 years, with standard deviation of 4.1 years.

a. Find an approximate 95% confidence interval for the average age difference for all married British couples. Write a sentence interpreting the interval.

b. Were the samples of husband ages and wife ages collected as paired data or as two independent samples? Briefly explain.

12.67 A study was conducted on pregnant women and subsequent development of their children (Olds et al., 1994). One of the questions of interest was whether the IQ of children would differ for mothers who smoked at least 10 cigarettes a day during pregnancy and those who did not smoke at all. The mean IQ at age 4 for the children of the 66 nonsmokers was 113.28 points, while for the children of the 47 smokers it was 103.12 points. The standard deviations were not reported, but from other information provided, the pooled standard deviation was about 13.5 points.

a. Assuming these women and children are a representative sample, find a 95% confidence interval (use t^*) for the difference in mean IQ scores at age 4 for children of mothers who do not smoke during pregnancy and those whose mothers smoke at least 10 cigarettes a day during pregnancy. Write a sentence interpreting the interval.

b. One of the statements in the article was "after control for confounding background variables . . . the average difference observed at 36 and 48 months was reduced to 4.35 points (95% CI: 0.02, 8.68) [Olds et al., p. 224]." Explain the purpose of this statement and interpret the result given.

12.68 In Example 12.8 we compared the time in seconds to cross an intersection when someone was or was not staring at the driver. We did not assume equal population variances. Here's the Minitab (Version 12) output that resulted from clicking "Assume Equal Variance":

```
Two-sample T for CrossTime

Group     N     Mean    StDev    SE Mean
NoStare   14    6.63    1.36     0.36
Stare     13    5.592   0.822    0.23

95% CI for mu (NoStare) − mu (Stare) : (0.14, 1.94)
T-Test mu (NoStare) = mu (Stare) (vs >): T = 2.37
P = 0.013    DF = 25
Both use Pooled StDev = 1.14
```

a. What is the value of s_p?

b. Compare the 95% confidence interval with the one in Example 12.8, and discuss whether it is reasonable to use the "equal variance" assumption for this study.

12.69 Suppose 100 researchers each plan to independently gather data and construct a 90% confidence interval for a population mean. If X = the number of those intervals that actually cover the population means, then X is a binomial random variable.

a. What is a "success" for this binomial random variable?

b. What is the numerical value of p, the probability of a success?

c. What is the expected number of intervals that will cover their population means?

d. Will each researcher know whether he or she has a "successful" interval? Explain.

e. What is the probability that all 100 researchers actually get intervals that cover their population means of interest?

12.70 Fatty acids present in fish oil may be useful for treating some psychiatric disorders. An article reported September 3, 1998 at the Yahoo Health website described a randomized experiment done by a Harvard Medical School researcher in which 14 bipolar patients received fish oil daily, while 16 other patients received a placebo daily. After four months, 9 of the 14 patients who received fish oil had responded favorably, but only 3 of the 16 placebo patients had done so.

a. Calculate the difference between the proportions showing favorable response in the two groups, and also calculate a standard error for this difference.

b. Discuss whether the method described in Section 12.6 should be used to calculate a confidence interval for the difference. If you believe it should, calculate the confidence interval.

c. In Chapter 3, blind and double-blind procedures were discussed. Explain how those concepts would be applied to this experiment.

Dataset Exercises

12.71 Use the dataset **pennstate1** for this exercise. Case Study 1.1 presented data given by college students in response to the question, "What is the fastest you have ever driven a car? _____ mph" The variable **fastest** contains the responses to this question. Determine a 95% confidence interval for the difference in the mean response of college men and college women. Interpret the interval, state any assumptions about the sample, and verify necessary conditions.

12.72 Use the dataset **pennstate1** for this exercise. The variable **RtSpan** has the raw data for measurements of the stretched right handspans of $n = 190$ Penn State students.

a. Calculate a 95% confidence interval for the mean stretched handspan in the population of men represented by this sample. Write a sentence that interprets the interval.

b. Calculate a 95% confidence interval for the mean stretched handspan in the population of women represented by this sample. Write a sentence that interprets the interval.

c. Compute a 95% confidence interval to estimate the difference in the mean handspans of college men and women. Write a sentence that interprets the interval.

12.73 Use the dataset **UCDavis2** for this exercise. In the survey, students reported ideal height (**IdealHt**) and actual height (**height**).

a. For the men in the sample, calculate a 90% confidence interval to estimate the mean difference between ideal and actual height. Be certain to graphically analyze the data in order to verify necessary conditions.

b. Repeat part (a) for the women in the sample.

c. Based on the intervals computed in parts (a) and (b), is there a difference between men and women with regard to mean difference between ideal and actual height? Explain.

12.74 Use the dataset **GSS-93.** The variable **owngun** indicates whether the respondent owns a gun or not, and the variable **polparty** contains the respondent's political party preference. Determine a 95% confidence interval for the difference in proportions of Republicans and Democrats who owned a gun that year.

12.75 Use the dataset **cholest.** The variable **control** contains cholesterol levels for individuals who have not had a heart attack, while the variable **2-day** contains cholesterol levels for heart attack patients two days after the attack.

a. Calculate a 98% confidence interval for the mean cholesterol level of individuals who have not had a heart attack. Interpret the interval, state any assumptions, and verify necessary conditions.

b. Calculate a 98% confidence interval for the mean cholesterol level of heart attack patients two days after the attack. Interpret the interval, state any assumptions, and verify necessary conditions.

c. Calculate a 98% confidence interval for the mean difference in the cholesterol levels of the populations represented by the two samples described in this problem. Interpret the interval, state any assumptions, and verify necessary conditions.

12.76 Use the **deprived** dataset for this exercise. Students in a statistics class were asked whether they generally felt sleep deprived (variable name = **Deprived**) and also were asked how many hours they usually slept per night (variable name = **SleepHrs**).

a. For the sample, find the mean hours of sleep per night for students who said they are sleep deprived and for students who said they are not sleep deprived. What is the difference in the two sample means?

b. Find a 95% confidence interval for the difference in population mean of hours of sleep for students who would say they are not sleep deprived versus students who would say they are sleep deprived. Interpret the interval in the context of this situation.

Turn On Your Computer

*For these exercises, use the **ConfidenceLevel** applet described in Section 12.3. It is on the CD for this text.*

12.77 Set the confidence level at 68%.

   a. Use the ***sample!*** button to generate a new random sample. Write the confidence interval displayed, and indicate whether it captures the population mean (170) or not.

   b. Generate 150 different samples. How many times did the confidence interval capture the population mean? What percent of the samples gave a confidence interval that captured the population mean?

   c. Refer to the previous part. Compare the number and percent of times that the confidence interval cap-tured the population mean to what might be expected for the 68% confidence level.

12.78 Change the confidence level to 95%. Repeat the parts of exercise 12.77 using the 95% confidence level.

12.79 Change the confidence level to 99%. Repeat the parts of exercise 12.77 using the 99% confidence level.

12.80 For the same random sample, give the 90%, 95%, and 99% confidence intervals for the population mean. (This can be done by observing what occurs when the scroll-bar is used to change the confidence level.) What is in-dicated about the relationship between the width of the confidence interval and the confidence level?

12.81 Write a definition of the confidence level for a con-fidence interval procedure.

REFERENCES

Cole, C. R., J. M. Foody, E. H. Blackstone, and M. S. Lauer (2000). "Heart rate recovery after submaximal exercise as a predictor of mortality in a cardiovascularly healthy cohort," *Annals of Internal Medicine,* Vol. 132, pp. 552–555.

Dabbs, J. M., D. de La Rue, and P. Williams (1990). "Testosterone and occupational choice: Actors, ministers, and other men," *Journal of Personality and Social Psychology.* Vol. 59, pp. 1261–1265.

Ellsworth, P. C., J. M. Carlsmith, and A. Henson, (1972). "The stare as a stimulus to flight in human subjects: A series of field experiments," *Journal of Personality and Social Psychology,* Vol. 21, pp. 302–311.

Gustafson, R., and H. Kallmen (1990). "Alcohol and Color Word Test," *Perceptual and Motor Skills,* Vol. 71, pp. 99–105.

Iman, R. L. (1994). *A Data-Based Approach to Statistics,* Belmont, CA: Wadsworth.

Jorenby, D. E., S. J. Leischow, M. A. Nides, S. I. Rennard, J. A. Johnston, A. R. Hughes, S. S. Smith, M. L. Muramoto, D. M. Daughton, K. Doan, M. C. Fiore, and T. B. Baker (1999). "A controlled trial of sustained-release bupropion, a nicotine patch, or both for smoking cessation," *New England Journal of Medicine,* Vol. 340, No. 9, pp. 685–691.

Lesko, S. M., L. Rosenberg, and S. Shapiro (1993). "A case-control study of baldness in relation to myocardial infarction in men," *Journal of the American Medical Association,* Vol. 269, No. 8, pp. 998–1003.

Maugh, Thomas H., II. "Study says calcium can help ease PMS," *Sacramento Bee,* 26 August 1998, pp. A1 and A9.

Norton, P. G., and E. V. Dunn (1985). "Snoring as a risk factor for disease," *British Medical Journal,* Vol. 291, pp. 630–632.

Olds, D. L., C. R. Henderson, Jr., and R. Tatelbaum (1994). "Intellectual impairment in children of women who smoke cigarettes during pregnancy," *Pediatrics,* Vol. 93(2), Feb. 1994, pp. 221–227.

Prasad, A. S., J. T. Fitzgerald, B. Bao, F. W. J. Beck, and P. H. Chandrasekar (2000). "Duration of symptoms and plasma cykotine levels in patients with the common cold treated with zinc acetate," *Annals of Internal Medicine,* Vol. 133, pp. 245–252.

Schwartz, B. S., W. F. Stewart, D. Simon, and R. B. Lipton (1998). "Epidemiology of tension-type headache," *Journal of the American Medical Association,* Vol. 279, pp. 381–383.

Thys-Jacobs, S., P. Starkey, D. Bernstein, J. Tian, and the Premenstrual Syndrome Study Group (1998). "Calcium carbonate and the premenstrual syndrome: Effects on premenstrual and menstrual symptoms," *American Journal of Obstetrics and Gynecology,* Vol. 179, No. 2, pp. 444–452.

Wood, R. D., M. L. Stefanick, D. M. Dreon, B. Frey-Hewitt, S. C. Garay, P. T. Williams, H. R. Superko, S. P. Fortmann, J. J. Albers, K. M. Vranizan, N. M. Ellsworth, R. B. Terry, and W. L. Haskell (1988). "Changes in plasma lipids and lipoproteins in overweight men during weight loss through dieting as compared with exercise," *New England Journal of Medicine,* Vol. 319, No. 18, pp. 1173–1179.

More About Significance Tests

Would your behavior change if she were staring at you? See Example 13.3.

To evaluate the quality of a research study, consider such features as sampling procedure, methods used to measure individuals, wordings of questions, and possible lurking variables. When conclusions are based on hypothesis tests, additional issues and cautions discussed in this chapter should be considered.

I n Chapter 11, we learned the logic and basic steps of hypothesis testing and applied them to *z*-tests for one proportion. In this chapter, we'll look at the details of hypothesis tests for the same situations for which we developed confidence intervals in Chapter 12:

♦ Hypotheses about a population mean or the mean difference for paired data. For example, we might wish to determine whether or not the mean body temperature for humans is 98.6 degrees Fahrenheit.

♦ Hypotheses about the difference between the means of two populations. For example, we might ask whether or not the mean time of television watching per week is the same for men and women.

♦ Hypotheses about the difference between two population proportions. For example, we might question whether or not the proportion favoring the death penalty is the same for teenagers as it is for adults.

Before we delve into the details, it is worth reviewing three cautions we have already encountered concerning the use of data to make inferences beyond the sample:

1. Inference is only valid if the sample is representative of the population for the question of interest.
2. Hypotheses and conclusions apply to the larger population(s) represented by the sample(s).
3. If the distribution of a quantitative variable is highly skewed, we should consider analyzing the median rather than the mean. Methods for testing hypotheses about medians are a special case of **nonparametric methods,** covered briefly in Section 16.3 and more thoroughly in Chapter 18. ♦

13.1 | THE GENERAL IDEAS OF SIGNIFICANCE TESTING

The terms **hypothesis testing** and **significance testing** are synonymous. The term *significance testing* arises because the conclusion about whether to reject a null hypothesis is based on whether a sample statistic is "significantly" far from what would be expected if the null hypothesis were true in the population. Equivalently, we declare **statistical significance** and reject the null hypothesis if there is a relatively small probability that an observed difference or relationship in the sample would have occurred if the null hypothesis holds in the population. One major principle for all significance tests is that we declare "statistical significance" when the *p*-value of the test is "small."

The significance tests for the three situations that we examine in this chapter all have the same general format because the test statistic is a standardized score that measures the difference between an observed statistic and the null value of a parameter. This was also the case in Chapter 11 where we used a *z*-statistic to test hypotheses about a population proportion. The basic format that we followed in Chapter 11 and will also follow in this chapter is as follows:

◆ *The null hypothesis defines a specific value of a population parameter, called the null value.* In Example 13.1 below, for instance, the null hypothesis is that the mean normal body temperature of humans is $\mu = 98.6$ degrees Fahrenheit. Sometimes the null value is defined for the difference between two population parameters. In Example 13.4, the null hypothesis is that the difference between the mean amounts of time college men and women watch television in a typical day is $\mu_1 - \mu_2 = 0$.

◆ *A relevant statistic is calculated from sample information and summarized into a "test statistic."* In each case, we begin by determining the sample equivalent of the parameter of interest. For instance, we use $\bar{x} = $ mean of a sample of temperature measurements to estimate $\mu = $ population mean temperature. We then measure the difference between the sample statistic and the null value using the **standardized statistic:**

$$\frac{\text{Sample statistic} - \text{Null value}}{\text{Null standard error}}$$

The Distribution of the Standardized Statistic

For hypotheses about means, the *standardized statistic* is called a *t*-statistic and the *t*-distribution is used to find the *p*-value.

For hypotheses about proportions, the *standardized statistic* is called a *z*-statistic and the standard normal distribution is used to find the *p*-value.

◆ *A p-value is computed based on the standardized "test statistic."* The *p*-value is calculated by temporarily assuming the null hypothesis to be true and then calculating the probability that the test statistic could be as large in magnitude as it is (or larger) in the direction(s) specified by the alternative hypothesis.

◆ *Based on the p-value, we either reject or fail to reject the null hypothesis.* The most commonly used criterion (level of significance) is that we reject the null hypothesis when the *p*-value is less than 0.05. In many research articles, *p*-values are simply reported and readers are left to draw their own conclusions. Remember that a *p*-value measures the strength of the evidence against the null hypothesis, and the smaller the *p*-value, the stronger the evidence against the null (and for the alternative).

These general ideas are consistent with the five steps outlined in Chapters 6 and 11 that are used for *any* hypothesis test. Let's review those five steps:

Step 1: Determine the *null* and *alternative* hypotheses.

Step 2: Verify necessary data conditions, and if met, summarize the data into an appropriate *test statistic*.

Step 3: Assuming the null hypothesis is true, find the *p-value*.

Step 4: Decide whether or not the result is *statistically significant* based on the *p-value*.

Step 5: Report the conclusion in the context of the situation.

13.2 | TESTING HYPOTHESES ABOUT ONE MEAN OR PAIRED DATA

For questions about **the mean** of a quantitative variable **for one population,** the null hypothesis typically has the form

$$H_0: \mu = \mu_0 \quad \text{(mean is a specified value)}$$

The alternative may either be one-sided ($H_a: \mu < \mu_0$ or $H_a: \mu > \mu_0$) or two-sided ($H_a: \mu \neq \mu_0$), depending on the research question of interest. The usual procedure for testing these hypotheses is called a **one-sample *t*-test** because the *t*-distribution is used to determine the *p*-value. Let's look at an example of a situation where a one-sample *t*-test would be used.

EXAMPLE 13.1
Normal Human Body Temperature

What is normal body temperature? A paper published in the *Journal of the American Medical Association* presented evidence that normal body temperature may be less than 98.6 degrees Fahrenheit, the long-held standard (Mackowiak et al., 1992). The value 98.6 degrees seems to have come from determining the mean in degrees Celsius, rounding up to the nearest whole degree (37 degrees), and then converting that number to Fahrenheit using 32 + (1.8)(37) = 98.6. Rounding up may have produced a result higher than the actual average, which may therefore be lower than 98.6 degrees. To test this, the null hypothesis is $\mu = 98.6$, and we are only interested in rejecting in favor of lower values, so the alternative is $\mu < 98.6$. We can write these hypotheses as

$$H_0: \mu = 98.6$$
$$H_a: \mu < 98.6$$

Suppose that a random sample of $n = 18$ normal body temperatures is

98.2 97.8 99.0 98.6 98.2 97.8 98.4 99.7 98.2
97.4 97.6 98.4 98.0 99.2 98.6 97.1 97.2 98.5

For these data, the sample mean is $\bar{x} = 98.217$. This is lower than 98.6, but to determine whether this is a statistically significant difference from 98.6, the question we must answer is: What is the probability that the sample mean could be 98.217 or less if the population mean is actually 98.6 (the null value)?

The answer is the *p*-value that we will use to decide between the two hypotheses. We will return to this example after we provide some details for the five steps of a one-sample *t*-test, but we won't keep you in suspense. The *p*-value for this example is 0.015, so the data support the alternative hypothesis. ◆

For a one-sample *t*-test, details of the five steps in the hypothesis test are

Step 1: Determine the Null and Alternative Hypotheses

Remember that in general the null hypothesis states the status quo and the alternative hypothesis states the research question of interest. For a one-sample *t*-test, the possible null and alternative hypotheses are one of these three choices, depending on the research question:

$$(1)\ H_0: \mu = \mu_0 \quad \text{versus} \quad H_a: \mu \neq \mu_0 \quad \text{(two-sided)}$$
$$(2)\ H_0: \mu = \mu_0 \quad \text{versus} \quad H_a: \mu < \mu_0 \quad \text{(one-sided)}$$
$$(3)\ H_0: \mu = \mu_0 \quad \text{versus} \quad H_a: \mu > \mu_0 \quad \text{(one-sided)}$$

Sometimes, when the alternative hypothesis is one-sided, the null hypothesis is written to include values in the other direction from the alternative hypotheses as well as equality, for instance $H_0: \mu \leq \mu_0$, but the "null value" used when computing the test statistic is always μ_0.

Step 2: Verify Necessary Data Conditions, and If Met, Summarize the Data into an Appropriate Test Statistic

Recall from the previous two chapters that whenever we carry out a statistical inference procedure, we must check to see if certain data conditions are met. These conditions are connected to the assumptions that were made when the theory was derived for the particular situation. Carrying out a *t*-test for one population mean requires that one of the same two situations holds as for constructing a confidence interval for a mean, given in Chapter 12. Here are those situations, slightly modified from the way they were stated in Chapter 12:

Situations for Which a One-Sample t-Test Is Valid

Situation 1: The population of the measurements of interest is approximately normal, and a random sample of any size is measured. In practice, the method is used as long as there is no evidence that the shape is notably skewed or that there are extreme outliers.

Situation 2: The population of measurements of interest is not approximately normal, but a *large* random sample is measured. Thirty is usually used as an arbitrary demarcation of "large," but if there are extreme outliers or extreme skewness, it is better to have a larger sample.

For small or moderate sample sizes, use a boxplot, histogram, dotplot, or stem-and-leaf plot to check for notable skewness or extreme outliers. Checking the relative sizes of the sample mean and sample median might also be useful. Skewness and outliers both cause these two statistics to differ from each other. If either skewness or outliers are present, a one-sample *t*-test should not be used. In this event, a statistical test called the *sign test* can be used to analyze hypotheses about the *median*. We provide the details of the sign test in Supplemental Topic 2 on the CD. The procedure is available in most statistical software, including Minitab.

For a large sample size, the *t*-test will work well for testing hypotheses about the mean even if there is some skewness. However, when the data are obviously skewed, you should question whether the mean is the appropriate parameter to analyze. For skewed data, the median may be a better measure of location, and test procedures for analyzing the median might be more appropriate than the *t*-test.

Continuing Step 2: The Test Statistic for a One-Sample t-Test

The next part of step 2 is to compute a test statistic. The statistic $\bar{x}$, the sample mean, estimates the population mean μ. The "null standard error" for the denominator is the usual standard error of $\bar{x}$, s.e.$(\bar{x}) = s/\sqrt{n}$. Because this standard error does not depend on the null value μ_0, the word *null* can be omitted from the description of the denominator standard error. The test statistic, called the *t*-statistic, is a standardized score for measuring the difference between the sample mean and the null hypothesis value of the population mean and is written as

$$t = \frac{\text{Sample mean} - \text{Null value}}{\text{Standard error}} = \frac{\bar{x} - \mu_0}{s/\sqrt{n}}$$

This particular *t*-statistic has approximately a *t*-distribution with df $= n - 1$.

Step 3: Assuming the Null Hypothesis Is True, Find the p-Value

Using the *t*-distribution with df $= n - 1$, the *p*-value is the area in the tail(s) beyond the test statistic *t*, as follows:

- For H$_a$: $\mu < \mu_0$ (a one-sided test), the *p*-value is the area below *t*, even if *t* is positive.
- For H$_a$: $\mu > \mu_0$ (a one-sided test), the *p*-value is the area above *t*, even if *t* is negative.
- For H$_a$: $\mu \neq \mu_0$ (a two-sided test), the *p*-value is $2 \times$ area above $|t|$.

Table 13.1 summarizes and illustrates this information. While some specifics will differ for the other hypothesis-testing situations considered in this chapter, the

TABLE 13.1 ■ **Alternative Hypothesis Regions and *p*-Value Areas**

Statement of H$_a$		*p*-Value Area	*t*-Curve Region
$\mu < \mu_0$	(less than)	Area to the left of *t* (even if $t > 0$)	
$\mu > \mu_0$	(greater than)	Area to the right of *t* (even if $t < 0$)	
$\mu \neq \mu_0$	(not equal)	$2 \times$ area to the right of $|t|$	

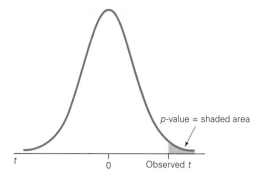

FIGURE 13.1 **The *p*-value for a one-tailed *t*-test with H$_a$: $\mu > \mu_0$**

general connection between the *p*-value and the computed area under the appropriate distribution holds in all cases that we consider.

Most software programs, including Minitab and Excel will do the work of finding the *p*-value for you. You can also find a range for the *p*-value using tables of the Student *t*-distribution. Table A.3 in the Appendix gives the areas to the right of eight different values of *t* for various values of degrees of freedom. The type of area given in Table A.3 is shown in Figure 13.1, which shows the area comprising the *p*-value for a one-tailed *t*-test with H$_a$: $\mu > \mu_0$. For instance, in Example 13.2 (p. 446), the *p*-value is the area to the right of $t = 2.68$ and the degrees of freedom are df = 9. In Table A.3, the *t*-value $t = 2.68$ is between the column headings 2.5 and 3.0, so the *p*-value is somewhere between 0.015 and 0.007. A *t*-distribution is symmetric, so Table A.3 can also be used to estimate *p*-values for a "less than" alternative hypothesis. This will be demonstrated when we complete Example 13.1.

Steps 4 and 5: Decide Whether or Not the Result Is Statistically Significant Based on the p-Value and Report the Conclusion in the Context of the Situation

These two steps remain the same for all of the hypothesis tests considered in this book. Choose a level of significance α, and reject the null hypothesis if the *p*-value is less than α. Otherwise, conclude that there is not enough evidence to support the alternative hypothesis.

EXAMPLE 13.1—*CONTINUED*
The Five Steps for Testing the Hypotheses About Normal Body Temperature

Step 1: The appropriate null and alternative hypotheses for this situation are

H$_0$: $\mu = 98.6$ degrees

H$_a$: $\mu < 98.6$ degrees

The parameter μ is the mean body temperature in the human population. The research question is whether μ is smaller than the long-held standard, so the alternative hypothesis displays a one-sided test.

Step 2: Verify necessary data conditions, and if met, summarize the data into an appropriate test statistic. The boxplot in Figure 13.2 illustrates that there are no outliers and the shape does not appear to be notably skewed. Although not shown on the boxplot, the sample mean = $\bar{x}$ = 98.217 is almost identical to the median of 98.2. This comparison pro-

97.0 97.4 97.8 98.2 98.6 99.0 99.4 99.8
Temperature

FIGURE 13.2 **Boxplot of the sample for Example 13.1**

vides additional evidence that skewness and outliers are not a problem, so the one-sample *t*-test can be used.

We used Minitab to carry out the test, and we can identify the elements necessary for constructing the *t*-statistic using the following Minitab output:

```
Test of mu = 98.600 vs mu < 98.600
Variable        N     Mean    StDev    SE Mean        T        P
Temperature    18    98.217    0.684      0.161    −2.38    0.015
```

The key elements are

- The sample statistic is $\bar{x} = 98.217$ (under "Mean").
- The standard error is $\text{s.e.}(\bar{x}) = \dfrac{s}{\sqrt{n}} = \dfrac{0.684}{\sqrt{18}} = 0.161$ (under "SE Mean").
- $t = \dfrac{\text{Sample statistic} - \text{Null value}}{\text{Standard error}} = \dfrac{98.217 - 98.6}{0.161} = \dfrac{-0.383}{0.161}$
  $= -2.38$ (under "T").

Step 3: The *p*-value is given as 0.015 in the last column of output. It was calculated as the area to the left of $t = -2.38$ in a *t*-distribution with $n - 1 = 18 - 1 = 17$ degrees of freedom, and Figure 13.3 illustrates this area. If we use Table A.3, we can determine that the *p*-value is between 0.016 and 0.010. Because the *t*-distribution is symmetric, the area to the left of $t = -2.38$ equals the area to the right of $t = +2.38$. The value $t = 2.38$ is between the column headings 2.33 and 2.58 in the table, and for 17 degrees of freedom, the corresponding one-sided *p*-values shown in the table are 0.016 and 0.010.

Step 4: Using $\alpha = 0.05$ as the level of significance criterion, the results are statistically significant because 0.015, the *p*-value of the test, is less than 0.05. In other words, we can reject the null hypothesis.

Step 5: We can conclude, based on these data, that the mean temperature in the human population is actually less than 98.6 degrees. ◆

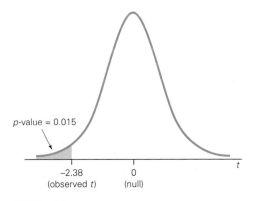

p-value = 0.015

−2.38   0   *t*
(observed *t*)   (null)

FIGURE 13.3 ***p*-Value region for Example 13.1**

Using Excel to Find the *p*-Value

tech note   In Example 13.1, the *p*-value can also be found in Excel by typing @TDIST(2.38,17,1) in any cell. Notice that Excel requires you to give the absolute value of the test statistic. The "1" indicates that you want the area in one tail only, since this is a one-tailed test.

Caution about using Excel for one-sided alternative hypotheses: If the sample mean happens to be in the direction opposite to that specified by the alternative hypothesis, then *p*-value > 0.5 and Excel gives $1 - p$-value. For instance, in Example 13.1, if the sample mean had been above 98.6 and had resulted in $t = +2.38$, then the appropriate *p*-value would *still* be the area to the *left* of *t*, or to the *left* of 2.38, which is $1 - 0.015 = 0.985$. Excel only gives areas to the *right* of the specified value of *t* for one-tailed tests, and only allows positive values of *t* to be specified.

---

in summary   The Steps for Testing a Single Mean with a "One-Sample *t*-Test"

Step 1: Determine the *null* and *alternative* hypotheses.

Null hypothesis: $H_0: \mu = \mu_0$

Alternative hypothesis: $H_a: \mu \neq \mu_0$ or $H_a: \mu > \mu_0$ or $H_a: \mu < \mu_0$

where the format of the alternative hypothesis depends on the research question of interest.

Step 2: Verify necessary data conditions, and if met, summarize the data into an appropriate *test statistic*.

If *n* is large, or if there are no extreme outliers or skewness, compute

$$t = \frac{\text{Sample mean} - \text{Null value}}{\text{Standard error}} = \frac{\bar{x} - \mu_0}{s/\sqrt{n}}$$

Step 3: Assuming the null hypothesis is true, find the *p-value*.

Using the *t*-distribution with df $= n - 1$, the *p*-value is the area in the tail(s) beyond the test statistic *t*, as follows:

For $H_a: \mu \neq \mu_0$, the *p*-value is $2 \times$ area above $|t|$ (a two-tailed test).

For $H_a: \mu > \mu_0$, the *p*-value is the area above *t* (a one-tailed test) even if *t* is negative.

For $H_a: \mu < \mu_0$, the *p*-value is the area below *t* (a one-tailed test) even if *t* is positive.

Step 4: Decide whether or not the result is *statistically significant* based on the *p-value*.

Choose a significance level ("alpha"); the standard is $\alpha = 0.05$. The result is statistically significant if the *p*-value $\leq \alpha$.

Step 5: Report the conclusion in the context of the situation.

turn on your mind

Consider the two possible one-sided alternative hypotheses in a one-sample *t*-test:

$$H_a: \mu > \mu_0 \quad \text{and} \quad H_a: \mu < \mu_0$$

In each case, what range of values of the *t*-statistic would result in a *p*-value > 0.5? What range of values of the sample mean would result in a *p*-value > 0.5? Write a general rule for one-sided tests explaining conditions under which the *p*-value > 0.5.*

Paired Data and the Paired *t*-Test

Remember from Chapter 12 that the term **paired data** is used to describe data collected in natural pairs. In Chapter 3 we learned about *matched-pair* designs in which units are paired and one in each pair receives each treatment. Data are collected in pairs in other types of studies as well, such as when cases and controls are matched in case-control studies. Often, paired data occur when the researcher collects two measurements from each observational unit. For instance, if we record weights both before and after a diet program for each person in a sample, we have paired data. Or, we might record performance on a college entrance exam for each individual before and after a training program designed to boost performance on that type of exam.

In most cases paired data are collected because the researchers want to know about the differences, and not about the original observations. In particular, it is often of interest to know if the mean difference in the population is different from 0. A one-sample *t*-test can be used on the sample of differences to examine whether the sample mean difference is significantly different from 0. When this is done, the test is called a **paired *t*-test**. It is nothing more than a one-sample *t*-test conducted on the *n* differences. To emphasize that the data used in the test are differences, it is commonplace to use d_i instead of x_i to denote the original data values, $\overline{d}$ instead of $\overline{x}$ for the sample mean of the differences, s_d instead of s for the sample standard deviation of the differences, and μ_d instead of μ for the population mean of the differences. Other than those notational differences, once the difference value has been computed for each pair, the test proceeds exactly like a one-sample *t*-test.

---

in summary   The Steps for a Paired *t*-Test

Step 1: Determine the *null* and *alternative* hypotheses.

Null hypothesis:	$H_0: \mu_d = 0$
Alternative hypothesis:	$H_a: \mu_d \neq 0 \quad or \quad H_a: \mu_d > 0 \quad or \quad H_a: \mu_d < 0$

---

*HINT: Remember that for $H_a: \mu < \mu_0$, the p-value is the area below the test statistic t. What values of t have more than half of the area below them? What relationship between the sample mean and μ_0 would result in those values of t?

where the format of the alternative hypothesis depends on the research question of interest and the order in which the differences are taken.

Step 2: Verify necessary data conditions, and if met, summarize the data into an appropriate *test statistic*.

Before proceeding with the details of this step, calculate the difference *d* between the two observations for each pair. Then verify that *n* is large or that there are no extreme outliers or skewness in the *differences*. Compute the sample mean and standard deviation of the differences, $\bar{d}$ and s_d, and then compute the test statistic:

$$t = \frac{\text{Sample mean} - \text{Null value}}{\text{Standard error}} = \frac{\bar{d} - 0}{s_d/\sqrt{n}}$$

Steps 3, 4, and 5: Proceed exactly as instructed in the In Summary "The Steps for Testing a Single Mean with a One-Sample *t*-Test" on page 444, except in the instructions for Step 3, replace μ with μ_d and μ_0 with 0.

EXAMPLE 13.2
The Effect of Alcohol on Useful Consciousness

Ten pilots performed tasks at a simulated altitude of 25,000 feet. Each pilot performed the tasks in a completely sober condition and, three days later, after drinking alcohol. The response variable is the time (in seconds) of useful performance of the tasks for each condition. The longer a pilot spends on useful performance, the better. The research hypothesis is that useful performance time decreases with alcohol use, so the data of interest is the decrease (or increase) in performance with alcohol compared to when sober. The data (Devore and Peck, 1993, p. 575) are as follows:

Pilot	No Alcohol	Alcohol	Difference = No Alcohol − Alcohol
1	261	185	76
2	565	375	190
3	900	310	590
4	630	240	390
5	280	215	65
6	365	420	−55
7	400	405	−5
8	735	205	530
9	430	255	175
10	900	900	0

Step 1: This is a paired-data design. Let μ_d = population mean difference between no alcohol and alcohol measurements, if all pilots were to take these tests. Null and alternative hypotheses about μ_d are

H_0: $\mu_d = 0$ seconds

H_a: $\mu_d > 0$ seconds   (i.e., no alcohol > alcohol)

The alternative hypothesis is one-sided because we hope to show that if performance does change, there is longer useful performance in the "No Alcohol" condition.

FIGURE 13.4   Boxplot of differences for Example 13.2

Step 2: Figure 13.4 displays a boxplot of the differences for the ten participants. A check of the necessary conditions for doing a *t*-test reveals that while the sample size is small and the dataset of differences for the sample does have some skewness, outliers and extreme skewness do not appear to be serious problems. So, a paired *t*-test will be used to examine this question.

Minitab output for analyzing the mean difference follows. Notice that the analysis involves only the sample of $n = 10$ differences and does not otherwise utilize the original performance scores.

```
Test of mu = 0.0 vs mu > 0.0
Variable      N     Mean    StDev    SE Mean       T       P
Diff         10    195.6    230.5       72.9    2.68   0.013
```

The *t*-statistic shown in the output is $t = 2.68$ (under "T" in the output). The calculation of the test statistic is

- The sample statistic is the observed mean difference, $\bar{d} = 195.6$.
- The standard error is s.e.$(\bar{d}) = \dfrac{s_d}{\sqrt{n}} = \dfrac{230.5}{\sqrt{10}} = 72.9$.

$$t = \frac{\text{Sample statistic} - \text{Null value}}{\text{Standard error}} = \frac{195.6 - 0}{72.9} = 2.68$$

Step 3: At the right side of the output, we see that the *p*-value is 0.013. Because the alternative hypothesis was "greater than," this *p*-value was computed as the area to the right of 2.68 in a *t*-distribution with df $= 10 - 1 = 9$, an area illustrated in Figure 13.5. If we had used Table A.3, we would have found that $0.007 < p\text{-value} < 0.015$. Finding the exact *p*-value of 0.013 requires a computer or calculator that gives areas for the *t*-distribution. For instance, in Excel if you type @TDIST(2.68,9,1) into any cell, the result will be 0.013 (rounded to three decimal places).

Steps 4 and 5: The *p*-value is 0.013, so we can reject the null hypothesis using the standard significance level of 0.05. Even with such a small experiment, we can declare that alcohol has a statistically significant effect and decreases useful performance time. ◆

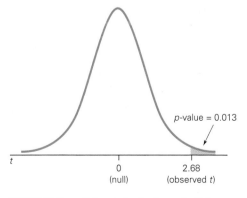

FIGURE 13.5   *p*-Value region for Example 13.2

13.2

turn on your mind

Suppose in Example 13.2 the purpose of the study was to determine whether pilots should be allowed to consume alcohol the evening prior to their flights and the alcohol consumption occurred 12 hours before the measurement of time of useful performance. Refer to the discussion of type 1 and type 2 errors in Chapter 11. Explain what the consequences of each type of error would be in this example. Which would be more serious? Given the data and results of the study, which type of error could have been made?*

The Rejection Region Approach for *t*-Tests

In Chapter 11, Section 11.4, we learned that the decision in hypothesis testing can be made by finding a **rejection region,** which is the region of possible values for the test statistic that would lead to rejection of the null hypothesis. The method is based on finding the **critical value,** which is a value such that if the test statistic is beyond it in one or both tails (depending on H_a), the result is statistically significant. Using this approach, the null hypothesis is rejected if the test statistic falls into the rejection region.

This method consists of replacing Steps 3 and 4 with:

Substitute Step 3: Find the critical value and rejection region for the test.

Substitute Step 4: If the test statistic is in the rejection region, conclude that the result is *statistically significant* and reject the null hypothesis. Otherwise, do not reject the null hypothesis.

In Chapter 11 we learned what values are in the rejection region for *z*-tests with a level of significance of 0.05 or 0.01. This approach can be used for *t*-tests as well, but the rejection region changes depending on the degrees of freedom accompanying the test. Table A.2 can be used to find the critical value and rejection region.

Steps for Finding the Critical Value and Rejection Region for a t-Test

1. Determine the degrees of freedom for the test.
2. Decide on α, the level of significance for the test; 0.05 is the most common but sometimes 0.10, 0.01, or some other value is used.
3. In Table A.2, read the critical value from the row with appropriate degrees of freedom for the test. Use the critical value to construct the rejection region, as follows:

   ◆ For H_a: *Parameter > null value,* read the critical value from the column with confidence level $1 - 2\alpha$. For instance, for $\alpha = 0.05$, use the 0.90 column, because $1 - 2\alpha = 1 - 2(0.05) = 1 - 0.10 = 0.90$. Reject the null hypothesis if $t \geq$ critical value. For example, to test H_a: $\mu > 0$ with df = 10 and $\alpha = 0.05$, the critical value is 1.81. The rejection region is $t \geq 1.81$. If the test statistic *t* falls in that region, reject the null hypothesis.

*HINT: Remember that a type 1 error can be made only when the null hypothesis is true. In this case, the null hypothesis is that on average, alcohol has no effect on pilots' performance.

♦ For H_a: *Parameter* < *null value,* also read the critical value from the column with confidence level $1 - 2\alpha$, but add a negative sign. Reject the null hypothesis if $t \leq$ critical value. For example, to test H_a: $\mu < 0$ with df = 10 and $\alpha = 0.05$, the critical value is -1.81. The rejection region is $t \leq -1.81$. If the test statistic t falls in that region, reject the null hypothesis.

♦ If the test is two-sided (H_a: *Parameter* ≠ *null value*), use the column with confidence level $1 - \alpha$. For instance, for $\alpha = 0.05$, use the column headed 0.95. Reject the null hypothesis if $|t| \geq$ critical value. For example, to test H_a: $\mu \neq 0$ with df = 10 and $\alpha = 0.05$, the critical value is 2.23. The rejection region is $t \leq -2.23$ and $t \geq 2.23$. If the test statistic falls in that region, reject the null hypothesis.

Here is a summary and examples of the procedure for finding the critical value and rejection region for any *t*-test. Always use the row of Table A.2 corresponding to the degrees of freedom for the test.

Alternative Hypothesis	Column Heading in Table A.2	Column in Table A.2 for $\alpha = 0.05$	Example: Rejection Region for df = 10, $\alpha = 0.05$
H_a: Parameter ≠ null value	$1 - \alpha$	0.95	$t \leq -2.23$ and $t \geq 2.23$
H_a: Parameter > null value	$1 - 2\alpha$	0.90	$t \geq 1.81$
H_a: Parameter < null value	$1 - 2\alpha$	0.90	$t \leq -1.81$

EXAMPLE 13.1—*CONTINUED*
The Critical Value and
Rejection Region for Testing
Normal Body Temperature

In Example 13.1, the alternative hypothesis was H_a: $\mu < 98.6$ degrees, the degrees of freedom = 17, and $\alpha = 0.05$. From Table A.2, the critical value is found to be -1.74, using the df = 17 row and confidence level = 0.90 column. The rejection region is $t \leq -1.74$. The test statistic $t = -2.38$ is in the rejection region, so the null hypothesis is rejected. This is the same conclusion reached by finding the *p*-value and comparing it to 0.05. ♦

EXAMPLE 13.2—*CONTINUED*
The Critical Value and
Rejection Region for Alcohol
and Useful Consciousness

To find the critical value for Example 13.2, use Table A.2 with df = 9. Because the alternative hypothesis is H_a: $\mu_d > 0$, for $\alpha = 0.05$ use the 0.90 column. The critical value is 1.83 and the rejection region is $t \geq 1.83$. Because the test statistic $t = 2.68$ is in the rejection region, we reject the null hypothesis. This is the same conclusion we reached in step 4 of the example, based on the *p*-value. ♦

Comparing the p-Value Approach to the Rejection Region Approach

Once the level of significance α has been decided, the rejection region method and the *p*-value method will always arrive at the same conclusion about statistical significance. It is always true that if the *p*-value $\leq \alpha$ then the test statistic falls into the rejection region, and vice versa. However, the *p*-value method provides additional information because it allows us to determine what decision would be made for every possible value of α. The rejection region method only provides a decision for the specified α. For that reason, and because statistical software programs and research journals generally report *p*-values, we will continue to emphasize the *p*-value approach.

Minitab Tip

Minitab v. 14 update on CD

tech note   **Computing a One-Sample *t*-Test or a Paired *t*-Test**

◆ To carry out a one-sample *t*-test, use **Stat**>**Basic Statistics**> **1-Sample t.** Specify the column that contains the raw data, and specify the *null value* of the mean in the "Test mean:" box. To specify the *alternative hypothesis*, use the **Options** button.

◆ For *paired data,* first calculate a column of differences using **Calc**>**Calculator,** then use the *1-sample t* procedure to test that the mean difference is 0 (the null value).

◆ Alternatively, for *paired data,* use **Stat**>**Basic Stats**>**Paired t.** Specify the two columns that contain the raw data for the pair of measurements.

◆ In Version 12 a confidence interval is not automatically provided. Beginning with Version 13, a confidence interval is provided, and if the alternative hypothesis selected is one-sided, the reported confidence interval will also be one-sided.

Note: For both the *1-sample* and *Paired t,* the **Graphs** button can be used to create visual displays of the data.

13.3 │ TESTING THE DIFFERENCE BETWEEN TWO MEANS (INDEPENDENT SAMPLES)

It is often of interest to determine whether the means of populations represented by two independent samples of a quantitative variable differ. The two populations may be represented by two categories of a categorical variable, such as males and females, or they may be two hypothetical populations represented by different treatment groups in an experiment. In most cases, when comparing two means, the null hypothesis is that they are equal:

$$H_0: \mu_1 - \mu_2 = 0 \quad (\text{or } \mu_1 = \mu_2)$$

The procedure for testing this null hypothesis is called the **two-sample *t*-test.** Let's look at an example of a situation where a two-sample *t*-test would be used.

EXAMPLE 13.3
The Effect of a Stare on Driving Behavior

In Example 12.8, we discussed an experiment done by social psychologists at the University of California at Berkeley. The researchers either did not stare or did stare at automobile drivers stopped at a campus stop sign. In the experiment, the response variable was the time (in seconds) it took the drivers to drive from the stop sign to a mark on the other side of the intersection. The two populations represented by the observed times are the hypothetical ones that would consist of the times drivers like these would take to move through a similar intersection, either under normal conditions (no stare) or the experimental condition (stare). An hypothesis the researchers wished to test was that the stare would speed up the crossing times, so the mean crossing time would be greater (slower) for those who did not experience the stare than it would be for those who did. So, the null and alternative hypotheses are

$$H_0: \mu_1 - \mu_2 = 0 \quad (\text{or } \mu_1 = \mu_2)$$
$$H_a: \mu_1 - \mu_2 > 0 \quad (\text{or } \mu_1 > \mu_2)$$

In these hypotheses, the subscript 1 is used to denote the No Stare population and the subscript 2 is used to denote the Stare population. For the data given in Chapter 12, the mean crossing time was $\bar{x}_1 = 6.63$ seconds for $n = 14$ drivers who crossed under normal conditions (No Stare) and $\bar{x}_2 = 5.59$ seconds for $n = 13$ who crossed in the experimental condition (Stare). The difference between the sample means is 1.04 seconds. Is this difference large enough to be statistically significant evidence against the null hypothesis? We will answer this question after describing the details of the five hypothesis-testing steps for a two-sample t-test. ◆

Step 1: Determine the Null and Alternative Hypotheses

The *null* hypothesis is usually that the difference in means is 0:

$$H_0: \mu_1 - \mu_2 = 0 \quad (\text{or } \mu_1 = \mu_2)$$

The *alternative* hypothesis may either be one-sided or two-sided:

$$H_a: \mu_1 - \mu_2 > 0 \quad or \quad H_a: \mu_1 - \mu_2 < 0 \quad or \quad H_a: \mu_1 - \mu_2 \neq 0$$

In one-sided tests, the null hypothesis may be written to include an inequality in the opposite direction from the alternative hypothesis.

Step 2: Verify Necessary Data Conditions, and If Met, Summarize the Data into an Appropriate Test Statistic

The situations in which the t-test for the difference in two means is valid are similar to those given in Chapter 12 for constructing a confidence interval for the difference in two means.

Situations for Which a t-Test for the Difference in Two Population Means Is Valid

Situation 1: Both populations represented by the measurements of interest are approximately normal, and random samples of any size are measured. In practice, the method is used as long as there is no evidence that the shapes are notably skewed or that there are extreme outliers.

Situation 2: The populations of measurements of interest are not approximately normal, but a *large* random sample is measured from each one. Thirty is usually used as an arbitrary demarcation of "large," but if there are extreme outliers or extreme skewness, it is better to have larger samples.

Independence: The samples must be independent. In other words, they must not have been measured as paired or blocked data.

To verify the necessary conditions, examine histograms, stem-and-leaf plots, boxplots, or dotplots for each sample. Within each sample, a comparison of the mean and median can also provide information about possible skewness and outliers. If both of the two independent samples are large, we are in "Situation 2" and the two-sample t-test generally is valid for comparing the means unless there are remarkably extreme outliers. Keep in mind, however, that for skewed data the median may be the better measure of location.

Continuing Step 2: The Test Statistic for a Two-Sample t-Test

The relevant sample statistic is $\bar{x}_1 - \bar{x}_2$ = the difference between sample means. As with a one-sample t-test, the standard error of $\bar{x}_1 - \bar{x}_2$ does not depend on whether the null hypothesis is true, so the term *null* is dropped from the denominator standard error. The "standardized" test statistic is

$$t = \frac{\text{Sample statistic} - \text{Null value}}{\text{Standard error}} = \frac{\bar{x}_1 - \bar{x}_2 - 0}{\sqrt{\dfrac{s_1^2}{n_1} + \dfrac{s_2^2}{n_2}}}$$

Step 3: Assuming the Null Hypothesis Is True, Find the p-Value

To find the p-value, use the t-distribution with appropriate degrees of freedom:

◆ For H_a: $\mu_1 - \mu_2 < 0$, the p-value is the area below t, even if t is positive.

◆ For H_a: $\mu_1 - \mu_2 > 0$, the p-value is the area above t, even if t is negative.

◆ For H_a: $\mu_1 - \mu_2 \neq 0$, the p-value is $2 \times$ area above $|t|$.

As discussed in Chapter 12, the t-distribution is only an approximation for the distribution of the t-statistic, and appropriate degrees of freedom are found by a complicated formula called "Welch's approximation." The formula is given in Section 12.5 and will not be repeated here. In practice, if computer software such as Minitab is used, it will provide the numerical value of degrees of freedom found from the approximation. If software is not available, a conservative approach is to use the smaller of $n_1 - 1$ and $n_2 - 1$ as the degrees of freedom.

Equal Variances and Pooled Standard Error

tech note If we are willing to make the assumption that the two populations we are comparing have the same (or similar) variance, then there is a procedure for which the t-distribution *is* the correct distribution of the t-statistic when the null hypothesis is true. It involves "pooling" both samples to estimate the common variance. The procedure is called a *pooled two-sample t-test*, and is discussed at the end of this section.

Steps 4 and 5: Decide Whether or Not the Result Is Statistically Significant Based on the p-Value and Report the Conclusion in the Context of the Situation

These steps proceed as with any hypothesis test. Choose a significance level ("alpha"), usually $\alpha = 0.05$, and declare the difference in means to be statistically significant if the p-value $\leq \alpha$. Report a conclusion appropriate for the situation. When drawing a conclusion about differences between two populations, it also is important to consider the way in which the data were gathered. Remember from Chapter 3 that data collected in an experiment can provide stronger evidence of causation than data from an observational study.

EXAMPLE 13.3—CONTINUED
The Effect of a Stare on Driving Behavior

Step 1: *State the hypotheses.* The subscript 1 is used to represent the No Stare population and the subscript 2 to represent the Stare population. As stated above, the appropriate null and alternative hypotheses are

$$H_0: \mu_1 - \mu_2 = 0 \qquad (\text{or } \mu_1 = \mu_2)$$
$$H_a: \mu_1 - \mu_2 > 0 \qquad (\text{or } \mu_1 > \mu_2)$$

The alternative hypothesis reflects the fact that the researchers thought staring would cause drivers to speed up, so the mean time should be slower (larger values) for the No Stare group and faster (smaller values) for the Stare group.

Step 2: *Verify necessary data conditions, and if met, determine the test statistic.* The data conditions were checked in Example 12.8. We saw that the conditions for conducting the *t*-test appeared to be valid, as there were no extreme outliers and no extreme skewness in the data. There were $n_1 = 14$ observations in the No Stare group and $n_2 = 13$ observations in the Stare group. The following Minitab (Release 12) output provides a confidence interval for the difference between the population means and results for a one-tailed test:

```
Two-sample T for CrossTime

Group        N       Mean     StDev    SE Mean
NoStare      14      6.63     1.36     0.36
Stare        13      5.59     0.822    0.23

95% CI for mu (NoStare) − mu (Stare   ): (0.14, 1.93)
T-Test mu (NoStare) = mu (Stare   ) (vs >): T = 2.41   P = 0.013   DF = 21
```

The last line of output shows us that the *t*-statistic is T = 2.41. The elements of the calculation are

- The sample statistic is $\bar{x}_1 - \bar{x}_2 = 6.63 - 5.59 = 1.04$ seconds.

- $\text{s.e.}(\bar{x}_1 - \bar{x}_2) = \sqrt{\dfrac{s_1^2}{n_1} + \dfrac{s_2^2}{n_2}} = \sqrt{\dfrac{1.36^2}{14} + \dfrac{0.822^2}{13}} = 0.43.$

- $t = \dfrac{\text{Sample statistic} - \text{Null value}}{\text{Standard error}} = \dfrac{\bar{x}_1 - \bar{x}_2 - 0}{\text{s.e.}(\bar{x}_1 - \bar{x}_2)} = \dfrac{1.04 - 0}{0.43} = 2.41.$

Steps 3, 4, and 5: *Determine the p-value and make a conclusion in context.* Note from the output that the *p*-value of the test is 0.013, so we can reject the null hypothesis and attach the label "statistically significant" to the results. The *p*-value is the probability that a sample difference could be as far or further from 0 as 1.04 seconds is *in the positive direction,* if the population difference is actually 0. This probability is determined using a *t*-distribution with df = 21 (the approximate degrees of freedom computed by Minitab using the Welch approximation formula described in Chapter 12) and finding the area to the right of the test statistic value, $t = 2.41$. From Table A.3, we would learn that the *p*-value is between 0.009 and 0.015. We can conclude that if all drivers were stared at, the mean crossing times at an intersection would be faster than under normal conditions. It can also be noted that these data were collected in a randomized experiment, so it's reasonable to conclude that the staring or not staring caused the difference between the mean crossing times.

Substitute Steps 3 and 4: As is always the case, the same conclusion is reached using the *rejection region* approach. To find the *critical value* for this example, use Table A.2 with df = 21. (If Minitab had not provided the degrees of freedom, we would use the conservative value of smaller sample size − 1, which is 13 − 1 = 12, but we will assume we know df = 21.) Because the alternative hypothesis is one-sided, H_a: $\mu_1 - \mu_2 > 0$, for $\alpha = 0.05$ use the 0.90 column. The critical value is 1.72 and the rejection region is $t \geq 1.72$. Because the test statistic $t = 2.41$ is in the rejection region, we reject the null hypothesis. This is the same conclusion we reached based on the *p*-value of 0.013, and because the *p*-value is more informative than the rejection region approach, it is always better to report it if it is known. ◆

EXAMPLE 13.4
A Two-Tailed Test of Television Watching for Men and Women

In Example 12.2 we used information from $n = 175$ Penn State students to estimate the mean amount of time that college students watch television in a typical day. Do you think that the mean time watching television is *different* for the population of college men than it is for the population of college women?

Step 1: A two-sided alternative hypothesis is appropriate because there is no prior knowledge that one specific sex might have a higher mean. Therefore, if μ_1 and μ_2 are the mean population daily television viewing hours for men and women, respectively, for the populations represented by these students, then the appropriate hypotheses are

$$H_0: \mu_1 - \mu_2 = 0$$
$$H_a: \mu_1 - \mu_2 \neq 0$$

Step 2: The conditions are met for a two-sample *t*-test to be valid because television viewing hours are available for large independent samples of men and women. Minitab provides the test statistic and furnishes the information necessary to show how to compute it by hand:

```
Sex        N      Mean    StDev    SE Mean
Male      59      2.37    1.87     0.24
Female   116      1.95    1.51     0.14

95% CI for mu (Male) − mu (Female): (−0.14, 0.98)
T-Test of mu (Male) − mu (Female) (vs not =): T = 1.49   P = 0.140   DF = 97
```

In the last line of output, we see that the test statistic is T = 1.49. The calculation is

- ◆ $\bar{x}_1 - \bar{x}_2 = 2.37 - 1.95 = 0.42$ hours

- ◆ $\text{s.e.}(\bar{x}_1 - \bar{x}_2) = \sqrt{\dfrac{s_1^2}{n_1} + \dfrac{s_2^2}{n_2}} = \sqrt{\dfrac{1.87^2}{59} + \dfrac{1.51^2}{116}} = 0.281$

- ◆ $t = \dfrac{\text{Sample statistic} - \text{Null value}}{\text{Standard error}} = \dfrac{0.42 - 0}{0.281} = 1.49$

Steps 3, 4, and 5: *p-value and conclusion.* The *p*-value is the probability that a sample mean difference could be as far or further from 0 as 0.42 hours is, *in either direction,* if the population mean difference is actually 0. This is determined by finding the probability that the value of a Student's *t*-distribution is above 1.49 or below −1.49. For this test, use approximate df = 97 (see Minitab output above) or if necessary, use df = 59 − 1 = 58 (minimum sample size − 1). In either case, the *p-value* is 2 times

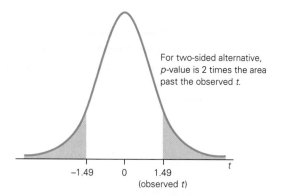

FIGURE 13.6   *p*-Value region for Example 13.4

the area to the right of 1.49 in a *t*-distribution with the stated degrees of freedom (97 or 58), which is 0.14 for either degrees of freedom. Figure 13.6 illustrates this area and demonstrates that it is actually the area to the left of -1.49 plus the area to the right of 1.49. It is simpler computationally to find $2 \times$ the area to the right of 1.49.

Notice that although the sample mean of the daily viewing hours is higher for the men, the *p*-value of 0.14 is greater than 0.05, so we cannot reject the null hypothesis that $\mu_1 - \mu_2 = 0$. On the basis of these *samples,* there is insufficient evidence to conclude that the mean *population* television viewing hours for college men and women are different.

Substitute Steps 3 and 4: As is always the case, the same conclusion is reached using the *rejection region* approach. The conclusion in this case is that we cannot reject the null hypothesis. To find the critical value for this example, use Table A.2 with df = 58. (If Minitab had not provided the degrees of freedom, we would have to use the conservative value of smaller sample size -1, which is $59 - 1 = 58$, so we will use it here.) Because the alternative hypothesis is two-sided, H_a: $\mu_1 - \mu_2 \neq 0$, for $\alpha = 0.05$ use the 0.95 column. The critical value falls between 2.01 and 2.00; we will use the more conservative 2.01 as the critical value. The rejection region is $t \leq -2.01$ and $t \geq 2.01$. Because the test statistic $t = 1.49$ is *not* in the rejection region, we *cannot* reject the null hypothesis. This is the same conclusion we reached based on the *p*-value of 0.14. ◆

---

in summary   The Steps for a Two-Sample
t-Test (Unpooled)

Step 1: Determine the *null* and *alternative* hypotheses.

Null hypothesis: H_0: $\mu_1 - \mu_2 = 0$

Alternative hypothesis: H_a: $\mu_1 - \mu_2 \neq 0$ or H_a: $\mu_1 - \mu_2 > 0$ or
H_a: $\mu_1 - \mu_2 < 0$

where the format of the alternative hypothesis depends on the research question of interest and the order in which population 1 and population 2 are defined.

Step 2: Verify necessary data conditions, and if met, summarize the data into an appropriate *test statistic*.

Verify that both n's are large *or* that there are no extreme outliers or skewness in either sample. The samples must also be independent. Compute the test statistic:

$$t = \frac{\text{Sample statistic} - \text{Null value}}{\text{Standard error}} = \frac{\bar{x}_1 - \bar{x}_2 - 0}{\sqrt{\frac{s_1^2}{n_1} + \frac{s_2^2}{n_2}}}$$

Steps 3, 4, and 5: Proceed exactly as instructed in the In Summary "The Steps for Testing a Single Mean with a One-Sample *t*-Test" on page 444, except in the instructions for step 3, use df = (smaller of $n_1 - 1$, $n_2 - 1$) and replace μ with $(\mu_1 - \mu_2)$ and μ_0 with 0.

Pooled Two-Sample *t*-Test

The use of the *t*-distribution for finding *p*-values for the two-sample *t*-test is an approximation. However, if the population standard deviations are the same, then there is a more precise method available. The **pooled two-sample *t*-test** is based on the assumption that $\sigma_1 = \sigma_2 = \sigma$, the common standard deviation for both populations. The sample variances are combined to provide a **pooled sample variance, s_p^2,** and **pooled standard deviation, s_p.** The formula for the pooled sample variance is

$$\text{Pooled sample variance} = s_p^2 = \frac{(n_1 - 1)s_1^2 + (n_2 - 1)s_2^2}{n_1 + n_2 - 2}$$

The computations for the *t*-statistic and *p*-value remain the same as with the regular two-sample *t*-test except for two small changes. First, the pooled sample variance is used in place of the individual sample variances when computing the standard error. Second, the degrees of freedom are df $= n_1 + n_2 - 2$.

in summary   Pooled Two-Sample *t*-Test

The test statistic is

$$\text{Pooled } t = \frac{\text{Sample statistic} - \text{Null value}}{\text{Pooled standard error}}$$

where

$$\text{Pooled standard error} = \text{Pooled s.e.}(\bar{x}_1 - \bar{x}_2) = \sqrt{\frac{s_p^2}{n_1} + \frac{s_p^2}{n_2}} = s_p\sqrt{\frac{1}{n_1} + \frac{1}{n_2}}$$

Under the null hypothesis, the pooled *t*-statistic has a *t*-distribution with df $= n_1 + n_2 - 2$. All other steps are the same as for the unpooled two-sample *t*-test.

Although the pooled t-test provides an exact solution and the "unpooled" version provides only an approximate solution, we do not recommend that the pooled t-test be used except in very specific circumstances. It is better to have an approximate solution that is known to work well than to use an exact solution based on an assumption that cannot be verified.

Guidelines for Using the Pooled Two-Sample t-Test

◆ When $n_1 = n_2$, the pooled and unpooled standard errors are equal and so the t-statistic also is the same for both the unpooled and pooled procedures. The pooled procedure, however, is not exact unless the population standard deviations are equal, so generally it is preferable to use Welch's approximate df. If this value is not available, when the two sample sizes are equal it is an acceptable substitute to use df $= n_1 + n_2 - 2$. This recommendation can be extended to include situations where the sample sizes are close.

◆ When n_1 and n_2 are very different, the pooled test can be quite misleading unless the sample standard deviations are similar. Some books recommend using the pooled test as long as the ratio of sample standard deviations (larger s/smaller s) is ≤ 2, but this cutoff is completely arbitrary and we do not recommend it.

◆ It can be shown using algebra that

$$\text{Pooled s.e.}(\bar{x}_1 - \bar{x}_2) \approx \sqrt{\frac{s_1^2}{n_2} + \frac{s_2^2}{n_1}}$$

where the symbol "$\approx$" means "is approximately equal to." Therefore, using the pooled standard error is essentially equivalent to reversing the roles of n_1 and n_2 in the standard error. If the larger sample produces the smaller sample standard deviation, the standard error will tend to be underestimated, the t-statistic will tend to be too large, and the null hypothesis will be rejected too easily. Therefore, if the sample sizes are very different and the smaller standard deviation accompanies the larger sample size, we do not recommend using the pooled procedure.

◆ If the sample sizes are very different, the standard deviations are similar, and the larger sample size produced the larger standard deviation, the pooled t-test is acceptable because it will be conservative.

EXAMPLE 13.4—*CONTINUED*
Misleading
Pooled t-Test for
Television Watching
for Men and Women

In Example 13.4, we compared television viewing hours for male and female college students and could not reject the hypothesis that they had the same population means. The sample size for males was only about half of what it was for females, but the standard deviation was larger for males. Here are the summary statistics:

	Sample Size	Mean	Standard Deviation	Variance
Male	$n_1 = 59$	2.37	1.87	3.50
Female	$n_2 = 116$	1.95	1.51	2.28

The difference in means is 0.42 hour, the unpooled standard error is 0.281, the test statistic is 1.49, and the resulting *p*-value is 0.14. Here are the computations for the *pooled t-test:*

$$s_p^2 = \frac{(n_1 - 1)s_1^2 + (n_2 - 1)s_2^2}{n_1 + n_2 - 2} = 2.69, \qquad s_p = 1.64$$

$$\text{Pooled s.e.}(\bar{x}_1 - \bar{x}_2) = \sqrt{\frac{s_p^2}{n_1} + \frac{s_p^2}{n_2}} = s_p\sqrt{\frac{1}{n_1} + \frac{1}{n_2}} = 1.64\sqrt{\frac{1}{59} + \frac{1}{116}} = 0.262$$

$$t = \frac{\text{Sample statistic} - \text{Null value}}{\text{Pooled standard error}} = 0.42/0.262 = 1.60$$

$$\text{df} = 59 + 116 - 2 = 173; \qquad p\text{-value} = 0.11$$

Although the *p*-value for the pooled test still would not lead to rejection of the null hypothesis, it is noticeably smaller than the *p*-value for the unpooled test. In similar situations, in which sample sizes are very different and the larger sample has the smaller variance, the pooled *t*-test and inflated test statistic could provide a misleading conclusion to reject the null hypothesis. ◆

EXAMPLE 13.5
Legitimate Pooled t-Test for Comparing Male and Female Sleep Time

In Example 12.9, we compared the mean sleep time for male and female college students by finding a confidence interval for the difference. The data were collected in statistics classes at the University of California, Davis, at the beginning of the Spring 2000 quarter. Here is a summary of the results from Minitab, displaying both the unpooled and pooled versions:

Two Sample T-Test and Confidence Interval

Two sample T for sleep [without "Assume Equal Variance" option]

sex	N	Mean	StDev	SE Mean
f	83	7.02	1.75	0.19
m	65	6.55	1.68	0.21

95% CI for mu (f) − mu (m): (−0.10, 1.02)
T-Test mu (f) = mu (m) (vs not =): T = 1.62   P = 0.11   DF = 140

Two sample T for sleep [with "Assume Equal Variance" option]

sex	N	Mean	StDev	SE Mean
f	83	7.02	1.75	0.19
m	65	6.55	1.68	0.21

95% CI for mu (f) − mu (m): (−0.10, 1.03)
T-Test mu (f) = mu (m) (vs not =): T = 1.62   P = 0.11   DF = 146
Both use Pooled StDev = 1.72

Notice that in this case, the sample sizes are similar, as are the sample standard deviations. Consequently, the pooled standard deviation of 1.72 is similar to the separate standard deviations of 1.75 and 1.68. The test statistic of 1.62 is identical under both procedures, and the *p*-value of 0.11 is the same. The use of the pooled procedure was warranted in this case. ◆

Minitab Tip

Minitab v. 14 update on CD

 tech note **Computing a Two-Sample *t*-Test to Compare Means**

♦ To carry out a two-sample *t*-test, use **Stat**>**Basic Statistics**> **2-sample t.** Specify the location of the data. There are two ways the data can be organized. The raw data for the response may be in one column (*Samples*) and the raw data for group categories (*Subscripts*) in a second column, or the raw data for the two independent groups may be in two separate columns.

♦ To compute a *pooled* two-sample *t*-test, click on ***Assume Equal Variances.***

♦ To specify the alternative hypothesis, use the ***Options*** button.

Note: Use the ***Graphs*** button to create a comparative dotplot or a comparative boxplot.

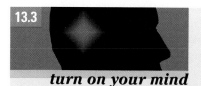

13.3

turn on your mind

The paired *t*-test introduced in Section 13.2 and the two-sample *t*-test introduced in this section are both used to compare two sets of measurements, and the null hypothesis in both cases is usually that the mean population difference is 0. Explain the difference in the situations for which they are used. Suppose researchers wanted to know if college students spend more time watching TV or exercising. Explain how they could collect data appropriate for a paired *t*-test and how they could collect data appropriate for a two-sample *t*-test.*

13.4 | TESTING THE DIFFERENCE BETWEEN TWO POPULATION PROPORTIONS

Researchers often wish to examine the difference between two populations with regard to the proportions falling into a particular category of a response variable. It is almost always of interest to test whether the two population proportions are equal, and that is the only case we will consider. The two populations may be represented by two categories of a categorical variable, such as when we want to compare the proportions of Republicans and Democrats who support a certain political issue. Or, the two populations may be hypothetical, represented by different treatments in an experiment. For instance, we may want to compare the proportions of smokers who would quit smoking if wearing a nicotine patch versus if wearing a placebo patch.

When two proportions are equal, the standard errors of the sample statistics (the sample proportions) are equal as well. This fact will be used to estimate the

*HINT: If you were in the study and they asked you (and other participants) about both time watching TV and time exercising, would the data be paired or independent samples?

null standard error for the test statistic. So, unlike in tests for means, the null standard error formula used for the test differs from the standard error formula used in Chapter 12 to find confidence intervals for the difference in two proportions. Also, it is appropriate to use the phrase "null standard error" rather than simply "standard error" because the calculation is connected to assuming the null hypothesis to be true.

EXAMPLE 13.6
The Prevention of Ear Infections

Based on its biochemical properties, Finnish researchers hypothesized that regular use of the sweetener xylitol might be useful for preventing ear infections in preschool children and carried out a study to test this hypothesis (Uhari, 1998). In a randomized experiment, $n_1 = 165$ children took five daily doses of placebo syrup, and 68 of these children got an ear infection during the three months of the study. Another $n_2 = 159$ children took five daily doses of xylitol, and 46 of these children got an ear infection during the study. The sample proportions getting an ear infection are $\hat{p}_1 = 68/165 = .412$ for the placebo group and $\hat{p}_2 = 46/159 = .289$ for the xylitol group, and the difference between these two proportions is .123 (12.3%). Is this observed difference in proportions large enough to conclude in general that using xylitol reduces the risk of ear infection? We will return to this example after summarizing the steps for the **hypothesis test for comparing two population proportions.** ◆

Step 1: Determine the Null and Alternative Hypotheses

The *null* hypothesis is that there is no difference in population proportions:

$$H_0: p_1 - p_2 = 0 \quad (\text{or } p_1 = p_2)$$

The *alternative* hypothesis may be either two-sided or one-sided, depending on the research question of interest:

$$H_a: p_1 - p_2 \neq 0 \quad or \quad H_a: p_1 - p_2 > 0 \quad or \quad H_a: p_1 - p_2 < 0$$

The alternative hypothesis may also be written as

$$H_a: p_1 \neq p_2 \quad or \quad H_a: p_1 > p_2 \quad or \quad H_a: p_1 < p_2$$

The direction of the alternative hypothesis for a one-sided test depends on which group is defined as population 1 and which is defined as population 2. It is often easier to clarify the correct direction by writing the alternative hypothesis in the second format.

Step 2: Verify Necessary Data Conditions, and If Met, Summarize the Data into an Appropriate Test Statistic

Conditions Necessary for a z-Test of the Difference in Two Proportions to Be Valid

◆ Independent samples are available from the two populations.

◆ The number with the trait or response of interest and the number without the trait or response of interest is at least 5 in each sample, and preferably at least 10.

Make sure the data collection was done so as to ensure independent samples, then simply make sure there are at least 5 (or 10) with and without the trait or response of interest in each of the two samples.

Continuing Step 2: Computing the Test Statistic for a z-Test for Two Proportions

The sample statistic $\hat{p}_1 - \hat{p}_2$ (difference between sample proportions) estimates the parameter $p_1 - p_2$ (difference between population proportions). To find the null standard error, assume the null hypothesis is true, so that $p_1 = p_2 = p$. Estimate the common population proportion p using all of the data:

$$\hat{p} = \frac{n_1\hat{p}_1 + n_2\hat{p}_2}{n_1 + n_2}$$

Despite the complicated formula, this combined estimate is simply the sample proportion for both samples combined. It is used instead of separate estimates of p_1 and p_2 to find the null standard error. The rationale for this substitution is that the test is based on the distribution of the test statistic *assuming the null hypothesis is true.* If $p_1 = p_2$, then $\hat{p}$ is the best estimate for each of the population proportions because it makes use of all available sample data. The standardized test statistic is

$$z = \frac{\text{Sample statistic} - \text{Null value}}{\text{Null standard error}} = \frac{\hat{p}_1 - \hat{p}_2 - 0}{\sqrt{\dfrac{\hat{p}(1 - \hat{p})}{n_1} + \dfrac{\hat{p}(1 - \hat{p})}{n_2}}}$$

$$= \frac{\hat{p}_1 - \hat{p}_2 - 0}{\sqrt{\hat{p}(1 - \hat{p})\left(\dfrac{1}{n_1} + \dfrac{1}{n_2}\right)}}$$

If the null hypothesis is true, the sampling distribution of this z-statistic is approximately the standard normal curve.

Step 3: Assuming the Null Hypothesis is True, Find the p-Value

The standard normal curve is used to find the *p*-value:

◆ For $H_a: p_1 - p_2 > 0$, the *p*-value is the area above z (a one-tailed test), even if z is negative.

◆ For $H_a: p_1 - p_2 < 0$, the *p*-value is the area below z (a one-tailed test), even if z is positive.

◆ For $H_a: p_1 - p_2 \neq 0$, the *p*-value is $2 \times$ area above $|z|$ (a two-tailed test).

Do not confuse the *p-value* with the population or sample *proportions,* which also use the letter *p*.

Steps 4 and 5: Decide Whether or Not the Result Is Statistically Significant Based on the p-Value and Make a Conclusion in the Context of the Situation

Choose a significance level ("alpha"); standard practice is $\alpha = 0.05$. The result is statistically significant if the *p*-value $\leq \alpha$. Interpret the conclusion in the context of the situation. Whenever two populations are compared, the manner in which the data were collected should also be considered. Remember from Chapter 3 that an experiment is generally a stronger proof of causation than an observational study.

EXAMPLE 13.6—*CONTINUED*
The Prevention of Ear Infections

Step 1: *Determine the null and alternative hypotheses.* The researchers hoped to show that xylitol reduces ear infections. The parameter p_1 is the proportion who would get an ear infection in the population of children similar to those in the study if taking the placebo. The parameter p_2 is the proportion who would get an ear infection in that population if taking the xylitol. Therefore, the null and alternative hypotheses of interest are

$$H_0: p_1 - p_2 = 0 \quad \text{(or } p_1 = p_2\text{)}$$
$$H_a: p_1 - p_2 > 0 \quad \text{(or } p_1 > p_2\text{)}$$

A one-sided alternative is used because the researchers want to show that the proportion getting an ear infection is significantly lower in the xylitol group than in the placebo group.

Step 2: *Verify necessary data conditions, and if met, summarize the data into an appropriate test statistic.* There are at least 10 children in each sample who did and did not get ear infections, so the conditions are met.

♦ $\hat{p}_1 = \dfrac{68}{165} = .412 \quad \text{and} \quad \hat{p}_2 = \dfrac{46}{159} = .289.$

♦ The sample statistic is $\hat{p}_1 - \hat{p}_2 = .412 - .289 = .123.$

♦ The combined proportion is $\hat{p} = \dfrac{68 + 46}{165 + 159} = \dfrac{114}{324} = .35.$

♦ null s.e.$(\hat{p}_1 - \hat{p}_2) = \sqrt{\hat{p}(1-\hat{p})\left(\dfrac{1}{n_1} + \dfrac{1}{n_2}\right)} = \sqrt{(.35)(.65)\left(\dfrac{1}{165} + \dfrac{1}{159}\right)} = .053.$

♦ $z = \dfrac{\text{Sample statistic} - \text{Null value}}{\text{Null standard error}} = \dfrac{.123}{.053} = 2.32.$

Steps 3, 4, and 5: Because the alternative is $H_a: p_1 - p_2 > 0$, the *p*-value is the area above $z = 2.32$. Figure 13.7 illustrates this probability. From the normal probability table (Appendix Table A.1), we learn that the probability is 0.9898 that z is less than 2.32, so *p*-value $= 1 - 0.9898 = 0.0102$. We can reject the null hypothesis and attach the label "statistically significant" to the result because the *p*-value is small. Based on this experiment, we can conclude that taking xylitol would reduce the proportion of ear infections in the population of similar preschool children in comparison to taking a placebo. ♦

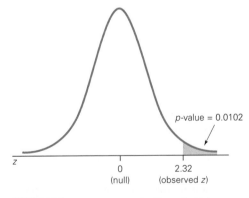

z

0
(null)

2.32
(observed z)

p-value = 0.0102

FIGURE 13.7   *p*-Value region for Example 13.6

> **in summary** The Steps for a z-Test for the Difference in Two Proportions
>
> **Step 1:** Determine the *null* and *alternative* hypotheses.
>
> | *Null hypothesis:* | $H_0: p_1 - p_2 = 0$ |
> |---|---|
> | *Alternative hypothesis:* | $H_a: p_1 - p_2 \neq 0$ *or* $H_a: p_1 - p_2 > 0$ *or* |
> | | $H_a: p_1 - p_2 < 0$ |
>
> where the format of the alternative hypothesis depends on the research question of interest and the order in which population 1 and population 2 are defined.
>
> **Step 2:** Verify necessary data conditions, and if met, summarize the data into an appropriate *test statistic*.
>
> Verify that samples are large enough so that for each sample, $n\hat{p}$ and $n(1 - \hat{p})$ are at least 5, preferably 10. The samples also must be independent. Compute the test statistic:
>
> $$z = \frac{\text{Sample statistic} - \text{Null value}}{\text{Null standard error}} = \frac{(\hat{p}_1 - \hat{p}_2) - 0}{\sqrt{\hat{p}(1 - \hat{p})\left(\dfrac{1}{n_1} + \dfrac{1}{n_2}\right)}}$$
>
> **Step 3:** Assuming the null hypothesis is true, find the *p-value*.
>
> Using the z (standard normal) distribution, the *p*-value is the area in the tail(s) beyond the test statistic z as follows:
>
> For $H_a: p_1 - p_2 \neq 0$ the *p*-value is $2 \times$ area above $|z|$ (a two-tailed test).
> For $H_a: p_1 - p_2 > 0$ the *p*-value is the area above z even if z is negative.
> For $H_a: p_1 - p_2 < 0$ the *p*-value is the area below z even if z is positive.
>
> **Steps 4 and 5:** Proceed exactly as instructed in the In Summary "The Steps for Testing a Single Mean with a One-Sample *t*-Test" on page 444.

Using Minitab for a Test of Two Proportions

Output from the Minitab (Release 12) procedure for examining two proportions for Example 13.6 is shown below:

```
Sample          X          N         Sample p
Placebo        68         165          0.412
Xylitol        46         159          0.289

Estimate for p(1) − p(2): 0.123
95% CI for p(1) − p(2): (0.020, 0.226)
Test for p(1) − p(2) = 0 (vs > 0): Z = 2.34   P-Value = 0.010
```

The default in Minitab is *not* to use the combined sample proportion in the formula for the standard error, because the Minitab procedure for two proportions is used to compute a confidence interval as well as to carry out a test of hypotheses. In other words, Minitab is using the standard error formula in Chapter 12 in the denominator of the test statistic, rather than the null standard error formula given in this chapter. However, the results are just slightly different and the conclusion is the same. This will generally, but not always, be the case.

Minitab Tip

 tech note   **Computing a Two-Sample *z*-Test for the Difference in Proportions**

- ◆ To test hypotheses about the difference in two proportions, use **Stat>Basic Statistics>2 Proportions.** See note below for information about inputting data.
- ◆ To specify the alternative hypothesis, use the **Options** button.
- ◆ To compute the *z*-statistic described in this section, use **Options,** and then click **Use pooled estimate of p for test.**

Note: There are three possibilities for inputting data. The raw data for the response may be in one column and the raw data for group categories (*Subscripts*) may be in a second column. Or, the raw data for the two independent groups may be in two separate columns. Or, the data may already be summarized. If so, click **Summarized Data,** then specify the sample size and the *number* of successes for each group.

13.4

turn on your mind

The raw data for Example 13.6 can be thought of as two categorical variables (with two possible categories each) for each participant in the experiment. What are the two variables and their categories? Construct a contingency table in the format learned in Chapter 6, with the explanatory variable as rows. In Chapter 6, you learned how to look for relationships by finding a chi-square statistic and associated *p*-value. In the context of this example, how do you think the null and alternative hypotheses would be stated if the problem is viewed from that perspective? (You will learn how to do this in Chapter 15.)*

13.5 | THE RELATIONSHIP BETWEEN SIGNIFICANCE TESTS AND CONFIDENCE INTERVALS

In the situations covered in Chapters 10–13, you may have noticed a direct correspondence between the values covered by a confidence interval and the results of two-sided hypothesis tests. The correspondence is precise for tests involving one or two population means and a two-sided alternative.

*HINT: One variable is treatment (xylitol or placebo). What else did the researchers measure for each child?

Confidence Intervals and Tests with Two-Sided Alternatives

When testing one population mean or the difference in two population means with the hypotheses

$$H_0: \text{parameter} = \text{null value} \quad and \quad H_a: \text{parameter} \neq \text{null value}$$

- If the null value is covered by a $(1 - \alpha)100\%$ confidence interval, the null hypothesis is not rejected and the test is not statistically significant at level α.
- If the null value is not covered by a $(1 - \alpha)100\%$ confidence interval, the null hypothesis is rejected and the test is statistically significant at level α.

Don't be confused by the cumbersome notation $(1 - \alpha)100\%$. The correspondence is between a 95% confidence interval and a significance level of 0.05, a 99% confidence interval and a significance level of 0.01, and so on.

VIDEO

EXAMPLE 13.4—
CONTINUED
Mean Daily Television Hours of Men and Women

In Example 13.4, we tested whether the population mean daily television viewing hours differed for male and female college students. The Minitab output is reprinted here:

Sex	N	Mean	StDev	SE Mean
Male	59	2.37	1.87	0.24
Female	116	1.95	1.51	0.14

95% CI for mu (Male) − mu (Female): (−0.14, 0.98)
T-Test of mu (Male) − mu (Female) (vs not =): T = 1.49   P = 0.140   DF = 97

Notice that a 95% confidence interval for the difference in population means is −0.14 to +0.98 hours. In other words, with 95% confidence we can say that $-0.14 \leq \mu_1 - \mu_2 \leq 0.98$ hours. The null value for the test was 0 hours, which is covered by this interval. Therefore, the difference in sample means, which is 0.42 hours, is not significantly different from the population null value of 0.

The correspondence also holds for one or two proportions, with one caveat. In those situations, the standard error used for the confidence interval is slightly different from the null standard error used for the hypothesis tests. Therefore, the correspondence may not hold if the null value is close to one of the boundaries of the confidence interval. In most cases, however, this minor technical detail will not interfere with the correspondence.

This correspondence means that a confidence interval can be used as an alternative way to conduct a two-sided significance test. Some researchers think that whenever possible, a confidence interval should be used to test hypotheses because an interval provides information about the value of a parameter as well. ◆

Confidence Intervals and One-Sided Tests

It is possible to construct one-sided confidence intervals by allowing one end of the interval to extend to the minimum or maximum possible value of the parameter, then adjusting the confidence level appropriately. In that case, the same correspondence would hold between one-sided confidence intervals and one-sided tests. It is also possible to use a standard two-sided confidence interval to test a one-sided alternative by appropriately adjusting the significance level of the test or the confidence level for the interval. We present the general result, but the caveat just given in the two-sided case for one or two proportions holds in this case as well.

When testing the hypotheses H_0: parameter = null value versus a one-sided alternative, compare the null value to a $(1 - 2\alpha)100\%$ confidence interval:

- If the null value is covered by the interval, the test is not statistically significant at level α.
- For the alternative H_a: parameter > null value, the test is statistically significant at level α if the entire interval falls *above* the null value.
- For the alternative H_a: parameter < null value, the test is statistically significant at level α if the entire interval falls *below* the null value.

For instance, if a 90% confidence interval falls completely above the null value for a test and the alternative hypothesis is H_a: parameter > null value, then the null hypothesis can be rejected at a significance level of $\alpha = 0.05$. As an illustration, a 90% confidence interval for $\mu_1 - \mu_2$ that contained only positive values would allow researchers to accept the alternative hypothesis H_a: $\mu_1 - \mu_2 > 0$ using $\alpha = 0.05$.

EXAMPLE 13.6—*CONTINUED*
Ear Infections and Xylitol

Consider Example 13.6 in which a 95% confidence interval for $p_1 - p_2$, the difference in proportions of children who would get ear infections with placebo and xylitol, is .020 to .226. This interval tells us that we can reject H_0: $p_1 - p_2 = 0$ and accept H_a: $p_1 - p_2 > 0$ with $\alpha = 0.025$, because the entire 95% confidence interval falls above the null value of 0. The *p*-value for the test was 0.01, and since $0.01 < 0.025$, this conclusion is confirmed using the standard approach as well. Notice that if the alternative hypothesis had been in the opposite direction, that $p_1 - p_2 < 0$, the null hypothesis would *not* be rejected even though the null value of 0 is not contained in the interval. This conclusion fits with our common sense. The interval indicates that the possible values for the difference are all *above* 0 (.02 to .226), so the alternative hypothesis that the true difference is *less* than 0 is certainly not supported by the interval. ◆

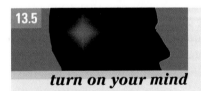

13.5

turn on your mind

Refer to the continuation of Example 13.6. A 95% confidence interval for the difference in proportions who would get ear infections with placebo compared to with xylitol was .02 to .226. Based on this information, specify a one-sided 97.5% confidence interval and explain how you would use it to test H_0: $p_1 - p_2 = 0$ versus H_a: $p_1 - p_2 > 0$ with $\alpha = 0.025$.*

13.6 | CHOOSING AN APPROPRIATE INFERENCE PROCEDURE

In this chapter and the previous three, we have learned how to construct confidence intervals and conduct hypothesis tests to make inferences for several different population parameters. The focus has been on how to carry out and interpret these inference procedures after the procedure and population parameter of interest have already been identified. In this section, we provide guidance on how to identify the appropriate procedure based on the research question(s) of interest. Necessary decisions include what parameter to investigate and whether to construct a confidence interval, conduct a hypothesis test, or do both.

Confidence Interval or Hypothesis Test?

When using sample data to answer a question about a larger population, how can you decide whether to construct a confidence interval, conduct a hypothesis test, or do both? The key point to consider is whether the main purpose is to estimate the numerical value of a parameter or to make a "maybe not" or "maybe yes" type of conclusion about a specific hypothesized value. If the purpose is to estimate a numerical value, a confidence interval is appropriate. If the purpose is to make a conclusion about a specific value, hypothesis testing is appropriate. Because these procedures provide different types of information about the population, it may be appropriate to use both in many situations.

As an example, think about the difference between the types of information asked for by the following two questions:

> *Question 1*: What is the mean number of hours that students study each week?
>
> *Question 2*: Is there a difference between mean hours of study per week for females versus males?

Notice that question 1 asks for a numerical value. The answer might be something like "about 18 hours per week." A confidence interval can be used to give an estimate that reflects the uncertainty due to using a sample to estimate the population parameter. For example, we might answer the question about mean number of hours studying with a statement like "With 95% confidence, we estimate that the mean hours of study per week for all students at our school is between 17.2 and 18.8 hours." In general, a confidence interval is used if the main purpose is to estimate the value of a parameter.

How would you answer question 2? It doesn't ask us to estimate numerical values of mean hours of study for men and women, but instead simply asks if there is a difference. Given the question wording, we only have to answer with something like "Yes, there appears to be a difference" or "We cannot say there is a difference." We can create an answer by conducting a hypothesis test (also called a significance test) of the null hypothesis that the difference in means for men and women is 0 versus the alternative hypothesis that the difference in means is not 0. In general, a hypothesis test is used if the main purpose is to provide a "maybe not" or "maybe yes" type of answer for a research question.

In many situations, it may be informative to conduct a significance test and construct a confidence interval as well. For question 2 above, for instance, it would not only be useful to know whether there was a difference, it would also be informative to know the magnitude of a possible difference. A confidence interval for the difference in means for the populations of men and women would give us that information. It is not always the case that both procedures should be used. When the primary purpose is to estimate the numerical value of a single parameter, it usually doesn't make sense to conduct a hypothesis test because there's no specific value of interest. As an example, there's no specific or hypothesized value for the parameter of interest for question 1 above.

Let's summarize what can be learned from constructing a confidence interval or conducting a hypothesis test.

- ◆ A confidence interval estimate provides a range of values that is likely to contain the truth about the population parameter, so it gives information about the *magnitude* of the parameter. The confidence level, typically 95%, gives information about how often the procedure will result in an interval that successfully captures its prey (the parameter) in the long run.

- ◆ The *p*-value and conclusion from hypothesis testing provide a "maybe not" or "maybe yes" answer to the question, "Is the parameter equal to (or different from) a certain value?" They do not give information about the magnitude of a parameter.

- ◆ In many cases where hypothesis testing is the primary goal, it makes sense to construct a confidence interval to accompany the results of the hypothesis test.

- ◆ In most cases where a confidence interval estimate is the primary goal, it does not make sense to conduct a hypothesis test to accompany the interval because there is no specific parameter value to be tested.

Determining the Appropriate Parameter

In Chapters 10–13 we learned inference procedures for five specific population parameters:

- p: One proportion = population proportion in a category of a categorical variable, or long-run probability of a particular outcome.
- μ: One mean = mean for a quantitative variable in a population, or a quantitative outcome in a random circumstance.
- μ_d: Paired difference mean = population mean for the difference in two paired variables or quantitative outcomes; use when paired samples are available.
- $p_1 - p_2$: Difference in two population proportions or long-run probabilities; use when independent samples are available.
- $\mu_1 - \mu_2$: Difference in two means = difference in population means for two populations, use when independent samples are available.

Determining which of these is the parameter of interest depends on the answers to these questions:

- ◆ Is there one sample or two? If there are two, are they independent samples or are they paired?
- ◆ For one sample, is the variable measured on each unit quantitative or categorical?
- ◆ For two independent samples, one variable identifies the two samples and is categorical. Is the other variable also categorical?

TABLE 13.2 ■ The Parameters and Types of Data Used for Inferences About Them

Variable Type (Parameter Type)	One Sample (No Pairing)	Paired Data	Two Independent Samples
Categorical (Proportions)	p	none	$p_1 - p_2$
Quantitative (Means)	μ	μ_d	$\mu_1 - \mu_2$

Based on the answers to these questions, you can identify which of the five parameters is appropriate. Table 13.2 illustrates the process.

When you are first learning to distinguish among the situations, it helps to have some clues. Here are tips for helping you determine which parameter to use, assuming the choices are from the five covered in Chapters 10–13.

- ◆ *Are two different sample sizes given, or is it clear that two independent samples were taken?* If so, the parameter of interest must be either $p_1 - p_2$ or $\mu_1 - \mu_2$.

- ◆ *Are one or more sample means given?* If so, the parameter is probably a mean or difference in two means.

- ◆ *Are sample counts or proportions given?* If so, the parameter is probably a proportion or a difference in two proportions.

After you determine the parameter of interest and whether to conduct a hypothesis test and/or construct a confidence interval, you can use the summary tables at the end of this chapter and at the end of Chapter 12 to review the computational details of the analysis.

Examples of Choosing an Inference Procedure

The following examples are quotes about studies from other chapters in this textbook. In each case, think about what type of inference is appropriate.

EXAMPLE 13.7
Kids and Weight Lifting

In Case Study 3.2, a randomized experiment was presented in which children were randomly assigned to one of two weight-lifting conditions or a control condition. Muscular strength and endurance were measured after eight weeks. One of the results reported in Case Study 3.2 was "leg extension strength significantly increased in both exercise groups compared with that in the control subjects."

Notice that the researchers are comparing each of the two weight-lifting conditions to the control condition. Let's focus on comparing the heavy-lifting group to the control group for the reported response variable, leg extension strength.

Decision 1: Confidence interval, hypothesis test or both? Based on the statement about the results, the researchers were clearly interested in knowing *if* there is a significant difference in leg extension strength after heavy lifting compared with no weight lifting. They were also most likely interested in the *magnitude* of the difference. Therefore, both inference procedures would be appropriate.

Decision 2: What is the appropriate parameter to investigate? Let's investigate the three questions used to determine which parameter is appropriate.

♦ *Is there one sample or two? If there are two, are they independent samples or are they paired?* It is clear that there are two samples and that they are independent. One group of children did heavy lifting and the other group did no weight training.

♦ *For one sample, is the variable measured on each unit quantitative or categorical?* This question is not relevant because there are two samples.

♦ *For two independent samples, one variable identifies the two samples and is categorical. Is the other variable also categorical?* The training (heavy or none) identifies the two samples. The other variable measured is the leg extension strength, which is not categorical. Presumably, a quantitative measurement was made for each child.

Based on the answers to these questions, the appropriate column of Table 13.2 is "Two Independent Samples" and the appropriate row is "Quantitative (Means)." Therefore, the parameter of interest is $\mu_1 - \mu_2$, the difference in population means for leg extension strength if all similar children were to participate in the heavy-lifting or control conditions. ♦

EXAMPLE 13.8
Loss of Cognitive Functioning

Example 3.5 reported on an observational study of 6000 older people, which found that 70% of the participants did not lose functioning over time. What inference procedure can be used for this result?

Decision 1: Confidence interval, hypothesis test, or both? There is no information provided about the study that would lead to a natural hypothesis-testing question. The reported result, that 70% of the *sample* lost cognitive functioning over time, leads to a consideration of the margin of error and confidence interval estimate for the proportion of the *population* in this age group that loses cognitive functioning over time. Therefore, hypothesis testing is not appropriate, but a confidence interval is.

Decision 2: What is the appropriate parameter to investigate? The sample statistic reported is a sample proportion for a single sample. People in the study were categorized by whether or not they lost cognitive functioning. Therefore, the parameter of interest is p = proportion that loses cognitive functioning over time for the population of older people represented by this sample. ♦

13.7 | THE TWO TYPES OF ERRORS AND THEIR PROBABILITIES

In Chapter 11, we learned that if the null hypothesis is true but it is rejected, a type 1 error has been made, while if the null hypothesis is false but we fail to reject it, a type 2 error has been made. In most situations, a type 1 error is more serious because it means that the status quo is overturned, and some consequential action is likely. There are situations where it is more serious to make a type 2 error because failure to reject the null hypothesis when it is false could result in the end of a promising line of research.

Determining which type of error is more serious in a given situation is important, especially since researchers have some control over the associated probabilities. In this section, we will learn how decisions can be made to minimize the more serious type of error.

When we describe the probability of making an error, we can only talk in conditional terms. For instance, by definition a type 1 error can only occur when the null hypothesis is true. This means that any description of type 1 error has to include the condition "if the null hypothesis is true." The difficulty is that in practice we never know what is actually true. If we did, we wouldn't need statistical methods. Therefore, it is necessary to consider both possible types of error in any situation.

Probability of a Type 1 Error and the Level of Significance

Recall that a type 1 error occurs when the decision is for the alternative hypothesis but the null hypothesis is actually true. Also, remember that the *level of significance,* or α-level, is the borderline used to decide if a p-value is small enough to justify a decision in favor of the alternative. The level of significance that we use for a hypothesis test also provides the (conditional) **probability of a type 1 error.**

definition When the null hypothesis is true, the **probability of a type 1 error,** the level of significance, and the α-level are all equivalent. When the null hypothesis is not true, a type 1 error cannot be made.

To see the correspondence between the level of significance and the probability of a type 1 error, suppose the null value is the correct population parameter value. In that case, the distribution of the test statistic (z or t) is known and is either the standard normal or the Student's t-distribution. The p-value is the tail area at and beyond the observed test statistic in the tail(s) specified by the alternative hypothesis. Suppose the 0.05 level of significance is being used. By chance alone, the test statistic will fall into the most extreme 5% of the distribution just 5% of the time. When that happens, the p-value will be <0.05, the null hypothesis will be rejected, and a type 1 error will have been made. The same holds for any specified value for α.

This correspondence means that with the standard level of significance of 0.05 or 5%, about 5% of all hypothesis tests done *when the null hypothesis is really true* will result in a false claim of statistical significance. Notice that this is not the same thing as saying that 5% of all rejected null hypotheses should not have been rejected. In fact, some statisticians and researchers argue that in the "real world," the null hypothesis is rarely true. If this argument is correct, the overall number of type 1 errors in practice may be quite small.

One caution to keep in mind is that most research projects involve many hypothesis tests, each done using a significance level of 0.05 or 0.01. We should be cautious when we read that a statistical analysis involved many different hypothesis tests but only a small number of the results were found to be statistically significant. A possible explanation is that all of the various null hypotheses are actually true, but with the 0.05 standard for the level of significance, about 5% of the tests resulted in a conclusion for the alternative hypothesis by chance.

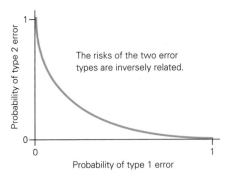

FIGURE 13.8 **The general relationship between the probabilities of the two errors**

The media often fails to report this **multiple-testing** phenomenon, focusing instead on the one or two tests that were statistically significant out of many that were conducted in a study.

The Trade-Off in Probability for the Two Types of Error

If we wanted to be sure that we rarely made a type 1 error, we could simply set the level of significance at a ridiculously low value, making it almost impossible to reject the null hypothesis when it was true. However, if we did this, we would obviously be at great risk of making a type 2 error, because our stringent criterion would make it difficult to reject the null hypothesis even when it was not true.

There is an inverse relationship between the probabilities of the two types of error, illustrated in Figure 13.8. As you can see, an increase in the probability of a type 1 error leads to a decrease in the probability of a type 2 error (and vice versa). The figure is only intended to illustrate the general pattern of the trade-off between the two risks. The exact probabilities will depend on the sample size and the type of test statistic used in a particular situation. Notice that it is *not* the case that the probabilities of the two types of errors sum to 1.0, or to any other specified value.

Type 2 Errors and Power

The probability of making a type 1 error is easy to specify because it involves the assumption that the population parameter is equal to a specific null value. With that assumption, we know precisely what distribution the test statistic has and can thus find probabilities associated with possible outcomes.

It is not so simple to find the **probability of a type 2 error.** Remember that a type 2 error is made when the alternative hypothesis is true. But the alternative hypothesis covers a wide range of possibilities. If the truth is close to the null value, it may be difficult for sample data to provide adequate evidence to reject the null hypothesis, and the probability of making a type 2 error will be high. If the truth is far from the null value, then the sample data will likely provide enough evidence to reject the null value, and the probability of a type 2 error will be low.

There are three factors that affect the probability of making a type 2 error:

1. The sample size; larger n reduces the probability of a type 2 error without affecting the probability of a type 1 error.

2. The level of significance used for the test; larger α reduces the probability of a type 2 error by increasing the probability of a type 1 error.
3. The actual value of the population parameter, which is not in the researcher's control. The farther the truth falls from the null value (as long as it is in the alternative hypothesis), the lower the probability of a type 2 error.

Rather than focusing on the risk of making a mistake, many investigators prefer to focus on the chance that their sample will provide the evidence necessary to make the right choice. A common question when a study is planned is "If something is really going on in the population, what is the probability that we'll be able to detect it?" This question leads to the statistical meaning of the word *power:*

> *definition* When the alternative hypothesis is true, the probability of making the correct decision is called the **power** of a test.

Power is simply the flip side of the risk of a type 2 error, and the same factors that influence the risk of a type 2 error also affect the power.

The concept of power is the same for all statistical hypothesis tests, but the technical formulas and details differ depending on the situation. The "by hand" calculations can be difficult, but fortunately some statistical software programs (including Minitab) can be used to easily determine the power of a test under specified conditions.

VIDEO

EXAMPLE 13.9
Planning a Student Survey

A university is considering a plan to offer regular classes year-round by making the summer session a regular term but will only do so if there is sufficient student interest. A university administrator is planning a student survey to determine if a majority of all students at the school would attend summer session under that structure. The plan will only be implemented if a majority would attend, so the status quo is to assume that the proportion who would attend is not a majority. With p representing the proportion of all students who would attend in the summer, a hypothesis testing structure is

$H_0: p \le .50$ (the proportion who would attend is not a majority)
$H_a: p > .50$ (a majority would attend)

Notice that the alternative hypothesis includes a broad range of possibilities and does not specify an exact value. If the true population proportion who would attend is any value greater than .50, the correct decision is to pick the alternative hypothesis.

Table 13.3 shows the *power* for three different "true" proportions and for three different sample sizes. As you look at the table, keep in mind that the power is the probability that the sample evidence leads us to conclude that a majority of the student population would attend a regular summer term. For example, suppose that in truth 60% of students think they would attend a regular summer term and $n = 100$ students are surveyed. The probability that the sample proportion would be large enough to conclude that $p > .50$ is given in Table 13.3 as .64. That means the probability is .36 that a type 2 error would be

TABLE 13.3 ■ Power for Selected Sample Sizes and True Proportions

		True Population Proportion		
		.52	.60	.65
Sample Size	$n = 50$	.09	.41	.69
	$n = 100$	.11	.64	.92
	$n = 400$	.20	.99	nearly 1

made, and the null hypothesis would not be rejected. Similarly, if the truth is that 52% think they would attend, and a sample of 400 is chosen, the power is only .20 and the probability of a type 2 error is .80.

These power values were determined for tests done with a 0.05 level of significance. Don't worry about how to do the calculations (Minitab was used to get these numbers). ◆

Example 13.9 illustrates that power is a function of multiple factors. What patterns can we see in Table 13.3? You should be able to detect two important relationships that apply to power for all hypothesis tests:

- *The **power** increases when the **sample size** is increased.* We can see this by looking down any column of the table. This makes sense because when the sample size is increased, the standard error is decreased, leading to larger values of the test statistic. Also, the sample statistic is a more accurate estimate of the population value, making it easier to detect a difference between the true population value and the null value.

- *The power increases when the difference between the true population value and the null hypothesis value increases.* We can see this looking across any row of the table. This makes sense because the probability of detecting a large difference is higher than the probability of detecting a small difference. However, remember that the truth about the population is not something the researcher can control or change.

Researchers should evaluate power before they collect data that will be used to do hypothesis tests to make sure they have sufficient power to make the study worthwhile. In Example 13.9, we see that a sample of 50 students would have low power even if the true proportion who would attend in the summer were as high as .60 or even .65, a definite majority. If this hypothesis test is important to the administrator for making a decision, he or she should sample more than 50 students. We also see that if the true population proportion is .52, only a slight majority, there is little chance that the administrator will be able to decide in favor of the alternative, even if 400 students are surveyed.

Occasionally, news reports, especially in science magazines, will insightfully note that a study may have failed to find a relationship between two variables because the test had low power. This is a common consequence of conducting research with samples that are too small, but it is one that is often overlooked in media reports.

13.6

turn on your mind

Draw a tree diagram (Chapter 7) illustrating the possible errors in hypothesis testing by making the first set of branches "the truth" (null or alternative) and the second set of branches the decision (do not reject the null, reject the null). Show where the α-level, the probability of a type 2 error, and the power fit on the tree, and how they relate to each other. For any branches on which you cannot assign probabilities, explain why you can't assign them.*

****HINT:** Remember that conditional probabilities go on the second set of branches—in this case, the probability of a specific decision, given that the null (or alternative) hypothesis is true.

13.8 | EFFECT SIZE

In most of the hypothesis-testing situations in Chapters 11 and 13, we are interested in comparing a population mean or proportion to a specific null value, or to another population mean or proportion. In many research situations, we would like to know something about the magnitude of the comparison. The test statistic and p-value for a test are not useful for this purpose because they depend on the size of the sample. A confidence interval may not be useful because it depends on the units of measurement, and we want a measure that is independent of the units of measurement. This is particularly true if we want to compare research results across studies, where different measurement units may have been used or slightly different treatments may have been assigned.

In general, the **effect size** for a research question is a measure of how much the truth differs from chance or from a control condition. The most common effect-size measure used for a single mean is

$$d = \frac{\mu_1 - \mu_0}{\sigma}$$

where μ_1 is the true population mean, μ_0 is the null value, and σ is the population standard deviation. Notice that d measures the distance between the true mean and the null mean value in terms of number of standard deviations.

As an example, suppose it is known that the mean IQ in a certain population under usual conditions is 110, with a standard deviation of 15. Researchers speculate that listening to classical music temporarily boosts IQ. If indeed the mean IQ after listening to classical music is 115, then the effect size is $(115 - 110)/15 = 1/3$. In other words, listening to classical music boosts the mean IQ by one-third of a standard deviation.

The most common effect-size measure for comparing two means is

$$d = \frac{\mu_1 - \mu_2}{\sigma}$$

where μ_1 and μ_2 are the two population means and σ is the population standard deviation, assumed to be the same for both populations. If this assumption is not reasonable, and if one population corresponds to a control group condition, then the standard deviation for the control condition is used. If that is not possible either, then a pooled standard deviation is used. Notice that d measures the distance between μ_1 and μ_2 in terms of number of standard deviations, using the standard deviation for the individual populations as the yardstick. In most situations, the order of the difference is not important to the magnitude of the effect, so the absolute value of d is reported.

As an example, suppose pulse rates for those who don't exercise have a mean of 72 and standard deviation of 4 beats per minute, while pulse rates for those who do exercise have a mean of 68 with the same standard deviation. Then the effect size for the difference is $(72 - 68)/4 = 1$. In other words, average pulse rates for those who exercise and those who don't exercise differ by 1 standard deviation.

There are many uses for effect sizes, some of which require d to be estimated from data. There are various suggestions for estimating d based on complicated statistical principles (see, e.g., Hedges and Olkin, 1985), but the simplest method is to substitute sample values for the population parameters in d.

Estimating Effect Sizes for One and Two Samples

formula   For a single sample, the estimated effect size is $\hat{d} = \dfrac{\bar{x} - \mu_0}{s}$, where $\bar{x}$ and s are the sample mean and standard deviation, respectively.

For two samples, the estimated effect size is $\hat{d} = \dfrac{\bar{x}_1 - \bar{x}_2}{s}$, where $\bar{x}_1$ and $\bar{x}_2$ are the two sample means and s is either the standard deviation of the control condition measurements (if there is one) or the pooled sample standard deviation.

Notice that there is a relationship between the test statistic t and the estimated effect size. The formula for the estimated effect size includes everything in the formula for the t statistic *except* the influence of the sample size(s).

Relationship Between Test Statistic t and Effect Size

formula   For one sample: $t = \dfrac{\bar{x} - \mu_0}{\dfrac{s}{\sqrt{n}}} = \dfrac{\sqrt{n}(\bar{x} - \mu_0)}{s} = \sqrt{n}\,\hat{d}$, so $\hat{d} = \dfrac{t}{\sqrt{n}}$

For two samples: *Pooled* $t = \dfrac{\bar{x}_1 - \bar{x}_2}{s\sqrt{\dfrac{1}{n_1} + \dfrac{1}{n_2}}} = \dfrac{\hat{d}}{\sqrt{\dfrac{1}{n_1} + \dfrac{1}{n_2}}}$

Although we won't cover other effect-size measures, the relationship for *t*-tests holds in most hypothesis-testing situations for which effect sizes are used. Rosenthal (1991) provides a summary of many of these procedures, and notes that in most cases,

Test statistic = Size of effect × Size of study

as we have seen in the tests for means.

Interpreting an Effect Size

Effect sizes provide information about how strong a difference or effect is in the population, relative to another population or a hypothesized value. The interpretation is similar to the interpretation of a standardized score for a specific individual value. In fact, you should notice that the effect size d for a single mean of a normal population is simply the standardized score for μ_1 relative to the normal distribution with mean μ_0 and standard deviation σ.

Cohen (1988) defined small, medium, and large effect sizes for $|d|$ somewhat arbitrarily to be 0.2, 0.5, and 0.8, respectively. In other words, if the true mean μ_1 is half of a standard deviation away from the null mean value ($|d| = 0.5$), Cohen called this a medium effect. If the true mean μ_1 is only 1/5 of a standard deviation away from the null mean ($|d| = 0.2$), Cohen called it a small effect; similarly for 0.8 as a large effect. These are arbitrary, but he said that a small effect should be detectable only through statistics, a medium effect should be obvious to a careful observer, and a large effect should be obvious to any observer.

EXAMPLE 13.10
Could Aliens Tell
That Women Are Shorter?

Suppose an alien were to visit Earth and observe humans. Would it be obvious that men are taller, on average, than women? The mean heights of adult men and women are about 70 inches and 65 inches, respectively, with standard deviations of about 2.5 inches within each population. Therefore, the effect size for the difference in heights is approximately $d = (70 - 65)/2.5 = 2.0$. Such a large difference should be immediately obvious to anyone.

However, suppose the difference in men's and women's heights was smaller. The following table shows what differences correspond to small, medium, and large effect sizes. Think about whether you would notice the difference in mean heights for two groups whose means differed by these amounts. ◆

Effect Size Magnitude	Difference in Heights when $\sigma = 2.5$	Interpretation
Small ($d = 0.2$)	$2.5 \times .2 = 0.50$ inch	Not obvious without statistics
Medium ($d = 0.5$)	$2.5 \times .5 = 1.25$ inches	Obvious to careful observer
Large ($d = 0.8$)	$2.5 \times .8 = 2.00$ inches	Obvious to most observers

EXAMPLE 13.1—CONTINUED
Normal Body Temperature

Have you noticed that many people comment on how their normal body temperature seems to be lower than the previously held standard of 98.6? If so, then you have detected a medium effect size. In Example 13.1, data were presented for normal body temperature for 18 people. The sample mean and standard deviation were 98.217 and .684, respectively. The estimated effect size for comparing normal body temperature to the previously held standard of 98.6 is

$$\hat{d} = \frac{\bar{x} - \mu_0}{s} = \frac{98.217 - 98.6}{.684} = -0.56$$

This is a medium effect size, which should be evident to a careful observer such as a health care worker who routinely measures the temperature of healthy individuals. ◆

Comparing Effect Sizes Across Studies

Effect sizes have become increasingly important in the past few decades with the increase in the use of **meta-analysis,** which is the statistical combination of results on the same topic across studies. Studies are all somewhat different, and may use different units of measurement, different sample sizes and so on. It is useful to have a method of comparison across studies that isn't influenced by sample size or units of measurement, and effect sizes serve that purpose.

In early attempts to review many studies on the same topic, researchers made the mistake of using p-values. In fact, they often used a **vote count,** in which they simply counted the proportion of studies that achieved statistical significance. As we have learned, the problem with this method is that statistical significance is heavily dependent on sample size. To illustrate the problem, consider the scenario in the following example.

EXAMPLE 13.11
The Hypothesis-Testing Paradox

Suppose a researcher conducts a one-sample t-test in a study with $n = 100$ and finds that the test statistic $t = 2.5$, so with a two-tailed test the p-value is 0.012. The results of the study are clearly statistically significant. Just to be sure, the researcher decides to repeat the study, but this is just for confirmation, so a smaller sample size of $n = 25$ is used. The researcher is disappointed to find that in the replication study, $t = 1.25$, so the p-value is only 0.21, and the result is not statistically significant. Perplexed about why the effect disappeared, the researcher decides to combine the data. For the combined data, the researcher is very surprised to find that $t = 2.795$ and the p-value is 0.0052! The paradox is that when the researcher considered the second study alone, it seemed to detract from the evidence for an effect because it was not statistically significant. When the researcher considered the second study in conjunction with the first study, it seemed to strengthen the evidence for an effect.

Now let's look at the effect size for each study and for the combined studies. Here is a summary of the results:

	Sample Size	Test Statistic t	p-Value	Effect Size $\hat{d} = t / \sqrt{n}$
Study #1	100	2.50	0.012	0.25
Study #2	25	1.25	0.21	0.25
Combined	125	2.795	0.0052	0.25

Notice that the effect size is exactly the same for all three studies! The problem is that the researcher was using statistical significance as the basis for the conclusion about whether the first study replicated the second one. The different sample sizes rendered that comparison meaningless. ◆

Thus, effect sizes are quite useful for comparing the results of studies on the same topic. Replication of results requires that similar effect sizes be obtained across studies, and not that statistical significance or the lack of it be obtained. Comparing p-values or statistical significance across studies is valid only if the sample sizes are similar.

Effect Size and Power

It is obvious from Table 13.3 that the *power* for a test depends on the difference between the null value and the true value of the parameter. However, the difference alone is not sufficient. For means, the relevant difference is the effect size, d.

Suppose a researcher wants to find a statistically significant result as long as the truth is a certain size effect, say D. All that he or she needs to do in addition to specifying D is to choose a significance level, usually 0.05 or 0.01. There is then a trade-off between sample size(s) and power that can be quantified. Many researchers aim for power of 0.80, meaning that there is an 80% chance of achieving statistical significance if the true effect size is D. Many software packages will compute the sample size(s) needed to achieve the specified power, or the power that will result from a specified sample size.

EXAMPLE 13.12
Planning a Weight-
Loss Study

Suppose you were convinced that your new weight-loss plan worked, and you wanted to conduct a study to verify that. How would you decide how many participants to include? You need to specify four things: (1) the effect size you hope to detect, (2) the planned level of significance α, (3) whether the test is one- or two-tailed, and (4) the desired power. Let's suppose you hope to detect a medium effect size of 0.5, using $\alpha = 0.05$ for a one-tailed test, and want power =

0.80. Using Minitab, we find that a sample of $n = 27$ participants is required. To detect a small effect size of 0.2 with power of 0.80, a sample of $n = 156$ is required.

The interpretation of an effect size of 0.2 or 0.5 in terms of weight lost depends on the standard deviation of the weight losses. If the standard deviation is 10 pounds, then effect sizes of 0.2 and 0.5 correspond to average weight losses of 2 pounds and 5 pounds, respectively.

Table 13.4 shows some additional power and sample size results for this situation. Notice that if indeed the effect is very small ($d = 0.1$), it will be very difficult to achieve statistical significance. Even with $n = 100$ participants, the probability of detecting a statistically significant effect is only .26. On the other hand, if the true effect is a medium effect size, then even with only 20 participants there is a fairly high probability (.695) of achieving a statistically significant effect. ◆

TABLE 13.4 ■ **Power Based on True Effect Size Values for One-Sided, One-Sample *t*-Test, $\alpha = 0.05$**

	True Effect Size			
Sample Size	*0.1*	*0.2*	*0.5*	*0.8*
$n = 20$	.11	.22	.695	.964
$n = 50$	.17	.40	.967	nearly 1
$n = 100$	.26	.63	.9996	nearly 1

Minitab Tip

tech note **Computing Sample Size for Specified Power or Power for Specified Sample Size**

- ◆ To compute power for a one-sample *t*-test, use **Stat>Power and Sample Size>1-Sample t.** Use the default sigma = 1.0, and for the differences specify the effect sizes for which you want power information.

There are three choices for what to compute:

- ◆ **Calculate power for each sample size.** If you want to use this option, fill in the sample sizes for which you want to find power, and for "difference" specify the effect size you hope to detect.

- ◆ **Calculate power for each difference.** If you want to use this option, fill in a list of effect sizes you hope to detect as the "differences." Then specify the sample size.

- ◆ **Calculate sample size for each power value.** If you want to use this option, fill in a list of power values and the effect size you hope to detect as the "difference."

- ◆ Use **Options** to specify the form of the alternative hypothesis (not equal to, greater than, less than), the level of significance, and optional storage locations.

13.9 | EVALUATING SIGNIFICANCE IN RESEARCH REPORTS

To evaluate the quality of a research study, you should always consider features like the sampling procedure, the methods used to measure individuals, the wording of questions, and the possibility of lurking variables. When the conclusions are based on hypothesis tests, several additional items should be considered:

1. Is the *p-value* reported? If you know the *p*-value, you can make your own decision, based on the severity of a type 1 error and the size of the *p*-value.
2. If the word *significant* is used to describe a result, determine if the word is being used in the everyday sense or in the statistical sense only. The phrase *statistically significant* just means that a null hypothesis has been rejected, which is no guarantee that the result has "real-world" importance.
3. If you read that "no difference" or "no relationship" has been found in a study, try to determine if the sample size was small. The hypothesis test may have had very low power because not enough data were collected to be able to make a firm conclusion.
4. Think carefully about conclusions based on extremely large samples. If a study is based on a very large sample size, even a weak relationship or a small difference can be statistically significant.
5. If possible, determine what confidence interval should accompany the hypothesis test. This interval will provide better information about the magnitude of the effect as well as information about the margin of error in the sample estimate.
6. Determine how many hypothesis tests were conducted in the study. Sometimes researchers perform a multitude of tests, but only a few of the tests achieved statistical significance. If all of the null hypotheses tested are actually true, then about 1 in 20 tests will achieve statistical significance just by chance at the 0.05 level of significance.

SUMMARY OF PROCEDURES FOR HYPOTHESIS TESTS

The basic structure of hypothesis testing for the parameters in the table below is

Null hypothesis: H_0: parameter = null value

Alternative hypothesis:
$$\left.\begin{cases} H_a: \text{parameter} \neq \text{null value} \\ H_a: \text{parameter} > \text{null value} \\ H_a: \text{parameter} < \text{null value} \end{cases}\right\} \text{Choose one.}$$

Test statistic: $t \text{ or } z = \dfrac{\text{Sample statistic} - \text{Null value}}{\text{Null standard error}}$

p-value   For H_a: parameter $\neq$ null value, *p*-value is $2 \times$ area above $|t|$ or $|z|$.
For H_a: parameter $>$ null value, *p*-value is the area *above t* or *z*, even if negative.
For H_a: parameter $<$ null value, *p*-value is the area *below t* or *z*, even if positive.

If *p*-value $>$ significance level, do not reject H_0.
If *p*-value $\leq$ significance level, reject H_0:, accept H_a, and conclude result is statistically significant.

	Parameter	Statistic	Null Value	Null Standard Error	t or z; df for t	Example
One Mean	μ	$\bar{x}$	μ_0	$\dfrac{s}{\sqrt{n}}$	t df $= n - 1$	13.1
Mean Difference in Paired Variables	μ_d	$\bar{d}$	Usually 0	$\dfrac{s_d}{\sqrt{n}}$	t df $= n - 1$	13.2
Difference Between Means (unpooled)	$\mu_1 - \mu_2$	$\bar{x}_1 - \bar{x}_2$	Usually 0	$\sqrt{\dfrac{s_1^2}{n_1} + \dfrac{s_2^2}{n_2}}$	t df = Welch's approximation or smaller of $n_1 - 1, n_2 - 1$	13.3
Difference Between Means (pooled)	$\mu_1 - \mu_2$	$\bar{x}_1 - \bar{x}_2$	Usually 0	$s_p\sqrt{\dfrac{1}{n_1} + \dfrac{1}{n_2}}$	t df $= n_1 + n_2 - 2$	13.5
One Proportion	p	$\hat{p}$	p_0	$\sqrt{\dfrac{p_0(1 - p_0)}{n}}$	z	11.7
Difference Between Proportions	$p_1 - p_2$	$\hat{p}_1 - \hat{p}_2$	Usually 0	$\sqrt{\hat{p}(1 - \hat{p})\left(\dfrac{1}{n_1} + \dfrac{1}{n_2}\right)}$	z	13.6

KEY TERMS

EXERCISES   Blue = Basics   **Bold, Bold** = Answer in back

For all exercises requiring hypothesis tests, make sure you check conditions, define all parameters, and state your conclusion in words in the context of the problem.

Sections 13.1 and 13.2

13.1 For each of the following research questions specify the "parameter" and the "specific value" that constitute the null hypothesis of "parameter = specific value." In other words, define the population parameter of interest and specify the null value that is being tested.

a. Do a majority of Americans between the ages of 18 and 30 think the use of marijuana should be legalized?

b. Is the mean of the Math SAT scores in California in a given year different from the target mean of 500 set by the test developers?

c. Is the mean age of death for left-handed people lower than it is for right-handed people?

d. Is there a difference in the proportions of male and female college students who smoke cigarettes?

13.2 Refer to the previous exercise. In each case, specify whether the alternative hypothesis would be one-sided or two-sided.

13.3 Explain whether or not each of the following statements is true.
   a. Hypotheses and conclusions from hypothesis testing apply only to the samples on which they are based.
   b. The *p*-value is calculated assuming that the null hypothesis is true.
   c. One of the two possible conclusions in hypothesis testing is to accept the null hypothesis.
   d. The statements "reject the null hypothesis" and "accept the alternative hypothesis" are equivalent.

13.4 Give the value of the test statistic *t* in each of the following situations.
   a. $H_0: \mu = 50, \bar{x} = 60, s = 90, n = 100$.
   b. $H_0: \mu_d = 0, \bar{d} = -4, s_d = 15, n = 50$.
   c. Null value $= 100$, sample mean $= 98$, $s = 15$, sample size $= 40$.
   d. $H_0: \mu = 250, \bar{x} = 270$, standard error $= 5, n = 100$.

13.5 Find the *p*-value and draw a sketch showing the *p*-value area for each of the following situations in which the value of *t* is the test statistic for the hypotheses given:
   a. $H_0: \mu = \mu_0, H_a: \mu > \mu_0, n = 28, t = 2.00$.
   b. $H_0: \mu = \mu_0, H_a: \mu > \mu_0, n = 28, t = -2.00$.
   c. $H_0: \mu = \mu_0, H_a: \mu \neq \mu_0, n = 81, t = 2.00$.
   d. $H_0: \mu = \mu_0, H_a: \mu \neq \mu_0, n = 81, t = -2.00$.

13.6 Use Table A.2 to find the critical value and rejection region in each of the following situations. Then determine whether the null hypothesis would be rejected. In each case the null hypothesis is $H_0: \mu = 100$.
   a. $H_a: \mu > 100, n = 21, \alpha = 0.05$, test statistic $t = 2.30$.
   b. $H_a: \mu > 100, n = 21, \alpha = 0.01$, test statistic $t = 2.30$.
   c. $H_a: \mu \neq 100, n = 21, \alpha = 0.05$, test statistic $t = 2.30$.
   d. $H_a: \mu \neq 100, n = 21, \alpha = 0.01$, test statistic $t = 2.30$.
   e. $H_a: \mu > 100, n = 10, \alpha = 0.05$, test statistic $t = 1.95$.
   f. $H_a: \mu < 100, n = 10, \alpha = 0.05$, test statistic $t = -1.95$.
   g. $H_a: \mu < 100, n = 10, \alpha = 0.05$, test statistic $t = 1.95$.
   h. $H_a: \mu \neq 100, n = 10, \alpha = 0.05$, test statistic $t = 1.95$.

13.7 The dataset **cholest** on the CD accompanying this book includes cholesterol levels for heart attack patients and for a group of control patients. It is recommended that people try to keep their cholesterol level below 200. The following Minitab output is for the control patients.

```
Test of mu = 200.00 vs mu < 200.00

Variable    N     Mean    StDev   SE Mean      T      P
control    30    193.13   22.30    4.07    -1.69   0.051
```

a. What are the null and alternative hypotheses being tested? Write them in symbols.
b. What is the mean cholesterol level for the sample of control patients?
c. How many patients were in the sample?
d. Use the formula for the standard error of the mean to show how to compute the value of 4.07 reported by Minitab.

e. What values does Minitab report for the test statistic and the *p*-value?
f. Identify the numbers that were used to compute the *t*-statistic, and verify that the reported value is correct.
g. What conclusion would be made in this situation, using a 0.05 level of significance?

13.8 Suppose a study is done to test the null hypothesis $H_0: \mu = 100$. A random sample of $n = 50$ observations results in $\bar{x} = 102$ and $s = 15$.
   a. What is the null standard error in this case?
   b. Plug numbers into the formula

$$\frac{\text{Sample statistic} - \text{Null value}}{\text{Null standard error}}$$

   c. Based on the information given, can the *p*-value for this test be found? If so, find it. If not, explain what additional information would be needed.

13.9 It has been hypothesized that the mean pulse rate for college students is about 72 beats per minute. A sample of Penn State students recorded their sexes and pulse rates. Assume the samples are representative of all Penn State men and women for pulse rate measurements. The summary statistics were as follows:

Sex	*n*	Mean	StDev
Female	35	76.9	11.6
Male	57	70.42	9.95

a. Test whether the pulse rates of all Penn State men have a mean of 72.
b. Test whether the pulse rates of all Penn State women have a mean of 72.
c. Write a sentence or two summarizing the results of parts (a) and (b) in words that would be understood by someone with no training in statistics.

13.10 Most people complain that they gain weight during the December holidays, and Yanovski et al. (2000) wanted to determine if that was the case. They sampled the weights of 195 adults in mid-November and again in early to mid-January. The mean weight change for the sample was a gain of 0.37 kg, with a standard deviation of 1.52 kg. State and test the appropriate hypotheses. Be sure to carefully define the population parameter(s) you are testing.

13.11 In Exercise 12.65 a study was reported in which students were asked to place as many dried beans into a cup as possible in 15 seconds with their dominant hand, and again with their nondominant hand (in randomized order). The differences in number of beans (dominant hand–nondominant hand) for 15 students were as follows:

4, 4, 5, 1, −2, 0, 2, 4, −3, 0, 0, 0, −2, 2, 1

The data also are given in the CD dataset **beans.**
a. The research question was whether students have better manual dexterity with their dominant hand than with their nondominant hand. Write the null and alternative hypotheses.

b. Check the necessary conditions for doing a one sample *t*-test.

c. Carry out the test using $\alpha = 0.05$.

d. Carry out the test using $\alpha = 0.10$.

e. Write a conclusion about this situation that would be understood by other students of statistics.

13.12 Data from the CD dataset **UCDavis1** included information on height (**height**) and mother's height (**momheight**) for 93 female students. Here is the output from the Minitab paired *t* procedure comparing these heights:

```
Paired T for height − momheight

                N     Mean    StDev    SE Mean
height          93    64.342  2.862    0.297
momheigh        93    63.057  2.945    0.305
Difference      93    1.285   3.136    0.325

95% CI for mean difference: (0.639, 1.931)
T-Test of mean difference = 0 (vs > 0): T-Value = 3.95
P-Value = 0.000
```

a. It has been hypothesized that college students are taller than they were a generation ago, and therefore that college women should be significantly taller than their mothers. State the null and alternative hypotheses to test this claim. Be sure to define any parameters you use.

b. Using the information in the Minitab output, the test statistic is $t = 3.95$. Identify the numbers that were used to compute the *t*-statistic, and verify that the stated value is correct.

c. What are the degrees of freedom for the test statistic?

d. Carry out the remaining steps of the hypothesis test.

e. Draw a sketch that illustrates the connection between the *t*-statistic and the *p*-value in this problem.

Section 13.3

13.13 In each of the following situations, determine whether the alternative hypothesis was Ha: $\mu_1 - \mu_2 > 0$, or Ha: $\mu_1 - \mu_2 < 0$, or Ha: $\mu_1 - \mu_2 \neq 0$.

a. $H_0: \mu_1 - \mu_2 = 0$, $t = 2.33$, df = 8, *p*-value = 0.048.

b. $H_0: \mu_1 - \mu_2 = 0$, $t = -2.33$, df = 8, *p*-value = 0.024.

c. $H_0: \mu_1 - \mu_2 = 0$, $t = 2.33$, df = 8, *p*-value = 0.976.

d. $H_0: \mu_1 - \mu_2 = 0$, $t = -2.33$, df = 8, *p*-value = 0.976.

13.14 For each of the following situations, identify whether a paired *t*-test or a two-sample *t*-test is appropriate:

a. The weights of a sample of 15 marathon runners were taken before and after a training run to test if marathon runners lose dangerous levels of fluids during a run.

b. Random samples of 200 new freshmen and 200 new transfer students at a university were given a 50-question test on current events to test if the level of knowledge of current events differs for new freshmen and transfer students.

c. Sixty students were matched by initial pulse rate, with the two with the highest pulse forming a pair,

and so on. Within each pair, one student was randomly chosen to drink a caffeinated beverage, while the other one drank an equivalent amount of water. Their pulse rates were measured 10 minutes later, to test if caffeine consumption elevates pulse rates.

13.15 Calculate the value of the test statistic *t* in each of the following situations. In each case, assume the null hypothesis is $H_0: \mu_1 - \mu_2 = 0$.

a. $\bar{x}_1 = 35$, $s_1 = 10$, $n_1 = 100$; $\bar{x}_2 = 33$, $s_2 = 9$, $n_2 = 81$.

b. The difference in sample means is 48, s.e. $(\bar{x}_1 - \bar{x}_1) = 22$.

c. Minitab output:

```
              N    Mean    StDev    SE Mean
Sample 1     68    80.58   4.22     0.51
Sample 2     68    78.55   3.31     0.40
```

13.16 Do hard- and softcover books likely to be found on a professor's shelf have the same average number of pages? Data on the number of pages for eight hardcover and seven softcover books from a professor's shelf were presented in Example 3.4 and are in the file **ProfBooks** on the CD accompanying this book. Here is the Minitab output from the "2-sample t" procedure. Assume the books are equivalent to a random sample.

```
Cover    N    Mean    StDev    SE Mean
Hard     8    307     134      47
Soft     7    429.4   80.9     31

95% CI for mu (Hard) − mu (Soft) : ( −246, 2)
T-Test mu (Hard) = mu (Soft)(vs not =) : T = −2.16
P = 0.054 DF = 11
```

a. Give the null and alternative hypotheses using symbols.

b. What is the value of the test statistic *t*?

c. Identify the numbers that were used to compute the *t*-statistic, and verify that the reported value is correct.

d. What conclusion would be made using a 0.05 level of significance? Write the conclusion in statistical terms and in the context of the problem.

13.17 Example 2.7 gave the weights of eight rowers on each of the Cambridge and Oxford crew teams. The weights are shown again here. Assuming these men represent appropriate random samples, test the hypothesis that the mean weight of rowers on the Cambridge and Oxford crew teams are equal versus the alternative that they are not equal.

Cambridge: 188.5, 183.0, 194.5, 185.0, 214.0, 203.5, 186.0, 178.5

Oxford: 186.0, 184.5, 204.0, 184.5, 195.5, 202.5, 174.0, 183.0

13.18 Case Study 1.1 presented data given in response to the question "What is the fastest you have ever driven a car? _____ mph." The summary statistics are:

Females: $n = 102$, mean = 88.4, standard deviation = 14.4

Males: $n = 87$, mean $= 107.4$, standard deviation $= 17.4$

Assuming these students represent a random sample of college students, test whether the mean fastest speed driven by college men and college women is equal versus the alternative that it is higher for men.

13.19 Example 12.6 presented results for the number of hours slept the previous night from a survey given in two statistics classes. One class was a liberal arts class, while the other class was a general introductory class. The survey was given following a Sunday night after classes had started. For simplicity, let's assume these classes represent a random sample of sleep hours for college students in liberal arts and non-liberal arts majors. The data are given next.

	n	Mean	St. Dev.
Lib. arts	25	7.66	1.34
Non-Lib. arts	148	6.81	1.73

a. Test the hypothesis that the mean number of hours of sleep for the two populations of students are equal versus the alternative that they are not equal. Use the unpooled *t*-test. (*Note:* The approximate df = 38 for the unpooled test.)

b. The figure below displays a dotplot of the data. Briefly explain what is indicated about the necessary conditions for doing a two-sample *t*-test.

c. Repeat the hypothesis test using the pooled procedure. Compare the results to those in part (a), and discuss which procedure you think is more appropriate in this situation.

EXERCISE 13.19(b)   Dotplot of sleep values for exercise 13.19

13.20 Example 12.4 presented data from a study in which sedentary men were randomly assigned to be placed on a diet or exercise for a year to lose weight. Forty-two men were placed on a diet, while the remaining 47 were put on an exercise routine. The group on a diet lost an average of 7.2 kg, with a standard deviation of 3.7 kg. The men who exercised lost an average of 4.0 kg, with a standard deviation of 3.9 kg.

a. State and test appropriate null and alternative hypotheses to determine if the mean weight loss would

be different under the two routines for the population of men similar to those in this study.

b. Explain how you decided whether to do a pooled or unpooled test in part (a).

13.21 Do students sleep more in Pennsylvania or in California? Data from surveys in elementary statistics classes at Penn State University and the University of California at Davis resulted in the following summary statistics for the number of hours students sleep:

	n	Mean	St. Dev.	S.E. Mean
UC Davis	173	6.93	1.71	0.13
Penn State	190	7.11	1.95	0.14

Assume these students are representative of all students at those two schools. Is there sufficient evidence to conclude that the mean hours of sleep are different at the two schools? Carry out all steps of the hypothesis test, and define all parameters.

13.22 Students in a statistics class at Penn State were asked: About how many minutes do you typically exercise in a week? Responses from the *women* in the class were:

60, 240, 0, 360, 450, 200, 100, 70, 240, 0, 60, 360, 180, 300, 0, 270

Responses from the *men* in the class were:

180, 300, 60, 480, 0, 90, 300, 14, 600, 360, 120, 0, 240

a. Draw appropriate graphs to check if the conditions for conducting a two-sample *t*-test are met. Discuss the results of your graphs.

b. What additional assumption or condition is required if conclusions are to be made about amount of exercise for the population of all Penn State students based on these sample results?

c. Assume that the conditions are met, and conduct a test to determine if the mean amount of exercise differs for men and women.

Section 13.4

13.23 Find the *p*-value and draw a sketch of the *p*-value area for each of the following situations in which the value of *z* is the test statistic for the hypotheses given:

a. $H_0: p_1 - p_2 = 0$, $H_a: p_1 - p_2 > 0$; $z = 1.75$.

b. $H_0: p_1 - p_2 = 0$, $H_a: p_1 - p_2 > 0$; $z = -1.75$.

c. $H_0: p_1 - p_2 = 0$, $H_a: p_1 - p_2 \neq 0$; $z = 1.75$.

d. $H_0: p_1 - p_2 = 0$, $H_a: p_1 - p_2 \neq 0$; $z = -1.75$.

13.24 State a conclusion for each of the following situations, using $\alpha = 0.05$:

a. $H_0: p_1 - p_2 = 0$, $H_a: p_1 - p_2 < 0$; $z = -1.99$.

b. $H_0: p_1 - p_2 = 0$, $H_a: p_1 - p_2 > 0$; $z = 1.78$.

c. $H_0: p_1 - p_2 = 0$, $H_a: p_1 - p_2 < 0$; $z = 0.33$.

d. $H_0: p_1 - p_2 = 0$, $H_a: p_1 - p_2 > 0$; $z = 0.33$.

13.25 Refer to the previous exercise and to the discussion of the rejection region approach to hypothesis testing (for *z*-tests) in Section 11.4. In each part (a to d), find the rejection region that would be used for $\alpha = 0.05$, then state a conclusion for the *z*-statistic given.

13.26 Is the proportion of men who write a shopping list before going shopping less than the proportion of women who do so? Let p_1 and p_2 represent the population proportions of men and women, respectively, who use shopping lists. Suppose a random sample of 50 men and 50 women was asked if they use a shopping list when they have more than ten items to buy. Of the 50 men, 20 said yes and of the 50 women 30 said yes.

a. Write the null and alternative hypotheses in terms of p_1 and p_2, taking into account the initial question asked at the beginning of this exercise.

b. Provide numerical values for $\hat{p}_1$ and $\hat{p}_2$.

c To find the null standard error, the combined sample proportion $\hat{p}$ is required. Compute $\hat{p}$ for this situation.

d. Find the test statistic for this situation.

e. Make a conclusion using the 0.05 level of significance.

13.27 Exercises 6.6 and 6.56 presented data on grumpy old men and heart disease. In Chapter 6, the goal was to determine if there was a relationship between the two variables. A one-tailed test may also make sense. We might hypothesize that if there is a difference in the probability of developing heart disease for men with no anger and men with lots of anger, the latter group would have a higher probability. Minitab output for testing this claim follows. Note that p_1 is the probability of developing heart disease for men with no anger and p_2 is the probability of developing heart disease for men with the most anger.

Sample	X	N	Sample p
1	8	199	0.040201
2	59	559	0.105546

Estimate for p(1) − p(2): 0.0653446
95% CI for p(1) − p(2): (−0.102675, −0.0280137)
Test for p(1) − p(2) = 0 (vs < 0): Z = −2.79 P-Value = 0.003

a. Write the null and alternative hypotheses in terms of p_1 and p_2, taking into account the claim being tested.

b. What is the value of the test statistic?

c. What conclusion would you make? Write your conclusion in statistical terms and in the context of the problem.

13.28 Exercise 12.57 presented results of a survey asking college students if they would return the money if they found a wallet on the street. Of the 93 women, 84 said they would, and of the 75 men, 53 said they would. Assume these students represent all college students. Test the hypothesis that equal proportions of college men and women would say they would return the money versus the alternative hypothesis that a higher proportion of women would do so.

13.29 Refer to exercise 12.64, in which men were classified according to whether they had a pierced ear and a tattoo. Of the 424 men with no ear pierce, 43 had a tattoo. Of the 141 men with at least one ear pierce, 42 had a tattoo. Assuming these men are a random sample of college men,

test the hypothesis that college men with at least one ear pierce are more likely to have a tattoo than college men with no ear pierces.

13.30 A Gallup Poll taken in May 2000 asked the question "In general, do you feel that the laws covering the sale of firearms should be made: more strict, less strict, or kept as they are now?" Of the $n = 493$ men who responded, 52% said "more strict," while of the $n = 538$ women who responded, 72% said "more strict." Assuming these respondents constitute random samples of U.S. men and women, is there sufficient evidence to conclude that a higher proportion of women than men in the population think these laws should be made more strict? Justify your answer.

13.31 Refer to the previous question. The same poll asked the question, "Which of the following do you think is the primary cause of gun violence in America—the availability of guns, the way parents raise their children, or the influences of popular culture such as movies, television, and the Internet?" 51% of the $n = 493$ men and 38% of the $n = 538$ women responded, "Way parents raise kids."

a. Is there sufficient evidence to conclude that a higher proportion of men than women in the population think the "way parents raise kids" is the primary cause? Justify your answer.

b. Refer to the data in this exercise as well as the previous one. Notice that 52% of the men thought laws for firearm sales should be more strict, and 51% of men thought the way parents raise kids is the primary cause of gun violence. Can the two-sample test for proportions covered in Section 13.4 be used with these data to test whether the corresponding population proportions differ? If so, carry out the test. If not, explain why not.

13.32 Case Study 10.3 reported on a survey of Penn State students asking the question, "Would you date someone with a great personality even though you did not find them attractive?" Of the $n = 131$ women, 61.1% said yes, and of the $n = 61$ men, 42.6% said yes. Is there sufficient evidence to conclude that for the populations represented by these students, a higher proportion of women than men would answer yes to this question?

Sections 13.5 and 13.6

13.33 In each of the following cases, explain whether the null hypothesis H_0: $\mu = 25$ can be rejected. Use $\alpha = 0.05$.

a. 95% confidence interval for μ is (10 to 30), H_a: $\mu \neq 25$.

b. 95% confidence interval for μ is (26 to 50), H_a: $\mu \neq 25$.

c. 90% confidence interval for μ is (10 to 30), H_a: $\mu > 25$.

d. 90% confidence interval for μ is (10 to 30), H_a: $\mu < 25$.

e. 90% confidence interval for μ is (26 to 50), H_a: $\mu > 25$.

f. 90% confidence interval for μ is (26 to 50), H_a: $\mu < 25$.

13.34 Refer to the rules for the relationship between confidence intervals and two-sided alternatives, given in the two bullets on page 465.

a. Rewrite the rules specifically for $\alpha = 0.05$.

b. Rewrite the rules specifically for $\alpha = 0.01$.

13.35 Refer to the rules for the relationship between confidence intervals and one-sided tests, given in the three bullets and sentence preceding them on page 466.
  a. Rewrite the rules specifically for $\alpha = 0.05$.
  b. Rewrite the rules specifically for $\alpha = 0.01$.

13.36 Refer to the five parameters given in Table 13.2. For each of the following situations identify which of the five is or are appropriate. Give the parameter(s) in symbols, and define what the symbols represent in the context of the situation.
  a. Researchers want to compare the mean running times for the 50-yard dash for first grade boys and girls. They select a random sample of 35 boys and 32 girls in first grade and time them running the 50-yard dash.
  b. Researchers want to know what proportion of a certain type of tree growing in a national forest suffers from a disease. They test a representative sample of 200 of the trees from around the forest and find that 15 of them have the disease.
  c. Researchers want to know what percent of adults have a fear of going to the dentist. They also want to know the average number of visits made to a dentist in the past 10 years for adults who have that fear. They ask a random sample of adults whether or not they fear going to the dentist and also how many times they have gone in the past 10 years.
  d. Refer to part (c). Researchers also want to compare the average number of visits made by adults who fear going to the dentist with the average number of visits for those who don't have the fear.

13.37 Refer to the previous exercise. In each case, explain whether a confidence interval, a hypothesis test, or both would be more appropriate.

13.38 Each of the following presents a two-sided 95% confidence interval and the alternative hypothesis of a corresponding hypothesis test. In each case, state a conclusion for the test, including the level of significance you are using.
  a. C.I. for μ is (101 to 105), H_a: $\mu \neq 100$.
  b. C.I. for p is (.12 to .28), H_a: $p < .10$.
  c. C.I. for $\mu_1 - \mu_2$ is (3 to 15), H_a: $\mu_1 - \mu_2 > 0$.
  d. C.I. for $p_1 - p_2$ is ($-.15$ to .07), H_a: $p_1 - p_2 \neq 0$.

13.39 As stated in Section 13.5, "a confidence interval can be used as an alternative way to conduct a two-sided significance test." If a test were conducted using this method, would the p-value for the test be available? Explain.

13.40 For each of the following situations, can you conclude whether a 90% confidence interval for μ would include the value 10? If so, make the conclusion. If not, explain why you can't tell.
  a. H_0: $\mu = 10$, H_a: $\mu < 10$, do not reject the null hypothesis for $\alpha = 0.05$.
  b. H_0: $\mu = 10$, H_a: $\mu < 10$, reject the null hypothesis for $\alpha = 0.05$.
  c. H_0: $\mu = 10$, H_a: $\mu \neq 10$, do not reject the null hypothesis for $\alpha = 0.10$.
  d. H_0: $\mu = 10$, H_a: $\mu \neq 10$, reject the null hypothesis for $\alpha = 0.10$.

13.41 Give an example of a situation for which the appropriate inference procedure would be each of the following:

  a. A hypothesis test for one proportion.
  b. A hypothesis test and confidence interval for a paired difference mean.
  c. A confidence interval for one mean.
  d. A confidence interval for the difference in two means for independent samples.

13.42 Refer to each of the following scenarios from exercises in various chapters of this book. In each case, determine the most appropriate inference procedure(s), including the appropriate parameter. If you think inference about more than one parameter may be of interest, answer the question for all parameters of interest. Explain your choices.
  a. A sample of college students was asked if they would return the money if they found a wallet on the street. Of the 93 women, 84 said "yes," and of the 75 men, 53 said "yes." Assume these students represent all college students.
  b. A study was conducted on pregnant women and subsequent development of their children (Olds et al., 1994). One of the questions of interest was whether the IQ of children would differ for mothers who smoked at least 10 cigarettes a day during pregnancy and those who did not smoke at all.
  c. Max likes to keep track of birthdays of people he meets. He has 170 birthdays on his birthday calendar. One cold January night, he comes up with the theory that people are more likely to be born in October than they would be if all 365 days were equally likely. He consults his birthday calendar and finds that 22 of the 170 birthdays are in October.
  d. The dataset **cholest** reports cholesterol levels of heart attack patients 2, 4, and 14 days after the heart attack. Data for days 2 and 4 were available for 28 patients. The mean difference (2 day–4 day) was 23.29, and the standard deviation of the differences was 38.28.

Sections 13.7, 13.8, and 13.9

13.43 A researcher is deciding whether to use a sample size of 100 or whether to increase the sample size to 200. Explain how this choice will affect the power of any hypothesis tests done using data from the resulting sample.

13.44 A researcher is deciding whether to use a level of significance equal to 0.05 or a level of significance equal to 0.01. Explain how this choice will affect the power of the hypothesis test.

13.45 Explain which type of error (1 or 2) could be made in each of the following situations. If both are possible, state that as well.
  a. The null hypothesis is true.
  b. The alternative hypothesis is true.
  c. The null hypothesis is rejected
  d. The null hypothesis is not rejected.

13.46 Compute the effect size for each of the following situations, and state whether it would be considered closer to a small, medium, or large effect:
  a. In a one-sample test with $n = 100$, the test statistic is $t = 2.24$.

b. In a one-sample test with $n = 50$, the test statistic is $t = -2.83$.

c. In a paired-difference test with $n = 30$ pairs, the test statistic is $t = 1.48$.

d. In a test for the difference in two means with independent samples, with $n_1 = 40$ and $n_2 = 50$, the test statistic is $t = -2.33$.

13.47 Refer to Table 13.4, which presents power for a one-sided, one-sample t-test. In a test of $H_0: \mu_d = 0$ versus $H_a: \mu_d > 0$, suppose the truth is that the population of differences is a normal distribution with mean $\mu_d = 2$ and standard deviation $\sigma = 4$.

   a. Recalling the Empirical Rule from Chapter 2, draw a picture of this distribution, showing the ranges into which 68% and 95% of the differences fall.

   b. On your picture, indicate where the null value of 0 falls.

   c. If a sample of size 20 is taken, what is the power for the test, assuming a 0.05 level of significance will be used?

   d. Explain in words what probability the power of the test represents.

13.48 In Table 13.4, it is shown that the power is 0.40 for a one-sided, one-sample t-test with 0.05 level of significance, $n = 50$, and true effect size of 0.2. Would the power be higher or lower for each of the following changes?

   a. The true effect size is 0.4.

   b. The sample size used is $n = 75$.

   c. The level of significance used is 0.01.

13.49 Refer to the list of items in Section 13.9. Explain which ones should be of concern if the sample size(s) for a test are large.

13.50 Refer to the list of items in Section 13.9. Explain which ones should be of concern if the sample size(s) for a test are small.

13.51 Suppose a one-sample t-test of $H_0: \mu = 0$ versus $H_a: \mu \neq 0$ results in a test statistic of $t = 0.65$ with df $= 14$. Suppose a new study is done with $n = 150$, and the sample mean and standard deviation turn out to be exactly the same as in the first study.

   a. What conclusion would you make in the original study, using $\alpha = 0.05$?

   b. What conclusion would you make in the new study, using $\alpha = 0.05$?

   c. Compare your results in parts (a) and (b) and comment.

13.52 Minitab can provide power for a variety of situations. Suppose a test for ESP has four choices, and the probability of a correct guess by chance on each trial is .25. A researcher believes the true probability of a correct guess is .33. The following output shows the power of the one-sided test for this situation for three possible sample sizes:

```
Test for One Proportion
Testing proportion = 0.25 (versus > 0.25)
Calculating power for proportion = 0.33
Alpha = 0.05   Difference = 0.08
Sample
Size     Power
  50     0.3776
 100     0.5740
 400     0.9705
```

   a. What is the power of the test if $n = 50$ trials are used?

   b. Write a sentence providing the power of the test for $n = 100$ and explaining its meaning.

   c. If the researcher wants to have at least a .95 probability of detecting ESP in the study and is correct that the true probability of a success is .33, would a sample of size 400 be sufficient? Explain.

   d. If the true probability of success is actually .40 on each trial, would the power for each sample size be higher or lower than that shown in the output? Explain.

13.53 Refer to Case Study 1.6, comparing heart attack rates for men who had taken aspirin or placebo. Suppose the observed proportions of .017 and .0094 are actually the correct population proportions who would have heart attacks with placebo and with aspirin. The following Minitab output shows the power of a one-sided test for two proportions for this situation, for three sample sizes. The samples are the number of participants in *each* group (aspirin and placebo). Suppose you are the statistician advising a research team about conducting a new study to confirm the results of the old study. The researchers comment that samples of size 500 in each condition should be sufficient, since the effect is obviously so strong, based on the small p-value for the previous study. What would you advise? Explain.

```
Testing proportion 1 = proportion 2 (versus >)
Calculating power for proportion 1 = 0.017 and
proportion 2 = 0.0094
Alpha = 0.05   Difference = 0.0076

Sample
Size      Power
 500      0.2768
1000      0.4380
3000      0.8250
```

13.54 A study is conducted to see if results of an IQ test are significantly higher after listening to Mozart than after sitting in silence. Explain what has happened in each of the following scenarios:

   a. A type 1 error was committed.

   b. A type 2 error was committed.

   c. The power of the test was too low to detect the difference that actually exists.

   d. The power of the test was so high that a very small difference resulted in a statistically significant finding.

13.55 (Computer software required.) Find the power for the following one-sample t-test situations. In each case, assume a 0.05 level of significance will be used.

   a. Effect size $= 0.3$, sample size $= 45$, $H_a: \mu > \mu_0$.

   b. Sample size $= 30$, $H_a: \mu \neq 10$, true mean $= 13$, $\sigma = 4$.

   c. Effect size $= -1.0$, sample size $= 15$, $H_a: \mu < \mu_0$.

13.56 (Computer software required.) In parts (a) to (d), find the sample size necessary to achieve 80% power for a one-sample t-test with $H_a: \mu > \mu_0$ and level of significance of 0.05 for each of the following effect sizes:

   a. 0.2

   b. 0.4

   c. 0.6

   d. 0.8

e. Make a scatterplot of the sample size (vertical axis) versus the effect size for the effect sizes in parts (a) to (d). Does the relationship between the effect size and the sample size required to achieve 80% power appear to be linear? If not, what is the nature of the relationship?

13.57   For a z-test for one proportion, a possible effect size measure is $(p_1-p_0)/\sqrt{p_0(1-p_0)}$ where p_0 is the null value and p_1 is the true population proportion.

a. What is the relationship between this effect size and the z-test statistic for this situation?

b. Does the effect size fit the relationship "Test statistic = Size of effect × Size of study?" If not, explain why not. If so, show how it fits.

c. What would be a reasonable way to estimate this effect size?

13.58   Refer to the effect-size measure in the previous exercise. For parts (a) to (c), compute the effect size.

a. $p_1 = .35, p_0 = .25$.

b. $p_1 = .15, p_0 = .05$.

c. $p_1 = .95, p_0 = .85$.

d. Based on the results in parts (a) to (c), does this effect size stay the same when $p_1 - p_0$ stays the same? Explain.

e. In statistical software for computing power for a test for a single proportion, unlike for a single mean, both p_1 and p_0 must be specified, rather than just the difference between them. Explain why it is not enough to specify the difference, using the results of the previous parts of this exercise.

13.59   Explain why it is more useful to compare effect sizes than p-values when trying to determine if many studies about the same topic have found similar results.

13.60   Refer to item 6 in the list of concerns in Section 13.9. Is that statement the same thing as saying that the null hypothesis is likely to be true in about 1 out of 20 tests that have achieved statistical significance? Explain.

13.61   Refer to the list of items in Section 13.9.

a. For which of the concerns would the p-value for a test be useful to have? Explain why in each case.

b. For which of the concerns would a confidence interval estimate for the parameter be useful to have? Explain why in each case.

c. For which of the concerns would the sample size(s) be useful to know? Explain why in each case.

Chapter Exercises

13.62   In a random sample of 170 married British couples, the difference between the husband's and wife's ages had a mean of 2.24 years and a standard deviation of 4.1 years.

a. Test the hypothesis that British men are significantly older than their wives, on average.

b. Explain what is meant by the use of "significant" in part (a), and discuss how it compares with the everyday use of the word.

13.63   A Gallup Poll taken on a random sample of Canadian adults in February 2000 asked the question, "Do you favour or oppose marriages between people of the same sex?" A similar poll was taken in April 1999 (Edwards and

Mazzuca, 2000). The Minitab output for the "two proportions" procedure (with the "pooled estimate for the test") is as follows, where Sample 1 is the February 2000 poll and Sample 2 is the April 1999 poll, and X is the number who answered "favour":

Sample	X	N	Sample p
1	431	1003	0.429711
2	360	1000	0.360000
Estimate for p(1) − p(2): 0.0697109			
95% CI for p(1) − p(2): (0.0270067, 0.112415)			
Test for p(1) − p(2) = 0 (vs not = 0): Z = 3.19			
P-Value = 0.001			

a. Define appropriate notation and write the null and alternative hypotheses to test whether the proportion of adult Canadians who favor marriages between people of the same sex was different in April 1999 and February 2000.

b. Using the Minitab output, go through the remaining four steps of hypothesis testing to test the hypotheses you defined in part (a).

c. Use the confidence interval given in the Minitab output to test the hypotheses in part (a).

13.64   Refer to Example 13.6, comparing the proportion of children with ear infections while taking xylitol or a placebo. Read the discussion of type 1 and type 2 errors, and explain the consequences of each type of error if it were to have been made in this study.

13.65   Suppose a highway safety researcher makes modifications to the design of a highway sign. The researcher believes that the modifications will make the mean maximum distance at which drivers are able to read the sign greater than 450 feet. The maximum distances (in feet) at which $n = 16$ drivers can read the sign are

440	490	600	540	540	600	240	440
360	600	490	400	490	540	440	490

Use this sample to determine whether the mean maximum sign-reading distance at which drivers can read the sign is greater than 450. Show all five steps of a hypothesis test, and be sure to state a conclusion. Also, describe any assumptions that you make.

13.66   Case Study 1.6 presented data on 22,071 physicians who were randomly assigned to take aspirin or a placebo every other day for five years. Of the 11,037 taking aspirin, 104 had a heart attack, while of the 11,034 taking placebo, 189 had a heart attack.

a. Test whether the proportion having heart attacks in the population of men similar to the ones in this study would be smaller if taking aspirin than if taking a placebo.

b. What would be the consequences of type 1 and type 2 errors for this study? Which do you think would be more serious?

13.67   Refer to the previous exercise. The test showed that the difference in proportions that had heart attacks after taking aspirin and after taking a placebo was highly statistically significant. Suppose the study had 2,200 participants instead of over 22,000, but that the proportions having heart attacks in the two groups were about the same as they were in the original study. In other words, suppose 10 out of 1100 in the aspirin group had a heart attack and 19 out of 1100 in the placebo group had a heart attack. For a two-sided test of the hypothesis that the proportions likely to have a heart attack are equal after taking aspirin and taking placebo, the test statistic is $z = 1.68$.

a. Using $\alpha = 0.05$, what would you conclude?

b. Discuss this example in the context of the material presented in Section 13.9.

13.68 In exercise 12.67 a study is described in which the mean IQs at age 4 for children of smokers and nonsmokers were compared. The mean for the children of the 66 nonsmokers was 113.28 points, while for the 47 smokers it was 103.12 points. Assume the pooled standard deviation for the two samples is 13.5 points and that a pooled *t*-test is appropriate. Test the hypothesis that the population mean IQ at age 4 is the same for children of smokers and nonsmokers versus the alternative that it is higher for children of nonsmokers.

13.69 Refer to the previous exercise. One of the statements made in the research article and reported in exercise 12.67 was: "After control for confounding background variables . . . the average difference observed at 36 and 48 months was reduced to 4.35 points (95% CI: 0.02, 8.68)." Use this statement to test the null hypothesis that the difference in the mean IQ scores for the two populations is 0 versus the alternative that it is greater than 0. (Notice that the results have now been adjusted for confounding variables such as parents' IQ, whereas the data given in the previous exercise had not.)

13.70 It is believed that regular physical exercise leads to a lower resting pulse rate. Following are data for $n = 20$ individuals on resting pulse rate and whether the individual regularly exercises or not. Assuming this is a random sample from a larger population, use this sample to determine whether the mean pulse is lower for those who exercise. Clearly show all five steps of the hypothesis test.

Person	Pulse	Regularly Exercises
1	72	No
2	62	Yes
3	72	Yes
4	84	No
5	60	Yes
6	63	Yes
7	66	No
8	72	No
9	75	Yes
10	64	Yes
11	62	No
12	84	No
13	76	No
14	60	Yes
15	52	Yes
16	60	No
17	64	Yes
18	80	Yes
19	68	Yes
20	64	Yes

13.71 In Case Study 5.1, results were presented for a sample of 63 men who were asked to report their actual weight and their ideal weight. The mean difference between actual and ideal weight was 2.48 pounds, and the standard deviation of the differences was 13.77 pounds. Is there sufficient evidence to conclude that for the population of men represented by this sample the actual and ideal weights differ, on average? Justify your answer by showing all steps of a hypothesis test.

13.72 Perry et al. (1999) wanted to test the folklore that women who have not been given information about the sex of their unborn child can guess it at better than chance levels. They asked a sample of 104 pregnant women to guess the sex of their babies, and 57 guessed correctly. Assuming chance guessing would result in 50% correct guesses, test the hypothesis that women can guess at a better than chance level. Carry out the test using $\alpha = 0.05$, and make sure to state a conclusion. Note that the hypothesis involves only one proportion, so you may need to refer to an earlier chapter for guidance.

13.73 Refer to the previous exercise. The authors also wanted to test whether women with different levels of education would differ in their ability to guess the sex of their babies correctly. Of the $n = 45$ women with more than 12 years of education, 32 guessed correctly. Of the $n = 57$ women with 12 or fewer years of education, 24 guessed correctly. (Two women did not report their level of education.) Is there sufficient evidence to conclude that women in the populations with these two different levels of education differ in their ability to guess the sex of their baby? Carry out the test using $\alpha = 0.05$, and make sure to state a conclusion.

13.74 Exercise 6.9 described a case control study comparing short to not-short English secondary school students. Of $n = 92$ short students, 42 said they had been bullied in school. Of $n = 117$ not-short students, 30 said they had been bullied in school. Here is Minitab output for comparing two proportions:

Sample	Bullied	N	Sample p
Short	42	92	0.456522
Not Short	30	117	0.256410

Estimate for p(1) − p(2): 0.200111
95% lower bound for p(1) − p(2): 0.0919200
Test for p(1) − p(2) = 0 (vs > 0): Z = 3.02
P-Value = 0.001

a. The researchers wanted to know if short students are bullied more often. Write the null and alternative hypotheses for this question, defining and using appropriate notation.

b. Use the Minitab output to carry out the remaining steps of the hypothesis test.

c. The Minitab procedure found a one-sided 95% confidence interval for the difference in the two proportions, and the lower end of the interval is given in the output. What is the upper end of the interval? Write the complete interval.

d. Use the one-sided 95% confidence interval to test the hypotheses stated in part (a).

13.75 A Gallup Poll taken in August 2000 (Chambers, 2000) asked U.S. adults in a random sample of $n = 1019$ about their satisfaction with K–12 education. One question was: "Overall, how satisfied are you with the quality of education students receive in grades kindergarten through grade twelve in the U.S. today—would you say: completely satisfied, somewhat satisfied, somewhat dissatisfied or completely dissatisfied?" There were 622 respondents who said that they were somewhat or completely dissatisfied. Is this sufficient evidence to conclude that a majority of U.S. adults were dissatisfied in August 2000? Carry out the test using $\alpha = 0.05$, and make sure to state a conclusion.

13.76 Refer to the previous exercise. A Gallup Poll the previous August (1999) asked $n = 1028$ U.S. adults the same question, and 524 responded that they were somewhat or completely dissatisfied.
   a. Repeat the test in the previous exercise and make a conclusion for August 1999.
   b. Referring to the data presented in the previous exercise and in this one, and assuming the samples were taken independently, is there sufficient evidence to conclude that the proportion of U.S. adults who were dissatisfied changed from August 1999 to August 2000? Justify your answer.

13.77 In an experiment conducted by one of the authors, ten students in a graduate-level statistics course were given this question about the population of Canada: "The population of the U.S. is about 270 million. To the nearest million, what do you think is the population of Canada?" (The population of Canada at the time was slightly over 30 million.) The responses were

20, 90, 1.5, 100, 132, 150, 130, 40, 200, 20

Eleven other students in the same class were given the same question with different introductory information: "The population of Australia is about 18 million. To the nearest million, what do you think is the population of Canada?" The responses were

12, 20, 10, 81, 15, 20, 30, 20, 9, 10, 20

The experiment was done to demonstrate the *anchoring effect,* which is that responses to a survey question may be "anchored" to information provided to introduce the question. In this experiment, the research hypothesis was that the individuals who saw the U.S. population figure would generally give higher estimates of Canada's population than the individuals who saw the Australia population figure.
   a. Write null and alternative hypotheses for this experiment. Use proper notation.
   b. Test the hypotheses stated in part (a). Be sure to state a conclusion in the context of the experiment.

   c. As a step in part (b), you should have created a graphical summary to verify necessary conditions. Do you think any possible violations of the necessary conditions have affected the results of part (b) in a way that produced a misleading conclusion? Explain why or why not.

13.78 The dataset **cholest** reports cholesterol levels of heart attack patients 2, 4, and 14 days after the heart attack. Data for days 2 and 4 were available for 28 patients. The mean difference (2 day − 4 day) was 23.29, and the standard deviation of the differences was 38.28. A histogram of the differences was approximately bell-shaped. Is there sufficient evidence to indicate that the mean cholesterol level of heart attack patients decreases from the 2nd to the 4th day after the heart attack?

13.79 Refer to the previous exercise. Suppose physicians will use the answer to that question to decide whether to retest patients' cholesterol levels on day 4. If there is no conclusive evidence that cholesterol goes down, they will use the day 2 level to decide whether to prescribe drugs for high cholesterol. If there is evidence that cholesterol goes down between days 2 and 4, patients will all be retested on day 4 and the prescription drug decision will be made then.
   a. What are the consequences of type 1 and type 2 errors for this setting?
   b. Which type of error do you think is more serious?

Dataset Exercises

13.80 Refer to the previous two exercises. Using the dataset **cholest,** determine whether there is sufficient evidence to conclude that the cholesterol level drops, on average, from day 2 to day 14 after a heart attack.

13.81 The dataset **UCDavis2** includes information on *Sex, Height,* and *Dadheight.* Use the data to test the hypothesis that college men are taller, on average, than their fathers. Assume the male students in the survey represent a random sample of college men.

13.82 The dataset **GSS-93** includes information on *sex* and on whether respondents think marijuana should be legal (*marijuan*). Use the dataset to test the hypothesis that the proportions of males and females who thought marijuana should be legal differed for the 1993 population.

13.83 The dataset **UCDavis2** includes grade point average *GPA* and answers to the question: "Where do you typically sit in a classroom (circle one): Front, Middle, Back." The answer to this question is coded as F, M, B for the variable *Seat.* Assuming these students represent all college students, test whether there is a difference in mean GPA for students who sit in the front versus the back of the classroom.

REFERENCES

Chambers, Chris (2000). "Americans dissatisfied with U.S. education in general, but parents satisfied with their kids' schools," Gallup Poll Release, Sept. 5, 2000 (www.gallup.com).

Cohen, J. (1988). *Statistical Power Analysis for the Behavioral Sciences,* 2nd ed., Hillsdale, NJ: Lawrence Erlbaum Associates.

Devore, J., and R. Peck (1993). *Statistics: The Exploration and Analysis of Data,* 2nd ed., Belmont, CA: Duxbury Press.

Edwards, Gary, and Josephine Mazzuca (2000). "About four-in-ten Canadians accepting of same sex marriages, adoption," Gallup Poll Release, March 7, 2000 (www.gallup.com).

Faigenbaum, Avery D., Wayne L. Westcott, Rita LaRosa Loud, and Cindy Long (1999). "The effects of different resistance training protocols on muscular strength and endurance development in children," *Pediatrics Electronic Article,* Vol. 104, No. 1, p. e5.

Hedges, L. V., and I. Olkin (1985). *Statistical Methods for Meta-Analysis,* New York: Academic Press.

Mackowiak, P. A., S. S. Wasserman, and M. M. Levine (1992). "A critical appraisal of 98.6 degrees F, the upper limit of the normal body temperature, and other legacies of Carl Reinhold August Wunderlich," *Journal of the American Medical Association,* Vol. 268, No. 12 (Sep. 23), pp. 1578–1580.

Olds, D. L., C. R. Henderson, Jr., and R. Tatelbaum (1994). "Intellectual impairment in children of women who smoke cigarettes during pregnancy," *Pediatrics,* Vol. 93, No. 2, pp. 221–227.

Perry, D. F., J. DiPetro, and K. Costigan (1999). "Are women carrying 'basketballs' really having boys? Testing pregnancy folklore," *Birth,* Vol. 26, pp. 172–177.

Rosenthal, R. (1991). *Meta-Analytic Procedures for Social Research,* revised ed., Newbury Park, CA: Sage Publications.

Uhari, M., T. Kenteerkar, and M. Niemala (1998). "A novel use of Xylitol sugar in preventing acute Otitis Media," *Pediatrics,* Vol. 102, pp. 879–884.

Yanovski, J. A., S. Z. Yanovski, K. N. Sovik, T. T. Nguyen, P. M. O'Neil, and N. G. Sebring (2000). "A prospective study of holiday weight gain," *The New England Journal of Medicine,* March 23, 2000, Vol. 342, No. 12, pp. 861–867.

More About Regression

When you get older, how much closer will you have to get to read highway signs? See Example 14.3.

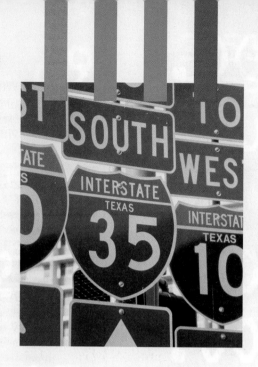

In Chapter 5 we used regression to describe a relationship in a sample. Now we make inferences about the population represented by the sample. What is the relationship between handspan and height in the population? What is the mean handspan for people 65 inches tall? What interval covers the handspans of most individuals of that height?

We learned in Chapter 5 that often a straight line describes the pattern of a relationship between two quantitative variables. For instance, in Example 5.1 we explored the relationship between the handspans (cm) and heights (in.) of 167 college students and found that the pattern of the relationship in this sample could be described by the equation

Average handspan = −3 + 0.35 Height

An equation like the one relating handspan to height is called a *regression equation,* and the term **simple regression** is sometimes used to describe the analysis of a straight-line relationship (*linear relationship*) between a response variable (*y* variable) and an explanatory variable (*x* variable).

In Chapter 5, we only used regression methods to describe a sample and did not make statistical inferences about the larger population. Now, we consider how to make inferences about a relationship in the population represented by the sample. Some questions involving the population that we might ask when analyzing a relationship are

1. Does the observed relationship also occur in the population? For example, is the observed relationship between handspan and height strong enough to conclude that the relationship also holds in the population?
2. For a linear relationship, what is the slope of the regression line in the population? For example, in the larger population, what is the slope of the regression line that connects handspans to heights?
3. What is the mean value of the response variable (*y*) for individuals with a specific value of the explanatory variable (*x*)? For example, what is the mean handspan in a population of people 65 inches tall?

4. What interval of values predicts the value of the response variable (*y*) for an individual with a specific value of the explanatory variable (*x*)? For example, what interval predicts the handspan of an individual 65 inches tall? ◆

14.1 | SAMPLE AND POPULATION REGRESSION MODELS

A **regression model** describes the relationship between a quantitative response variable (the *y* variable) and one or more explanatory variables (*x* variables). The *y* variable is sometimes called the *dependent variable,* and because regression models may be used to make predictions, the *x* variables may be called the *predictor variables.* The labels *response variable* and *explanatory variable* may be used for the variables on the *y* axis and *x* axis, respectively, even if there is not an obvious way to assign these labels in the usual sense.

Any regression model has two important components. The most obvious component is the equation that describes how the mean value of the *y* variable is connected to specific values of the *x* variable. The equation for the connection between handspan and height, Average handspan $= -3 + 0.35$ (Height), is an example. In this chapter, we focus on *linear relationships,* so a straight-line equation will be used, but it is important to note that some relationships are *curvilinear.*

The second component of a regression model describes how individuals vary from the regression line. Figure 14.1, which is identical to Figure 5.6, displays the raw data for the sample of $n = 167$ handspans and heights along with the regression line that estimates how the mean handspan is connected to specific heights. Notice that most individuals vary from the line. When we examine sample data, we will find it useful to estimate the general size of the deviations from the line. When we consider a model for the relationship within the population represented by a sample, we will state assumptions about the distribution of deviations from the line.

If the sample represents a larger population, we need to distinguish between the **regression line for the sample** and the **regression line for the population.** The observed data can be used to determine the regression line for the sample, but the regression line for the population can only be imagined. Because we do not observe the whole population, we will not know numerical values for the intercept and slope of the regression line in the population. As in nearly every sta-

FIGURE 14.1   **Regression line linking handspan and height for a sample of college students**

tistical problem, the statistics from a sample are used to estimate the unknown population parameters, which in this case are the slope and intercept of the regression line.

The Regression Line for the Sample

In Chapter 5, we introduced this notation for the regression line that describes sample data:

$$\hat{y} = b_0 + b_1 x$$

In any given situation, the sample is used to determine values for b_0 and b_1.

- $\hat{y}$ is spoken as "y-hat" and it is also referred to either as *predicted y* or *estimated y.*
- b_0 is the **intercept** of the straight line. The *intercept* is the value of $\hat{y}$ when $x = 0$.
- b_1 is the **slope** of the straight line. The *slope* tells us how much of an increase (or decrease) there is for $\hat{y}$ when the x variable increases by one unit. The sign of the slope tells us whether $\hat{y}$ increases or decreases when x increases. If the slope is 0, there is no linear relationship between x and y because $\hat{y}$ is the same for all values of x.

The equation describing the relationship between handspan and height for the sample of college students can be written as

$$\hat{y} = -3 + 0.35x$$

In this equation,

- $\hat{y}$ estimates the average handspan for any specific height x. If height = 70 in., for instance, $\hat{y} = -3 + 0.35(70) = 21.5$ cm.
- The *intercept* is $b_0 = -3$. While necessary for the line, this value does not have a useful statistical interpretation in this example. It estimates the average handspan for individuals who have height = 0 in., an impossible height far from the range of the observed heights. It also is an impossible handspan.
- The *slope* is $b_1 = 0.35$. This value tells us that the *average increase* in handspan is 0.35 cm for every 1-inch increase in height.

Reminder: The Least-Squares Criterion

In Chapter 5, we described the least-squares criterion. This mathematical criterion is used to determine numerical values of the intercept and slope of a sample regression line. The least-squares line is the line, among all possible lines, that has the smallest sum of squared differences between the sample values of y and the corresponding values of $\hat{y}$.

Deviations from the Regression Line in the Sample

The terms *random error, residual variation,* and *residual error* all are used as synonyms for the term **deviation.** Most commonly, the word **residual** is used to describe the deviation of an observed y value from the sample regression line. A *residual* is easy to compute. It simply is the difference between the observed y value for an individual and the value of $\hat{y}$ determined from the x value for that individual.

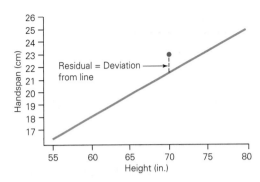

FIGURE 14.2 **Residual for a person 70 inches tall with a handspan = 23 cm. The residual is the difference between observed $y = 23$ and $\hat{y} = 21.5$, the predicted value for a person 70 inches tall.**

EXAMPLE 14.1
Residuals in
the Handspan
and Height Regression

Consider a person 70 inches tall whose handspan is 23 centimeters. The sample regression line is $\hat{y} = -3 + 0.35x$, so $\hat{y} = -3 + 0.35(70) = 21.5$ cm for this person. The *residual* = observed y − predicted $y = y - \hat{y} = 23 - 21.5 = 1.5$ cm. Figure 14.2 illustrates this residual. ◆

definition For an observation y_i in the sample, the **residual** is

$$e_i = y_i - \hat{y}_i$$

y_i = the value of the response variable for the observation

$\hat{y}_i = b_0 + b_1 x_i$, where x_i is the value of the explanatory variable for the observation

tech note The sum of the residuals is 0 for any least-squares regression line. The "least-squares" formulas for determining the equation always result in $\sum y_i = \sum \hat{y}_i$, so $\sum e_i = 0$.

The Regression Line for the Population

The regression equation for a simple linear relationship in a population can be written as

$$E(Y) = \beta_0 + \beta_1 x$$

◆ $E(Y)$ represents the mean or expected value of y for individuals in the population who all have the same particular value of x. Note that $\hat{y}$ is an estimate of $E(Y)$.

◆ β_0 is the **intercept** of the straight line in the **population.**

◆ β_1 is the **slope** of the line in the **population.** Note that if the slope $\beta_1 = 0$, there is no linear relationship in the population.

Unless we measure the entire population, we cannot know the numerical values of β_0 and β_1. These are population parameters that we estimate using the corresponding sample statistics. In the handspan and height example, $b_1 = 0.35$ is a sample statistic that estimates the population parameter β_1, and $b_0 = -3$ is a sample statistic that estimates the population parameter β_0.

Multiple Regression

> **tech note** In **multiple regression,** the mean of the response variable is a function of two or more explanatory variables. Put another way, in multiple regression we use the values of more than one explanatory (predictor) variable to predict the value of a response variable. For example, a college admissions committee might predict college GPA for an applicant based on Verbal SAT, Math SAT, high school GPA, and class rank. The general structure of an equation for doing this might be
>
> $$\text{College GPA} = \beta_0 + \beta_1 \text{ Verbal SAT} + \beta_2 \text{ Math SAT} + \beta_3 \text{ HS GPA} + \beta_4 \text{ Class Rank}$$
>
> As in simple regression, numerical estimates of the parameters $\beta_0, \beta_1, \beta_2, \beta_3$, and β_4 would be determined from a sample. Multiple regression is covered in Chapter 19.

Assumptions About Deviations from the Regression Line in the Population

To make statistical inferences about the population, two assumptions about how the y values vary from the population regression line are necessary. First, we assume that the general size of the deviation of y values from the line is the same for all values of the explanatory variable (x), an assumption called the **constant variance** assumption. This assumption may or may not be correct in any particular situation, and a scatterplot should be examined to see if it is reasonable or not. In Figure 14.1, the constant variance assumption looks reasonable because the magnitude of the deviation from the line appears to be about the same across the range of observed heights.

The second assumption about the population is that for any specific value of x, the distribution of y values is a normal distribution. Equivalently, this assumption is that deviations from the population regression line have a normal curve distribution. Figure 14.3 illustrates this assumption along with the other

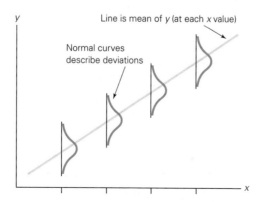

FIGURE 14.3 **Regression model for population**

elements of the population regression model for a linear relationship. The line $E(Y) = \beta_0 + \beta_1 x$ describes the mean of y, and the normal curves describe deviations from the mean.

The Simple Regression Model for a Population

A useful format for expressing the components of the population regression model is

$$y = \text{Mean} + \text{Deviation}$$

This conceptual equation states that for any individual, the value of the response variable (y) can be constructed by combining two components:

1. The **mean,** which in the population is the line $E(Y) = \beta_0 + \beta_1 x$ if the relationship is linear. There are other possible relationships, such as curvilinear, a special case of which is a quadratic relationship, $E(Y) = \beta_0 + \beta_1 x + \beta_2 x^2$. Relationships that are not linear will not be discussed in this book.
2. The individual's **deviation = y − mean,** which is what is left unexplained after accounting for the mean y value at that individual's x value.

This format also applies to the sample, although technically we should use the term *estimated mean* when referring to the sample regression line.

EXAMPLE 14.1—*CONTINUED*
Mean and Deviation
for Height and
Handspan Regression

Recall that the sample regression line for handspans and heights is $\hat{y} = -3 + 0.35x$. Although it is not likely to be true, let's assume for convenience that this equation also holds in the population. If your height is $x = 70$ in. and your handspan is $y = 23$ cm, then

$$\text{Mean} = -3 + 0.35(70) = 21.5,$$
$$\text{Deviation} = y - \text{Mean} = 23 - 21.5 = 1.5, \text{ and}$$
$$y = 23 = \text{Mean} + \text{Deviation} = 21.5 + 1.5$$

In other words, your handspan is 1.5 cm above the mean for people with your height. ◆

In the theoretical development of procedures for making statistical inferences for a regression model, the collection of all *deviations* in the population is assumed to have a normal distribution with mean 0 and standard deviation σ (so the variance is σ^2). The value of the standard deviation σ is an unknown population parameter that is estimated using the sample. This standard deviation can be interpreted in the usual way that we interpret a standard deviation. It is, roughly, the average distance between individual values of y and the mean of y as described by the regression line. In other words, it is roughly the size of the average deviation across all individuals in the range of x values.

Keeping the regression notation straight for populations and samples can be confusing. Although we have not yet introduced all relevant notation, a summary at this stage will help you keep it straight.

in summary Simple Linear Regression Model: Population and Sample Versions

For (x_1, y_1), (x_2, y_2), . . . , (x_n, y_n), a sample of n observations of the explanatory variable x and the response variable y from a large population, the simple linear regression model describing the relationship between y and x follows.

Population Version

$$\text{Mean:} \quad E(Y) = \beta_0 + \beta_1 x$$

$$\text{Individual:} \quad y_i = \beta_0 + \beta_1 x_i + \varepsilon_i = E(Y_i) + \varepsilon_i$$

The deviations ε_i are assumed to follow a normal distribution with mean 0 and standard deviation σ.

Sample Version

$$\text{Mean:} \quad \hat{y} = b_0 + b_1 x$$

$$\text{Individual:} \quad y_i = b_0 + b_1 x_i + e_i = \hat{y}_i + e_i$$

where e_i is the *residual* for individual i. The sample statistics b_0 and b_1 *estimate* the population parameters β_0 and β_1. The mean of the residuals is 0, and the residuals can be used to estimate the population standard deviation σ.

14.2 | ESTIMATING THE STANDARD DEVIATION FOR REGRESSION

Recall that the standard deviation in the regression model measures, roughly, the average deviation of y values from the mean (the regression line). Expressed another way, the **standard deviation for regression** measures the general size of the residuals. This is an important and useful statistic for describing individual variation in a regression problem, and it also provides information about how accurately the regression equation might predict y values for individuals. A relatively small standard deviation from the regression line indicates that individual data points generally fall close to the line, so predictions based on the line will be close to the actual values.

The calculation of the estimate of standard deviation is based on the sum of the squared residuals for the sample. This quantity is called the **sum of squared errors** and is denoted by **SSE.** Synonyms for "sum of squared errors" are **residual sum of squares** or **sum of squared residuals.** To find the SSE, residuals are calculated for all observations, then the residuals are squared and summed. The *standard deviation from the regression line* is

$$s = \sqrt{\frac{\text{Sum of Squared Residuals}}{n - 2}} = \sqrt{\frac{\text{SSE}}{n - 2}}$$

and this sample statistic estimates the population standard deviation σ.

Estimating the Standard Deviation for a Simple Regression Model

formula   The formula for estimating the standard deviation for a simple regression model is

$$\text{SSE} = \sum(y_i - \hat{y}_i)^2 = \sum e_i^2$$

$$s = \sqrt{\frac{\text{SSE}}{n-2}} = \sqrt{\frac{\sum(y_i - \hat{y}_i)^2}{n-2}}$$

The statistic s is an estimate of the population standard deviation σ.

Technical Note: Notice the difference between the estimate of σ in the regression situation compared to if we simply had a random sample of the y_i's without information about the x_i's:

$$\text{Sample of } y\text{'s only:} \quad s = \sqrt{\frac{\sum(y_i - \bar{y})^2}{n-1}}$$

$$\text{Sample of } (x,y) \text{ pairs, linear regression:} \quad s = \sqrt{\frac{\sum(y_i - \hat{y}_i)^2}{n-2}}$$

Remember that in the regression context, σ is the standard deviation of the y values at *each x*, not the standard deviation of the whole population of y values.

EXAMPLE 14.2
Relationship Between Height and Weight for College Men

Figure 14.4 displays regression results from the Minitab program and a scatterplot for the relationship between y = weight (pounds) and x = height (inches) in a sample of n = 43 men in a Penn State statistics class. The regression line for the sample is $\hat{y} = -318 + 7x$, and this line is drawn onto the plot. We see from

FIGURE 14.4   **The relationship between weight and height for n = 43 college men**

the plot that there is considerable variation from the line at any given height. The standard deviation, shown in the row of computer output immediately above the plot, is "S = 24.00." This value roughly measures, for any given height, the general size of the deviations of individual weights from the mean weight for the height.

The standard deviation from the regression line can be interpreted in conjunction with the Empirical Rule for bell-shaped data stated in Section 2.7. Recall, for instance, that about 95% of individuals will fall within 2 standard deviations of the mean. As an example, consider men who are 72 inches tall. For men with this height, the estimated average weight determined from the regression equation is $-318 + 7.00(72) = 186$ pounds. The estimated standard deviation from the regression line is $s = 24$ pounds, so we can estimate that about 95% of men 72 inches tall have weights within $2 \times 24 = 48$ pounds of 186 pounds, which is 186 ± 48, or 138 to 234 pounds. Think about whether this makes sense for all the men you know who are 72 inches (6 feet) tall. ◆

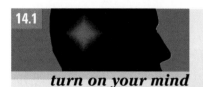

14.1

turn on your mind

Regression equations can be used to predict the value of a response variable for an individual. What is the connection between the accuracy of predictions based on a particular regression line and the value of the standard deviation from the line? If you were deciding between two different regression models for predicting the same response variable, how would your decision be affected by the relative values of the standard deviations for the two models?*

The Proportion of Variation Explained by x

In Chapter 5, we learned that a statistic denoted as r^2 is used to measure how well the explanatory variable actually does explain the variation in the response variable. This statistic is also denoted as R^2 (rather than r^2), and the value is commonly expressed as a percent. Researchers typically use the phrase **"proportion of variation explained by x"** in conjunction with the value of r^2. For example, if $r^2 = 0.60$ (or 60%), the researcher may write that the explanatory variable explains 60% of the variation in the response variable.

The formula for r^2 presented in Chapter 5 was

$$r^2 = \frac{\text{SSTO} - \text{SSE}}{\text{SSTO}}$$

The quantity SSTO is the sum of squared differences between observed y values and the sample mean $\bar{y}$. It measures the size of the deviations of the y values from the overall mean of y, whereas SSE measures the deviations of the y values from the predicted values y.

EXAMPLE 14.2—CONTINUED
R^2 for Heights and
Weights of College Men

In Figure 14.4, we can find the information "R-sq = 32.3%" for the relationship between weight and height. A researcher might write "the variable height explains 32.3% of the variation in the weights of college men." This isn't a particularly impressive statistic. As we noted before, there is substantial deviation

*HINT: Read the first paragraph of this section.

of individual weights from the regression line, so a prediction of a college man's weight based on height may not be particularly accurate. ◆

14.2

turn on your mind

Look at the formula for SSE and explain in words under what condition SSE = 0. Now explain what happens to r^2 when SSE = 0, and explain whether that makes sense according to the definition of r^2 as "proportion of variation in y explained by x."*

EXAMPLE 14.3
Driver Age and Highway Sign-Reading Distance

In Example 5.2, we examined data for the relationship between y = maximum distance (feet) at which a driver can read a highway sign and x = the age of the driver. There were n = 30 observations in the dataset. Figure 14.5 displays Minitab regression output for these data. The equation describing the linear relationship in the sample is

$$\text{Average distance} = 577 - 3.01 \times \text{Age}$$

From the output, we learn that the standard deviation from the regression line is s = 49.76 and R-sq = 64.2%. Roughly, the average deviation from the regres-

```
The regression equation is
Distance = 577 - 3.01 Age

Predictor        Coef      SE Coef        T         P
Constant       576.68        23.47    24.57     0.000
Age            -3.0068       0.4243   -7.09     0.000

S = 49.76      R-Sq = 64.2%      R-Sq (adj) = 62.9%

Analysis of Variance

Source           DF         SS         MS        F        P
Regression        1     124333     124333    50.21    0.000
Residual Error    28      69334       2476
Total             29     193667

Unusual Observations
Obs     Age    Distance      Fit     SE Fit    Residual    St Resid
 27    75.0      460.00   351.17      13.65      108.83       2.27R

R denotes an observation with a large standardized residual
```

FIGURE 14.5   **Minitab output: Sign reading distance and driver age**

*HINT: Remember that SSE stands for "sum of squared errors." The formula for r^2 is given just before Example 14.2—*CONTINUED*.

sion line is about 50 feet, and the proportion of variation in sign reading distances explained by age is .642, or 64.2%.

The "Analysis of Variance" table provides the pieces needed to compute r^2 and s:

$$SSE = 69,334$$

$$s = \sqrt{\frac{SSE}{n-2}} = \sqrt{\frac{69,334}{28}} = 49.76$$

$$SSTO = 193,667$$

$$SSTO - SSE = 193,667 - 69,334 = 124,333$$

$$r^2 = \frac{124,333}{193,667} = .642, \text{ or } 64.2\% \blacklozenge$$

14.3 | INFERENCE ABOUT THE LINEAR REGRESSION RELATIONSHIP

When researchers do a regression analysis, they occasionally know based on past research or common sense that the variables are indeed related. In some instances, however, it may be necessary to do a hypothesis test in order to make the generalization that two variables are related in the population represented by the sample. The **statistical significance of a linear relationship** can be evaluated by testing whether or not the slope is 0. Recall that if the slope is 0 in a simple regression model, the two variables are not related because changes in the x variable will not lead to changes in the y variable. The usual null hypothesis and alternative hypothesis about β_1, the slope of the population regression line $E(Y) = \beta_0 + \beta_1 x$, are

$H_0: \beta_1 = 0$ (the population slope is 0, so y and x are *not linearly related*)
$H_a: \beta_1 \neq 0$ (the population slope is not 0, so y and x *are linearly related*)

The alternative hypothesis may be one-sided or two-sided, although most statistical software uses the two-sided alternative.

The test statistic used to do the hypothesis test is a t-statistic with the same general format that we saw in Chapter 13. That format, and its application to this situation, is

$$t = \frac{\text{Sample statistic} - \text{Null value}}{\text{Standard error}} = \frac{b_1 - 0}{\text{s.e.}(b_1)}$$

This is a standardized statistic for the difference between the sample slope and 0, the null value. Notice that a large value of the sample slope (either positive or negative) relative to its standard error will give a large value of t. If the mathematical assumptions about the population model described in Section 14.1 are correct, the statistic has a t-distribution with $n - 2$ degrees of freedom. The p-value for the test is determined using that distribution.

"By hand" calculations of the sample slope and its standard error are cumbersome. Fortunately, the regression analysis of most statistical software includes a t-statistic and a p-value for this significance test.

Formula for the Sample Slope and Its Standard Error

formula   In case you ever need to compute the values by hand, here are the formulas for the sample slope and its standard error:

$$b_1 = r\frac{s_y}{s_x}$$

$$\text{s.e.}(b_1) = \frac{s}{\sqrt{\Sigma(x_i - \bar{x})^2}} \qquad \text{where } s = \sqrt{\frac{\text{SSE}}{n-2}}$$

In the formula for the sample slope, s_x and s_y are the sample standard deviations of the x and y values, respectively, and r is the correlation between x and y.

EXAMPLE 14.3—CONTINUED
Driver Age and Highway Sign-Reading Distance

Figure 14.5 (p. 502) presents the Minitab output for the regression of sign-reading distance and driver age. The sample estimate of the slope is $b_1 = -3.01$. This sample slope is different from 0, but is it different enough to enable us to generalize that a linear relationship exists in the population represented by this sample?

The part of the Minitab output that can be used to test the statistical significance of the relationship is shown in bold in Figure 14.5, and the relevant *p*-value is underlined (by the authors of this text, not by Minitab). This line of the output provides information about the sample slope, the standard error of the sample slope, the *t*-statistic for testing statistical significance and the *p*-value for the test of

H_0: $\beta_1 = 0$ (the population slope is 0, so y and x are *not linearly related*)
H_a: $\beta_1 \neq 0$ (the population slope is not 0, so y and x are *linearly related*)

The test statistic is

$$t = \frac{\text{Sample statistic} - \text{Null value}}{\text{Standard error}} = \frac{b_1 - 0}{\text{s.e.}(b_1)} = \frac{-3.0068 - 0}{0.4243} = -7.09$$

The *p*-value is, to three decimal places, 0.000. This means the probability is virtually 0 that the observed slope could be as far from 0 or farther than it is if there is no linear relationship in the population. So, as we might expect for these variables, we can conclude that the relationship between the two variables in the sample represents a real relationship in the population. ◆

tech note   Most statistical software reports a *p*-value for a two-sided alternative hypothesis when doing a test for whether the slope in the population is 0. It may sometimes make sense to use a one-sided alternative hypothesis instead. In that case, the *p*-value for the one-sided alternative is (reported $p/2$) if the sign of b_1 is consistent with H_a, but is $1 - $ (reported $p/2$) if it is not.

Confidence Interval for the Population Slope

The significance test of whether or not the population slope is 0 only tells us if we can declare the relationship to be statistically significant. If we decide that the true slope is not 0, we might ask, "What is the value of the slope?" We can answer this question with a confidence interval for β_1, the population slope.

The format for this confidence interval is the same as the general format used in Chapters 10 and 12, which is

$$\text{Sample estimate} \pm \text{Multiplier} \times \text{Standard error}$$

The estimate of the population slope β_1 is b_1, the slope of the least-squares regression line for the sample. As shown already, the standard error formula is complicated and we'll usually rely on statistical software to determine this value. The "multiplier" will be labeled t^* and is determined using a t-distribution with df $= n - 2$. Table A.2 can be used to find the multiplier for the desired confidence level.

Formula for Confidence Interval for β_1, the Population Slope

formula A confidence interval for β_1 is

$$b_1 \pm t^* \text{ s.e.}(b_1)$$

The multiplier t^* is found using a t-distribution with $n - 2$ degrees of freedom and is such that the probability between $-t^*$ and $+t^*$ equals the confidence level for the interval.

EXAMPLE 14.3—CONTINUED
95% Confidence Interval for Slope Between Age and Sign-Reading Distance

In Figure 14.5, we see that the estimated slope is $b_1 = -3.01$ and s.e.$(b_1) = 0.4243$. There are $n = 30$ observations so df $= 28$ for finding t^*. For a 95% confidence level, $t^* = 2.05$ (see Table A.2). The 95% confidence interval for the population slope is

$$-3.01 \pm 2.05 \times 0.4243$$
$$-3.01 \pm 0.87$$
$$-3.88 \text{ to } -2.14$$

With 95% confidence, we can estimate that in the population of drivers represented by this sample, the *mean* sign-reading distance decreases somewhere between 3.88 and 2.14 feet for each one-year increase in age. ◆

14.3

turn on your mind

In previous chapters, we learned that a confidence interval can be used to determine whether a hypothesized value for a parameter can be rejected. How would you use a confidence interval for the population slope to determine if there was a statistically significant relationship between x and y? For example, why is the interval just computed for the sign-reading example evidence that sign-reading distance and age are related?*

Testing Hypotheses About the Correlation Coefficient

In Chapter 5, we learned that the correlation coefficient is 0 when the regression line is horizontal. In other words, if the slope of the regression line is 0, the correlation is 0. This means that the results of a hypothesis test for the population slope can also be interpreted as applying to equivalent hypotheses about the correlation between x and y in the population.

As we did for the regression model, we use different notation to distinguish between a correlation computed for a sample and a correlation within a population. It is commonplace to use the symbol ρ (pronounced "rho") to represent the correlation between two variables within a population. Using this notation, null and alternative hypotheses of interest are

H_0: $\rho = 0$ (x and y are not correlated)
H_a: $\rho \neq 0$ (x and y are correlated)

The results of the hypothesis test described before for the population slope β_1 can be used for these hypotheses as well. If we reject H_0: $\beta_1 = 0$, we also reject H_0: $\rho = 0$. If we decide in favor of H_a: $\beta_1 \neq 0$, we also decide in favor of H_a: $\rho \neq 0$.

Many statistical software programs, including Minitab, will give a p-value for testing whether the population correlation is 0 or not. This p-value will be the same as the p-value given for testing whether the population slope is 0 or not.

In the following Minitab output for the relationship between pulse rate and weight in a sample of 35 college women, notice that 0.292 is given as the p-value for testing that the slope is 0 (look under P in the regression results) and for testing that the correlation is 0. Because this is not a small p-value, we cannot reject the null hypotheses for the slope and the correlation.

Regression Analysis: Pulse Versus Weight

The regression equation is
Pulse = 57.2 + 0.159 Weight

Predictor	Coef	SE Coef	T	P
Constant	57.17	18.51	3.09	0.004
Weight	0.1591	0.1487	1.07	0.292

Correlations: Pulse, Weight

Pearson correlation of Pulse and Weight = 0.183
P-Value = 0.292

The Effect of Sample Size on Significance

The size of a sample always affects whether a specific observed result achieves statistical significance. For example, $r = 0.183$ is not a statistically significant correlation for a sample size of $n = 35$, as in the pulse and weight example, but it would be statistically significant if $n = 1000$. With very large sample sizes, weak relationships with low correlation values can be statistically significant. The "moral of the story" here is that with a large sample size, it may not be saying much to say that two variables are significantly related. This only means that we think the correlation is not precisely 0. To assess the practical significance of the result, we should carefully examine the observed strength of the relationship.

tech note   The usual t-statistic for testing whether the population slope is 0 in a linear regression could also be found using a formula that involves only n = sample size and r = correlation between x and y. The algebraic equivalence is

$$t = \frac{b_1}{\text{s.e.}(b_1)} = \sqrt{n-2}\,\frac{r}{\sqrt{1-r^2}}$$

In the output for the pulse rate and body weight example just given, notice that the t-statistic for testing whether the slope $\beta_1 = 0$ is $t = 1.07$. This was calculated as

$$t = \frac{b_1}{\text{s.e.}(b_1)} = \frac{0.1591}{0.1487} = 1.07$$

The sample size is $n = 35$ and the correlation is $r = 0.183$, so an equivalent calculation of the t-statistic is

$$t = \sqrt{n-2}\,\frac{r}{\sqrt{1-r^2}} = \sqrt{35-2}\,\frac{0.183}{\sqrt{1-0.183^2}} = 1.07$$

This second method for calculating the t-statistic illustrates two ideas. First, there is a direct link between the correlation value and the t-statistic used to test whether the slope is 0. Second, notice that for any fixed value of r, increasing the sample size n will increase the size of the t-statistic. And, the larger the value of the t-statistic, the stronger the evidence against the null hypothesis.

14.4 | PREDICTING THE VALUE OF y FOR AN INDIVIDUAL

An important use of a regression equation is to estimate or predict the unknown value of a response variable for an *individual* with a known specific value of the explanatory variable. Using the data described in Example 14.3, for instance, we can predict the maximum distance at which an individual can read a highway sign by substituting his or her age for x in the sample regression equation. Consider a person 21 years old. The predicted distance for such a person is approximately $\hat{y} = 577 - 3(21) = 514$ feet.

There will be variation among 21-year-olds with regard to the sign-reading distance, so the predicted distance of 514 feet is not likely to be the exact distance for the next 21-year-old who views the sign. Rather than predicting that the distance will be exactly 514 feet, we should instead predict that the distance will be within a particular interval of values. A **95% prediction interval** for the value of the response variable (y) accounts for the variation among individuals with a particular value of x. This interval can be interpreted in two equivalent ways:

1. The 95% prediction interval estimates the central 95% of the values of y for members of the population with a specified value of x.
2. The probability is .95 that a randomly selected individual from the population with a specified value of x falls into the corresponding 95% prediction interval.

Notice that a prediction interval differs conceptually from a confidence interval. A confidence interval estimates an unknown population parameter, which is a numerical characteristic or summary of the population. An example in this chapter is a confidence interval for the slope of the population line. A prediction interval, however, does not estimate a parameter; instead, it estimates the potential data value for an individual. Equivalently, it describes an interval into which a specified percentage of the population may fall.

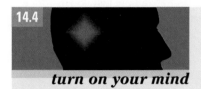

14.4

turn on your mind

If we knew the population parameters β_0, β_1, and σ, under the usual regression assumptions, we would know that the population of y values at a specific x value was normal with mean $\beta_0 + \beta_1 x$ and standard deviation σ. In that case, what interval would cover the central 95% of the y values for that x value? Use your answer to explain why a prediction interval would not have zero width even with complete population details.*

As with most regression calculations, the "by hand" formulas for prediction intervals are formidable. Statistical software can be used to create the interval. Figure 14.6 shows Minitab output that includes the 95% prediction intervals for three different ages (21, 30, and 45). The intervals are toward the bottom-right side of the display in a column labeled "95% PI" and are highlighted with bold type. (**Note:** The term "Fit" is a synonym for $\hat{y}$, the estimate of the average response at the specific x value.) Here is what we can conclude:

♦ The probability is .95 that a randomly selected 21-year-old will read the sign at somewhere between roughly 407 and 620 feet.

♦ The probability is .95 that a randomly selected 30-year-old will read the sign at somewhere between roughly 381 and 592 feet.

♦ The probability is .95 that a randomly selected 45-year-old will read the sign at somewhere between roughly 338 and 545 feet.

We can also interpret each interval as an estimate of the sign-reading distances for the central 95% of a population of drivers with a specified age. For instance, about 95% of all drivers 21 years old will be able to read the sign at a distance somewhere between 407 and 620 feet.

We're not limited to using only 95% prediction intervals. With Minitab, we can describe any central percentage of the population that we wish. For example, here are 50% prediction intervals for the sign-reading distance at the three specific ages we considered above.

Age	Fit	50.0% PI
21	513.54	(477.89, 549.18)
30	486.48	(451.38, 521.58)
45	441.37	(406.76, 475.98)

*The regression equation gives mean y for a specific x.

HINT: Generally, standard deviation measures how individuals vary from the mean.

```
The regression equation is
Distance = 577 − 3.01 Age

Predictor       Coef      SE Coef        T        P
Constant      576.68        23.47     24.57    0.000
Age          −3.0068       0.4243     −7.09    0.000

S = 49.76       R-Sq = 64.2%      R-Sq(adj) = 62.9%

Analysis of Variance

Source          DF         SS        MS        F        P
Regression       1     124333    124333    50.21    0.000
Residual Error   28      69334      2476
Total            29     193667

Unusual Observations
Obs    Age    Distance      Fit    SE Fit    Residual    St Resid
27    75.0      460.00   351.17     13.65      108.83        2.27R

R denotes an observation with a large standardized residual

Predicted Values for New Observations

New Obs      Fit    SE Fit        95.0% CI              95.0% PI
1         513.54     15.64   ( 481.50, 545.57)   ( 406.69, 620.39)
2         486.48     12.73   ( 460.41, 512.54)   ( 381.26, 591.69)
3         441.37      9.44   ( 422.05, 460.70)   ( 337.63, 545.12)

Values of Predictors for New Observations
New Obs      Age
1           21.0
2           30.0
3           45.0
```

FIGURE 14.6   Minitab output showing prediction intervals of distance

For each specific age, the 50% prediction interval estimates the central 50% of the maximum sign-reading distances in a population of drivers with that age. For example, we can estimate that 50% of drivers 21 years old would have a maximum sign-reading distance somewhere between about 478 feet and 549 feet. The distances for the other 50% of 21-year-old drivers would be predicted to be outside this range, with 25% above 549 feet and 25% below 478 feet.

> *definition* A **prediction interval** estimates the value of y for an individual with a particular value of x, or equivalently, the range of values of the response variable for a specified central percentage of a population with a particular value of x.

tech note   The formula for the prediction interval for y at a specific x is

$$\hat{y} \pm t^*\sqrt{s^2 + [\text{s.e.(fit)}]^2}$$

where

$$\text{s.e.(fit)} = s\sqrt{\frac{1}{n} + \frac{(x - \bar{x})^2}{\Sigma(x_i - \bar{x})^2}}$$

The multiplier t^* is found using a t-distribution with $n - 2$ degrees of freedom and is such that the probability between $-t^*$ and $+t^*$ equals the desired level for the interval.

Note:

- The s.e.(fit), and thus the width of the interval, depends upon how far the specified x value is from $\bar{x}$. The further the specific x is from the mean, the wider the interval.

- When n is large, s.e.(fit) will be small, and the prediction interval will be approximately $\hat{y} \pm t^*s$.

14.5 | ESTIMATING THE MEAN y AT A SPECIFIED x

In the previous section, we focused on the estimation of the values of the response variable for individuals. A researcher may instead want to estimate the *mean* value of the response variable for individuals with a particular value of the explanatory variable. We might ask, "What is the mean weight for college men who are 6 feet tall?" This question only asks about the mean weight in a group with a common height, and it is not concerned with the deviations of individuals from that mean.

In technical terms, we wish to estimate the population mean $E(Y)$ for a specific value of x that is of interest to us. To make this estimate, we use a confidence interval. The format for this confidence interval is again:

Sample estimate $\pm$ Multiplier $\times$ Standard error

The *sample estimate* of $E(Y)$ is the value of $\hat{y}$ determined by substituting the x value of interest into $\hat{y} = b_0 + b_1 x$, the least-squares regression line for the sample. The *standard error* of $\hat{y}$ is the s.e.(fit) shown in the Tech Note box in the previous section, and its value is usually provided by statistical software. The *multiplier* is found using a t-distribution with df $= n - 2$, and Appendix A.2 can be used to determine its value.

EXAMPLE 14.2—
CONTINUED
Estimating Mean
Weight of College
Men at Various Heights

Based on the sample of $n = 43$ college men in Example 14.2, let's estimate the mean weight in the population of college men for each of three different heights: 68 inches, 70 inches, and 72 inches. Figure 14.7 shows Minitab output that includes the three different confidence intervals for these three different heights. These intervals are toward the bottom of the display in a column labeled "**95% CI**." The first entry in that column is the estimate of the population mean weight for men who are 68 inches tall. With 95% confidence, we can estimate that the mean weight of college men 68 inches tall is somewhere between 147.78 and

The regression equation is
Weight = − 318 + 7.00 Height

Predictor	Coef	SE Coef	T	P
Constant	−317.9	110.9	−2.87	0.007
Height	6.996	1.581	4.42	0.000

S = 24.00   R-Sq = 32.3%   R-Sq(adj) = 30.7%

--- **Some Output Omitted** ---

Predicted Values for New Observations

New Obs	Fit	SE Fit	95.0% CI	95.0% PI
1	157.80	4.96	(147.78, 167.81)	(108.31, 207.29)
2	171.79	3.66	(164.39, 179.19)	(122.76, 220.82)
3	185.78	4.72	(176.25, 195.31)	(136.38, 235.18)

Values of Predictors for New Observations

New Obs	Height
1	68.0
2	70.0
3	72.0

FIGURE 14.7   **Minitab output with confidence intervals for mean weight**

167.81 pounds. The second row under "**95% CI**" contains the information that the 95% confidence interval for the mean weight of college men 70 inches tall is 164.39 to 179.19 pounds. The 95% confidence interval for the mean weight for men 72 inches tall is 176.25 to 195.31 pounds.

Again, it is important to realize that the confidence intervals for $E(Y)$ do not describe the variation among *individuals*. They only are estimates of the *mean* weights for specific heights. The prediction intervals for individual responses describe the variation among individuals. You may have noticed that 95% prediction intervals, labeled "95% PI," are next to the confidence intervals in the output. Among men 70 inches tall, for instance, we would estimate that 95% of the individual weights would be in the interval from about 123 to about 221 pounds. ◆

Minitab Tip

tech note   **Creating Prediction Intervals for *y***
and Confidence Intervals for *E*(*Y*)

◆ Use **Stat**>**Regression**>**Regression,** then use the ***Options*** button of the dialog box.

◆ In the box labeled "Prediction intervals for new observations," either specify the numerical value of a single *x* value of interest or specify a worksheet column that contains two or more different *x* values of interest. The confidence level can be specified in the box labeled "Confidence Level."

Note: Minitab computes both a prediction interval for *y* and a confidence interval for $E(Y)$ for each specified value of *x*.

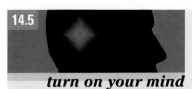

14.5

turn on your mind

Draw a picture similar to the one in Figure 14.3, illustrating the regression line and the normal curves for the *y* values at several values of *x*. Use it to illustrate the difference between a prediction interval for *y* and a confidence interval for the mean of the *y*'s, at a specific value of *x*. *

14.6 | CHECKING CONDITIONS FOR USING REGRESSION MODELS FOR INFERENCE

There are a few conditions that should be at least approximately true when we use a regression model to make an inference about a population. Of the five conditions that follow, the first two are particularly crucial.

Conditions for Linear Regression

1. The form of the equation that links the mean value of *y* to *x* must be correct. For instance, we won't make proper inferences if we use a straight line to describe a curved relationship.
2. There should not be any extreme outliers that influence the results unduly.
3. The standard deviation of the values of *y* from the mean *y* is the same regardless of the value of the *x* variable. In other words, *y* values are similarly spread out at all values of *x*.
4. For individuals in the population with the same particular value of *x*, the distribution of the values of *y* is a normal distribution. Equivalently, the distribution of deviations from the mean value of *y* is a normal distribution. This condition can be relaxed if the sample size is large.
5. Observations in the sample are independent of each other.

Checking the Conditions with Plots

A scatterplot of the raw data and **plots of the residuals** provide information about the validity of the assumptions. Remember that a residual is the difference between an observed value and the predicted value for that observation and that some assumptions made for a linear regression model have to do with how *y* values deviate from the regression line. If the properties of the residuals for the sample appear to be consistent with the mathematical assumptions made about deviations within the population, we can use the model to make statistical inferences.

Conditions 1, 2, and 3 can be checked using two useful plots:

- A scatterplot of *y* versus *x* for the sample
- A scatterplot of the residuals versus *x* for the sample

If Condition 1 holds for a linear relationship, then:

- The plot of *y* versus *x* should show points randomly scattered around an imaginary straight line.
- The plot of residuals versus *x* should show points randomly scattered around a horizontal line at residual = 0.

*HINT: Each bell curve in Figure 14.3 describes individual variation at a specific *x* value. How does this relate to the purpose of a prediction interval?

If *Condition 2 holds,* extreme outliers should not be evident in either plot. If *Condition 3 holds,* neither plot should show increasing or decreasing spread in the points as x increases.

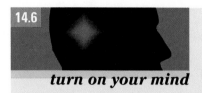

14.6

turn on your mind

A residual is the difference between an observed value of y and the predicted value of y for that observation. Based on the size of a residual for an observation, how would you decide whether an observation was an outlier? Is it enough to know the value of the residual, or do you need to know other information to make this judgment? How could you apply the methods for detecting outliers described in Chapter 2?*

VIDEO

**EXAMPLE 14.2 —
CONTINUED
*Checking the
Conditions for the
Weight and Height Problem***

Figure 14.4 (p. 500) displayed a scatterplot of the weights and heights of $n = 43$ college men. In that plot, it appears that a straight line is a suitable model for how mean weight is linked to height: In Figure 14.8 there is a plot of the residuals (e_i) versus the corresponding values of height for these 43 men. This plot is further evidence that the right model has been used. If the right model has been used, the way in which individuals deviate from the line (residuals) will not be affected by the value of the explanatory variable. The somewhat random-looking blob of points in Figure 14.8 is the way a plot of residuals versus x should look if the right equation for the mean has been used. Both plots (Figures 14.4 and 14.8) also show that there are no extreme outliers and that the heights have approximately the same variance across the range of heights in the sample. Therefore, Conditions 2 and 3 appear to be met. ◆

Condition 4, which is that deviations from the regression line are normally distributed, is difficult to verify but it is also the least important of the conditions because the inference procedures for regression are *robust.* This means that if there are no major outliers or extreme skewness, the inference procedures work well even if the distribution of y values is not a normal distribution. In Chapters 12 and 13, we saw that confidence intervals and hypothesis tests for a mean or a difference between two means also were robust.

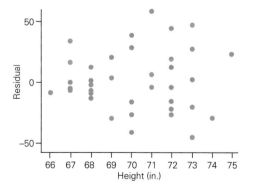

FIGURE 14.8   **Plot of residuals versus *x* for Example 14.2; the absence of a pattern indicates the right model has been used**

* **HINT:** We must know the general size of the residuals to judge the size of a single residual. Section 2.6 discusses how to use the five-number summary to identify outliers.

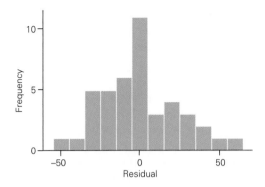

FIGURE 14.9   **Histogram of residuals for Example 14.2**

To examine the distribution of the deviations from the line, a *histogram of the residuals* is useful, although for small samples a histogram may not be informative. A more advanced plot called a *normal probability plot* can also be used to check whether the residuals are normally distributed, but we do not provide the details in this text. Figure 14.9 displays a histogram of the residuals for Example 14.2. It appears that the residuals are approximately normally distributed, so Condition 4 is met.

Condition 5 follows from the data collection process. It is met as long as the units are measured independently. It would not be met if the same individuals were measured across the range of x values, such as if x = average speed and y = gas mileage were to be measured for multiple tanks of gas on the same cars. More complicated models are needed for *dependent* observations, and those models will not be discussed in this book.

Corrections When Conditions Are Not Met

There are some steps that can be taken if Conditions 1, 2, or 3 are not met. If Condition 1 is not met, more complicated models can be used. For instance, Figure 14.10 shows a typical plot of residuals that occurs when a straight-line model is used to describe data that are curvilinear. It may help to think of the residuals as prediction errors that would occur if we use the regression line to predict the value of y for the individuals in the sample. In the plot shown in Figure 14.10, the "prediction errors" are all negative in the central region of x and nearly all positive for outer values of x. This occurs because the wrong model is being used to make the predictions. A curvilinear model, such as the quadratic model discussed earlier, may be more appropriate.

Condition 2, that there are no influential outliers, can be checked graphically with the scatterplot of y versus x and the plot of residuals versus x. The appropriate correction if there are outliers depends on the reason for the outliers. The same considerations and corrective action discussed in Chapter 2 would be taken, depending on the cause of the outlier. For instance, Figure 14.11 shows a scatterplot and a residual plot for the data of exercise 5.55. A potential outlier is seen in both plots.

In this example, the x variable is weight and the y variable is time to chug a beverage. The outlier probably represents a legitimate data value. The relationship appears to be linear for weights ranging up to about 210 pounds, but then it appears to change. It could either become quadratic, or it could level off. We do not have enough data to determine what happens for higher weights. The so-

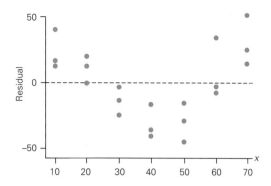

FIGURE 14.10   **A residual plot indicating the wrong model has been used**

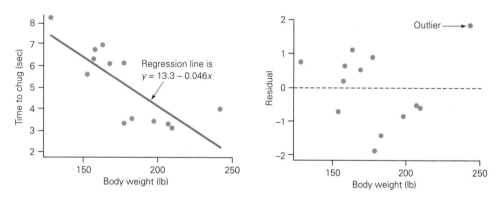

FIGURE 14.11   **Scatterplot and corresponding residual plot with an outlier**

lution in this case would be to remove the outlier and use the linear regression relationship only for body weights under about 210 pounds. Determining the relationship for higher body weights would require a larger sample of individuals in that range.

If either Condition 1 or Condition 3 is not met, a **transformation** may be required. This is equivalent to using a different model. Fortunately, often the same transformation will correct problems with Conditions 1, 3, and 4. For instance, when the response variable is monetary, such as salaries, it is often more appropriate to use the relationship

$$\ln(y) = b_0 + b_1 x + e$$

In other words, to assume that there is a linear relationship between the natural log of y and the x values. This is called a *log transformation on the y's*. We will not pursue transformations further in this book.

Minitab Tip

> *tech note*   **Plotting and Storing Residuals**
>
> ◆ Use **Stat**>**Regression**>**Regression**. The *Graphs* button of the dialog box offers options for several possible residual plots for the regression.
>
> ◆ To store residuals in a column of the worksheet, use the *Storage* button of the dialog box for regression.

CASE STUDY 14.1
A Contested Election

Case Study 11.1 discussed allegations of fraud in a 1993 Philadelphia special election for state senator. The Democratic candidate, William Stinson, beat the Republican candidate, Bruce Marks, by a difference of 461 votes, but Judge Clarence Newcomer overturned the results in a decision issued April 24, 1994. Judge Newcomer wrote, "The court will order the certification of Bruce Marks as the winner of the 1993 Special Election in the Second Senatorial District, because this court finds that the record of evidence overwhelmingly supports the finding that Bruce Marks would have won the election but for wrongdoing" (Newcomer, 1994).

Judge Newcomer partially based his decision on statistical evidence presented by expert witnesses. Other evidence suggested that a number of absentee ballots had been filed illegally, and a statistical analysis indicated that the absentee count was indeed unusual when compared to the machine vote count.

The data used for the statistical analysis consisted of two variables for each of the 21 senatorial elections held in the previous 11 years (1982–1992) in the Philadelphia area. The variables were:

y = *Diff absentee* = (Democrat candidate − Republican candidate) votes by absentee ballot

x = *Diff machine* = (Democrat candidate − Republican candidate) votes by machine ballot

The data for the 21 prior elections were given by economist and expert witness Orley Ashenfelter in a report to Judge Newcomer (Ashenfelter, 1994), and they are presented in Table 14.1 along with the contested 1993 results.

Figure 14.12 shows a plot of the data, with the contested election marked in red. In the contested election, the observed difference (Democrat − Republican) appears well above what might be expected based on the overall connection between absentee and machine votes. Notice also that there is an additional outlier, for the election held in District 1 in 1992. No explanation or discussion of that outlier occurred in the court records.

Is the combination of absentee and machine differences in the contested election at all reasonable, if all other factors remain similar to circumstances of the other elections? The output shown on the next page are for a simple linear regression computed using data from the 21 prior elections. Included in the results are 95% confidence and prediction intervals for the difference in absentee ballots when the

TABLE 14.1 ■ Differences in Democratic and Republican Votes in Philadelphia Elections

Year	District	Diff Absentee	Diff Machine
82	2	346	26427
82	4	282	15904
82	8	223	42448
84	1	593	19444
84	3	572	71797
84	5	−229	−1017
84	7	671	63406
86	2	293	15671
86	4	360	36276
86	8	306	36710
88	1	401	21848
88	3	378	65862
88	5	−829	−13194
88	7	394	56100
90	2	151	700
90	4	−349	11529
90	8	160	26047
92	1	1329	44425
92	3	368	45512
92	5	−434	−5700
92	7	391	51206
93 (Contested)	2	1025	−564

machine ballot difference is −564, the value observed in the contested election. In particular, notice that the 95% prediction interval for the absentee ballot difference for this machine ballot difference is −854.7 to 588.6. The actual ab-

FIGURE 14.12 **Plot of absentee difference versus voting machine difference**

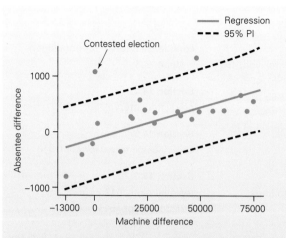

FIGURE 14.13 **Prediction limits for Diff Absentee, given machine difference**

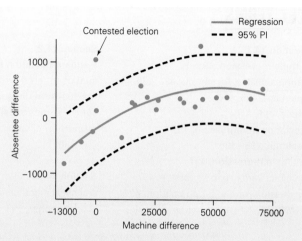

FIGURE 14.14 **Prediction limits for quadratic model**

sentee difference in the contested election was +1025, a difference far outside this range. Figure 14.13 displays this result graphically, giving "prediction limits" for the entire range of possible values of machine differences covered by the data.

```
The regression equation is
Diff Absentee = − 126 + 0.0127 Diff Machine
Predictor       Coef        StDev        T        P
Constant       −125.9       114.3      −1.10    0.284
Diff Mac       0.012703     0.002980    4.26    0.000

S = 324.8      R-Sq = 48.9%       R-Sq(adj) = 46.2%

Predicted Values

   Fit     StDev Fit      95.0% CI          95.0% PI
 −133.1     115.6     ( −375.0, 108.8)   ( −854.7, 588.6)
```

The analysis reported in the *New York Times,* and quoted in Case Study 11.1, used the linear relationship we have just used. However, an examination of the plot of the data in Figure 14.12 indicates that the relationship may actually be curvilinear. A plot of the residuals versus *x* (not shown) confirms this effect. Therefore, another analysis was done by including a "quadratic" term, $(\text{Diff Machine})^2$, in the regression equation. Results are shown in the output in the next column, including 95% confidence and prediction intervals for the contested election. Notice that the prediction interval is even more disparate with the observed result of +1025 than it was with the simple linear regression equation.

```
The regression equation is
Diff Absentee = − 219 + 0.0297 Diff Machine −0.000000
Quadratic

Predictor        Coef          StDev          T        P
Constant        −219.0        105.4        −2.08    0.052
Diff Mac        0.029709      0.006887      4.31    0.000
Quadrati       − 0.00000028   0.00000011   −2.67    0.016

S = 282.6      R-Sq = 63.3%       R-Sq(adj) = 59.3%

Predicted Values

   Fit     StDev Fit        95.0% CI           95.0% PI
 −235.9     107.7       ( −462.1, −9.6)    ( −871.2, 399.5)
```

The *p*-value reported for the "quadratic term" is 0.016 and it is for a test of the null hypothesis that the quadratic term is not needed. Therefore, it is clear that the new model, including the quadratic term, is better than the simple linear model. Figure 14.14 shows the "prediction limits" based on the quadratic model over the range of machine differences for the model including the quadratic term.

The regression analysis and accompanying figures cast doubt on the results of the contested election, *assuming* all other factors were the same as in the past. There may have been other factors that led more people who cast votes for the Democratic candidate to do so by absentee ballot in this election. That issue cannot be addressed using the statistical evidence available, but Judge Newcomer was convinced that other potential differences could not account for the statistical anomaly displayed by the regression analysis.

KEY TERMS

Section 14.1

simple regression, 493

regression line, for sample, 494, 495, 499

intercept, 495, 496

slope, 495, 496

deviation in regression, 495

residual, 495–496

multiple regression, 497

constant variance in regression, 497

regression model, for population, 498, 499

Section 14.2

standard deviation for regression, 499

sum of squared errors (SSE), 499

residual sum of squares, 499

sum of squared residuals, 499

proportion of variation explained by x, R^2, 501

Section 14.3

statistical significance of a linear relationship, 503

confidence interval for population slope, 505

Section 14.4

prediction interval, 507, 509

Section 14.5

confidence interval for $E(Y)$, 510

Section 14.6

residual plots, 512

transformation, 515

EXERCISES   Blue = Basics   Bold, **Bold** = Answer in back

Sections 14.1, 14.2, and 14.3

14.1 The **temperature** dataset on this book's CD gives y = mean April temperature (Fahrenheit) and x = geographic latitude for 20 U.S. cities. The simple linear regression equation for the sample is $\hat{y} = 119 - 1.64\,x$.
   a. The value of latitude for Pittsburgh is 40. What is the predicted April temperature for Pittsburgh?
   b. The mean April temperature for Pittsburgh is 50. What is the residual for Pittsburgh?

14.2 Refer to the previous exercise. Minitab output for the linear regression is

```
The regression equation is
AprTemp = 119 − 1.64 latitude

Predictor      Coef    SE Coef        T       P
Constant    118.776      4.467    26.59   0.000
latitude    −1.6436     0.1165   −14.11   0.000

S = 2.837       R-Sq = 91.7%
```

   a. Write null and alternative hypotheses for testing whether there is a linear relationship between mean April temperature and latitude. Use proper statistical notation.
   b. Use information given in the output to test the hypotheses written in part (a). State a conclusion and give a reason for the conclusion.
   c. Using values given in the output, show how to calculate the t-statistic for testing the hypotheses stated in part (a).

14.3 Refer to the previous two exercises.
   a. In the output, what value is given for the standard deviation from the regression line? Write a sentence that interprets this value.
   b. In the output, the value of R^2 is given as 91.7%. Write a sentence that interprets this value.

14.4 The **UCDavis1** dataset on the text CD includes grade point averages and self-reported hours per week of watching television for students in a statistics class. In a simple linear regression for the relationship between y = grade point average and x = hours per week of television watching, the p-value is 0.552 for testing H_0: $\beta_1 = 0$ versus H_a: $\beta_1 \neq 0$.
   a. Is the observed relationship statistically significant or not? Explain.
   b. What would be the p-value for testing the null hypothesis that correlation between the variables is 0 in the population represented by the sample? Assume a two-sided alternative hypothesis.

14.5 For men over the age of 40, a linear regression is done to examine the relationship between y = systolic blood pressure and x = age. A 95% confidence interval for the population slope β_1 is 0.3 to 0.7. In the context of this situation, write a sentence that interprets this confidence interval.

14.6 The population model for simple linear regression can be defined as

$$y_i = \beta_0 + \beta_1 x_i + \varepsilon_i$$

   a. Explain what the parameters β_0 and β_1 represent.
   b. Write the equation that represents how the mean of the y variable relates to the x variable.

14.7 Find r^2 in each of the following situations:
   a. SSTO = 500, SSE = 300.
   b. SSTO = 200, SSE = 40.
   c. SSTO = 80, SSE = 0.
   d. SSTO = 100, SSE = 95.

14.8 Example 14.3 in Section 14.2 gave the regression equation $y = 577 - 3.01x$ for the relationship between y = maximum distance at which a driver can read a highway sign and x = driver age.
   a. What is the predicted value of the maximum sign-reading distance for a driver 21 years old?
   b. Compute the residual for a 21-year-old driver who can read the sign at a maximum distance of 525 feet.

c. The standard deviation from the regression line in this example is approximately $s = 50$ feet. Use the Empirical Rule to give an interval that describes the maximum sign-reading distances for about 95% of all drivers who are 21 years old.

d. Would it be unusual for a 21-year-old driver to be able to read the sign from 650 feet? Justify your answer.

14.9 In the simple linear regression model for a population, the relationship between the x variable and the mean of the y variable is $E(Y) = \beta_0 + \beta_1 x$.

a. Suppose $\beta_1 = 0$. What would be the appearance of this line? Draw a sketch illustrating your answer.

b. Explain how the answer to part (a) illustrates the lack of a linear relationship between x and y.

14.10 Data for $y =$ hours of sleep the previous day and $x =$ hours of studying the previous day for $n = 116$ college students were shown in Figure 5.12 and described in Example 5.9. Some regression results for those data are

The regression equation is
Sleep = 7.56 − 0.269 Study

Predictor	Coef	SE Coef	T	P
Constant	7.5555	0.2239	33.74	0.000
Study	−0.26917	0.06616		0.000

S = 1.509   R-Sq = 12.7%   R-Sq(adj) = 11.9%

a. What is the estimated mean decrease in hours of sleep per 1-hour increase in hours of studying? What notation is used for this value?

b. Calculate an approximate 95% confidence interval for β_1. Write a sentence that interprets this interval.

c. Using proper statistical notation, write null and alternative hypotheses for assessing the statistical significance of the relationship.

d. We omitted from the output the t-statistic for testing the hypotheses of part (c). Compute the value of this t-statistic using other information shown in the output. What are the degrees of freedom for this t-statistic?

e. Is there a statistically significant relationship between hours of sleep and hours of studying? Justify your answer based on information shown in the output.

14.11 Refer to the previous problem about hours of sleep and hours of study.

a. What is the value of the standard deviation from the regression line? Write a sentence that interprets this value.

b. Calculate the predicted value of hours of sleep the previous day for a student who studied 4 hours the previous day.

c. Using the Empirical Rule, determine an interval that describes hours of sleep for approximately 95% of students who studied 4 hours the previous day.

d. The value of R^2 is given as 12.7%. Write a sentence that interprets this value.

14.12 Refer to the previous two exercises about hours of sleep and hours spent studying. What is the intercept of the regression line? Does this value have a useful interpretation in the context of this problem? If so, what is the interpretation? If not, why not?

14.13 The relationship between $y =$ hours of watching television in a typical day and $x =$ age was examined in Example 5.8. The data were gathered in the 1996 General Social Survey done by the National Opinion Research Center at the University of Chicago, and there were $n = 1913$ observations. The correlation between the variables is $r = .12$, a value indicative of a weak connection between hours of watching television and age.

The regression equation is
tvhours = 2.19 + 0.0173 age

Predictor	Coef	SE Coef	T	P
Constant	2.1899	0.1577	13.89	0.000
age	0.017255	0.003348	5.15	0.000

S = 2.371   R-Sq = 1.4%   R-Sq(adj) = 1.3%

a. What is the estimated increase in average daily hours of television watching for each 1-year increase in age?

b. Using proper statistical notation, write null and alternative hypotheses for testing the statistical significance of the relationship between the two variables.

c. Use information in the output to determine if there is a statistically significant relationship between age and daily hours of watching television. Include relevant information in your answer.

d. Explain whether or not you think the observed relationship has practical significance.

14.14 The least-squares regression line is $\hat{y} = 9 + 2x$ for these six observations of x and y:

x	1	1	3	3	5	5
y	10	12	13	17	17	21

a. For each observation, calculate the residual. Verify that the sum of the residuals is 0.

b. Compute SSE, the sum of squared errors (residuals).

c. For the sample, calculate the standard deviation from the regression line.

14.15 In Example 14.2, a sample of $n = 43$ college men was used to determine the regression equation, Weight = −318 + 7 (Height). Weights were measured in pounds and heights were measured in inches.

a. The slope of the line is 7. Explain whether the proper notation for this slope is b_1 or β_1.

b. The standard error for the slope was given as 1.581 in Figure 14.4. Calculate an approximate 95% confidence interval for the slope in the population. Write a sentence that interprets this interval.

c. It has been stated that for each 1-inch increase in height, average weight increases by 5 pounds. Explain whether the interval computed in part (b) supports this statement or not.

d. Explain why it would not be useful to compute a confidence interval for the population intercept in this example.

14.16 Refer to the previous exercise in which a regression equation is given for the relationship between weight and height for college men. What is the value of the residual for a college man 72 inches tall who weighs 160 pounds?

Sections 14.4 and 14.5

14.17 For each part of this question, explain whether a confidence interval for the mean or a prediction interval for a value of y should be used.
   a. Estimate the first-year college GPA for a student whose high school GPA was 3.5.
   b. Estimate the mean first-year college GPA for students whose high school GPA was 3.5.
   c. Estimate the average income for individuals who have 20 years of experience within a particular occupation.
   d. Estimate the adult height of a daughter of a woman who is 64 inches tall.

14.18 Suppose a regression model is used to analyze the relationship between y = grade point average and x = number of classes missed in a typical week for college students.
   a. In the context of this situation, explain what would be predicted by a prediction interval for y when x = 2 classes missed per week.
   b. In the context of this situation, explain what would be estimated by a confidence interval for $E(Y)$ when x = 2 classes missed per week.

14.19 A college finds that among students who had a total SAT score of approximately 1200, about 90% have a first-year GPA in the range 2.7 to 3.7. Explain whether this interval is a prediction interval or a confidence interval for the mean.

14.20 Forty students measure their resting pulse rates, then march in place for one minute and measure their pulse rates after the marching. The regression line for the sample is $\hat{y}$ = 17.79 + 0.894x, where y = pulse after marching and x = pulse before marching. The following Minitab output gives a confidence interval and a prediction interval for pulse after marching when pulse before marching is 70 (*Data source:* **pulsemarch** dataset on the text CD).

Fit	SE Fit	95.0% CI	95.0% PI
80.35	1.15	(78.02, 82.67)	(65.50, 95.19)

   a. Write the 95% confidence interval for $E(Y)$. In the context of this situation, write a sentence that interprets this interval.
   b. Write the 95% prediction interval for y. In the context of this situation, write a sentence that interprets this interval.

14.21 Example 2.11 described the heights of a sample of married British women. The ages of the women and their husbands were available in the same dataset for n = 170

couples. For the sample, the linear regression line relating y = husband's age and x = wife's age is $\hat{y}$ = 3.59 + 0.967x. The following results computed with Minitab give a confidence interval and prediction interval for husband's age when wife's age is 40:

Fit	SE Fit	95.0% CI	95.0% PI
42.258	0.313	(41.641, 42.875)	(34.200, 50.316)

   a. Verify that the "Fit" is consistent with the predicted value that would be given by the regression equation.
   b. Interpret the "95% CI" given by Minitab. Be specific about what the interval estimates.
   c. Interpret the "95% PI" given by Minitab. Be specific about what the interval estimates.
   d. Explain why the "95% PI" is much wider than the "95% CI."

14.22 Refer to the previous exercise. Using the general format for a 95% confidence interval, verify the confidence interval for the mean given by Minitab. Notice that the standard error of the "Fit" is given.

14.23 Exercise 14.10 gave linear regression results for the relationship between y = hours of sleep the previous day and x = hours spent studying the previous day. Following is Minitab output showing a confidence interval and a prediction interval for hours of sleep when hours of studying = 3 hours:

Fit	SE Fit	95.0% CI	95.0% PI
6.748	0.142	(6.466, 7.029)	(3.746, 9.750)

Write down the interval given for predicting the hours of sleep when hours of studying is 3 hours. Give two different interpretations of this interval.

14.24 Suppose a linear regression analysis of the relationship between y = systolic blood pressure and x = age is done for women between 40 and 60 years old. For women who are 45 years old, a 90% confidence interval for $E(Y)$ is determined to be 128.2 to 131.3. Explain why it is incorrect to conclude that about 90% of women who are 45 years old have a systolic blood pressure in this range. Write a sentence that correctly interprets the interval.

Section 14.6

14.25 There are five conditions listed at the beginning of Section 14.6 that should be at least approximately true for linear regression. Which of the conditions can be checked using each of the following methods? In each case, list all of them that can be checked.
   a. Drawing a histogram of the residuals.
   b. Drawing a scatterplot of the residuals versus the x values.

c. Learning how the data were collected.

d. Drawing a scatterplot of the raw data, y versus x.

14.26 Refer to exercise 14.20 about a linear regression for $y =$ pulse after marching in place and $x =$ pulse before marching in place. The figure for this exercise is a plot of residuals versus the pulse before marching for a sample of 40 students. Discuss what the plot indicates about Conditions 1, 2, and 3 for linear regression listed at the beginning of Section 14.6.

EXERCISE 14.26   **Residuals versus pulse before marching**

14.27 The figure for this exercise is a histogram of the residuals for a linear regression relating $y =$ height (inches) and $x =$ foot length (cm) for a sample of college men. Discuss what the histogram indicates about Conditions 2 and 4 for linear regression listed at the beginning of Section 14.6 (*Data source:* **heightfoot** dataset on the text CD).

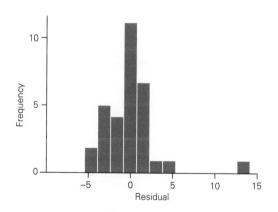

EXERCISE 14.27   **Histograms of residuals**

14.28 Regression results for the relationship between $y =$ hours of sleep the previous day and $x =$ hours spent studying the previous day were given in exercise 14.10.

The figure accompanying this exercise is a plot of residuals versus hours spent studying. What does the plot indicate about the necessary conditions for doing a linear regression? Be specific about which of the five conditions given in Section 14.6 are verified by this plot.

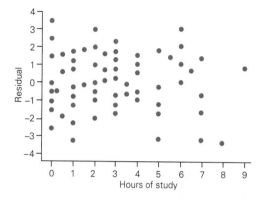

EXERCISE 14.28   **Residuals versus hours spent studying**

14.29 Refer to the previous exercise. The figure accompanying this exercise is a histogram of the residuals for the regression relating hours of sleep and hours spent studying. What does this plot indicate about the necessary conditions for doing a linear regression? Be specific about which of the five necessary conditions are verified in this figure.

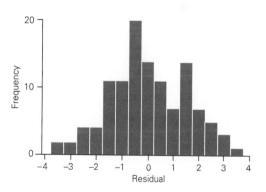

EXERCISE 14.29   **Histogram of residuals**

14.30 Plot residuals versus x for the data given in exercise 14.14.

14.31 Observed data along with the sample regression line for the relationship between body weight (lbs) and neck girth (in.) in 19 female bears of various ages are shown in the figure on the next page. (*Note:* The data are in the dataset **bears-female** on the data CD.)

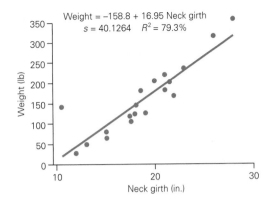

EXERCISE 14.31   Data and regression line for the body weight and neck girth of $n = 19$ female bears

a. Which of the necessary conditions for linear regression appears to be violated in this dataset?

b. What corrective actions would you consider in order to properly estimate the relationship between body weight and neck girth?

c. Sketch a histogram that illustrates the pattern of the distribution of the residuals for this problem. Your sketch does not have to be accurate in the numerical details, but should correctly show the pattern.

14.32 The accompanying figure shows data for the relationship between the average stopping distance (ft) of a car when the brakes are applied and vehicle speed (mph). The regression line for these data is also shown on the plot. (*Note:* The raw data were given in exercise 5.7.)

EXERCISE 14.32   Stopping distance and vehicle speed for automobiles

a. Which of the five necessary conditions for linear regression appears to be violated in this situation?

b. What corrective action would you take in order to correctly estimate the connection between stopping distance and vehicle speed?

c. Make a sketch that illustrates the pattern of a plot of residuals versus speed for the data shown in the figure in this exercise. Your sketch does not need to be numerically accurate, but should correctly show the pattern of the plot.

Chapter Exercises

14.33 Sketch sample data for which the standard deviation from the regression line is $s = 0$.

14.34 An example in Sections 14.1 and 14.2 described the relationship between $y =$ handspan (cm) and $x =$ height (in.). The sample consisted of measurements of both variables for $n = 167$ college students. Computer output for this example is

```
The regression equation is
HandSpan = − 3.00 + 0.351 Height

Predictor      Coef      SE Coef        T        P
Constant     −3.002       1.694     −1.77    0.078
Height       0.35057     0.02484    14.11    0.000


S = 1.301            R-Sq = 54.7%
```

a. What is the value of the standard deviation from the regression line? Write a sentence that interprets this value.

b. What is the value of R^2 given in the output? Write a sentence that interprets this value.

14.35 Refer to the previous exercise. Compute a 90% confidence interval for the population slope. Write a sentence that interprets this interval.

14.36 This output is for a linear regression relating $y =$ GPA and $x =$ hours per week spent using the computer for $n = 162$ college students (*Note:* The data are in the **UCDavis1** dataset).

```
The regression equation is
GPA = 3.01 − 0.00555 computer

162 cases used 11 cases contain missing values

Predictor      Coef       SE Coef        T        P
Constant     3.00588      0.07335     40.98    0.000
computer    −0.005548     0.003979    −1.39    0.165
```

a. Is there a statistically significant relationship between the two variables? Justify your answer using information from the output. Be specific about the null and alternative hypotheses that you're evaluating.

b. Suppose that ρ denotes the correlation between the two variables in the population represented by the sample. What is the p-value for testing $H_0: \rho = 0$ versus $H_a: \rho \neq 0$?

14.37 These data are $x =$ average on five quizzes prior to midterm exam and $y =$ score on midterm exam for $n = 11$ students randomly selected from a statistics class of about 950 students:

x = Quizzes	80	68	94	72	74	83	56	68	65	75	88
y = Exam	72	71	96	77	82	72	58	83	78	80	92

a. Plot the data, and describe the important features of this plot.

b. Using statistical software, calculate the regression line for this sample.

c. What is the predicted midterm exam score for a student with a quiz average equal to 75?

d. With statistical software, determine a 50% prediction interval for the midterm exam score of a student whose quiz average was 75.

14.38 This exercise refers to the following Minitab output, relating y = son's height to x = father's height for a sample of n = 76 college males. (*Note:* The data are in the dataset **UCDavis1** on the data CD.)

The regression equation is
Height = 30.0 + 0.576 dadheight

76 cases used 3 cases contain missing values

Predictor	Coef	SE Coef	T	P
Constant	29.981	5.129	5.85	0.000
dadheigh	0.57568	0.07445	7.73	0.000

S = 2.657   R-Sq = 44.7%

Predicted Values [for dad's heights of 65, 70 and 74]

Fit	SE Fit	95.0% CI	95.0% PI
67.400	0.415	(66.574, 68.226)	(62.041, 72.759)
70.279	0.318	(69.645, 70.913)	(64.946, 75.612)
75.581	0.494	(71.596, 73.566)	(67.195, 77.967)

---

a. What is the equation for the regression line?

b. Identify the value of the *t*-statistic for testing whether or not the slope is 0. Verify that the value is correct, using the formula for the *t*-statistic and the information provided by Minitab for the parts that go into the formula.

c. State and test the hypotheses about whether or not the population slope is 0. Use relevant information provided in the output.

d. Compute a 95% confidence interval for β_1, the slope of the relationship in the population. Write a sentence that interprets this interval.

14.39 Refer to the previous exercise.

a. What is the value of R^2 for the observed linear relationship between height and father's height? Write a sentence that interprets this value.

b. What is the value of the correlation coefficient r?

14.40 Refer to the previous two exercises. Notice that Minitab provided prediction intervals and confidence intervals for father's heights of 65, 70, and 74 inches.

a. Verify that the "Fit" given by Minitab for dad's height of 65 inches is consistent with the predicted height that would be given by the regression equation.

b. Write down the interval Minitab provided for predicting an individual son's height if his dad's height is 70

inches. Provide two different interpretations for the interval.

c. Write a sentence that interprets the "95% CI" given for son's height when father's height is 74 inches.

d. Explain why the prediction interval is much wider than the corresponding confidence interval for each father's height provided.

e. The accompanying figure is a plot of the residuals versus father's height. Which of the five necessary conditions for regression listed in Section 14.6 appears to be violated? What corrective actions should you consider?

EXERCISE 14.40(e)   **Residuals versus father's height**

14.41 The five steps for hypothesis testing were given in Chapters 11 and 13. Describe those steps as they apply to testing whether there is a relationship between two variables in the simple regression model.

14.42 A sample of 837 ninth-grade girls was asked how much time they spent watching music videos each week. The girls also were asked to rate how concerned they were about their weight on a scale of 0–100 (100 = extremely concerned) as well as to rate their perception of how important appearance is on a scale of 1–6 (6 = very important). The researchers (Borzekowski et al., 2000) wrote the following about the results: "When media use was separated into distinct media genres, only hours of watching music videos was related to perceived importance of appearance and weight concerns (r = .12, $p < 0.001$, and r = .08, $p < 0.05$, respectively)."

a. What are the null and alternative hypotheses for which *p*-values are given in the researchers' statement? What conclusions have the researchers made about these hypotheses?

b. Do the correlation values given by the researchers indicate weak or strong relationships between the variables? Support your answer with a sketch illustrating the appearance of a scatterplot for the given correlation values.

14.43 Explain why rejecting H_0: β_1 = 0 in a simple linear regression model does not prove that the relationship is linear. To answer this question, you might find it helpful to consider the figure in exercise 14.32, which shows stopping distance and vehicle speed for automobiles.

Dataset Exercises

14.44 Use the dataset **letters** for this exercise. A sample of 63 students wrote as many letters of the alphabet, in order as capital letters, as they could in 15 seconds using their dominant hand, and then repeated this task using their nondominant hand. The variables **dom** and **nondom** contain the raw data for the results.

a. Plot $y = $ **dom** versus $x = $ **nondom.** Describe the important features of the plot.

b. Compute the simple regression equation for the relationship. What is the equation?

c. What are the values of the standard deviation from the regression line and R^2? Interpret these values in the context of this problem.

d. Determine a 95% confidence interval for the population slope. Write a sentence that interprets this interval.

e. Consider the statement, "On average, a student can write about 23 more letters in 15 seconds with the dominant hand than with the nondominant hand." What regression equation would accompany this statement? Based on your answers to parts (b) and (d), explain why this statement is reasonable.

14.45 Refer to the previous exercise about letters written with the dominant (y) and nondominant (x) hands.

a. Plot residuals versus $x = $ **nondom.** What does this plot indicate about conditions for using the linear regression model?

b. Create a histogram of the residuals. What does this plot indicate about the conditions for using the linear regression model?

14.46 Use the **bears-female** dataset for this exercise. Weights (lbs) and chest girths (in.) are given for $n = 19$ female wild bears. The corresponding variable names are **Weight** and **Chest.**

a. Plot $y = $ **Weight** versus $x = $ **Chest.** Describe the important features of the plot.

b. Compute a simple linear regression equation for $y = $ **Weight** and $x = $ **Chest.**

c. What is the value of R^2 for this relationship? Write a sentence interpreting this value.

d. What is the predicted weight for a bear with a chest girth of 40 inches?

e. Compute a 95% prediction interval for the weight of a bear with a chest girth of 40 inches. Write a sentence interpreting this interval.

f. Compute a 95% confidence interval for the mean weight of bears with a chest girth of 40 inches. Write a sentence interpreting this interval.

14.47 Use the dataset **heightfoot.** Heights (in.) and foot lengths (cm) are given for 33 men.

a. Plot $y = $ height versus $x = $ foot length. What important features are evident in the plot?

b. Omit any outliers evident in the plot in part (a), and compute the linear regression line for predicting $y = $ height from $x = $ foot length.

c. Determine a 90% prediction interval for the height of a man whose foot length is 28 centimeters.

d. Discuss whether height can be accurately predicted from foot length. Use regression results to justify your answer.

e. For the data used for part (b), plot residuals versus $x = $ foot length. What does this plot indicate about the necessary conditions for doing a linear regression?

14.48 Refer to the previous exercise about the relationship between height and foot length.

a. Do not omit any outliers. Use the complete dataset to determine a 90% prediction interval for the height of a man whose foot length is 28 centimeters.

b. Explain why the interval computed in part (a) is wider than the interval computed for part (c) of the previous exercise.

c. For the regression based on all data points, plot residuals versus $x = $ foot length. What does this plot indicate about the necessary conditions for doing a linear regression?

REFERENCES

Ashenfelter, Orley (1994). "Report on expected absentee ballot," revised March 29, 1994. Unpublished report submitted to Judge Clarence Newcomer, U.S. District Court, Eastern District of Pennsylvania, March 30, 1994.

Borzekowski, D. L. G, T. N. Robinson, and J. D. Killen (2000). "Does the camera add 10 pounds? Media use, perceived importance of appearance, and weight concerns among teenage girls," *Journal of Adolescent Health,* Vol. 26, pp. 36–41.

Newcomer, Clarence (1994). Judgment in the Civil Action No. 93–6157, Bruce S. Marks et al. v. William Stinson et al., United States District Court for the Eastern District Court of Pennsylvania, entered April 26, 1994.

CHAPTER 15

More About Categorical Variables

With whom is it easiest to make friends? See Example 15.2.

Are the digits drawn in lotteries actually equally likely, as claimed? Are belief and performance on ESP tests related? Do males and females have the same opinions about whether it's easier to make friends with the same sex or the opposite sex? Chi-square tests can be used to answer these kinds of questions about categorical variables.

n Chapter 6, we learned techniques for analyzing a fundamental question that researchers often ask about two categorical variables, which is

Is there a relationship between the variables so that the chance that an individual falls into a particular category for one variable depends upon the particular category they fall into for the other variable?

For instance, in Example 6.1 we saw that the likelihood of divorce is greater for smokers than it is for nonsmokers. A procedure for assessing the statistical significance of a relationship between categorical variables is the chi-square test. We introduced the basic ideas of this statistical method in Chapter 6, and in this chapter we examine the chi-square test more fully. We will use the chi-square procedure to test hypotheses about two-way tables, and also we will learn how the chi-square procedure is used to test hypotheses about a single categorical variable. ◆

15.1 | THE CHI-SQUARE TEST FOR TWO-WAY TABLES

In Chapter 6, we learned these terms:

- ◆ The raw data from a **categorical variable** consist of group or category names that don't necessarily have any logical ordering. Your sex (male or female) and your handedness (left or right), for instance, are both categorical variables.

- ◆ A **two-way table** displays the counts of how many individuals fall into each possible combination of categories of two categorical variables. A two-way table may also be called a **contingency table**.

◆ Each combination of a row and a column is referred to as a **cell** of the contingency table.

◆ **Row percents** are the percents across a row of a contingency table. They are based on the total number of observations in the row.

◆ **Column percents** are the percents down a column of a contingency table. They are based on the total number of observations in the column.

As usual, there are five steps required to assess statistical significance. In this case, the question of interest is whether a relationship observed in a sample can be used to infer that there is a relationship in the population(s) from which the sample(s) was drawn.

Step 1: The Null and Alternative Hypotheses

The **null** and **alternative hypotheses** about the categorical variables that form a two-way table are

H₀: The two variables are not related.
Hₐ: The two variables are related.

It is commonplace to use the word *associated* in place of the word *related*. Within a specific context, we may be able to write more informative versions of the general statements about the relationship.

EXAMPLE 15.1
*Ear Infections
and Xylitol Sweetener*

Xylitol is a food sweetener that may also have antibacterial properties. In an experiment conducted in Finland (and discussed in Example 13.6), researchers investigated whether the regular use of chewing gum containing Xylitol could reduce the risk of a middle ear infection for children in daycare centers (Uhari et al., 1998). The investigators randomly divided 533 children in daycare centers into three groups. One group regularly chewed gum that contained Xylitol, another group regularly took Xylitol lozenges, and the third group regularly chewed gum that did not contain Xylitol. The experiment lasted for three months and, for each child, the researchers recorded whether the child had an ear infection during that period. (In Example 13.6, the data concerned Xylitol administered in a syrup form rather than in gum.)

Table 15.1 displays the data along with row percents. As usual, the explanatory variable, which is the type of gum or lozenge each child took, is displayed as rows. The response variable, which is whether the child had an ear infection, is displayed as columns. The results of the test would be the same if the vari-

TABLE 15.1 ■ **Ear Infections and the Use of Xylitol**

Group	Ear Infection in Three Months?		Total
	No	*Yes*	*Total*
Placebo Gum	129 (72.5%)	49 (27.5%)	178
Xylitol Gum	150 (83.8%)	29 (16.2%)	179
Xylitol Lozenge	137 (77.8%)	39 (22.2%)	176
Total	416 (78%)	117 (22%)	533

ables in the rows and columns were reversed, with the explanatory variable as columns. However, displaying the categories of the explanatory variable as rows makes it easier to find and interpret the conditional percents that are usually of interest, the row percents. The display in Table 15.1 readily allows us to see that only 16.2% of the children in the Xylitol gum group got an ear infection during the experiment, compared to 22.2% of the Xylitol lozenge group and 27.5% of the placebo gum group.

The purpose of the experiment is to compare the three groups with regard to the proportion of children who experienced an ear infection. The goal is to see if the sample results allow us to infer that differences would exist in the population of all children if they were to be administered these treatments. For this purpose, there are three relevant parameters.

> p_1 = proportion who would get an ear infection in a population given placebo gum
>
> p_2 = proportion who would get an ear infection in a population given Xylitol gum
>
> p_3 = proportion who would get an ear infection in a population given Xylitol lozenges

If the chance of falling into the ear infection category is not related to treatment method, the risk of an ear infection would be the same for the three treatments. So, null and alternative hypotheses about the parameters are

> H_0: $p_1 = p_2 = p_3$. (No relationship between treatment and outcome.)
> H_a: p_1, p_2, and p_3 are not all the same. (There is a relationship.)

Notice that the alternative hypothesis simply states that the three proportions are not all the same. A limitation of chi-square tests is that no particular direction for the relationship can be stated in the alternative hypothesis. In contrast, if we were comparing only two proportions, as in Chapter 13, we would be able to specify a one-sided alternative hypothesis if it were appropriate. ◆

EXAMPLE 15.2
With Whom Do You Find It Easiest to Make Friends?

Students in a Penn State statistics class were asked, "With whom do you find it easiest to make friends?" Possible responses were "opposite sex," "same sex," and "no difference." Students were also categorized as male or female. We would like to determine whether there is a statistically significant relationship between the sex of the respondent and response. To emphasize the comparison of men and women, we might express the null and alternative hypotheses as

> H_0: There would be no difference in the distribution of responses of men and women if the populations of them were asked this question. (No relationship between gender and response.)
> H_a: There would be a difference in the distribution of responses of men and women if the populations of them were asked this question. (There is a relationship between gender and response.)

As is always the case in hypothesis testing, these hypotheses are statements about the larger populations represented by the samples of men and women. Again, notice that the alternative hypothesis does not specify any particular way in which the men and women might differ.

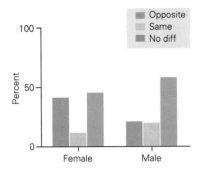

TABLE 15.2 ■ **Data For Example 15.2**

| | With Whom Is It Easiest to Make Friends? | | | |
	Opposite Sex	*Same Sex*	*No Difference*	*Total*
Females	58 (42%)	16 (12%)	63 (46%)	137
Males	15 (22%)	13 (19%)	40 (59%)	68
Total	73 (36%)	29 (14%)	103 (50%)	205

Table 15.2 compares the responses of males and females, and Figure 15.1 displays the relevant conditional percentage distributions. We see that there appears to be a relationship. For instance, a decidedly higher percentage of females than males responded "opposite sex." ◆

Homogeneity and Independence

tech note   It is generally sufficient to state the null and alternative hypotheses as H_0: No relationship and H_a: There is a relationship. At the theoretical level, however, there are two specific variations of these general statements, and the method of sampling dictates which variation is appropriate in a particular situation. If samples have been taken from separate populations, the null hypothesis statement is a statement of **homogeneity** (sameness) among the populations. Example 15.1 is such an instance. If a sample has been taken from a single population, and two categorical variables measured for each individual, the statement of no relationship is a statement of **independence** between the two variables. Example 15.2 is such an instance. (*Note:* Independence was discussed in Chapter 7.) Fortunately, for sufficiently large samples, the same chi-square test is appropriate for both situations, so generally we will not be concerned about whether the null hypothesis is about homogeneity or about independence.

Step 2: The Chi-square Statistic for Two-Way Tables and Necessary Conditions

The chi-square test described in Section 6.5 can be used to assess statistical significance for tables with any numbers of rows and columns. This procedure is an approximate method that requires a "large" sample. The larger the sample, the better the approximation. Commonly used guidelines for the term *large sample* are

1. All expected counts should be greater than 1.
2. At least 80% of the cells should have an expected count greater than 5.

We will provide more details about the chi-square test shortly, but first let's review the basic ideas from Section 6.5.

- ◆ The counts in the cells of a two-way table of the sample data are called the **observed counts.**
- ◆ The chi-square statistic measures the difference between the observed counts and corresponding **expected counts.** The *expected counts* are hy-

pothetical counts that would occur if the null hypothesis were true. The expected counts can be calculated as

$$\text{Expected} = \frac{\text{Row total} \times \text{Column total}}{\text{Total } n}$$

- The **chi-square statistic** is

$$\chi^2 = \sum_{\text{all cells}} \frac{(\text{Observed} - \text{Expected})^2}{\text{Expected}}$$

The symbol Σ stands for "sum," and the sum is over all cells of the two-way table. A large chi-square value occurs when there is a large difference between the observed and expected counts.

- The **p-value** of the chi-square test is the probability that the chi-square statistic, χ^2, could be as large as or larger than it is if the null hypothesis is true. A chi-square probability distribution is used to determine this probability. The degrees of freedom for the chi-square distribution are given by df = (Rows − 1)(Columns − 1), where "Rows" indicates the number of rows in the table (excluding the row of totals), and "Columns" indicates the number of columns in the table (excluding the column of totals).

Notation for Chi-square Statistic for Two-Way Tables

tech note   The following mathematical notation can be used for the elements of the chi-square test:

O_{ij} = observed count for the cell in the ith row and jth column of the table

E_{ij} = expected count for the cell in the ith row and jth column of the table

R_i = total count of observations in the ith row

C_j = total count of observations in the jth column

n = total count

$E_{ij} = \dfrac{R_i C_j}{n}$ = expected count for the cell in the ith row and jth column

$\chi^2 = \sum_{i,j} \dfrac{(O_{ij} - E_{ij})^2}{E_{ij}}$

r = number of rows; c = number of columns

EXAMPLE 15.1—*CONTINUED*
Chi-square Results for the Xylitol and Ear Infection Data

Minitab output for testing the significance of the Xylitol and ear infection relationship is displayed in Figure 15.2 (see next page). Toward the bottom of the output, we see that the *p*-value is 0.035, which is below the usual significance level standard of 0.05. As a result, we can conclude that there is a statistically significant relationship between the risk of an ear infection and the preventative treatment used. Notice that expected counts are shown below observed counts in the results and that values for the chi-square statistic and the degrees of freedom are also given. There are three rows and two columns, so the degrees of freedom are df = (3 − 1)(2 − 1) = 2. ◆

```
Rows:      Treatment    Columns:    Infectn

             No           Yes         All
Placebo     129           49          178
            138.93        39.07       178.00

Xyl Gum     150           29          179
            139.71        39.29       179.00

Xyl Loz     137           39          176
            137.37        38.63       176.00

All         416           117         533
            416.00        117.00      533.00

Chi-Square = 6.690,   DF = 2,   P-Value = 0.035

Cell Contents:   Observed
                 Expected
```

FIGURE 15.2   **Minitab output for Example 15.1**

Interpreting and Calculating Expected Counts

The expected counts in a chi-square analysis are the counts that would be expected to occur if the null hypotheses were true. Three important properties of the expected counts for a two-way table are

- The expected counts have the same row and column totals as the observed counts.
- The pattern of row percents is identical for all rows of expected counts.
- The pattern of column percents is identical for all columns of expected counts.

Calculating Expected Counts

formula   For each cell in a two-way table,

$$\text{Expected count} = \frac{\text{Row total} \times \text{Column total}}{\text{Total } n}$$

"Row total" is the total sample size in the row, "Column total" is the total sample size in the column, and "Total n" is the total sample size over the entire table.

There are several equivalent ways to derive the formula for calculating the expected counts. For instance, we could utilize the property that the row percents would be identical for all rows if the null hypothesis were true. In Table 15.1 for the Xylitol and ear infection example, the proportion that experienced an ear infection in the total sample was 117/533. Note that this is the sample size in the second column divided by the total sample size. Under the null hypothesis, the proportion experiencing an ear infection is the same for all treatments, so the quantity 117/533 estimates the *proportion* with an ear infection for any treatment. For any treatment, the expected *count* of those who would get an ear infection can be determined by multiplying the sample size for the treatment (a row total) by the overall proportion with an ear infection (column total/total n). The same argument could be made for the expected counts of those who do not get an ear infection.

EXAMPLE 15.1—*CONTINUED*
Calculation of Expected Counts and Chi-square for the Xylitol and Ear Infection Data

Consider the "Placebo Gum, Yes Ear Infection" cell. There are 178 observations in the "Placebo Gum" row, and 117 observations in the "Yes Ear Infection" column. For the cell in question, the expected count is

$$\frac{\text{Row total} \times \text{Column total}}{\text{Total } n} = \frac{178 \times 117}{533} = 39.07$$

The calculation of all expected counts for this example is shown in Table 15.3. The calculation of the chi-square statistic is

$$\chi^2 = \sum_{\text{all cells}} \frac{(\text{Observed} - \text{Expected})^2}{\text{Expected}}$$

$$= \frac{(129 - 138.93)^2}{138.93} + \frac{(49 - 39.07)^2}{39.07} + \frac{(150 - 139.71)^2}{139.71} + \frac{(29 - 39.29)^2}{39.29}$$

$$+ \frac{(137 - 137.37)^2}{137.37} + \frac{(39 - 38.63)^2}{38.63} = 6.69 \; \blacklozenge$$

Step 3: The *p*-Value for the Chi-square Test

The *p*-value for a chi-square test is the probability that the chi-square statistic could have been as large or larger if the null hypotheses were true. This probability is determined using a chi-square distribution with degrees of freedom = (Rows − 1)(Columns − 1) = $(r - 1)(c - 1)$, where r = Rows is the number of rows in the contingency table and c = Columns is the number of columns. This number is equivalent to the minimum number of cells for which expected counts need to be computed by the formula. The remaining expected counts are then determined by the fact that the row and column totals are fixed in advance. For instance, in the 3 × 2 table in Example 15.1, once you compute (3 − 1) × (2 − 1) = 2 expected counts, the rest are determined by the row and column totals. So you are only "free" to fill in two cells before the others are determined. In Table 15.3, notice that after the counts of 138.93 and 139.71 had been computed, the rest of the counts could be found by subtraction from knowing what the row and column totals must be.

TABLE 15.3 ■ Calculation of Expected Counts for Example 15.1

	No Ear Infection	Yes Ear Infection	Row Total
Placebo Gum	$\frac{178 \times 416}{533} = 138.93$	$\frac{178 \times 117}{533} = 39.07$	178
Xylitol Gum	$\frac{179 \times 416}{533} = 139.71$	$\frac{179 \times 117}{533} = 39.29$	179
Xylitol Lozenge	$\frac{176 \times 416}{533} = 137.37$	$\frac{176 \times 117}{533} = 38.63$	176
Column Total	416	117	533

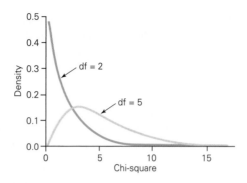

FIGURE 15.3   Two different chi-square distributions

> ***formula***   For the chi-square statistic for a two-way table,
>
> $$\text{Degrees of freedom} = (r - 1)(c - 1)$$
>
> r = number of rows in the table, and c = number of columns in the table.

The Chi-square Family of Distributions

A **chi-square distribution** is a skewed probability distribution that stretches to the right. The minimum value is 0 for a variable with a chi-square distribution. A single parameter, called *degrees of freedom,* completely determines the precise characteristics of a specific chi-square distribution. Figure 15.3 shows a chi-square distribution with 2 degrees of freedom and also shows a chi-square distribution with 5 degrees of freedom.

EXAMPLE 15.1—*CONTINUED*
p-Value for the
Xylitol Example

Figure 15.4 illustrates the *p*-value area for the Xylitol and ear infection problem. The chi-square statistic was 6.69, degrees of freedom = $(3 - 1)(2 - 1) = 2$, and the *p*-value is the probability that a chi-square statistic with df = 2 would be 6.69 or larger. This is the area to the right of 6.69 under a chi-square probability density curve (df = 2). ◆

Three Ways to Determine the *p*-Value

The chi-square test for two-way tables is used so frequently that any statistical software program will include this procedure and provide the *p*-value necessary

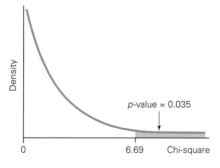

FIGURE 15.4   *p*-Value for Example 15.1

for deciding between the null and alternative hypotheses about the table. Some of you might not be using statistical software, however, so we consider three different ways to determine the *p*-value:

- Use statistical software to do the chi-square test. The *p*-value will be part of the output (as in Figure 15.1 of this chapter).
- Use the Excel command CHIDIST(chi-square value, df). This command will return the area to the right of the value you specify. For instance, CHIDIST (6.69,2) returns 0.035, the *p*-value for Example 15.1.
- Use Table A.5 in the Appendix to give a range for the *p*-value.

Using Table A.5 to Approximate the p-Value

First, determine the degrees of freedom for your two-way table. Then, look in the corresponding "df" row of Table A.5. Scan across that row until you locate approximately where the calculated chi-square statistic falls.

- If the value of the chi-square statistic falls between two table entries, the *p*-value is between the values of *p* (column headings) for these two entries.
- If the value of the chi-square statistic is larger than the entry in the rightmost column (labeled $p = 0.001$), the *p*-value is less than 0.001 (written as $p < 0.001$).
- If the value of the chi-square statistic is smaller than the entry in the leftmost column (labeled $p = 0.50$), the *p*-value is greater than 0.50 ($p > 0.50$).

EXAMPLE 15.3
A Moderate p-Value

Suppose that a two-way table has three rows and three columns and the chi-square statistic is $\chi^2 = 8.12$. The degrees of freedom are df $= (3 - 1)(3 - 1) = 4$. To find the *p*-value, we scan the df = 4 row in Table A.5 and learn that 8.12 is between the entries 7.78 ($p = 0.10$) and 8.50 ($p = 0.075$). Thus, the *p*-value is between 0.075 and 0.10. We would write this as $0.075 < p\text{-value} < 0.10$. Or, often it is written simply as $0.075 < p < 0.10$. In Excel, the command CHIDIST (8.12,4) tells us that the *p*-value is 0.087. ◆

EXAMPLE 15.4
A Tiny p-Value

Suppose that a two-way table has four rows and two columns and the chi-square statistic is $\chi^2 = 17.67$. The degrees of freedom are $(4 - 1)(2 - 1) = 3$. In Table A.5, we see that 17.67 is larger than the rightmost entry in the df = 3 row, so the *p*-value is less than 0.001 ($p < 0.001$). In Excel, the command CHIDIST (17.67,3) tells us that the *p*-value is 0.00051. ◆

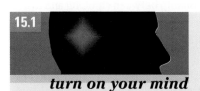

15.1

turn on your mind

Consider Example 15.2 about gender and the question about with whom it's easiest to make friends. What are the degrees of freedom for the chi-square statistic for these data? To be statistically significant at the 0.05 level, how large would the calculated chi-square have to be?*

*HINT: You should use Table A.5 to determine this.

Steps 4 and 5: Making a Decision and Reporting a Conclusion

An approximate p-value, or even a range for the p-value, is enough information to make a decision about the null hypothesis. If we use the 0.05 significance level, we simply have to determine whether the p-value is less than 0.05. In Example 15.3, for instance, we learned from Table A.5 that the p-value is between 0.075 and 0.10. Thus, we know that it is not less than 0.05 and we would fail to reject the null hypothesis.

The value in the 0.05 column of Table A.5 is referred to as the **critical value** for the 0.05 significance level. The region of chi-square values greater than the critical value is called the **rejection region.** The p-value will be less than 0.05 when the chi-square statistic is greater than this critical value, in other words, if the chi-square statistic is in the rejection region. Consequently, if the level of significance is $\alpha = 0.05$, two equivalent rules for declaring statistical significance based on a chi-square statistic are

- Reject the null hypothesis when the p-value is less than 0.05.
- Reject the null hypothesis when the chi-square statistic is greater than the entry in the 0.05 column of Table A.5 (the critical value), in other words, when the chi-square statistic is in the rejection region.

Reporting a Conclusion in Context

Remember that the two possible conclusions for hypothesis tests are "Do not reject the null hypothesis" or "Reject the null hypothesis." The latter conclusion is equivalent to "accept the alternative hypothesis" or "conclude that the results are statistically significant."

In chi-square tests for two-way tables, it is important to write these conclusions in context. Suppose we were testing whether there is a relationship between smoking (yes or no) and drinking alcohol (never, occasionally, often). Here are various ways to write the possible conclusions.

Ways to write the conclusion "do not reject the null hypothesis"

- The relationship between smoking and drinking alcohol is not statistically significant.
- The proportions of smokers who never drink, drink occasionally, and drink often are not significantly different from the proportions of nonsmokers who do so.
- There is insufficient evidence to conclude that there is a relationship in the population between smoking and drinking alcohol.

Ways to write the conclusion "reject the null hypothesis"

- There is a statistically significant relationship between smoking and drinking alcohol.
- The proportions of smokers in the population who never drink, drink occasionally, and drink often are not the same as the proportions of nonsmokers who do so.
- Smokers have significantly different drinking behavior than nonsmokers.

EXAMPLE 15.2—*CONTINUED*
Making Friends

Figure 15.5 shows partial Minitab output for a chi-square test of the relationship between sex and response to the question "With whom is it easiest to make

Minitab v. 14 update on CD

Expected counts are printed below observed counts

	Opposite	Same	No Diff	Total
Female	58	16	63	137
	48.79	19.38	68.83	
Male	15	13	40	68
	24.21	9.62	34.17	
Total	73	29	103	205

Chi-Sq =   1.740 + 0.590 + 0.494 +
            3.507 + 1.188 + 0.996 = 8.515

FIGURE 15.5   **Minitab output for Example 15.2**

friends?" Recall that the possible responses concerned the sex of the friends. In the last line of output, we see that the chi-square statistic is 8.515. Although Minitab provides the p-value, this value has been omitted to provide practice using Table A.5, the chi-square distribution table. The contingency table has two rows and three columns, so df $= (2 - 1)(3 - 1) = 2$. In the df $= 2$ row of Table A.5, the computed chi-square falls between the entries in the 0.025 column (7.38) and the 0.01 column (9.21). The p-value is between 0.01 and 0.025, so the result is statistically significant if we use the standard 0.05 significance level.

Equivalently, find the critical value in Table A.5. Using the df $= 2$ row and the 0.05 column, the critical value is 5.99. Reject the null hypothesis because the computed chi-square statistic of 8.515 is greater than the critical value of 5.99.

In the context of the situation, the conclusion is that there is a statistically significant relationship between sex and response to the question asked. Equivalently, conclude that there would be a difference in the distribution of responses of men and women if the populations of them were asked this question. Using the bar chart of row percents shown in Figure 15.1 (p. 530), we can describe the nature of the difference between the females and males. Compared to men, women are more likely to say "opposite sex" and less likely to say "same sex." ◆

Supporting Analyses

We don't learn anything about the specific nature of a relationship from a chi-square test. The results of a statistically significant chi-square test should be accompanied by some other descriptive or inferential statistics that elaborate on the connection between the variables. Some or all of the following procedures may be useful for this purpose:

◆ Description of row (or column) percents.

◆ Bar chart of counts or percents.

◆ Examination of $\dfrac{(\text{Observed} - \text{Expected})^2}{\text{Expected}}$ within each cell. These values are called "contributions to chi-square." The cells with the largest values have contributed the most to the statistical significance of the relationship, so they deserve attention in any description of the relationship.

◆ Confidence intervals for important proportions or for differences between proportions. In Example 15.1, for instance, confidence intervals for the proportions getting an ear infection in the three treatment groups would be informative.

---

in summary The Five Steps in a Chi-square Test of a Relationship Between Two Variables

Step 1: *Determine the null and alternative hypotheses.*

Null hypothesis: H_0: The two variables are not related.

Alternative hypothesis: H_a: The two variables are related.

Step 2: *Verify necessary data conditions, and if met, summarize the data into an appropriate test statistic.* If at least 80% of the expected counts are greater than 5 and none are less than 1, compute

$$\chi^2 = \sum_{\text{all cells}} \frac{(\text{Observed} - \text{Expected})^2}{\text{Expected}}$$

The *expected count* for a cell is computed as $\dfrac{\text{Row total} \times \text{Column total}}{\text{Total } n}$.

Step 3: *Assuming the null hypothesis is true, find the p-value.* Using the χ^2 distribution with df = $(r - 1)(c - 1)$, where r = number of rows and c = number of columns in the two-way table, the *p*-value is the area in the tail to the right of the calculated test statistic χ^2.

Step 4: *Decide whether or not the result is statistically significant based on the p-value.* Choose a significance level ("alpha"); the standard is $\alpha = 0.05$. The result is statistically significant if the *p*-value $\leq \alpha$.

Step 5: *Report the conclusion in the context of the situation.*

Minitab Tip

Minitab v. 14 update on CD

tech note **Computing a Chi-square Test for a Two-Way Table**

♦ If the raw data are stored in columns of the worksheet, use **Stat>Tables>Cross Tabulation.** In the dialog box, specify the two columns of interest and also click on ***Chi-square analysis.***

♦ If the data are already summarized into counts, enter the table of counts into columns of the worksheet and then use **Stat>Tables>Chi-Square Test.** In the dialog box, specify the columns containing the counts.

15.2 | ANALYZING 2 × 2 TABLES

There are a number of special circumstances that apply to analyzing 2 × 2 tables. First, there is a handy computational shortcut for computing the chi-square statistic for a 2 × 2 table that does not require computing expected counts. It can be accomplished with a calculator, without needing to write down any intermediate results.

Shortcut Formula for a 2 × 2 Table

Suppose we label a 2 × 2 table as follows:

	Column 1	Column 2	Total
Row 1	A	B	R_1
Row 2	C	D	R_2
Total	C_1	C_2	N

Then the test statistic is

$$\chi^2 = \frac{N(AD - BC)^2}{R_1 R_2 C_1 C_2}$$

with df = 1.

EXAMPLE 6.10—*CONTINUED*
Order and Choice of S or Q   In Example 6.10 of Chapter 6, we examined the results of a survey in which students were asked to "Randomly choose one of the letters *S* or *Q*" or "Randomly choose one of the letters *Q* or *S*." The Minitab output for examining the relationship between the form of the question and the letter picked is shown in Example 6.10 and again here.

Minitab v. 14 update on CD

> **Minitab Output for the *S* or *Q* Experiment (Example 6.10)**
>
> Expected counts are printed below observed counts
>
> | Form | Letter Picked S | Q | Total |
> |------|------|------|-------|
> | S first | 61 | 31 | 92 |
> | | 51.33 | 40.67 | |
> | Q first | 45 | 53 | 98 |
> | | 54.67 | 43.33 | |
> | Total | 106 | 84 | 190 |
>
> Chi-Sq =   1.823 + 2.301 +
>   1.712 + 2.160 = 7.995
> DF = 1,   P-Value = 0.005

The value of the chi-square statistic is 7.995. The shortcut formula results in precisely the same value and is found as follows:

$$\chi^2 = \frac{N(AD - BC)^2}{R_1 R_2 C_1 C_2} = \frac{190(61 \times 53 - 31 \times 45)^2}{92 \times 98 \times 106 \times 84} = 7.995$$

The remaining steps of the test are the same as always. ◆

Chi-square Test or *z*-Test for Difference in Two Proportions?

In Section 13.4, we learned that a *z*-test can be used to compare two proportions. In such situations, we also can create a 2 × 2 table of the observed counts. The two groups would constitute the two rows, and whether or not the trait of interest is present would constitute the two columns. When the data are presented as a contingency table, we may naturally consider using a chi-square test. Does it make a difference whether we choose to use the z-test of Section 13.4 or the chi-square test to compare two proportions? The answer depends upon the nature of the alternative hypothesis. If the desired alternative hypothesis has no specific direction (is two-sided), the two tests give exactly the same p-value. If the desired alternative hypothesis has a direction (is one-sided), the z-test should be used because in a chi-square test a specific direction cannot be specified in the alternative hypothesis.

The only essential difference between the two tests is the ability to specify a one-sided alternative hypothesis. In fact, there is an algebraic connection between the values of the test statistics for the two procedures. For any 2×2 table, the squared value of the z-statistic equals the chi-square statistic ($z^2 = \chi^2$). Or, put another way, $|z| = \sqrt{\chi^2}$. This connection between the test statistics is also the connection between the theoretical standard normal distribution and the theoretical chi-square distribution with df $= 1$.

15.2

turn on your mind

Suppose that you read that men are more likely to be left-handed than women are. To investigate this claim, you survey your class and find that 11 of 84 men and 7 of 78 women are left-handed. Should you compare the men and women using a z-test or a chi-square test? Or does it matter?*

VIDEO

EXAMPLE 15.5
Age and Tension Headaches

Exercise 12.46 was about a study for which the response variable was whether or not the respondent had experienced an episodic tension-type headache during the previous year (Schwartz et al., 1998). Table 15.4 is a contingency table that compares the incidence of tension headaches for women 18 to 29 years old and women 30 to 39 years old.

Let p_1 and p_2 represent the proportions experiencing tension headaches in the populations of women aged 18 to 29 and women aged 30 to 39, respectively. Because the researchers did not speculate about the possible relative sizes of p_1 and p_2, the alternative hypothesis is two-sided. Null and alternative hypotheses are

$H_0: p_1 = p_2$
$H_a: p_1 \neq p_2$

TABLE 15.4 ■ **Age Group and the Incidence of Tension Headaches**

	Episodic Tension Headaches		
Age	Yes	No	Total
18 to 29	653 (40.8%)	947 (59.2%)	1600
30 to 39	995 (46.9%)	1127 (53.1%)	2122
Total	1648 (44.3%)	2074 (57.7%)	3722

The null hypothesis is equivalent to stating that there isn't a relationship between age group and the risk of episodic tension headaches. In this example, we can use either a chi-square test or a two-sample z-test because no specific direction is specified in the alternative hypothesis. Using statistical software (Minitab) to do a chi-square test, we found that $\chi^2 = 13.66$ and the p-value was reported to be 0.000 (only three decimal places were given). The observed relationship between age group and the incidence of tension headaches is statistically significant, and the row percents show us that the older women are more likely to experience tension headaches. If we had done a z-test for comparing two proportions, the value of z would have been $z = \sqrt{13.66} \approx 3.70$, and the p-value would have been identical to the p-value for the chi-square test. ◆

EXAMPLE 15.6
Sheep, Goats, and ESP

In parapsychology, the sheep–goat effect is that people who believe in extrasensory perception (sheep) tend to be more successful in ESP experiments than people who do not (goats). The effect is slight, but it has been observed in many forced-choice ESP experiments (Lawrence, 1993; Schmeidler, 1988, pp. 56–58). In a forced-choice experiment, the participant "guesses" which of several known possibilities has been (or will be) selected by the experimenter.

Students in a statistics class at the University of California at Davis classified themselves as sheep (believers) or goats (nonbelievers). Among 192 students, there were 112 sheep and 80 goats. Each student was paired with another student in the same belief group. One member of each pair flipped a coin ten times while the other member guessed the outcome of each flip. Students then reversed their roles and repeated the flipping, so each person was able to make ten guesses. The guessing process was uncontrolled, as each student could decide whether he or she would guess in advance of the flip, during the flip, or after the flip. Table 15.5 displays data for the numbers of successful and unsuccessful guesses in each group. The sheep did slightly better than the goats, guessing correctly about 52% of the time compared to 49% for the goats.

The activity was done to determine whether sheep would do better, so a one-sided alternative hypothesis is appropriate. Define p_1 and p_2 to be the population proportions of successful guesses for sheep and goats. The null and alternative hypotheses of interest are

$H_0: p_1 \le p_2$
$H_a: p_1 > p_2$

A z-test for comparing two proportions should be done rather than a chi-square test because the alternative hypothesis describes a relationship in a specific direction. Figure 15.6 displays Minitab output for this test. From the bottom line of output, we learn that the p-value is 0.075. The null hypothesis cannot be rejected at the 0.05 significance level. ◆

TABLE 15.5 ■ ESP Belief and Success at Guessing Coin Flips

Group	Successful Guess		Total
	Yes	No	
Sheep (Believers)	582 (52%)	538 (48%)	1120
Goats (Nonbelievers)	389 (49%)	411 (51%)	800
Total	971 (51%)	949 (49%)	1920

```
Sample       X          N       Sample p
1           582       1120      0.519643
2           389        800      0.486250

Estimate for p(1) − p(2): 0.0333929
Test for p(1) − p(2) = 0 (vs > 0): Z = 1.44   P-Value = 0.075
```

FIGURE 15.6   **Minitab output for Example 15.6**

Fisher's Exact Test for 2 × 2 Tables

For 2 × 2 tables, **Fisher's Exact Test** is another test that can be used to analyze the statistical significance of the relationship. In theory, this test can be used for any 2 × 2 table, but most commonly it is used when necessary sample size conditions for using the z-test or the chi-square test are violated. In large part, this has to do with computational difficulties that occur for large sample sizes.

Although the computations are cumbersome, the idea behind Fisher's Exact Test is easy. Suppose, for example, that a randomized experiment is done to see if taking the herb echinacea reduces the risk of getting a cold, and the observed data are 1 of 10 people taking echinacea get a cold during the study while 4 of 10 taking a placebo get a cold. In all, 5 of the 20 participants got a cold, but only 1 was in the echinacea group. The p-value for a one-sided Fisher's Exact Test is the answer to this question:

Given that 5 of 20 participants get a cold regardless of treatment method, what is the probability that just 1 or fewer would be in the echinacea group?

Notice that the test statistic is simply the count of how many got a cold in the echinacea group. As always, the p-value question addresses how likely it is that the test statistic would be as extreme or more extreme as it is in the direction of the alternative hypothesis if the null hypothesis is true. The answer (it's 0.152) is determined using a probability distribution called the *hypergeometric distribution*. This distribution is discussed further in Chapter 17, which covers three different classes of discrete random variables.

Minitab v. 14 update on CD

Some, but not all, statistical software programs include the Fisher's Exact Test. The Statistical Package for the Social Sciences (SPSS) does, and Minitab just recently began offering it, with Version 14. Several websites provide a calculator designed to give the p-value for Fisher's Exact Test.

EXAMPLE 15.7
Butterfly Ballots   The winner of the 2000 United States presidential election (George W. Bush) was not determined until about a month after election day. There were several court hearings in Florida about the legality of vote counting and voting procedures in that state. One of the Florida controversies concerned a ballot format called the butterfly ballot used in Palm Beach County. Candidates were listed in two columns, and to indicate a choice, a voter punched a hole in the appropriate place in an area between the two columns of names. The first three presidential candidates were listed as follows:

After the election, several Palm Beach County voters said they meant to vote for Al Gore, but mistakenly voted for Pat Buchanan, the Reform Party candidate. These voters said they punched a hole in the second punch location because they saw Al Gore's name listed second in the left column. That location, however, was for Pat Buchanan who was listed in the right column. Perhaps not coincidentally, Buchanan's vote total in Palm Beach County was decidedly higher than expected.

A brief communication published in *Nature* (Sinclair et al., 2000) gave results from an experiment done in Canada to study whether the butterfly ballot increases the likelihood of voter mistakes. Shortly after the United States election, the researchers recruited people in an Edmonton, Alberta, shopping mall to vote in a mock election for the prime minister of Canada. Participants were randomly assigned to use either the double-column butterfly format or the more conventional single-column format. After they voted, the participants were asked for whom they voted. Of 55 voters using the butterfly format, four voted for a different candidate than they intended. Of 52 voters using the single-column format, none made a mistake.

Is the observed result statistically significant? Define p_1 = population proportion that makes a mistake with the butterfly ballot and p_2 = population proportion that makes a mistake with the single-column ballot. The null and alternative hypotheses of interest are

$H_0: p_1 \leq p_2$
$H_a: p_1 > p_2$

The observed counts don't satisfy the necessary conditions for doing a z-test, which are that at least five observations with and without the trait of interest should be observed in each group (see Section 13.4). Fisher's Exact Test is more appropriate for these data, and the p-value can be determined by answering this question:

> *Given that 4 of 107 participants will make a mistake, what is the probability that all 4 are among the 55 voters randomly selected to use the butterfly ballot?*

The answer is 0.0661, a p-value not quite below the usual 0.05 standard for statistical significance. It is important to note, however, that given that 4 participants will make a mistake, the evidence for the alternative is as strong as possible since all 4 were using the butterfly format, so it may be reasonable to use a higher level of significance and reject the null hypothesis. Also, the researchers asked all participants to rate the confusion level of the ballot they used on a 7-point scale, with 7 representing the most confusing. The mean confusion score for the butterfly ballot was significantly higher than for the single-column format. ◆

Significance Levels and *p*-Values Close to 0.05

tech note   Although we have used the common standard of rejecting the null hypothesis when the *p*-value ≤ 0.05, there is nothing magic about 0.05. In Examples 15.6 and 15.7, the *p*-values were 0.075 and 0.0661, respectively. In these and similar cases it is up to you as the reader of the research to decide whether or not to reject the null hypothesis. Remember that the level of significance used corresponds to the conditional probability of rejecting a true null hypothesis. Sometimes there may be reasons to set that at a higher (or lower) level than 0.05. Where to set the cutoff is something that you as a reader should always decide, which is why it is important for researchers to report *p*-values rather than to simply report a decision about whether or not to reject the null hypothesis.

15.3 | TESTING HYPOTHESES ABOUT ONE CATEGORICAL VARIABLE: GOODNESS OF FIT

15.3

turn on your mind

Imagine tossing a six-sided die 60 times. How many times would you expect each side to occur? What did you assume when you calculated these *expected counts*? How would you measure the difference between the observed and expected counts of how often each side occurs in the 60 tosses?*

We sometimes wish to test whether the probabilities of falling into the possible categories of a single categorical variable are given by a specified set of values. For example, if a lottery drawing involves randomly drawing digits between 0 and 9, we may wish to analyze observed data to confirm that the probability over the long run is 1/10 for drawing each of the ten digits from 0 to 9. With two modifications from how it is applied to two-way tables, the chi-square statistic can be used to test hypotheses about the probability distribution of a single categorical variable. In such instances, the significance test is called the **chi-square goodness of fit test.**

The first modification of the chi-square procedure necessary for the goodness of fit situation involves the calculation of the expected counts. This calculation is straightforward and intuitively obvious. Suppose we hypothesize, for example, that 45% of a population has brown eyes, 35% has blue eyes, and 20% has an eye color other than brown or blue. If we randomly select $n = 200$ persons from this population, it probably makes sense to you that we would expect about 45% of 200 = $.45 \times 200 = 90$ persons to have brown eyes if our hypothesis is correct. We also would expect about $.35 \times 200 = 70$ persons to have blue eyes, and $.20 \times 200 = 40$ persons to have an eye color other than brown or blue.

The other modification of the chi-square procedure concerns the degrees of freedom. For a goodness of fit test, the degrees of freedom are computed as $df = k - 1$, where k is the number of categories for the variable of interest. In the eye color example of the previous paragraph, there were three eye color categories. Consequently, the degrees of freedom for a chi-square goodness of fit test would be $df = 3 - 1 = 2$.

15.4

turn on your mind

Remember that the "degrees of freedom" for the chi-square test for a two-way table represents the largest number of cells for which you were "free" to find expected counts. The remaining expected counts were determined because the row and column totals had to be the same as they were for the observed counts. Explain how the same principle applies in specifying the degrees of freedom for a chi-square goodness of fit test, which are $k - 1$ when there are k categories.*

---

in summary   The Five Steps in a Chi-square Goodness of Fit Test

Step 1: *Determine the null and alternative hypotheses.*

Null hypothesis: H_0: The probabilities for the k categories of a categorical variable are given by $p_1, p_2, \ldots, p_k$.

Alternative hypothesis: H_a: Not all probabilities specified in H_0 are correct.

Note: The probabilities specified in the null hypothesis must sum to 1.

Step 2: *Verify necessary data conditions, and if met, summarize the data into an appropriate test statistic.* If at least 80% of the expected counts are greater than 5 and none are less than 1, compute

$$\chi^2 = \sum_{\text{all categories}} \frac{(\text{Observed} - \text{Expected})^2}{\text{Expected}}$$

where the expected count for the ith category is computed as np_i.

Step 3: *Assuming the null hypothesis is true, find the p-value.* Using the χ^2 distribution with $df = k - 1$, the p-value is the area in the tail beyond the test statistic χ^2.

Step 4: *Decide whether or not the result is statistically significant based on the p-value.* Choose a significance level ("alpha"); the standard is $\alpha = 0.05$. The result is statistically significant if the p-value $\leq \alpha$. Equivalently, the result is statistically significant if the test statistic χ^2 is at least as large as the value in the α column and $df = k - 1$ row of Table A.5.

Step 5: *Report the conclusion in the context of the situation.*

---

*HINT: The total expected count must equal the total observed count. If you know $k - 1$ expected counts, what must the remaining one be?

EXAMPLE 15.8
The Pennsylvania
Daily Number

The Pennsylvania Daily Number is a state lottery game in which the state constructs a three-digit number by drawing a digit between 0 and 9 from each of three different containers. If the digits drawn, in order, were 3, 6, and 3, for example, then the daily number would be 363. In this example, we focus only on draws from the first container. If numbers are randomly selected, each value between 0 and 9 would be equally likely to occur. This leads to the following null hypothesis:

$$H_0: p = \frac{1}{10} \text{ for each of the 10 possible digits in first container}$$

Simply stated, the alternative hypothesis is that the null hypothesis is false. In this setting, that would mean that the probability of selection for some digits is different from 1/10.

Figure 15.7 shows the frequency distribution of draws from the first container for the $n = 500$ days between July 19, 1999 and November 29, 2000. The observed data were downloaded from the Pennsylvania lottery website. The expected count for each of the 10 possible outcomes is $1/10 \times 500 = 50$, and the red dashed line in the figure marks this value. The greatest difference between observed and expected occurs for the number 5, which was drawn from the first container only 39 times over these 500 days.

The chi-square goodness of fit statistic for these data is

$$\chi^2 = \sum_{\text{categories}} \frac{(\text{Observed} - \text{Expected})^2}{\text{Expected}}$$

$$= \frac{(47-50)^2}{50} + \frac{(50-50)^2}{50} + \frac{(55-50)^2}{50} + \frac{(46-50)^2}{50} + \frac{(53-50)^2}{50}$$

$$+ \frac{(39-50)^2}{50} + \frac{(55-50)^2}{50} + \frac{(55-50)^2}{50} + \frac{(44-50)^2}{50} + \frac{(56-50)^2}{50} = 6.04$$

There are ten possible outcomes for the number drawn, so df = 10 − 1 = 9. To find the *p*-value, we find the probability that a chi-square statistic would be as large as or larger than 6.04 using a chi-square distribution with df = 9. From Table A.5, the chi-square distribution table, we learn that the *p*-value is greater than 0.50 because 6.04 is smaller than (to the left of) the entry of 8.34 under 0.50 in the row for df = 9. The Excel command CHIDIST(6.04,9)

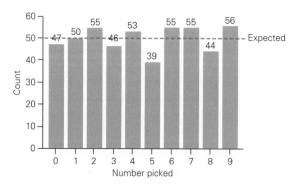

FIGURE 15.7   Observed and expected counts for first digit drawn in the Pennsylvania Daily Number Game, July 19, 1999 through November 29, 2000

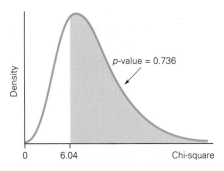

FIGURE 15.8 *p*-Value for Example 15.8 is area to right of 6.04 in chi-square distribution with df = 9

provides the information that the *p*-value is 0.736. This *p*-value is illustrated in Figure 15.8. The result is not statistically significant; the null hypothesis is not rejected. ◆

Mathematical Notation for Chi-square Goodness of Fit Statistic

tech note The following mathematical notation can be used for the chi-square goodness of fit test:

O_i = observed count in the *i*th category

p_i = hypothesized probability for the *i*th category

$E_i = np_i$ = expected count in *i*th category

$$\chi^2 = \sum_i \frac{(O_i - E_i)^2}{E_i}$$

k = number of categories; df = $k - 1$

Sample Size and the Goodness of Fit Test

The null hypothesis often is the desired hypothesis for the investigators when they do a goodness of fit test. In Example 15.8 about the Pennsylvania Daily Number, lottery officials would hope to "prove" that all numbers have an equal chance to be selected. This creates a technical and logical difficulty because with a small sample, a null hypothesis might be retained simply because there is not enough data to reject it. This was the basis for the advice given in Chapter 11 to use the phrase "cannot reject the null" rather than "accept the null." If the goal is to not reject the null hypothesis, the most scientifically valid approach is to use as large a sample as possible. This prevents the possible criticism that the null hypothesis was retained only because the sample was not large enough to provide conclusive evidence. On the other hand, the risk of a large sample size is that small, relatively unimportant departures from the null hypothesis can achieve statistical significance. It is often useful to include confidence intervals for the parameters if possible, to determine the magnitude of departures from the null hypothesis.

Minitab Tip

tech note   **Calculating the Chi-square Statistic for Goodness of Fit**
Minitab does not have a chi-square goodness of fit procedure, but it is not difficult to compute the chi-square statistic using the Minitab calculator. The basic steps are

- Enter the observed counts into a column. Suppose it is C1.
- Enter the null hypothesis proportions into a second column. Suppose it is C2.
- Use **Calc**>**Calculator.** Create a column of expected counts by multiplying C2 by the total sample size. Suppose it is C3.
- Again use **Calc**>**Calculator.** Create (by computation) a column (say, C4) that contains a chi-square value. Given the column designations of the previous steps, the expression is SUM((C1−C3)**2/C3). Look carefully at the parenthesis locations. You can also compute (C1−C3)**2/C3 and store the results in C4, then sum C4. That will allow you to determine which categories are contributing the most to the value of the test statistic.

CASE STUDY 15.1
Do You Mind If I Eat the Blue Ones?

You'll find six different colors of M&Ms in a bag of plain M&Ms: brown, red, yellow, blue, orange, and green. If you're like some people, you may have a color preference (or dislike) when it comes to eating M&Ms and might have fleeting thoughts about what the color distribution is or should be. At the M&Ms website (http://www.m-ms.com/factory/history/faq1.html), the distribution of colors for plain M&Ms is given as: brown 30%, red 20%, yellow 20%, blue 10%, orange 10%, and green 10%.

Students in a University of California at Davis statistics class (Fall 2000 term) counted the frequency of each color among 6918 plain M&Ms from about 130 small (1.69 ounces) bags. All bags were purchased at the same store in California. The observed distribution of colors in this sample was

	Color					
	Brown	Red	Yellow	Blue	Orange	Green
Count	1911	1072	1308	804	821	1002
Percent	27.6%	15.5%	18.9%	11.6%	11.9%	14.5%
Expected %	30%	20%	20%	10%	10%	10%

Generally, the observed percents in the UC Davis data are not far from the percents advertised by the manufacturer, but some differences are evident. Most notably (especially for those who like red M&Ms), there were fewer red and more green candies than expected.

We can use a chi-square goodness of fit test to determine if the difference between the expected and observed distributions is statistically significant. If we assign numbers to colors using the order in which the colors are listed above (1 = brown, 2 = red, and so on), the null hypothesis of interest is

$$H_0: p_1 = .3, p_2 = .2, p_3 = .2, p_4 = .1, p_5 = .1, p_6 = .1$$

The first step in computing the chi-square statistic is to calculate the expected counts for each color. To do so, multiply the null hypothesis proportion for a color by the total sample size, $n = 6918$. For example, the expected count for brown is $.3 \times 6918 = 2075.4$. Figure 15.9 displays expected counts and observed counts for each color. As noted before, the largest differences between observed and expected counts occur for red and green.

FIGURE 15.9   **Observed and expected counts for a sample of 6918 plain M&Ms**

The chi-square goodness of fit statistic for these data is $\chi^2 = 268.75$, with degrees of freedom = number of categories $- 1 = 6 - 1 = 5$. In Table A.5, the entry under 0.001 for df = 5 is 20.51. The computed chi-square statistic, 268.75, is so much larger than 20.51 that it's reasonable to say that the p-value is essentially 0. The result is highly significant in that it would be nearly impossible for a random sample of 6918 M&Ms to have a color distribution that differed from the null hypothesis distribution by as much (or more) as the observed sample does.

This experiment provided an opportunity to examine the data in another interesting way. Twenty-five student teams sorted and counted the M&Ms; for most teams, the sample size was approximately 275 (five bags). A separate chi-square statistic was computed for each team's sample, and statistical significance was achieved for most, but not all, samples. For 16 of the 25 samples, the p-value was less than 0.05, and in most of these cases, the p-value was much less than 0.05.

Figure 15.10 shows the chi-square distribution for 5 degrees of freedom along with the 25 calculated chi-square values, indicated by the blue circles along the horizontal axis. If many, many chi-square tests are performed and the null hypothesis is always true, Figure 15.10 illustrates that when df = 5, the test statistic should generally fall between 0 and about 10. In fact, when df = 5, the *critical value* for the 0.05 significance level is 11.07 (look under 0.05 in Table A.5). Computed chi-square values of 11.07 or more are statistically significant. We see that the majority of the computed chi-square values were greater than 11.07, and several were much greater than 11.07. Such large chi-square values offer strong evidence against the null hypothesis. The results summarized in Figure 15.10 also show us why it's useful to replicate studies. Although the results were not significant for a few teams, the consistency of the significance over most teams helps us believe that there is a real difference.

After obtaining these results, one of the authors posted a message about them to an internet "listserv" for teachers of Advanced Placement statistics courses. She received several replies from other instructors who had done this activity. Nearly all described statistically significant departures from the color distribution given at the M&Ms website, but the nature of the difference varied considerably. In some cases, for example, the number of red candies was much more than expected, while in other cases the number of red candies was much less than expected.

To further investigate sample-to-sample variation, one of the authors counted the frequency of the six colors in four one-pound bags of plain M&Ms purchased at four different stores in Pennsylvania. The color distribution in the total sample was reasonably close to the expected distribution, but there were substantial differences among the bags. For instance, the observed percent of brown M&Ms ranged from 23% in one bag to 40% in another. And the null hypothesis was rejected for three of the four bags. See exercise 15.36 for details.

Why do so many different samples provide statistically significant evidence against the distribution given by the manufacturer? The answer probably has to do with the manufacturing and distribution process. A bag of M&Ms, or a group of bags purchased at the same store, may not be a *random sample* from a larger population. The color composition of the bags produced in a manufacturing plant on a given day could be affected by the available quantities of the various colors. Also, the process used to put candies into the bags may not produce a well-mixed random sampling of the colors. In other words, students counting M&M colors most likely are not using random or representative samples of the larger population.

FIGURE 15.10   **Chi-square values for 25 samples of M&Ms and the chi-square distribution for df = 5**

KEY TERMS

Section 15.1
categorical variable, 527
two-way table, 527
contingency table, 527
cell, 528
row percent, 528
column percent, 528
null hypothesis for two-way tables, 528
alternative hypothesis for two-way
   tables, 528

chi-square test for homogeneity, 530
chi-square test for independence, 530
observed counts for chi-square test,
   530
expected counts for chi-square test,
   530–531, 532
chi-square test statistic, 531
p-value, 531, 533
degrees of freedom, 534
chi-square distribution, 534

critical value, 536
rejection region, 536

Section 15.2
Fisher's Exact Test, 542

Section 15.3
chi-square goodness of fit test, 544

EXERCISES   Blue = Basics   Bold, **Bold** = Answer in back

Sections 15.1 and 15.2

15.1 For each pair of variables, indicate whether or not a two-way table would be appropriate for summarizing the relationship. In each case, briefly explain why or why not.
  a. Satisfaction with quality of local K–12 schools (satisfied or not satisfied) and political party (Republican, Democrat, etc.).
  b. Height (centimeters) and foot length (centimeters).
  c. Gender and amount willing to spend on a home theater system.
  d. Age group (under 20, 21–29, etc.) and handedness (right-handed, left-handed, or ambidextrous).

15.2 In each of the following situations, give the p-value for the given chi-square statistic. Either use the information in Table A.5 to estimate a range for the p-value or use software to determine an exact value.
  a. $\chi^2 = 3.84$, df = 1.
  b. $\chi^2 = 6.7$ for a table with 3 rows and 3 columns.
  c. $\chi^2 = 26.23$ for a table with 2 rows and 3 columns.
  d. $\chi^2 = 2.28$, df = 9.

15.3 Recall that the critical value for a chi-square test is the chi-square value for which the area to its right equals the level of significance. An observed relationship is statistically significant if the calculated chi-square statistic is greater than or equal to the critical value. Use Table A.5 to determine the critical value in each of the following situations.
  a. Level of significance is $\alpha = 0.05$; df = 1.
  b. Level of significance is $\alpha = 0.01$; table has 3 rows and 4 columns.
  c. Level of significance is $\alpha = 0.05$; df = 6.

15.4 For each of the following situations determine whether the result is statistically significant at the 0.05 level of significance.
  a. $\chi^2 = 2.89$, df = 1.
  b. $\chi^2 = 5.00$, df = 1.
  c. $\chi^2 = 23.60$, df = 4.
  d. $\chi^2 = 23.60$, df = 15.

15.5 Refer to the previous exercise. For each part determine whether the result is statistically significant at the 0.01 level of significance.

15.6 Suppose investigators conduct a study on the relationship between birth order (first or only child, not first or only child) and activity preference (indoor or outdoor).
  a. Suppose the p-value of the study is not small enough to reject the null hypothesis. Write this conclusion in the context of the situation.
  b. Now suppose the p-value of the study is small enough to reject the null hypothesis. In the context of the situation, express the conclusion in two different ways.

15.7 Students from two different statistics classes at UC Davis reported their gender (male or female) and recorded which class they were taking. One class is for liberal arts students and the other is for non-liberal-arts students. (These data are in the file **UCDavis1** on the CD accompanying the text.) The results for the 173 students are given below. In each cell the observed counts are printed first, with the expected counts below them. Assume these students are a representative sample of all UC Davis students who would take either class.

	NonLib	LibArt	All
Female	83	11	94
	80.42	13.58	94.00
Male	65	14	79
	67.58	11.42	79.00
All	148	25	173
	148.00	25.00	173.00

  a. State the null and alternative hypotheses that could be tested from these data.
  b. What are the degrees of freedom for the chi-square test statistic?

c. Compute the value of the chi-square test statistic.

d. Find the *p*-value or *p*-value range for the chi-square statistic found in part (c).

e. Make a conclusion and write it in the context of the situation. Be sure to state the level of significance you are using.

15.8 The data for this exercise are from a sample of 12th graders, collected as part of the 2001 Youth Risk Behavior Surveillance System, in the dataset **YouthRisk** on the CD accompanying this book. The students were asked how often they wear a seat belt while driving, and for this exercise we combine the responses for "never" and "rarely" and for "most times" and "always" and ignore the "sometimes" responses. Here is a contingency table of the results for males and females.

	Seat Belt Use		
	Most Times or Always	*Rarely or Never*	*Total*
Female	964	97	1061
Male	924	254	1178
Total	1888	351	2239

Data source: http://www.cdc.gov/nccdphp/dash/yrbs/

a. Calculate the row percents for females and the row percents for males. Compare them.

b. Specify the null and alternative hypotheses that could be tested using a chi-square statistic in this situation.

c. Calculate the expected count for the cell "Male, Most times or Always."

d. Use the expected count you calculated in part (c) and obtain the expected counts for the remaining cells by subtraction.

e. Compute the chi-square statistic.

f. Use the chi-square statistic computed in part (e) to find a *p*-value or *p*-value range.

g. Make a conclusion and write it in the context of the situation. Be sure to state the level of significance you are using.

15.9 Refer to the previous exercise, in which the relationship between sex and seat belt use was examined.

a. Compute the chi-square statistic using the shortcut formula given in Section 15.2. Show your work.

b. What are the degrees of freedom for this situation?

c. Find the *p*-value or *p*-value range for the chi-square statistic found in part (a).

d. Make a conclusion and write it in the context of the situation. Be sure to state the level of significance you are using.

15.10 In the 1993 General Social Survey conducted by the National Opinion Research Center at the University of Chicago, participants were asked:

Do you favor or oppose the death penalty for persons convicted of murder?

Do you think the use of marijuana should be made legal or not?

A two-way table of counts for the responses to these two questions is

		Marijuana?		
		Legal	*Not Legal*	*Total*
	Favor	152	561	713
Death Penalty?	**Oppose**	61	159	220
	Total	213	720	933

Data source: http://csa.berkeley.edu:7502/archive.htm

a. Is there an obvious choice for which variable should be the explanatory variable in this situation? Explain.

b. Calculate row and/or column percents for the table. Use these percents to describe the relationship (or absence of a relationship).

c. State null and alternative hypotheses about the two variables that have been used to create the contingency table.

d. Do a chi-square test of the hypotheses stated in part (c). State a conclusion and support your conclusion with a *p*-value.

e. If you did not already do so in part (d), use the "shortcut" computational formula to find the value of the test statistic. Show how the numbers fit into the formula.

15.11 The following two contingency tables separate the data of exercise 15.10 by sex of the respondent. The first table shows the responses of males while the second table displays the responses of females.

Males Only

		Marijuana?		
		Legal	*Not Legal*	*Total*
	Favor	80	238	318
Death Penalty?	**Oppose**	30	41	71
	Total	110	279	389

Females Only

		Marijuana?		
		Legal	*Not Legal*	*Total*
	Favor	72	323	395
Death Penalty?	**Oppose**	31	118	149
	Total	103	441	544

a. For each table, compute relevant conditional percents. Based on these percents, briefly discuss, for each gender, the nature of the relationship between the variables.

b. For males, do a chi-square test to assess the statistical significance of the relationship between opinions about marijuana and the death penalty. State a conclusion within the context of the problem.

c. For females, do a chi-square test to assess the statistical significance of the relationship between opinions about marijuana and the death penalty. State a conclusion within the context of the problem.

15.12 Refer to the previous two exercises. Based on the results from exercise 15.10, a reporter wrote, "There is a significant relationship between opinions on the death penalty and legalization of marijuana. While about 28% of those who oppose the death penalty think marijuana should be legal, only about 21% of those who favor the death penalty think marijuana should be legal."

a. Referring to the data in exercise 15.11, make appropriate changes and rewrite the reporter's statement based on men only.

b. Make appropriate changes and rewrite the reporter's statement based on women only.

c. Discuss the validity of the reporter's statement based on your results from parts (a) and (b). Write a paragraph explaining the results taking the sex of the respondents into account.

15.13 For the expected counts shown in Table 15.3 for Xylitol and ear infection data, verify that the null hypothesis "expected" proportion getting an ear infection is the same for the three treatment groups.

15.14 An article at the Gallup Poll website entitled "SOCIAL AUDIT, Gambling in America" included the following comparison of the responses of teenagers and adults to the question, "Generally speaking, do you approve or disapprove of legal gambling or betting?" The survey was conducted in April 1999.

	Opinion of Legal Gambling			
	Approve	*Disapprove*	*No Opinion*	*Total*
Adults	959 (63%)	488 (32%)	76 (5%)	1523
Teens	261 (52%)	235 (47%)	5 (1%)	501
Total	1220	723	81	2024

Source: http://www.gallup.com/poll/socialaudits/gambling2.asp

a. State null and alternative hypotheses for this contingency table.

b. Here's Minitab output for a chi-square test. What conclusion can we make about the hypotheses stated in part (a)? Justify this conclusion using information from the output.

```
Expected Counts are shown below observed counts

Rows: Age Group    Columns: Opinion

          Approve   Disappro   No Opini     All

Adults       959        488         76    1523
          918.01     544.04      60.95 1523.00

Teens        261        235          5     501
          301.99     178.96      20.05  501.00

All         1220        723         81    2024
         1220.00     723.00      81.00 2024.00

Chi-Square = 45.723,   DF = 2,   P-Value = 0.000
```

c. What is the expected count for the "Teens, Approve" cell? Show how to calculate this value.

15.15 Refer to the previous exercise. Would a z-test for the difference in two proportions have been appropriate instead of a chi-square test? If your answer is yes, explain whether the results would have been identical to the results of the chi-square test. If your answer is no, explain why not.

15.16 In each of the following situations, explain which test would be most appropriate: a chi-square test, a one-sided z-test for the difference in two proportions, or a Fisher's Exact Test.

a. The manufacturer of a safety seal used in cars wants to know if the safety seals perform as well in extreme cold temperatures as they do in normal temperatures. Testing the seals is expensive, so the manufacturer tests only six seals at an extreme cold temperature and six other seals at a normal temperature. Two of the six seals tested at the extreme cold fail and none of the six seals tested at a normal temperature fails.

b. In June, before an upcoming November presidential election, a random sample of 1000 voters was asked which candidate they would vote for if the election were held then. In September, a new random sample of 1000 voters is asked the same question. The polling agency wants to know if the population proportions planning to vote for each candidate are different in September than they were in June.

c. Frequent computer use may be a cause of carpal tunnel syndrome, a painful condition that affects peoples' hands and wrists. Researchers hypothesized that people who took typing classes in school would be less likely to suffer from carpal tunnel syndrome than those who did not take typing classes. To examine this hypothesis, the researchers asked a random sample of 3000 people who frequently used computers in their jobs if they had ever taken typing classes and if they had ever suffered from carpal tunnel syndrome.

15.17 Refer to the previous exercise. In each case, specify the null and alternative hypotheses.

15.18 The use of magnets has been proposed as a cure for various illnesses. Suppose researchers conduct a study with ten participants to determine whether using magnets as therapy reduces pain from migraine headaches. Five participants are randomly assigned to receive the magnet treatment and of those, two report that their pain is reduced. The remaining five participants are given an indistinguishable sham treatment and of those, none report that their pain is reduced. Fisher's Exact Test would be appropriate in this situation.

a. Write in words the question that the p-value would answer in this situation.

b. (Appropriate software required.) Carry out Fisher's Exact Test.

15.19 In exercise 2.15, data were presented on seat belt use and grades for a sample of 2530 12th graders, collected as part of the 2001 Youth Risk Behavior Surveillance System. Students were asked how often they wear a seat belt while driving, and possible choices were never, rarely, sometimes, most times, and always. For this exercise, we combine the responses for "never" and "rarely" and for "most times" and "always" and ignore the "sometimes" responses. The contingency table on the next page summarizes the results.

Seat Belt Use			
Typical Grades	Most Times or Always	Rarely or Never	Total
A or B	1354	180	1534
C	428	125	553
D or F	65	40	105
Total	1847	345	2192

Data source: http://www.cdc.gov/nccdphp/dash/yrbs/

Carry out a chi-square test for the relationship between seat belt use and typical grades for 12th graders. Be sure to cover all five steps.

15.20 The following drawing illustrates the water-level task. Several developmental psychologists have investigated performance on this task. The figure to the left shows the water level in a glass of water that's half full (or is it half empty?). The figure on the right shows the same glass tipped to the right, but the water level is not shown. The task is to correctly draw the water line in this glass.

Water line is shown   Draw the water line

A University of South Carolina psychologist examined the water-level task success rate of 50 students from each of five different colleges within the university (Kalichman, 1986). Within each college, the sample included 25 men and 25 women. The investigators showed each participant eight different drawings of tipped glasses. If the participant correctly drew the water line on four or more of the drawings, he or she was given a "passing" result. Here's a contingency table of the results.

Result on Task			
College	Pass	Fail	Total
Business	33	17	50
Language Arts	32	18	50
Social Sciences	25	25	50
Natural Sciences	38	12	50
Engineering	43	7	50
Total	171	79	250

a. Compare the colleges using row percents. What differences are apparent from these percents?
b. Write null and alternative hypotheses for this table.
c. What are the degrees of freedom for a chi-square test?
d. The chi-square statistic for this table is 16.915. Determine a p-value, and state a conclusion about the null and alternative hypotheses.
e. For each college, what are the expected counts for "Pass" and "Fail" outcomes?

f. (Obviously, optional) What is the correct water line in the glass at the right above?

15.21 The data in the table below first appeared in exercise 6.9. The variables are height (short or not) and whether or not the student had ever been bullied in school for 209 secondary school students in England. The researchers gathered the data to test their hypothesis that short students are more likely to be bullied in school than other students are.

Height and Bullying in School

	Ever Bullied		
Height	Yes	No	Total
Short	42	50	92
Not Short	30	87	117
Total	72	137	209

a. Should these data be analyzed using a chi-square test or a z-test for comparing two proportions? Explain.
b. Explain why it makes sense to designate height as the explanatory variable.
c. Carry out an appropriate hypothesis test. Be sure to state null and alternative hypotheses, and clearly indicate your conclusion about these hypotheses.
d. Calculate separate 95% confidence intervals for the proportion ever bullied for short students and for students who are not short. Compare these two intervals. What does this comparison indicate about the relationship between height and the risk of being bullied?

Section 15.3

15.22 In the following situations, give the expected count for each of the k categories.
a. $k = 3$, H_0: $p_1 = p_2 = p_3 = 1/3$, and $n = 300$.
b. $k = 3$, H_0: $p_1 = 1/4$, $p_2 = 1/4$, $p_3 = 1/2$, and $n = 1000$.
c. $k = 4$, H_0: $p_1 = .2$, $p_2 = .4$, $p_3 = .1$, $p_4 = .3$, and $n = 2000$.
d. $k = 5$, H_0: all p_i are the same, and $n = 2500$.

15.23 In a chi-square goodness of fit test, is it possible for all of the expected counts to be larger than the corresponding observed counts? Explain.

15.24 Explain whether each of these is possible in a chi-square goodness of fit test.
a. The chi-square statistic is negative.
b. The chi-square statistic is 0.
c. The expected counts are not whole numbers.
d. The observed counts are not whole numbers.
e. The probabilities specified in the null hypothesis sum to less than 1.
f. The degrees of freedom are larger than the number of categories.

15.25 The following table below shows the results from 190 students at Penn State (dataset **pennstate1** on the CD) "randomly" choosing among the numbers from 1 to 10. For example, there were 2 students who chose the digit 1, and 9 students who chose the digit 2.

Digit	1	2	3	4	5	6	7	8	9	10
Frequency	2	9	22	21	18	23	56	19	14	6

Assuming that the students were actually choosing digits at random, the expected number of students choosing any digit is 19 (19 = 190/10), so all expected counts are 19.

a. State the null and alternative hypotheses. (*Hint:* Refer to Example 15.8.)
b. Calculate the chi-square goodness of fit test statistic.
c. What are the degrees of freedom for this test statistic?
d. Find the *p*-value or *p*-value range.
e. Make a conclusion and write it in the context of the situation. Be sure to state the level of significance you are using.

15.26 Suppose that on a typical day the proportion of students who drive to campus is .3 (30%), the proportion who bike is .6 (60%), and the remaining .1 (10%) come to campus in some other way (for example, walk, take the bus, get a ride). The campus sponsors a "spare the air" day to encourage people not to drive to campus on that day. They want to know if the proportions using each mode of transportation on that day differ from the norm. To test this hypothesis, a random sample of 300 students that day was asked how they got to campus, with the following results:

Method of Transportation	Drive	Bike	Other	Total
Frequency	80	200	20	300

a. State the null and alternative hypotheses.
b. Find the expected counts for the three modes of transportation.
c. Calculate the chi-square goodness of fit statistic.
d. Find the *p*-value or *p*-value range.
e. Make a conclusion and write it in the context of the situation. Be sure to state the level of significance you are using.

15.27 In a class survey done in a statistics class at Penn State, students were asked, "Suppose you are buying a new car and the model you are buying is available in three colors: silver, blue, or green. Which color would you pick?" Of the *n* = 111 students who responded, 59 picked silver, 27 picked green, and 25 picked blue. Is there sufficient evidence to conclude that the colors are not equally preferred? Carry out a significance test. Be sure to state the null hypothesis and specify the population to which your conclusion applies.

15.28 Refer to the previous exercise. Suppose a car manufacturer had hypothesized that 50% would prefer silver, 30% would prefer blue, and 20% would prefer green. Test the manufacturer's hypothesis.

15.29 The following table contains the observed distribution of the last digit of the forecasted high temperature on December 10, 2000 for *n* = 150 United States and international cities on December 10, 2000 (*Source: New York Times*, Dec. 10, 2000, p. 47).

Last Digit	0	1	2	3	4	5	6	7	8	9
Count	11	21	11	23	10	17	11	15	13	18

a. Compute expected counts for the null hypothesis that all digits 0, 1, ..., 9 are equally likely to be the last digit of the forecasted high temperature.
b. Calculate the chi-square goodness of fit statistic for these data. What are the degrees of freedom for this statistic?
c. State a conclusion about the null hypothesis that all digits are equally likely to be the last digit selected. Explain.

15.30 Suppose a statistics teacher assigns her class of 60 students the homework task of flipping a coin two times, counting the number of heads, and submitting that number to her during the next class period.

a. List the possible sequences of results (in order) for two flips of a coin. Using this list, determine the probability distribution for *X* = number of heads in two flips if the coin is fair.
b. Using proper notation, express the probabilities determined in part (a) as a null hypothesis about the number of heads that may occur in two flips of a fair coin.
c. The distribution of the number of heads reported by the 60 students in the class was

Number of Heads	0	1	2
Number of Students	8	40	12

Using these data, do a chi-square goodness of fit test to test the null hypothesis stated in part (b). Clearly show all steps, and state a conclusion about the null hypothesis.
d. If you were the teacher, what would you conclude about how the class may have done this assignment? Briefly justify your answer.

15.31 The California Daily 3 lottery game is identical to the Pennsylvania Daily Number game described in Example 15.8. The following table contains the observed distribution of all digits drawn in the California Daily 3 on the 200 days between May 14, 2000 and November 29, 2000. Three digits are drawn each day, so the sample size is *n* = 600 for this table.

Last Digit	0	1	2	3	4	5	6	7	8	9
Count	49	61	64	62	50	64	59	65	63	63

Carry out a significance test of the null hypothesis that all digits 0, 1, ..., 9 are equally likely to be drawn.

Chapter Exercises

15.32 In a survey reported in a special issue of *Newsweek* magazine (Special Edition: Health for Life, Spring/Summer 1999), *n* = 747 randomly selected women were asked, "How satisfied are you with your overall appearance?" There were four possible responses to this question, and the following table shows the distribution of counts for the possible responses for each of three age groups. (*Note:* The counts were estimated from percents given in *Newsweek*.)

How Satisfied Are You with Your Overall Appearance?					
Age	Very	Somewhat	Not Too	Not at All	Total
Under 30	45	82	10	4	141
30 – 49	73	168	47	6	294
Over 50	106	153	41	12	312
Total	224	403	98	22	747

a. Determine conditional percents within each age group. Summarize these data using a bar chart, and describe what is shown about the relationship between age and satisfaction with appearance.

b. Minitab output for a chi-square test of association between age group and satisfaction with appearance follows. Verify that the degrees of freedom shown are correct.

c. Is there a statistically significant association between the two variables for this problem? Justify your answer with information from the output.

d. Refer to the output for part (b). Draw a sketch that illustrates the *p*-value for this problem, and write a sentence interpreting the *p*-value.

e. Refer to the output for part (b). What is the expected count for the "50+, Not Too Satisfied" cell? Show how to calculate this expected count.

```
Expected counts are printed below observed counts

         Very    Somewhat   Not Too   Not at All   Total
18-29     45        82        10          4         141
         42.28     76.07     18.50       4.15

30-49     73       168        47          6         294
         88.16    158.61     38.57       8.66

50+      106       153        41         12         312
         93.56    168.32     40.93       9.19

Total    224       403        98         22         747

Chi-Sq = 0.175 + 0.463 + 3.904 + 0.006 +
         2.607 + 0.556 + 1.842 + 0.816 +
         1.655 + 1.395 + 0.000 + 0.860 = 14.278
DF = 6, P-Value = 0.027
1 cells with expected counts less than 5.0
```

15.33 Refer to the previous exercise. Notice that the "contributions to chi-square" are given in the output for all cells. For instance, the "contribution to chi-square" for the row "18–29" and the column "Very" is 0.175, and for the row "18–29" and the column "Not at All" it is 0.006.

a. Identify the two cells with the highest "contributions to chi-square." Specify the numerical value of the "contribution" and the row and column categories for each of the two cells.

b. For each of the two cells identified in part (a), determine whether the expected count is higher or lower than the observed count.

c. Using the information in parts (a) and (b), explain how the women in those category combinations contribute to the overall conclusion for this study.

15.34 Exercise 6.45 gave data on ear pierces and tattoos for a sample of 678 college women. The data are presented again in the following table. The ear pierce response is

the total number of ear pierces for a woman, and this has been categorized.

Ear Pierces and Tattoos, 678 College Women

Pierces	No Tattoo	Have Tattoo
2 or less	245	19
3 or 4	210	26
5 or 6	91	32
7 or more	25	30

Data source: One of the authors.

a. Describe the relationship with relevant percents and an appropriate graph.

b. Write null and alternative hypotheses for the relationship between ear pierces and tattoos.

c. Test the null and alternative hypotheses written for part (b). Use $\alpha = 0.05$ for the level of significance. State a conclusion.

15.35 In a study reported in the *Annals of Internal Medicine* (Lotufo et al., 2000), the investigators examined the possible relationship in men between baldness and the risk of coronary heart disease. Other researchers have reported a possible link between these two variables (see Case Study 3.4, for example). In this study, 19,112 men aged 40 to 84 years old were followed for 11 years. All men were free of coronary heart disease at the beginning of the study. During the study, the men were asked about any baldness pattern they may have had at the age of 45. The following table shows counts for the cross-classification of hair loss pattern at age 45 and whether a participant developed coronary heart disease during the study period or not. (*Note:* Vertex baldness occurs on top in the area of the crown or peak of the head.)

Hair Loss Pattern and Coronary Heart Disease

	Coronary Heart Disease		
Hair Pattern	Yes	No	Total
No Baldness	548	7611	8159
Baldness			
Frontal	333	4075	4408
Mild Vertex	275	3148	3423
Moderate Vertex	163	1608	1771
Severe Vertex	127	1224	1351
Total	1446	17666	19112

a. Analyze the relationship between hair loss pattern and the risk of coronary heart disease. Write a short report in which you provide relevant descriptive statistics as well the results of a test for statistical significance. If there is an association, describe the nature of the association.

b. Is this an observational study or an experiment? Briefly explain how the answer affects the interpretation of any observed link between hair loss pattern and coronary heart disease in men.

15.36 One of the authors of this book purchased four 1-pound bags of plain M&Ms at different stores in Pennsylvania to compare the color distribution to the one stated at the manufacturer's website. The observed results for the combined bags, and the proportions alleged by the manufacturer, are given in the following table. State and test

Minitab v. 14 update on CD

the appropriate hypotheses to determine if it is reasonable to assume that the observed colors are a random sample from a population with the manufacturer's alleged proportions.

	Brown	Red	Yellow	Blue	Orange	Green	Total
Observed	602	396	379	227	242	235	2081
	(.289)	(.190)	(.182)	(.109)	(.116)	(.113)	
Alleged Proportions	.30	.20	.20	.10	.10	.10	

15.37 Refer to Example 15.6, describing an experiment in which students at the University of California at Davis were classified as "sheep" who believe in ESP or as "goats" who do not. Each student then guessed the results of 10 coin tosses. Classify students as "Stars" if they guessed 5 or more correctly, and as "Duds" if they did not. The results are shown in the following table.

	Stars	Duds	Total
Sheep	79	33	112
Goats	48	32	80
Total	127	65	192

a. Describe the relationship between belief and performance with relevant percents and an appropriate graph.
b. Write null and alternative hypotheses for the relationship between belief and performance. Be sure to state the population to which your hypotheses apply.
c. Test the null and alternative hypotheses written for part (b). Use $\alpha = 0.05$ for the level of significance. State a conclusion.

15.38 Refer to the previous exercise in which each student guessed the results of 10 coin flips. If all students are just guessing, and if the coins are fair, then the number of correct guesses for each student should follow a binomial distribution.
   a. What are the parameters n and p for the binomial distribution, assuming the coins are fair and students were just guessing?
   b. Specify the probabilities of getting 0 correct, 1 correct, ..., 10 correct for this experiment if students were just guessing. (Hint: These are the probabilities in the pdf for a binomial distribution with parameters specified in part (a).)
   c. The following table shows how many students got ≤2 right, 3 right, 4 right, and so on, separately, for students classified as sheep (believe in ESP) and classified as goats (don't believe in ESP). Using your results from part (b), fill in the "null probabilities" that correspond to the hypothesis that students are just guessing.

Number Correct	Sheep	Goats	Null Probabilities
≤2	6	5	
3	11	12	
4	16	15	
5	29	19	
6	28	16	
7	14	10	
≥8	8	3	
Total	112	80	

d. Test the hypothesis that the population of sheep represented by these students guess according to the null probabilities specified in part (c). Give a p-value, and state a conclusion.
e. Repeat part (d) for the population of goats represented by these students.

15.39 Household sizes for the households participating in the 1996 General Social Survey conducted by the National Opinion Research Center at the University of Chicago are shown in the following table. The table also shows the proportion of households of each size in the survey and the corresponding proportions of all U.S. households of each size.

Household Size	Number of Cases	Sample Proportion	U.S. Proportion
1	744	0.256	0.257
2	988	0.340	0.322
3	454	0.156	0.169
4	453	0.156	0.149
5 or more	265	0.091	0.103
Total	2904		

Sources: *Sample data:* SDA data archive, http://csa.berkeley.edu:7502; *Population data:* www.census.gov/populatio/socdemo/hh-fam/98ppla.txt (Table A. Households by Type and Selected Characteristics, 1998).

a. Carefully state the null and alternative hypotheses for testing whether the sizes of households in the General Social Survey reflect the size of households for the United States as a whole.
b. Carry out the appropriate test for your hypotheses in part (a). State your conclusion in the context of the situation.

15.40 The data in this exercise were first presented in exercise 6.48. In the 1996 General Social Survey, religious preference and opinion about when premarital sex might be wrong were among the measured variables. The contingency table of counts for these variables is shown in exercise 6.48 and again here.

Religious Preference and Opinion About Premarital Sex

	When Is Premarital Sex Wrong?				
Religion	Always	Almost Always	Sometimes	Never	Total
Protestant	355	117	227	384	1083
Catholic	62	37	120	226	445
Jewish	0	3	14	34	51
None	20	13	45	147	225
Other	15	13	23	40	91
Total	452	183	429	831	1895

Source: SDA archive at http://csa.berkeley.edu:7502.

a. Are the conditions necessary for carrying out a chi-square test met? Explain.
b. Test whether or not there is a statistically significant relationship between these two variables. Show all five steps for the hypothesis test.

15.41 Case Study 10.3 described a survey in which students in a statistics class at Penn State were asked, "Would you date someone with a great personality even though you did not find them attractive?" The results were that 80 of

131 women answered "yes" while 26 of 61 men answered "yes."

a. Construct a contingency table for this situation, being careful to identify the explanatory and response variables.

b. State the appropriate hypotheses for this situation.

c. Given the hypotheses you specified in part (b), would it be more appropriate to conduct a chi-square test, a z-test, or does it not matter?

d. Conduct the appropriate test of the hypotheses you specified in part (b). Carry out all five steps.

15.42 Refer to the previous exercise. Use the shortcut formula for 2 × 2 tables to compute the chi-square statistic. Show the formula with all numbers entered.

15.43 Weindling et al. (1986; also in Hand et al., 1994, p. 15) were interested in the health of juvenile delinquents. They classified 16 boys who failed a vision test by whether or not they wore glasses and whether or not they were a juvenile delinquent. They were interested in knowing if the likelihood of wearing glasses differed for juvenile delinquents compared to nondelinquents. The results are shown in the following table.

Data for Exercise 15.43

	Wears Glasses	Does Not Wear Glasses	Total
Delinquent	1	8	9
Nondelinquent	5	2	7
Total	6	10	16

a. Specify the appropriate null and alternative hypotheses.

b. Explain why a chi-square test is not appropriate in this situation.

c. Find statistical software or a website that will conduct a Fisher Exact Test, and carry out the test.

d. Explain in words what the p-value for this test represents.

15.44 Wilding and Cook (2000) asked 352 males and 376 females to listen to a male voice and a female voice. One week later they attempted to identify the voices they had heard in lineups of six male and six female voices. The results are shown in the following table. Using appropriate subsets of the data, conduct a test of the null hypotheses in parts (a) to (d).

	Identified Male Voice?		Identified Female Voice?	
Listener's Sex	Yes	No	Yes	No
Male ($n = 352$)	145	207	132	220
Female ($n = 376$)	162	214	191	185
Total ($n = 728$)	307	421	323	405

a. H_0: Sex of listener and ability to identify a male voice are not related.

b. H_0: Sex of listener and ability to identify a female voice are not related.

c. H_0: Sex of listener and ability to identify a voice of the same sex are not related.

d. H_0: Sex of listener and ability to identify a voice of the opposite sex are not related.

e. Using the results from parts (a) to (d), write a short paragraph about the relationship between listener's sex, speaker's sex, and ability to subsequently identify a voice.

15.45 In a class survey, Penn State statistics students were asked, "Which one of these choices describes your perception of your weight: about right, overweight, underweight?" The table in exercise 6.7 displayed the results by sex. Displayed below are these results along with the Minitab output for a chi-square test. (*Source: The authors.*)

Minitab v. 14 update on CD

```
Expected counts are printed below observed counts

           About right    Overwt    Underwt    Total
F                   87        39          3      129
                 91.88     25.56      11.56

M                   64         3         16       83
                 59.12     16.44       7.44

Total              151        42         19      212

Chi-Sq = 0.259 +  7.072 + 6.340 +
         0.403 + 10.991 + 9.853 = 34.918
DF = 2,   P-Value = 0.000
```

a. State the null and alternative hypotheses being tested in this situation.

b. Make a conclusion for the test using $\alpha = 0.05$.

c. Using the results shown in the Minitab output, identify the cells with large "contributions to chi-square." Explain in words that would be understood by someone with no training in statistics why the contributions are so large for those cells.

15.46 In the 1996 General Social Survey, participants were asked, "Should divorce in this country be easier or more difficult to obtain than it is now?" The results are shown in the following table and Minitab output.

Data for Exercise 15.46

Sex	Easier	More Difficult	Stay Same	Total
Male	247	413	151	811
Female	280	547	206	1033
Total	527	960	357	1844

Data source: 1996 SDA archive at http://csa.berkeley.edu:7502/.

Minitab v. 14 update on CD

Minitab Results for Exercise 15.46

```
           Easier    More Diff.    Same    Total
M             247         413       151      811
           231.78      422.21    157.01

F             280         547       206     1033
           295.22      537.79    199.99

Total         527         960       357     1844

Chi-Sq = 1.000 + 0.201 + 0.230 +
         0.785 + 0.158 + 0.181 = 2.554
DF = 2,   P-Value = 0.279
```

a. Carry out the five steps for a chi-square test for this situation.

b. Draw a picture of the appropriate chi-square distribution and shade the region for the p-value for this test.

15.47 Refer to the previous exercise. Suppose the same question were to be asked in many independent surveys. Assuming the null hypothesis is true, into what range should the test statistic fall about 95% of the time, where 0 is at the lower end of the range?

15.48 Explain why a chi-square test statistic cannot be negative.

15.49 Students ($n = 183$) were asked to identify their own eye color as well as the eye color to which they are most attracted. The results are shown here.

Eye Color and Attraction

	Eyes Attracted To				
Own Eyes	*Brown*	*Blue*	*Hazel*	*Green*	*Total*
Brown	30	22	6	13	71
Blue	15	37	3	11	66
Hazel	4	12	7	7	30
Green	4	8	1	3	16
Total	53	79	17	34	183

a. The Minitab results of a chi-square analysis follow. Based on these results, explain why a chi-square analysis is not appropriate.

	Brown	Blue	Hazel	Green	Total
Brown	30	22	6	13	71
	20.56	30.65	6.60	13.19	
Blue	15	37	3	11	66
	19.11	28.49	6.13	12.26	
Hazel	4	12	7	7	30
	8.69	12.95	2.79	5.57	
Green	4	8	1	3	16
	4.63	6.91	1.49	2.97	
Total	53	79	17	34	183

Chi-Sq = 4.331 + 2.441 + 0.054 + 0.003 +
       0.886 + 2.541 + 1.599 + 0.130 +
       2.530 + 0.070 + 6.369 + 0.365 +
       0.087 + 0.173 + 0.159 + 0.000 = 21.738
DF = 9, P-Value = 0.010
4 cells with expected counts less than 5.0

Minitab v. 14 update on CD

b. Combine the categories for "hazel" and "green" and rerun the analysis. Carry out all five steps of the hypothesis test.

c. Refer to the results from part (b). Identify the cells that have the largest "contributions to chi-square," and explain why the results in those cells support the alternative hypothesis.

15.50 Refer to the previous exercise. Using the original data (not combining green and hazel) construct a contingency table for the two variables:

Explanatory variable: Eye color.
Response variable: Finds own eye color most attractive, yes or no.

a. Conduct a chi-square test to determine if these two variables are related. Specify all five steps of the test.

b. Determine which eye color(s) contributes the most to the chi-square statistic. Explain whether that eye color(s) is more attractive to people who have it as their own color than would be expected, or less so.

15.51 Gillespie (1999) and Chambers (2000) reported on two Gallup polls, taken in August 1999 and August 2000 using independent samples, which asked parents the question, "How satisfied are you with the quality of education your oldest child is receiving? Would you say completely satisfied, somewhat satisfied, somewhat dissatisfied or completely dissatisfied?" The results of the two polls are shown in the following table.

Results of Gallup Polls Taken in 1999 and 2000

	August 24–26, 1999	August 24–27, 2000
Completely Satisfied	125	87
Somewhat Satisfied	155	133
Somewhat Dissatisfied	41	34
Completely Dissatisfied	7	17
Just Starting School (Volunteered Answer)	7	11
No Opinion	3	0
Total	338	282

a. Refer to the discussion of the technical difference between tests for homogeneity and tests for independence. Which is more appropriate for this situation, for comparing the results of the survey for the two years? Specify the appropriate null and alternative hypotheses.

b. Ignore the last two categories (Just starting school; no opinion). Carry out the five steps for a chi-square test to determine if opinions differed in 1999 and 2000. Be sure to identify the appropriate populations represented by these data.

Dataset Exercises

15.52 Use the dataset **GSS-93** for this exercise. The variable *owngun* indicates whether the respondent owns a gun or not, and the variable *polparty* contains the respondent's political party preference. Is there a significant relationship between *owngun* and *polparty?*

15.53 Use the dataset **UCDavis2**. Respondents were categorized as male or female (*Sex*) and were asked if they typically sit in the front, middle, or back of the classroom (*Seat*).

a. Carry out the five steps to determine whether there is a significant relationship between these two variables.

b. The p-value for your test in part (a) should be about 0.02. Explain in words which cells contribute the most to the chi-square statistic and how those cells support the alternative hypothesis.

15.54 Use the dataset **UCDavis2.** Two variables measured were whether the respondent was left- or right-handed

(*Hand*) and whether they find it easier to make friends with people of the same or opposite sex (*Friends*). Carry out the five steps to determine if there is a significant relationship between those variables.

15.55 Use the dataset **UCDavis2.** Identify two variables for which it would be of interest to you to test whether or not there is a relationship. Carry out the five steps of the chi-square test. If one or both of the variables are quantitative, create reasonable categories. For instance, you could classify students as nondrinkers, moderate drinkers, or heavy drinkers using the variable **Alcohol.** Make sure you explain what variables you used and any recoding you did.

REFERENCES

Chambers, Chris (2000). "Americans dissatisfied with U.S. education in general, but parents satisfied with their kids' schools," Gallup Poll Release, Sept. 5, 2000.

Gillespie, Mark (1999). "Local schools get passing grades: Americans support higher pay for teachers," Gallup Poll Release, Sept. 8, 1999.

Hand, D. J., E. Daly, A. D. Lunn, K. J. McConway and E. Ostrowski (1994). *A Handbook of Small Data Sets,* London: Chapman & Hall.

Kalichman, S. C. (1986). "Horizontality as a function of sex and academic major," *Perceptual and Motor Skills,* Vol. 63, pp. 903–908.

Lawrence, T. R. (1993). "Gathering in the sheep and goats. A Meta-analysis of forced choice sheep–goat ESP studies, 1947–1993," *Proceedings of Presented Papers: The 36th Annual Convention of the Parapsychological Association,* pp. 75–86.

Lotufo, P. A., C. U. Chae, U. A. Ajani, C. H. Henneken, and J. E. Manson (2000). "Male pattern baldness and coronary heart disease: The physicians' health study," *Arch. Intern. Med.,* Vol. 160, pp. 165–171.

Schmeidler, Gertrude R. (1988). *Parapsychology and Psychology: Matches and Mismatches,* Jefferson, NC: McFarland.

Schwartz, B. S., W. F., Stewart, D. Simon, and R. B. Lipton (1998). "Epidemiology of tension-type headache," *Journal of the American Medical Association,* Vol. 279, pp. 381–383.

Sinclair, R. C., M. M. Mark, S. E. Moore, L. A. Lavis, and A. S. Soldat (2000). "An electoral butterfly effect," *Nature,* Vol. 408, pp. 665–666.

Uhari, M., T. Kenteerkar, and M. Niemala (1998). "A novel use of Xylitol sugar in preventing acute Otitis Media," *Pediatrics,* Vol. 102, pp. 879–884.

Weindling, A. M., F. N. Bamford, and R. A. Whittall (1986). "Health of juvenile delinquents," *British Medical Journal,* Vol. 292, p. 447.

Wilding, J., and S. Cook (2000). "Sex differences and individual consistency in voice identification," *Perceptual and Motor Skills,* Vol. 91, pp. 535–538.

Analysis of Variance

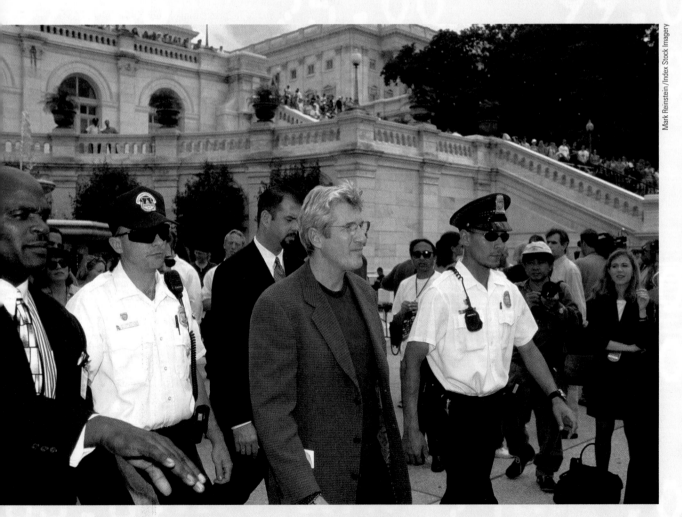

Do actors have higher testosterone levels than other men? See Example 16.2.

Analysis of variance is a versatile tool for analyzing how the mean value of a quantitative response variable is affected by one or more categorical explanatory factors. For instance, it can be used to compare the mean weight loss for three different weight-loss programs, or the mean testosterone levels of men in seven different occupations.

Suppose that a researcher wants to compare the mean weight loss for three different weight-loss programs. Or, that another researcher wants to compare the mean testosterone levels of men in seven different occupations. What statistical methods can be used to make the desired comparisons? In the data analysis, the researchers might, at some point, use confidence intervals and significance tests to compare two means at a time (for example, compare mean testosterone levels of occupations 1 and 2, occupations 1 and 3, and so on). Usually, however, an important first step in the analysis of more than two means is to do a significance test to determine if there are any differences at all among the population means being compared. The significance test for doing this is part of a procedure called the **analysis of variance,** which is also sometimes referred to as **ANOVA.**

In this chapter, we focus on the comparison of the means of more than two populations. When different values or levels of a single categorical explanatory variable (weight-loss programs, for instance) define the populations being compared, the ANOVA procedure is called **one-way analysis of variance.** In general, *analysis of variance* is a versatile tool for analyzing how the mean value of a quantitative response variable is related to one or more categorical explanatory factors. While most of the chapter is about one-way analysis of variance, the final section gives an overview of the concepts of *two-way analysis of variance,* which is used to examine the effects of two categorical explanatory variables on the mean value of a quantitative response variable. ◆

16.1 | COMPARING MEANS WITH AN ANOVA *F*-TEST

When we compare the means of populations represented by independent samples of a quantitative variable, a null hypothesis of interest is that all means have the same value. An alternative hypothesis is that the means are not all equal.

Notice that this alternative hypothesis does not require that all means must differ from each other. The alternative would be true, for example, if only one of the means were different from the others. If k = the number of populations, the null and alternative hypotheses for comparing population means can be written as

H_0: $\mu_1 = \mu_2 = \ldots = \mu_k$
H_a: The means are not all equal.

An **F-statistic** that arises from a *one-way analysis of variance* of the sample data is used to test the hypotheses about the population means, and the significance test is called an **F-test.** The F-statistic is sensitive to differences among a set of sample means. The greater the variation among the sample means, the larger the value of the test statistic. The smaller the variation among the observed means, the smaller the value of the test statistic. In this section, we are concerned with the general ideas of the F-test. The specific details and formulas are given in the next section.

Conceptually, the F-statistic can be viewed as follows:

$$F = \frac{\text{Variation among sample means}}{\text{Natural variation within groups}}$$

The "variation among sample means" is 0 if all k of the sample means are exactly equal and gets larger the more spread out they are. If that variation is large enough, it provides evidence that at least one of the k population means is different from the others, so the null hypothesis would be rejected. The denominator of the F-statistic, the "natural variation within groups," provides a yardstick for determining whether the numerator is large enough to reject the null hypothesis. Much like the standard error in a z-statistic, it "standardizes" the numerator so that the p-value can be found from common tables. In fact, the denominator of the F-statistic is simply a pooled estimate of the variance within each group, a fact that will be discussed in more detail in the next section.

To find the p-value, which is the probability the computed F-statistic would be as large as it is (or larger) if the null hypothesis is true, a probability distribution called the **F-distribution** is used. As with all other significance tests, the null hypothesis is rejected if the p-value is as small or smaller than the desired level of significance (usually $\alpha = 0.05$). When the null hypothesis is rejected, the conclusion is that the population means do not all have the same value.

EXAMPLE 16.1
Classroom Seat Location and Grade Point Average

Is it true that the best students sit in the front of a classroom, or is that a false stereotype? In surveys done in two statistics classes at the University of California at Davis, students reported their grade point averages and also answered the question, "Where do you typically sit in a classroom (front, middle, back)?" In all, 384 students gave valid responses to both questions, and among these students, 88 said they typically sit in the front, 218 said they typically sit in the middle, and 78 typically sit in the back. Figure 16.1 shows a boxplot comparing the GPAs of the students for the three seat locations and the analysis of variance results from Minitab with an F-test for comparing the mean grade point averages. In the boxplot, we see that students sitting in the front generally have slightly higher GPAs than the others; toward the bottom of the computer output the sample means are given as: Front = 3.2029,

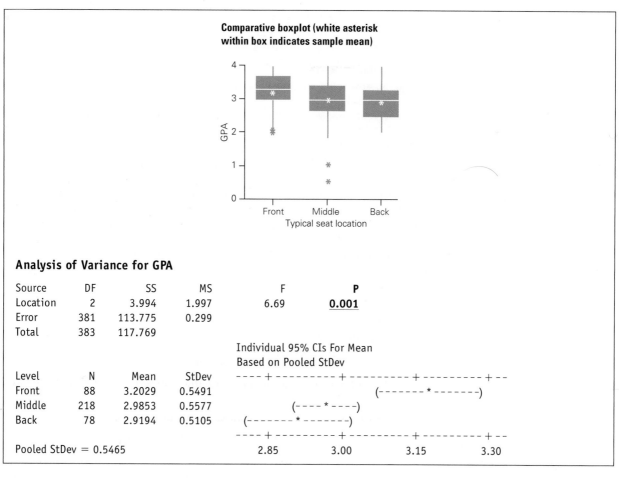

FIGURE 16.1   **Comparison of GPA for three classroom seat locations**

Middle = 2.9853, Back = 2.9194. The analysis of variance results can be used to test

$H_0: \mu_1 = \mu_2 = \mu_3$
H_a: The three means are not all equal.

where μ_1, μ_2, and μ_3 are the *population* mean GPAs for the populations of students who would typically sit in the front, middle, and back of the classroom, respectively.

The details of the analysis of variance will be shown later, but it is relatively easy to interpret the result. The *p*-value for the *F*-test is under "P" in the rightmost column of the table entitled "Analysis of Variance for GPA," and its value is 0.001 (bold, underlined). With a *p*-value this small, we can reject the null hypothesis and thus conclude that there are differences among the means in the populations represented by the samples. Notice that the output in Figure 16.1 also shows 95% confidence intervals for the three population means. The location of the interval for the "Front" mean does not overlap with the other two intervals, which indicates a significant difference between the mean GPA for the front-row sitters and the mean GPA for the other students. It is not clear whether there is a significant difference between the "Middle" and "Back" groups since their confidence intervals overlap. ◆

Notation for Summary Statistics

Useful notation for summarizing statistics from the observed samples is

k = number of groups

$\bar{x}_i$, s_i, and n_i are the mean, standard deviation, and sample size for the ith sample group

N = total sample size ($N = n_1 + n_2 + \cdots + n_k$)

VIDEO

EXAMPLE 16.1— **CONTINUED** **Application of** **Notation to the GPA** **and Classroom Seat Sample**

Three seat locations (1 = front, 2 = middle, 3 = back) are compared, so $k = 3$. The group sample sizes are $n_1 = 88$, $n_2 = 218$, $n_3 = 78$. The total sample size is $N = 88 + 218 + 78 = 384$. Toward the bottom of the output shown in Figure 16.1, the sample means given are $\bar{x}_1 = 3.2029$, $\bar{x}_2 = 2.9853$, $\bar{x}_3 = 2.9194$. The sample standard deviations given in Figure 16.1 are $s_1 = 0.5491$, $s_2 = 0.5577$, $s_3 = 0.5105$. ◆

Assumptions and Necessary Conditions for the F-Test

In the derivation of the F-statistic, the assumptions are similar to those for the pooled two-sample t-procedures described in Chapters 12 and 13. The assumptions about the populations and samples representing them are

- The samples are independent random samples.
- The distribution of the response variable is a normal curve within each population.
- The different populations may have different means.
- All populations have the same standard deviation, σ.

These assumptions make up a model for the probability distributions in the populations being compared. Figure 16.2 illustrates, for three populations, how this model might look if the alternative hypothesis were true, and how it would look if the null hypothesis were true. Notice that in the alternative hypothesis case, the only difference among the populations is the difference among their means, μ_1, μ_2, and μ_3. The spread of the normal curve is assumed to be the same for each population.

When we compare two means, it is not imperative to assume equal population standard deviations because an "unpooled" procedure is available that does not require this assumption. Unfortunately, when we do a one-way analysis of variance, the equal standard deviations assumption is necessary. In practice, equality of the observed standard deviations does not have to hold exactly for the F-test to provide a valid significance test of differences among population means. One commonly used criterion is that the F-test can be used if the largest of the sample standard deviations is not more than twice as large as the smallest of the sample standard deviations. There are significance tests for the equality of several standard deviations, but those tests tend not to work well when the data are not normally distributed. They can also have low power for small or moderate sample sizes, leading to false acceptance of equal variances when they are actually quite different. Therefore, we do not recommend using those tests.

The assumption of normally distributed data is not strictly necessary in practice either. Unless there are extreme outliers, or extreme skewness in the data, the F-test works well even for relatively small sample sizes. When the raw

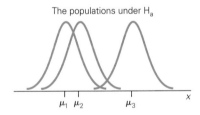

The populations under H_a

The populations under H_0

FIGURE 16.2 **The model for one-way** **analysis of variance under the two** **different hypotheses**

data are available, a visual display should be created and examined in order to evaluate whether skewness or outliers are present.

Necessary Conditions for Using the *F*-Statistic to Compare Means

- ◆ The *F*-statistic can be used if the data are not extremely skewed, there are no extreme outliers, and the group standard deviations are not markedly different.
- ◆ As with the *t*-tests and confidence intervals in Chapters 12 and 13, tests based on the *F*-statistic are valid for data with skewness or outliers if the sample sizes are large.
- ◆ A rough criterion for standard deviations is that the largest of the sample standard deviations should not be more than twice as large as the smallest of the sample standard deviations.

EXAMPLE 16.1—
CONTINUED
Assessing the Necessary
Conditions for the GPA
and Seat Location Data

The boxplot in Figure 16.1 (p. 563) shows two outliers in the group of students who typically sit in the middle of a classroom, but there are 218 students in that group so these outliers don't have much influence on the results. The standard deviations for the three groups are nearly the same, and the data do not appear to be skewed. The necessary conditions for doing an *F*-test are satisfied in this example. ◆

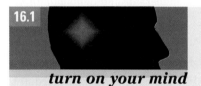

turn on your mind

To what populations do the conclusions of Example 16.1 apply? Do you think it matters that the data were collected at a single university? Does it matter that the surveys were done only in statistics classes?*

EXAMPLE 16.2
Occupational Choice
and Testosterone Level

Many research studies have been done to examine how testosterone affects human behavior, and, conversely, some studies have considered how behavior affects testosterone level. In a study done at Georgia State University, psychologists compared the salivary testosterone concentrations of men who were ministers, salesmen, firemen, professors, physicians, professional football players, and actors. There were 66 men in the overall sample, and the sample sizes in the seven occupational groups ranged from 6 (actors) to 16 (physicians).

The researchers reported in the *Journal of Personality and Social Psychology* (Dabbs et al., 1990, p. 1262) that "analysis of variance showed an overall difference among the groups, $F(6,59) = 2.50$, $p < 0.05$, and a Newman-Keuls test indicated that actors and football players were significantly higher than ministers."

In this quote, the information given is that the p-value for an F-test was less than 0.05, so the null hypothesis that all population means are the same was rejected. Additional information is given about the pattern of the difference, which is that actors and football players had higher testosterone levels than ministers. Presumably, all other differences involving the other occupational groups were not statistically significant. The researchers focused on the difference between actors and ministers and reported on two additional studies they did that replicated the finding. In their discussion, they related the observed testosterone differences to the social, competitive, and dominance behaviors of ministers and actors. ◆

The Family of F-Distributions

An F-distribution is used to find the p-value for an ANOVA F-test of the null hypothesis that several population means are equal. The family of F-distributions is a family of skewed distributions, each with a minimum value of 0. A specific F-distribution is indicated by two parameters called *degrees of freedom*. The first of the two parameters is called the **numerator degrees of freedom;** the second is called the **denominator degrees of freedom.** The values of the two degrees of freedom parameters always are given in the order *numerator df, denominator df.* In one-way ANOVA, the numerator df $= k - 1$ (number of groups $- 1$), and the denominator df $= N - k$ (total sample size $-$ number of groups).

Figure 16.3 shows an F-distribution with 2 and 30 degrees of freedom along with an F-distribution with 6 and 30 degrees of freedom. Notice that both curves are skewed and the minimum possible value is 0. These are properties of all F-distributions, regardless of the values for the numerator and denominator degrees of freedom.

Determining the p-Value

All statistical software programs incorporate analysis of variance procedures, and the p-value will be reported as part of the output. In situations where an F-statistic has been computed without statistical software, the Excel command FDIST(*value, numerator df, denominator df*) can be used to find the p-value. When software is not available a range for the p-value can be found using tabled values for the appropriate F-distribution. Table A.4 in the Appendix gives **critical values** for two different levels of significance, $\alpha = 0.05$ and $\alpha = 0.01$. If the observed

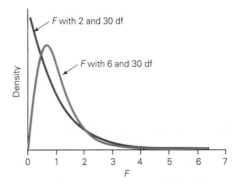

FIGURE 16.3   **An F-distribution with 2 and 30 df and an F-distribution with 6 and 30 df**

F-statistic is greater than the critical value for a particular level of significance, the result is statistically significant at that level. The region of F-statistic values greater than the critical value is called the **rejection region** for the test.

Using Table A.4 to Judge Statistical Significance

First, determine values for the numerator degrees of freedom and the denominator degrees of freedom. Possible values for the numerator degrees of freedom are given as column headings across the top of Table A.4. Possible values for the denominator degrees of freedom are given as row labels in the leftmost column.

- If the value of the F-statistic is greater than the critical value in the $\alpha = 0.05$ portion of the table, the result is statistically significant at the 0.05 significance level. In this case, the p-value is less than 0.05, which may be written $p < 0.05$.

- If the value of the F-statistic is greater than the critical value in the $\alpha = 0.01$ portion of the table, the result is statistically significant at the 0.01 significance level. In this case, the p-value is less than 0.01, which may be written $p < 0.01$.

- If the value of the F-statistic is between the critical values given for $\alpha = 0.05$ and $\alpha = 0.01$, the p-value is between 0.05 and 0.01 ($0.01 < p < 0.05$). In this case, the result is statistically significant for $\alpha = 0.05$ but not for $\alpha = 0.01$.

As an example, suppose the F-statistic for comparing four means is $F = 4.26$ and numerator df = 3 and denominator df = 20. In Table A.4, for $\alpha = 0.05$, the critical value is given as 3.10 (in the column labeled "3" and row labeled "20"). Because the observed $F = 4.26$ is greater than the critical value = 3.10, the result is statistically significant at the $\alpha = 0.05$ level. The critical value for the $\alpha = 0.01$ is given as 4.94 in Table A.4. Because 4.26 is not greater than this critical value, the result is not significant at the 0.01 level. Thus, the p-value is between 0.01 and 0.05.

EXAMPLE 16.2—CONTINUED
The p-Value for the Testosterone and Occupational Choice Example

The authors reported that the value of the F-statistic for comparing the seven occupations was $F = 2.5$ and the p-value was less than 0.05. With $k = 7$ occupational groups and $N = 66$ observations in the total sample, the degrees of freedom are $7 - 1 = 6$ and $66 - 7 = 59$. Figure 16.4 illustrates the p-value, which is the area to the right of 2.5 under an F-distribution with 6 and 59 df. In Excel, FDIST(*value, numerator df, denominator df*) can be used to find the probability of an F-statistic as large as or larger than the specified value. In this case, FDIST(2.5,6,59) provides the information that the p-value is 0.032. Using Table A.4, we learn that the approximate critical value is 2.25 for $\alpha = 0.05$. This is the value given for numerator df = 6 and denominator df = 60. Because 2.5 is larger than this critical value, the result is statistically significant for the 0.05 significance level, and the p-value is smaller than 0.05. ◆

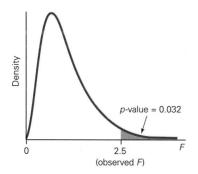

FIGURE 16.4 **The *p*-value for Example 16.2 is the area to the right of *F* = 2.5 under an *F*-distribution with 6 and 59 degrees of freedom**

Multiple Comparisons

The term **multiple comparisons** is used when two or more comparisons are made to examine the specific pattern of differences among means. The most commonly analyzed set of multiple comparisons is the set of all pairwise comparisons among population means. In Example 16.1 about GPA and classroom seat location, the possible pairwise comparisons are Front versus Middle,

Middle versus Back, and Front versus Back. In Example 16.2 about occupation and testosterone level, the mean testosterone levels for 21 different pairs of occupations can be compared (actor versus minister, football player versus professor, and so on).

Either of two equivalent approaches can be used to make inferences about pairs of means. For each pair, a significance test could be done to determine if the two means significantly differ. Or, a confidence interval for the difference in each pair of means could be computed, and interpreted in terms of statistical significance. If a confidence interval for a difference does not include the value 0, there is a statistically significant difference. In its one-way analysis of variance procedure, Minitab uses the confidence interval approach.

When many comparisons are made within the same study, some statistically significant results may occur just by chance. Remember that when the $\alpha = 0.05$ significance level is used, about 1 in 20 independent statistical tests will achieve statistical significance even if all null hypotheses are true. In other words, when many statistical tests are done there is an increased risk of making at least one type I error (erroneously rejecting a null hypothesis). Consequently, several procedures have been developed to control the overall **family type I error rate** or the overall **family confidence level** when inferences for a set (family) of multiple comparisons are done. **Tukey's procedure** is one such procedure for the family of pairwise comparisons. If the family error rate is not a concern, **Fisher's procedure** is used.

definition ◆ A **family error rate** for a set of significance tests is the probability of making one or more type I errors (erroneously rejecting a null hypothesis) when more than one significance test is done.

◆ A **family confidence level** for a procedure used to create a set of confidence intervals is the proportion of times that all intervals in the set capture their true parameter values.

EXAMPLE 16.1—
CONTINUED
Pairwise Comparisons
of GPAs Based
on Seat Locations

Figure 16.5 shows Minitab output for pairwise comparisons of the mean GPA by seat location. The family confidence level used for the Tukey procedure was 0.95 (95%), while for Fisher's procedure a 0.95 (95%) confidence level was used for each individual interval. For both methods, the output gives confidence intervals for the difference in mean GPAs for each of the three possible comparisons of seat locations (Front versus Middle, Front versus Back, and Middle versus Back). The two numbers given for each combination of two seat locations are the lower and upper limits of a confidence interval for the difference in mean GPAs for the corresponding populations. In the results for the Tukey procedure, for instance, the values given for the "Front" (column label) versus "Back" (row label) combination are 0.0846 and 0.4824. Thus, a confidence interval for the difference in means ($\mu_{\text{Front}} - \mu_{\text{Back}}$) is 0.0846 to 0.4824. In the results given for the Fisher procedure, the confidence interval for ($\mu_{\text{Front}} - \mu_{\text{Back}}$) is 0.1164 to 0.4506.

Minitab v. 14 update on CD

Tukey's pairwise comparisons
Intervals for (column level mean) − (row level mean)

	Front	Middle
Middle	0.0561	
	0.3791	
Back	0.0846	−0.1028
	0.4824	0.2347

Fisher's pairwise comparisons
Intervals for (column level mean) − (row level mean)

	Front	Middle
Middle	0.0819	
	0.3533	
Back	0.1164	−0.0758
	0.4506	0.2077

FIGURE 16.5   Confidence intervals for pairwise differences in the GPA and seat location example

In general, two means are significantly different if the confidence interval for the difference does not cover 0. Notice that the two procedures lead to the same conclusions about statistical significance in this example, but that will not always be the case. In both procedures, the only confidence interval that covers 0 is the one for $(\mu_{Middle} - \mu_{Back})$. So, we can conclude that the population mean GPAs differ for students who sit in the front and middle of the room and for those who sit in the front and back of the room, but not for those who sit in the middle and back.

Figure 16.6 is a graph of the confidence intervals found using the Tukey method. Each interval estimates the difference between the population mean GPAs for two different seat locations. A vertical line is drawn at 0, the value that would indicate no difference between two means. Notice that the confidence interval for $(\mu_{Middle} - \mu_{Back})$ covers the line drawn at 0, whereas confidence intervals for $(\mu_{Front} - \mu_{Back})$ and $(\mu_{Front} - \mu_{Middle})$ do not. ◆

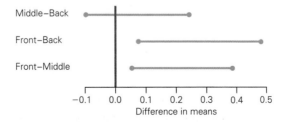

FIGURE 16.6   Confidence intervals for differences between pairs of means in the GPA and seat location example (Tukey procedure)

Minitab Tip

tech note   **One-Way Analysis of Variance**

- Use **Stat>ANOVA>One-way** if values of the response variable are in one column (Response) and values or names for groups are in a second column (Factor). In the dialog box, the ***Comparisons*** button is used for pairwise comparisons.

- Use **Stat>ANOVA>One-way (unstacked)** if the data for different groups are stored in separate columns. Specify the separate columns for the groups in the dialog box. There is not a pairwise comparisons capability in this procedure (Version 13 or before).

16.2 | DETAILS OF ONE-WAY ANALYSIS OF VARIANCE

In this section, we provide some computational details for the *F*-statistic and give the necessary conditions for using the statistic. In practice, statistical software is nearly always used to do a one-way analysis of variance, so it is not necessary to learn formulas in order to do the computations "by hand." The formulas, however, provide useful insights into the concepts of analysis of variance and how the *F*-test is constructed.

The Analysis of Variance Table

A fundamental concept in one-way analysis of variance is that the variation among the data values in the overall sample can be separated into (1) differences between group means, and (2) natural variation among observations within a group. The calculations and theory for analysis of variance stem from the fact that for a particular way of measuring variation, the sum of the variation between group means and the variation among observations within groups equals the total variation. Expressed as an equation, the relationship is

Total variation = Variation between groups + Variation within groups

An **analysis of variance table,** like the one shown in Figure 16.1 for Example 16.1 (p. 563), is used to display information about the sources of variation in the response variable. The *F*-statistic used to compare population means measures the relative size of the variation between group means and the natural variation within groups.

EXAMPLE 16.3
***Comparison of
Weight-Loss Programs***

Suppose that a researcher does a randomized experiment to compare the mean weight loss for three different programs for losing weight, and the observed weight losses after three months are as follows:

Program 1	Program 2	Program 3
7	9	15
9	11	12
5	7	18
7		

FIGURE 16.7 **Dotplot for Example 16.3**

A look at the dotplot of the data shown in Figure 16.7 shows that over the total dataset, the weight losses ranged from 5 to 18 pounds. The sample means for the three programs are $\bar{x}_1 = 7, \bar{x}_2 = 9, \bar{x}_3 = 15$. While there is some variation among individuals within each program, generally it appears that differences *between* weight-loss programs account for much of the variation in the total dataset. The weight losses for program 3 in particular are noticeably greater than the weight losses for the other two programs. ◆

Measuring Variation Between Groups

The variation between group means is measured with a weighted sum of squared differences between the sample means (the $\bar{x}_i$) and $\bar{x}$, the overall mean of all data. Each squared difference is multiplied by the appropriate group sample size, n_i, in this sum. This quantity is called **sum of squares for groups** or **SS Groups.** A formula is

$$\text{SS Groups} = n_1(\bar{x}_1 - \bar{x})^2 + n_2(\bar{x}_2 - \bar{x})^2 + \cdots + n_k(\bar{x}_k - \bar{x})^2 = \sum_{\text{groups}} n_i(\bar{x}_i - \bar{x})^2$$

The numerator of the F-statistic for comparing means is called the **mean square for groups** or **MS Groups,** and it is calculated as

$$\text{MS Groups} = \frac{\text{SS Groups}}{k - 1}$$

Measuring Variation Within Groups

To measure the variation among individuals within groups, find the sum of squared deviations between data values and the sample mean in each group, and then add these quantities. This is called the **sum of squared errors** or **SSE.** A formula in terms of sample standard deviations is

$$\text{SS Error} = (n_1 - 1)s_1^2 + (n_2 - 1)s_2^2 + \cdots + (n_k - 1)s_k^2 = \sum_{\text{groups}} (n_i - 1)s_i^2$$

The denominator of the F-statistic is called the **mean square error** or **MSE,** and it is calculated as

$$\text{MSE} = \frac{\text{SSE}}{N - k} = \frac{(n_1 - 1)s_1^2 + (n_2 - 1)s_2^2 + \cdots + (n_k - 1)s_k^2}{n_1 + n_2 + \cdots + n_k - k}$$

Notice that MSE is just a weighted average of the sample variances for the k groups. In fact, if all n_i are equal, MSE is simply the average of the k sample variances. The square root of MSE, denoted by s_p and called the **pooled standard deviation,** estimates the population standard deviation of the response variable.

Recall that the populations being compared are assumed to have equal standard deviations.

> **definition** The **pooled standard deviation** $s_p = \sqrt{\text{MSE}}$ estimates the population standard deviation, σ. This assumes that each population has the same value of σ.

Measuring Total Variation

The total variation in all samples combined is measured by computing the sum of squared deviations between data values and the mean of all the data. This quantity is referred to as the **total sum of squares** or **SS Total.** The total sum of squares may also be referred to as SSTO. A formula for the sum of squared differences from the overall mean is

$$\text{SS Total} = \sum_{\text{values}} (x_{ij} - \overline{x})^2$$

where x_{ij} represents the jth observation within the ith group, and $\overline{x}$ is the mean of all observed data values. Notice that SS Total would be the numerator of the sample variance (with denominator $N - 1$) if all the data were combined and treated as a single sample.

The relationship between SS Total, SS Groups, and SS Error is

$$\text{SS Total} = \text{SS Groups} + \text{SS Error}$$

This additive relationship is useful computationally. For example, if the values of SS Total and SS Groups are known, then SS Error can be calculated as

$$\text{SS Error} = \text{SS Total} - \text{SS Groups}$$

EXAMPLE 16.3 — CONTINUED
Analysis of Variation
Among Weight Losses

For Example 16.3, the $k = 3$ sample means are $\overline{x}_1 = 7$, $\overline{x}_2 = 9$, $\overline{x}_3 = 15$, and the mean of all $N = 10$ observations is $\overline{x} = 10$.

$$\text{SS Groups} = n_1(\overline{x}_1 - \overline{x})^2 + n_2(\overline{x}_2 - \overline{x})^2 + n_3(\overline{x}_3 - \overline{x})^2$$
$$= 4(7 - 10)^2 + 3(9 - 10)^2 + 3(15 - 10)^2 = 114$$

$$\text{MS Groups} = \frac{114}{3 - 1} = 57$$

$$\begin{aligned}
\text{SS Total} = &(7 - 10)^2 + (9 - 10)^2 + (5 - 10)^2 + (7 - 10)^2 + (9 - 10)^2 \\
&+ (11 - 10)^2 + (7 - 10)^2 + (15 - 10)^2 + (12 - 10)^2 \\
&+ (18 - 10)^2 = 148
\end{aligned}$$

$$\text{SS Error} = \text{SS Total} - \text{SS Groups} = 148 - 114 = 34$$

$$\text{MSE} = \frac{34}{10 - 3} = 4.857$$

$$F = \frac{57}{4.857} = 11.74 \qquad \text{with 2 and 7 df}$$

Figure 16.8 shows Minitab output for Example 16.3. The analysis of variance table format used is standard for most software. The line labeled "Programs" gives information about SS Groups and MS Groups, and also includes the

Analysis of Variance

Source	DF	SS	MS	F	P
Programs	2	114.00	57.00	11.74	0.006
Error	7	34.00	4.86		
Total	9	148.00			

```
                              Individual 95% CIs For Mean
                              Based on Pooled StDev
Level       N    Mean   StDev  ----------+---------+---------+-----
Program1    4    7.000  1.633    (------*-----)
Program2    3    9.000  2.000       (-------*------)
Program3    3   15.000  3.000                  (------*-------)
                                 ----------+---------+---------+-----
Pooled StDev =    2.204                    8.0      12.0      16.0
```

FIGURE 16.8   **Minitab output for Example 16.3**

F-statistic and the p-value. The label for this line is the name the user gives the explanatory variable when the dataset is created. In some instances, "Factor" may be used as a generic label for the explanatory variable. The line labeled "Error" gives information about SS Error and MSE. At the bottom of the output, the pooled standard deviation is given as 2.204. This can be computed as $\sqrt{\text{MSE}} = \sqrt{4.86}$. ◆

General Format of a One-Way ANOVA Table

In summary, the general format of the one-way analysis of variance table is

Source	df	SS	MS	F
Between groups (due to factor)	$k - 1$	$\text{SS Groups} = \sum_{\text{groups}} n_i(\bar{x}_i - \bar{x})^2$	$\dfrac{\text{SS Groups}}{k-1}$	$F = \dfrac{\text{MS Groups}}{\text{MSE}}$
Error (within groups)	$N - k$	$\text{SSE} = \sum_{\text{groups}} (n_i - 1)s_i^2$	$\dfrac{\text{SSE}}{N-k}$	
Total	$N - 1$	$\text{SSTO} = \sum_{\text{values}} (x_{ij} - \bar{x})^2$		

EXAMPLE 16.4
Top Speeds of Supercars

Kitchens (1998, p. 783) gives data gathered by *Car and Driver* magazine on the top speeds of five supercars from five different countries—Acura NSX-T from Japan, Ferrari F355 from Italy, Lotus Esprit S4S from Great Britain, Porsche 911 Turbo from Germany, and Dodge Viper RT/10 from the United States. The data represent the top speeds for six runs on each car, using as much distance as necessary without exceeding the engine's redline. There were three runs in each direction on the test facility, to cancel grade or wind effects. Figure 16.9 on the next page includes a dotplot of the data, as well as output for a one-way analysis of variance that compares the five cars. Some important features of the results shown in the figure are

◆ The p-value is 0.000, so we can reject the null hypothesis that the population mean speeds are the same for all five cars.

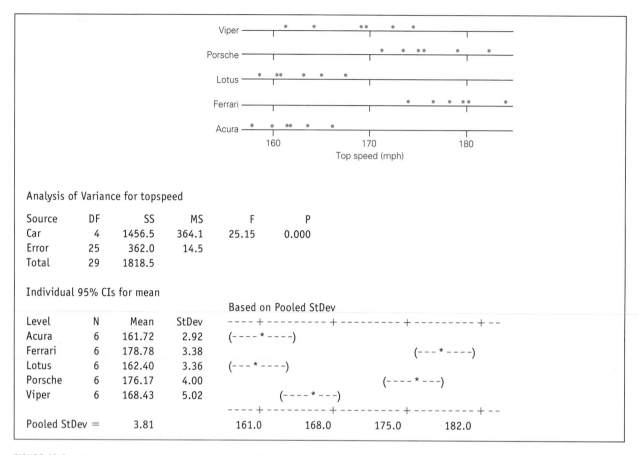

FIGURE 16.9 **Dotplot and analysis of variance output for Example 16.4**

◆ $F = 25.15$. The calculation was $F = $ MS Cars/MS Error $= 364.1/14.5$. An F-distribution with 4 and 25 df was used to determine the p-value.

◆ The necessary conditions for doing an ANOVA F-test are present. The data are not skewed and there are no extreme outliers. The largest sample standard deviation (5.02 for Viper) is not more than twice as large as the smallest standard deviation (2.92 for Acura).

◆ MS Error $= 14.5$ is an estimate of the variance of the top speed for the hypothetical distribution of all possible runs with one car. The estimated standard deviation for each car is $\sqrt{\text{MS Error}} = \sqrt{14.5} = 3.81$. This value is given as the pooled standard deviation at the bottom of the output.

◆ Based on the sample means and corresponding confidence intervals for population means, the Porsche and Ferrari seem to be significantly faster than the other three cars. ◆

Computation of 95% Confidence Intervals for the Population Means

It is informative to examine and compare confidence intervals for the population means. In Examples 16.1 and 16.4, we saw that the Minitab output for one-way analysis of variance includes a graph showing a 95% confidence interval for each population mean. These intervals are computed using the same general format we used in Chapter 12 for a confidence interval for a mean:

$$\text{Sample mean} \pm \text{Multiplier} \times \text{Standard error}$$

Because it is assumed that all populations have the same standard deviation, $s_p = \sqrt{\text{MSE}}$ is used to estimate the standard deviation within each group. This affects the standard error and the degrees of freedom. For the ith sample mean, s.e.$(\bar{x}_i) = s_p/\sqrt{n_i}$. Notice that the pooled standard deviation is used regardless of group, but the sample size is specific to the group. The "multiplier" is determined using a t-distribution as it was for a single mean in Section 12.4, but in this situation, df $= N - k$. These degrees of freedom are used because they are associated with the *mean squared error,* which is used to estimate the standard deviation.

Confidence Interval for a Population Mean

formula   In one-way analysis of variance, a **confidence interval for a population mean** μ_i is

$$\bar{x}_i \pm t^* \frac{s_p}{\sqrt{n_i}}$$

where $s_p = \sqrt{\text{MSE}}$ and t^* is such that the confidence level is the probability between $-t^*$ and t^* in a t-distribution with $N - k$ degrees of freedom.

EXAMPLE 16.4—*CONTINUED*
95% *Confidence Intervals for Mean Car Speeds*

The pooled standard deviation for Example 16.4 is $s_p = 3.81$ mph, a value displayed at the bottom of Figure 16.9. For each car, $n_i = 6$, so the standard error for any car is s.e.$(\bar{x}_i) = 3.81/\sqrt{6}$. The df for the t^* multiplier are $N - k = 30 - 5 = 25$. From the table of t^* multipliers (Table A.2), we learn that for a 0.95 confidence level and df $= 25$, the multiplier is $t^* = 2.06$. A 95% confidence interval for any mean is $\bar{x}_i \pm 2.06(3.81/\sqrt{6})$, or $\bar{x}_i \pm 3.02$. For example, the 95% confidence interval for the Acura is 161.72 ± 3.02 mph, for the Ferrari it is 178.78 ± 3.02 mph, and so on. ◆

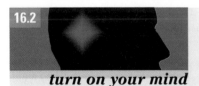

16.2

turn on your mind

In Example 16.4, each 95% confidence interval had the same width. Why did this happen? When would the 95% confidence intervals have different widths?*

in summary   **Steps for the *F*-Test for Comparing Several Means**

Step 1: *Determine the* null *and* alternative *hypotheses.*

Null hypothesis: $H_0: \mu_1 = \mu_2 = \ldots = \mu_k$
Alternative hypothesis: H_a: The μ's are not all equal.

*HINT: In the formula given just before Example 16.4—*CONTINUED,* which of the quantities affecting interval width may or may not differ from group to group?

Step 2: *Verify necessary data conditions, and if met, summarize the data into an appropriate* test statistic. The *F*-statistic can be used if the data are not extremely skewed, there are no extreme outliers, and the group standard deviations are not markedly different. A criterion for standard deviations is that the largest of the sample deviations should not be more than twice as large as the smallest of the sample standard deviations. The test statistic is

$$F = \frac{\text{MS Groups}}{\text{MS Error}} = \frac{\dfrac{\Sigma n_i(\overline{x}_i - \overline{x})^2}{k - 1}}{\dfrac{\Sigma(n_i - 1)s_i^2}{N - k}}$$

Step 3: *Assuming the null hypothesis is true, find the p-value.* Using the *F*-distribution with numerator df $= k - 1$ and denominator df $= N - k$, the *p*-value is the area in the tail to the right of the test statistic *F*.

Step 4: *Decide whether or not the result is* statistically significant *based on the p-value.* Choose a significance level ("alpha"); standard is $\alpha = 0.05$. The result is statistically significant if the *p*-value $\leq \alpha$.

Step 5: *Report the conclusion in the context of the situation.* In analysis of variance, when the null hypothesis is rejected these steps are usually followed by a multiple comparison procedure to determine which group means are significantly different from each other.

16.3 | OTHER METHODS FOR COMPARING POPULATIONS

You probably won't be surprised to learn that the necessary conditions for using an analysis of variance *F*-test don't hold for all datasets. In this section, we discuss methods that can be used when one or both of the assumptions about equal population standard deviations and normal distributions are violated. It is important to remember that no inference method is appropriate if the observed data do not represent the population for the question of interest.

EXAMPLE 16.5
Drinks per Week and Seat Location

In the two surveys described for Example 16.1, students were asked, "How many alcoholic beverages do you consume each week?" Let's compare responses for the same three groups compared in Example 16.1: students who typically sit in the front, middle, or back of a classroom. Figure 16.10 displays a boxplot of the data and Table 16.1 contains summary statistics for the three seat locations. The sample sizes are a bit different here than in Example 16.1 because some students did not give a response to every question.

The information in Figure 16.10 and Table 16.1 reveals that the students who typically sit in the back report drinking more alcohol than the other students do. The information in the figure and table also shows that the group standard deviations differ, and the data appear to be skewed. The boxplot shows much more variation in the responses of the back of the classroom group than in the other two groups. In Table 16.1, we see that the standard deviation in the "back" group (10.5) is more than twice as large as the standard deviation in the "front" group

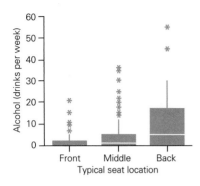

FIGURE 16.10 **Boxplot by seat location for number of alcoholic beverages per week**

TABLE 16.1 ■ **Summary Statistics by Seat Location for Number of Alcoholic Beverages per Week**

Location	n	Mean	Median	s.d.
Front	87	1.6	0	3.4
Middle	207	3.9	1	6.5
Back	79	8.5	5	10.5

(3.4). Also, the mean is greater than the median in each group, which is evidence of skewness in the data. The necessary conditions for doing an analysis of variance are violated in this dataset. ◆

Hypotheses About Medians

When the observed data are skewed, or when extreme outliers are present, it usually is better to analyze the median rather than the mean. This is also what should usually be done if the response variable is an ordinal variable. When several population medians are compared, null and alternative hypotheses of interest are

H_0: Population medians are equal.
H_a: Population medians are not all equal.

The notation used for a population median varies from author to author. The most commonly used symbol is η (pronounced "eta"). With this notation, the null hypothesis about the medians of k populations is H_0: $\eta_1 = \eta_2 = \ldots = \eta_k$. The letter M is generally used to denote a sample median. When considering several samples, the median of the ith sample would be written M_i.

Kruskal—Wallis Test for Comparing Medians

One test for comparing medians is the **Kruskal–Wallis Test.** It is based on a comparison of the relative rankings (sizes) of the data in the observed samples, and for this reason is called a **rank test.** The term **nonparametric test** also is used to describe this test because there are no assumptions made about a specific distribution for the population of measurements.

The general idea is that values in the total data set of N observations are ranked from lowest to highest (lowest = 1, highest = N). The ranks of the values are averaged for each group, and the test statistic measures the variation among the average ranks of the groups. If most of the small data values were in one particular group, for example, that group would have a lower average rank than the other groups. A p-value can be determined by finding the probability that the variation among the set of rank averages for the groups would be as large (or larger) as it is if the null hypothesis is true. More information about the Kruskal–Wallis Test as well as material about other nonparametric methods is given in Chapter 18.

EXAMPLE 16.5—*CONTINUED*
Kruskal–Wallis Test for Alcoholic Beverages per Week by Seat Location

Figure 16.11 contains output from Minitab for a Kruskal–Wallis Test of the null hypothesis that the *median* number of alcoholic beverages consumed per week is the same for the three classroom seat locations. The p-value shown at the bottom of the output is P = 0.000, which indicates strong evidence against the null hypothesis. You'll see that two p-values are shown; the second is followed by the phrase "adjusted for ties." There is a "tie" in the rankings when two or more observations have the same value. For example, all students who reported 0 drinks per week are tied with each other. It is possible to determine an adjustment to the p-value based on the number of ties that occur. In practice, the adjusted and unadjusted values rarely differ by much.

The output in Figure 16.11 on the next page includes the sample median and the average rank within each group. Oddly, the medians are given by Minitab in

Kruskal–Wallis Test on alcohol

Seat	N	Median	Ave Rank	Z
Front	87	0.00E+00	143.7	−4.27
Middle	207	1.00E+00	183.6	−0.68
Back	79	5.00E+00	243.6	5.25
Overall	373		187.0	

$H = 35.98$ $DF = 2$ $P = 0.000$
$H = 40.18$ $DF = 2$ $P = 0.000$ (adjusted for ties)

FIGURE 16.11   **Kruskal–Wallis Test for comparing median number of alcoholic beverages per week in three classroom locations**

scientific notation. Notice that the median number of alcoholic beverages per week for the "back" group is 5, which is clearly higher than the medians for the other two groups. Because the reported data values tended to be higher in the back group, the average rank is highest for this group. The data values tended to be lowest for the front group (median = 0), so the average rank is lowest for that group. The Kruskal–Wallis statistic measures the variation among the average ranks shown for the three groups much like the F-statistic measures the variation among the sample means in analysis of variance. ◆

Mood's Median Test for Comparing Medians

TABLE 16.2 ■ **Format of the Two-Way Table for Mood's Median Test. M = Median of All Data in Overall Dataset**

Group	Number ≤ M	Number > M
1		
2		
⋮		
k		

Another *nonparametric* test used to compare population medians is **Mood's Median Test.** The idea is easy to grasp. First, the median (M) of the total data set with all groups combined is determined. Then, within each group, the number of observations less than or equal to M and the number of observations greater than M are counted. These counts can be displayed in the two-way table shown in Table 16.2. A chi-square statistic for two-way tables (Chapter 15) is used to test the null hypothesis that the population medians are the same. The test is equivalent to testing that the two variables creating the rows and columns in Table 16.2 are not related. A statistically significant result indicates that the medians are not all the same.

16.3

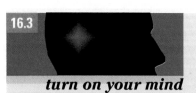

turn on your mind

Show why the null hypothesis of equal population medians is equivalent to the null hypothesis that the two variables in Table 16.2 are not related as follows. Assume for simplicity that no values are tied with the sample median M. In that case, fill in the marginal totals for Table 16.2 using the notation for group sizes of $n_1, n_2, \ldots, n_k$; $N = n_1 + n_2 + \cdots + n_k$. What would be the expected cell counts for the test of the null hypothesis that the two variables are unrelated? Explain why those are the same cell counts that would be expected under the null hypothesis that all population medians are equal.*

*****HINT:** If the null were true, what fraction of each group would be on either side of the overall median? See Section 15.1 to review expected counts for a two-way table.

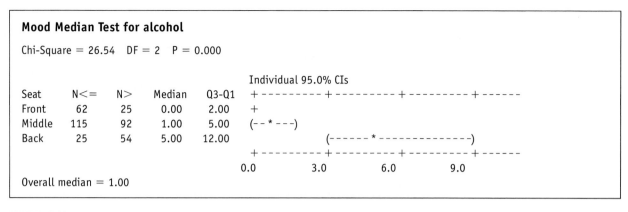

FIGURE 16.12 Mood's Median Test for the alcoholic beverages per week by seat location example

EXAMPLE 16.5—CONTINUED
Mood's Median Test for
the Alcoholic Beverages
and Seat Location Example

Output from Minitab for Mood's Median Test in the alcoholic beverages consumed per week by classroom seat location example is shown in Figure 16.12. The first line shows that the *p*-value for the chi-square test is 0.000, indicating that the null hypothesis of equal population medians can be rejected. The overall median of all responses is shown at the bottom of the output as "Overall median = 1.00." In the central portion of the output, information is given about how many responses in each seat location group were less than or equal to this value and how many were greater than this value. In the "front" group, for instance, 62 students reported consuming one or fewer alcoholic beverages per week and 25 reported drinking more than this. A much different pattern is evident in the "back" group. In that group, 25 reported consuming one or fewer drinks per week compared to 54 who reported drinking more than one drink per week. Notice also that sample medians and interquartile ranges are given for each group along with a 95% confidence interval for the median of the population represented by each group. ◆

Minitab Tip

tech note **Nonparametric Procedures for Comparing Several Medians**
The Kruskal–Wallis and Mood's Median tests are listed in the menu that results from using **Stat>Nonparametrics.**

Advanced Technical Note

tech note **Transforming the Response Variable for Analysis of Variance**
When the group standard deviations and the group means are related, it may be possible to transform the values of the response variable so that the necessary conditions for ANOVA are satisfied on the transformed scale. An analysis of variance is then done on the transformed data. Three transformations that frequently work for this purpose are square root, logarithm, and reciprocal. For instance, if the ratio $s_i/\bar{x}_i$ is about the same for all groups, then the values of log(response) tend to have equal standard deviations for all groups. Using this "log transformation" frequently works for financial data, such as salaries, for which the variability increases as the mean increases.

The type of pattern for which a transformation might work is present in Example 16.5. Notice that sample means and sample standard deviations are ordered in the same way over the three groups. In this example, if $x =$ alcoholic beverages per week, the group standard deviations are nearly equal for the transformed response $1/(0.5 + x)$. (Adding 0.5 to x avoids dividing by 0.) It is difficult, however, to interpret the transformed responses, so a comparison of medians may be preferable.

16.4 | TWO-WAY ANALYSIS OF VARIANCE

A **two-way analysis of variance** is used to examine how two categorical explanatory variables affect the mean of a quantitative response variable. For example, an industrial psychologist may want to see how type of background music and loudness of background music affect productivity in a workplace. Or an economist may be interested to learn the effects of gender and race on mean income. In problems like these, there is interest in the effect of each separate explanatory factor, and there is interest in the combined effect of the two explanatory factors.

The **interaction** between two variables is usually the most interesting feature of a two-way analysis of variance problem. When two variables interact, the effect on the response variable of one explanatory variable depends upon the specific value or level present for the other explanatory variable. A statement like "being overweight caused greater increases in blood pressure for men than for women" is a statement describing an interaction. The effect on blood pressure of weight (overweight or not) depends upon gender (male or female).

The term **main effect** is used to describe the overall effect of a single explanatory variable. In the type of background music and loudness of music example just described, the main effect of the explanatory variable "music loudness" is the effect on productivity of loudness averaged over all music types. A main effect may not be meaningful if an interaction is present. For instance, it may not be sensible or useful to report the effect on blood pressure of being overweight without reporting that the effect is different for men than it is for women.

EXAMPLE 16.6
Happy Faces and Restaurant Tips

When a restaurant server writes a friendly note or draws a "happy face" on your restaurant check, is this just a friendly act or is there a financial incentive? Temple University psychologists conducted a randomized experiment to investigate whether drawing a happy face on the back of a restaurant bill increased the average tip given to the server (Rind and Bordia, 1996). One female server and one male server in a Philadelphia restaurant either did or did not draw a happy face on checks during the experiment. In all, they drew happy faces on 45 checks (22 for the female, 23 for the male) and did not draw happy faces on 44 other checks (23 for the female and 21 for the male). The sequence of drawing the happy face or not was randomized in advance.

Figure 16.13 is a graph of the mean tip percent for each of the four combinations of server gender (female or male) and check message (none or happy face drawing). The type of graph shown is called an **interaction plot.** The mean

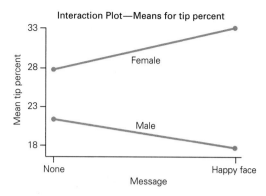

FIGURE 16.13   **Mean tip percent related to sex of server and message on check (none or happy face)**

response (tip percent) is graphed on the vertical axis and message type (none or happy face) is graphed on the horizontal axis. The higher line in the figure is for the female server (she got better tips than the male). We see that her mean tip percent was about 5% higher for the "happy face" checks. The lower line in the figure shows the results for the male server, and for him, drawing a happy face decreased the mean tip percent by about 4%. The effect of drawing a happy face depended upon gender, so there is an interaction between sex of server and message type. The researchers speculated that customers might have felt that drawing a happy face was not gender appropriate for males. ◆

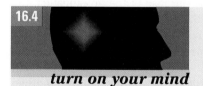

16.4

turn on your mind

In Example 16.6, there was only one server of each sex. What problem does this cause in the interpretation and generalization of the results? How would you have designed the experiment to better examine the interaction between sex of the server and drawing a happy face (or not)?*

EXAMPLE 16.7
You've Got to Have Heart

Psychologist Lee Salk (*Scientific American,* May 1973) observed 287 mothers within four days after giving birth and found that 83% of the right-handed mothers and 78% of the left-handed mothers held their babies on the left side. When asked why they chose the left side, the right-handed mothers said it was so their right hand would be free. The left-handed mothers said it was because they could hold the baby better with their dominant hand. Salk speculated that "it is not in the nature of nature to provide living organisms with biological tendencies unless such tendencies have survival value." He surmised that there must be survival value to placing a newborn infant close to the sound of the mother's heart.

To test his conjecture, Salk arranged for a baby nursery at a New York City hospital to have the continuous sound of a human heartbeat played over a loudspeaker. At the end of four days, he measured how much weight $n = 102$ babies

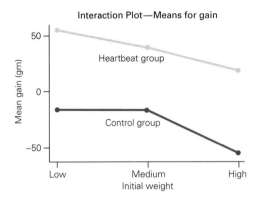

FIGURE 16.14   **Mean weight gain (or loss) by initial weight and experimental conditions (heartbeat or not)**

in the nursery had gained or lost. Later, with a new group of $n = 112$ babies in the nursery, no sound was played and weight gains (or losses) were again measured after four days. Because initial birth weight affects weight gain or loss in the early stages of life, Salk also categorized the babies into three birth-weight categories.

In all, there are six possible combinations of experimental group (heartbeat or not) and birth weight (low, medium, high). Figure 16.14 is an interaction plot of the sample mean weight gain or loss (grams) for these combinations. (Salk did not give means; we have estimated them from the data shown in a dotplot in his *Scientific American* report.) Considering the purpose of the investigation, the most important finding revealed by Figure 16.14 is that the weight gain was generally greater for the heartbeat group. This confirmed what Salk had suspected. Although they did not eat more than the control group, the infants treated to the sound of the heartbeat gained more weight (or lost less). Further, they spent much less time crying. Salk's conclusion was that "newborn infants are soothed by the sound of the normal adult heartbeat." Somehow, mothers intuitively know that it is important to hold their babies on the left side.

The difference in weight gain evident in Figure 16.14 for all three birth-weight groups indicates that there is a *main effect* for the explanatory variable of "heartbeat or control group." The difference between the mean weight gain or loss in the heartbeat and control groups is roughly the same in each of the birth-weight groups. This is evident from the approximately equal distance between the two lines (heartbeat and control) shown in Figure 16.14 regardless of initial weight. This similarity across groups indicates that there may not be an *interaction* between the experimental condition variable and initial birth-weight group. In this example, it's sensible to report the main effect of heartbeat versus control averaged over birth-weight groups. ◆

Two-Way Analysis of Variance and *F*-Tests

In a two-way analysis of variance, three *F*-statistics are constructed. One is used to test the statistical significance of the interaction, while the other two are used to test the significance of the two separate main effects. The computational details for these tests and a more extensive discussion of two-way ANOVA are in Chapter 20.

Source	DF	Adj SS	Adj MS	F	P
Message	1	14.7	14.7	0.13	0.715
Sex	1	2602.0	2602.0	23.69	0.000
Interaction	1	438.7	438.7	3.99	0.049
Error	85	9335.5	109.8		
Total	88	12407.9			

FIGURE 16.15   **Two-way analysis of variance of restaurant tipping data**

EXAMPLE 16.6—CONTINUED
Two-Way Analysis of Variance for Happy Face Example

The Minitab output in Figure 16.15 is for a two-way analysis of variance of the data for the happy face and restaurant tip example. (Professor Bruce Rind of Temple University provided the raw data). Only consider the *p*-values shown in the last column of the table, and don't worry about the other details.

The significance test of the effect of "Message" (happy face or none) has a *p*-value of 0.715, indicating a nonsignificant result. That result, however, is nearly meaningless because, as we saw before, there is an interaction between message type and sex of server. The interaction, which is significant with a *p*-value of 0.049, was that drawing a happy face increased the tip percent for the woman, but decreased it for the man. For the main effect of sex of server, the *p*-value is 0.000, indicating a statistically significant difference in average tips for the female and male.

In the last Turn On Your Mind, you were asked to think about the problem caused by having only one server of each sex. The answer is that we can't be certain whether the observed effects have to do with a gender difference or with the difference between these particular individuals. Multiple servers of each sex should be included in the study in order to determine whether there really is a gender difference. ◆

KEY TERMS

Section 16.1
analysis of variance (ANOVA), 561
one-way analysis of variance, 561
F-statistic for comparing means, 562
F-test for comparing means, 562, 575–576
F-distribution, 562, 566
assumptions and conditions for *F*-test, 564–565
numerator degrees of freedom, 566
denominator degrees of freedom, 566
critical values, 566–567
rejection region, 567
multiple comparisons, 567
family type I error rate, 568

family confidence level, 568
Tukey's multiple comparison procedure, 568
Fisher's multiple comparison procedure, 568

Section 16.2
analysis of variance table, 570, 573
sum of squares (SS) for groups, 571
mean square (MS) for groups, 571
sum of squared errors (SSE), 571
mean squared error (MSE), 571
pooled standard deviation in ANOVA, 571–572
total sum of squares (SS Total, SSTO), 572

confidence interval for a population mean in ANOVA, 575

Section 16.3
Kruskal–Wallis test, 577
rank test, 577
nonparametric test, 577
Mood's median test, 578

Section 16.4
two-way analysis of variance, 580
interaction, 580
main effect, 580
interaction plot, 580–581

EXERCISES   Blue = Basics   Bold, **Bold** = Answer in back

Section 16.1

16.1 In each situation, determine whether one-way analysis of variance is an appropriate method for analyzing the data described. Briefly explain why or why not for each part.
  a. A researcher compares the mean blood pressures of men over 50 years old for three different ethnic groups. He samples 100 men in each ethnic group, and measures their blood pressures.
  b. A psychiatrist compares four treatment programs for clinical depression. The response variable is whether a patient shows improvement after two months of treatment or not. A randomized experiment is done in which 20 patients are randomly allocated to each treatment.
  c. Three methods for memorizing information are compared. Forty-five participants are randomly divided into three groups, and each group memorizes information with a different method. All participants then take a test on the memorized information, and the scores are used to compare the methods.
  d. Fifty individuals all listen to five songs, and rate each song on a scale of 0 to 100. The mean scores for the five songs are compared.

16.2 Researchers studying the connection between body weight and age report the following finding: "The p-value was 0.003 for a one-way analysis of variance done to compare body mass index values for the three age groups."
  a. What null hypothesis did the researchers test? Write the null hypothesis in words, and also write it using proper statistical notation.
  b. Explain what conclusion can be made about the comparison of body mass index values for the three age groups.

16.3 The output for this exercise is an analysis of variance comparing hours spent watching television in a typical day for five categories of highest educational degree attained for $n = 1591$ respondents in the 1993 General Social Survey. Educational degree groups are: graduate, bachelor's, junior college, high school, and less than high school (*Data source:* **GSS-93** dataset on the CD for this book).

  a. Write null and alternative hypotheses for this situation. Write the hypotheses in words, and also write the null hypothesis using proper notation.
  b. What are the values of the F-statistic and the p-value for testing the hypotheses written in part (a)? What conclusion can be made about the hypotheses?
  c. Compare the sample means given in the output below the analysis of variance table. Describe how mean hours of watching television per day relates to highest degree attained.

16.4 For $n = 153$ female students in the **UCDavis2** dataset on the CD accompanying this book, mean height (inches) by student's preferred seat location in a classroom is as follows:

Location	Mean
Front ($n = 38$)	63.9
Middle ($n = 93$)	64.8
Back ($n = 22$)	66.5

  a. The p-value is 0.001 for an F-test that compares the mean heights of female students in the three seating locations. In the context of this situation, what conclusion can be made?
  b. Describe how mean height of female students in the dataset relates to preferred seating location.

16.5 Refer to the previous exercise about female height and preferred classroom seating location. The following output from Minitab gives confidence intervals for all pairwise comparisons of seating locations. A Tukey procedure with 95% family confidence interval was used.

Minitab v. 14 update on CD

Intervals for (column level mean) − (row level mean)

	Back	Front
Front	1.018	
	4.362	
Middle	0.270	−2.142
	3.230	0.261

Analysis of Variance for tvhours

Source	DF	SS	MS	F	P
degree	4	639.70	159.92	35.98	0.000
Error	1586	7050.14	4.45		
Total	1590	7689.84			

Level	N	Mean
Bachelor	252	2.060
Graduate	118	1.797
HighScho	834	2.969
JunColl	96	2.635
NotHS	291	3.928

a. What confidence interval is given for the difference between the mean heights of female students who prefer to sit in the back versus those who prefer to sit in the front? Explain whether the interval gives evidence that the population means differ for these two locations.

b. What confidence interval is given for the difference between the mean heights of female students who prefer to sit in the back versus those who prefer to sit in the middle? Explain whether the interval gives evidence that the population means differ for these two locations.

c. What confidence interval is given for the difference between the mean heights of female students who prefer to sit in the front versus those who prefer to sit in the middle? Explain whether the interval gives evidence that the population means differ for these two locations.

16.6 In each situation use Table A.4 to find a critical value and then state a conclusion for an F-test of the null hypothesis of equal population means:

a. F-statistic = 6.27 with 2 and 12 degrees of freedom; $\alpha = 0.05$.

b. F-statistic = 1.63 with 5 and 70 degrees of freedom; $\alpha = 0.05$.

c. F-statistic = 3.27 with 3 and 20 degrees of freedom; $\alpha = 0.05$.

d. F-statistic = 3.27 with 3 and 20 degrees of freedom; $\alpha = 0.01$.

16.7 In the 1993 General Social Survey, a randomly selected sample of U.S. adults was asked what they think is the ideal number of children for a couple to have. The accompanying output is for a one-way analysis of variance that compares the mean response in four age groups: 18–29, 30–44, 45–59, and 60+ (*Source:* SDA archive at csa.berkeley.edu:7502/).

a. Write null and alternative hypotheses for this problem. State the hypotheses in words and also state them using mathematical notation. To what populations do these hypotheses apply?

b. What values are given in the output for the F-statistic and the p-value? What conclusion can be made about the hypotheses?

c. Based on the 95% confidence intervals given for the population means, describe the differences among the age groups.

d. Is the assumption of equal population standard deviations reasonable for these data? Briefly explain.

16.8 Refer to the previous exercise about the ideal number of children for a couple to have. The following output is for a Tukey procedure done to examine all pairwise comparisons of the four age groups. A 95% family confidence level was used.

Minitab v. 14 update on CD

```
Intervals for (column level mean) − (row level mean)

                18−29      30−39      40−59

30−39   −0.0854
         0.3023

40−59   −0.1650    −0.2507
         0.2675     0.1363

60+     −0.5099    −0.5952    −0.5608
        −0.0854    −0.2170    −0.1369
```

a. Which pairs of age groups exhibit statistically significant differences? In one or two sentences, summarize the differences in mean response for the four age groups.

b. Briefly explain what it means to say that a 95% family confidence level was used for the six confidence intervals shown in the output for this problem.

16.9 To evaluate a standard test of the flammability of fabric for children's sleepwear, the American Society for Testing Materials had five laboratories each test 11 pieces of the same type of fabric. The response variable is the length of the char mark made when the fabric sample is held over a flame for a specified time period. Ideally, all labs using the standardized test to test the same fabric should observe about the same value for this response variable. Ryan and Joiner (2001, p. 269) give the following data for the observed char lengths (the measurement unit was not specified by Ryan and Joiner).

```
Analysis of Variance for chldidel
Source    DF      SS       MS      F       P
agegrp     3    23.519   7.840   10.91   0.000
Error    955   686.128   0.718
Total    958   709.648

                                Individual 95% CIs For Mean
                                Based on Pooled StDev
Level    N     Mean    StDev    +---------+---------+---------+------
1       202   2.4010   0.8654              (-----*-----)
2       335   2.2925   0.7409    (----*---)
3       203   2.3498   0.8150       (----*-----)
4       219   2.6986   1.0003                        (-----*-----)
                                 +---------+---------+---------+------
Pooled StDev =   0.8476              2.40      2.60      2.80
```

```
Lab 1:  2.9  3.1  3.1  3.7  3.1  4.2  3.7  3.9  3.1  3.0  2.9
Lab 2:  2.7  3.4  3.6  3.2  4.0  4.1  3.8  3.8  4.3  3.4  3.3
Lab 3:  3.3  3.3  3.5  3.5  2.8  2.8  3.2  2.8  3.8  3.5  3.8
Lab 4:  3.3  3.2  3.4  2.7  2.7  3.3  2.9  3.2  2.9  2.6  2.8
Lab 5:  4.1  4.1  3.7  4.2  3.1  3.5  2.8  3.5  3.7  3.5  3.9
```

a. Graph the data in a way that is useful for comparing the laboratories. Describe any differences among laboratories that can be identified in your graph.

b. Use the graph drawn for part (a) to assess whether the necessary conditions are present for using one-way analysis of variance to compare the five laboratories.

16.10 Refer to the previous exercise about testing the flammability of children's sleepwear. Minitab output for a one-way analysis of variance of the data is shown below.

a. Write null and alternative hypotheses for comparing the mean char lengths for the five labs. State the hypotheses in words, and also state them using mathematical notation. To what populations do these hypotheses apply?

b. Are there statistically significant differences among the labs? Justify your answer using the F-statistic and p-value found in the output.

c. Draw a sketch illustrating how the p-value was determined for this problem. See Figure 16.4 (p. 567) for guidance.

d. Based on sample means and the 95% confidence intervals given for the population means, describe the differences among the five laboratories. Be specific about which laboratories appear to have different mean values from the others.

16.11 Koopmans (1987, p. 93) gave data on the testosterone levels (measured as milligrams per 100 milliliters of blood samples) of 46 women classified into three occupational groups: (1) not employed, (2) employed in job not requiring an advanced degree, and (3) employed in job requiring advanced degree. The data are displayed in the accompanying dotplot.

a. Write null and alternative hypotheses for comparing the mean testosterone levels in the three groups. Use proper notation.

EXERCISE 16.11   Dotplot of testosterone levels of women by degree required for profession

b. What does the dotplot indicate about the necessary conditions for doing a one-way analysis of variance of the data? Do any conditions appear to be violated? If so, which conditions?

Section 16.2

16.12 Give a value for each of the missing elements in the following analysis of variance table:

Source	df	SS	MS	F
Between Groups	5	40	—	—
Error	10	60	—	
Total	—		—	

16.13 Give a value for each of the missing elements in the following analysis of variance table:

Source	df	SS	MS	F
Between Groups	2	10	—	—
Error	—	—	—	
Total	30	300		

```
Analysis of Variance for CharLng
Source    DF      SS      MS       F        P
Lab        4    2.987   0.747    4.53    0.003
Error     50    8.233   0.165
Total     54   11.219

                          Individual 95% CIs For Mean
                          Based on Pooled StDev
Level    N     Mean    StDev   -- + --------- + --------- + --------- + ----
1        11   3.3364   0.4523          (------ * ------)
2        11   3.6000   0.4604              (------ * ------)
3        11   3.3000   0.3715         (------ * ------)
4        11   3.0000   0.2864   (------ * ------)
5        11   3.6455   0.4321              (------ * ------)
                               -- + --------- + --------- + --------- + ----
Pooled StDev = 0.4058           2.80      3.15      3.50      3.85
```

EXERCISE 16.10

16.14 Refer to exercises 16.4 and 16.5 about female student height and preferred classroom seating location. Sample sizes, means, and standard deviations by seating location are as follows:

Location	N	Mean	Std. Dev.
Front	38	63.9	2.09
Middle	93	64.8	2.80
Back	22	66.6	2.76
All	153	64.8	

  a. Show calculations verifying that SS Groups is approximately 102.

  b. Use the formula $\text{SS Error} = \sum_{\text{groups}} (n_i - 1)s_i^2$ to verify that SS Error is approximately 1043.

  c. The value of df for error is 150. Find the value of MSE, and find the value of the pooled standard deviation s_p.

16.15 Suppose that SS Groups = 0 in an analysis of variance. What would this indicate about the sample means?

16.16 Thirty male college students were randomly divided into three groups of 10, and the groups received different doses of caffeine (0, 100, and 200 mg). Two hours after consuming the caffeine, each participant tapped a finger as rapidly as possible, and the number of taps per minute was recorded. The data are displayed in a comparative boxplot in the following figure (*Data source:* Hand et. al., 1994, dataset 50).

  a. Write null and alternative hypotheses for comparing the mean taps per minute in the three caffeine amount groups. Use proper notation.

  b. Using the figure, evaluate the necessary conditions for doing a one-way analysis of variance to compare the mean taps per minute for the three caffeine amount groups.

EXERCISE 16.16 **Boxplot of finger taps per minute by caffeine amount**

16.17 Refer to the previous exercise about finger tapping and caffeine amount. Here is a partial analysis of variance table for the problem with some elements intentionally omitted:

Source	df	SS	MS	F	p
Caffeine	—	61.40	—	—	0.006
Error	=	—	—		
Total	29	195.50			

  a. Write out the completed analysis of variance table. Indicate how you determined a value for each element not shown in the table.

  b. Calculate the pooled standard deviation, s_p. Describe what parameter is estimated by this statistic.

  c. Based on the p-value given in the ANOVA table, what conclusion can be made about the mean taps per minute in the population for the three caffeine amounts?

16.18 Refer to exercises 16.9 and 16.10 about testing the flammability of children's sleepwear. Assuming that the equal standard deviations assumption is valid, calculate 95% confidence intervals for each of μ_1, μ_2, μ_3, μ_4, and μ_5. The output for exercise 16.10 shows the sample means and the pooled standard deviation.

16.19 Suppose that four soil treatments that might be useful for improving the yield of alfalfa grown in fields are compared in an experiment. Each treatment is used in six fields, and the response variable is the crop yield per acre. Suppose further that SS Groups = 150 and SS Error = 200.

  a. Write null and alternative hypotheses for comparing the mean crop yields in the four groups. Use proper notation.

  b. Calculate the F-statistic for comparing the mean yield of the four treatments.

  c. What are the degrees of freedom for the F-statistic?

  d. The correct p-value is about 0.01. Draw a sketch illustrating how the p-value is found.

  e. Based on the p-value given in part (d), what conclusion can be made about the mean yields for the four soil treatments?

16.20 Refer to the previous exercise. Use the information given or determined in that exercise to write out a completed analysis of variance table.

16.21 Suppose that three drugs used to reduce cholesterol are compared in a randomized experiment in which three people use each drug for a month. The data for the reductions in cholesterol level for the $N = 9$ participants are

Drug 1	Drug 2	Drug 3
6	10	9
4	14	12
2	9	6

  a. Calculate the overall mean, $\bar{x}$, and the group means, $\bar{x}_1$, $\bar{x}_2$, and $\bar{x}_3$.

  b. Calculate SS Groups, the sum of squares for groups.

  c. Calculate SS Total, the total sum of squares.

  d. Determine SS Error, the sum of squared errors.

  e. Calculate the F-statistic for comparing the mean cholesterol reductions for the three drugs. What are the degrees of freedom for this statistic?

Section 16.3

16.22 Refer to exercise 16.3 about hours of watching television in a typical day and the highest educational degree attained for respondents in the 1993 General Social Survey. Minitab output for a Mood's Median Test of the same data is as follows:

```
Mood median test for tvhours

Chi-Square = 138.16    DF = 4    P = 0.000

degree      N<=     N>    Median
Bachelor    182     70    2.00
Graduate     94     24    2.00
HighScho    409    425    3.00
JunCol       62     34    2.00
NotHS        90    201    4.00

Overall median = 2.00
```

a. In the context of this situation, write the null and alternative hypotheses that are tested by the Mood's Median Test.
b. What is the p-value for testing the hypotheses written in part (a)? Explain what conclusion can be made about the hypotheses.
c. Which group had the highest median hours of television watched in a typical day? What was the median value for that group?
d. The overall median for the sample is 2 hours of watching television in a typical day. What percent of respondents with a bachelor's degree said they watch television more than 2 hours in a typical day? What percent of those who did not get a high school degree watch more than 2 hours of television in a typical day?

16.23 A sample of $n = 729$ college students is asked to rate how much they like various types of music on a scale of 1 to 6, where 1 = don't like at all and 6 = like a lot. The students also were asked whether their hometown was a big city, rural, a small town, or suburban. Results for a Kruskal–Wallis test comparing ratings of reggae music for the four types of hometown are given in the output for this exercise.
a. Write null and alternative hypotheses for comparing the four hometowns with regard to the rating of reggae music.
b. What p-value is given in the output? Explain what conclusion can be made based on this p-value.
c. For which type of hometown was the median rating of reggae the highest? What was the median rating in that type?

```
Kruskal–Wallis Test on Reggae

Hometown      N    Median   Ave Rank      Z
Big city     89    5.000     476.2      5.32
Rural        96    3.000     335.2     -1.49
Small town  176    3.000     360.3     -0.34
Suburban    368    3.000     348.2     -2.18
Overall     729               365.0

H = 29.17    DF = 3    P = 0.000
```

16.24 In the same survey described in the previous exercise, students were asked how many times they pray per week and what they felt was the importance of religion in their life (very important, fairly important, or not very). Descriptive statistics comparing number of times praying per week for the three religious importance groups are

Importance	N	Mean	Median	StDev
Very	169	9.57	7	8.64
Fairly	312	4.76	3	2.15
Not very	252	0.77	0	1.83

Why would it be preferable to compare the prayer frequency for three religious importance groups using a nonparametric technique rather than analysis of variance? (*Hint:* Which of the necessary conditions for using an F-statistic to compare means are violated by these data?)

16.25 A sample of $n = 235$ college students was asked, "Consider how important a person's personality and looks (attractiveness) are to you. Rate this importance on a scale of 1 (personality is most important) to 25 (looks are most important)."
a. Explain whether the response is a quantitative variable, a categorical variable, or an ordinal variable.
b. Results for a Mood's Median Test that compares the responses of men and women are shown in the Minitab output given on the next page. What are the null and alternative hypotheses of the Mood's Median Test in this instance? Make your answer specific to this situation.
c. What is the overall median for all responses? What percent of the women gave an answer that was less than or equal to the overall median? What percent of the men gave an answer that was less than or equal to the overall median?
d. What conclusion can be made about the difference between men and women with regard to the relative importance of looks versus personality? Explain how you made this decision, and be specific about the populations to which you think the conclusion applies.

16.26 Refer to the previous exercise. The p-value is given as P = 0.000, and the value of the chi-square statistic is given as Chi-Square = 15.05. Explain the connection between the p-value and the chi-square statistic.

16.27 Refer to exercise 16.11 about women's testosterone levels. Results are shown in the accompanying output for a Kruskal–Wallis test done to compare the testosterone levels of the three groups of women.

```
group       N    Median   Ave Rank      Z
Adv Degr   24    3.400     30.1      3.46
No Adv D   11    2.200     17.6     -1.67
Not Empl   11    2.000     15.1     -2.38
Overall    46               23.5

H = 12.19   DF = 2   P = 0.002
H = 12.21   DF = 2   P = 0.002 (adjusted for ties)
```

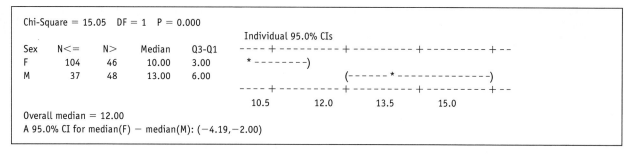

```
Chi-Square = 15.05   DF = 1   P = 0.000

                                         Individual 95.0% CIs
Sex   N<=    N>    Median   Q3-Q1   ---- + --------- + --------- + --------- + --
F     104    46    10.00    3.00    * --------)
M     37     48    13.00    6.00                      (------ * -------------)
                                    ---- + --------- + --------- + --------- + --
                                        10.5       12.0      13.5      15.0

Overall median = 12.00
A 95.0% CI for median(F) − median(M): (−4.19,−2.00)
```

EXERCISE 16.25

a. Write null and alternative hypotheses for the Kruskal–Wallis test. Make your answer specific to this situation.

b. Based on the output, what conclusion can be made about testosterone levels in the three groups? Justify your answer using information given in the output.

c. What are the sample medians for the three groups? Based on these medians, and the appearance of the dotplot in the figure for Exercise 16.11, describe the difference among the three groups.

Section 16.4

16.28 For each scenario explain whether an interaction is described or not. If there is an interaction, what are the two variables that interact?

a. For women, there was no difference between the mean hours spent studying per week for members and nonmembers of Greek organizations. For men, there was a five-hour difference between mean hours spent studying per week for members of Greek organizations versus nonmembers.

b. The difference between the mean survival times for the two treatments was about six months in each patient age group.

16.29 Explain whether the following statement is an example of a main effect or an interaction: The mean number of classes missed per week was significantly lower for students who think religion is very important than it was for students who think religion is either fairly or not very important.

16.30 Exercise 16.23 described a survey in which college students were asked to rate various types of music on a 1 to 6 scale where 1 = don't like and 6 = like a lot. The figure for this exercise is an interaction plot showing how the mean rating of rock music relates to hometown and gender.

a. Which gender liked rock music better?

b. What is the evidence of a possible main effect of hometown?

c. Discuss whether you think there is evidence of a possible interaction between gender and hometown.

16.31 The figure for this exercise shows how mean hours slept the previous night is related to gender and preferred classroom seat location for a sample of $n = 173$ college students (*Data source:* **UCDavis** dataset).

a. For males, describe how mean hours slept the previous night is related to preferred classroom seat location.

b. For females, describe how mean hours slept the previous night is related to preferred classroom seat location.

c. Explain why the plot is evidence of a possible two-way interaction between gender and seat location.

16.32 Give an example not given in this chapter in which two-way analysis of variance would be used to make a comparison of means. Be specific about what the response variable of interest is, what the explanatory variables are, and what specific levels or categories of the explanatory variables are used.

16.33 In Example 16.1 about mean GPA and typical classroom seat location, the finding was that there were differences among the mean grade point averages for the three lo-

EXERCISE 16.30

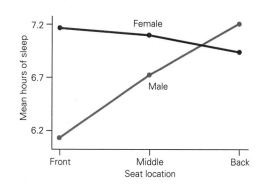

EXERCISE 16.31

cations. Suppose that the sex of the students is also considered along with the seat location.

a. What would be indicated about the difference in the mean GPA of men and women if there were an interaction between student gender and seat location?

b. What would be indicated about the difference in the mean GPA of men and women if there were not an interaction between student gender and seat location?

16.34 Students in a statistics class gave information on their height and weight and also reported their perception of their weight (about right, underweight, overweight). The mean body mass index, BMI, is shown in the following table, along with sample size, for each combination of student gender and perception of weight category. (BMI is calculated as Weight in kilograms/(Height in meters)^2.)

Mean Body Mass Index by Gender and Perception of Weight

	Perception of Weight		
	Underweight	*About Right*	*Overweight*
Females	19.3 ($n = 7$)	21.1 ($n = 107$)	24.4 ($n = 31$)
Males	20.1 ($n = 14$)	23.9 ($n = 56$)	27.8 ($n = 13$)

a. Draw an interaction plot of the means. Figures 16.13 (p. 581) and 16.14 (p. 582) may provide guidance for how to do this.

b. Based on the given data, do you think gender and perception of weight are interacting variables? Briefly explain why you think this.

16.35 Ryan and Joiner (2001) give data collected in an experiment done to examine the effects of storage temperature on the rot that occurs in stored potatoes. Potatoes were injected with bacteria known to cause potato rot, and three different bacteria amounts (low, medium, and high) were used. For each bacteria amount, half of the injected potatoes were stored at 10°C, while the others were stored at 16°C. The response variable was the diameter (mm) of the rot in a potato after being stored. The mean rot diameters for all combinations of bacteria amount and temperature are shown in the following table. For each combination, the sample size is 9.

Mean Diameter of Potato Rot by Storage Temperature and Bacteria Amount

	Temperature	
Bacteria	*10°C*	*16°C*
Low	3.6	7.0
Medium	4.8	13.6
High	8.0	19.6

a. Draw an interaction plot of the means. Figures 16.13 (p. 581) and 16.14 (p. 582) may provide guidance for how to do this.

b. What is the evidence that there may be an interaction between bacteria amount and storage tempera-

ture? Briefly describe the pattern of this possible interaction.

Chapter Exercises

16.36 Give an example not already given anywhere in this chapter in which one-way analysis of variance could be used to make a comparison of means. Be specific about what the response variable of interest is, and what groups are compared.

16.37 Refer to Figure 16.1 for Example 16.1 about GPA by classroom seat location. Explain why the 95% confidence intervals for the mean GPA in the three seat locations have unequal widths.

16.38 Students in grades 3 to 8 in 11 public schools in Ohio were asked how often they engage in violent behaviors and how often they watch television (Singer et al., 1998). A violent behaviors score was calculated by adding frequency scores (0 = never, 3 = almost every day) for five different violent acts: threatening, hitting before being hit, hitting after being hit, beating up, and attacking with knife. For five groups determined by self-reported hours of television, the sample size, sample mean, and sample standard deviation are

Daily Hours of Television	n_i	$\bar{x}_i$	s_i
<1	227	2.94	2.73
1–2	526	2.59	2.44
3–4	666	2.87	2.36
5–6	310	3.10	2.45
>6	488	4.03	2.81

a. Write null and alternative hypotheses for comparing the mean violent behaviors score for the five television-watching groups. State the hypotheses in words, and also state them using mathematical notation.

b. What do you think are the populations to which the results of an analysis of variance F-test would apply?

c. The researchers did a one-way analysis of variance and reported the results as $F = 22.95$, $p < 0.001$. What does this result indicate about the null and alternative hypotheses for this problem?

d. Notice that the highest mean violent behaviors score is in the group that watches the most television. Explain why this cannot necessarily be interpreted to mean that watching a lot of television causes violent behavior in children. What is another explanation for the observed result?

16.39 Refer to the previous exercise.

a. Based on the information given, which of the necessary conditions for doing a one-way analysis of variance appear to be satisfied? What is the evidence of this?

b. For children who watch television less than 1 hour per day, use the Empirical Rule (see Section 2.7) to estimate the interval in which the middle 95% of scores would fall if the data were bell-shaped. Considering that 0 is the lowest possible violent behaviors score, explain why the result of your calculation is evidence that the data either are skewed or contain outliers.

c. Is the difficulty noted in part (b) also present in the other television-watching groups? Show calculations that support your answer.

16.40 Refer to the previous two exercises. The investigators reported results separately for boys and girls in the five television-watching groups. In this two-way classification, the approximate observed mean violent behaviors scores were

	Hours of Daily Television Watching				
	<1	*1–2*	*3–4*	*5–6*	*6+*
Boys	3.7	3.0	3.4	3.4	4.1
Girls	2.3	2.3	2.4	2.8	3.9

Explain whether there is an interaction between sex and amount of television watching or not. If there is an interaction, describe the pattern of the interaction.

16.41 Refer to Example 16.4 about the top speeds of supercars. The accompanying output shows Fisher's procedure for pairwise comparisons, with a 95% confidence level for each individual confidence interval. Interpret the output. Specifically, use the output to describe the statistically significant differences among the mean top speeds of the cars. Look back at Example 16.4 to see which were the fastest and slowest cars.

Minitab v. 14 update on CD

```
Intervals for (column level mean) − (row level mean)

              Acura    Ferrari    Lotus    Porsche

Ferrari     −21.593
            −12.541

Lotus        −5.209    11.857
              3.843    20.909

Porsche     −18.976    −1.909   −18.293
             −9.924     7.143    −9.241

Viper       −11.243     5.824   −10.559    3.207
             −2.191    14.876    −1.507   12.259
```

REFERENCES

16.42 At a private university in southern California, 334 freshman (197 males and 137 females) reported their frequency of drinking alcoholic beverages and also completed a test of personality and psychological characteristics (Ichimaya and Kruse, 1998). One response variable was the score on a 12-item test of conscientiousness for which a high score indicates dependability. Summary statistics for three groups defined by frequency of binge drinking (five or more drinks in one session) are

Drinking Group	n_i	$\bar{x}_i$	s_i
Nonbinge	162	32.59	7.15
Occasional Binge	67	29.93	6.43
Frequent Binge	105	30.10	6.45

a. Write null and alternative hypotheses for comparing the mean conscientiousness score for the three groups. Write the hypotheses in words, and also write the null hypothesis using proper statistical notation.

b. What do you think are the populations to which the results of an analysis of variance F-test would apply?

c. The researchers did a one-way analysis of variance and reported the results as $F = 5.95$, $p < 0.0005$. What does this result indicate about the null and alternative hypotheses for this problem? (*Note:* The p-value reported by the researchers is incorrect. The correct p-value is $p = 0.003$).

d. Using the reported sample means, describe the differences among the mean conscientiousness scores for the three groups. Explain whether you think any observed differences have practical significance or not.

e. Is it reasonable to assume that population standard deviations are equal? Why or why not?

16.43 Refer to the previous exercise. Use formulas in Section 16.2 to compute SS Error, MSE, and the pooled standard deviation, s_p.

16.44 Refer to exercise 16.16 about finger tapping and amount of caffeine consumed. Notice that caffeine amount is a quantity, so it could be analyzed as a quantitative variable. What statistical technique other than one-way analysis of variance could be used to analyze the relationship between taps per minute and caffeine amount? What would be an advantage of using that technique to analyze this relationship? (*Tip:* It might help to look again at the boxplot for exercise 16.16.)

Dabbs, J. M., D. de LaRue, and P. Williams (1990). "Testosterone and occupational choice," *Journal of Personality and Social Psychology,* Vol. 59, No. 6, pp. 1261–1265.

Hand, D. J., F. Daly, A. D. Lunn, K. J. McConway, and E. Ostrowski (1994). *A Handbook of Small Data Sets,* London: Chapman and Hall.

Ichimaya, M. A., and M. I. Kruse (1998). "The social contexts of binge drinking among university freshmen," *Journal of Alcohol and Drug Education,* Vol. 44, pp. 18–33.

Kitchens, L. J. (1998). *Exploring Statistics: A Modern Introduction to Data Analysis and Inference,* 2nd ed., Boston: Duxbury Press.

Koopmans, L. H. (1987). *An Introduction to Contemporary Statistical Methods,* 2nd ed., Boston: Duxbury Press.

Rind, B., and P. Bordia (1996). "Effect on restaurant tipping of male and female servers drawing a happy face on the backs of customers' checks," *Journal of Social Psychology,* Vol. 26 (3), pp. 215–225.

Ryan, B., and B. L. Joiner (2001). *MINITAB Handbook,* Fourth Edition., Belmont: Duxbury Press.

Salk, L. (1973). "The role of the heartbeat in the relations between mother and infant," *Scientific American,* May, pp. 262–264.

Singer, M. J., K. Slovak, T. Frierson, and P. York (1998). "Viewing preferences, symptoms of psychological trauma, and violent behaviors among children who watch television," *J. Am. Acad. Child Adolesc. Psychiatry,* Vol. 37, No. 10, pp. 1041–1049.

CHAPTER 17

Additional Discrete Random Variables

Will the teacher ever call on you back there? See Example 17.7.

Andrew Lichtenstein/CORBIS

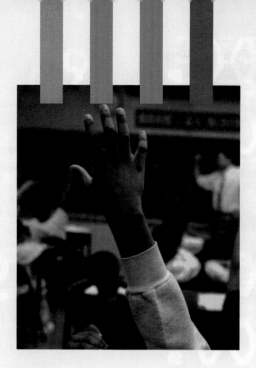

What does the number of women giving birth in a hospital in a week have to do with the number of customers entering a store in a different town that same week? They are both examples of *Poisson random variables*. Recognizing when a random variable fits a known profile allows complex probabilities to be computed easily.

n Chapter 8 we learned about discrete random variables, including the special case of binomial random variables. Remember that a *random variable* assigns a number to each possible outcome of a random circumstance or to each unit in a population. A *discrete random variable* is one for which there are a countable number of possibilities, rather than a continuous set.

We will use the notation used in Chapter 8 for discrete random variables, in which X is the random variable, k is a number the random variable could take, and $P(X = k)$ is the probability that X does indeed take the value k. There is an additional complication for the random variables introduced in Section 17.3, because multiple outcomes are possible. ◆

17.1 | HYPERGEOMETRIC DISTRIBUTION

Suppose there are 100 passengers on a flight from New York to Los Angeles. Among the 100 passengers there are 15 members of a certain ethnic group, say group E. If five passengers are randomly selected for additional security screening, what is the probability that none of them is of ethnicity E? What is the probability that all of them are of ethnicity E? One? Two? If the random variable X = number of screened passengers who are of ethnicity E, then X is called a *hypergeometric random variable*.

> **definition** Consider a population consisting of N individual units in which a units are of Type A and $b = N - a$ units are of Type B. A random sample of n individual units is selected. The random variable
>
> X = number of Type A in sample
>
> is called a **hypergeometric random variable**. X can take any value from 0 to the smaller of a and n. We also say that X has a **hypergeometric distribution**.

In general, consider a situation in which there are two types of individuals in a population and a random sample of size n is taken. We are interested in the probability associated with the number of one type that ends up in the sample.

Survey and Poll Results Are Hypergeometric Random Variables
In Chapter 8 we discussed surveys from a large population and defined the random variable X = number in the sample with a certain trait or opinion. We treated X as a binomial random variable but noted that if the sample was taken without replacement, then this treatment was an approximation. Technically, X is a hypergeometric random variable. If the sample size n is small compared to the population size N, then probabilities derived from the two approaches will be very similar.

EXAMPLE 17.1
***Random Security
Screening***

Continuing the example at the beginning of this section, suppose a flight consists of 100 passengers of whom 15 are of ethnicity E and the remaining 85 are not. If five passengers are randomly selected for screening, and X = number of ethnicity E selected for screening, then X is a hypergeometric random variable. The possible values of X are 0, 1, 2, 3, 4, and 5. Using the notation in the definition, $N = 100$, $n = 5$, $a = 15$, and $b = 85$. X can take any value from 0 to the smaller of a and n, which in this instance is $n = 5$. ◆

EXAMPLE 17.2
***Betting Birthdays
for the Lottery***

In Example 9.6 in Chapter 9 we discussed the Pennsylvania Cash 5 lottery game and the fact that some people use a strategy of betting numbers based on birthdays. Players select five numbers from the integers 1 to 39, and the grand prize is won by anyone who selects the same five numbers randomly drawn for that play of the lottery. Suppose someone selects only from the numbers 1 to 31, which are possible birthdays. What is the probability that at least one of the five winning numbers will be outside that range? Define X = number of winning numbers that are 32 or higher. Notice that this scenario fits the hypergeometric framework, with $N = 39$, $n = 5$, $a = 8$ (the numbers 32 to 39), and $b = 31$. ◆

The Hypergeometric Probability Distribution

Remember that the *probability distribution function* (pdf) for a discrete random variable X is a table or rule that assigns probabilities to the possible values of the random variable. The pdf for a hypergeometric random variable is a rule that depends on the population size, the sample size, and the number of each type in the population. It is based on counting the total number of possible samples of size n that can be constructed out of the N units in the population and then determining the number of those for which $X = k$ for each number k. The formula is long and involves factorials, but it is not difficult to compute. Remember that for any integer m, "m-factorial" is $m! = m \times (m - 1) \times (m - 2) \cdots \times 1$. By convention, $0! = 1$.

The notation for the number of ways to select n items out of N is $\binom{N}{n}$. The formula is $\binom{N}{n} = N!/[n!(N - n)!]$. Rather than explain the general formula, let's consider a simple example. Suppose a customer at a coffee bar has five coins in his pocket—a penny, nickel, dime, quarter, and dollar—and chooses two at random without replacement to put into a tip jar. How many combinations of two coins are possible? The formula is written and solved as follows:

$$\binom{5}{2} = \frac{5!}{2!(5-2)!} = \frac{5 \times 4 \times 3 \times 2 \times 1}{(2 \times 1)(3 \times 2 \times 1)} = \frac{5 \times 4}{2 \times 1} = 10$$

Notice that in computing the answer from the formula, you can cancel out part of the numerator with one of the two parts of the denominator. You can easily list the ten possibilities in this case as {penny and nickel, penny and dime, etc.} to convince yourself that there are indeed ten combinations.

Hypergeometric Probability Distribution Function

formula   A population consists of N units, with a of Type A and b of Type B. A random sample of n units is selected, and X = number of Type A in the sample. Then X has a **hypergeometric probability distribution** and the probability distribution function is

$$P(X = k) = \frac{\binom{a}{k}\binom{b}{n-k}}{\binom{N}{n}}$$

$$= \frac{\left(\dfrac{a!}{k!(a-k)!}\right)\left(\dfrac{b!}{(n-k)!(b-n+k)!}\right)}{\left(\dfrac{N!}{n!(N-n)!}\right)} \quad \text{for } k = 0, 1, \ldots, \text{ smaller of } (a, n)$$

EXAMPLE 17.1—
CONTINUED
Random Security Screening

What is the probability that if five passengers out of 100 are selected for security screening, they will all be of the ethnic group that has only 15 members? Remember that in this example of a hypergeometric random variable, $N = 100$, $n = 5$, $a = 15$, and $b = 85$. Therefore, the pdf of X = number of ethnic group E in the sample is

$$P(X = k) = \frac{\binom{15}{k}\binom{85}{5-k}}{\binom{100}{5}}$$

$$= \frac{\left(\dfrac{15!}{k!(15-k)!}\right)\left(\dfrac{85!}{(5-k)!(80+k)!}\right)}{\left(\dfrac{100!}{5!(95)!}\right)} \quad \text{for } k = 0, 1, 2, 3, 4 \text{ or } 5$$

Specific values are shown in the table below:

k	0	1	2	3	4	5
$P(X = k)$	.43568	.40341	.13775	.02158	.00154	.00004

Notice that the probability is only .00004 that all five screened passengers would be from this ethnic group if the sample were randomly selected. Thus, if that were to happen, the passengers might legitimately consider that the selection was not random. ◆

EXAMPLE 17.2—CONTINUED
Betting Birthdays
for the Lottery

In Chapter 9 we examined data from the Pennsylvania Cash 5 lottery game. We found that for 1560 games played between April 23, 1992, and May 18, 2000, there were 444 games, which is 28.46%, for which all five winning numbers were contained in the integers from 1 to 31. We can now find the theoretical probability that all five winning numbers will be in that range. Define X = number of winning numbers that are 32 or higher. Then X is a hypergeometric random variable with $N = 39$, $n = 5$, $a = 8$, and $b = 31$. So we can find $P(X = 0)$, the probability that none of the winning numbers are 32 or higher, as

$$P(X = 0) = \frac{\left(\dfrac{8!}{0!(8!)}\right)\left(\dfrac{31!}{(5)!(26)!}\right)}{\left(\dfrac{39!}{5!(34)!}\right)} = .295$$

Notice that this differs slightly from the observed proportion of .285, but the observed proportion of .285 is well within the range of sampling variability we would expect in a sample of 1560 games. Thus, for people who select their choices of numbers from the integers 1 to 31, the probability is only .295 that the winning numbers selected by the state will make it possible for them to win. The probability is $1 - .295 = .705$ that at least one of the winning numbers will be greater than 31. ◆

Using Minitab to Compute Hypergeometric Probabilities

 tech note Hypergeometric probabilities can be computed using Minitab in Version 13 or higher.

- Use **Calc > Probability Distributions > Hypergeometric.**
- Specify values for the population size (N), the number of Type A in the population (labeled "Number of Successes in Population" in Minitab), and the sample size (n).
- If the values of k are in a column, specify the column; otherwise, click **Input Constant** and enter the desired value of k.
- To find $P(X \leq k)$ select **Cumulative Probability.**

Using Excel to Compute Hypergeometric Probabilities

tech note To use Excel to find the probability that $X = k$ for a hypergeometric random variable with notation as described in this Section, use HYPGEOMDIST(k,n,a,N). For example, if a population of 100 includes 15 members of ethnic group E, then the probability that a random sample of five will include two members of that ethnic group is found as HYPGEOMDIST(2,5,15,100) = .13775.

Mean, Variance, and Standard Deviation for a Hypergeometric Random Variable

Remember that the mean of a discrete random variable is not necessarily a realizable value, but it is what the long-run average would be if values for the random variable were chosen over and over again. A synonym for "mean" for a random variable is "expected value."

As we will see, the mean for a hypergeometric random variable makes logical sense. For example, suppose that 40% of a population opposes the death penalty, and a jury of 12 people is selected at random from this population. If X = number on the jury who oppose the death penalty, then X is a hypergeometric random variable with $n = 12$, N = population size, and a = 40% of the population size. What is the expected value of X in this situation?

If the jury members were selected with replacement or from a very large population, then X would be a binomial random variable. Remember that the "expected value" for a binomial random variable is np. So the expected value for the number on the jury who oppose the death penalty would be $np = 12(.4) = 4.8$, close to 5 out of the 12. But clearly, for a jury, the individuals are selected without replacement, so X is a hypergeometric random variable. It makes sense that the expected value would still be close to $12(.4)$, and indeed it is. We will return to this example after stating the general formula for the mean (expected value) of a hypergeometric random variable. We also give formulas for the variance and standard deviation.

Mean (Expected Value), Variance, and Standard Deviation for a Hypergeometric Random Variable	**formula** A population of size N has a units of Type A and b units of Type B. Define $p = a/N$ to be the proportion of Type A units in the population. A random sample of size n is selected and X = number of Type A in the sample. The mean (expected value), variance, and standard deviation for the hypergeometric random variable X are

$$Mean: E(X) = \mu = \frac{na}{N} = np$$

$$Variance: \sigma^2 = \frac{nab(N - n)}{N^2(N - 1)} = np(1 - p)\frac{(N - n)}{(N - 1)}$$

$$Standard\ deviation: \sigma = \sqrt{np(1 - p)\frac{(N - n)}{(N - 1)}}$$

Notice that the mean is the same as it would be for a binomial random variable in this setting. Thus, the expected number of jurors on a jury of 12 who oppose the death penalty, when 40% of the population do so, is $np = 12(.4) = 4.8$, the same as it would be if jurors were chosen from the population with replacement.

Let's compare what happens in general when sampling with and without replacement. Consider a population of size N in which $p = a/N$ is the proportion with a certain trait (or opinion). Suppose we take a random sample of n units and define the random variable X = number in the sample with the trait. Then, depending on whether sampling is done with or without replacement, the characteristics of X are as follows:

Sampling	X	Mean	Variance
With Replacement	binomial	np	$np(1 - p)$
Without Replacement	hypergeometric	np	$np(1 - p)\dfrac{(N - n)}{(N - 1)}$

Notice that the means are identical. The variance is slightly smaller if sampling is without replacement. This is obvious because the hypergeometric variance includes an extra term that must be less than one because $(N - n) < (N - 1)$, so $(N - n)/(N - 1) < 1$. If N is large compared with n, then the variances are essentially equal as well. That is the situation for most surveys, where we ignore the fact that sampling is done without replacement and treat X as if it were a binomial random variable instead of a hypergeometric random variable. For instance, for a population of 1,000,000 and a sample of 1,000, the extra factor $(N - n)/(N - 1)$ is 0.9990. The square root, the factor by which the standard deviation is multiplied, is even closer to 1.0, at 0.9995.

17.2 | POISSON DISTRIBUTION

For binomial and hypergeometric random variables, interest is in the number of one type of outcome in a fixed number of trials or individuals. In contrast, a **Poisson random variable** X describes the number of a particular type of random occurrence in fixed time or space. Here are some examples of random variables X that are approximately Poisson random variables:

1. $X =$ the number of defects in the paint job on a new car.
2. $X =$ the number of phone calls received by a help line for a product in a day.
3. $X =$ the number of women giving birth in a given hospital in a week.
4. $X =$ the number of typographical errors in a new book.
5. $X =$ the number of customers entering a small gift shop from 9 A.M. to 10 A.M.

A few explanations are in order. The events in question should be independent; thus for the third example we count a "birth" as a single birth, twins, triplets, and so on. In other words, twins do not count as two births. For the last example, we define a "customer" as a single customer or a group of customers who enter the store as a unit.

Finding Probabilities for Poisson Random Variables

Probabilities associated with a Poisson random variable are completely determined by a single parameter, which is usually denoted by μ because the parameter also happens to be the mean for a Poisson random variable. The only requirement for μ is that it be greater than 0. For example, consider the random variable $X =$ number of customers entering a small gift shop from 9 A.M. to 10 A.M. on a randomly selected day. Suppose that the long-run average number of customers during that hour is 3 per day. Then we can use the probability distribution function for a Poisson random variable with parameter $\mu = 3$ to find probabilities for $X = 0, 1, 2$, etc.

> *definition* The **probability distribution function (pdf)** for a Poisson random variable X is
>
> $$P(X = k) = \frac{\mu^k e^{-\mu}}{k!} \qquad \text{for } k = 0, 1, 2, \ldots$$
>
> where $\mu > 0$ is the long-run average for X. If X is a random variable with those probabilities, we say that X has a **Poisson distribution.**

The letter e represents a number for which the natural log is 1; the value of e is approximately 2.7183. The "exponential function" for a number x is e^x and is sometimes written as $\exp(x)$. It is available on most scientific calculators, with a button that may be written as e^x, meaning that if the number x is entered in the calculator and the button is pressed, the number returned is $(2.7183)^x$. Therefore, to find $e^{-\mu}$ using a calculator, simply enter the value for $-\mu$ and press the exponential function button. As an example, for $x = 3$, $\exp(-3) = e^{-3} = .0498$.

An Infinite Number of Possibilities

tech note   Notice that there is no upper limit on the possible values for a Poisson random variable X. In theory, X can be any integer from 0 to infinity. However, the probability of actually getting very large values of X is small, where "very large values" depends on the value of μ. For instance, when $\mu = 3$, as in the next example, $P(X \geq 11) = .0003$.

EXAMPLE 17.3
Customers Entering a Small Shop

Suppose that a gift shop opens at 9 A.M. and that, on average, three customers visit the shop during the first hour it's open. What is the probability that no customers will come in during the first hour? What is the probability that five customers will come in during the first hour? What is the probability that at least one customer visits the store in the first hour? If X is the number of customers entering during that hour, then X is a Poisson random variable with parameter $\mu = 3$, and the pdf for X is

$$P(X = k) = \frac{3^k e^{-3}}{k!} = \frac{3^k (.0498)}{k!} \qquad \text{for } k = 0, 1, 2, \ldots$$

It is easy to see that $P(X = 0) = .0498$, because $3^0 = 1$ and $0! = 1$. (Any number raised to the 0 power is one.) This means there is about a 5% chance that no customers will enter the store during the first hour. The probability that five customers come during the first hour is $P(X = 5) = 3^5 (.0498)/5! = .1008$, or about 10%. We can find the probability that at least one customer comes into the store during the first hour by using the complement rule (Rule 1) from Chapter 7:

$$P(X \geq 1) = 1 - P(X = 0) = 1 - .0498 = .9502$$

Thus, there is about a 95% chance that someone (at least one customer) will visit the store during the first hour. ◆

Finding Poisson Probabilities Using Minitab

tech note   ◆ Use **Calc>Probability Distributions>Poisson.**

◆ Specify the mean, which is the parameter μ.

◆ If the values of k are in a column, specify the column; otherwise, click **Input Constant** and enter the desired value of k.

◆ To find $P(X \leq k)$ select **Cumulative Probability.**

Finding Poisson Probabilities Using Excel

tech note ◆ $P(X = k)$ is calculated with the command POISSON(k,μ,false) where μ is the parameter for the Poisson random variable. The "false" indicates that we want the probability of exactly k occurrences.

◆ $P(X \le k)$ is calculated with the command POISSON(k,μ,true). The "true" indicates that we want the cumulative probability of k or fewer occurrences.

Mean, Variance, and Standard Deviation for a Poisson Random Variable

The mean of a Poisson random variable X is the parameter μ used in the formula for the pdf. This result can be verified using the definition of the mean (expected value) for a discrete random variable. If you revisit the definitions of expected value and variance for discrete random variables in Chapter 8, you will find that the summations required to find the mean and variance for a Poisson random variable are easy to write down. The calculations are beyond the level of treatment provided here, but if you are experienced at solving sums of infinite series, you should have little difficulty verifying these results.

Mean, Variance, and Standard Deviation for Poisson Random Variables

formula If X is a Poisson random variable with parameter μ, then the mean, variance, and standard deviation are

$$\textit{Mean:}\quad E(X) = \sum_{k=0}^{\infty} kP(X = k) = \sum_{k=0}^{\infty} k\frac{\mu^k e^{-\mu}}{k!} = \mu$$

$$\textit{Variance:}\ V(X) = \sum_{k=0}^{\infty} (k - \mu)^2 P(X = k) = \sum_{k=0}^{\infty} (k - \mu)^2 \frac{\mu^k e^{-\mu}}{k!} = \mu$$

$$\textit{Standard deviation} = \sqrt{\text{Variance}} = \sqrt{\mu}$$

Note that the mean and variance are equal—and are equal to the parameter μ.

Poisson Processes

One way in which a Poisson random variable occurs is as the result of a **Poisson process.** A Poisson process with rate λ is a series of events occurring at random over time, with the following properties:

1. The probability that exactly one event occurs in a time period of duration d, where d is small, is λd and is the same for all time periods of that length.
2. The probability that two or more events occur in a small time period of duration d is essentially 0.
3. The numbers of events that occur in nonoverlapping time intervals are independent of one another.

These conditions are somewhat simplified versions of the precise mathematical assumptions required. A more precise treatment is given in most books covering the theory of probability, for example, Casella and Berger (2002).

If events follow a Poisson process with rate λ, and if X is the number of events in a time period t, then X is a Poisson random variable with parameter $\mu = \lambda t$. In other words, once we know the rate at which events occur, we can find the average μ for any time period of interest.

EXAMPLE 17.4
Earthquakes in the Coming Year

Suppose the probability is .006 that an earthquake of at least magnitude 4.0 occurs in a particular region on any given day, the probability of two or more earthquakes on a given day is essentially 0, and occurrences are independent from one time period to the next. How many earthquakes are likely to occur over a one-year period? The occurrence of earthquakes is approximated by a Poisson process with rate of .006 per day, or (.006)(365) = 2.2 per year. Therefore, the random variable X = number of earthquakes in the next year is a Poisson random variable with $\mu = 2.2$. ◆

EXAMPLE 17.5
Emergency Calls to a Small Town Police Department

Suppose that the emergency calls for help to a small-town police department occur at the rate of one every three days, on average. What is the probability that there will be no calls on a given day? Two or more calls in a given day? The calls can be approximated by a Poisson process with rate $\lambda = 1$ call per 3 days. Therefore, the random variable X = number of calls in one day is a Poisson random variable with parameter $\mu = 1/3$, because the average number of calls per day is 1/3. The pdf for X is

$$P(X = k) = \frac{(1/3)^k e^{-1/3}}{k!} = \frac{(1/3)^k(.7165)}{k!} \qquad k = 0, 1, 2, \ldots$$

It is then easy to find probabilities for specific values. For instance, $P(X = 0)$ = .7165, so on about 71.65% of all days there will be no emergency calls. $P(X = 1)$ = (1/3)(.7165) = .2388, so on about 24% of the days there will be one call. $P(X \geq 2) = 1 - P(X = 0 \text{ or } 1) = 1 - (.7165 + .2388) = 1 - .9553 = .0447$, so there is about a 4.5% chance that the department will receive two or more calls on any given day. ◆

Using Poisson Random Variables to Approximate Binomial Random Variables

In Chapter 8 we learned that the normal distribution could be used to approximate binomial random variables. The approximation required that the binomial parameters n and p be such that both np and $n(1 - p)$ are at least 10. It was most useful for finding cumulative probabilities over a range of values of interest, such as the probability of passing a multiple-choice exam when just guessing.

A different approximation is available if n is large, p is small, and np is no more than about 7, which is useful for finding $P(X = k)$ for a binomial random variable X. For instance, suppose the border patrol between two countries knows that the probability that any given car will be hiding illegal drugs is 1/10,000. What is the probability that none of the next 5,000 cars to cross the border will be hiding illegal drugs?

Poisson Approximation for Binomial Random Variables

> ***formula***   If X is a binomial random variable based on n trials with probability of success p, and n is large, p is small, and np is less than 7, then X is approximately a Poisson random variable with parameter $\mu = np$. Therefore,
>
> $$P(X = k) \approx \frac{(np)^k e^{-np}}{k!} \qquad \text{for } k = 0, 1, 2, \ldots, n$$

EXAMPLE 17.6
Are There Illegal Drugs in the Next 5,000 Cars?

Consider the scenario in which there are illegal drugs in 1/10,000 cars that cross at a border. What is the probability that there will be no drugs hidden in the next 5,000 cars that cross? Let X = number of cars with illegal drugs out of the next 5,000 cars. Then X is a binomial random variable with $n = 5000$ and $p = 1/10,000$. What is $P(X = 0)$? Using the binomial pdf, the probability is $(9,999/10,000)^{5000} = .6065$, an answer many calculators could provide. However, X can also be approximated as a Poisson random variable with parameter $\mu = np = (5,000)(1/10,000) = 1/2$. Thus, $P(X = 0)$ can be approximated as $e^{-1/2}$, which also is .6065. The approximation is very accurate. The approximation is particularly useful for larger values of k. For instance,

$$P(X = 4) = [5000!/(4! \times 4996!)](.9999)^{4996}(.0001)^4 \approx (.5)^4 e^{-.5}/4! = .0016$$

The probability that exactly four cars out of the next 5,000 will be carrying illegal drugs is .0016. ◆

17.3 | MULTINOMIAL DISTRIBUTION

In Chapter 8 we defined a binomial experiment to be one in which there were n independent trials with two possible outcomes on each trial. In this section, we consider an extension of that setting in which there are more than two possible outcomes on each trial.

> ***definition***   A **multinomial experiment** is defined by the following conditions:
>
> 1. There are n "trials" where n is determined in advance and is not a random value.
> 2. There are k possible outcomes on each trial.
> 3. The outcomes are independent from one trial to the next.
> 4. The probabilities associated with outcomes 1, 2, . . . , k remain the same from one trial to the next. The probabilities are denoted by p_1, $p_2, \ldots, p_k$ where $p_1 + p_2 + \cdots + p_k = 1$.

Notice that these are the same conditions used to define a binomial experiment in Chapter 8, except that the number of possible outcomes is k instead of two. In fact, a binomial experiment is the special case of a multinomial experiment with $k = 2$.

definition Consider a multinomial experiment with n trials and k possible outcomes. Let $X_1, X_2, \ldots, X_k$ be the numbers of trials in the experiment with outcomes $1, 2, \ldots, k$, respectively. Then the random variables $X_1, X_2, \ldots, X_k$ have a **multinomial probability distribution.** The joint probability distribution function is

$$P(X_1 = a_1, X_2 = a_2, \ldots, X_k = a_k) = \frac{n!}{a_1! a_2! \cdots a_k!} p_1^{a_1} p_2^{a_2} \cdots p_k^{a_k}$$

where $a_1 + a_2 + \cdots + a_k = n$.

Notice that the multinomial probability distribution function is called a **joint probability distribution function,** which in general is a rule, table, or formula that gives probabilities for a specific combination of several discrete random variables.

Relationship Between Binomial and Multinomial Distributions

tech note As noted earlier, a multinomial experiment with $k = 2$ is simply a binomial experiment. But any multinomial experiment can be reduced to a binomial experiment when interest is in just one particular outcome, say outcome m. In that case, a "success" is outcome m and a "failure" is any other outcome. The random variable X_m is a binomial random variable with success probability p_m. In fact, each X_i for $i = 1, 2, \ldots, k$ is a binomial random variable. Thus, we know from Chapter 8 that

$E(X_i) = np_i$

Variance $(X_i) = np_i(1 - p_i)$

Standard deviation $= \sqrt{np_i(1 - p_i)}$

EXAMPLE 17.7
Calling On the Back of the Class

Suppose that in a large class, 20% of the students are characterized as sitting in the front of the class, 60% in the middle, and 20% in the back. The students in the back are reluctant to answer questions, so the instructor decides to randomly call on students to answer. He has a computerized list of students, and before class begins he randomly selects five names, with replacement, to call on that day. What is the probability that none of the students called upon will be in the front, two will be in the middle, and three will be in the back of the class?

This situation fits the description of a multinomial experiment with $n = 5$, $p_1 = .2, p_2 = .6$, and $p_3 = .2$. The desired probability is

$$P(X_1 = 0, X_2 = 2, X_3 = 3) = \frac{5!}{(0!)(2!)(3!)} (.2)^0 (.6)^2 (.2)^3 = \frac{5 \times 4}{2} (.36)(.008) = .0288$$

Notice that if interest is only in the probability that three students called upon will be in the back of the class, without regard for where the other two sit, then the situation is a binomial experiment. The desired probability is $P(X = 3)$ where X is a binomial random variable with $n = 5$ and $p = .2$. The solution in that case is

$$P(X = 3) = \frac{5!}{(2!)(3!)} (.8)^2 (.2)^3 = .0512$$

This solution could also be found using the multinomial formulation, but it would require adding probabilities for all combinations of X_1 and X_2 that summed to 2. This idea is covered in the exercises for Section 17.3. ◆

KEY TERMS

Section 17.1

hypergeometric random variable, 593

hypergeometric distribution, 593

hypergeometric probability
distribution, 595

Section 17.2

Poisson random variable, 598

Poisson distribution, 598

Poisson process, 600

Poisson approximation for binomial
random variables, 601–602

Section 17.3

multinomial experiment, 602

multinomial probability
distribution, 602

joint probability distribution
function, 603

EXERCISES   Blue = Basics   **Bold, Bold** = Answer in back

Section 17.1

17.1 Five hundred new students are moving into the dormitories at a university. The 300 women and 200 men will be randomly assigned to dorms. Grace is a continuing student, and 60 of the new students will be assigned to her dorm. She is interested in X = number of women among the 60 new students assigned to her dorm. Is X a hypergeometric random variable? If so, specify values for N, n, and a. If not, explain why not.

17.2 A club has sold 250 raffle tickets, with 100 sold to members and 150 to nonmembers. Five winning tickets will be drawn without replacement. X = the number of winning tickets held by club members. Is X a hypergeometric random variable? If so, specify values for N, n, and a. If not, explain why not.

17.3 A police officer observes cars arriving at a stop sign at a rural intersection in which there is little traffic. The probability is .7 that a randomly selected car will stop at the stop sign, independent from one car to the next. The officer observes 50 cars and X = the number that stop. Is X a hypergeometric random variable? If so, specify values for N, n, and a. If not, explain why not.

17.4 In a state with an adult population of 10 million people, 40% of the adults would answer "yes" if asked whether they support the current governor for re-election. A poll is taken in which 1,000 people are randomly selected without replacement. X = the number in the poll who say they would support the governor. Is X a hypergeometric random variable? If so, specify values for N, n, and a. If not, explain why not.

17.5 A jury of 12 is randomly selected from 100 potential jurors, of whom 60 are men and 40 are women. Specify N, n, and a for the hypergeometric random variable X = number of women on the jury.

17.6 Refer to the previous exercise. What are the possible values for X?

17.7 An experiment is conducted by randomly assigning 16 volunteers to receive either a vitamin or a placebo, with 8 assigned to each treatment. The goal is to see if the vitamin helps prevent the flu. Suppose the vitamin is not effective, and 5 of the 16 volunteers will get the flu in any case. Specify N, n, and a for the hypergeometric random variable X = number of volunteers in the vitamin group who get the flu. (*Hint:* Define Type A units to be the individuals who will get the flu in any case.)

17.8 Refer to the previous exercise. What are the possible values for X?

17.9 A lottery game is played in which players select six numbers from the integers from 1 to 51. The grand prize is won by anyone who selects the same six numbers randomly drawn for that play of the lottery. Suppose someone only selects his or her choices from the integers 1 to 31. What is the probability that all of the six winning numbers will be in the range 1 to 31?

17.10 Refer to the description of the Pennsylvania Cash 5 lottery game in Example 17.2. A player selects five numbers from the integers 1 to 39. Define the player's five numbers as "Type A" numbers and the remaining 34 numbers as "Type B." Define X = number of winning numbers among the five Type A numbers.

  a. X is a hypergeometric random variable. Specify N, n, a, and b.

  b. Find $P(X = 0)$, the probability that the player does not match any of the winning numbers.

  c. Find $P(X = 1)$.

  d. If a player matches two numbers he or she wins $1.00, which is the cost of the ticket. Find $P(X = 2)$.

  e. A player needs to match at least two numbers to win any prize. Find the probability that the player wins any prize.

17.11 A class has 10 girls and 12 boys. The teacher selects four students to participate in a special activity and claims she selected them at random. All four are girls. What is the probability that all four students selected would be girls if they were selected at random?

17.12 Give an example of an actual situation (such as those specified in Examples 17.1 and 17.2) for which a random variable X would be hypergeometric with $N = 25$, $n = 5$, and $a = 15$.

17.13 For each of the following scenarios, find the mean and variance for X. Note that X may be a binomial random variable or a hypergeometric random variable.

  a. A survey is done by selecting 1000 adults from a state with a population of 8 million, in which 38% believe in life after death. X = number in the sample who believe in life after death.

  b. Customers entering a store buy something with probability .25, and this behavior is independent from one customer to the next. X = number out of the next 100 customers who will buy something.

  c. A candy company claims that 20% of the candies it manufactures are blue. A customer buys a bag containing 200 pieces of candy and X = number of blue candies in the bag.

d. Matthew and his younger sister Maria are given a bag of candy to divide. There are five colors of candy in the bag of 200, including 40 red pieces. Matthew, who likes the red candy, offers to divide the 200 pieces by giving them 100 each. X = number of red pieces Maria will receive if Matthew randomly divides the candy without regard for color.

17.14 Refer to part (d) of the previous exercise. Maria notices that there are only five red candies in her 100 pieces. Find the probability that Maria would receive five or fewer red pieces if Matthew divided them randomly. (*Note:* You should use computer software to do the computations for this exercise.)

Section 17.2

17.15 The number of accidents occurring on a particular road in a month is approximately a Poisson random variable with $\mu = 1$. Find probabilities for the following numbers of accidents in the next month.
  a. None.
  b. At least one.
  c. Exactly two.

17.16 The number of pieces of mail Debbie receives in a day is a Poisson random variable with $\mu = 5$. One day she arrives home and finds that she has not received any mail. She suspects that the mail was not delivered that day. What is the probability that she would receive no mail by chance?

17.17 Refer to the previous exercise. Mail is delivered six days per week. Let X = number of pieces of mail Debbie receives in a week. What is the distribution of X, including the value(s) of any parameters?

17.18 A Poisson random variable can take on any integer value from 0 to infinity, which is an infinite number of values. Is a Poisson random variable discrete or continuous? Explain.

17.19 In the past, an average of ten new cases of cancer per year has been diagnosed among employees of a large organization. The organization moves to a large new office building. In the next year, 20 new cases of cancer are diagnosed. An employee organization suspects that there is a problem with the building. If chance alone is responsible, and the rate has not changed from the past, what is the probability that 20 or more new cases of cancer would be diagnosed in a year? (*Hint:* Use software to find an appropriate cumulative probability, and then use the complement rule.)

17.20 The number of people who will be bitten by a poisonous snake during the summer in a national park has a Poisson distribution with $\mu = 3$. Find probabilities for the following numbers of snake bites in a summer.
  a. None.
  b. At least one.
  c. More than three.

17.21 Refer to the previous exercise. There are three months in the summer—June, July, and August. Assume the distribution of snakebites is similar across the three months. Define the random variable X to be the number of bites that will occur in June.

  a. What is the distribution of X? Include value(s) for any parameter(s).
  b. Suppose antivenom is in short supply and the park has only three doses for the month of June. What is the probability that there will be more than three snakebites that month?

17.22 For each of the following situations, X is a Poisson random variable. Give the value of the parameter μ in each case.
  a. The number of births per day in a hospital is a Poisson random variable with $\mu = 2$. X = number of births in April.
  b. Alan regularly takes a 2-hour hike in the woods. The number of mosquito bites he gets during the hike is a Poisson random variable with $\mu = 4$. X = number of bites he gets during the first hour of his next hike.
  c. The number of calls for appointments a massage therapist gets in a day is a Poisson random variable with $\mu = 3$. The therapist will be gone on vacation for four days. X = number of calls she gets during the time she is gone.
  d. The number of junk e-mail messages Lauren gets during the work week (Monday to Friday) is a Poisson random variable with $\mu = 7$. X = number of junk e-mail messages she will get next Tuesday.

17.23 Give the mean, variance, and standard deviation for a Poisson random variable X with each of the following values of μ.
  a. $\mu = 4$.
  b. $\mu = 0.25$.
  c. $\mu = 25$.

17.24 Refer to the previous exercise. In each case, give $P(X = 0)$.

17.25 Suppose a random variable X is observed 100 times and the sample mean and variance of the 100 observations are $\bar{x} = 15$ and $s^2 = 1$. Explain how you know that it is unlikely that X is a Poisson random variable.

17.26 For each of the following binomial random variables, specify whether the "Poisson approximation for binomial random variables" can be used. If not, explain why not. If so, specify the value of the parameter μ.
  a. $n = 100, p = .05$.
  b. $n = 10, p = .5$.
  c. $n = 1000, p = .1$.
  d. $n = 1000, p = .001$.

17.27 A candy manufacturer produces bags of 200 pieces of candy. The firm knows that about 1% of all candies are malformed. Assume the 200 pieces that are packaged in each bag constitute a random sample of pieces. Let X = number of malformed candies in a randomly chosen bag.
  a. X is a binomial random variable. What are the values of n and p?
  b. Explain how the Poisson approximation for binomial random variables would be used in this situation.
  c. Using the binomial distribution, find $P(X = 0)$.
  d. Using the Poisson distribution, find the approximate $P(X = 0)$ and compare it to your answer in part (c).

17.28 A large hotel has 400 rooms. The hotel knows that there is a .01 chance that a customer who has reserved a room and guaranteed it for payment will not show up. They book 403 rooms for an upcoming convention.

Define X = number of booked customers who do not show up.

a. Assume that whether or not one booked customer shows up is independent of whether any others show up. X has a binomial distribution. What are the values of n and p?

b. X is approximately a Poisson random variable. What is the value of the parameter μ?

c. For what range of values of X will the hotel *not* be oversold?

d. Use the Poisson approximation to find the probability that the hotel will not be oversold.

Section 17.3

17.29 A fair die is rolled 30 times. Define the random variable X_i to be the number of times i dots show on the top face, $i = 1, 2, \ldots, 6$. Go through the conditions for a multinomial experiment and explain how this scenario fits. Specify values for the p_i.

17.30 Refer to the previous exercise. Write out the formula (but do not solve it) for the probability that each face lands as the top face five times.

17.31 Refer to the previous two exercises. Give the mean and variance for each of the random variables X_i. (They all have the same mean and variance.)

17.32 A fair die is rolled until each possible face (1 dot, 2 dots, etc.) shows as the top face at least five times. Is this a multinomial experiment? Explain why or why not.

17.33 A sociologist has observed that drivers approaching a set of multiple tollbooths tend to cluster at the ones in the center. To test this theory, she observes 200 cars approach a set of seven tollbooths and she observes how many cars line up at each one. Is this a multinomial experiment? Explain why or why not.

17.34 There are three viable candidates in an upcoming primary election. A polling agency asks a random sample of 1000 adults in the United States which of the three candidates they would vote for if the election were held today. Is this a multinomial experiment? Explain why or why not. (Assume the population is large enough that it is as if sampling is done with replacement.)

17.35 Refer to the previous exercise. Candidate A wants to know what proportion of voters support her. Define the random variable X to be the number of voters who say they would vote for her. What distribution does the random variable X have?

17.36 Refer to Example 17.7, in which five students are randomly selected, with replacement, to answer questions during class. The random variables X_1, X_2, and X_3 were defined as the number of students selected from the front, middle, and back of the class, respectively. One probability, $P(X_1 = 0, X_2 = 2, X_3 = 3) = .0288$, was found. It was also shown that if X = number of students selected from the back is defined, then X is a binomial random variable and $P(X = 3) = .0512$.

a. Find $P(X_1 = 1, X_2 = 1, X_3 = 3)$.

b. Find $P(X_1 = 2, X_2 = 0, X_3 = 3)$.

c. Show how the result $P(X = 3) = .0512$ can be found using the multinomial probabilities given or found in this exercise.

REFERENCE

Casella, George, and Roger L. Berger (2002). *Statistical Inference,* 2nd ed., Belmont, CA: Brooks/Cole.

Nonparametric Tests of Hypotheses

Will the highest ranked wines come from here? See Example 18.6.

Robert Glusic/Photodisc Red/GettyImages

Most nonparametric tests are based on simple counting and ranking procedures. Generally, nonparametric procedures are resistant to the influence of outliers and are robust, which means they are valid over a broad range of circumstances.

A distinguishing feature of a **nonparametric test** is that it can be used without having to assume a specific type of distribution for the measurements in the population(s) being studied. The validity of a *nonparametric* method, for instance, does not depend on the assumption that the response variable has a normal distribution, as is the case for the *t*-test procedures described in Sections 13.2 and 13.3 for testing hypotheses about either one or two population means. Thus, almost by definition, nonparametric methods are **robust,** which means they are valid over a broad range of circumstances.

Most nonparametric tests are based on simple counting and ranking procedures. As an example, suppose we want to know whether the *median* amount spent on textbooks last year by students at a university was $700 (a null value) or greater than $700 (an alternative hypothesis). We could simply count how many students in a random sample spent more than $700. If the population median actually was $700, then by the definition of a median, roughly half of the sample would have spent more than $700 (and the other half would have spent less). If "significantly" more than half of the sample spent over $700, it would be evidence that the population median was greater than $700. We'll learn how to determine a *p*-value for this type of situation in Section 18.1, where we describe the *sign test.* An important point to note is that the procedure we use to find the *p*-value is appropriate for any continuous random variable, not just normal random variables.

Generally, nonparametric procedures are **resistant** to the influence of outliers. The number of students who spent over $700 won't be unduly influenced by outliers. If somebody had the misfortune to have spent $2,000, they are merely counted as a student who spent more than $700, just as somebody who spent $750 would be. Because the test statistic (how many spent over $700) does not use the specific amounts, it is resistant to the influence of outliers.

The term *nonparametric test* originated because the test statistic in such a test does not depend upon sample estimate(s) of parameter(s) in a population distribution. In contrast, recall the one-sample *t*-test procedure described in Section 13.2 for testing hypotheses about a population mean. The one-sample *t*-test is a *parametric* test. The *t*-statistic involves $\bar{x}$ and *s*, the sample mean and sample standard deviation. These are estimates of the parameters μ and σ, the mean and standard deviation of the population distribution, which is assumed to be a normal curve in the one-sample *t*-test problem. ◆

18.1 | THE SIGN TEST

The **sign test** can be used to test hypotheses about the *population median* when the response is a *continuous* variable. (Recall from Chapter 8 that a continuous random variable is one for which any value within some interval is a possible outcome.) We will use the symbol η ("eta") to represent the population median. Thus the null hypothesis for a sign test could be written as

$$H_0: \eta = \eta_0 \quad \text{(population median equals a specified value)}$$

The alternative hypothesis can be one-sided (either $H_a: \eta < \eta_0$ or $H_a: \eta > \eta_0$) or two-sided ($H_a: \eta \neq \eta_0$). When the alternative hypothesis is one-sided, the null hypothesis can be written to include an inequality in the opposite direction.

If the response of interest is a discrete random variable (a variable with a countable set of possible outcomes), the hypotheses must be written differently, particularly when the number of possible outcomes is small. We will cover that situation later in this section. For now we assume that the response variable is continuous. In that case, we are using the *sign test* to test hypotheses about the value of the population median.

As an example, we could use the sign test to test whether the median amount spent on textbooks last year at a university was $700 (or less) or greater than $700. The null and alternative hypotheses in this situation are

$$H_0: \eta = 700 \quad \text{(population median equals 700)}$$
$$H_a: \eta > 700 \quad \text{(population median is greater than 700)}$$

The null hypothesis could also be written as $H_0: \eta \leq 700$.

Most commonly, the sign test is used to analyze paired data, and this application gives the test its name. The usual null hypothesis for the difference between the paired measurements is that the median difference is 0 ($H_0: \eta = 0$). Suppose, for example, that a sample of college men reports their actual and desired weights. For each man, we could find *difference = actual − desired*, the difference between the two responses. This difference will be greater than 0 (positive) when a man wants to lose weight; it will be less than 0 (negative) when a man wants to gain weight. To consider the null hypothesis that the population median difference is 0, we could count the number of positive differences and the number of negative differences in the sample. If the null hypothesis were true, these counts should each be roughly equal to one-half of the sample size.

Finding the *p*-Value for a Sign Test

In a sign test of the null hypothesis that a population median equals a specified value (the null value), S^+ = number of observations in the sample greater than the null value can be used as a **test statistic.** If the null hypothesis is true, from the definition of a median, it follows that $p = .5$ would be the probability that any randomly selected value is greater than the null value. For a random sample of n values, the statistic S^+ = number of values greater than the null value has a binomial distribution (covered in Section 8.4) with parameters n and $p = .5$ when the null hypothesis is true. Thus a p-value for the sign test can be determined using the binomial distribution to find the probability that S^+ would be as "extreme" as it is (or more extreme) in the direction of the alternative hypothesis.

**Finding the
p-Value for a Sign Test**

formula   Suppose the response variable is continuous, η = population median, the null hypothesis is $H_0: \eta = \eta_0$, and a random sample of n observations is available. Let

S^+ = number of values in the sample greater than η_0

S^- = number of values in the sample less than η_0

$n_U = S^+ + S^-$, the number of values in the sample not equal to η_0

Finding an Exact p-Value
Define Y to be a binomial random variable with parameters n_U and $p = .5$. Using S^+ as the test statistic, we can find an exact p-value for the sign test as follows:

◆ For $H_a: \eta < \eta_0$, the p-value $= P(Y \le S^+)$.
◆ For $H_a: \eta > \eta_0$, the p-value $= P(Y \ge S^+)$.
◆ For $H_a: \eta \ne \eta_0$, the p-value $= 2 \times$ [smaller of $P(Y \le S^+)$ and $P(Y \ge S^+)$].

Finding an Approximate p-Value
The normal approximation for finding binomial probabilities can be used to approximate the p-value for sufficiently large sample sizes. When the null hypothesis is true, the following z-statistic has approximately a standard normal distribution.

$$z = \frac{S^+ - (n_U/2)}{\sqrt{n_U/4}}$$

With this z-statistic, an approximate p-value for a sign test can be found using the standard normal curve (Table A.1) as follows:

◆ For $H_a: \eta < \eta_0$, the p-value $= P(Z < z)$ = the area below z.
◆ For $H_a: \eta > \eta_0$, the p-value $= P(Z > z)$ = the area above z.
◆ For $H_a: \eta \ne \eta_0$, the p-value $= 2 \times P(Z > |z|) = 2 \times$ area above $|z|$.

**EXAMPLE 18.1
*Normal Human
Body Temperature***

Example 13.1 of Chapter 13 described a paper published in the *Journal of the American Medical Association* that presented evidence that normal human body temperature might actually be less than 98.6 degrees Fahrenheit, the long-held standard. Figure 18.1 is a dotplot of the (imaginary) random sample of $n = 18$ normal body temperatures given in Example 13.1.

FIGURE 18.1   Normal body temperature for $n = 18$ individuals

Letting η = population median normal human body temperature, we'll use the sample shown in Figure 18.1 to test

H_0: $\eta = 98.6$
H_a: $\eta < 98.6$

From the dotplot, it's easy to determine that only three data values are above 98.6, so $S^+ = 3$, whereas 13 data values are below 98.6 degrees, so $S^- = 13$. Notice that relatively few observations are above 98.6, compared to the number below 98.6. This is evidence that the population median might actually be less than 98.6. Minitab can be used to do a sign test, and the output for this example follows:

Sign test of median = 98.60 versus < 98.60

	N	Below	Equal	Above	P	Median
Bodytemp	18	13	2	3	0.0106	98.20

The *p*-value is 0.0106 (given under "P" in the output). This is below the usual 0.05 standard for significance, so we can reject the null hypothesis and conclude that the population median normal human body temperature is below 98.6 degrees Fahrenheit.

The *p*-value is the probability that S^+ would be as small as 3 (or smaller), computed assuming the population median actually is 98.6. The number of observations used for the sign test in this situation is $n_U = S^+ + S^- = 16$. Two observations equal 98.6, but a convention of the sign test is that values tied with the null value are not used. If the null hypothesis is true, the probability is $p = .5$ that any randomly selected value will be above 98.6 (by definition of a median). Thus the *p*-value is the probability that a binomial random variable with parameters $n_U = 16$ and $p = .5$ will have a value less than or equal to 3 (the observed value of S^+). We learned from the output that the answer is 0.0106. ◆

Using Minitab to Do a Sign Test

 tech note   For paired data it will be necessary to first calculate a column of differences between the paired measurements.

- Use **Stat>Nonparametrics>1-Sample Sign.**
- In the box labeled "Variables," specify the column that contains the data.
- Click the *Test Median* radio button and enter the null value in the adjacent box.
- Select the desired alternative hypothesis.

EXAMPLE 18.2
Heights of Male Students and Their Fathers

Do male college students tend to be taller than their fathers? The **UCDavis2** dataset on the CD for this text includes student height (inches) and father's height (reported by the student) for $n = 76$ male students in an elementary statistics class. To compare the heights of sons and fathers, we can compute *difference = student height − father's height* for each of the 76 men in the sample. The sample of differences is graphed in Figure 18.2. One notable feature of the plot is the outlier. One student reported that he is 37 inches taller than his father! A positive difference (>0) occurs when the student is taller than his father, and a negative difference occurs when the student is shorter. Notice in Figure 18.2 that more values fall above 0 than below.

0   10   20   30
Difference in heights (in.)

FIGURE 18.2   **Difference between student height and father's height for** $n = 76$ **college men**

It's believed that humans are gradually getting taller from one generation to the next, so let's use the sign test to test the following:

H_0: $\eta = 0$ (population median difference = 0)
H_a: $\eta > 0$ (population median difference is greater than 0)

Minitab output for this situation follows:

```
Sign test of median = 0.00000 versus > 0.00000

                 N    N*    Below    Equal    Above         P    Median
Difference      76    10       21       11       44    0.0032     1.000
```

The *p*-value given in the output is 0.0032 (under "P"), so we can reject the null hypothesis and conclude that the population median difference is greater than 0. In particular, this indicates that students in the population represented by the sample tend to be taller than their fathers. The sample median is given in the output; the median difference in the sample was 1 inch. We also see in the output that $S^+ = 44$ (number of differences above 0) and $S^- = 21$ (number of differences below 0). The difference was equal to 0 for 11 students. The alternative hypothesis is a "greater than" hypothesis, so the *p*-value was computed as the probability that S^+ would be greater than or equal to 44, assuming the null hypothesis is true. The exact *p*-value can be found as $P(Y \geq 44) = 1 - P(Y \leq 43) = 1 - .997 = .003$, where Y is a binomial random variable with parameters $n_U = 65$ and $p = .5$. A z-statistic could be also be used to find the *p*-value. In this example,

$$z = \frac{S^+ - (n_U/2)}{\sqrt{n_U/4}} = \frac{44 - (65/2)}{\sqrt{65/4}} = 2.85$$

Using this z-statistic, we find the approximate *p*-value to be $P(Z > 2.85) = .0022$. ◆

**Hypotheses for
Discrete Variables**

tech note When the response variable is discrete, the null hypothesis should be written as H_0: $P(X < \eta_0) = P(X > \eta_0)$ where X denotes the response variable. For continuous variables, this is the same as writing that η_0 is the population median, but this is not the case for discrete variables due to the nonzero probability for $P(X = \eta_0)$. For discrete variables, the two possible one-sided alternative hypotheses are

H_a: $P(X < \eta_0) > P(X > \eta_0)$ (X below η_0 is more probable than X above η_0.)

or

H_a: $P(X < \eta_0) < P(X > \eta_0)$ (X below η_0 is less probable than X above η_0.)

The two-sided alternative hypothesis can be written as

H_a: $P(X < \eta_0) \neq P(X > \eta_0)$ (X below η_0 and X above η_0 are not equally probable.)

18.2 | THE TWO-SAMPLE RANK-SUM TEST

The **Wilcoxon rank-sum test,** also known as the **Mann–Whitney test** and sometimes called the Mann–Whitney–Wilcoxon test, is a nonparametric alternative to the two-sample *t*-test for comparing two means described in Section 13.3 of Chapter 13. Usually, we'll refer to this test simply as the two-sample rank-sum test. It can be used to compare two populations when the variable of interest is either quantitative or ordinal and the data are from two independent samples. The hypotheses of interest concern whether or not values in one population tend to be larger than values in the other population. Some examples of research questions that could be addressed using a two-sample rank-sum test follow:

1. Do the resting pulse rates of women tend to be greater than the resting pulse rates of men?
2. Do students who say religion is very important in their life tend to miss fewer classes than students who say religion is not very important?
3. When people rate how much they like rap music on a scale of 1 (don't like) to 6 (like a lot), do individuals from big cities tend to give rap higher ratings than individuals from small towns?

The Null and Alternative Hypotheses for the Two-Sample Rank-Sum Test

The null and alternative hypotheses for the two sample rank-sum test can be stated as

H_0: No difference in the distribution of values in the two populations.
H_a: The values in one population tend to be larger than values in the other.

The alternative will be one-sided if we specify which particular population might tend to have larger values. For instance, a statement that women tend to have higher pulse rates than men would be a one-sided alternative hypothesis.

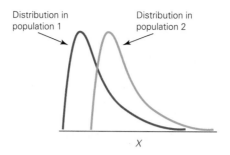

X

FIGURE 18.3   **An example of the assumption of same shape, but possibly different location**

The null and alternative hypotheses can be written as hypotheses about the two population medians if we assume that the response variable is continuous and the two population distributions have the same shape, differing possibly only by a shift of location. Figure 18.3 illustrates this assumption. With this assumption, the null hypothesis can be written as H_0: $\eta_1 = \eta_2$ where η_1 and η_2 denote the medians of the two populations being compared. As usual, the alternative hypothesis may be one-sided (either H_a: $\eta_1 > \eta_2$ or H_a: $\eta_1 < \eta_2$) or two-sided (H_a: $\eta_1 \neq \eta_2$).

The Rank-Sum Statistic

The two-sample rank-sum test is based on ranks assigned to the observed values in the two samples. The **rank** of an observation is its location in the ordered list of data, where the data are ordered from smallest to largest. The rank is 1 for the smallest data value, 2 for the second smallest value, and so on. For instance, the ranks for the values 65, 62, 67 are 2, 1, 3. When two or more observations have the same value, the rank for each observation is the **midrank,** or average, of the lowest and highest ranks that would have been given if the values had not been tied. For example, the ranks for the ordered list of values 60, 65, 65, 67, 70 are 1, 2.5, 2.5, 4, 5, respectively. The two observations equal to 65 are tied for second and third place in the ordered list, so each is given the rank 2.5, the average of 2 and 3.

The **test statistic for the two-sample rank-sum test** is W = sum of the ranks (within the overall dataset) for the observations in the first sample. The procedure for determining W is simple:

1. Combine the data from two samples and assign ranks to the values as 1 = smallest value, 2 = second smallest, and so on.
2. When observations have the same value, assign the *midrank* of the tied values to each of the tied values.
3. Find W = sum of ranks for the values in sample 1.

EXAMPLE 18.3
Estimating the Size
of Canada's Population

Exercise 2.76 in Chapter 2 described an experiment conducted by one of this book's authors in which students were asked to estimate (in millions) the population of Canada, which at that time was about 30 million. Before they made their estimates, ten students (sample 1) were told that the population of the United States was about 290 million at that time. Nine students (sample 2) were told

FIGURE 18.4 **Estimates of Canada's Population**

that the population of Australia was roughly 20 million at that time. Figure 18.4 is a comparative dotplot that compares the responses in the two groups. It's clear that students who were given the United States population tended to give higher estimates of Canada's population than students who were given Australia's population. For instance, the six highest estimates over the whole dataset were made by students who were given the United States population.

The table below shows the data, along with the rank for each data value. To make it easy to see the ranking pattern, we show the data sorted in order from low to high within each sample. The rank for a data value gives the value's location within the 19 values in the overall dataset. Notice, for example, that the largest overall value (200 in the USA sample) has rank 19. Also notice that an estimate of 35 million occurred twice (once in each group); each of those observations is assigned rank 7.5 because they are tied in the seventh and eighth places of the ordered data. A rank-sum statistic for comparing the two groups is the sum of ranks for data in the sample told the U.S. population. This rank sum is $W = 1 + 6 + 7.5 + 12 + 14 + 15 + 16 + 17 + 18 + 19 = 125.5$.

USA Data Value	Rank	Australia Data Value	Rank
2	1	8	2
30	6	12	3
35	7.5	16	4
70	12	29	5
100	14	35	7.5
120	15	40	9
135	16	45	10
150	17	46	11
190	18	95	13
200	19		

◆

Finding the p-Value for a Two-Sample Rank-Sum Test

A few textbooks, most notably those specializing in nonparametric methods, have tables that can be used to assess statistical significance for a two-sample rank-sum test. These tables, however, tend to be limited to small sample sizes, so usually an approximate p-value for the two-sample rank test is found using a standardized z-statistic for W.

Finding the *p*-Value for a Two-Sample Rank-Sum Test

formula Suppose either quantitative or ordinal data have been collected for two independent samples. Define:

n_1 = sample size for sample 1

n_2 = sample size for sample 2

$N = n_1 + n_2$, the total sample size for samples 1 and 2 combined

W = sum of ranks for observations in sample 1 when the overall dataset is ranked

When the population distribution is the same for both populations (the null hypothesis is true), the mean and standard deviation of W are

$$\text{Mean} = E(W) = \mu_W = \frac{n_1(1 + N)}{2}$$

$$\text{Standard deviation} = \sigma_W = \sqrt{\frac{n_1 n_2(1 + N)}{12}}$$

For sufficiently large samples, the following standardized statistic has approximately a standard normal distribution (when the null hypothesis is true):

$$z = \frac{W - \mu_W}{\sigma_W} = \frac{W - [n_1(1 + N)/2]}{\sqrt{(n_1 n_2(1 + N)/12}}$$

Using this z-statistic, an approximate p-value for a Wilcoxon–Mann–Whitney rank-sum test can be found as follows:

Alternative Hypothesis	*p*-Value
Values in population 1 tend to be *greater than* values in population 2.	$P(Z > z)$, area to the right of z
Values in population 1 tend to be *less than* values in population 2.	$P(Z < z)$, the area to the left of z
Values in one population tend to be greater than values in the other, but a specific ordering is not specified (*two-sided*).	$2 \times P(Z > |z|)$, two times the area to the right of absolute z

EXAMPLE 18.3 — *CONTINUED*
***p*-Value for Testing**
Whether Information
Given Affects Estimates
of Canada's Population

The experiment was done to determine whether the information given would influence the estimate of Canada's population. Specifically, it was believed that people who were told the U.S. population would tend to give higher estimates than people who were told the Australia population.

The null and alternative hypotheses about the populations represented by the two samples can be stated as

H_0: No difference in the population distribution of values guessed for the two conditions.

H$_a$: Estimates made by people told the U.S. population tend to be larger than estimates made by people told the Australia population.

To find the p-value we can use the normal curve approximation described above. We found that the sum of ranks for the U.S. sample is $W = 125.5$, and the sample sizes are $n_1 = 10$ (U.S. sample), $n_2 = 9$ (Australia sample), and $N = 10 + 9 = 19$ overall. Details for finding the z-statistic are

$$E(W) = \mu_W = [10(1 + 19)/2] = 100$$

$$\text{Standard deviation} = \sigma_W = \sqrt{\frac{(10)(9)(1 + 19)}{12}} = 12.25$$

$$z = (125.5 - 100)/12.25 = 2.08$$

Because the alternative is a "greater than" hypothesis, the p-value is $P(Z > 2.08)$, the area to the right of $z = 2.08$ under a standard normal curve. Utilizing Table A.1, we find that $P(Z > 2.08) = 1 - P(Z < 2.08) = 1 - .9812 = .0188 \approx .02$. The p-value is below the usual 0.05 standard used for significance, so we can conclude that the experiment is evidence that estimates of Canada's population by people who were told the U.S. population size will tend to be higher than estimates by people who were told the Australia population size. ◆

Using Minitab to Do a Two-Sample Rank-Sum Test

 tech note   The data for the two samples must be in two separate columns.

◆ Use **Stat>Nonparametrics>Mann–Whitney.**

◆ Specify the columns containing the data for samples 1 and 2 in the boxes labeled "Sample 1" and "Sample 2."

◆ Select the desired alternative hypothesis.

EXAMPLE 18.3—*CONTINUED*
Minitab Output for
the Population of
Canada Experiment

In Minitab, the two-sample rank-sum test is called the Mann–Whitney test. The Mann–Whitney test for our estimates of Canada's population follows:

```
USA               N = 10          Median =              110.00
Australi          N =  9          Median =               35.00
Point estimate for ETA1-ETA2 is                          72.50
95.5 Percent CI for ETA1-ETA2 is (1.02,123.01)
W = 125.5
Test of ETA1 = ETA2 vs ETA1 > ETA2 is significant at 0.0206
The test is significant at 0.0206 (adjusted for ties)
```

The last line gives a p-value adjusted for the presence of tied observations in the data, and the next-to-last line gives a p-value that is not adjusted for ties. Here, the two p-values are identical; they rarely will differ by very much. The phrase "is significant at" used in the last two lines can be interpreted as "the p-value is." Notice that the output contains references to the population medians "ETA1" and "ETA2." The next-to-last line includes information about the null hypothesis (ETA1 = ETA2) and alternative hypothesis (ETA1 > ETA2) for the test. ◆

18.3 | THE WILCOXON SIGNED-RANK TEST

The **Wilcoxon signed-rank test** (not to be confused with the Wilcoxon rank-sum test in the previous section) is used to test hypotheses about the median of one population. As with the *sign test* and the *one-sample t-test*, the signed-rank test can be used either to examine a single response variable or to examine the difference between paired measurements. Most often, in practice, the test is applied to paired data. An important necessary condition for using this test is that the response variable have a symmetric (but not necessarily bell-shaped) distribution in the population; the signed-rank test should not be used when the data are skewed.

The specific hypotheses tested are the same as those for the sign test when the response is continuous. Again, we will use the symbol η ("eta") to represent the population median. Thus the null hypothesis for a signed-rank test could be written as

$$H_0: \eta = \eta_0 \qquad \text{(population median equals a specified value)}$$

The alternative hypothesis can be one-sided (either $H_a: \eta < \eta_0$ or $H_a: \eta > \eta_0$) or two-sided ($H_a: \eta \neq \eta_0$). When the alternative hypothesis is one-sided, the null hypothesis can be written to include an inequality in the opposite direction. For paired differences, the null value of interest in most situations is $\eta_0 = 0$ ("no difference").

The Test Statistic for the Wilcoxon Signed-Rank Test

The procedure for determining the test statistic used in the Wilcoxon signed-rank test is as follows:

1. For each data value x_i, record whether the value is below η_0 (a negative difference) or above η_0 (a positive difference). Values equal to η_0 are not used in the test.
2. For each data value x_i, calculate $|x_i - \eta_0|$, the absolute difference between the data value and the null value.
3. Assign ranks to the absolute differences computed in the previous step. (Do not include observations where the difference = 0.)
4. The test statistic is $T^+ = $ sum of ranks for data values above η_0 (the positive differences).

Notice that T^+ will be influenced both by how many observations are above the null value (as in the sign test) and by how far the positive differences are from the null value (not a feature of the sign test).

EXAMPLE 18.4
Calculating T^+ for a Sample of Systolic Blood Pressures

Suppose that we want to use the Wilcoxon signed-rank procedure to test the null hypothesis that the median systolic blood pressure within a particular population is $\eta = 120$, and the available sample of systolic blood pressures is 125, 118, 123, 120, 135, 129, and 117. The following table shows necessary steps for determining the value of T^+ in this situation.

Data Value (x_i)	> 120 or < 120?	$\lvert x_i - \eta_0 \rvert = \lvert x_i - 120 \rvert$	Rank of $\lvert x_i - \eta_0 \rvert$
125	+	5	4
118	−	2	1
123	+	3	2.5
120	equal	0	n/a
135	+	15	6
129	+	9	5
117	−	3	2.5

There are two observations below 120, four observation above 120, and one observation equal to 120. Notice that only the $n_U = 6$ observations not equal to 120 are used in the ranking process. The value 118 is closest to the null value of 120, so it gets rank 1. The value 135 is furthest from 120 so it gets rank 6. $T^+ =$ sum of ranks for the four values above 120 (the positive differences), and its value is $T^+ = 4 + 2.5 + 6 + 5 = 17.5$. ◆

Finding the p-Value for a Wilcoxon Signed-Rank Test

In general, a relatively large value of T^+ may be evidence that $\eta > \eta_0$, and a relatively small value may be evidence that $\eta < \eta_0$. An approximate p-value for evaluating the statistical significance of T^+ can be found using a standardized z-statistic that has approximately a standard normal distribution.

Finding the p-Value for a Wilcoxon Signed-Rank Test

formula Suppose the response variable is continuous and symmetric, $\eta =$ population median, the null hypothesis is H_0: $\eta = \eta_0$, and a random sample of n observations $x_1, x_2, \ldots, x_n$ is available. Let

$T^+ =$ Wilcoxon signed-rank statistic

$n_U =$ The number of values in the sample not equal to η_0, the sample size used for the test

When the null hypothesis is true ($\eta = \eta_0$), the mean and standard deviation of T^+ are

$$\text{Mean} = E(T^+) = \mu_{T^+} = \frac{n_U(n_U + 1)}{4}$$

$$\text{Standard deviation} = \sigma_{T^+} = \sqrt{\frac{n_U(n_U + 1)(2n_U + 1)}{24}}$$

For sufficiently large samples, the following standardized statistic has approximately a standard normal distribution (when the null hypothesis is true):

$$z = \frac{T^+ - \mu_{T^+}}{\sigma_{T^+}} = \frac{T^+ - [n_U(n_U + 1)/4]}{\sqrt{n_U(n_U + 1)(2n_U + 1)/24}}$$

With this z-statistic, an approximate p-value for a Wilcoxon signed-rank test can be found using the standard normal curve (Table A.1) as follows:

◆ For H_a: $\eta < \eta_0$, the p-value = $P(Z < z)$ = the area below z.
◆ For H_a: $\eta > \eta_0$, the p-value = $P(Z > z)$ = the area above z.
◆ For H_a: $\eta \neq \eta_0$, the p-value = $2 \times P(Z > \lvert z \rvert)$ = 2 × area above $\lvert z \rvert$.

EXAMPLE 18.5
Difference Between Student Height and Mother's Height for College Women

The **UCDavis2** dataset on the CD includes student height (inches) and mother's height for $n = 132$ women in an elementary statistics class. For each woman, we can compute *difference = own height − mother's height,* and a histogram of those differences is shown in Figure 18.5. Notice that the majority of observations are above 0, so more students are taller than their mothers rather than are shorter. Also notice that the shape of the histogram is more or less symmetric, so the Wilcoxon signed-rank test could be used to test hypotheses about the population median difference between student height and mother's height.

We'll use the Wilcoxon signed-rank procedure to test

H_0: $\eta = 0$ (population median difference = 0)
H_a: $\eta > 0$ (population median difference is greater than 0)

Minitab output for this example follows:

```
Test of median = 0.000000 versus median > 0.000000

                         N for    Wilcoxon              Estimated
                  N      Test     Statistic     P       Median
Difference       132     120      5021.0       0.000    1.000
```

The *p*-value is given as 0.000 (under "P"). Thus we can reject the null hypothesis and conclude that in the population represented by this sample, college women tend to be taller than their mothers. That is, the population median difference is greater than 0. Notice that the median difference within the sample was 1 inch (given in the last column of output).

The value of T^+ (5021.0) is given under "Wilcoxon Statistic." The sample size used for the test is $n_U = 120$, given under "N for Test." (Twelve of the 132 women shown in the histogram reported no difference between their height and their mother's height.) The *z*-statistic for T^+ is not given in the output. Its value is

$$z = \frac{T^+ - [n_U(n_U + 1)/4]}{\sqrt{n_U(n_U + 1)(2n_U + 1)/24}} = \frac{5021 - [120(121)/4]}{\sqrt{120(121)(241)/24}} = \frac{5021 - 3630}{381.84} = 3.64$$

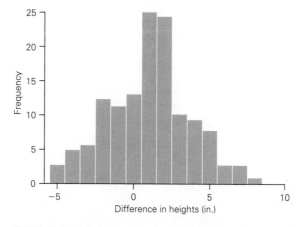

FIGURE 18.5   **Difference between student height and mother's height for $n = 132$ college women**

Because the alternative is a "greater than" hypothesis, the p-value is found as $P(Z > 3.64)$. Using the "In the Extreme" portion of Table A.1 we can determine that this probability is approximately .0001. ◆

Using Minitab to Do a Wilcoxon Signed-Rank Test

> **tech note**   For paired data it will be necessary to first calculate a column of differences between the paired measurements.
>
> ◆ Use **Stat>Nonparametrics>1-Sample Wilcoxon.**
>
> ◆ In the box labeled "Variables," specify the column that contains the data.
>
> ◆ Click the ***Test Median*** radio button and enter the null value in the adjacent box.
>
> ◆ Select the desired alternative hypothesis.

18.4 | THE KRUSKAL–WALLIS TEST

We briefly discussed the **Kruskal–Wallis test** in Section 16.3 of Chapter 16. Here, we provide more details of the test. The Kruskal–Wallis test is used to compare three or more populations when the response variable is quantitative and the data are from independent random samples from the populations being compared. Some examples of questions that could be examined using a Kruskal–Wallis test follow:

1. Are testosterone levels the same for men in four different occupations (teachers, doctors, firefighters, and lawyers)?
2. Is the number of classes missed per week the same for students who say religion is very important, students who say religion is fairly important, and students who say religion is not very important?
3. Students rate how much they like Rap music on a scale of 1 (don't like) to 6 (like a lot). Do the ratings differ by type of hometown (big city, suburban, small town, rural)?

The precise nature of hypotheses for the Kruskal–Wallis test depends on what assumptions, if any, are made about the distribution of the response variable in the populations. In all circumstances, the null hypothesis and alternative hypothesis could be written as

H_0: The distribution of values is the same for all populations.
H_a: Values in at least one population tend to be larger (or smaller) than values in the other populations.

If we can assume that the variable of interest is continuous and that the population distributions all have the same shape (see Figure 18.3 in Section 18.2 for an example), we can express the hypotheses as statements about population medians. In this case, the null and alternative hypotheses are

H_0: $\eta_1 = \eta_2 = \cdots = \eta_k$ (population medians are equal)
H_a: At least one population median differs from the others.

We wrote the hypotheses in this manner in Section 16.3, although we did not state the assumption that the population distributions all have the same shape.

Determining the Test Statistic for the Kruskal–Wallis Test

The value of the test statistic for the Kruskal–Wallis is found using these steps:

1. Assign ranks to all values within the overall dataset (all samples combined). The ranking procedure is the same as it was for the two-sample rank-sum test. The smallest data value is assigned rank 1, the second smallest is assigned rank 2, and so on. Use midranks for tied observations.
2. For each independent sample, find the average rank assigned to observations within that sample. Let $\bar{R}_i$ represent the average rank in sample i.
3. Let n_i = sample size in sample i, N = the total of the sample sizes for all groups, and k = number of independent samples. The Kruskal–Wallis statistic is

$$H = \frac{12}{N(N+1)} \sum_i n_i \left(\bar{R}_i - \frac{1+N}{2} \right)^2$$

$$= \frac{12}{N(N+1)} \left[n_1 \left(\bar{R}_1 - \frac{1+N}{2} \right)^2 + n_2 \left(\bar{R}_2 - \frac{1+N}{2} \right)^2 + \cdots + n_k \left(\bar{R}_k - \frac{1+N}{2} \right)^2 \right]$$

The average rank for all observations in the combined dataset is $(1+N)/2$. Notice that the test statistic is a function of the difference between the average rank in each sample and the overall average rank for the dataset, $(1+N)/2$.

Finding the p-Value for a Kruskal–Wallis Test

When the null hypothesis is true, the test statistic, H, has a chi-square distribution with degrees of freedom = number of samples $-1 = k - 1$. The p-value for a Kruskal–Wallis test is found as the probability that a chi-square variable with $k - 1$ degrees of freedom will have a value greater than the value of H. Table A.5 in the Appendix can be used to approximate the p-value.

EXAMPLE 18.6
Comparing the Quality of Wine Produced in Three Different Regions

A dataset included with the Minitab program gives data for the quality of 38 different Pinot Noir wines produced in three different regions. (The regions aren't specified, nor is the original source of the data.) Several brands of wine were sampled from each region, and each wine was evaluated by a panel of wine experts. The sample sizes for the three regions are 17, 9, and 12, respectively, and the total sample size is $N = 38$.

Based on the expert evaluation, a quality rating score was given to each brand. Figure 18.6 compares the quality ratings for the three regions. One notable feature is that all wines from region 3 were given higher ratings than all wines from region 2.

FIGURE 18.6   **Quality ratings for Pinot Noir wines from three regions**

The data, along with the ranks for data values, follow:

Region 1	Region 2	Region 3
9.1 (3)	7.9 (1)	12.7 (21)
9.8 (5)	8.5 (2)	13.5 (26.5)
10.3 (6)	9.5 (4)	13.6 (28)
11.1 (10)	10.7 (7)	13.8 (29.5)
11.3 (11)	10.8 (8.5)	13.8 (29.5)
11.7 (13)	10.8 (8.5)	14.4 (32)
11.9 (14.5)	11.6 (12)	14.9 (33)
12.0 (16)	11.9 (14.5)	15.1 (34)
12.1 (17)	12.3 (19)	15.5 (35.5)
12.2 (18)		15.5 (35.5)
12.6 (20)		16.1 (37.5)
12.8 (22.5)		16.1 (37.5)
12.8 (22.5)		
13.2 (24)		
13.3 (25)		
13.5 (26.5)		
13.9 (31)		

As an example of the ranking procedure, notice that the smallest data value overall (7.9 in region 2) is assigned rank 1. The smallest data value (9.2) in region 1 is assigned rank 3 because two values in region 2 are smaller. The sum of ranks for the 17 observations in region 1 is 285, so the average rank in region 1 is $\overline{R}_1 = 285/17 = 16.76$. You can verify that $\overline{R}_2 = 76.5/9 = 8.5$ and $\overline{R}_3 = 379.5/12 = 31.625$. Note that the average rank in region 3 is much higher than the average ranks for the other two regions.

The value of the Kruskal–Wallis test statistic is

$$H = \frac{12}{38(38 + 1)} [17(16.76 - 19.5)^2 + 9(8.5 - 19.5)^2 + 12(31.625 - 19.5)^2]$$

$$= 24.13$$

The *p*-value is the probability that a chi-square value with $3 - 1 = 2$ degrees of freedom has a value greater than 24.13. Using Table A.5 (chi-square distribution) in the Appendix, we can estimate that the *p*-value is less than 0.001 because 24.13 exceeds the value in the last column of the df $= 2$ row of the table. In the following Minitab output for this example, Minitab gives the *p*-value as 0.000.

```
Kruskal–Wallis Test on Quality

Region           N         Median       Ave Rank
1               17          12.10          16.8
2                9          10.80           8.5
3               12          14.65          31.6
Overall         38                         19.5

H = 24.13   DF = 2   P = 0.000
H = 24.15   DF = 2   P = 0.000 (adjusted for ties)
```

We barged through this example without explicitly stating the hypotheses, but the conclusion seems clear when we consider Figure 18.6 and the average ranks for three regions. It appears that wines from region 3 tend to have higher quality ratings than wines from the other two regions.

Using Minitab to Do a Kruskal–Wallis Test

 tech note   ◆ Use **Stat>Nonparametrics>Kruskal–Wallis.**

◆ Specify the column containing the response data in the box labeled "Response."

◆ Specify the column containing the group designations in the box labeled "Explanatory."

KEY TERMS

EXERCISES   Blue = Basics   **Bold, Bold** = Answer in back

Section 18.1

18.1 In each situation, explain whether it would be appropriate to use a sign test to analyze the question of interest. If so, write the null and alternative hypotheses in words, and again using proper statistical notation.
   a. Resting pulse rates are measured for $n = 25$ adult women. Is the median pulse rate of women equal to 72 or is it greater than 72?
   b. Are the median heights of 11-year-old girls and 11-year-old boys the same, or are they different?
   c. Blood pressures are measured in the morning and again at night for each individual in a random sample of $n = 100$ college students. On average, are the morning and nighttime blood pressure measurements about equal, or does blood pressure tend to be higher at night?

18.2 A sample of $n = 63$ college men reports their ***actual*** and ***ideal*** (desired) weights (pounds). The difference, computed as *actual* − *ideal*, was positive for 28 of the men, was negative for 19 others, and was equal to 0 for the remaining 16 men. (*Data source:* **idealwtmen** on the CD for the text.)

a. Suppose the data are used to test the null hypothesis that the population median difference is 0 versus the alternative hypothesis that the population median difference is greater than 0. What are the values of S^+, S^-, and n_U?
b. Explain how a *p*-value would be found in this situation. Specifically, what probability should be found, and what are the parameters of the binomial distribution that would be used to determine it?
c. Compute the value of the *z*-statistic that could be used to find an approximate *p*-value for this problem.
d. Find the *p*-value based on the *z*-statistic, and write a conclusion about the hypotheses.

18.3 Twenty individuals each place as many beans as they can into a container during a 30-second time period using their dominant hand and again using their nondominant hand. The order of using the dominant and nondominant hands is randomly determined for each person. Thirteen individuals in the sample placed more beans with their dominant hand, three people placed more beans with their nondominant hand, and four people placed the same number with each hand. Consider a sign test to determine whether the data are statistically significant evidence that, in general, people are able to place more beans with their dominant hand.

a. Write null and alternative hypotheses in words and using statistical notation.

b. What are the values of S^+, S^-, and n_U?

c. Refer to part (a). Carry out a sign test and state a conclusion.

18.4 Suppose that a sample of 16 students at a university is asked about how much they spent on textbooks for the present semester, and the responses (in dollars) are

250, 450, 300, 279, 360, 300, 670, 50, 430, 350, 220, 420, 375, 275, 360, 365

Consider a test of the null hypothesis that the population median amount spent is $400 (or more) versus the alternative hypothesis that the population median is less than $400.

a. Write the null and alternative hypotheses using statistical notation.

b. What is the value of the sample median amount spent?

c. What are the values of S^+, S^-, and n_U?

d. Explain how a p-value would be found in this situation. Specifically, what probability should be found, and what are the parameters of the binomial distribution that would be used to determine it?

e. Carry out a sign test and write a conclusion. Either find an exact p-value using the binomial distribution or use a z-statistic to find an approximate p-value.

18.5 In a survey about music interests done in a statistics class at Penn State, students were asked to rate how much they like various types of music on a scale of 1 (don't like) to 6 (like a lot). The Minitab output given for this exercise is for a sign test for the difference between student ratings of punk and country music (computed as *punk − country*). The output indicates that the test is of whether the median difference is 0 or not 0. Because the response is discrete with few possible values, it is more appropriate to write H_0 as P(difference below 0) = P(difference above 0) and H_a as P(difference below 0) ≠ P(difference above 0).

```
Sign test of median = 0.00000 versus not = 0.00000

                N  Below Equal Above   P    Median
punk − country 738  229   136   373 0.0000 1.000
```

a. How many students said they like punk music more than country music? How many said they like country better than punk? How many gave the two types of music the same rating?

b. What value is given in the output for the p-value of the test? Based on this p-value, what is the appropriate conclusion? Write a conclusion in the context of this situation.

c. Find the value of the z-statistic that could be used to find the p-value given in the output.

18.6 Twelve individuals have their systolic blood pressure measured and recorded at an appointment with a dentist and also have their systolic blood pressure measured at an appointment with a medical doctor. The data are

Person	BP at Dentist	BP at Doctor
1	137	123
2	130	142
3	144	120
4	154	145
5	128	144
6	124	112
7	126	121
8	137	133
9	133	127
10	137	129
11	124	120
12	122	108

Consider the difference in blood pressures measured at the dentist and those measured at the doctor. Carry out a sign test of the null hypothesis that the median difference is 0 versus the alternative hypothesis that the median difference is greater than 0, where the difference is computed as *dentist − doctor*. Show all details and write a conclusion in the context of this situation.

Section 18.2

18.7 The following data are the resting pulse rates of 8 individuals who say they do not exercise (sample 1) and 12 individuals who say they do exercise (sample 2).

Do not exercise: 72, 84, 66, 72, 62, 84, 76, 60
Exercise: 62, 72, 60, 63, 75, 64, 60, 52, 64, 80, 68, 64

a. It is believed that the pulse rates of people who exercise regularly tend to be lower than the pulse rates of people who do not exercise. Using this belief as the alternative hypothesis, write null and alternative hypotheses for a two-sample rank-sum test.

b. Rank the combined data and find W = sum of ranks for those who do not exercise.

c. Find values for μ_W and σ_W, the mean and standard deviation of W, assuming the null hypothesis is true.

d. Calculate the value of the z-statistic that would be used to determine a p-value in this problem, and then find the p-value.

e. State a conclusion about the null and alternative hypotheses in the context of this situation.

18.8 Suppose that 14 overweight individuals are divided randomly into two groups, with 7 in each group. The first group is assigned to use a diet intended to cause weight loss, and the second group is assigned to use an exercise plan to lose weight. After three months, weight losses (pounds) in the two groups are

Diet plan: 10, 8, 12, 16, 0, 7, 35
Exercise plan: 12, 7, 2, 3, 5, 4, 9

a. Find the value of W, the test statistic for a two-sample rank-sum test that would compare the two methods for losing weight. Give the details of how you found W.

b. Carry out a two-sample rank-sum test (Mann–Whitney–Wilcoxon test) in which the alternative hypothesis is that median weight loss is greater if dieting is the weight-loss method used. Show all details and state a conclusion in the context of this situation.

18.9 Refer to data given in the previous exercise. Explain why it would not be advisable to use those data to do a two-sample t-test to compare the mean weight losses for the two methods.

18.10 Case Study 1.1 in Chapter 1 gave data for responses by 189 Penn State students to the question "What's the fastest you have ever driven a car? ____ mph." The data showed that male respondents tended to report faster speeds than females. The following data are responses to the same question for 14 men and 10 women in a different statistics class at Penn State:

Males: 100, 95, 85, 130, 125, 90, 110, 120, 95, 92, 120, 85, 105, 75

Females: 95, 110, 87, 55, 95, 105, 80, 90, 80, 70

Carry out a two-sample rank-sum test in which the alternative hypothesis is that the fastest speeds ever driven reported by men tend to be faster than the fastest speeds ever driven reported by women. State the hypotheses, show all details of the test, and state a conclusion in the context of this situation.

18.11 Example 18.3 described an experiment in which individuals were asked to estimate the population of Canada. Some participants were given information about the population of the United States (condition 1), and others were given information about the population of Australia (condition 2). The same experiment was performed with a different group of participants. Those data were given in exercise 13.77 and are given again here:

Condition 1 ($n_1 = 11$): 20, 90, 1.5, 100, 132, 150, 130, 130, 40, 200, 20

Condition 2 ($n_2 = 10$): 12, 20, 10, 81, 15, 20, 30, 20, 9, 10, 20

a. Find the sample medians for condition 1 (given U.S. population) and condition 2 (given Australia population).

b. Rank the combined set of 21 values, and find $W = $ sum of ranks for condition 1.

c. Using the data given for this exercise, carry out a two-sample rank-sum test (Mann–Whitney–Wilcoxon test) of the same hypotheses tested in Example 18.3. State the hypotheses, show details of finding the p-value, and state a conclusion in the context of this situation.

18.12 The computer output (top of next column) for this exercise gives results of a two-sample rank-sum test (Mann–Whitney–Wilcoxon test) done to compare the median cholesterol level of heart attack patients (measured two days after the heart attack) to the median cholesterol of "control" patients who did not have a heart attack. (*Data source:* **cholest** dataset on the CD for this text.)

Heart Attack	N = 28	Median =	268.00
Control	N = 30	Median =	187.00

W = 1140.0
Test of ETA1 = ETA2 vs ETA1 > ETA2 is significant at 0.0000

a. State the null and alternative hypotheses that were tested. (A "shorthand" version of the hypotheses is given in the output.).

b. State a conclusion about the hypotheses. Justify your answer using information given in the output.

c. Find the values of μ_W and σ_W, the mean and standard deviation of W, assuming the null hypothesis is true.

18.13 In a survey about music interests done in a statistics class at Penn State, students were asked to rate how much they like Top 40 music on a scale of 1 (don't like) to 6 (like a lot). The output for this exercise gives results for a two-sample rank-sum test in which the alternative hypothesis is that women tend to give higher ratings to Top 40 music than men do. This one-sided alternative was used because it was a result observed in prior surveys of statistics classes.

Females	N =448	Median =	5.0000
Males	N =283	Median =	4.0000

W = 183726.5
Test of ETA1 = ETA2 vs ETA1 > ETA2 is significant at 0.0000

a. What were the sample median ratings for women and men, respectively?

b. State a conclusion for the hypothesis test in the context of this situation. Justify your answer using information given in the output.

c. Find the values of μ_W and σ_W, the mean and standard deviation of W, assuming the null hypothesis is true.

d. Calculate the value of a z-statistic that could be used to determine a p-value in this problem.

Section 18.3

18.14 Refer to exercise 18.4, where data are given for the amount spent on textbooks. Suppose a Wilcoxon signed-rank procedure will be used to test the null hypothesis that the population median amount spent is $400 (or more) versus the alternative hypothesis that the population median is less than $400.

a. Find the value of T^+, the Wilcoxon signed-rank test statistic. Show details.

b. Calculate the value of the z-statistic that would be used to determine a p-value in this problem, and then find the p-value.

c. Based on the p-value found in part (b), what conclusion can be made about the hypotheses?

18.15 Refer to exercise 18.6, in which data are given for blood pressures measured at a visit to a dentist and a visit to a doctor. Carry out a Wilcoxon signed-rank test of the null hypothesis that the population median difference is 0 versus the alternative hypothesis that the median difference is greater than 0, where the difference is computed as *dentist − doctor*. Show all details, and write a conclusion in the context of this situation.

18.16 Refer to Example 18.1 about normal human body temperature. The data are

98.2 97.8 99.0 98.6 98.2 97.8 98.4 99.7 98.2
97.4 97.6 98.4 98.0 99.2 98.6 97.1 97.2 98.5

Minitab output for a Wilcoxon signed-rank test of $H_0: \eta = 98.6$ versus $H_a: \eta < 98.6$ is

	N	Wilcoxon Test Statistic	P	Estimated Median
Bodytemp	18	27.0	0.018	98.2

a. Show the details for finding the value of T^+ (given in the output as 27.0).
b. Based on the p-value given in the output, what conclusion can be made about the hypotheses? Write a conclusion in the context of this situation.

18.17 Refer to the previous exercise. Find values for μ_{T^+} and σ_{T^+}, the mean and standard deviation of T^+, assuming the null hypothesis is true.

18.18 In humans, the length of the foot supposedly is about equal to the length of the forearm (from elbow to wrist). The output for this exercise is for a Wilcoxon signed-rank test to analyze the median difference between right foot length and right forearm length using data collected from $n = 55$ college students. (*Data source:* **physical** dataset on the CD for the text.)

Test of median = 0.0 versus median not = 0.0

	N	N for Test	Wilcoxon Statistic	P	Estimated Median
Diff	55	39	458.0	0.346	0.2500

a. Write the null and alternative hypotheses given in the output using appropriate statistical notation. Make sure you define any symbols you use for parameters.
b. Based on the p-value given in the output, what conclusion can be made about the hypotheses? Write a conclusion in the context of this situation.
c. Discuss whether or not the result in part (b) could be used as proof that the length of the foot and the length of the forearm are about the same for a human.
d. Calculate the value of a z-statistic that could be used to determine a p-value in this problem.

Section 18.4

Additional exercises for the Kruskal–Wallis test are in the exercises for Section 16.3 in Chapter 16.

18.19 Three different methods for memorizing information are compared in an experiment. Eighteen participants are randomized into three groups of six individuals. Each group uses a different method of memorization to memorize a list of 50 words, and then all participants are tested to determine how many words on the list they can recall. The data for number of words recalled for individuals in each group are

Method 1: 23, 27, 30, 32, 33, 36
Method 2: 28, 30, 31, 35, 37, 40
Method 3: 16, 18, 22, 24, 27, 37

a. Rank the combined dataset and find values for $\bar{R}_1$, $\bar{R}_2$, and $\bar{R}_3$, the mean ranks for the three groups.
b. Find the value of H, the Kruskal–Wallis test statistic.

18.20 Refer to the previous exercise. State null and alternative hypotheses, find a p-value, and write a conclusion.

18.21 In a survey, 391 college women who say they exercise regularly were asked how many times per week they typically exercise and also were asked about their main reason for exercising. Choices for possible reasons were enjoyment, health, reduce stress, and weight control. The output given for this problem is for a Kruskal–Wallis test that compares the distribution of times per week exercising in the four categories of reason for exercising.

Kruskal–Wallis Test on Reason

Reason	N	Median	Ave Rank
Enjoyment	31	5.000	282.2
Health	101	3.000	200.4
Reduce stress	48	2.000	139.2
Weight control	211	3.000	194.1
Overall	391		196.0

$H = 30.37$ $DF = 3$ $P = 0.000$

a. Write null and alternative hypotheses for the Kruskal–Wallis test in this situation.
b. Based on the p-value given in the output, what conclusion can be made about the hypotheses? Write a conclusion in the context of this situation.
c. Using the average ranks and sample medians given in the output, describe how times per week of exercising may be related to reason for exercising.
d. Explain the connection between the value of H (given as 30.37) and the p-value (given as 0.000).

18.22 Refer to the previous exercise. Using information given in the output, show how H was calculated by substituting values into the proper formula.

18.23 Suppose that a Kruskal–Wallis test is done to compare the distribution of a response variable in four groups, and the value of $H = 12.35$. Determine the p-value (or an interval for the p-value) using Table A.5.

18.24 For this exercise, use the **GSS-93** dataset on the CD for the text. The variable *degree* gives the highest educational degree achieved, and the variable *tvhours* gives responses for hours of watching television in a typical week for respondents in the 1993 General Social Survey. Use statistical software to carry out a Kruskal–Wallis test of the null hypotheses that the distribution of *tvhours* is the same for all educational degree levels. Write a conclusion and give a justification for the conclusion. If there is a statistically significant difference, describe how *tvhours* might be related to educational degree.

Multiple Regression

Randy Faris/CORBIS

Can you predict a hot August when you're cold in January? See Example 19.1 on pages 632 and 634.

In multiple regression, we use the values of more than one explanatory variable to predict or describe the values of a response variable. It can be difficult to interpret the values of the estimated regression coefficients in a multiple regression model.

A **regression equation** is used to describe the relationship between one or more explanatory variables (x variables) and a quantitative response variable (y). In Chapters 5 and 14 we learned about **simple regression,** in which a straight-line equation is used to describe the relationship between a single quantitative explanatory variable (x) and a quantitative response variable (y). Remember that the y variable is sometimes called the *dependent variable,* and because regression models may be used to make predictions, an x variable may also be called a *predictor variable.*

In **multiple regression,** we use the values of more than one explanatory variable to predict or describe the values of a response variable. Put another way, the equation for the mean of the response variable (y) is a function of two or more explanatory variables. As an example, suppose a researcher is studying the connection between behavioral variables and grade point average for college students. Predictor variables of interest to the researcher are

x_1 = Study hours = Hours spent studying per week

x_2 = Classes missed = Number of classes student misses in typical week

x_3 = Work hours = Hours per week that student works in a part-time or full-time job

The general form of a *linear multiple regression model* relating grade point average (GPA) to these three predictor variables is

$$\text{GPA} = \beta_0 + \beta_1 \text{ Study hours} + \beta_2 \text{ Classes missed} + \beta_3 \text{ Work hours}$$

Numerical estimates of the parameters β_0, β_1, β_2, and β_3 would be determined using data from a sample of students for whom information on grade point average and the three predictor variables is available. ◆

19.1 | THE MULTIPLE LINEAR REGRESSION MODEL

Suppose y is the response variable, and $x_1, x_2, \ldots, x_{p-1}$ is a set of $p-1$ explanatory variables. The **multiple linear regression equation** for the relationship between the mean value of y and the variables $x_1, x_2, \ldots, x_{p-1}$ is

$$E(Y) = \beta_0 + \beta_1 x_1 + \beta_2 x_2 + \cdots + \beta_{p-1} x_{p-1}$$

- ◆ $E(Y)$ represents the mean value of y for individuals in the population who all have the same particular values of $x_1, x_2, \ldots, x_{p-1}$.
- ◆ $\beta_0, \beta_1, \ldots, \beta_{p-1}$ are referred to as the **regression coefficients** or, alternatively, may be called either the β **coefficients** or the β **parameters.**

As in simple regression, a useful format for expressing the components of the overall population regression model is

$$y = \text{Mean} + \text{Deviation}$$

This conceptual equation states that for any individual, the value of the response variable (y) can be constructed by combining two components:

1. The **mean,** which is given by $E(Y) = \beta_0 + \beta_1 x_1 + \beta_2 x_2 + \cdots + \beta_{p-1} x_{p-1}$, is the mean y for individuals having the same specific set of x-values as this individual.
2. An individual's **deviation = y − mean,** which is the difference between an individual's actual value of y and the mean y for individuals having the same specific set of x-values as this individual.

When writing the regression model for an individual within the population, we must use additional notation in order to distinguish one individual from another. Let x_{ik} represent the value of x_k for the ith individual, and let y_i represent the value of the ith observation for variable y. The **multiple linear regression model** is

$$y_i = \text{Mean} + \text{Deviation} = \beta_0 + \beta_1 x_{i1} + \beta_2 x_{i2} + \cdots + \beta_{p-1} x_{i,\,p-1} + \epsilon_i$$

The deviations ϵ_i are assumed to follow a normal distribution with mean 0 and standard deviation σ. In multiple regression, we make the same assumptions about the deviations that we did in simple regression, and in fact you should notice that the multiple regression model with only one x variable gives the simple regression model.

EXAMPLE 19.1
Predicting Average
August Temperature

The **temperature** dataset on the CD for this text gives geographic latitude, mean January temperature, mean April temperature, and mean August temperature for 20 cities in the United States. The temperatures are recorded as degrees Fahrenheit. A multiple linear regression equation for the relationship between August temperature and the explanatory variables latitude, January temperature, and April temperature is

$$E(\text{August temp}) = \beta_0 + \beta_1 \, \text{Latitude} + \beta_2 \, \text{January temp} + \beta_3 \, \text{April temp}$$

Written more formally using statistical notation, the multiple regression model in this situation is

$$y_i = \beta_0 + \beta_1 x_{i1} + \beta_2 x_{i2} + \beta_3 x_{i3} + \epsilon_i$$

where the ϵ_i are assumed to follow a normal distribution with mean 0 and standard deviation σ.

We will see shortly that we may not need to use all three explanatory variables in the equation because they are predictable from each other. Data for the 20 cities in the temperature dataset can be used to estimate the β coefficients. An important first step when analyzing the data is to examine scatterplots for all pairs of variables in order to determine whether any influential outliers are present and whether any of the relationships between the response variable and an explanatory variable might be nonlinear. ◆

EXAMPLE 19.2
Blood Pressure
of Peruvian Indians

How would somebody's blood pressure be affected, in the long run, by migrating from a small village at a high altitude to an urban area at a low altitude? Anthropologists examined this question by studying $n = 39$ Peruvian Indian men who had made such a migration. The men in the sample ranged in age from 21 to 54 years old, and the number of years since migration to the urban area ranged from 2 to 40 years. In addition to information on age and years since migrating, the investigators collected data on systolic and diastolic blood pressure, age, body weight, and various skinfold measurements related to amount of body fat. The data are given by Ryan and Joiner (2001) and also are in the Minitab worksheet PERU.MTW (not on the text CD). The researchers also created and analyzed an additional variable not in the dataset, which was the percent of a man's lifetime that he had lived in the urban area.

In this example, we will use a multiple linear regression model in which the response variable is $y =$ systolic blood pressure and the explanatory variables are

$$x_1 = \text{Weight (kg)}$$
$$x_2 = \text{Age}$$
$$x_3 = \text{Years} = \text{Number of years since migrating}$$
$$x_4 = \text{Pct\_Life} = (\text{Years/Age}) \times 100 = \text{Percent of lifetime living in}$$
$$\text{urban area}$$

Although the principal focus of the study was on how blood pressure was related to the number of years since living at altitude, the explanatory variables weight and age are included because it is well known that blood pressure is related to these variables. Expressed somewhat loosely (in terms of notation), the model is

$$\text{Blood pressure} = \beta_0 + \beta_1 \text{ Weight} + \beta_2 \text{ Age} + \beta_3 \text{ Years} + \beta_4 \text{ Pct\_Life}$$

Written using statistical notation, the multiple regression model in this situation is

$$y_i = \beta_0 + \beta_1 x_{i1} + \beta_2 x_{i2} + \beta_3 x_{i3} + \beta_4 x_{i4} + \epsilon_i$$

where ϵ_i are assumed to follow a normal distribution with mean 0 and standard deviation σ. ◆

The Regression Equation for the Sample

The first step in a multiple regression problem is to estimate the β coefficients using sample data. Notation for the sample coefficients is $b_0, b_1, \ldots, b_{p-1}$. With this notation, the regression equation for the sample could be written as

$$\hat{y} = b_0 + b_1 x_1 + b_2 x_2 + \cdots + b_{p-1} x_{p-1}$$

For the ith observation in the sample,

- The **predicted value** (estimated mean) is $\hat{y}_i = b_0 + b_1 x_{i1} + b_2 x_{i2} + \cdots + b_{p-1} x_{i,\,p-1}$
- The **residual** (deviation or error) is $e_i = y_i - \hat{y}_i$

where y_i = the value of the response variable for the ith observation in the sample and $x_{i1}, x_{i2}, \ldots, x_{i,p-1}$ are the corresponding values of variables x_1, $x_2, \ldots, x_{p-1}$.

Numerical values for the sample regression coefficients $b_0, b_1, \ldots, b_{p-1}$ usually are found by applying the **least-squares** criterion described in Chapter 5. The values of $b_0, b_1, \ldots, b_{p-1}$ determined by this criterion give a smaller value of the sum of squared errors for the sample than would any other possibilities for the values $b_0, b_1, \ldots, b_{p-1}$. Put another way, the value of SSE = $\Sigma(y_i - \hat{y}_i)^2$ is minimized. In practice, statistical software is always used to estimate the coefficients of a regression equation, so we don't give formulas for calculating the sample regression coefficients "by hand."

EXAMPLE 19.1—*CONTINUED*
Sample Regression
Equation for Predicting
August Temperature

For the **temperature** dataset, separate plots of y = August temperature versus each of x_1 = geographic latitude, x_2 = January temperature, and x_3 = April temperature show linear associations with no obviously influential observations. Thus we can use the multiple linear regression model. With statistical software, it can be determined that the sample regression equation is

$$\hat{y} = -29 + 0.70x_1 - 0.44x_2 + 1.66x_3$$

Two oddities are evident when we consider the values of the sample regression coefficients, and they illustrate how difficult it is to interpret the values of the regression coefficients in a multiple regression. Notice that the coefficient multiplying x_1 = latitude is positive, indicating that August temperature goes up as latitude increases, which does not seem reasonable. The value of the coefficient multiplying x_2 = January temperature is negative, indicating that the warmer it is in January, the colder it is in August. Again, this is not reasonable. Regardless of month, average temperature tends to decrease as latitude increases in the Northern hemisphere. For instance, in this sample, the correlation between latitude and August temperature is $r = -0.781$, indicating a strong negative correlation. Thus we might expect that the regression coefficient multiplying latitude would be negative, but instead it's positive ($+0.70$). Similarly, we would expect a positive value for the regression coefficient multiplying x_2 = January temperature because, in general, relatively warm locations in January tend also to be relatively warm places in August. For our sample, the correlation between January and August temperatures is $+0.622$, so the sign of the corresponding regression coefficient (-0.44) is in contrast to the direction of the association.

The difficulty is caused by the degree of correlation among the explanatory variables, which is quite high in this example. Specifically, the correlation between x_1 = latitude and x_2 = January temperature is $r = -0.86$, the correlation between latitude and x_3 = April temperature is $r = -0.96$, and the correlation between x_2 = January temperature and x_3 = April temperature is $r = 0.91$.

With such strong associations among the explanatory variables, it becomes nearly impossible to separate their individual effects on the response variable. Each coefficient actually represents the *additional* contribution made by that variable, *given* that all of the others are already being used as predictors. The moral of the story is that caution must be employed when interpreting the individual regression coefficients in a multiple regression analysis.

The sample regression can be used to determine a predicted value of August temperature, even when it's difficult to interpret the sample regression coefficients. As an example, the city of Phoenix has latitude = 33, January temperature = 54, and April temperature = 70. The predicted value of August temperature for Phoenix is found as

$$\hat{y} = -29 + 0.70(33) - 0.44(54) + 1.66(70) = 86.5 \text{ degrees}$$

Phoenix is hotter in August than we would predict based on information about its geographic latitude and January and April temperatures. The actual average August temperature in Phoenix is 92 degrees Fahrenheit. The *residual* for Phoenix is $e = y - \hat{y}$ = actual − predicted = 92 − 86.5 = 5.5 degrees. ◆

VIDEO

EXAMPLE 19.2 —
CONTINUED
Sample Regression
Equation for Peruvian
Indian Blood Pressure Study

Using statistical software (Minitab, for instance), we can determine that the sample regression equation for the Peruvian Indian blood example is

$$\hat{y} = 116.84 + 0.832 \text{ Weight} - 0.951 \text{ Age} + 2.34 \text{ Years} - 1.08 \text{ Pct\_Life}$$

where $\hat{y}$ = predicted systolic blood pressure based on the four explanatory variables. The value of $\hat{y}$ is the estimated mean systolic blood pressure for a population of individuals with the same particular set of values for the explanatory variables, and it can also be used as the predicted value of systolic blood pressure for an individual with a specified set of x values.

In multiple regression, it may be difficult to interpret the values of the sample regression coefficients; that most definitely is the case in this example. In fact, the values of the individual coefficients should not be interpreted here. Notice, for instance, that the coefficient multiplying Age in the equation is equal to −0.951. Does blood pressure actually decrease as age increases? A naïve interpretation would be that average systolic blood pressure decreases 0.951 point per each one-year increase in age, given that all other variables in the equation are held constant (not allowed to change). But it's not possible to change the value of Age while holding Years and Pct_Life constant. The variable Pct_Life is a function of the variable Age; it must necessarily change if Age is increased while Years is held constant.

The following table gives the values of $\hat{y}$, found using the sample regression equation, for four different sets of specific values for explanatory variables:

Set	x_1, Weight (kg)	x_2, Age	x_3, Years	x_4, Pct_Life	$\hat{y}$
1	65	30	15	50	123.5
2	65	50	15	30	126.1
3	65	50	25	50	127.9
4	65	50	35	70	129.6

Notice that the weight is 65 kg for all four sets and that age equals 30 in the first set and equals 50 in the other three sets. For sets 2, 3, and 4, in which age is 50, the predicted values of systolic blood pressure are greater than the predicted value for set 1 where age is 30. This is consistent with the usual medical finding that blood pressure, on average, increases with age. By comparing the predicted values for sets 2, 3, and 4, we see that for men of the same age and weight, increasing the number of years since migration apparently increases average systolic blood pressure. ◆

Minitab Tip

> **tech note**   **Estimating the Sample Regression Equation**
>
> ◆ Use **Stat>Regression>Regression.**
>
> ◆ In the box labeled "Response," specify the worksheet column with the data for the response variable.
>
> ◆ In the box labeled "Predictors," specify the worksheet columns containing the data for the explanatory variables.

Estimating the Standard Deviation σ

Remember from Section 14.2 that the standard deviation in the regression model measures, roughly, the average deviation of y-values from the mean (the regression equation). Expressed another way, the standard deviation for regression measures the general size of the residuals. In Section 14.2 we gave a formula for calculating an estimate of the standard deviation for a simple regression model and discussed how to interpret its value. The interpretation is essentially the same in multiple regression as it is in simple regression, and the calculation for multiple regression involves only a minor (but important) modification to the procedure for simple regression.

Estimating the Standard Deviation for a Multiple Regression Model

> **formula**   Let p = the number of β coefficients in the multiple regression model and n = sample size. The formula for estimating the standard deviation for a multiple regression model is
>
> $$s = \sqrt{\text{MSE}} = \sqrt{\frac{\text{SSE}}{n-p}} = \sqrt{\frac{\Sigma(y_i - \hat{y}_i)^2}{n-p}}$$
>
> The statistic $s = \sqrt{\text{MSE}}$ is an estimate of the population standard deviation σ.

The Proportion of Variation Explained by the Explanatory Variables (R^2)

The statistic denoted by R^2 measures what researchers often term the "proportion of variation in the response explained by the predictor variables (the x variables)." This statistic may be called the **multiple squared correlation** or, synonymously, the **multiple coefficient of determination.** It is often expressed as a percent, and if so, its value is between 0 and 100%. A value of 100% would indicate that the x variables completely explain the observed variation in responses for y. A value of 0% would indicate that the explanatory variables explain absolutely nothing about the variation in the responses for y. The procedure described in Section 5.3 for simple regression also applies for multiple regression.

Calculating R^2 for a Multiple Regression Model

formula   The formula for the squared multiple correlation is

$$R^2 = \frac{\text{SSTO} - \text{SSE}}{\text{SSTO}}$$

where

$\text{SSTO} = \Sigma(y_i - \bar{y})^2$, the sum of squared differences between observed y values and the sample mean of the y-values.

$\text{SSE} = \Sigma(y_i - \hat{y}_i)^2$, the sum of squared errors (residuals); that is, the sum of squared differences between observed and predicted y-values.

R^2 may be expressed as a percent, and it can also be found as the square of the correlation between the values of y_i and $\hat{y}_i$.

EXAMPLE 19.1—CONTINUED
R^2 and s for the August Temperature Equation

Values for $s = \sqrt{\text{MSE}}$ and R^2 are given in the multiple regression output of nearly all statistical software packages. For the prediction of August temperature, the value of $R^2 = 0.871$, or 87.1%, and the value of the standard deviation is $s = 2.81$. The value of R^2 is high, indicating a generally strong linear relationship between August temperature and the three predictor variables. The standard deviation gives, roughly, the average absolute size of the deviations (residuals) between actual and predicted values in the dataset. On average, actual August temperatures might be off by about ± 2.81 degrees Fahrenheit from predicted August temperatures. ◆

EXAMPLE 19.2—CONTINUED
R^2 and s for the Peruvian Indian Blood Pressure Equation

In the Peruvian Indian blood pressure example, the value of $R^2 = 0.597$, or 59.7%, and the value of the standard deviation is $s = 8.795$. This is a moderately strong value of R^2, indicating that the four explanatory variables explain some, but not all, of the variation in individual systolic blood pressures. The interpretation of the standard deviation is that, roughly, the average absolute size of the deviations (residuals) between actual and predicted systolic blood pressure is about 8.795. ◆

---

in summary   In Summary: Multiple Linear Regression Model: Population and Sample Versions

Suppose there is a random sample of n observations of the explanatory variables $x_1, x_2, \ldots, x_{p-1}$ and the response variable y from a large population. Let x_{ik} represent the value of the ith observation for variable x_k, and let y_i represent the value of the ith observation for variable y. The multiple linear regression model follows:

Population Version

Mean: $E(Y) = \beta_0 + \beta_1 x_1 + \beta_2 x_2 + \cdots + \beta_{p-1} x_{p-1}$

Individual: $y_i = \beta_0 + \beta_1 x_{i1} + B_2 x_{i2} + \cdots + \beta_{p-1} x_{i, p-1} + \epsilon_i = E(Y) + \epsilon_i$

The deviations ϵ_i are assumed to follow a normal distribution with mean 0 and standard deviation σ.

Sample Version

$$\textit{Mean: } \hat{y} = b_0 + b_1x_1 + b_2x_2 + \cdots + b_{p-1}x_{p-1}$$
$$\textit{Individual: } y_i = b_0 + b_1x_{i1} + b_2x_{i2} + \cdots + b_{p-1}x_{i,p-1} + e_i = \hat{y}_i + e_i$$

where $e_i = y_i - \hat{y}_i$ is the **residual** for individual i. The sample statistics b_0, $b_1, \ldots, b_{p-1}$ estimate the population parameters $\beta_0, \beta_1, \ldots, \beta_{p-1}$, and they are determined using the criterion of least sum of squared errors. The mean of the residuals is 0, and the residuals can be used to estimate the population standard deviation σ.

19.2 | INFERENCE ABOUT MULTIPLE REGRESSION MODELS

In this section, we'll discuss four commonly used inference procedures in multiple regression.

1. A hypothesis test of the null hypothesis that an individual population regression coefficient β_k is equal to 0.
2. A hypothesis test of the null hypothesis that $\beta_1 = \beta_2 = \cdots = \beta_{p-1} = 0$, which is equivalent to stating that none of the x variables is related to y.
3. A prediction interval for the value of y for an individual with a specified set of x-values.
4. A confidence interval for the mean y for a population of individuals, all with the same particular set of values for the x variables.

Inference for a Coefficient β_k

The multiple regression output of nearly every statistical software program includes results for testing the null hypothesis $H_0: \beta_k = 0$ versus the alternative hypothesis $H_a: \beta_k \neq 0$ for each separate β_k in the model. Notice that if $\beta_k = 0$ for a β coefficient that multiplies an explanatory variable, then the variable x_k is not influencing the calculation of predicted y-values because $\beta_kx_k = 0$, no matter what value x_k might have. Thus these tests are used to make decisions about which explanatory variables may or may not be helpful predictors of the response variable. Great care must be taken, however, when interpreting these tests, because they are affected by the magnitude of the correlations between the x variables. We will discuss this issue in an example later in this section.

Testing Null and Alternative Hypotheses for a Regression Coefficient

formula   In the multiple regression model, a test of $H_0: \beta_k = 0$ versus $H_a: \beta_k \neq 0$ is done using the test statistic

$$t = \frac{b_k - 0}{\text{s.e.}(b_k)}$$

where b_k = value of the sample estimate of β_k and s.e.(b_k) is the standard error of b_k.

> The p-value is found in the usual way, using a t-distribution with $n - p$ degrees of freedom where n = sample size and p = number of β coefficients in the model. Most statistical software uses the two-sided alternative hypothesis.

If a confidence interval for any of the β_k is desired, it is computed as $b_k \pm t^*$ s.e.(b_k). The multiplier t^* is found using a t-distribution with $n - p$ degrees of freedom and is such that the probability between $-t^*$ and $+t^*$ equals the confidence level for the interval. Table A.2 can be used to find the value of t^*.

Inference for the Overall Model

Suppose that each of the parameters $\beta_1, \beta_2, \ldots, \beta_{p-1}$ in the multiple regression model were equal to 0. If that were so, then the regression equation would reduce to $E(Y) = \beta_0$. This would imply that the response variable y was in fact unrelated to any of the explanatory variables being studied. Although this occurs rarely in practice, it must still be considered as a possibility. Thus all statistical software provides results for a test of

$H_0: \beta_1 = \beta_2 = \cdots = \beta_{p-1} = 0$
$H_a:$ At least one of $\beta_1, \beta_2, \ldots, \beta_{p-1}$ is not 0.

The interpretation of the null hypothesis is that if it is true, then none of the explanatory variables are useful predictors of the values of y. The interpretation of the alternative hypothesis is that at least one of the explanatory variables is related to the response variable.

The test statistic in this situation is called an F-statistic. In computer output for multiple regression, it is usually given in a table called an **analysis of variance table.** The general format of the analysis of variance table for multiple regression follows:

Source	Degrees of Freedom	Sum of Squares	Mean Square = SS/df	F
Regression model	$p - 1$	SSR = SSTO − SSE = $\Sigma(\hat{y}_i - \bar{y})^2$	$MSR = \dfrac{SSR}{p - 1}$	$\dfrac{MSR}{MSE}$
Residual error	$n - p$	SSE = $\Sigma(y_i - \hat{y}_i)^2$	$MSE = \dfrac{SSE}{n - p}$	
Total	$n - 1$	SSTO = $\Sigma(y_i - \bar{y})^2$		

The p-value is found using an F-distribution. An F-distribution is a skewed distribution with a minimum value of 0, and it has two parameters called *degrees of freedom*. The first of the two parameters is called the **numerator degrees of freedom,** and the second is called the **denominator degrees of freedom.** In multiple regression, the numerator df = $p - 1$ = number of x variables in the equation, and the denominator df = $n - p$ (sample size − number of groups). Thus the p-value could be found as the probability that an F-distribution with degrees of freedom $p - 1$ and $n - p$ would have a value larger than the observed F-statistic. Table A.4 provides specific numerical values for F-distributions, corresponding to the upper 5% and 1% tails of the distributions.

EXAMPLE 19.1—*CONTINUED*
Inference for the
August Temperature Model

Figure 19.1 displays Minitab output for the multiple linear regression model in which y = August temperature and the three x variables are geographic latitude, January temperature, and April temperature. At the far right of the analysis of variance table given toward the bottom of the output, a p-value is given for a test of H_0: $\beta_1 = \beta_2 = \beta_3 = 0$. Its value is 0.000 (bold, underlined), so the null hypothesis is rejected, which means that at least one of the explanatory variables in the model is linearly related to August temperature (no surprise).

Toward the top of the output there is a display of five columns with information about the individual regression coefficients. Remember that in the multiple regression model, a test of H_0: $\beta_k = 0$ versus H_a: $\beta_k \neq 0$ is done using the test statistic

$$t = \frac{b_k - 0}{\text{s.e.}(b_k)} = \frac{b_k}{\text{s.e.}(b_k)}$$

The t-statistic for each of the coefficients can be seen in the column headed "T" and is formed by taking the ratio of the values in the columns headed "Coef" and "SE Coef." For instance, for testing H_0: $\beta_1 = 0$, the test statistic is $t = 0.7009/0.4053 = 1.73$.

The p-values corresponding to these tests are given in the last column (under "P") and are used for testing the null hypothesis $\beta_k = 0$ separately for each coefficient in the model. We focus on the p-values in the rows labeled with the names of the explanatory variables. In order starting with the latitude row, these are p-values for testing $\beta_1 = 0$, $\beta_2 = 0$, and $\beta_3 = 0$. The tests for each of $\beta_2 = 0$ and $\beta_3 = 0$ (last two rows) are statistically significant because they have the p-values 0.001 and 0.000, respectively. On the other hand, the p-value (in the "Latitude" row) for testing $\beta_1 = 0$ is 0.103, which exceeds the usual 0.05 level of significance, so the test is not statistically significant. In fact, there is a strong connection between latitude and August temperature, but January and April temperatures, which themselves are related to latitude, provide sufficient information to pre-

```
The regression equation is
AugTemp = −29.0 + 0.701 latitude − 0.440 JanTemp + 1.66 AprTemp
```

Predictor	Coef	SE Coef	T	**P**
Constant	−28.99	29.42	−0.99	0.339
Latitude	0.7009	0.4053	1.73	**0.103**
JanTemp	−0.4397	0.1036	−4.24	**0.001**
AprTemp	1.6649	0.2964	5.62	**0.000**

```
S = 2.814    R-Sq = 87.1%    R-Sq(adj) = 84.7%
```

Analysis of Variance

Source	DF	SS	MS	F	P
Regression	3	858.07	286.02	36.11	**0.000**
Residual Error	16	126.73	7.92		
Total	19	984.80			

FIGURE 19.1   **Minitab output for the August temperature model**

```
The regression equation is
Systol = 117 + 0.832 Weight − 0.951 Age + 2.34 Years − 1.08 Pct_life

Predictor          Coef      SE Coef          T          P
Constant         116.84        21.98       5.32      0.000
Weight           0.8324       0.2754       3.02      0.005
Age             −0.9507       0.3164      −3.00      0.005
Years            2.3393       0.7714       3.03      0.005
Pct_life        −1.0807       0.2833      −3.81      0.001

S = 8.795     R-Sq = 59.7%     R-Sq(adj) = 55.0%

Analysis of Variance

Source            DF          SS         MS          F        P
Regression         4     3901.73     975.43      12.61    0.000
Residual Error    34     2629.71      77.34
Total             38     6531.44
```

FIGURE 19.2   Minitab output for the Peruvian Indian blood pressure model

dict August temperature. The consequence is that we could consider removing the latitude variable from the model because it is not making a "significant" additional contribution to the prediction of y. If we do so, however, we must again use statistical software to estimate the regression coefficients for the model with only two predictor variables.

The test for $\beta_0 = 0$ (information given in the row labeled "Constant") is not useful in this situation. The coefficient β_0 corresponds to the August temperature when the value of each explanatory variable is 0. That's not a realistic combination of x-values. Latitude = 0 occurs at the equator; the January and April temperatures there certainly are not 0 degrees Fahrenheit. ◆

EXAMPLE 19.2—*CONTINUED*
Inference for the Peruvian Indian Blood Pressure Model

Multiple regression output for the Peruvian Indian example is given in Figure 19.2. The interpretation of the results for significance tests represented in the output is easy. All of the p-values are small and are well below any reasonable standard for level of significance. Thus we can conclude that all four explanatory variables are making "significant" contributions to the explanation of variation in systolic blood pressure. ◆

Predicting an Individual Response and Estimating $E(Y)$

In the context of the simple regression model, Sections 14.4 and 14.5 described methods for predicting the value of the response variable for an individual and for estimating the mean value of y for a population of individuals all with the same particular value of x. The discussion in those sections applies to the multiple regression model as well; the only technical change is that in multiple regression, the degrees of freedom = $n - p$ for finding a t^* multiplier, rather than $n - 2$ as in simple regression. There were two key definitions in Sections 14.4 and 14.5, and they can be modified for multiple regression as follows:

> *definition* A **prediction interval** is an interval that estimates the value of *y* for an individual with a specific set of values for the explanatory variables. Equivalently, it is an interval that gives a range of values of the response variable for a specified central percentage of a population with the specified set of values for the explanatory variables.
>
> A **confidence interval for $E(Y)$** is an interval that estimates the population mean $E(Y)$ for a specific set of values for the explanatory variables of interest.

EXAMPLE 19.2—*CONTINUED*
Prediction and Confidence
Intervals for Peruvian
Indian Blood Pressure

Figure 19.3 (opposite) gives 95% confidence intervals for mean systolic blood pressure as well as 95% prediction intervals for individual blood pressure for the four sets of specific values of the *x* variables described earlier in the discussion of this example. The specific values are listed in the bottom portion of the output, and the confidence intervals and prediction intervals are given in the top portion. For example, the first set of *x*-values is Weight = 65, Age = 30, Years = 15, and Pct_Life = 50.0. At the top of the output, we see that a 95% confidence interval is 117.55 to 129.4 for the *mean systolic* blood pressure of a population of individuals with these values. A 95% prediction interval is 104.65 to 142.30 for the systolic blood pressure of an *individual* with these values for the *x* variables. This prediction interval is wide, reflecting the fact that there is substantial variation among blood pressures of individuals with similar characteristics. The column labeled "Fit" gives the predicted values for the cases, and "SE Fit" gives the standard error of the predicted value for each case. ◆

Minitab Tip

> *tech note* **Creating Prediction Intervals for *Y***
> **and Confidence Intervals for $E(Y)$**
>
> ◆ Use **Stat>Regression>Regression,** and then use the ***Options*** button of the dialog box.
>
> ◆ In the box labeled "Prediction intervals for new observations," either specify the numerical values for each explanatory variable (in order of appearance in the model) or specify a set of columns that give multiple values for the explanatory variables.
>
> ◆ The confidence level can be specified in the box labeled "Confidence Level."

19.3 | CHECKING CONDITIONS FOR MULTIPLE LINEAR REGRESSION

In Section 14.6 we listed these five conditions that should be present in order for us to use a regression model for inference:

1. The form of the regression equation must be correct.
2. There should not be any extreme outliers that influence the results unduly.

```
New Obs        Fit      SE Fit          95.0% CI                    95.0% PI
  1           123.47     2.91    (   117.55, 129.40)    (   104.65, 142.30)
  2           126.08     3.41    (   119.14, 133.02)    (   106.90, 145.25)
  3           127.85     3.27    (   121.21, 134.50)    (   108.79, 146.92)
  4           129.63     4.81    (   119.85, 139.41)    (   109.26, 150.00)


Values of Predictors for New Observations

New Obs    Weight      Age     Years    Pct_life
  1          65.0      30.0     15.0       50.0
  2          65.0      50.0     15.0       30.0
  3          65.0      50.0     25.0       50.0
  4          65.0      50.0     35.0       70.0
```

FIGURE 19.3   **Prediction and confidence intervals**

3. The standard deviation of the y variable (equivalently, the deviations) is the same regardless of the values of the x variables.
4. The distribution of y (equivalently, the deviations) is a normal distribution. This condition can be relaxed if the sample size is large.
5. Observations in the sample should be independent of one another. Essentially, this means the sample should be a random sample.

Chapter 14 covered simple regression, but these five conditions should be present also for multiple linear regression. Graphical methods for checking conditions 1 through 4 were covered in Section 14.6, and those techniques can be used to verify the conditions for multiple regression.

An additional plot not mentioned in Section 14.6 is informative about conditions 1, 2, and 3 in a multiple regression analysis. In this plot, the values of e_i (the residuals) are plotted on the vertical axis against the values of $\hat{y}_i$ (the predicted values of y) on the horizontal axis. Often this plot is referred to as **residuals versus fits** because the term *fit* is a synonym for "predicted value." Ideally, this plot should appear to be a band of points randomly spread around a horizontal line at *residual* = 0. A curved pattern in this plot indicates that the form of the regression equation may be wrong (condition 1 is violated). Outliers (extreme residuals) will show up as points that are noticeably distant in a vertical direction from the other points in the plot, in which case condition 2 may be violated. Condition 3 may be violated if the vertical range of the residuals is distinctly different in one region of predicted values than in another.

EXAMPLE 19.1—*CONTINUED*
Residuals Versus
Fits for Model Used to
Predict August Temperature

Figure 19.4 (on the next page) is a plot of residuals versus fits for a multiple regression equation in which January and April temperatures are used to predict mean August temperature, and latitude has not been used as an x variable. (The sample regression equation is $\hat{y} = 20.8 - 0.415$ January temp + 1.24 April temp.) The plot is roughly a horizontal band, as it should be if conditions 1, 2, and 3 are met. There are two potential outliers, however. One is at 85 along the horizontal axis, with a value for the residual that is somewhat larger than +5. This observation is Phoenix, which is hotter in August than would be predicted by its January and April temperatures. The other possible outlier is at about 70 along the horizontal axis, with a residual of about −6. This observation is San Francisco,

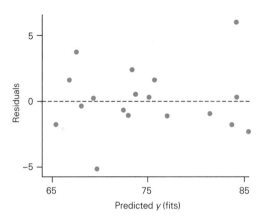

FIGURE 19.4   **Residuals versus fits for August temperature example**

which is cooler in August than would be predicted by its January and April temperatures. ◆

Minitab Tip

tech note   Graphing Residuals Versus Fits

- Use **Stat>Regression>Regression,** and then use the *Graphs* button of the dialog box.
- Click on the box labeled "Residuals versus Fits."

KEY TERMS

Section 19.1
multiple linear regression model, 632, 637–638
regression equation for sample, 633, 638
predicted value, 634
residual, 634

estimating standard deviation, 636
proportion of variation explained, R^2, 636–637

Section 19.2
testing hypotheses about β_k, 638–639
analysis of variance table, 639

prediction interval, 642
confidence interval for $E(Y)$, 642

Section 19.3
necessary conditions for regression, 642–643
residuals versus fits, 643

EXERCISES   Blue = Basics   **Bold, Bold** = Answer in back

19.1 Refer to Example 19.2—*CONTINUED*, Sample Regression Equation for Peruvian Indian Blood Pressure Study in Section 19.1 (p. 635). Show how the four values of $\hat{y}$ given in that example were calculated, and verify that the values are correct.

19.2 Suppose that a college admissions committee plans to use data for total SAT score, high school grade point average, and high school class rank to predict college freshman-year grade point averages for applicants. Write the population multiple linear regression model for this situation, and state any necessary assumptions. Specify what is the response variable and what are the explanatory variables.

19.3 Describe an example of a multiple linear regression model for a topic or subject matter area that interests you. Specify the response variable and the explanatory variables, and write the multiple regression model for your example.

19.4 The figure for this exercise (see next page) is a plot of residuals versus fits for the sample regression model used in Example 19.2 (the Peruvian Indian example). Are the necessary conditions for using a multiple regression model for statistical inference met in this situation? Explain.

19.5 Consider the Peruvian Indian blood pressure example. The sample regression equation was $\hat{y} = 116.84 + 0.832$ Weight $- 0.951$ Age $+ 2.34$ Years $- 1.08$ Pct_Life. Suppose the statement is made that if the men had not migrated to an urban area, their blood pressures would have decreased about 0.951 points per year of aging because Years and Pct_Life would both be 0, making the regression equation become $\hat{y} = 116.84 + 0.832$ Weight $- 0.951$ Age. Discuss why this statement is not valid.

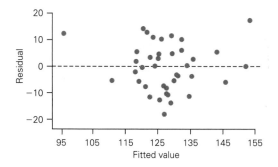

EXERCISE 19.4

19.6 Refer to Example 19.1, in which multiple regression is used to predict August temperature for locations in the United States. If latitude is not used as a predictor variable, the sample regression equation for predicting August temperature is

$$\hat{y} = 20.8 - 0.415 \text{ January temp} + 1.24 \text{ April temp}$$

a. For San Francisco, the average January temperature is 49 degrees Fahrenheit and the average April temperature is 56 degrees. Use the sample regression equation to estimate the average August temperature for San Francisco.

b. The actual average August temperature for San Francisco is 64 degrees. What is the value of the residual for San Francisco?

*Exercises 19.7 to 19.14 refer to the following: The **UCDavis1** dataset on the CD for this text includes student height, mother's height, and father's height for 76 male students in an elementary statistics class at UC Davis. All heights were recorded in inches. The output for these exercises is for a multiple regression with y = student height, x_1 = mother's height (momheigh in output), and x_2 = father's height (dadheigh in output).*

19.7 Locate the regression equation in the output.
  a. Give a numerical value for each of the sample regression coefficients b_0, b_1, and b_2.
  b. Calculate a predicted height for a student whose mother's height = 65 inches and father's height = 74 inches.
  c. Refer to part(b). Suppose student height is 71 inches for someone whose mother is 65 inches tall and whose father is 74 inches tall. What is the value of the residual for this student?

19.8 Explain why the value of b_0 does not have a meaningful interpretation in the context of the relationship between student height and parents' heights.

19.9 Suppose two students have mothers whose heights are the same and fathers whose heights differ by one inch. What is the difference in estimated heights of these two students? The answer is given by one of the regression coefficients. Which one?

19.10 In the output, locate the result given for a test of $H_0: \beta_1 = 0$.
  a. What is the p-value for the test? Based on this p-value, what conclusion can be made? Explain what this conclusion means about whether mother's height is useful for predicting student height.

```
The regression equation is
Height = 20.2 + 0.193 momheigh + 0.539 dadheigh

Predictor        Coef       SE Coef       T        P
Constant        20.248       6.680      3.03     0.003
momheigh       0.19294      0.08777     2.20     0.031
dadheigh       0.53897      0.07449     7.24     0.000

S = 2.591     R-Sq = 48.1%      R-Sq(adj) = 46.7%

Analysis of Variance

Source          DF       SS        MS        F       P
Regression       2     454.71    227.35    33.86   0.000
Residual Error  73     490.14      6.71
Total           75     944.85

New Obs     Fit    SE Fit     95.0% CI          95.0% PI
1        70.670   0.349  ( 69.976, 71.365)  ( 65.460, 75.881)

Values of Predictors for New Observations

New Obs      momheigh     dadheigh
1              63.0         71.0
```

b. What is the value of the t-statistic for a test of $H_0: \beta_1 = 0$? Show how this value is calculated from other information in the output.

19.11 Locate the "Analysis of Variance" in the output.

   a. Write the null and alternative hypotheses that are tested, using the p-value given in the analysis of variance.

   b. In this situation, explain what conclusion can be made about the hypotheses written in part (a). What does the conclusion indicate about the relationship between student height and parents' heights?

19.12 Locate the values of s and R^2 in the output.

   a. What is the value of s? Write a sentence that interprets the value in the context of this situation.

   b. What is the value of R^2? Write a sentence that interprets the value in the context of this situation.

19.13 The output gives a confidence interval and a prediction interval for the case where mother's height = 63 inches and father's height = 71 inches.

   a. State the 95% confidence interval that's given, and write a sentence that explains what is estimated by this interval.

   b. State the 95% prediction interval that's given, and write a sentence that explains what is predicted by this interval.

   c. The value of the "Fit" for this combination of parents' heights is reported to equal 70.67. Explain what this value estimates, and show how it was calculated.

19.14 The figure for this exercise (top of next column) is a plot of residuals versus fits for the regression model in which student height is the response variable and mother's and father's heights are two explanatory variables. Are the necessary conditions for using the regression model for statistical inference met in this situation? Explain.

Exercises 19.15 to 19.17. The **physical** *dataset on the CD for the text includes measurements of forearm and foot lengths (cm) for $n = 55$ college students. The output for this exercise is for a multiple regression model with y = left forearm length, x_1 = left foot length, and x_2 = right foot length.*

19.15 Locate the "Analysis of Variance" in the output below.

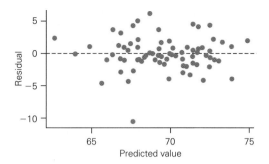

EXERCISE 19.14

a. Write the null and alternative hypotheses that are tested using the p-value given in the analysis of variance.

b. In this situation, explain what conclusion can be made about the hypotheses you wrote in part (a). What does this conclusion indicate about the relationship between left forearm length and the lengths of the right and left feet?

19.16 Locate the results for the (separate) tests of $H_0: \beta_1 = 0$ and $H_0: \beta_2 = 0$.

   a. Write a conclusion for each test. Justify your answers using information in the output.

   b. Explain why the conclusions for part (a) of this exercise do not contradict the conclusion for part (b) of the previous exercise.

19.17 Locate the values of s and R^2 in the output.

   a. What is the value of s? Write a sentence that interprets this value in the context of this situation.

   b. What is the value of R^2? Write a sentence that interprets this value in the context of this situation.

19.18 In the Peruvian Indian dataset (not on the CD), the variable **Calf** gives calf skinfold measurements for the 39 men in the sample. The output for this exercise gives the results for t-tests of $H_0: \beta_k = 0$ when **Calf** is included in the regression model as an explanatory variable along with the four explanatory variables used in Example 19.2.

The regression equation is
LeftArm = 11.7 + 0.352 LeftFoot + 0.185 RtFoot

Predictor	Coef	SE Coef	T	P
Constant	11.710	2.518	4.65	0.000
LeftFoot	0.3519	0.2961	1.19	0.240
RtFoot	0.1850	0.2816	0.66	0.514

S = 1.796   R-Sq = 36.9%   R-Sq(adj) = 34.4%

Analysis of Variance

Source	DF	SS	MS	F	P
Regression	2	98.019	49.010	15.19	0.000
Residual Error	52	167.785	3.227		
Total	54	265.804			

Predictor	Coef	SE Coef	T	P
Constant	119.58	21.57	5.54	0.000
Weight	0.9636	0.2817	3.42	0.002
Age	-1.0903	0.3217	-3.39	0.002
Years	2.6295	0.7762	3.39	0.002
Pct_life	-1.2080	0.2884	-4.19	0.000
Calf	-0.6413	0.4022	-1.59	0.120

a. Give numerical values for each of the sample regression coefficients b_0, b_1, b_2, b_3, b_4, and b_5.

b. Use information in the output to explain whether or not it is useful to include the variable **Calf** as an explanatory variable in the model.

c. What is the value of the t-statistic for testing $H_0: \beta_5 = 0$? Show how this value is calculated from other values reported in the output.

19.19 Refer to the previous exercise. The sample regression coefficient for the variable *Calf* is a negative value (-0.6413). Explain why this should not be interpreted to mean that as calf skinfold increases, average systolic blood pressure decreases.

19.20 The data source for this exercise is the CD dataset **UC-Davis1,** which gives data collected from college students. The output below is for a multiple linear regression analysis of the connection between y = grade point average and three predictor variables: x_1 = weekly time spent using the computer, x_2 = weekly time spent watching television, and x_3 = weekly time spent exercising.

a. Locate the value of R^2 reported in the output. What is the value? Write a sentence that interprets the value in the context of this situation.

b. Using the results of the significance tests reported in the output, discuss whether you think there is any relationship between grade point average and the three x variables used in the multiple regression model.

19.21 Suppose that a multiple linear regression model includes two explanatory variables.

a. Write the population regression model, using appropriate statistical notation.

b. Explain the difference between what is represented by the symbol b_2 and what is represented by the symbol β_2.

19.22 In a regression analysis, what are the two different interpretations for a predicted value $\hat{y}$?

19.23 Use the **bears-female** dataset on the CD and statistical software for this exercise. The variable *Weight* is the weight of a bear (pounds), *Neck* is the bear's neck girth (inches), and *Chest* is the bear's chest girth (inches). In this exercise, *Weight* will be the response variable (y).

a. Draw scatterplots of *Weight* versus *Neck* and *Weight* versus *Chest*, using *Weight* as the y variable in each instance. Write a short description of each plot. Do the relationships appear to be linear? Are there any outliers?

b. Omit the sixth observation in the dataset. It's an outlier with values that don't seem legitimate. Then find the estimate of the multiple linear regression equation, using *Weight* as the response variable and using *Neck* and *Chest* as the explanatory variables. What is the sample regression equation?

c. Based on the relevant t-tests, explain whether both explanatory variables are "significant" predictors of *Weight.*

d. What is the value of R^2? Write a sentence that interprets the value in the context of this situation.

Predictor	Coef	SE Coef	T	P
Constant	3.0424	0.1013	30.04	0.000
Computer	-0.005594	0.004042	-1.38	0.168
TV	-0.002773	0.004507	-0.62	0.539
Exercise	-0.00201	0.01073	-0.19	0.851

$S = 0.5924$ R-Sq = 1.5% R-Sq(adj) = 0.0%

Analysis of Variance

Source	DF	SS	MS	F	P
Regression	3	0.8120	0.2707	0.77	0.512
Residual Error	157	55.0912	0.3509		
Total	160	55.9032			

EXERCISE 19.20

REFERENCE

Ryan, B., and B. L. Joiner (2001). *MINITAB Handbook,* 4th ed., Belmont, CA: Duxbury Press.

Two-Way Analysis of Variance

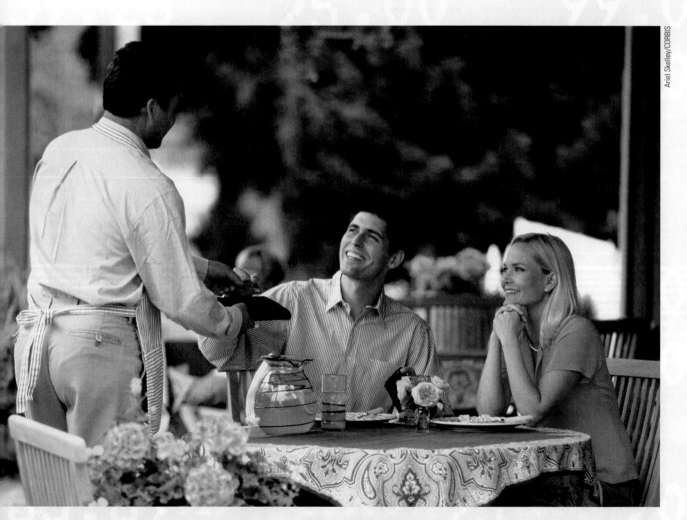

Ariel Skelley/CORBIS

What simple gesture might decrease his tip? See Example 20.4.

In Chapter 16 we learned how to compare means of three or more groups using analysis of variance. Now we learn how to compare means for a continuous variable when individuals are grouped by combinations of two categorical variables. The procedure is called *two-way analysis of variance*.

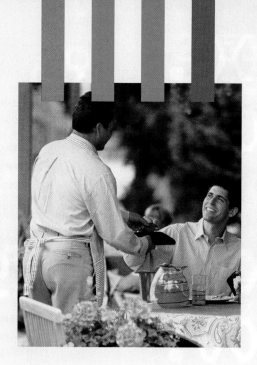

n Chapter 16 we introduced two-way analysis of variance (ANOVA) and described some of the relevant ideas. However, we did not furnish details about the models and analyses. In this chapter we provide some of those details. ◆

20.1 | ASSUMPTIONS AND MODELS FOR TWO-WAY ANOVA

Two-way analysis of variance is appropriate when the response variable is continuous and there are two categorical explanatory variables. For instance, we might want to compare grade point averages for students categorized by whether they attend school full-time or part-time and whether they are working in addition to attending school. Or we might want to compare pulse rates for adults on the basis of gender (male, female) and smoking behavior (yes, no).

EXAMPLE 20.1
Sleep Hours Based on Gender and Seat Location

In Exercise 16.31 we showed the relationship between responses to the survey question "hours of sleep the previous night" (continuous) and the two categorical explanatory variables "gender" and "seat location" (using the **UCDavis1** dataset on the CD). Figure 20.1 shows the mean responses to the question "How many hours did you sleep last night?" categorized by gender and preferred seat location. Notice that the mean hours of sleep for students in the front of the class differ considerably for males and females, with females reporting more than a full hour more sleep than males. The differences between males and females are smaller for those who prefer the middle and back of the class, but there are still differences. Questions of interest include the following:

◆ Is there an overall difference in mean hours of sleep for males and females?

◆ Are there overall differences in mean hours of sleep for those who prefer to sit in the front, middle, and back of the class?

◆ Is there an interaction between gender and seat location, so that the difference in mean hours of sleep for males and for females changes on the basis of seat location preference? This certainly appears to be the case, but the question of interest is about the population these students represent, and the difference observed in the sample data may or may not reflect a real difference in the population. ◆

Main Effects and Interaction Effects

In Chapter 16 we defined the terms *main effect* and *interaction*. A **main effect** describes the individual relationship between one of the categorical variables and the response variable. In Example 20.1, one main effect is the relationship between gender and average hours of sleep. The other main effect is the relationship between preferred seat location and average hours of sleep.

An **interaction effect** describes the extent to which the relationship between one explanatory variable and the response variable depends on the other explanatory variable. In Example 20.1, if the difference in mean hours of sleep for males and females is similar for all three preferred seat location categories, then there is little or no interaction between gender and seat location in their relationship with hours of sleep. On the other hand, if the male–female sleep difference changes for the seat location groups, as appears to be the case, then there is an interaction effect.

Assumptions and Conditions for Two-Way ANOVA

The assumptions and conditions required for two-way ANOVA are the same as those listed for one-way analysis of variance in Section 16.1. The only difference is in how the populations of interest are defined. In one-way ANOVA we are interested in comparing the means of k populations. In two-way ANOVA we are interested in comparing the means of ab populations, where a is the number of categories of the first explanatory variable and b is the number of categories of the second explanatory variable. For instance, in Example 20.1, the first explanatory variable had $a = 2$ categories (male, female) and the second explanatory variable had $b = 3$ (front, middle, back) categories. We are interested in

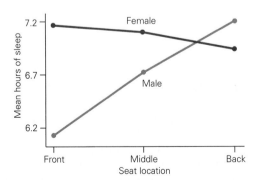

FIGURE 20.1   **Main effects and interaction effects**

comparing the mean hours of sleep for the $ab = 6$ populations defined by the combinations of the two explanatory variables.

The comparisons in two-way ANOVA are more complicated than those made in one-way ANOVA, but the assumptions are not. Here are the assumptions about the populations and the samples representing them, repeated from Section 16.1:

- The samples are independent random samples.
- The distribution of the response variable is a normal curve within each population.
- The different populations may have different means.
- All populations have the same standard deviation, σ.

As in Chapter 16, these assumptions may be relaxed somewhat in practice. For instance, it is rarely the case that all of the population standard deviations are identical. As long as they are similar, the procedures will work. Similarly, the assumption of normal curves can be relaxed if the sample sizes are moderate and the samples don't have major outliers. The assumption about independent random samples is stricter than is usual in practice. In general, separate representative samples or one combined sample that can be split into separate samples using the two categorical variables will suffice.

Model for One-Way ANOVA

In Chapter 16 we did not explicitly state the model for observations in one-way ANOVA. We do so now to facilitate understanding of the two-way ANOVA model. In one-way ANOVA we assume that there is a random sample from each of k normal populations and that population i has mean μ_i and standard deviation σ.

---

definition Let x_{ij} be the notation for observation j from population i, where $i = 1, 2, \ldots, k$ and $j = 1, 2, \ldots, n_i$. The **one-way ANOVA model** is

$$x_{ij} = \mu_i + \epsilon_{ij}$$

where the collection of ϵ_{ij} are independent normal random variables with mean 0 and standard deviation σ.

This one-way ANOVA model is less complicated than it may appear. It says that each observation can be thought of as being made up of the population mean and an additional term that may be positive or negative. For instance, suppose a woman's height is $67''$ and she belongs to a population of women whose mean height is $65''$. Then her height can be written as $67'' = 65'' + 2''$. In the format of the one-way ANOVA, her height would be written as $67'' = x_{ij} = \mu_i + \epsilon_{ij} = 65'' + 2''$. Another woman's height might be $64''$, in which case for her, $\epsilon_{ij} = -1''$. As in the regression model, ϵ is sometimes called the *error term,* which is a misnomer because the amount by which an individual differs from the population mean is certainly not an "error" in most cases.

Models for Two-Way ANOVA

Before defining a model for two-way ANOVA, let's define some additional terms and notation. The categorical explanatory variables in ANOVA are called **factors.** In two-way ANOVA there are two factors, and it is standard to call them **Factor A** and **Factor B.** In Example 20.1, Factor A is gender and Factor B is seat location. (There is nothing specific about this order; we could define Factor A to be seat location and Factor B to be gender.) Factor A has a categories and Factor B has b categories. Therefore, there are ab populations, with a random sample from each one for which the response variable is measured.

We use the letter i to indicate the category level for Factor A and the letter j to indicate the category level for Factor B. Thus μ_{ij} is the population mean for level i of Factor A and level j of Factor B. For instance, in Example 20.1, Factor A (gender) has level 1 = male and level 2 = female. Factor B (seat location) has level 1 = front, level 2 = middle, and level 3 = back. Thus the mean hours of sleep for the population of male students who prefer to sit in the front of the room is μ_{11}. The notation μ_{23} represents the mean hours of sleep for the population of female students ($i = 2$) who prefer to sit in the back of the room ($j = 3$).

We also need notation and a definition for the combined mean for each level of one factor averaged across the levels of the other factor. We use a dot (.) to represent the factor over which the average has been taken. The definitions follow.

> *definition* The **population mean for Factor A, level i** is
>
> $$\mu_{i.} = \frac{1}{b} \sum_{j=1}^{b} \mu_{ij}$$
>
> The **population mean for Factor B, level j** is
>
> $$\mu_{.j} = \frac{1}{a} \sum_{i=1}^{a} \mu_{ij}$$
>
> The **overall population mean** is
>
> $$\mu_{..} = \frac{1}{ab} \sum_{i=1}^{a} \sum_{j=1}^{b} \mu_{ij}$$

We use the letter k to indicate which of the individual observations we are writing. The sample observations are thus denoted as x_{ijk} where i and j specify the levels of Factors A and B, respectively, and k specifies the individual observations. For instance, in Example 20.1, observation x_{125} represents the hours of sleep reported by the fifth student ($k = 5$) in the sample of male ($i = 1$) students who prefer to sit in the middle of the room ($j = 2$).

The size of the sample from the population with level i of Factor A and level j of Factor B is n_{ij}. In many cases, all of the ab sample sizes will be identical, and we will simply use the letter n for each one. The combined sample size is denoted by N. If all of the sample sizes are the same, then $N = nab$. If not, then N is the sum of all of the individual n_{ij}.

> ### *in summary*   Notation for Two-Way ANOVA
>
> There are a levels of Factor A and b levels of Factor B. Individual observations are denoted by x_{ijk} where i, j, and k represent the level of Factor A, the level of Factor B, and the number of the individual observation within the sample, respectively. Notation for means and sample sizes is as follows:
>
> *Population means:*
> μ_{ij} = Population mean for Factor A level i and Factor B level j.
> $\mu_{i.}$ = Population mean for Factor A level i.
> $\mu_{.j}$ = Population mean for Factor B level j.
> $\mu_{..}$ = Overall population mean for all ab combinations of Factor A and Factor B.
>
> *Sample sizes:*
> n_{ij} = Sample size for Factor A level i and Factor B level j.
> $n_{i.}$ = Combined sample size for Factor A level i.
> $n_{.j}$ = Combined sample size for Factor B level j.
> N = Total sample size
>
> *Sample means:*
>
> The notation for sample means is the same as for population means, but with μ replaced by $\bar{x}$. For instance, the sample mean for Factor A level i and Factor B level j is $\bar{x}_{ij}$.

> **definition** Let x_{ijk} represent observation k from population ij, where $i = 1, 2, \ldots, a; j = 1, 2, \ldots, b$, and $k = 1, 2, \ldots, n_{ij}$. The **general two-way ANOVA model** is
>
> $$x_{ijk} = \mu_{ij} + \epsilon_{ijk}$$
>
> where the collection of ϵ_{ijk} are independent normal random variables with mean 0 and standard deviation σ.

If we were simply to study the general two-way ANOVA model, it would provide no more information than the one-way model. They are identical except for the additional subscript required for the notation in the general two-way model. What makes the two-way ANOVA situation more useful is that we break up the population mean into pieces. There are four pieces—the overall mean $\mu_{..}$ and an adjustment for each of Factor A, Factor B, and the possible interaction between Factor A and Factor B. Before introducing notation for these pieces, let's consider a simple example.

EXAMPLE 20.2
Pulse Rates,
Gender, and Smoking

How are resting pulse rates different for males and females? Does it depend on whether people smoke? On the basis of data collected in various statistics classes, the following are reasonable guesses for the population mean pulse rates, in beats per minute, for male and female smokers and nonsmokers.

	Smoker	Nonsmoker	Combined
Male	$\mu_{11} = 72$	$\mu_{12} = 70$	$\mu_{1.} = 71$
Female	$\mu_{21} = 80$	$\mu_{22} = 74$	$\mu_{2.} = 77$
Combined	$\mu_{.1} = 76$	$\mu_{.2} = 72$	$\mu_{..} = 74$

Notice that the average of 71 for male pulse rates is 3 beats lower than the overall mean of 74, and the average for females is 3 beats higher than the overall mean. Also, the mean of 76 for smokers is 2 beats above the overall mean of 74, and for nonsmokers it's 2 beats below the overall mean. Thus we might think we can define the mean for male smokers as overall mean of 74 beats −3 beats (for males) + 2 beats (for smokers) = 73 beats. But notice that in fact the mean for male smokers isn't completely explained by adding those components. The mean is 72 beats, one beat less than the added components produced. The additional −1 is due to the interaction effect. Even after accounting for the fact that males have lower pulse rates and smokers have higher pulse rates, an adjustment was needed for the combination of male and smoker. ◆

Trying to describe the individual means using only the overall mean and a contribution for each factor is called the *additive model.* Including an interaction term produces the *full model.* If the additive model describes the data almost as well as the full model, then there is little or no interaction between the two factors in their relationship to the response variable.

> *definition*   The **main effects for Factor A** are defined as $\alpha_i = \mu_{i.} - \mu_{..}$ for $i = 1, 2, \ldots, a$.
> The **main effects for Factor B** are defined as $\beta_j = \mu_{.j} - \mu_{..}$ for $j = 1, 2, \ldots, b$.
> The **interaction effects** are defined as $\alpha\beta_{ij} = \mu_{ij} - (\mu_{..} + \alpha_i + \beta_j)$ for all pairs i, j.
> The **additive model for two-way ANOVA** is defined as
>
> $$\mu_{ij} = \mu_{..} + \alpha_i + \beta_j$$
>
> The **full model for two-way ANOVA** is defined as
>
> $$\mu_{ij} = \mu_{..} + \alpha_i + \beta_j + \alpha\beta_{ij}$$
>
> Notice that the *interaction effects* are equivalent to the differences between the full model and the additive model for each mean.

It is easy to show using algebra that the sum of the $\alpha_i = 0$, the sum of the $\beta_j = 0$, and for *each* fixed value of i, the sum over j of $\alpha\beta_{ij} = 0$. Similarly, for *each* fixed value of j, the sum over i of $\alpha\beta_{ij} = 0$.

The additive model and full model can be written for individual observations as well by replacing the mean μ_{ij} in the general two-way ANOVA model with the additive or full model version. The **full model for each observation** in two-way ANOVA is

$$x_{ijk} = \mu_{ij} + \epsilon_{ijk} = \mu_{..} + \alpha_i + \beta_j + \alpha\beta_{ij} + \epsilon_{ijk}$$

where the ϵ_{ijk} are independent normal random variables with mean 0 and standard deviation σ. The **additive model for each observation** is

$$x_{ijk} = \mu_{ij} + \epsilon_{ijk} = \mu_{..} + \alpha_i + \beta_j + \epsilon_{ijk}$$

where the ϵ_{ijk} are independent normal random variables with mean 0 and standard deviation σ.

EXAMPLE 20.2—*CONTINUED*
Models for Pulse Rates

Let's find the main effects and the interaction terms for the mean pulse rates in Example 20.2. The main effects for Factor A are

$$\alpha_1 = \mu_{1.} - \mu_{..} = 71 - 74 = -3 \qquad \text{for males}$$
$$\alpha_2 = \mu_{2.} - \mu_{..} = 77 - 74 = +3 \qquad \text{for females}$$

The main effects for Factor B are

$$\beta_1 = \mu_{.1} - \mu_{..} = 76 - 74 = +2 \qquad \text{for smokers}$$
$$\beta_2 = \mu_{.2} - \mu_{..} = 72 - 74 = -2 \qquad \text{for nonsmokers}$$

Notice that in both cases, the sum of the main effects across the levels of the factor is 0. This makes sense both intuitively and algebraically based on the definition of the main effects. These values represent the amount by which each mean differs from the overall mean, either as a positive or a negative amount.

These main effects tell us that, averaged over smoking behaviors, males have an average pulse rate about 3 beats lower than the overall average, whereas females have an average pulse rate about 3 beats higher. Similarly, smokers have average pulse rates about 2 beats higher than the overall average, and nonsmokers about 2 beats lower.

As noted in Example 20.2, the additive model is not adequate to describe the individual population means. For instance, using the additive model for male smokers would result in $\mu_{11} = \mu_{..} + \alpha_1 + \beta_1 = 74 - 3 + 2 = 73$, which is not correct because $\mu_{11} = 72$. The remainder is the *interaction effect*. The interaction effect for male smokers is

$$\alpha\beta_{11} = \mu_{11} - (\mu_{..} + \alpha_i + \beta_j) = 72 - 73 = -1$$

You can verify that $\alpha\beta_{12} = \alpha\beta_{21} = +1$ and $\alpha\beta_{22} = -1$. In fact, the interaction effects must sum to 0 over each row and column.

Figure 20.2 displays an *interaction plot* for this situation, as described in Chapter 16. If there were no interaction effect, the lines would be parallel. Notice

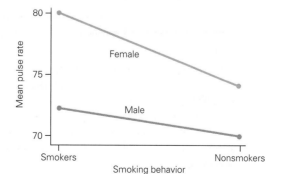

Interaction plot—Mean pulse rate

FIGURE 20.2   Interaction plot for Example 20.2

that the lines do not cross, but they are not parallel either. The plot indicates an interaction between gender and smoking behavior in their relationship with pulse rate. This can be seen by noticing that the difference between the means for males and females is larger for smokers than it is for nonsmokers. ◆

Interpreting Main Effects and Interactions

The main effects for each factor measure how much the means differ for the levels of that factor, averaged over all levels of the other factor. The interaction effects measure how much the mean for each combination of the two explanatory variables differs from what it would be if the variables were considered separately.

If there are interaction effects, it may be misleading to average over one factor when comparing the means for the other factor. An important point to keep in mind when interpreting main effects is that the interpretation is different when interactions are present than when they are not. For instance, in Example 20.2, the mean pulse rate of 76 for smokers is 4 beats per minute higher than the mean of 72 for nonsmokers. However, the increase in means for smokers over nonsmokers is only 2 beats per minute for males, whereas it is 6 beats per minute for females. Thus it would be misleading to say that smokers have a mean pulse rate 4 beats higher than nonsmokers. The effect of smoking (Factor B) should be reported separately for males and females, the categories of Factor A. Similarly, the effect of Factor A should be reported separately for the categories of Factor B. It is necessary to report the means for one factor separately for the levels of the other factor only when interactions are present.

Sometimes the interaction is so strong that the differences in means for one factor actually occur in the opposite direction for some levels of the other factor than for others. We saw this in Example 20.1 and Figure 20.1. In that example, among students who prefer to sit in the front of the room, women slept almost an hour more than men, on average. But for students who preferred the back of the room, the men slept more on average than the women. This example again illustrates that it is misleading to compare the means, in this case mean hours of sleep, for the categories of one of the factors without taking into account the other factor when interaction is present.

20.2 | TESTING FOR MAIN EFFECTS AND INTERACTIONS

There are three sets of hypotheses of interest in two-way ANOVA: one set for each main effect and one set for the interactions. Each set of hypotheses is tested using an F-test. We will cover the intermediate details only for the situation where all n_{ij} are equal because the formulas are more complicated when the sample sizes differ. When all sample sizes are equal in ANOVA, we say we have a **balanced ANOVA** or a **balanced design.** When the sample sizes are not equal, we say we have an **unbalanced design.**

In fact there are multiple approaches for testing the hypotheses in two-way ANOVA with an unbalanced design, and there is some debate about which approach is best. The details are beyond the scope of this book, and we will simply use the most common approach.

Hypotheses in Two-Way ANOVA

In one-way ANOVA we were interested in whether the population means were all equal. In two-way ANOVA we are also interested in whether the ab population means are equal. But in addition, if they do differ, we want to know whether it's Factor A, Factor B, the interaction, or some combination that is responsible for the difference. Therefore, we usually forego the overall test for a difference in means (which can be done using one-way ANOVA on the ab groups) and conduct separate tests for Factor A, Factor B, and the interaction.

In each case we can write the hypotheses in words or in symbols. We will show how to write them in words and then illustrate two ways to write them in symbols.

Hypotheses for Factor A

H_0: The population means are equal for all levels of Factor A.
H_a: The population means are not all equal for all levels of Factor A.

or

H_0: $\mu_{1.} = \mu_{2.} = \cdots = \mu_a$
H_a: Not all $\mu_{i.}$ are equal.

or

H_0: $\alpha_1 = \alpha_2 = \cdots = \alpha_a = 0$
H_a: Not all $\alpha_i = 0$.

Hypotheses for Factor B

H_0: The population means are equal for all levels of Factor B.
H_a: The population means are not all equal for all levels of Factor B.

or

H_0: $\mu_{.1} = \mu_{.2} = \cdots = \mu_{.b}$
H_a: Not all $\mu_{.j}$ are equal.

or

H_0: $\beta_1 = \beta_2 = \cdots = \beta_b = 0$
H_a: Not all $\beta_j = 0$.

Hypotheses for Interaction

H_0: The population means fit the additive model.
H_a: The population means do not fit the additive model.

or

H_0: $\mu_{ij} = \mu_{..} + \alpha_i + \beta_j$ for all i, j.
H_a: $\mu_{ij} \neq \mu_{..} + \alpha_i + \beta_j$ for at least one i, j pair.

or

H_0: All $\alpha\beta_{ij} = 0$.
H_a: Not all $\alpha\beta_{ij} = 0$.

Sums of Squares, Degrees of Freedom, and Mean Squares

Recall from Chapter 16 that in one-way ANOVA the total variability among sample observations, SS Total, can be divided into two parts, SS Groups and SS Error. SS Groups measures the amount by which the group means differ from the overall mean, and SS Error measures the amount by which individual observations differ from their own group mean.

In two-way ANOVA we can still divide SS Total into the same two pieces. But now SS Groups, which represents the amount by which the ab group means differ from the overall mean, is further divided into pieces reflecting mean differences for Factor A, Factor B, and the AB interaction.

We need some additional notation before explaining how to divide SS Groups into pieces. The sample observations can be used to estimate the main effect and interaction terms in the model, and those estimates are then used to construct sums of squares for the two factors and the interaction. The estimates are constructed by replacing the population means that define the terms with their corresponding sample versions.

Estimating Main Effect and Interaction Terms

formula **Note:** These estimates are appropriate when the sample sizes are all equal—that is, for a balanced design.

The estimate of the overall mean μ is $\hat{\mu} = \bar{x}_{..}$.
The estimates of the Factor A effects α_i are $\hat{\alpha}_i = \bar{x}_{i.} - \bar{x}_{..}$ for $i = 1, 2, \ldots, a$.
The estimates of the Factor B effects β_j are $\hat{\beta}_j = \bar{x}_{.j} - \bar{x}_{..}$ for $j = 1, 2, \ldots, b$.
The estimates of interaction effects $(\alpha\beta)_{ij}$ are $\widehat{\alpha\beta}_{ij} = \bar{x}_{ij} - (\hat{\mu} + \hat{\alpha}_i + \hat{\beta}_j)$ for all i, j pairs.

These estimates are used to allocate the variability among the sample means to sources for the two factors and interaction. Using two-way ANOVA notation, SS Groups, which measures the variability among the sample means, is written as follows:

$$\text{SS Groups} = n \sum_i \sum_j (\bar{x}_{ij} - \bar{x}_{..})^2$$

The summand, which is the amount by which the sample mean for group ij differs from the overall sample mean, can be rewritten by adding together a contribution for each factor and for interaction:

$$(\bar{x}_{ij} - \bar{x}_{..}) = (\bar{x}_{i.} - \bar{x}_{..}) + (\bar{x}_{.j} - \bar{x}_{..}) + [(\bar{x}_{ij} - (\hat{\mu} + \hat{\alpha}_i + \hat{\beta}_j)] = \hat{\alpha}_i + \hat{\beta}_j + (\widehat{\alpha\beta})_{ij}$$

If the group mean and overall sample mean are equal, then this summand will be 0. This could happen if all three of the contributions on the right are 0 or if they somehow cancel each other out. If the latter is the case, then when the terms are squared to form sums of squares, it will become evident that they are not all 0.

The SS Groups can be rewritten as follows, using the above formulation for the summand and some algebra.

Sums of Squares for Two-Way ANOVA

formula $\text{SS Groups} = n \sum_i \sum_j (\bar{x}_{ij} - \bar{x}_{..})^2 = nb \sum_i \hat{\alpha}_i^2 + na \sum_j \hat{\beta}_j^2 + n \sum_{i,j} (\widehat{\alpha\beta})_{ij}^2$

$$= \text{SSA} + \text{SSB} + \text{SSAB}$$

$$\text{SS Error} = \sum_{i,j} (n_{ij} - 1)s_{ij}^2 = \sum_i \sum_j \sum_k (x_{ijk} - \bar{x}_{ij})^2$$

$$\text{SS Total} = \text{SS Groups} + \text{SS Error}$$

Notice that SSA is defined as the sum of the squares of the individual estimated Factor A terms, and similarly for SSB and SSAB. The sum is actually taken over all *nab* individual observations, but because the term is the same for all *nb* individuals at level *i*, the sum over the subscripts *j* and *k* is summarized in the multiple *nb*. A similar simplification occurs for SSB and SSAB.

The degrees of freedom can be divided in a similar way. The technical details are beyond the level of discussion here, but the results are as follows:

Degrees of Freedom for Two-Way ANOVA

formula   **Note:** These formulas apply whether or not the sample sizes are equal for all groups

- Total degrees of freedom = $N - 1$ = df for groups + df for error = $(ab - 1) + (N - ab)$
- df for groups = $(ab - 1) = (a - 1) + (b - 1) + (a - 1)(b - 1) =$ (df for A) + (df for B) + (df for AB)

When the sample sizes are equal, df for error can be written as $(N - ab) = (nab - ab) = ab(n - 1)$, and the degrees of freedom for all of the other sources remain the same.

As in one-way ANOVA, the "mean square" for each source is MS = SS/df. Thus MSA = SSA/$(a - 1)$, MSB = SSB/$(b - 1)$, MSAB = SSAB/$(a - 1)(b - 1)$, and MS Error = SS Error/$(N - ab)$.

F-Tests for Two-Way ANOVA

The mean squares for Factor A, Factor B, and the AB interaction represent the total variability among the response variable measurements that can be attributed to each of those sources. If there is really no effect in the population for A, B, or AB, then the mean square for that term should be small relative to the error mean square (MSE). In that case, we would not want to reject the null hypothesis for that source. For instance, if H_0: all $\alpha_i = 0$ is true, then it is reasonable to expect SSA = $nb\Sigma_i\hat{\alpha}_i^2$ to be small (relative to MSE). On the other hand, if some of the α_i are large, then SSA should also be large, and in that case the correct decision would be to reject the null hypothesis and conclude that at least some of the Factor A means do differ.

To determine whether the Factor A, Factor B, and AB interaction mean squares are large enough for us to reject the corresponding null hypotheses, we construct an *F*-test statistic for each one. The denominator in each case is MSE, and the numerator is the MS for A, B, or AB. Table 20.1 summarizes the relevant information and is called a **two-way analysis of variance table.** In practice, the formulas in the table are replaced with their numerical values, and the terms *Factor A* and *Factor B* are replaced by descriptive names for the variables they represent.

In each case, test the hypotheses about the effect (A, B, AB) using the corresponding *F*-test statistic. Computer software will provide *p*-values, or Table A.4 in the Appendix can be used to find the critical values for $\alpha = 0.05$ or 0.01. (Note that we are using the symbol α here to represent the significance level as we have throughout the book, not in the two-way ANOVA context of a Factor A

TABLE 20.1 ■ **Two-Way ANOVA Table**

Source	Degrees of Freedom	Sum of Squares	Mean Square	F
Factor A	$a - 1$	SSA	$MSA = SSA/(a - 1)$	MSA/MSE
Factor B	$b - 1$	SSB	$MSB = SSB/(b - 1)$	MSB/MSE
AB interaction	$(a - 1)(b - 1)$	SSAB	$MSAB = SSAB/(a - 1)(b - 1)$	MSAB/MSE
Error	$N - ab$	SSE	$MSE = SSE/(N - ab)$	
Total	$N - 1$	SSTO		

effect. The distinction should be clear from the context.) In each case, the denominator degrees of freedom = $(N - ab)$, corresponding to MSE, and the numerator degrees of freedom are found from the appropriate row of the ANOVA table. The details of the hypotheses, test statistics, and p-values are shown in Table 20.2.

Minitab for Two-Way ANOVA

tech note
  ◆ Use **Stat>ANOVA>Two-way** if the sample sizes are all equal. Identify the response variable and the two factors (labeled as "row factor" and "column factor") in the dialog box.

  ◆ You can use **Stat>ANOVA>General linear model** whether or not the sample sizes are equal, and there are some features not available in the "Two-way" procedure. Identify the response variable. For the model, write "A B A*B" where A and B are the names of the columns containing the levels of Factors A and B, respectively. In the **Options** box, click "Adjusted (Type III)" for Sums of Squares (the default).

  ◆ If you would like to compare the means for one or both factors, or for the *ab* individual means, use **Stat>ANOVA>General linear model** and open the **Comparisons** box. Enter the terms for which you want comparisons in the box labeled "Terms" using the names of the columns for Factor A and Factor B and using the names with "*" between them for the *ab* means. Click the "Tukey" box, and the display will show pairwise comparisons using the Tukey method, as described in Chapter 16 for one-way ANOVA.

TABLE 20.2 ■ **Hypotheses and Test Statistics for Two-Way ANOVA**

Effect	Null Hypothesis	F-Test Statistic	p-value = Area Above Value of F-Test Statistic for F-Distribution with df of:
Factor A	H_0: All $\mu_{i.}$ equal	MSA/MSE	$(a - 1), N - ab$
Factor B	H_0: All $\mu_{.j}$ equal	MSB/MSE	$(b - 1), N - ab$
AB interaction	H_0: All $(\alpha\beta)_{ij} = 0$	MSAB/MSE	$(a - 1)(b - 1), N - ab$

Examples and Interpretation of Two-Way ANOVA Results

It is instructive to consider three types of examples because the results are interpreted differently in each case. The interpretation depends on whether there are significant interactions and whether there are significant main effects for one or both variables. We cover examples for each of these three situations:

1. Significant main effects but no significant interaction
2. One significant main effect, one nonsignificant main effect, and significant interaction
3. Significant main effects and significant interaction

Be very careful about interpreting main effects if interaction is present. In that case, the differences in means for the levels of Factor A should be examined separately for each level of Factor B, and vice versa. We have already seen this concept in Examples 20.1 and 20.2. In both cases there appeared to be an interaction between the two factors, and we examined the difference in means for one factor separately for the levels of the other factor.

EXAMPLE 20.3
Nature Versus
Nurture in IQ Scores

Are the IQs of children related to the socio-economic status (SES) of their parents? If children are adopted, are their IQs related to the SES of their biological parents, their adoptive parents, or some combination? These are the questions French researchers set out to investigate in a study. They identified children from adoption registers in France whose biological and adoptive parents were from the highest and the lowest SES categories. SES was determined by occupation and years of education. Factor A is the SES of the adoptive parents (high, low) and Factor B is the SES of the biological parents (high, low). They attempted to create a balanced design with ten children for all groups but found only eight for the combination with high SES for biological parents but low SES for the adoptive parents. The IQs and categories for the 38 children are shown in Table 20.3. (*Source:* Data reported in Ramsey and Schafer, 1997, p. 396; adapted from Capron and Duyme, 1991.)

The null hypotheses for this example can be written in context as follows:

For Factor A (adoptive parents' SES):
H_0: The mean IQs for the populations of children with high-SES and low-SES adoptive parents are equal.

For Factor B (biological parents' SES):
H_0: The mean IQs for the populations of adopted children with high-SES and low-SES biological parents are equal.

TABLE 20.3 ■ Data for Example 20.3

Adoptive Parents' SES (Factor A)	Biological Parents' SES (Factor B)	
	High	**Low**
High	136, 99, 121, 133, 125, 131, 103, 115, 116, 117	94, 103, 99, 125, 111, 93, 101, 94, 125, 91
Low	98, 99, 91, 124, 100, 116, 113, 119	92, 91, 98, 83, 99, 68, 76, 115, 86, 116

For the AB interaction:

H_0: The relationship between mean IQ and adoptive parents' SES status does not depend on the biological parents' SES.

Figure 20.3 displays an interaction plot for this example, showing the means for the four categories of the two factors. Notice that the lines are parallel. This indicates that there is little or no interaction. The difference between mean IQ of children of high- and low-SES adoptive parents is about the same, regardless of the SES of their biological parents.

The plot also indicates that both main effects are present. The gap between the lines for low and high adoptive parents' SES indicates that the means for the two levels of Factor A differ (by the amount shown by the gap, about 12 points). The slope of the lines indicates that the means for the two levels of Factor B differ. Notice that for both lines, the mean IQ for children of high-SES biological parents is about 16 points higher than the mean IQ for children of low-SES biological parents.

The individual means, analysis of variance table, and other results produced by Minitab for this example are shown in Figure 20.4. The two different SS columns, "Seq SS" and "Adj SS," represent two methods for analyzing an unbalanced design. The results are based on the "Adj SS," as is common practice. When the design is not balanced, the three hypothesis tests of interest are not independent. The "Adj SS" method tests for each effect after adjusting for the presence of the other effects. The "Seq SS" method tests the effects in order, starting with Factor A, and thus the results depend on which factor is designated as A and which is designated as B.

The *p*-value for the test of the Factor A effect is 0.011, so the null hypothesis can be rejected. Remember that the null hypothesis for Factor A stated that the population mean IQs are equal for children of high-SES and low-SES adoptive parents. This is clearly not the case. In fact, the sample means for the children of high- and low-SES adoptive parents are 111.60 and 99.11, respectively, a difference of over 12 IQ points. The *p*-value confirms that the difference seen in the sample results reflects a real difference in the population means.

The *p*-value for the test of Factor B effect is 0.001, so the null hypothesis can be rejected. The null hypothesis for Factor B stated that the population mean IQs are equal for children of high-SES and low-SES biological parents. This is also clearly not true; the sample means are 114.22 and 98.00, respectively, and the low *p*-value confirms that the difference in the sample means reflects a real difference in the population means.

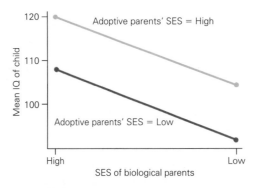

FIGURE 20.3 **Interaction plot for SES and IQ in Example 20.3**

Mean IQ for Adopted Children
Rows: ADOPTIVE   Columns: BIOLOGIC

	High	Low	All
High	119.60	103.60	111.60
Low	107.50	92.40	99.11
All	114.22	98.00	105.68

Two-Way ANOVA Using General Linear Model
General Linear Model: IQ versus ADOPTIVE, BIOLOGIC

Factor	Type	Levels	Values
ADOPTIVE	fixed	2	High Low
BIOLOGIC	fixed	2	High Low

Analysis of Variance for IQ, Using Adjusted SS for Tests

Source	DF	Seq SS	Adj SS	Adj MS	F	P
ADOPTIVE	1	1477.6	1277.4	1277.4	7.31	0.011
BIOLOGIC	1	2291.5	2275.8	2275.8	13.02	0.001
ADOPTIVE*BIOLOGIC	1	1.9	1.9	1.9	0.01	0.917
Error	34	5941.2	5941.2	174.7		
Total	37	9712.2				

FIGURE 20.4   **Minitab results for Example 20.3**

The *p*-value for the test of AB interaction is 0.917, so the null hypothesis cannot be rejected. Consistent with the interaction plot, this *p*-value confirms that the difference in mean IQs for the levels of one factor does not depend much on the level of the other factor.

In summary, the mean IQs of children are related to the SES of both their biological and adoptive parents. Children whose adoptive parents have high SES have average IQs about 6.6 points higher than the overall average, and children whose biological parents have high SES have average IQs about 7.7 points higher than the overall average. The combined effect is additive, so that children for whom both sets of parents have high SES have average IQs about 14 points above the overall average. This can be seen in Figure 20.4; the mean for the children with high SES for both sets of parents is 119.6, whereas the overall mean is 105.68. ◆

EXAMPLE 20.4
Happy Faces and Restaurant Tips Revisited

In Chapter 16 we presented an example in which a male and a female server in a restaurant drew a "happy face" on some checks presented to customers and not on others, randomly determined. The percent tip accompanying each check was recorded. The researcher was interested in knowing if the happy face would increase or decrease the tip percent, and if it depended on whether the server was male or female. One shortcoming of the study was that only one server of each sex was used, so the results may apply to those two servers only. Nonetheless, it is a useful example for illustrating a somewhat unusual situation. There is a significant interaction effect, but only one of the main effects was significant.

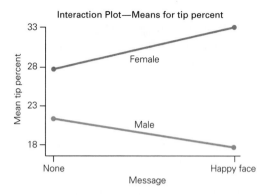

FIGURE 20.5   **Mean tip percent related to sex of server and message on check (none or happy face)**

The response variable for this example is tip percent. Factor A is the message on the check, with level 1 = none and level 2 = happy face. Factor B is sex, with level 1 = female and level 2 = male.

Figure 20.5, reprinted from Chapter 16, shows the interaction plot. Notice that the lines are not parallel. For the female, tips were higher with the happy face message, whereas for the male, tips were lower with the happy face message. This difference indicates that there is an interaction effect. The plot also demonstrates that there is a sex effect. Overall, the mean tip for the female is much higher than that for the male. However, it appears from the plot that there is not a "message" effect. Notice that the mean tip percent for no message is about the same as the mean tip percent for the happy face message, when averaged across the two sexes.

The means and analysis of variance results are shown in Figure 20.6. The numerical results confirm what the plot illustrated. The p-value for testing for the

Means for Tip Percent

Message	Female	Male	All
Happy Face	33.041	17.776	25.239
None	27.784	21.406	24.740
All	30.354	19.508	24.992

Two-Way ANOVA Using General Linear Model

General Linear Model

Factor	Type	Levels	Values	
Message	fixed	2	Happy Face	None
Sex	fixed	2	Female	Male

Analysis of Variance for TipPct, Using Adjusted SS for Tests

Source	DF	Seq SS	Adj SS	Adj MS	F	P
Message	1	5.5	14.7	14.7	0.13	0.715
Sex	1	2628.1	2602.0	2602.0	23.69	0.000
Message*Sex	1	438.7	438.7	438.7	3.99	0.049
Error	85	9335.5	9335.5	109.8		
Total	88	12407.9				

FIGURE 20.6   **Means and analysis of variance table for Example 20.4**

AB interaction is 0.049, so the null hypothesis of "no interaction" can be rejected using $\alpha = 0.05$. The p-value for the Factor B (sex) effect is 0.000, indicating a strong difference in average tip percents for the male and female servers.

However, the p-value for testing for a Factor A effect is 0.715, so the null hypothesis cannot be rejected. Remember that the null hypothesis for Factor A in this example is that the population mean tip percent is the same when no message is written on the check as when the happy face is written. We have seen that this is not true for the female, whose tips went up with the happy face, nor is it true for the male, whose tips went down for the happy face. It is true only when averaged across them. This illustrates the danger of interpreting a main effect overall, in the presence of interaction. When interaction is present, it is very important to examine the means for the different levels of one factor separately for the levels of the other factor. In this example, it is important to examine the difference in means for the two messages separately for the male and the female. ◆

EXAMPLE 20.5
Does Smoking Lead to More Errors?

This example is adapted from a study of the relationship between smoking and performance on three types of tasks. The study was done by Spilich, June, and Renner (1992). Howell (1997, p. 416) simplified the data for illustrative purposes; the data presented here are a slight modification based on communication with Professor Howell, who thought that this adaptation was more similar to the original results.

Participants were asked to perform one of three types of tasks. Factor A is the type of task performed. The first type (level 1) is pattern recognition, in which participants had to locate a target on a computer screen. The second type (level 2) is a cognitive task involving recall of a passage read earlier. The third type (level 3) is a driving simulation computer game. For Factor B, participants were categorized by their smoking behavior. Level 1 is nonsmoker, level 2 is for smokers who did not smoke within 3 hours of the task, and level 3 is for participants who were smoking during the task. The response variable is the number of errors made while performing the task.

The data are presented in Table 20.4. An interaction plot is shown in Figure 20.7, and analysis of variance results are shown in Figure 20.8. From the interaction plot, the most obvious result is that there are many more errors in the

TABLE 20.4 ■ Data for Example 20.5

Pattern Recognition														
Nonsmoker	9,	8,	12,	10,	7,	10,	9,	11,	8,	10,	8,	10,	8,	11,	10
>3 Hours	12,	7,	14,	4,	8,	11,	16,	17,	5,	6,	9,	6,	6,	7,	16
Active Smoker	8,	8,	9,	1,	9,	7,	16,	19,	1,	1,	22,	12,	18,	8,	10

Cognitive Task														
Nonsmoker	27,	34,	19,	20,	56,	35,	23,	37,	4,	30,	4,	42,	34,	19,	49
>3 Hours	48,	29,	34,	6,	18,	63,	9,	54,	28,	71,	60,	54,	51,	25,	49
Active Smoker	34,	65,	55,	33,	42,	54,	21,	44,	61,	38,	75,	61,	51,	32,	47

Driving Simulation														
Nonsmoker	3,	2,	0,	0,	6,	2,	0,	6,	4,	1,	0,	0,	6,	2,	3
>3 Hours	7,	0,	6,	0,	12,	17,	1,	11,	4,	4,	3,	5,	16,	5,	11
Active Smoker	15,	2,	2,	14,	5,	0,	16,	14,	9,	17,	15,	9,	3,	15,	13

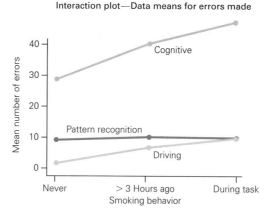

FIGURE 20.7   Interaction plot for Example 20.5

cognitive task than in the other two tasks. Is there an interaction effect? It appears that the mean number of errors increases as smoking behavior gets more immediate for the cognitive task and the driving task, but does change much for the pattern recognition task. If that pattern is statistically significant, it is interesting because it indicates that smoking is related to performance in memory (the cognitive task) and in simulated driving, but not in pattern recognition.

The analysis of variance results in Figure 20.8 confirm that there is indeed a significant interaction (p-value = 0.023) and that there are significant Factor A and Factor B effects. Because of the significant interaction effect, interpretation of the differences in factor level means should be done within the levels of the other factor. Notice that the means shown in Figure 20.8 confirm that for the sample, the mean number of errors for the pattern recognition task remained relatively constant across smoking behaviors. The mean number of errors for the other two tasks was highest for active smokers and lowest for nonsmokers, with those who had not smoked recently falling between them. ◆

Mean Number of Errors Made

	Nonsmoker	>3hrs	During	All
Pattern Rec	9.400	9.600	9.933	9.644
Cognitive	28.867	39.933	47.533	38.778
Driving	2.333	6.800	9.933	6.356
All	13.533	18.778	22.467	18.259

Analysis of Variance for Errors Made, Using Adjusted SS for Tests

Source	DF	Seq SS	Adj SS	Adj MS	F	P
Task	2	28661.5	28661.5	14330.8	132.90	0.000
SmokeGrp	2	1813.7	1813.7	906.9	8.41	0.000
Task*SmokeGrp	4	1269.5	1269.5	17.4	2.94	0.023
Error	126	13587.2	13587.2	107.8		
Total	134	45331.9				

FIGURE 20.8   Means and analysis of variance results for Example 20.5

Cautions About Nonsignificant Results

We end this discussion with a reminder that applies to all hypothesis-testing situations. Remember that a nonsignificant result does not mean that the null hypothesis is true. It simply means that the sample data didn't provide enough evidence for us to conclude that there is an effect in the population.

EXAMPLE 20.1—*CONTINUED*
Seat Location,
Gender, and Hours of Sleep

We began our discussion of two-way ANOVA with an example for which the interaction plot indicated that there would be a strong interaction as well as main effects. Figure 20.1 showed the mean hours of sleep for male and female students who prefer to sit in the front, middle, and back of the class. The ANOVA results for this example are presented in Figure 20.9. Notice that none of the effects is statistically significant, as reflected by the *p*-values, which are all much greater than 0.05. Why do the ANOVA results appear to be contradictory to the interaction plot? The answer is that there is a large amount of variability in the data. The interaction plot does not take the standard deviations into account; it is a plot of the means.

If you look at the standard deviations in Figure 20.9, you will see that they range from about 1.3 hours to 2.0 hours of sleep. The natural variability in sleep times, reflected in the denominator of the *F*-test through MSE, is so large that any difference that might exist in the population means cannot be detected with this sample.

The lesson is that it is not enough to look at either an interaction plot or an analysis of variance table alone. The combination provides a picture of the main effects and interactions in the sample, as well as a gauge for whether we can conclude that they exist in the population from which the sample was drawn. ◆

Means and Standard Deviations

	Front	Middle	Back	All
Female	7.1731	7.1085	6.9667	7.1037
S.D.	1.7085	1.6623	2.0219	1.7174
Male	6.1333	6.7250	7.1818	6.7403
S.D.	1.2743	1.9080	1.4601	1.6987
All	6.7927	6.9435	7.0946	6.9401
S.D.	1.6278	1.7724	1.6867	1.7136

Analysis of Variance for Sleep, Using Adjusted SS for Tests

Source	DF	Seq SS	Adj SS	Adj MS	F	P
Sex	1	5.592	5.589	5.589	1.91	0.169
Seat	2	2.959	3.410	1.705	0.58	0.560
Sex*Seat	2	7.273	7.273	3.637	1.24	0.292
Error	165	483.374	483.374	2.930		
Total	170	499.198				

FIGURE 20.9   **Means, standard deviations, and analysis of variance for sleep by gender and seat location**

KEY TERMS

Section 20.1

main effect, 650, 654

interaction effect, 650, 654

one-way ANOVA model, 651

factor, 652

general two-way ANOVA
  model, 653

additive model for two-way
  ANOVA, 654, 655

full model for two-way ANOVA, 654

Section 20.2

balanced ANOVA, 656

balanced design, 656

unbalanced design, 656

two-way analysis of variance table,
  659–660

EXERCISES   Blue = Basics   **Bold, Bold** = Answer in back

20.1 For each of the following two-way ANOVA situations specify the response variable, the two factors and values for *a* and *b*, and the number of levels for Factors A and B, respectively.

    a. One hundred overweight men are classified by whether they smoke or not. They are randomly assigned to participate in a swimming program, a jogging program, or a yoga program. Weight loss after two months is measured.

    b. A random sample of kindergarten children in a certain state is given a reading test. The children are categorized by whether they attended preschool (regularly, some, not at all) and whether they have older siblings at home (yes, no).

20.2 Researchers measure the salaries and years of experience for a random sample of male and female professors at a university. They want to know whether there is a difference in the mean salaries of males and females, adjusted for years of experience. Is this a situation for which two-way ANOVA is appropriate? If so, specify the response variable and the two factors. If not, explain why not.

20.3 A plant biologist randomly assigns seedlings to one of four conditions: sunlight and drip watering, artificial light and drip watering, sunlight and hand watering, artificial light and hand watering. He measures the biomass of each plant after three weeks. Is this a situation for which two-way ANOVA is appropriate? If so, specify the response variable and the two factors. If not, explain why not.

Use the following scenario for Exercises 20.4 to 20.9.

    Suppose that the population mean hours of sleep for college students, categorized by whether they are coffee drinkers and whether they are over 21, are as follows:

	Coffee Drinker?	
	Yes	No
18–21	6.0	8.0
Over 21	7.0	7.4

20.4 Draw an interaction plot of these means. Comment on whether there appear to be effects for Factor A (age) and Factor B (coffee-drinking behavior), and whether there appears to be an interaction between them.

20.5 Give numerical values for each of the following:

    a. μ_{12}

    b. $\mu_{..}$

    c. $\mu_{i.}$ for $i = 1, 2$

    d. $\mu_{.j}$ for $j = 1, 2$

20.6 Give numerical values for each of the following:

    a. α_i for $i = 1, 2$

    b. β_j for $j = 1, 2$

    c. $\alpha\beta_{ij}$ for all i, j pairs

20.7 Based on relevant numerical values found in the previous two exercises, write each of the means μ_{ij} using the additive model. Is the additive model adequate in this situation?

20.8 Write each of the means μ_{ij} using the full model, showing the numerical value of each of the terms in the model.

20.9 Explain in words what the table of means indicates in terms of main effects and interaction in this situation.

Use the following scenario for Exercises 20.10 to 20.14.

    A study was done to investigate the effect of the pesticide DDT in falcons. The response variable was the amount of DDT measured in each bird. Factor A was the nesting site of the bird (1 = United States, 2 = Canada, 3 = Arctic region) and Factor B was the age of the bird (1 = Young, 2 = Middle-aged, 3 = Old). The data, means, and analysis of variance table are shown on the next page, and the interaction plot is shown in the accompanying figure. (*Source:* Falcon.MTW dataset in the Student1 Directory of Minitab files.)

20.10 Based on the interaction plot, comment on whether there appears to be

    a. A Factor A (nesting site) main effect.

    b. A Factor B (age) main effect.

    c. An interaction between the factors.

20.11 State the null and alternative hypotheses in the context of this problem for

    a. A Factor A effect.

    b. A Factor B effect.

    c. An interaction between the factors.

20.12 Conduct and interpret the hypothesis test for this situation for

    a. A Factor A effect.

    b. A Factor B effect.

    c. An interaction between the factors.

20.13 Based on examining the sample means and the analysis of variance table, explain in words

    a. The effect of nesting site on DDT in falcons.

    b. The effect of age on DDT in falcons.

    c. In this example, does the interpretation of the effects of one factor depend on the levels of the other factor? Explain why or why not.

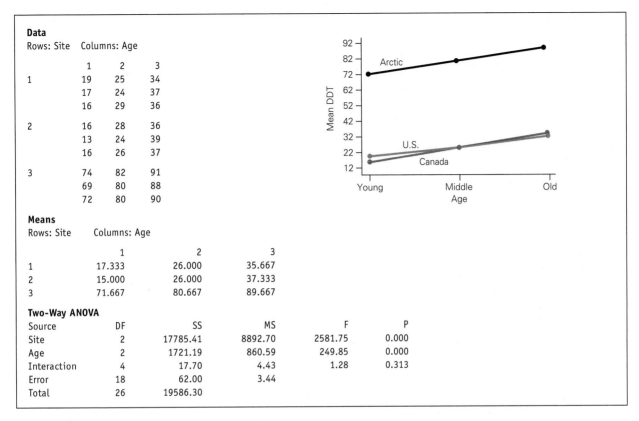

Data

Rows: Site   Columns: Age

	1	2	3
1	19	25	34
	17	24	37
	16	29	36
2	16	28	36
	13	24	39
	16	26	37
3	74	82	91
	69	80	88
	72	80	90

Means

Rows: Site   Columns: Age

	1	2	3
1	17.333	26.000	35.667
2	15.000	26.000	37.333
3	71.667	80.667	89.667

Two-Way ANOVA

Source	DF	SS	MS	F	P
Site	2	17785.41	8892.70	2581.75	0.000
Age	2	1721.19	860.59	249.85	0.000
Interaction	4	17.70	4.43	1.28	0.313
Error	18	62.00	3.44		
Total	26	19586.30			

EXERCISES 20.10 TO 20.14

20.14 Show numerically how the F-test statistic was created for the A, B, and AB interaction tests by plugging in the appropriate mean squares.

Use the following scenario for Exercises 20.15 to 20.20.

Jennifer lives in the San Francisco Bay Area and drives to work. She has a choice of three routes (Factor A) and wants to know if any of them is faster than the others. She suspects that it may depend on whether it's Monday, Friday, or a mid-week day (Factor B), due to vacationers leaving the area on Fridays and returning on Mondays. She randomly chooses which route to take each day until she has taken each route 10 times for each of Monday, mid-week day, and Friday, for a total of 90 measurements (3 routes × 3 types of day × 10 observations). The response variable is commute time. The means are shown in the following table.

	Monday	Mid-week	Friday
Route 1	34.2	30.8	32.1
Route 2	22.7	24.6	26.0
Route 3	38.6	34.1	32.9

She computes these sums of squares: SSA = 1868.9, SSB = 65, SSAB = 229.33, SSE = 552.8.

20.15 Construct the analysis of variance table for this problem.

20.16 Construct an interaction plot for this situation.

20.17 State the null and alternative hypotheses in the context of this problem for
   a. A Factor A effect.
   b. A Factor B effect.
   c. An interaction between the factors.

20.18 Conduct and interpret the hypothesis test for this situation for
   a. A Factor A effect.
   b. A Factor B effect.
   c. An interaction between the factors.

20.19 Based on examining the sample means and the analysis of variance table, explain in words
   a. Whether there is a significant difference in commute time for different routes.
   b. Whether there is a significant difference in commute times for different days.
   c. In this example, does the interpretation of the effects of one factor depend on the levels of the other factor? Explain why or why not.

20.20 What route would you recommend that Jennifer take to work? Does it depend on the day of the week? Explain.

20.21 Explain why an interaction plot may show lines that are not parallel, or even that cross, yet the analysis of variance results show a nonsignificant interaction effect.

20.22 Fill in the missing values in the following two-way analysis of variance table, where the letters occupying the space indicate the parts to this exercise. For instance, for part (a) give the degrees of freedom for Factor B, and for part (b) give MSA.

Analysis of Variance				
Source	DF	SS	MS	F
Factor A	2	450	b	<u>f</u>
Factor B	a	840	c	<u>g</u>
Interaction	2	130	<u>d</u>	<u>h</u>
Error	48	1050	<u>e</u>	
Total	53	2470		

20.23 Refer to the previous exercise. Carry out the hypothesis test for each of the following:
a. Factor A.
b. Factor B.
c. AB interaction.

20.24 Refer to the previous two exercises. The results came from a balanced design. Based on the analysis of variance table, give a numerical value for each of the following:
a. The number of levels of Factor A, a.
b. The number of levels of Factor B, b.
c. The sample size for each group, n.
d. The overall sample size, N.

20.25 Refer to the previous three exercises. If a one-way analysis of variance had been done using the ab groups, instead of a two-way analysis of variance, give a numerical value for each of the following:
a. SS Groups.
b. SS Error.
c. Degrees of freedom for groups.
d. Degrees of freedom for error.

REFERENCES

Capron, C., and M. Duyme (1991). "Children's IQs and SES of biological and adoptive parents in a balanced cross-fostering study," *European Bulletin of Cognitive Psychology,* Vol. 11, No. 3, pp. 323–348.

Howell, David C. (1997). *Statistical Methods for Psychology,* 4th ed., Belmont, CA: Duxbury.

Ramsey, Fred L., and Daniel W. Schafer (1997). *The Statistical Sleuth: A Course in Methods of Data Analysis,* Belmont, CA: Duxbury.

Spilich, G. J., L. June, and J. Renner (1992). "Cigarette smoking and cognitive performance," *British Journal of Addiction,* Vol. 87, 1313–1326.

Ethics

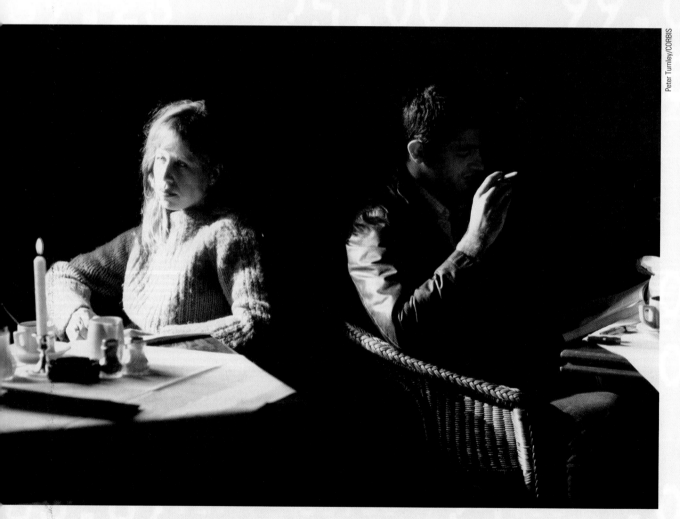

Peter Turnley/CORBIS

Is his smoking putting her at risk? See Example 21.4.

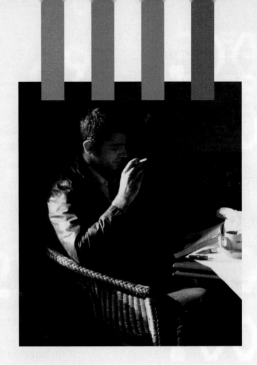

Ethical issues in experiments include ensuring proper treatment of human and animal participants, making sure data are of high quality, carrying out appropriate statistical analyses, and reporting the results in a fair and unbiased way. Many professional societies have a code of ethics for their members to follow when conducting research.

As with any human endeavor, in statistical studies there are ethical considerations that must be taken into account. These considerations fall into multiple categories, and we discuss the following issues here:

1. Ethical treatment of human and animal participants
2. Assurance of data quality
3. Appropriate statistical analyses
4. Fair reporting of results

Most professional societies have a code of ethics that their members are asked to follow. Guidelines about the conduct of research studies are often included. For instance, the American Psychological Association first published a code of ethics in 1953 and has updated it at least every ten years since then. In 1999 the American Statistical Association, one of the largest organizations of professional statisticians, published *Ethical Guidelines for Statistical Practice* listing 67 recommendations, divided into categories such as "Professionalism" and "Responsibilities to Research Subjects." ◆

21.1 | ETHICAL TREATMENT OF HUMAN AND ANIMAL PARTICIPANTS

There has been increasing attention paid to the ethics of experimental work, partly as a consequence of some deceptive experiments that led to unintended harm to participants. Here is a classic example of an experiment conducted in the early 1960s that would probably be considered unethical today.

Social psychologist Stanley Milgram was interested in the extent to which ordinary citizens would obey an authority figure, even if it meant injuring another human being. Through newspaper ads offering people money to participate in an experiment on learning, he recruited people in the area surrounding Yale University, where he was a member of the faculty. When participants arrived, they were greeted by an authoritative researcher in a white coat and introduced to another person who they were told was a participant like them but who was actually an actor. Lots were drawn to see who would be the "teacher" and who would be the "student," but in fact it had been predetermined that the actor would be the "student" and the local citizen would be the "teacher."

The student/actor was placed in a chair with restricted movement and hooked up to what was alleged to be an electrode that administered an electric shock. The teacher conducted a memory task with the student and was instructed to administer a shock when the student missed an answer. The shocking mechanism was shown to start at 15 volts and increase in intensity with each wrong answer, up to 450 volts. When the alleged voltage reached 375, it was labeled as "Danger/Severe" and when it reached 435 it was labeled "XXX." The experimenter sat at a nearby table, encouraging the teacher to continue to administer the shocks. The student/actor would respond with visible and increasing distress. The teacher was told that the experimenter would take responsibility for any harm that came to the student.

The disturbing outcome of the experiment was that 65% of the participants continued to administer the alleged shocks up to the full intensity, even though many of them were quite distressed and nervous about doing so. Even at "very strong" intensity, 80% of the participants were still administering the electric shocks. (*Sources:* http://www.vanderbilt.edu/AnS/Anthro/Anth101/stanley_milgram_experiment.htm and Milgram (1983).) ◆

This experiment would be considered unethical today because of the stress it caused the participants. The American Psychological Association continues to update its *Ethical Principles of Psychologists and Code of Conduct.* The latest version as of fall 2004 can be found at http://www.apa.org/ethics/code2002.html. One of the sections of the code is called "Deception in Research" and includes three instructions (http://www.apa.org/ethics/code2002.html#8_07); Milgram's experiment would most likely fail the criterion in part (b):

(a) Psychologists do not conduct a study involving deception unless they have determined that the use of deceptive techniques is justified by the study's significant prospective scientific, educational, or applied value and that effective nondeceptive alternative procedures are not feasible.

(b) Psychologists do not deceive prospective participants about research that is reasonably expected to cause physical pain or severe emotional distress.

(c) Psychologists explain any deception that is an integral feature of the design and conduct of an experiment to participants as early as is feasible, preferably at the conclusion of their participation, but no later than at the conclusion of the data collection, and permit participants to withdraw their data.

Informed Consent

Virtually all experiments with human participants require that the researchers obtain the **informed consent** of the participants. In other words, participants are to be told what the research is about and given an opportunity to make an informed choice about whether to participate. If you were a potential participant in a research study, what would you want to know in advance to make an informed choice about participation? Because of such issues as the need for a control group and the use of double-blinding, it is often the case that participants cannot be told everything in advance. For instance, it would be antithetical to good experimental procedure to tell participants in advance whether they were taking a drug or a placebo, or to tell them whether they were in the treatment group or the control group. Instead, the use of multiple groups is explained and participants are told that they will be randomly assigned to a group but will not know what it is until the conclusion of the experiment.

The information provided in this process is slightly different in such research as psychology experiments than it is in medical research. In both cases, participants are supposed to be told the nature and purpose of the research and to be informed about any risks or benefits. In medical research the participants generally are suffering from a disease or illness, and an additional requirement is that they be informed about alternative treatments. Of course, it is unethical to withhold a treatment known to work.

In section 8.02 of its code of ethics, the American Psychological Association provides these guidelines for informed consent in experiments in psychology (http://www.apa.org/ethics/code2002.html#8_02):

> (a) When obtaining informed consent as required in Standard 3.10, Informed Consent, psychologists inform participants about (1) the purpose of the research, expected duration, and procedures; (2) their right to decline to participate and to withdraw from the research once participation has begun; (3) the foreseeable consequences of declining or withdrawing; (4) reasonably foreseeable factors that may be expected to influence their willingness to participate such as potential risks, discomfort, or adverse effects; (5) any prospective research benefits; (6) limits of confidentiality; (7) incentives for participation; and (8) whom to contact for questions about the research and research participants' rights. They provide opportunity for the prospective participants to ask questions and receive answers.

> (b) Psychologists conducting intervention research involving the use of experimental treatments clarify to participants at the outset of the research (1) the experimental nature of the treatment; (2) the services that will or will not be available to the control group(s) if appropriate; (3) the means by which assignment to treatment and control groups will be made; (4) available treatment alternatives if an individual does not wish to participate in the research or wishes to withdraw once a study has begun; and (5) compensation for or monetary costs of participating including, if appropriate, whether reimbursement from the participant or a third-party payor will be sought.

Informed Consent in Medical Research

The United States Department of Health and Human Services has a detailed policy on informed consent practices in medical research, which can be accessed on the Internet at http://www.hhs.gov/ohrp/humansubjects/guidance/ictips.htm. A more "user-friendly" website is provided by the Department of Health and Human Services Office for Human Research Protections, Office for Protection from Research Risks, which has provided the following tips and checklist.

TIPS ON INFORMED CONSENT

The process of obtaining informed consent must comply with the requirements of 45 CFR 46.116. The documentation of informed consent must comply with 45 CFR 46.117. The following comments may help in the development of an approach and proposed language by investigators for obtaining consent and its approval by IRBs:[1]

◆ Informed consent is a process, not just a form. Information must be presented to enable persons to voluntarily decide whether or not to participate as a research subject. It is a fundamental mechanism to ensure respect for persons through provision of thoughtful consent for a voluntary act. The procedures used in obtaining informed consent should be designed to educate the subject population in terms that they can understand. Therefore, informed consent language and its documentation (especially explanation of the study's purpose, duration, experimental procedures, alternatives, risks, and benefits) must be written in "lay language" (i.e., understandable to the people being asked to participate). The written presentation of information is used to document the basis for consent and for the subjects' future reference. The consent document should be revised when deficiencies are noted or when additional information will improve the consent process.

◆ Use of the first person (e.g., "I understand that . . .") can be interpreted as suggestive, may be relied upon as a substitute for sufficient factual information, and can constitute coercive influence over a subject. Use of scientific jargon and legalese is not appropriate. Think of the document primarily as a teaching tool not as a legal instrument.

◆ **Describe the overall experience that will be encountered.** Explain the research activity, how it is experimental (e.g., a new drug, extra tests, separate research records, or nonstandard means of management, such as flipping a coin for random assignment or other design issues). Inform the human subjects of the reasonably foreseeable harms, discomforts, inconvenience, and risks that are associated with the research activity. If additional risks are identified during the course of the research, the consent process and documentation will require revisions to inform subjects as they are recontacted or newly contacted.

◆ **Describe the benefits that subjects may reasonably expect to encounter.** There may be none other than a sense of helping the public at large. If payment is given to defray the incurred expense for participation, it must not be coercive in amount or method of distribution.

◆ **Describe any alternatives to participating in the research project.** For example, in drug studies the medication(s) may be available through their family doctor or clinic without the need to volunteer for the research activity.

◆ **The regulations insist that the subjects be told the extent to which their personally identifiable private information will be held in confidence.** For example, some studies require disclosure of information to other parties. Some studies inherently are in need of a Certificate of Confidentiality which protects the investigator from involuntary release

[1] The IRB (Institutional Review Board) is a board that all research institutions are required to maintain for oversight of research involving human and animal participants.

(e.g., subpoena) of the names or other identifying characteristics of research subjects. The IRB will determine the level of adequate requirements for confidentiality in light of its mandate to ensure minimization of risk and determination that the residual risks warrant involvement of subjects.

◆ **If research-related injury (i.e., physical, psychological, social, financial, or otherwise) is possible in research that is more than minimal risk (see 45 CFR 46.102[g]), an explanation must be given of whatever voluntary compensation and treatment will be provided.** Note that the regulations do not limit injury to "physical injury." This is a common misinterpretation.

◆ **The regulations prohibit waiving or appearing to waive any legal rights of subjects.** Therefore, for example, consent language must be carefully selected that deals with what the institution is *voluntarily* willing to do under circumstances such as providing for compensation beyond the provision of immediate or therapeutic intervention in response to a research-related injury. In short, subjects should not be given the impression that they have agreed to and are without recourse to seek satisfaction beyond the institution's *voluntarily* chosen limits.

◆ **The regulations provide for the identification of contact persons who would be knowledgeable to answer questions of subjects about the *research, rights as a research subject,* and *research-related injuries.* These three areas must be explicitly stated and addressed in the consent process and documentation.** Furthermore, a single person is not likely to be appropriate to answer questions in all areas. This is because of potential conflicts of interest or the appearance of such. Questions about the research are frequently best answered by the investigator(s). However, questions about the rights of research subjects or research-related injuries (where applicable) may best be referred to those not on the research team. These questions could be addressed to the IRB, an ombudsman, an ethics committee, or other informed administrative body. Therefore, each consent document can be expected to have at least two names with local telephone numbers for contacts to answer questions in these specified areas.

◆ **The statement regarding voluntary participation and the right to withdraw at any time can be taken almost verbatim from the regulations** (45 CFR 46.116[a][8]). It is important not to overlook the need to point out that no penalty or loss of benefits will occur as a result of *both* not participating or withdrawing at any time. It is equally important to alert potential subjects to any foreseeable consequences to them should they unilaterally withdraw while dependent on some intervention to maintain normal function.

◆ Don't forget to ensure provision for appropriate **additional requirements** which concern consent. Some of these requirements can be found in sections 46.116(b), 46.205(a)(2), 46.207(b), 46.208(b), 46.209(d), 46.305(a) (5–6), 46.408(c), and 46.409(b). The IRB may impose additional requirements that are not specifically listed in the regulations to ensure that adequate information is presented in accordance with institutional policy and local law.

Source: http://www.hhs.gov/ohrp/humansubjects/guidance/ictips.htm

INFORMED CONSENT CHECKLIST

Basic and Additional Elements

☐   A statement that the study involves research

☐   An explanation of the purposes of the research

☐   The expected duration of the subject's participation

☐   A description of the procedures to be followed

☐   Identification of any procedures which are experimental

☐   A description of any reasonably foreseeable risks or discomforts to the subject

☐   A description of any benefits to the subject or to others which may reasonably be expected from the research

☐   A disclosure of appropriate alternative procedures or courses of treatment, if any, that might be advantageous to the subject

☐   A statement describing the extent, if any, to which confidentiality of records identifying the subject will be maintained

☐   For research involving more than minimal risk, an explanation as to whether any compensation, and an explanation as to whether any medical treatments are available, if injury occurs and, if so, what they consist of, or where further information may be obtained

☐   Research Qs

☐   Rights Qs

☐   Injury Qs

     An explanation of whom to contact for answers to pertinent questions about the research and research subjects' rights, and whom to contact in the event of a research-related injury to the subject

☐   A statement that participation is voluntary, refusal to participate will involve no penalty or loss of benefits to which the subject is otherwise entitled, and the subject may discontinue participation at any time without penalty or loss of benefits, to which the subject is otherwise entitled

Additional Elements, as Appropriate

☐   A statement that the particular treatment or procedure may involve risks to the subject (or to the embryo or fetus, if the subject is or may become pregnant), which are currently unforeseeable

☐   Anticipated circumstances under which the subject's participation may be terminated by the investigator without regard to the subject's consent

☐   Any additional costs to the subject that may result from participation in the research

☐   The consequences of a subject's decision to withdraw from the research and procedures for orderly termination of participation by the subject

☐   A statement that significant new findings developed during the course of the research, which may relate to the subject's willingness to continue participation, will be provided to the subject

☐   The approximate number of subjects involved in the study

Source: http://www.hhs.gov/ohrp/humansubjects/assurance/consentckls.htm

As you can see, the process of obtaining informed consent can be daunting, and some potential participants may opt out because of an excess of information. There are additional issues that arise when participants are in certain categories. For example, young children cannot be expected to fully understand the informed consent process. The regulations state that children should be involved in the decision-making process to whatever extent possible and that a parent must also be involved. There are special rules applying to research on prisoners and to research on "Pregnant Women, Human Fetuses and Neonates." These can be found on the Internet at http://ohrp.osophs.dhhs.gov/human subjects/guidance/45cfr46.htm.

Research on Animals

Perhaps nothing is as controversial in research as the use of animals. There are some who believe that any experimentation on animals is unethical, whereas others argue that humans have benefited immensely from such research and that justifies its use. There is no doubt that in the past, some research with animals, as with humans, has been clearly unethical. But again, the professions of medicine and behavioral sciences, which are the two arenas in which animal research is most prevalent, have developed ethical guidelines that their members are supposed to follow.

Nonetheless, animals do not have the rights of humans as participants in experiments. Here are two of the elements of the American Psychological Association's Code of Ethics section on "Humane Care and Use of Animals in Research." Clearly these statements would not be made regarding human subjects, and even these are stated ideals that many researchers may not follow. (*Source:* http://www.apa.org/ethics/code2002.html#8_09).

> (e) Psychologists use a procedure subjecting animals to pain, stress, or privation only when an alternative procedure is unavailable and the goal is justified by its prospective scientific, educational, or applied value.
>
> (g) When it is appropriate that an animal's life be terminated, psychologists proceed rapidly, with an effort to minimize pain and in accordance with accepted procedures.

The United States National Institutes of Health has also established the *Public Health Service Policy on Humane Care and Use of Laboratory Animals,* which can be found at http://www.grants.nih.gov/grants/olaw/references/phspol.htm. However, the guidelines for research on animals are not nearly as detailed as they are for human subjects. A research study of the approval of animal-use protocols found strong inconsistencies when protocols approved or disapproved by one institution were submitted for approval to another institution's review board. The decisions of the two boards to approve or disapprove the protocols agreed at no better than chance levels, suggesting that the guidelines for approval of animal research are not spelled out in sufficient detail (Plous and Herzog, 2001.) For more information and links to a number of websites on the topic of research with animals, visit http://www.socialpsychology.org/methods.htm#animals.

21.2 | ASSURANCE OF DATA QUALITY

When reading the results of a study, you should be able to have reasonable assurance that the researchers collected data of high quality. This is not as easy as it sounds, and we have explored many of the difficulties and disasters of collecting and interpreting data for research studies in Chapters 3 and 4.

The quality of data becomes an ethical issue when there are personal, political, or financial reasons motivating one or more of those involved in the research, and steps are not taken to assure the integrity of the data. Data quality is also an ethical issue when researchers knowingly fail to report problems with data collection that may distort the interpretation of the results. As a simple example, survey data should always be reported with an explanation of how the sample was selected, what questions were asked, who responded, and how they may differ from those who didn't respond.

United States Federal Statistical Agencies

The United States government spends large amounts of money to collect and disseminate statistical information on a wide variety of topics. In recent years there has been an increased focus on making sure the data are of high quality. The Federal Register Notices on June 4, 2002 (Vol. 67, No. 107), provided a report called *Federal Statistical Organizations' Guidelines for Ensuring and Maximizing the Quality, Objectivity, Utility, and Integrity of Disseminated Information.* (The report can be found on the Web at http://www.fedstats.gov/policy/Stat-Agency-FR-June4–2002.pdf.) The purpose of the report was to provide notice that each of over a dozen participating federal statistical agencies were making such guidelines available for public comment. (A list of federal statistical agencies, with links to their websites and the data they provide, can be found at www.fedstats .gov.)

Most of the guidelines proposed by the statistical agencies were common-sense, self-evident good practice. As an example, the following box includes data collection principles from the Bureau of Transportation Statistics website.

3.1 Data Collection Operations

Principles

◆ Forms, questionnaires, automated collection screens, and file layouts are the medium through which data are collected. They consist of sets of questions or annotated blanks on paper or computer that request information from data suppliers. They need to be designed to maximize communication to the data supplier.

◆ Data collection includes all the processes involved in carrying out the data collection design to acquire data. Data collection operations can have a high impact on the ultimate data quality, especially when they deviate from the design.

♦ The data collection method should be appropriate to the data complexity, collection size, data requirements, and amount of time available.

For example, a reporting system will often primarily rely on the required reporting mechanism, with follow-up for missing data. Similarly, a large survey requiring a high response rate will often start off with a mail out, followed by telephone contact, and finally by a personal visit.

♦ Specific data collection environmental choices can significantly affect error introduced at the collection stage.

For example, if the data collector is collecting as a collateral duty or is working in an uncomfortable environment, it may adversely affect the quality of the data collected. Also, if the data are particularly difficult to collect, it will affect the data quality.

♦ Conversion of data on paper to electronic form (e.g., key entry, scanning) introduces a certain amount of error which must be controlled.

♦ Third-party sources of data introduce error in their collection processes.

♦ Computer-assisted information collection can result in more timely and accurate information. Initial development costs will be higher, and much more lead time will be required to develop, program, and test the data collection system. However, the data can be checked and corrected when originally entered, key-entry error is eliminated, and the lag between data collection and data availability is reduced.

♦ The use of sensors for data can significantly reduce error.

Source: http://www.bts.gov/statpol/guide/chapter3.html

So where does this fit in a discussion of ethics? Although it is laudable that the federal statistical agencies are spelling out these principles, the ethical issues involved with ensuring the quality of government data are more subtle than can be evoked from general principles. There are always judgments to be made, and politics can sometimes enter into those judgments.

For instance, a report by the National Research Council (Citro and Norwood, 1997) examined many aspects of the functioning of the Bureau of Transportation Statistics (BTS) and formulated suggestions for improvement. The BTS was created in 1992 and thus had been in existence for only five years at the time of the report. One of the data-quality issues discussed in the report was "comparability [of statistics about transportation issues] across data systems and time" (p. 32). One example given was that the definition of a transportation fatality was not consistent across modes of transportation until 1994, when consistency was mandated by the Secretary of Transportation. Before that, a highway traffic fatality was counted if death resulted from the accident within 30 days. But for a railroad accident, the time limit was 365 days. Continuing to use different standards for different modes of transportation would make for unfair safety comparisons. In an era where various transportation modes compete for federal dollars, politics could easily enter into a decision to report fatality statistics one way or another.

Probably the most controversial federal data collection issue surrounds the decennial United States census, most recently conducted in 2000. In 1988 a lawsuit was filed, led by New York City, alleging that urban citizens are undercounted in the census process. The political ramifications are enormous, because redistricting of congressional seats results from shifting population counts. The lawsuit began a long saga of political and statistical twists and turns that was still unresolved at the taking of the 2000 census. For an interesting and readable account, see *Who Counts? The Politics of Census-Taking in Contemporary America,* by Anderson and Fienberg (2001).

A recent report by the National Research Council's Committee on National Statistics (Martin, Straf, and Citro, 2001), called *Principles and Practices for a Federal Statistical Agency,* enumerated three principles and eleven practices that such agencies should follow. One of the recommended practices was "A strong position of independence." The recommendation included a clear statement about separating politics from the role of statistical agencies:

> In essence, a statistical agency must be distinct from those parts of the department that carry out enforcement and policy-making activities. It must be impartial and avoid even the appearance that its collection, analysis, and reporting processes might be manipulated for political purposes or that individually identifiable data might be turned over for administrative, regulatory, or enforcement purposes.
>
> The circumstances of different agencies may govern the form that independence takes. In some cases, the legislation that establishes the agency may specify that the agency head be professionally qualified, be appointed by the President and confirmed by the Senate, serve for a specific term not coincident with that of the administration, and have direct access to the secretary of the department in which the agency is located. (p. 6)

It should be apparent to you that it is not easy to maintain complete independence from politics; for instance, many of the heads of these agencies are appointed by the President and confirmed by the Senate. Other steps taken to help maintain independence include prescheduled release of important statistical information such as unemployment rates and authority to release information without approval from the policy-making branches of the organization.

Experimenter Effects and Personal Bias

We learned in Chapter 3 that there are numerous ways in which *experimenter effects* can bias statistical studies. If a researcher has a desired outcome for a study and if conditions are not very carefully controlled, it is quite likely that the researcher will influence the outcome. Here are some of the precautions that it may be possible to take to help prevent this from happening:

- ◆ Randomization done by a third party with no vested interest in the experiment, or at least done by a well-tested computer randomization device

- ◆ Automated data recording without intervention from the researcher

♦ Double-blind procedures to ensure that no one who has contact with the participants knows which treatment or condition they are receiving

♦ An honest evaluation that what is being measured is appropriate for the research question of interest

♦ A standard protocol for the treatment of all participants that must be strictly followed

---

EXAMPLE 21.2
Janet's (Hypothetical) Dissertation Research

This is a hypothetical example to illustrate some of the subtle (and not so subtle) ways experimenter bias can alter the data collected for a study. Janet is a Ph.D student and is under tremendous pressure to complete her research successfully. For her study, she hypothesized that role-playing assertiveness training for women would help them learn to say "no" to telephone solicitors. She recruited 50 undergraduate women as volunteers for the study. The plan was that each volunteer would come to her office for half an hour. For 25 of the volunteers (the control group), she would simply talk with them for 30 minutes about a variety of topics, including their feelings about saying "no" to unwanted requests. For the other 25 volunteers, Janet would spend 15 minutes on similar discussion and the remaining 15 minutes on a prespecified role-playing scenario in which the volunteer got to practice saying "no" in various situations. Two weeks after each volunteer's visit, Brad, a colleague of Janet's, would phone them anonymously, pretending to be a telephone solicitor selling a magazine for a good price, and would record the conversation so that Janet could determine whether or not they were able to say "no."

It's the first day of the experiment, and the first volunteer is in Janet's office. Janet has the randomization list that was prepared by someone else, which randomly assigns each of the 50 volunteers to either Group 1 or Group 2. This first volunteer appears to be particularly timid, and Janet is sure she won't be able to learn to say "no" to anyone. The randomization list says that volunteer 1 is to be in Group 2. But what was Group 2? Did she say in advance? She can't remember. Oh well, Group 2 will be the control group.

The next volunteer comes in, and according to the randomization list, volunteer 2 is to be assigned to Group 1, which now is defined to be the role-playing group. Janet follows her predefined protocol for the half-hour. But when the half-hour is over, the student doesn't want to leave. Just then Brad comes by to say "hello," and the three of them spend another half-hour having an amiable conversation.

The second phase of the experiment begins and Brad starts phoning the volunteers. The conversations are recorded so that Janet can assess the results. When listening to volunteer 2's conversation, Janet notices that almost immediately she says to Brad, "Your voice sounds awfully familiar, do I know you?" When he assures her that she does not and asks her to buy the magazine, she says "I can't place my finger on it, but this is a trick, right? I'm sure I know your voice. No thanks, no magazine!" Janet records the data: a successful "no" to the solicitation.

Janet listens to another call, and although she is supposed to be blind to which group the person was in, she recognizes the voice as being one of the role-playing volunteers. Brad pitches the magazine to her and her response is "Oh, I already get that magazine. But if you are selling any others, I might be able to buy one." Janet records the data: a successful "no" to the question of whether she wants to buy that magazine.

The second phase is finally over and Janet has a list of the results. But now she notices a problem. There are 26 people listed in the control group and 24 listed in the role-playing group. She tries to resolve the discrepancy but can't figure it out. She notices that the last two people on the list are both in the control group and that they both said "no" to the solicitation. She figures she will just randomly choose one of them to move over to the role-playing data, so she flips a coin to decide which one to move.

This example illustrates just a few of the many ways in which experimenter bias can enter into a research study. Even when it appears that protocols are carefully in place, in the real world it is nearly impossible to place controls on all aspects of the research. In this case, notice that every decision Janet made benefited her desired conclusion that the role-playing group would learn to say "no." It is up to the researchers to take utmost care not to allow these kinds of unethical influences on the results. ◆

21.3 | APPROPRIATE STATISTICAL ANALYSES

There are a number of decisions that need to be made when analyzing the results of a study, and care must be taken not to allow biases to affect those decisions. Here are some examples of decisions that can influence the results:

- ◆ Should the alternative hypothesis be one-sided or two-sided? This decision must be made before examining the data.
- ◆ What level of confidence or level of significance should be used?
- ◆ How should outliers be handled?
- ◆ Should a nonparametric procedure be used?
- ◆ Have all appropriate conditions and assumptions been investigated and verified?

One of the easiest ethical blunders to make is to try different methods of analysis until one produces the desired results. Ideally, the planned analysis should be spelled out in advance. If various analysis methods are attempted, all analyses should be reported along with the results.

EXAMPLE 21.3
Jake's (Hypothetical) Fishing Expedition

Jake was doing a project for his statistics class, and he knew it was important to find a statistically significant result because all of the interesting examples in class had statistically significant results. He decided to compare the memory skills of males and females, but he did not have a preconceived idea of who should do better, so he planned to do a two-sided test. He constructed a memory test in which he presented people with a list of 100 words and allowed them to study it for 10 minutes. Then, the next day, he presented another list of 100 words to the participants and asked them to indicate for each word whether it had been on the list the day before. The answers were entered into a bubble sheet and scored by computer so that Jake didn't inadvertently misreport any scores. The data for number of correct answers were as follows:

Males: 69, 70, 71, 61, 73, 68, 70, 69, 67, 72, 64, 72, 65, 70, 100
Females: 64, 74, 72, 76, 64, 72, 76, 80, 72, 73, 71, 70, 64, 76, 70

Jake remembered that they had been taught two different tests for comparing independent samples: a two-sample *t*-test to compare means and a Mann–

Whitney rank-sum test to compare medians. He decided to try them both to see which one gave better results. He found the following:

Two-Sample T-Test and CI: Males, Females

Two-sample T for Males vs Females

	N	Mean	StDev	SE Mean
Males	15	70.73	8.73	2.3
Females	15	71.60	4.75	1.2

Difference = mu Males − mu Females
Estimate for difference: −0.87
95% CI for difference: (−6.20, 4.47)
T-Test of difference = 0 (vs not =): T-Value = −0.34 P-Value = 0.739 DF = 21

Mann−Whitney Test and CI: Males, Females

Males   N = 15   Median = 70.00
Females   N = 15   Median = 72.00
Point estimate for ETA1-ETA2 is   −3.00
95.4 Percent CI for ETA1-ETA2 is (−6.00,1.00)
W = 193.5
Test of ETA1 = ETA2 vs ETA1 not = ETA2 is significant at 0.1103
The test is significant at 0.1080 (adjusted for ties)

Cannot reject at alpha = 0.05

Jake was disappointed that neither test gave a small enough p-value to reject the null hypothesis. He was also surprised by how different the results were. The t-test produced a p-value of 0.739, whereas the Mann−Whitney test produced a p-value of 0.108, adjusted for ties. It dawned on him that maybe he should conduct a one-tailed test instead. After all, it was clear that the mean and median for the females were both higher, so he decided that the alternative hypothesis should be that females would do better. He reran both tests. This time, the p-value for the t-test was 0.369 and for the Mann−Whitney test was 0.054. Maybe he should simply round that one off to 0.05 and be done.

But Jake began to wonder why the tests produced such different results. He looked at the data and realized that there was a large outlier in the data for the males. Someone scored 100%! Jake thought that must be impossible. He knew that he shouldn't just remove an outlier, so he decided to replace it with the median for the males, 70. Just to be sure he was being fair, he reran the original two-sided hypothesis tests. This time, the p-value for the t-test was 0.066 and the p-value for the Mann−Whitney test (adjusted for ties) was 0.0385. Finally! Jake wrote his analysis, explaining that he did a two-sided test because he didn't have a preconceived idea of whether males or females should do better. He said that he decided to do a nonparametric test because he had small samples and he wanted to make sure none of the assumptions was violated. He didn't mention replacing the outlier because he didn't think it was a legitimate value anyway.

This hypothetical situation is obviously exaggerated to make a point, but it illustrates the dangers of **data-snooping.** If you manipulate the data and try enough different procedures, something will eventually produce desired results.

It is not ethical to keep trying different methods of analysis until one produces a desired result. ◆

EXAMPLE 21.4
The Debate Over Passive Smoking

On July 28, 1993, the *Wall Street Journal* featured an article with the headline "Statisticians occupy front lines in battle over passive smoking" (Jerry E. Bishop, pp. B-1 and B-4). The interesting feature of this article was that it highlighted a debate between the U.S. Environmental Protection Agency (EPA) and the tobacco industry over what level of confidence should prevail in a confidence interval. Here is the first side of the story, as reported in the article:

> The U.S. Environmental Protection Agency says there is a 90% probability that the risk of lung cancer for passive smokers is somewhere between 4% and 35% higher than for those who aren't exposed to environmental smoke. To statisticians, this calculation is called the "90% confidence interval."

Now for the other side of the story:

> And that, say tobacco-company statisticians, is the rub. "Ninety-nine percent of all epidemiological studies use a 95% confidence interval," says Gio B. Gori, director of the Health Policy Center in Bethesda, MD, who has frequently served as a consultant and an expert witness for the tobacco industry.

The problem underlying this controversy is that the amount of data available at the time did not allow an extremely accurate estimate of the true change in risk of lung cancer for passive smokers. The EPA statisticians were afraid that the public would not understand the interpretation of a confidence interval. If a 95% confidence interval actually went below zero percent, which it might do, the tobacco industry could argue that passive smoke might reduce the risk of lung cancer. As noted by one of the EPA's statisticians,

> Dr. Gori is correct in saying that using a 95% confidence interval would hint that passive smoking might reduce the risk of cancer. But, he says, this is exactly why it wasn't used. The EPA believes it is inconceivable that breathing in smoke containing cancer-causing substances could be healthy and any hint in the report that it might be would be meaningless and confusing. (p. B-4)

What do you think? Should the EPA have used the common standard and reported a 95% confidence interval? If they had done so, the interval would have included values indicating that the risk of lung cancer for those who are exposed to passive smoke may actually be lower than the risk for those who are not. The 90% confidence interval reported was narrower and did not include those values. (*Source: Seeing Through Statistics,* 3rd ed., Jessica M. Utts, Belmont, CA: Duxbury Press, 2005.) ◆

21.4 | FAIR REPORTING OF RESULTS

Research results are usually reported in articles published in professional journals. Most researchers are careful to report the details of their research, but there are more subtle issues that researchers and journalists sometimes ignore. There are also more blatant reporting biases that can mislead readers, usually in the direction of making stronger conclusions than are appropriate. We have discussed some of the problems with interpreting statistical inference results in earlier chapters, and some of those problems are related to how results are reported. Let's revisit some of them and discuss some other possible ethical issues that can arise when reporting the results of research.

Sample Size and Statistical Significance

Remember that whether a study achieves statistical significance depends not only on the magnitude of whatever effect or relationship may actually exist but also on the size of the study. In particular, if a study fails to find a statistically significant result, it is important to include a discussion of the sample size used and the power to detect an effect that would result from that sample size. Often, research is reported as having found no effect or no difference, when in fact the study had such low power that even if an effect existed, the study would have been unlikely to detect it.

One of the ethical responsibilities of a researcher is to collect enough data to have relatively high probability of finding an effect if indeed it really does exist. In other words, it is the responsibility of the researcher to determine an appropriate sample size in advance, as well as to discuss the issue of power when the results are presented. A study that is too small to have adequate power is a waste of everyone's time and money.

As we have discussed earlier in the book, the other side of the problem is the recognition that statistical significance does not necessarily imply practical importance. If possible, report the magnitude of an effect or difference through the use of a confidence interval, rather than just reporting a *p*-value.

Multiple Hypothesis Tests and Selective Reporting

In most research studies there are multiple outcomes measured. Many hypothesis tests may be done, looking for whatever statistically significant relationships may exist. For example, a study may ask people to record their dietary intake over a long time period so that the investigators can then look for foods that are correlated with certain health outcomes. The problem is that even if there are no legitimate relationships in the population, something in the sample may be statistically significant just by chance. It is unethical to report conclusions only about the results that were statistically significant without informing the reader about all of the tests that were done.

The *Ethical Guidelines of the American Statistical Association* (1999) lists the problem under "Professionalism" as follows:

> Recognize that any frequentist statistical test has a random chance of indicating significance when it is not really present. Running multiple tests on the same data set at the same stage of an analysis increases the chance of obtaining at least one invalid result. Selecting the one "significant" result from a multiplicity of parallel tests poses a grave risk of an incorrect conclusion. Failure to disclose the full extent of tests and their results in such a case would be highly misleading.

In many cases it is not the researcher who makes this mistake; it is the media. The media are naturally interested in surprising or important conclusions, not in results showing that there is nothing going on. For instance, the story of interest will be that a particular food is related to higher cancer incidence, not that 30 other foods did not show that relationship. It is unethical for the media to publicize such results without explaining the possibility that multiple testing may be responsible for uncovering a spurious chance relationship.

Even if there is fairly strong evidence that observed statistically significant relationships represent real relationships in the population, the media should mention other, less interesting results because they may be important for people making lifestyle decisions. For example, if the relationship between certain foods and disease are explored, it is interesting to know which foods do not appear to be related to the disease, as well as which foods appear to be related.

EXAMPLE 21.5
Helpful and Harmless Outcomes from Hormone Replacement Therapy

In July 2002 the results were released from a large clinical trial studying the effects of estrogen plus progestin hormone replacement for postmenopausal women. Some of these results were covered in earlier chapters. The trial was stopped early because of increased risk of breast cancer and coronary heart disease among the women taking the hormones. However, many news stories failed to report some of the other results of the study, which showed that the hormones actually decreased the risk of other adverse outcomes and were unresolved about others. The original article (Writing Group for the Women's Health Initiative Investigators, 2002) reported the results as follows:

> Absolute excess risks per 10,000 person-years attributable to estrogen plus progestin were 7 more CHD [coronary heart disease] events, 8 more strokes, 8 more PEs [pulmonary embolism], 8 more invasive breast cancers, while absolute risk reductions per 10,000 person-years were 6 fewer colorectal cancers and 5 fewer hip fractures.

These results show that in fact some outcomes were more favorable for those taking the hormones, specifically colorectal cancer and hip fractures. Because different people are at varying risk for certain diseases, it is important to report all of these outcomes so that an individual can make an informed choice about whether to take the hormones. In fact, overall, 231 out of 8506 women taking the hormones died of any cause during the study, which is 2.72%. Of the 8102 women taking the placebo, 218, or 2.69% died, a result virtually identical to that in the hormone group. In fact, when the results are adjusted for the time spent in the study, the death rate was slightly lower in the hormone group, with an annualized rate of 0.52% compared to 0.53% in the placebo group.

The purpose of this example is not to negate the serious and unexpected outcomes related to heart disease, which hormones were thought to protect against, or the serious breast cancer outcome. Instead, the purpose is to show that the results of a large and complex study such as this one are bound to be mixed and should be presented as such so that readers can make informed decisions. ◆

Making Stronger or Weaker Conclusions Than Are Justified

As we have learned throughout this book, for many reasons research studies cannot be expected to measure all possible influences on a particular outcome. They are also likely to have problems with ecological validity, in which the fact that someone is participating in a study is enough to change their behavior or produce a result that would not naturally occur. It is important that research results be presented with these issues in mind and that the case not be overstated even when an effect or relationship is found. An obvious example, often discussed in this book, is that a cause-and-effect conclusion cannot generally be made on the basis of an observational study.

However, there are more subtle ways in which conclusions stronger or weaker than is justified may be made. For example, often little attention is paid to how representative the sample is of a larger population, and results are presented as though they would apply to all men, or all women, or all adults. It is important to consider and report an accurate assessment of who the participants in the study are really likely to represent.

Sometimes there are financial or political pressures that can lead to stronger (or weaker) conclusions than are justified. Results of certain studies may be suppressed while others are published, if some studies support the desired outcome and others don't. As we have cautioned throughout this book, when you read the results of a study that has personal importance to you, try to gain access to as much information as possible about what was done and who was included and to examine all of the analyses and results.

CASE STUDY 21.1
Science Fair Project or Fair Science Project?

In 1998 a fourth-grade girl's science project received extensive media coverage after its publication in the *Journal of the American Medical Association* (Rosa et al., 1998). Later that year, in its December 9, 1998 issue, the journal published a series of letters criticizing the study and its conclusions on a wide variety of issues. There are a number of ethical issues related to this study, some of which were raised by the letters and others that have not been raised before.

The study was supposed to be examining "therapeutic touch" (TT), a procedure practiced by many nurses that involves working with patients through a five-step process, including a sensing and balancing of their "energy." The experiment proceeded as follows: Twenty-one self-described therapeutic touch practitioners participated. They were asked to sit behind a cardboard screen and place their hands through cutout holes, resting them on a table on the other side of the screen. The 9-year-old "experimenter" then flipped a coin and used the outcome to decide which of the practitioner's hands to hold her hand over. The practitioner was to guess which hand the girl was hovering over. Fourteen of the practitioners contributed 10 tries each, and the remaining 7 contributed 20 tries each, for a total of 280 tries.

The paper had four authors, including the child and her mother. It is clear from the affiliations of the authors of the paper, as well as from language throughout the paper, that the authors were biased against therapeutic touch before the experiment began. For example, the first line read "Therapeutic touch (TT) is a widely used nursing practice rooted in mysticism but alleged to have a scientific basis" (Rosa et al., 1998, p. 1005). The first author of the paper was the child's mother, and her affiliation is listed as "the Questionable Nurse Practices Task Force, National Council Against Health Fraud, Inc."

The paper concludes that "Twenty-one experienced TT practitioners were unable to detect the investigator's 'energy field.' Their failure to substantiate TT's most fundamental claim is unrefuted evidence that the claims of TT are groundless and that further professional use is unjustified." The conclusion was widely reported in the media, presumably at least in part because of the novelty of a child having done the research. That would have been cute if it hadn't been taken so seriously by people on both sides of the debate on the validity of therapeutic touch.

The letters responding to the study point out many problematic issues with how the study was done and with its conclusions. Here are several quotes:

> The experiments described are an artificial demonstration that some number of self-described mystics were unable to "sense the field" of the primary investigator's 9-year-old daughter. This hardly demonstrates or debunks the efficacy of TT. The vaguely described recruitment method does not ensure or even suggest that the subjects being tested were actually skilled practitioners. More important, the experiments described are not relevant to the clinical issue supposedly being researched. Therapeutic touch is not a parlor trick and should not be investigated as such. (Freinkel, 1998)

> To describe this child's homework as "research" is without foundation since it clearly fails to meet the criteria of randomization, control, and valid intervention. . . . Flagrant violations against TT include the fact that "sensing" an energy field is not TT but rather a nonessential element in the 5-step

process; inclusion of many misrepresentations of cited sources; use of inflammatory language that indicates significant author bias; and bias introduced by the child conducting the project being involved in the actual trials. (Carpenter et al., 1998)

I critiqued the study on TT and was amazed that a research study with so many flaws could be published. . . . The procedure was conducted in different settings with no control of environmental conditions. Even though the trials were repeated, the subjects did not change, thus claims of power based on possible repetitions of error are inappropriate. The true numbers in groups are 15 and 13, thus making a type II error highly probable with a study power of less than 30%. Another concern is whether participants signed informed consent documents or at least were truly informed as to the nature of this study and that publication of its results would be sought beyond a report to the fourth-grade teacher. (Schmidt, 1998)

As can be seen by these reader comments, many of the ethical issues covered in this chapter may cloud the results of this study. However, there are two additional points that were not raised in any of the published letters. First, it is very likely that the child knew that her mother wanted the results to show that the participants would not be able to detect which hand was being hovered over. And what child does not want to please her mother? That would not matter so much if there hadn't been so much room for "experimenter effects" to influence the results. One example is the randomization procedure. The young girl flipped a coin each time to determine which hand to hover over. Coin tosses are very easy to influence, and presumably even a nine-year-old child could pick up the response biases of the subjects. The fact that a proper randomization method wasn't used should have ended any chance that this experiment would be taken seriously.

Is there evidence that experimenter bias may have entered the experiment? Absolutely. Of the 280 tries, the correct hand was identified in 123 (44%) of them. The authors of the article conclude that this number is "close to what would be expected for random chance." In fact, that is not the case. Using the binomial distribution, the chance of getting 123 *or fewer* guesses by chance is only 0.0242. If a two-tailed test had been used instead of a one-tailed test, the *p*-value would have been 0.048, a statistically significant outcome. The 9-year-old did an excellent job of fulfilling her mother's expectations.

KEY TERMS

Section 21.1
informed consent, 674–675

Section 21.2
experimenter effects, 682

Section 21.3
data-snooping, 685

Section 21.4
selective reporting, 687–688

EXERCISES   Blue = Basics   Bold, **Bold** = Answer in back

21.1 Visit the websites of one or more professional organizations related to your major. (You may need to ask one of your professors for help with this.) Find one that has a code of ethics. Describe whether the code includes anything about research methods. If not, explain why you think nothing is included. If so, briefly summarize what is included. Include the website address with your answer.

21.2 A classic psychology experiment was conducted by psychology professor Philip Zimbardo in the summer of 1971 at Stanford University. The experiment is described at the website http://www.prisonexp.org/. Visit the website; if it is no longer operative, try an Internet search on "Zimbardo" and "prison" to find information about this study.
  a. Briefly describe the study and its findings.
  b. The website has a number of discussion questions related to the study. Here is one of them: "Was it ethical to do this study? Was it right to trade the suffering experienced by participants for the knowledge gained by the research?" Discuss these questions.
  c. Another question asked at the website is "How do the ethical dilemmas in this research compare with the ethical issues raised by Stanley Milgram's obedience experiments? Would it be better if these studies had never been done?" Discuss these questions.

21.3 In the report *Principles and Practices for a Federal Statistical Agency* (Martin, Straf, and Citro, 2001), one of the recommended practices is "a strong position of independence." Each of the following parts gives one of the characteristics that is recommended to help accomplish this. In each case, explain how the recommendation would help ensure a position of independence for the agency.
  a. "Authority for selection and promotion of professional, technical, and operational staff." (p. 6)

b. "Authority for statistical agency heads and qualified staff to speak about the agency's statistics before Congress, with congressional staff, and before public bodies." (p. 6)

21.4 Refer to Example 21.2 about Janet's dissertation research. Explain whether each of the following changes would have been a good idea or a bad idea.

    a. Have someone who is blind to conditions and has not heard the volunteers' voices before listen to the phone calls to assess whether a firm "no" was given.

    b. Have Janet flip a coin each time a volunteer comes to her office to decide whether that person should go into the role-playing group or the control group.

    c. Use a variety of different solicitors rather than Brad alone.

21.5 Refer to Example 21.3 about Jake's memory experiment. What do you think Jake should have done regarding the analysis and reporting of it?

21.6 Find an example of a statistical study reported in the news. Explain whether you think multiple tests were done and, if so, whether they were reported.

21.7 Marilyn is a statistician who works for a company that manufactures components for sound systems. Two development teams have each come up with a new method for producing one of the components, and management is to decide which one to adopt based on which produces components that last longer. Marilyn is given the data for both and is asked to make a recommendation about which one should be used. She computes 95% confidence intervals for the mean lifetime using each method and finds an interval from 96 hours to 104 hours, centered on 100 hours for one of them, and an interval from 92 hours to 112 hours, centered on 102 for the second one. What should she do? Should she make a clear recommendation? Explain.

Use the following scenario for Exercises 21.8 to 21.12: Based on the information in Case Study 21.1, describe the extent to which each of the following conditions for reducing the experimenter effect (listed in Section 21.2) was met. If you cannot tell from the description of the experiment, then explain what additional information you would need.

21.8 Randomization done by a third party with no vested interest in the experiment, or at least done by a well-tested computer randomization device.

21.9 Automated data recording without intervention from the researcher.

21.10 Double-blind procedures to ensure that no one who has contact with the participants knows which treatment or condition they are receiving.

21.11 An honest evaluation that what is being measured is appropriate for the research question of interest.

21.12 A standard protocol for the treatment of all participants that must be strictly followed.

Exercises 21.13 to 21.22: Explain the main ethical issue of concern in each of the following. Discuss what, if anything, should have been done differently to address that concern.

21.13 Example 21.1, Stanley Milgram's experiment.

21.14 In Example 21.2, Janet's decision to make Group 2 the control group after the first volunteer came to her office.

21.15 In Example 21.2, Janet's handling of the fact that her data showed 26 volunteers in the control group when there should have been only 25.

21.16 In Example 21.3, Jake's decision to replace the outlier with the median of 70.

21.17 In Example 21.4, the Environmental Protection Agency's decision to report a 90% confidence interval.

21.18 In Example 21.5, the fact that many media stories mentioned increased risk of breast cancer and coronary heart disease but not any of the other results.

21.19 In Case Study 21.1, the concerns raised in the letter from Freinkel.

21.20 In Case Study 21.1, the concerns raised in the letter from Carpenter et al.

21.21 In Case Study 21.1, the concerns raised in the letter from Schmidt.

21.22 In Case Study 21.1, the concerns raised about the experimenter effect.

REFERENCES

American Statistical Association (1999). Ethical guidelines for statistical practice; http://www.amstat.org/profession/ethicalstatistics.html.

Anderson, M. J., and S.E. Fienberg (2001). *Who Counts? The Politics of Census-Taking in Contemporary America*, New York: Russell Sage Foundation.

Carpenter, J., J. Hagemaster, and B. Joiner (1998). Letter to the editor, *Journal of the American Medical Association*, Vol. 280, No. 22, p. 1905.

Citro, C. F. and J. L. Norwood (eds.) (1997). *The Bureau of Transportation Statistics: Priorities for the Future*, Washington, D.C.: National Academy Press.

Martin, M. E., M. L. Straf, and C. F. Citro (eds.) (2001). *Principles and Practices for a Federal Statistical Agency*, 2nd ed., Washington, D.C.: National Academy Press.

Freinkel, A. (1998). Letter to the editor, *Journal of the American Medical Association*, Vol. 280, No. 22, p. 1905.

Milgram, Stanley (1983). *Obedience to Authority: An Experimental View*, New York: Harper/Collins.

Plous, S., and H. Herzog (2001, July 27). Reliability of protocol reviews for animal research. *Science*, Vol. 293, pp. 608–609; available at http://www.socialpsychology.org/articles/scipress.htm.

Rosa, L., E. Rosa, L. Sarner, and S. Barrett (1998). "A close look at therapeutic touch," *Journal of the American Medical Association*, Vol. 279, No. 13, pp. 1005–1010.

Schmidt, S. M. (1998). Letter to the editor, *Journal of the American Medical Association*, Vol. 280, No. 22, p. 1906.

Writing Group for the Women's Health Initiative Investigators (2002). "Risks and benefits of estrogen plus progestin in healthy postmenopausal women," *Journal of the American Medical Association*, Vol. 288, No. 3, pp. 321–333.

Turning Information Into Wisdom

Would you trust her less if she were selling cars? See Example 22.9.

Information developed through the use of statistics has enhanced our understanding of how life works, helped us learn about each other, allowed control over some societal issues, and helped individuals make informed decisions. Nearly every area of knowledge has been advanced by statistical studies.

I n the opening paragraph of Chapter 1, we said, "we hope to convince you that learning about this subject [statistics] will be interesting and useful." By the time you have read this far in the book, we hope you have seen some interesting examples and information. But are the methods in this book useful? Why would anyone care about what can be done with statistics?

In this chapter, we speculate on the wisdom to be gained from statistical studies and why people are interested in such studies. Certainly you can live without them. But information developed through the use of statistics has enhanced our understanding of how life works, helped us learn about each other, allowed control over some societal issues, and helped individuals make informed decisions. There is almost no area of knowledge that has not been advanced by statistical studies. As individuals, we learn from our own and others' experiences. But as a society, we learn through scientific studies—many of them based on statistical methods.

This chapter contains a number of examples that illustrate the wisdom to be gained from statistical studies. Before delving into examples, let's review the basic situations that allow us to turn data into information and information into wisdom. ◆

22.1 | BEYOND THE DATA

Much of the appeal of statistical methods is that they allow us to describe, predict, and sometimes control the world around us. This is possible because with the methods of statistical inference we can use relatively small amounts of data to provide information and sometimes make conclusions about immense populations. This process is one of the truly remarkable features of modern statistics.

However, it is important to remember that the extent to which data descriptions can be generalized to something more rests on two issues. These issues are *what group of individuals was measured,* and, if relevant, *whether or not randomization was used to assign conditions.* These are two different aspects of study design and lead to two different possible extensions of simple descriptive measures.

Random, Representative, or Restrictive Sample?

If pressed to be completely intellectually honest, most statisticians would have to agree with the statement made in *The Statistical Sleuth* (Ramsey and Schafer, 1997, p. 7): "Inferences to **populations** can be drawn from random sampling studies, but not otherwise." The mathematics supporting the material on statistical inference in this book demands simple random samples. However, most researchers, including statisticians, realize that true random samples are almost impossible to obtain. Methods of statistical inference would be of little use if they were restricted to situations in which true random samples were available. Instead, in practice most researchers follow the Fundamental Rule for Using Data for Inference we introduced in Chapter 3.

> **definition** The **Fundamental Rule for Using Data for Inference** is that available data can be used to make inferences about a much larger group *if the data can be considered to be representative with regard to the question(s) of interest.*

One of the key steps in interpreting a statistical study is the capacity to determine whether this rule holds for the questions of interest in the study. We will look at some examples in this chapter.

Randomized Experiments, Observational Studies, and Causal Conclusions

One of the most common errors made by the media in interpreting statistical studies is to conclude that a causal relationship has been established when that conclusion is not warranted by the way the study was conducted. If a statistically significant relationship between the explanatory and response variables is demonstrated, it is tempting to assume that the differences in the explanatory variable *caused* the differences in the response variable. But it is important to remember that even if a cause-and-effect conclusion is sensible, it is not justified by the results of an observational study.

> **definition** The **Rule for Concluding Cause and Effect** is that cause-and-effect relationships can be inferred from randomized experiments, but not from observational studies.

The reasoning is simple. When individuals are randomized into different treatment conditions, the effect of confounding variables should be similar in all treatment groups. Therefore, differences in the response variable(s) can be attributed to the one thing that is known to differ across groups—the explanatory variable, which is the set of treatment conditions. In observational studies, the groups defining the explanatory variable occur naturally and they may have other natural differences between them in addition to the explanatory variable. These other possible differences produce confounding variables, the effect of which cannot be separated from the effect of the explanatory variable. Therefore, differences in the response variable cannot be attributed to differences in the explanatory variable.

Here are some examples of headlines from various news services on the Internet, all based on observational studies. In each instance, the headline implies a cause-and-effect relationship. We're sure you can think of confounding variables that may have led to the observed relationship:

- *Emotional Support Helps Breast Cancer Survival* (Reuters News Service, http://dailynews.yahoo.com/headlines/hl/, Dec. 2, 2000)
- *Study: Even a Little* [of the drug] *Ecstasy can Lower IQ* (USA Today Health News, http://www.usatoday.com/life/health/archive.htm, May 16, 2000)
- *Study: Junk Food Raises Teens' Risk of Heart Disease* (CNN Health News, http://www.cnn.com/2000/HEALTH/03/14/teen.arteries/index.html, Mar. 14, 2000)
- *Lack of Folic Acid Causes Defects* (Associated Press, http://dailynews.yahoo.com/htx/ap/20001129/hl/folic_acid_1.html, Nov. 29, 2000)
- *City Living Increases Men's Death Risk* (ABC News, http://www.abcnews.go.com/sections/living/DailyNews/cityrisk001130.html, Nov. 30, 2000)

In most cases, neither the original researchers nor the authors of the news articles made the mistake made in the headline. The reporters who write the articles usually do not write the headlines, and the reporters may be better trained in how to interpret studies than are headline writers. Unfortunately, the public may not read the fine print beyond the headline.

Using Nonstatistical Considerations to Assess Cause and Effect

You might wonder why researchers conduct observational studies at all, given that the goal is probably to make a causal conclusion. There are some nonstatistical considerations that can be used to help assess whether a causal link is reasonable based on observational studies. Here are some features that lend evidence to a causal connection.

1. *There is a reasonable explanation of cause and effect.*

A potential causal connection will be more believable if an explanation exists for how the cause and effect occurs. For instance, in Example 3.4, we established that for hardcover books the number of pages is correlated with the price. We would probably not contend that higher prices cause more pages, but could reasonably argue that more pages cause higher prices. We can imagine that publishers set the price of a book based on the cost of producing it and that the more pages there are, the higher the cost of production. Thus, we have a reasonable explanation for how an increase in the length of a book could cause an increase in the price.

2. *The connection happens under varying conditions.*

If many observational studies conducted under different conditions all find the same link between two variables, that strengthens the evidence for a causal connection. This is especially true if the studies are not likely to have the same confounding variables. The evidence is also strengthened if the same type of relationship holds when the explanatory variable falls into different ranges.

For example, numerous observational studies have related cigarette smoking and lung cancer. Further, the studies have shown that the higher the number of cigarettes smoked, the greater the chances of developing lung cancer; similarly, a connection has been established with the age at which smoking began. These facts make it more plausible that smoking actually causes lung cancer.

3. *Potential confounding variables are ruled out.*

When a relationship appears in an observational study, potential confounding variables may immediately come to mind. For example, the researchers in Case Study 3.1, which showed a relationship between lead exposure and tooth decay in children, did consider that income level, carbohydrates, calcium, and lack of dental care may have been confounding variables. However, they were able to measure those variables and examine them in the analysis, and the relationship between lead exposure and tooth decay persisted. The greater the number of confounding factors that can be ruled out, the more convincing the evidence for a causal connection.

---

in summary   **Guidelines About Cause and Effect and Using Data for Inference**

- If the sample *does not* represent a larger population for the question of interest and *randomization* to treatments *was not used*, no inferences can be drawn.

- If the sample *does* represent a larger population for the question of interest but *randomization* to treatments *was not used*, statistically significant results *can* be extended to the larger population but generally not with a cause-and-effect explanation.

- If the sample *does not* represent a larger population for the question of interest but *randomization* to treatments *was used*, the relationship observed in the sample can be inferred to be causal, but cannot be extended to a larger population.

- If the sample *does* represent a larger population for the question of interest and *randomization* to treatments *was used*, causal inferences can be drawn about the larger population.

22.2 | TRANSFORMING UNCERTAINTY INTO WISDOM

We live in a world filled with uncertainty and variability. The field of statistics exists because this is so. Statistical methods allow us to gain information and control because we can use them to measure variability and find predictable patterns in the presence of uncertainty. Here are some reasons that statistical methods are popular and useful:

- As individuals, we need to *make personal decisions* about things like:

  What to eat to remain healthy.
  How to enhance our children's well-being.
  Which treatment to undergo in a medical situation.
  How to invest our time to enhance our enjoyment of life.

- As a society, we want to *have some control* over things like:

  Reducing rates of cancer, heart attacks, and other diseases.
  Growing better plants (pest-resistant, high-yield, and so on).
  Determining what public education strategies help reduce risky behavior
    for HIV infection.
  Deciding what new laws to enact to enhance driver safety.

- As intelligent and curious beings, we want *to understand* things like:

  Are psychic abilities real and if so, under what conditions?
  Do left-handed people die at a younger average age?
  Does listening to classical music really improve scores on intelligence
    tests?
  Are there differences in how men's and women's brains work?

- As social and curious beings, we want to *know about other people:*

  Who are they planning to vote for in the next election?
  What television programs are they watching?
  How much alcohol do they drink?
  How healthy are they?

22.3 | MAKING PERSONAL DECISIONS

Results of statistical studies and other statistical information can sometimes help when you are required to make a decision involving uncertainty. Such a decision usually involves both factual and emotional reasoning, so statistical information alone is often not sufficient. It is sometimes helpful to think about the decision in the framework of hypothesis testing and to consider the consequences of the two types of error we discussed. In many cases, the decision can be formulated as follows:

H_0: I will be better off if I take no action.
H_a: I will be better off if I do take action.

A type 1 error would correspond to taking action when you would have been better off not doing so. A type 2 error would correspond to taking no action when you would have been better off taking action.

EXAMPLE 22.1
Playing the Lottery

Suppose there is a lottery in your area and the prize money is many millions of dollars (or the equivalent in your currency). You have picked the numbers you would like to play but can't decide whether to buy a ticket. For simplicity, let's forget about the smaller possible prizes and consider only the decision about whether or not to spend money for a ticket based on the big prize. Here are your hypotheses:

H_0: The numbers I have chosen will not win the big prize.
H_a: The numbers I have chosen will win the big prize.

Let's consider the possible consequences of choosing each of these hypotheses and then analyze the choices.

	Likelihood	H_0 Is Chosen (Don't Buy)	H_a Is Chosen (Buy Ticket)
H_0 Is True (Won't Win)	Very high	No loss	Lose cost of ticket
H_a Is True (Would Win)	Very low	Lose millions, agony of seeing your numbers win	Win the big prize

The likelihood that the null hypothesis is true is very high, yet many people behave as if the alternative hypothesis is true. The reason becomes obvious by examining the consequences of each choice in the last two columns of the table. Although there is no loss for correctly choosing H_0, the possible error associated with that choice is devastating. The possible error associated with choosing H_a is minor for many people (lose cost of ticket) and the possible payoff is very high.

To summarize, if you choose H_a, the likelihood that you are making a type 1 error is very high but the consequences of it are minor. If you choose H_0, the likelihood that you are making a type 2 error is extremely small, but the consequences of it are not minor. By the way, the long-run average loss of 30 to 60 cents per ticket in most lotteries is apparently insignificant to most individual ticket buyers, who make their decision (consciously or not) based on the consequences shown in the table. ◆

EXAMPLE 22.2
Surgery or Uncertainty?

Sometimes life-or-death decisions must be made on the basis of little information. A somewhat common situation is as follows. Suppose your doctor discovers a lump and cannot tell whether it is a malignant or benign growth. You have two choices. You can have the affected organ removed with surgery, or you can wait and see if the lump continues to grow or to spread. What would you do? The hypotheses you must choose between are

H_0: The lump is benign.
H_a: The lump is malignant.

Let's examine the choices and their consequences:

	Likelihood	H_0 Is Chosen (No Surgery)	H_a Is Chosen (Surgery)
H_0 Is True (Benign)	??	No loss	Lose organ needlessly
H_a Is True (Malignant)	??	Possibly life-threatening	Stop possible spread

Statistical information may be available to assess the likelihood of each hypothesis, and presumably the medical practitioner could provide reasonable probabilities. You must then weigh the possible consequences of each choice based on nonstatistical issues. For instance, depending on which organ is involved, its loss may be minor or it may be extremely serious and require lifelong medical intervention. ◆

EXAMPLE 22.3
Fish Oil and
Psychiatric Disorders

A friend's child was recently diagnosed with a possible psychiatric disorder, and the child's physician recommended that the parents give the child fish oil. In an Internet search for more information, the parents found the following news story.

Thursday September 3, 1998

Fish Oil May Fight Psychiatric Disorders

NEW YORK, Sep 03 — The consumption of omega-3 polyunsaturated fatty acids found in fish and fish oil may reduce the symptoms of a variety of psychiatric illnesses, including schizophrenia, bipolar disorder, and depression, researchers report.

"Research suggests that (fatty acids) may have a role in psychiatric disorders," said Dr. Joseph Hibbeln of the National Institute on Alcohol Abuse and Alcoholism, part of the National Institutes of Health (NIH) in Bethesda, Maryland.

Hibbeln is one of a number of researchers attending an NIH-sponsored workshop on the issue in Bethesda this week.

The workshop was prompted in part by the results of three recent studies.

Findings from one study, conducted by Dr. Andrew Stoll of the Harvard Medical School in Boston, Massachusetts, suggest that fish oil supplementation could help alleviate the symptoms of bipolar (manic-depressive) disorder.

For a 4-month period, Stoll gave 14 bipolar patients daily supplements of either fish oil or a "dummy pill," or placebo. He found that "overall, 9 of 14 patients responded favorably to the addition of omega-3 fatty acids (to their diet), compared to only 3 of 16 patients receiving placebo."

Another study focused on the effects of one fish-oil fatty acid, eicosa-pentaenoic (EPA), in the treatment of schizophrenia. A 3-month trial conducted by Dr. Malcolm Peet of Northern General Hospital in Sheffield, England, concluded that [there was] "a 25% improvement (in schizophrenic symptoms) in the EPA treated group," compared with patients receiving either docosahexaenoic acid (DHA, another omega-3 fatty acid) or placebo.

A third study, conducted by Hibbeln, focused on levels of omega-3 fatty acids in the blood of 50 patients hospitalized after attempting suicide.

Hibbeln found that, among nondepressive (but not depressive) patients, high blood concentrations of EPA "predicted strikingly lower (better) scores in 6 different psychological rating scales which are related to suicidal risk." The NIH researcher says these findings suggest that "some subgroups of suicidal patients may reduce their suicidal risk with the consumption of EPA."

Hibbeln also noted that another study showed that dietary intake of EPA and DHA may influence serotonin function in the brain. "Such an alteration in serotinergic function may possibly reduce depressive, suicidal and violent behavior, but these changes have not yet been demonstrated in . . . clinical trials," he said in a statement.

Hibbeln explained that the brain's synaptic membranes, where much of the brain's neurological signaling takes place, "have a large proportion of essential fatty acids in them—fatty acids which are derived entirely from the diet."

He points out that "in the last century, (Western) diets have radically changed and we eat grossly fewer omega-3 fatty acids now. We also know that rates of depression have radically increased by perhaps a hundred-fold" over the same period of time.

Links between fish consumption and neurological health may be supported by the results of global studies. According to Hibbeln, those findings suggest that "rates of major depression are markedly different across countries, depending upon how much fish is consumed in those countries."

Source: "Fish Oil May Fight Psychiatric Disorders," Sept. 3, 1998. Yahoo News Archive © 1998.

What would you do if you were the parents of this child? Would you start giving fish oil to your child? The two hypotheses are

H_0: Fish oil will not help my child.
H_a: Fish oil will help my child.

The possible decisions and consequences are

	Likelihood	H_0 Is Chosen (No Fish Oil)	H_a Is Chosen (Fish Oil)
H_0 Is True (Fish Oil Won't Help)	??	No loss or gain	Child takes fish oil needlessly; presumably no harm
H_a Is True (Fish Oil Will Help)	??	Child does not receive a treatment that would help	Child is helped by fish oil

In this case, the parents must assess the likelihood that fish oil will help their child, and the consequences of making each decision. Some of the research discussed in the article is based on randomized experiments, so in those studies there appears to be a causal benefit from fish oil. The participants in those studies may or may not be representative of the same group as this particular child, and the parents need to assess that issue. Finally, the consequences of choosing H_a erroneously are minor, so with a moderate probability that it is true, the parents would be wise to decide in favor of giving fish oil to the child. ◆

22.4 | CONTROL OF SOCIETAL RISKS

Sometimes statistical studies can be used to guide policy decisions. In these cases, lawmakers, government regulatory agencies, and other decision makers must weigh decisions and their consequences.

EXAMPLE 22.4
Go, Granny, Go or Stop, Granny, Stop?

Efforts to improve driving safety are often based on results of statistical studies. For instance, studies have shown that wearing a seat belt is likely to reduce the chances of death or serious injury in an accident, that driving after consuming alcohol increases the risk of an accident, and so on. Laws have been passed to help protect drivers and passengers based on these findings. But there is clearly a trade-off between protection and personal freedom. If you were a lawmaker concerned about driver safety, would the following article convince you to take action?

Tuesday April 7, 1998 6:46 PM EDT   Yahoo 1998

Visual Field Loss Ups Elderly Car Crashes

NEW YORK — A 40% loss in range of vision among older drivers more than doubles their risk for a car accident, experts say. They believe around one third of all seniors suffer from such vision impairment.

   "Visual dysfunction and eye disease deserve further examination as causes of motor vehicle crashes and injury," say ophthalmology researchers at the University of Alabama at Birmingham, and the Western Kentucky University in Bowling Green, Kentucky.

   Their study, published in the April 8th issue of The Journal of the American Medical Association (JAMA), focused on the 3-year motor vehicle accident rates of 294 Alabama drivers between 55 and 87 years of age. Each of the individuals received various vision tests, including field of vision examinations.

The researchers discovered that, of all the vision-related factors they studied, "impaired useful field of view was the only one that demonstrated a marked elevation (in risk). Older drivers with a 40% or greater reduction in the useful field of view were 2.1 times more likely to have incurred a crash . . . compared with those with less than 40% reduction."

They found that an older driver's crash risk increased by 16% with every 10-point drop in their field of vision.

Since previous research has found that one in three older adults has a greater than 40% reduction in their field of vision, the study authors speculate that field of vision tests might "be a good candidate for inclusion in a functional test battery" aimed at determining driver safety. At the present time, most of the nation's drivers are only obligated to pass visual acuity tests that assess their sharpness of vision in order to obtain license renewals.

However, JAMA contributing editor Dr. Thomas Cole believes field of view tests are "probably too time-consuming and expensive to be used as a screening test in driver licensing offices, and should instead be considered a diagnostic test for drivers with suspected subtle (vision) impairment."

Source: "Visual Field Loss Ups Elderly Car Crashes." April 7, 1998. Yahoo News Archive © 1998.

Should additional vision screening be required for older drivers? There are statistical and nonstatistical issues involved in that decision. First, the research was based on an observational study; drivers cannot be randomly assigned to have vision impairment or not. There are obviously potential confounding factors that may affect accident rates. Also, the researchers mentioned that they studied many aspects of vision but that "impaired useful field of view" was the only one that was related to accident rates. It is possible that they made a type 1 error, which can easily happen for one or more hypotheses when a large number are tested. Finally, it is not clear how representative the participants in this study are of all older drivers, and the results may not apply to all older drivers.

Even if the relationship is real and causal, it may not be appropriate to restrict driving privileges based on visual field impairment. In addition to the statistical issues, there are issues of personal freedom and of the cost of vision testing. Policy makers must weigh all of these factors when making decisions based on statistical studies. ◆

EXAMPLE 22.5
When Smokers Butt Out, Does Society Benefit?

The debate over the health risks of breathing smoke from other peoples' cigarettes has been largely statistical in nature, and new information released at the end of 2000 added fuel to the fire. After more than ten years of active antismoking measures, the lung cancer rate in California was down substantially compared to the rest of the United States. The situation was summarized in this news article:

Thursday November 30, 2000 8:13 PM ET

Anti-Tobacco Measures Lessen Cancer

By JENNIFER COLEMAN, Associated Press Writer

SACRAMENTO, Calif. (AP) — California's tough anti-smoking measures and public health campaigns have resulted in a 14 percent decrease in lung cancer over the past 10 years, the government reported Thursday.

Other regions of the country reported only a 2.7 percent decrease over the same period, the Centers for Disease Control and Prevention said.

"Based on the California experience, we would hope to see similar effects in other states using similar programs," said Dr. Terry Pechacek, CDC associate director for science and public health.

Lung cancer develops slowly and the full benefits of quitting can take up to 15 years to be realized. However, Pechacek said, researchers can start seeing some results within five years.

Smoking rates in California began dropping in the late 1980s, helped in part by Proposition 99 in 1988. The voter-approved measure added a 25-cent-per-pack tax on tobacco products that paid for anti-smoking and education programs. Local governments also began restricting smoking in public buildings and workplaces.

Two years ago, voters bumped the price of cigarettes an additional 50 cents per pack, money also earmarked for education. And this year alone, the state will spend $136 million on smoking prevention, cessation and research—some $45 million of it on anti-tobacco advertising.

"California has the most comprehensive program for protecting nonsmokers from secondhand smoke," said Ken August, spokesman for the state health department. "Restaurants, bars and almost all indoor workplaces are smoke-free."

The effect of the anti-tobacco efforts has been fewer smokers and fewer deadly cases of cancer related to smoking, health officials said. August and Pechacek both said they expect the trend to continue.

August said that means there will be up to 4,000 fewer lung cancer cases in California this year and about 2,000 fewer deaths.

In its report, the CDC compared cancer registries in California, Connecticut, Hawaii, Iowa, New Mexico and Utah, as well as Seattle, Atlanta and Detroit.

In 1988, the lung cancer rate in California was 72 cases per 100,000 people, slightly higher than that of the other regions studied. By 1997, California's rate had dropped to about 60 per 100,000.

The CDC averages the statistics for the first two years and the last two years studied to arrive at an accurate representation, health officials said. The numbers the CDC used were 71.9 for 1988 and 70.3 for 1989, averaging to 71.1; and 62.2 for 1996 and 60.1 for 1997, averaging to 61.15. That computes to a 14 percent drop in lung cancer cases.

While lung cancer rates for women in the other regions rose 13 percent, the rate for California women dropped 4.8 percent. Among California men, lung cancer rates dropped 23 percent, compared with a 13 percent drop among men elsewhere.

Dr. David Burns, a volunteer with the American Lung Association in California, said: "This is an accomplishment of Proposition 99 money being invested wisely by the state to help people change their smoking behavior."

Source: "Anti-Tobacco Measures Lessen Cancer," Nov. 30, 2000. Copyright © 2000 Associated Press. Used with permission.

Suppose you were a lawmaker in another jurisdiction and read these results. Would you attempt to enact legislation similar to that in California? The evidence is certainly tempting, but there are some statistical issues that need to be considered. First, these results are based on observational data. California was not randomly chosen to apply tough antismoking laws. People in California may be more health conscious than people in other states. California has other strict air pollution legislation that was enacted in the same time period, such as tough maintenance and inspection laws for vehicle emissions. California industry has changed over the years, favoring high-tech industry likely to employ younger, healthier workers.

If you were a lawmaker, you would need to weigh consequences of possible decisions. For instance, as mentioned, California has had an active antismoking education program. The consequences of launching such a program would presumably not be harmful, so a decision to do so would have positive or neutral consequences. On the other hand, a decision to ban smoking in bars and restaurants would involve more controversial consequences. ◆

22.5 | UNDERSTANDING OUR WORLD

Sometimes statistical studies are done to help us understand ourselves and our world, without involving any decisions. There are thousands of academic journals containing the results of statistical studies, most of which will never be used to change personal or societal decisions or behavior. Scan any major news source for a few weeks and you will find reports of many interesting studies that are done to help us understand the world. Many of these studies are exploratory in nature and have results that are controversial. That's part of why they make interesting news.

EXAMPLE 22.6
Is It Wining or Dining That Helps French Hearts?

French citizens have lower rates of heart disease than people in neighboring countries, and some recent speculation has focused on red wine as a partial cause. The following article discusses some other possible explanations. Notice that no one is likely to change laws or behavior as a result of this research, it is simply interesting information.

Friday May 28 1:24 PM ET (1999)

Wine May Not Explain France's Lower Heart Disease Rate

NEW YORK, May 28 — Previous research has suggested that red wine consumption may explain why the French have a much lower rate of heart disease than other nations. But in a report published this week in the British Medical Journal, two UK experts dispute this theory, suggesting instead that the reduced risk lies in the fact that the French diet traditionally contains less animal fat.

And because the levels of animal fat and cholesterol in the French diet have risen in the last 15 years, their explanation suggests that it is only a matter of time before the heart disease rate in France catches up with the rate in Britain, where heart disease mortality is about four times higher.

Dr. Malcolm Law and Professor Nicholas Wald from the Wolfson Institute of Preventive Medicine in London, UK, suggest that the tendency of French doctors to attribute heart disease deaths to other causes "could account for about 20% of the difference," according to the report.

Law and Wald do give red wine some credit as a preventive factor, but say that this effect is small. The high consumption of red wine in France, they write, "explains (less than 5%) of the difference."

On the other hand, they provide considerable statistical evidence to support their "time lag" theory. "Mortality from . . . heart disease," the authors write, "was strongly associated with past animal fat consumption . . . and past (serum cholesterol) values, but not with recent values."

"Animal fat consumption and serum cholesterol concentration have been similar in France and Britain for a relatively short time—about 15 years," write Law and Wald. "Serum cholesterol concentration in 1970 was . . . lower in France than in Britain . . . and this explains most of its lower mortality from heart disease," they suggest.

Several related editorials question Law and Wald's hypothesis. Meir Stampfer and Eric Rimm from Harvard University, Boston, Massachusetts, write, "Obviously, other factors must play a role. . . . We think it more likely that the difference in coronary mortality rests on behavioral (especially dietary) differences that have not received adequate attention."

In another commentary, D.J.P. Barker from the University of Southampton in Southampton, UK, writes that "recent trends in coronary heart disease are only weakly related to trends in serum cholesterol" and argues that "coronary heart disease originates in utero, through adaptations that the fetus makes to undernutrition."

Finally, Johan Mackenbach and Anton Kunst from Erasmus University in Rotterdam, the Netherlands, contend that "heterogeneity of populations should be taken into account."

But Law and Wald dismiss these arguments. "We believe," they write in response, "that the time lag explanation is the major reason and that the alternative explanations offered in the commentaries are quantitatively unimportant."

Source: "Wine May Not Explain France's Lower Heart Disease Rate," May 28, 1999. Yahoo News Archive © 1999. ◆

EXAMPLE 22.7
Give Her the Car Keys

Memory is a fascinating ability, and most of us wish we had more of that ability. Studies about memory may eventually help us understand ways to improve it. In the meantime, the following study is simply interesting. If it leads to any decision at all, it would be related to who thinks they should get the car keys and who actually should. We'll let you figure it out.

Thursday March 26, 1998 1:30 PM EST   YAHOO Health News

Women Remember Item Location Better

NEW YORK — Women are better at remembering where objects are than men, but show less confidence in their memory, according to a study.

"When it comes to memory, women have more skill than confidence, and men have more confidence than skill," concludes Dr. Robin West, a University of Florida (UF) psychology professor and researcher.

West, along with UF psychology graduate student Duana Welch, compared the spatial memories of over 300 healthy men and women of various ages in a study funded by the National Institutes of Mental Health.

They used specially-designed computer tests that had each participant place 20 common household "objects" into one of 12 "rooms" (e.g., kitchen, bedroom, bathroom) depicted on the screen. After a 40-minute interval, each participant was asked to remember the location of each object.

According to Welch, the test focused on "something that has very important day-to-day meaning for people; that is, the ability to find where you put something."

The results? West and Welch say they found "younger adults performing better than older adults, and women performing better than men." In fact, women scored an average of 14.4 points on the test, compared with the men's average score of 13.5.

Those scores didn't meet male expectations of their own memory skills, the researchers say. In a statement issued by the university, West and Welch say that when questioned, "men tended to overestimate their ability to remember object locations," while women tended to underestimate their powers of recall. The study authors note these findings "are in keeping with other literature suggesting that women are not as confident about their cognitive (intellectual) abilities as they should be."

Those types of gender-based insecurities may be ending, however. In a previous study involving married couples, West discovered that gender-specific differences in memory-confidence disappeared among younger (ages 35 to 45) couples. West credits those results on "historical changes in the beliefs and attitudes of our culture. The older women grew up in a time when men were thought to have more mental ability than women. Such beliefs—learned in childhood and reinforced over time—are hard to change." ◆

22.6 | GETTING TO KNOW YOU

As social beings, we are curious about how others think and behave. What do they do with their time? Are we in the majority with our opinions on controversial issues? At what age do people typically get married? What proportion of the population is left-handed? What proportion is gay or lesbian? Are people basically honest? Many of these questions can be answered by surveying random or representative samples.

Most national governments have agencies that collect samples to answer some of these questions on a routine basis. The United States government collects diverse and numerous statistics on the behavior and opinions of its people. The website http://www.fedstats.gov is headlined "The gateway to statistics from over 100 U.S. Federal agencies" and provides links to all kinds of information about the U.S. population.

EXAMPLE 22.8
Lifestyle Statistics
from the Census Bureau

The United States Census Bureau collects voluminous data on many aspects of American life. The ongoing "Current Population Survey" polls a random sample of U.S. households on a wide variety of topics. From this survey, trends in lifestyle decisions can be tracked. Here are a few examples of trends reported in 1995 ("Marital Status and Living Arrangements," Arlene F. Saluter, http://www.census.gov/population/www/pop-profile/msla.html):

♦ "The estimated median age at first marriage is higher than ever before. In 1994, the median age at first marriage was 26.7 years for men and 24.5 years for women, approximately $3\frac{1}{2}$ years higher than the median age in 1970 (23.2 years for men and 20.8 years for women).

♦ An unmarried-couple household is composed of two unrelated adults of the opposite sex (one of whom is the householder) who share a housing unit with or without the presence of children under 15 years old. There were 7 unmarried couples for every 100 married couples in 1994, compared with only 1 for every 100 in 1970. About one-third had children under 15 years old present in the home.

♦ Children living with one parent (18.6 million) represented 27 percent of all children under 18 years old in 1994, up from 12 percent in 1970. The majority lived with their mother, but an increasing proportion lived with their father. In 1994, 12 percent of the children in a one-parent situation lived with their father, up from 9 percent in 1970."

Notice that these statistics are presented as if they were known for the population, but in fact, they are based on random samples of various sizes. Consequently, they have a margin of error that could be computed from knowing the sample size. ♦

EXAMPLE 22.9
In Whom Do We Trust?

If you want to know about the opinions of Americans on almost any topic, visit the website for the Gallup Organization (http://www.gallup.com) and type in the key word of your choice. (Unfortunately, Gallup now charges for this service.) For instance, a news release on November 27, 2000 was headlined "Nurses Remain at Top of Honesty and Ethics Poll, Car Salesmen Still Seen as Least Honest and Ethical." The accompanying story (Darren K. Carlson, http://

www.gallup.com/poll/releases/pr001127.asp) reported on a poll that has been taken yearly since 1977. Respondents were asked, "Please tell me how you would rate the honesty and ethical standards of people in these different fields—very high, high, average, low, or very low" and were then given a list of several dozen occupations, in random order. Some of the results highlighted in the news release were

- "For the second year in a row, Gallup's survey on honesty and ethics in professions finds that the American public rates nursing as the field with the highest standards of honesty and ethics. Almost eight in 10 Americans—79%—say nurses have 'very high' or 'high' ethical standards. Pharmacists finish second with 67%. Pharmacists had consistently finished first in the survey, until nurses were added to the list in 1999.

- Only one profession is rated as having 'low' or 'very low' standards by a majority of Americans: car sales.

- Americans remain skeptical of the honesty and ethics of their elected officials. As with every previous year, less than half the public said any elected official on Gallup's list had 'very high' or 'high' honesty and ethics.

- The long-term trends of Gallup's honesty and ethics survey reveal a growing skepticism among the American public regarding the ethics of news professionals and lawyers. The three news professions surveyed all remain significantly lower than when they were first placed on the list. Twenty-one percent of the public says journalists are honest and ethical, a percentage that has declined steadily from its debut of 33% in 1977.

- Lawyers are often the punch line to jokes regarding professional ethics and honesty. They debuted on Gallup's survey list in 1977 with 26%. Lawyers have not been rated as having high ethical standards by more than 20% of the public since 1991, and the latest poll shows a rating of 17%."

College teachers were rated highly by 59% and grade school and high school teachers by 62%. And, of course, to interpret the results correctly, you need to know that "the results reported here are based on telephone interviews with a randomly selected national sample of 1,028 adults, 18 years and older, conducted November 13–15, 2000. For results based on the whole sample, one can say with 95 percent confidence that the maximum error attributable to sampling and other random effects is plus or minus 3 percentage points." ◆

22.7 | WORDS TO THE WISE

We hope that by now you realize that you are likely to encounter statistical studies in your personal and professional life and that they can provide useful information. We also hope you realize that results from such studies can be misleading if you don't interpret them wisely.

Throughout this book we have presented warnings and examples of how statistical studies are often misinterpreted. The ten guiding principles presented here are a synthesis of those warnings. Keeping these principles in mind when you read statistical studies will help you gain wisdom about the world around you while maintaining a healthy dose of skepticism about legitimate conclusions.

in summary   Ten Guiding Principles

1. A representative sample can be used to make inferences about a larger population, but descriptive statistics are the only useful results for an unrepresentative sample.

2. Cause and effect can be inferred from randomized experiments, but generally not from observational studies, where confounding variables are likely to cloud the interpretation.

3. A conservative estimate of *sampling* error in a survey is the margin of error, $1/\sqrt{n}$. This provides a bound on the difference between the true proportion and the sample proportion that holds for at least 95% of properly conducted surveys.

4. The margin of error does *not* include nonsampling error, such as errors due to biased wording, nonresponse, and so on.

5. When the individuals measured constitute the whole population, there is no need for statistical inference because the truth is known.

6. A significance test based on a very large sample is likely to produce a statistically significant result even if the true value is close to the null value. In such cases, it is wise to examine the magnitude of the parameter with a confidence interval to determine if the result has practical importance.

7. A significance test based on a small sample may not produce a statistically significant result even if the true value differs substantially from the null value. It is important that the null hypothesis not be "accepted" on this basis.

8. When deciding how readily to reject the null hypothesis (what significance level to use), it is important to consider the consequences of type 1 and type 2 errors. If a type 1 error has serious consequences, the level of significance should be small. If a type 2 error is more serious, a higher level of significance should be used.

9. A study that examines many hypotheses could find one or more statistically significant results just by chance, so you should try to find out how many tests were conducted when you read about a significant result. For instance, with a level of significance of 0.05, about 1 test in 20 should result in statistical significance when all of the null hypotheses are correct and the tests are independent. It is common in large studies to find that one test attracts media attention, so it is important to know if that test was the only one out of many conducted that achieved statistical significance.

10. You will sometimes read that researchers were surprised to find "no effect" and that a study "failed to replicate" an earlier finding of statistical significance. In that case, consider two possible explanations. One is that the sample size was too small and the test had low power. The other possibility is that the result in the first study was based on a type 1 error. This is particularly likely if the effect was moderate and was part of a larger study that covered multiple hypotheses.

*If you still think statistics is a boring and useless subject,
check to make sure you are still breathing, your heart is beating,
and your mind has not turned off!*

EXERCISES   Blue = Basics   Bold, **Bold** = Answer in back

Chapter Exercises

Exercises 1 to 15 are based on the following reports of three studies taken from news organization websites. In each case, the headline and a few sentences from the news story are provided.

Study #1: "*Calcium Lowers Blood Pressure in Black Teens:* In the new study, researchers followed 116 African-American teens who were enrolled in grades 10 through 12 at three high schools in Los Angeles. Over the course of eight weeks, the study participants were given either a placebo or a calcium supplement. They were also asked to fill out food questionnaires. Researchers measured the teens' blood pressure at the outset of the study and two, four and eight weeks later." (Linda Carroll, Medical Tribune News Service, Sept. 7, 1998; study published in the *American Journal of Clinical Nutrition* (1998), Vol. 68, 648–655.)

Study #2: "*Daily Two-Mile Walk Halves Death Risk:* Between 1980 and 1982, multicenter researchers in the Honolulu Heart Program studied 707 nonsmoking, retired men, aged 61 to 81 years, and collected mortality data on these men over the following 12 years. During the study, 208 of the men died. The study results show that while 43.1% of men who walked less than one mile per day died, only half this figure—21.5%—of the men who walked more than two miles per day died." (Reuters News, Jan. 8, 1998; study published in *The New England Journal of Medicine* (1998), Vol. 338, 94–99.)

Study #3: "*Tea Doubles Chance Of Conception:* Tea for two might make three, according to a new study. The investigators asked 187 women, each of whom said they were trying to conceive, to record information concerning their daily dietary intake over a one-year period. According to the researchers, an analysis of the data suggests drinking one-half cup or more of tea daily approximately doubled the odds of conception per cycle, compared with non-tea drinkers. The mechanism behind tea's apparent influence on fertility remains unclear. But the investigators point out that, on average, tea drinking is associated with a 'preventive or healthier' lifestyle, which might encourage conception." (Reuters News, Feb. 27, 1998; study published in the *American Journal of Public Health* (1998), Vol. 88, No. 2: 270–274.)

22.1   Each of the three headlines implies a causal relationship between an explanatory and response variable. For each of the three studies, describe the explanatory and response variables.

22.2   Do you think Study #1 was a randomized experiment or an observational study? Explain. Based on your answer, explain whether or not the headline is justified.

22.3   Do you think Study #2 was a randomized experiment or an observational study? Explain. Based on your answer, explain whether or not the headline is justified.

22.4   Do you think Study #3 was a randomized experiment or an observational study? Explain. Based on your answer, explain whether or not the headline is justified.

22.5   Give an example of a possible confounding variable for Study #2. Explain why you think it fits the definition of a confounding variable.

22.6   Give an example of a possible confounding variable for Study #3. Explain why you think it fits the definition of a confounding variable.

22.7   Can the results of Study #1 be applied to a larger population? If so, to what population? If not, why not?

22.8   Can the results of Study #2 be applied to a larger population? If so, to what population? If not, why not?

22.9   Can the results of Study #3 be applied to a larger population? If so, to what population? If not, why not?

22.10   In Study #1, participants were also asked to fill out a food questionnaire. Another quote from the news article was: "Overall, diastolic blood pressure—the second of the two numbers given when blood pressure is measured—dropped about two points in the group given calcium supplements. But the drop was almost five points in teens with a particularly low intake of calcium-containing foods."

a. Why do you think the researchers wanted the information provided by the food questionnaire?

b. Is "intake of calcium-containing foods" a confounding variable or an interacting variable in this study? Explain.

22.11   Here are some additional quotes from the article for Study #2. Comment on each quote using issues covered in this chapter.

a. "The investigators also report that ' . . . when the distance walked is increased by one mile per day . . . the risk of death can be reduced by 19%.' "

b. "Cancer was the most common cause of death in the men studied. The research team found that 13.4% of the men who walked less than one mile per day died of cancer, but only 5.3% of those who walked more than two miles per day died of cancer. Even when age was included in the analysis, this result held."

c. "Walking also appeared to protect against dying from heart disease and stroke, but the findings were not statistically significant, meaning that they could have occurred by chance."

22.12   The article for Study #3 also noted: "The study results also show that the women who consumed the most tea ' . . . had significantly higher energy and fat intakes' than women who drank less tea."

a. Do you think the word "significantly" in this quote refers to statistical significance? If not, why not? If so, describe in words the null and alternative hypotheses you think the researchers tested to reach that conclusion.

b. Are "energy and fat intake" confounding variables or interacting variables in this study? Explain.

22.13   The article for Study #3 also said: "The California researchers say they found no significant association between coffee intake and fertility, but ' . . . in later cycles . . . there was a suggestion that both total caffeine and coffee were associated with a nonsignificant reduction in fertility,' a finding in keeping with previous research which pointed to the possibility that coffee or caffeine might lower conception rates."

a. The phrase "*no* significant association between coffee intake and fertility" seems to contradict the "suggestion that both total caffeine and coffee were associated with a nonsignificant reduction in fertility." Explain this apparent contradiction.

b. Refer to item 10 in Section 22.7 and discuss it in the context of this quote.

22.14 The two main variables measured in Study #1 were the treatment given to the participant and the participant's blood pressure.

    a. For each of the two variables, determine if it was recorded as a categorical variable or a quantitative variable. If it was a categorical variable, specify the categories.

    b. Researchers measured the participants' blood pressure at the beginning of the study and two, four, and eight weeks later. If they wanted to determine whether blood pressure had dropped significantly from the beginning of the study to eight weeks later for the participants in the calcium group, would they use a paired *t*-test or a test for the difference in means for independent samples? Explain.

    c. Refer to part (b). Write the null and alternative hypotheses the researchers would use.

    d. If the researchers wanted to compare the change in blood pressure after eight weeks of taking calcium compared to eight weeks of taking a placebo, what hypothesis-testing method would they use?

    e. Refer to part (d). Write the null and alternative hypotheses the researchers would use.

22.15 Refer to the headline in Study #2 and to the statement "While 43.1% of the men who walked less than one mile per day died, only half this figure—21.5%—of the men who walked more than two miles per day died."

    a. What two categorical variables measured on each participant are used in making that statement? What are the categories for each one?

    b. If the researchers wanted to determine if the result was statistically significant, what inference procedure would be appropriate? Be specific.

22.16 Find an example of an observational study in the news and answer these questions about it:

    a. Explain why it fits the definition of an observational study rather than a randomized experiment.

    b. Describe an explanatory and response variable considered in the study.

    c. Using the variables in part (b), give an example of a possible confounding variable and explain why it fits the definition of a confounding variable.

    d. Write two headlines to accompany the study—one that is correct and the other that implies a conclusion that is not justified. Explain which is which, and why. (You may use the actual headline accompanying the story as one of these.)

    e. Can the results of the study be extended to a larger population? Explain.

22.17 Find an example of a randomized experiment in the news and answer these questions about it:

    a. Explain why it fits the definition of a randomized experiment rather than an observational study.

    b. Describe an explanatory and response variable considered in the study.

    c. Using the variables in part (b), give an example of a possible interacting variable and explain why it fits the definition of an interacting variable. You may use one that is actually measured in the study and discussed in the news report.

    d. Give the headline that was provided with the news article and discuss whether it is justified based on the study.

    e. Can the results of the study be extended to a larger population? Explain.

22.18 In each of the following cases, explain whether the results can be extended to a larger population. If so, explain what the population is. If not, explain why not.

    a. A company measured the salaries of all of its employees and found that the average salary for men was 10% higher than the average salary for women.

    b. A geologist at Yellowstone National Park measured the length of time for each eruption and the waiting time until the next eruption for the Old Faithful geyser from Aug. 1–8 and 16–23, 1978 (data from Weisberg, 1985, p. 231). The correlation between the two measurements was 0.877, indicating that waiting time until the next eruption is strongly correlated with the duration of the previous eruption.

    c. A poll following the presidential debates of 2000 was conducted by using random-digit dialing. When someone answered the phone, they were asked if they had watched the debates. If the answer was yes, they were asked who won. If the answer was no, the call was terminated. The question of interest was the proportion of U.S. adults who thought each of the two candidates had won the debate.

22.19 Refer to Example 22.5 describing California's antismoking measures and reduction in lung cancer. An article in *The New England Journal of Medicine* (Fichtenberg and Glantz, 2000) studied mortality from heart disease in California during the same period. The conclusion the authors reached was: "A large and aggressive tobacco-control program is associated with a reduction in deaths from heart disease in the short run."

    a. The conclusion is based on results obtained through methods similar to those in Example 22.5. Is the wording of the conclusion justified? Explain.

    b. Write two headlines to accompany a news report on this study, one that presents the conclusion accurately and one that implies a conclusion that is not justified. Explain your reasoning.

22.20 Example 22.5 describes California's antismoking measures and subsequent reduction in lung cancer rates. Read the example and the material that follows it. Using the "nonstatistical considerations to assess cause and effect" given in Section 22.1, discuss the extent to which you think a cause-and-effect conclusion can be made about the relationship between the antismoking measures and reduction in lung cancer rates.

REFERENCES

Fichtenberg, C. M., and S. A. Glantz (2000). "Association of the California Tobacco Control program with declines in cigarette consumption and mortality from heart disease," *New England Journal of Medicine,* Vol. 343, No. 24, 1772–1777.

Ramsey, F., and D. Schafer (1997). *The Statistical Sleuth: A Course in Methods of Data Analysis,* Belmont, CA: Duxbury Press.

Weisberg, S. (1985). *Applied Linear Regression,* 2nd ed. New York: Wiley and Sons.

Appendix of Tables

TABLE A.1 ■ Standard Normal Probabilities (for $z < 0$)

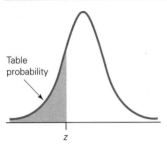

Table probability

z	.00	.01	.02	.03	.04	.05	.06	.07	.08	.09
−3.4	.0003	.0003	.0003	.0003	.0003	.0003	.0003	.0003	.0003	.0002
−3.3	.0005	.0005	.0005	.0004	.0004	.0004	.0004	.0004	.0004	.0003
−3.2	.0007	.0007	.0006	.0006	.0006	.0006	.0006	.0005	.0005	.0005
−3.1	.0010	.0009	.0009	.0009	.0008	.0008	.0008	.0008	.0007	.0007
−3.0	.0013	.0013	.0013	.0012	.0012	.0011	.0011	.0011	.0010	.0010
−2.9	.0019	.0018	.0018	.0017	.0016	.0016	.0015	.0015	.0014	.0014
−2.8	.0026	.0025	.0024	.0023	.0023	.0022	.0021	.0021	.0020	.0019
−2.7	.0035	.0034	.0033	.0032	.0031	.0030	.0029	.0028	.0027	.0026
−2.6	.0047	.0045	.0044	.0043	.0041	.0040	.0039	.0038	.0037	.0036
−2.5	.0062	.0060	.0059	.0057	.0055	.0054	.0052	.0051	.0049	.0048
−2.4	.0082	.0080	.0078	.0075	.0073	.0071	.0069	.0068	.0066	.0064
−2.3	.0107	.0104	.0102	.0099	.0096	.0094	.0091	.0089	.0087	.0084
−2.2	.0139	.0136	.0132	.0129	.0125	.0122	.0119	.0116	.0113	.0110
−2.1	.0179	.0174	.0170	.0166	.0162	.0158	.0154	.0150	.0146	.0143
−2.0	.0228	.0222	.0217	.0212	.0207	.0202	.0197	.0192	.0188	.0183
−1.9	.0287	.0281	.0274	.0268	.0262	.0256	.0250	.0244	.0239	.0233
−1.8	.0359	.0351	.0344	.0336	.0329	.0322	.0314	.0307	.0301	.0294
−1.7	.0446	.0436	.0427	.0418	.0409	.0401	.0392	.0384	.0375	.0367
−1.6	.0548	.0537	.0526	.0516	.0505	.0495	.0485	.0475	.0465	.0455
−1.5	.0668	.0655	.0643	.0630	.0618	.0606	.0594	.0582	.0571	.0559
−1.4	.0808	.0793	.0778	.0764	.0749	.0735	.0721	.0708	.0694	.0681
−1.3	.0968	.0951	.0934	.0918	.0901	.0885	.0869	.0853	.0838	.0823
−1.2	.1151	.1131	.1112	.1093	.1075	.1056	.1038	.1020	.1003	.0985
−1.1	.1357	.1335	.1314	.1292	.1271	.1251	.1230	.1210	.1190	.1170
−1.0	.1587	.1562	.1539	.1515	.1492	.1469	.1446	.1423	.1401	.1379
−0.9	.1841	.1814	.1788	.1762	.1736	.1711	.1685	.1660	.1635	.1611
−0.8	.2119	.2090	.2061	.2033	.2005	.1977	.1949	.1922	.1894	.1867
−0.7	.2420	.2389	.2358	.2327	.2296	.2266	.2236	.2206	.2177	.2148
−0.6	.2743	.2709	.2676	.2643	.2611	.2578	.2546	.2514	.2483	.2451
−0.5	.3085	.3050	.3015	.2981	.2946	.2912	.2877	.2843	.2810	.2776
−0.4	.3446	.3409	.3372	.3336	.3300	.3264	.3228	.3192	.3156	.3121
−0.3	.3821	.3783	.3745	.3707	.3669	.3632	.3594	.3557	.3520	.3483
−0.2	.4207	.4168	.4129	.4090	.4052	.4013	.3974	.3936	.3897	.3859
−0.1	.4602	.4562	.4522	.4483	.4443	.4404	.4364	.4325	.4286	.4247
−0.0	.5000	.4960	.4920	.4880	.4840	.4801	.4761	.4721	.4681	.4641

In the Extreme (for $z < 0$)

z	−3.09	−3.72	−4.26	−4.75	−5.20	−5.61	−6.00
Probability	.001	.0001	.00001	.000001	.0000001	.00000001	.000000001

S-PLUS was used to determine information for the "In the Extreme" portion of the table.

TABLE A.1 ■ Standard Normal Probabilities (for $z > 0$)

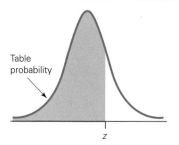

Table probability

z

z	.00	.01	.02	.03	.04	.05	.06	.07	.08	.09
0.0	.5000	.5040	.5080	.5120	.5160	.5199	.5239	.5279	.5319	.5359
0.1	.5398	.5438	.5478	.5517	.5557	.5596	.5636	.5675	.5714	.5753
0.2	.5793	.5832	.5871	.5910	.5948	.5987	.6026	.6064	.6103	.6141
0.3	.6179	.6217	.6255	.6293	.6331	.6368	.6406	.6443	.6480	.6517
0.4	.6554	.6591	.6628	.6664	.6700	.6736	.6772	.6808	.6844	.6879
0.5	.6915	.6950	.6985	.7019	.7054	.7088	.7123	.7157	.7190	.7224
0.6	.7257	.7291	.7324	.7357	.7389	.7422	.7454	.7486	.7517	.7549
0.7	.7580	.7611	.7642	.7673	.7704	.7734	.7764	.7794	.7823	.7852
0.8	.7881	.7910	.7939	.7967	.7995	.8023	.8051	.8078	.8106	.8133
0.9	.8159	.8186	.8212	.8238	.8264	.8289	.8315	.8340	.8365	.8389
1.0	.8413	.8438	.8461	.8485	.8508	.8531	.8554	.8577	.8599	.8621
1.1	.8643	.8665	.8686	.8708	.8729	.8749	.8770	.8790	.8810	.8830
1.2	.8849	.8869	.8888	.8907	.8925	.8944	.8962	.8980	.8997	.9015
1.3	.9032	.9049	.9066	.9082	.9099	.9115	.9131	.9147	.9162	.9177
1.4	.9192	.9207	.9222	.9236	.9251	.9265	.9279	.9292	.9306	.9319
1.5	.9332	.9345	.9357	.9370	.9382	.9394	.9406	.9418	.9429	.9441
1.6	.9452	.9463	.9474	.9484	.9495	.9505	.9515	.9525	.9535	.9545
1.7	.9554	.9564	.9573	.9582	.9591	.9599	.9608	.9616	.9625	.9633
1.8	.9641	.9649	.9656	.9664	.9671	.9678	.9686	.9693	.9699	.9706
1.9	.9713	.9719	.9726	.9732	.9738	.9744	.9750	.9756	.9761	.9767
2.0	.9772	.9778	.9783	.9788	.9793	.9798	.9803	.9808	.9812	.9817
2.1	.9821	.9826	.9830	.9834	.9838	.9842	.9846	.9850	.9854	.9857
2.2	.9861	.9864	.9868	.9871	.9875	.9878	.9881	.9884	.9887	.9890
2.3	.9893	.9896	.9898	.9901	.9904	.9906	.9909	.9911	.9913	.9916
2.4	.9918	.9920	.9922	.9925	.9927	.9929	.9931	.9932	.9934	.9936
2.5	.9938	.9940	.9941	.9943	.9945	.9946	.9948	.9949	.9951	.9952
2.6	.9953	.9955	.9956	.9957	.9959	.9960	.9961	.9962	.9963	.9964
2.7	.9965	.9966	.9967	.9968	.9969	.9970	.9971	.9972	.9973	.9974
2.8	.9974	.9975	.9976	.9977	.9977	.9978	.9979	.9979	.9980	.9981
2.9	.9981	.9982	.9982	.9983	.9984	.9984	.9985	.9985	.9986	.9986
3.0	.9987	.9987	.9987	.9988	.9988	.9989	.9989	.9989	.9990	.9990
3.1	.9990	.9991	.9991	.9991	.9992	.9992	.9992	.9992	.9993	.9993
3.2	.9993	.9993	.9994	.9994	.9994	.9994	.9994	.9995	.9995	.9995
3.3	.9995	.9995	.9995	.9996	.9996	.9996	.9996	.9996	.9996	.9997
3.4	.9997	.9997	.9997	.9997	.9997	.9997	.9997	.9997	.9997	.9998

In the Extreme (for $z > 0$)

z	3.09	3.72	4.26	4.75	5.20	5.61	6.00
Probability	.999	.9999	.99999	.999999	.9999999	.99999999	.999999999

S-PLUS was used to determine information for the "In the Extreme" portion of the table.

TABLE A.2 ■ The *t** Multipliers for Confidence Intervals for Means or Difference Between Means

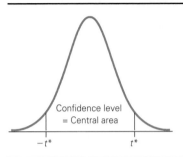

Confidence level
= Central area

$-t^*$ t^*

	Confidence Level			
df	0.90	0.95	0.98	0.99
1	6.31	12.71	31.82	63.66
2	2.92	4.30	6.96	9.92
3	2.35	3.18	4.54	5.84
4	2.13	2.78	3.75	4.60
5	2.02	2.57	3.36	4.03
6	1.94	2.45	3.14	3.71
7	1.89	2.36	3.00	3.50
8	1.86	2.31	2.90	3.36
9	1.83	2.26	2.82	3.25
10	1.81	2.23	2.76	3.17
11	1.80	2.20	2.72	3.11
12	1.78	2.18	2.68	3.05
13	1.77	2.16	2.65	3.01
14	1.76	2.14	2.62	2.98
15	1.75	2.13	2.60	2.95
16	1.75	2.12	2.58	2.92
17	1.74	2.11	2.57	2.90
18	1.73	2.10	2.55	2.88
19	1.73	2.09	2.54	2.86
20	1.72	2.09	2.53	2.85
21	1.72	2.08	2.52	2.83
22	1.72	2.07	2.51	2.82
23	1.71	2.07	2.50	2.81
24	1.71	2.06	2.49	2.80
25	1.71	2.06	2.49	2.79
26	1.71	2.06	2.48	2.78
27	1.70	2.05	2.47	2.77
28	1.70	2.05	2.47	2.76
29	1.70	2.05	2.46	2.76
30	1.70	2.04	2.46	2.75
40	1.68	2.02	2.42	2.70
50	1.68	2.01	2.40	2.68
60	1.67	2.00	2.39	2.66
70	1.67	1.99	2.38	2.65
80	1.66	1.99	2.37	2.64
90	1.66	1.99	2.37	2.63
100	1.66	1.98	2.36	2.63
1000	1.65	1.96	2.33	2.58
Infinite	1.645	1.960	2.326	2.576

Note that the *t*-distribution with infinite df is the standard normal distribution.

TABLE A.3 ■ One-Sided *p*-Values for Significance Tests Based on a *t*-Statistic

◆ A *p*-value in the table is the area to the right of *t*.
◆ Double the value if the alternative hypothesis is two-sided (not equal).

| df | \multicolumn{8}{c}{Absolute Value of *t*-Statistic} |
	1.28	1.50	1.65	1.80	2.00	2.33	2.58	3.00
1	0.211	0.187	0.173	0.161	0.148	0.129	0.118	0.102
2	0.164	0.136	0.120	0.107	0.092	0.073	0.062	0.048
3	0.145	0.115	0.099	0.085	0.070	0.051	0.041	0.029
4	0.135	0.104	0.087	0.073	0.058	0.040	0.031	0.020
5	0.128	0.097	0.080	0.066	0.051	0.034	0.025	0.015
6	0.124	0.092	0.075	0.061	0.046	0.029	0.021	0.012
7	0.121	0.089	0.071	0.057	0.043	0.026	0.018	0.010
8	0.118	0.086	0.069	0.055	0.040	0.024	0.016	0.009
9	0.116	0.084	0.067	0.053	0.038	0.022	0.015	0.007
10	0.115	0.082	0.065	0.051	0.037	0.021	0.014	0.007
11	0.113	0.081	0.064	0.050	0.035	0.020	0.013	0.006
12	0.112	0.080	0.062	0.049	0.034	0.019	0.012	0.006
13	0.111	0.079	0.061	0.048	0.033	0.018	0.011	0.005
14	0.111	0.078	0.061	0.047	0.033	0.018	0.011	0.005
15	0.110	0.077	0.060	0.046	0.032	0.017	0.010	0.004
16	0.109	0.077	0.059	0.045	0.031	0.017	0.010	0.004
17	0.109	0.076	0.059	0.045	0.031	0.016	0.010	0.004
18	0.108	0.075	0.058	0.044	0.030	0.016	0.009	0.004
19	0.108	0.075	0.058	0.044	0.030	0.015	0.009	0.004
20	0.108	0.075	0.057	0.043	0.030	0.015	0.009	0.004
21	0.107	0.074	0.057	0.043	0.029	0.015	0.009	0.003
22	0.107	0.074	0.057	0.043	0.029	0.015	0.009	0.003
23	0.107	0.074	0.056	0.042	0.029	0.014	0.008	0.003
24	0.106	0.073	0.056	0.042	0.028	0.014	0.008	0.003
25	0.106	0.073	0.056	0.042	0.028	0.014	0.008	0.003
26	0.106	0.073	0.055	0.042	0.028	0.014	0.008	0.003
27	0.106	0.073	0.055	0.042	0.028	0.014	0.008	0.003
28	0.106	0.072	0.055	0.041	0.028	0.014	0.008	0.003
29	0.105	0.072	0.055	0.041	0.027	0.013	0.008	0.003
30	0.105	0.072	0.055	0.041	0.027	0.013	0.008	0.003
40	0.104	0.071	0.053	0.040	0.026	0.012	0.007	0.002
50	0.103	0.070	0.053	0.039	0.025	0.012	0.006	0.002
60	0.103	0.069	0.052	0.038	0.025	0.012	0.006	0.002
70	0.102	0.069	0.052	0.038	0.025	0.011	0.006	0.002
80	0.102	0.069	0.051	0.038	0.024	0.011	0.006	0.002
90	0.102	0.069	0.051	0.038	0.024	0.011	0.006	0.002
100	0.102	0.068	0.051	0.037	0.024	0.011	0.006	0.002
1000	0.100	0.067	0.050	0.036	0.023	0.010	0.005	0.001
Infinite	0.1003	0.0668	0.0495	0.0359	0.0228	0.0099	0.0049	0.0013

Note that the *t*-distribution with infinite df is the standard normal distribution.

TABLE A.4 ▪ Critical Values for *F*-test ($\alpha = 0.05$)

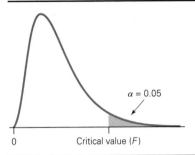

$\alpha = 0.05$

0   Critical value (*F*)

Denom df	Numerator df									
	1	2	3	4	5	6	7	8	9	10
1	161.45	199.50	215.71	224.58	230.16	233.99	236.77	238.88	240.54	241.88
2	18.51	19.00	19.16	19.25	19.30	19.33	19.35	19.37	19.38	19.40
3	10.13	9.55	9.28	9.12	9.01	8.94	8.89	8.85	8.81	8.79
4	7.71	6.94	6.59	6.39	6.26	6.16	6.09	6.04	6.00	5.96
5	6.61	5.79	5.41	5.19	5.05	4.95	4.88	4.82	4.77	4.74
6	5.99	5.14	4.76	4.53	4.39	4.28	4.21	4.15	4.10	4.06
7	5.59	4.74	4.35	4.12	3.97	3.87	3.79	3.73	3.68	3.64
8	5.32	4.46	4.07	3.84	3.69	3.58	3.50	3.44	3.39	3.35
9	5.12	4.26	3.86	3.63	3.48	3.37	3.29	3.23	3.18	3.14
10	4.96	4.10	3.71	3.48	3.33	3.22	3.14	3.07	3.02	2.98
11	4.84	3.98	3.59	3.36	3.20	3.09	3.01	2.95	2.90	2.85
12	4.75	3.89	3.49	3.26	3.11	3.00	2.91	2.85	2.80	2.75
13	4.67	3.81	3.41	3.18	3.03	2.92	2.83	2.77	2.71	2.67
14	4.60	3.74	3.34	3.11	2.96	2.85	2.76	2.70	2.65	2.60
15	4.54	3.68	3.29	3.06	2.90	2.79	2.71	2.64	2.59	2.54
16	4.49	3.63	3.24	3.01	2.85	2.74	2.66	2.59	2.54	2.49
17	4.45	3.59	3.20	2.96	2.81	2.70	2.61	2.55	2.49	2.45
18	4.41	3.55	3.16	2.93	2.77	2.66	2.58	2.51	2.46	2.41
19	4.38	3.52	3.13	2.90	2.74	2.63	2.54	2.48	2.42	2.38
20	4.35	3.49	3.10	2.87	2.71	2.60	2.51	2.45	2.39	2.35
21	4.32	3.47	3.07	2.84	2.68	2.57	2.49	2.42	2.37	2.32
22	4.30	3.44	3.05	2.82	2.66	2.55	2.46	2.40	2.34	2.30
23	4.28	3.42	3.03	2.80	2.64	2.53	2.44	2.37	2.32	2.27
24	4.26	3.40	3.01	2.78	2.62	2.51	2.42	2.36	2.30	2.25
25	4.24	3.39	2.99	2.76	2.60	2.49	2.40	2.34	2.28	2.24
26	4.23	3.37	2.98	2.74	2.59	2.47	2.39	2.32	2.27	2.22
27	4.21	3.35	2.96	2.73	2.57	2.46	2.37	2.31	2.25	2.20
28	4.20	3.34	2.95	2.71	2.56	2.45	2.36	2.29	2.24	2.19
29	4.18	3.33	2.93	2.70	2.55	2.43	2.35	2.28	2.22	2.18
30	4.17	3.32	2.92	2.69	2.53	2.42	2.33	2.27	2.21	2.16
40	4.08	3.23	2.84	2.61	2.45	2.34	2.25	2.18	2.12	2.08
50	4.03	3.18	2.79	2.56	2.40	2.29	2.20	2.13	2.07	2.03
60	4.00	3.15	2.76	2.53	2.37	2.25	2.17	2.10	2.04	1.99
70	3.98	3.13	2.74	2.50	2.35	2.23	2.14	2.07	2.02	1.97
80	3.96	3.11	2.72	2.49	2.33	2.21	2.13	2.06	2.00	1.95
90	3.95	3.10	2.71	2.47	2.32	2.20	2.11	2.04	1.99	1.94
100	3.94	3.09	2.70	2.46	2.31	2.19	2.10	2.03	1.97	1.93
200	3.89	3.04	2.65	2.42	2.26	2.14	2.06	1.98	1.93	1.88
500	3.86	3.01	2.62	2.39	2.23	2.12	2.03	1.96	1.90	1.85
1000	3.85	3.00	2.61	2.38	2.22	2.11	2.02	1.95	1.89	1.84

TABLE A.4 ■ Critical Values for *F*-test ($\alpha = 0.01$)

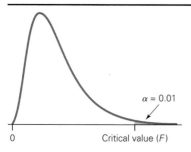

$\alpha = 0.01$

0   Critical value (*F*)

	Numerator df									
Denom df	1	2	3	4	5	6	7	8	9	10
1	4052	4999	5404	5624	5764	5859	5928	5981	6022	6056
2	98.50	99.00	99.16	99.25	99.30	99.33	99.36	99.38	99.39	99.40
3	34.12	30.82	29.46	28.71	28.24	27.91	27.67	27.49	27.34	27.23
4	21.20	18.00	16.69	15.98	15.52	15.21	14.98	14.80	14.66	14.55
5	16.26	13.27	12.06	11.39	10.97	10.67	10.46	10.29	10.16	10.05
6	13.75	10.92	9.78	9.15	8.75	8.47	8.26	8.10	7.98	7.87
7	12.25	9.55	8.45	7.85	7.46	7.19	6.99	6.84	6.72	6.62
8	11.26	8.65	7.59	7.01	6.63	6.37	6.18	6.03	5.91	5.81
9	10.56	8.02	6.99	6.42	6.06	5.80	5.61	5.47	5.35	5.26
10	10.04	7.56	6.55	5.99	5.64	5.39	5.20	5.06	4.94	4.85
11	9.65	7.21	6.22	5.67	5.32	5.07	4.89	4.74	4.63	4.54
12	9.33	6.93	5.95	5.41	5.06	4.82	4.64	4.50	4.39	4.30
13	9.07	6.70	5.74	5.21	4.86	4.62	4.44	4.30	4.19	4.10
14	8.86	6.51	5.56	5.04	4.69	4.46	4.28	4.14	4.03	3.94
15	8.68	6.36	5.42	4.89	4.56	4.32	4.14	4.00	3.89	3.80
16	8.53	6.23	5.29	4.77	4.44	4.20	4.03	3.89	3.78	3.69
17	8.40	6.11	5.19	4.67	4.34	4.10	3.93	3.79	3.68	3.59
18	8.29	6.01	5.09	4.58	4.25	4.01	3.84	3.71	3.60	3.51
19	8.18	5.93	5.01	4.50	4.17	3.94	3.77	3.63	3.52	3.43
20	8.10	5.85	4.94	4.43	4.10	3.87	3.70	3.56	3.46	3.37
21	8.02	5.78	4.87	4.37	4.04	3.81	3.64	3.51	3.40	3.31
22	7.95	5.72	4.82	4.31	3.99	3.76	3.59	3.45	3.35	3.26
23	7.88	5.66	4.76	4.26	3.94	3.71	3.54	3.41	3.30	3.21
24	7.82	5.61	4.72	4.22	3.90	3.67	3.50	3.36	3.26	3.17
25	7.77	5.57	4.68	4.18	3.85	3.63	3.46	3.32	3.22	3.13
26	7.72	5.53	4.64	4.14	3.82	3.59	3.42	3.29	3.18	3.09
27	7.68	5.49	4.60	4.11	3.78	3.56	3.39	3.26	3.15	3.06
28	7.64	5.45	4.57	4.07	3.75	3.53	3.36	3.23	3.12	3.03
29	7.60	5.42	4.54	4.04	3.73	3.50	3.33	3.20	3.09	3.00
30	7.56	5.39	4.51	4.02	3.70	3.47	3.30	3.17	3.07	2.98
40	7.31	5.18	4.31	3.83	3.51	3.29	3.12	2.99	2.89	2.80
50	7.17	5.06	4.20	3.72	3.41	3.19	3.02	2.89	2.78	2.70
60	7.08	4.98	4.13	3.65	3.34	3.12	2.95	2.82	2.72	2.63
70	7.01	4.92	4.07	3.60	3.29	3.07	2.91	2.78	2.67	2.59
80	6.96	4.88	4.04	3.56	3.26	3.04	2.87	2.74	2.64	2.55
90	6.93	4.85	4.01	3.53	3.23	3.01	2.84	2.72	2.61	2.52
100	6.90	4.82	3.98	3.51	3.21	2.99	2.82	2.69	2.59	2.50
200	6.76	4.71	3.88	3.41	3.11	2.89	2.73	2.60	2.50	2.41
500	6.69	4.65	3.82	3.36	3.05	2.84	2.68	2.55	2.44	2.36
1000	6.66	4.63	3.80	3.34	3.04	2.82	2.66	2.53	2.43	2.34

TABLE A.5 ■ Chi-square Distribution

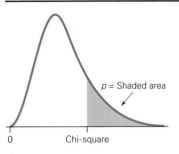

p = Shaded area

0   Chi-square

			p = Area to Right of Chi-square Value						
df	*0.50*	*0.25*	*0.10*	*0.075*	*0.05*	*0.025*	*0.01*	*0.005*	*0.001*
1	0.45	1.32	2.71	3.17	3.84	5.02	6.63	7.88	10.83
2	1.39	2.77	4.61	5.18	5.99	7.38	9.21	10.60	13.82
3	2.37	4.11	6.25	6.90	7.81	9.35	11.34	12.84	16.27
4	3.36	5.39	7.78	8.50	9.49	11.14	13.28	14.86	18.47
5	4.35	6.63	9.24	10.01	11.07	12.83	15.09	16.75	20.51
6	5.35	7.84	10.64	11.47	12.59	14.45	16.81	18.55	22.46
7	6.35	9.04	12.02	12.88	14.07	16.01	18.48	20.28	24.32
8	7.34	10.22	13.36	14.27	15.51	17.53	20.09	21.95	26.12
9	8.34	11.39	14.68	15.63	16.92	19.02	21.67	23.59	27.88
10	9.34	12.55	15.99	16.97	18.31	20.48	23.21	25.19	29.59
11	10.34	13.70	17.28	18.29	19.68	21.92	24.73	26.76	31.26
12	11.34	14.85	18.55	19.60	21.03	23.34	26.22	28.30	32.91
13	12.34	15.98	19.81	20.90	22.36	24.74	27.69	29.82	34.53
14	13.34	17.12	21.06	22.18	23.68	26.12	29.14	31.32	36.12
15	14.34	18.25	22.31	23.45	25.00	27.49	30.58	32.80	37.70
16	15.34	19.37	23.54	24.72	26.30	28.85	32.00	34.27	39.25
17	16.34	20.49	24.77	25.97	27.59	30.19	33.41	35.72	40.79
18	17.34	21.60	25.99	27.22	28.87	31.53	34.81	37.16	42.31
19	18.34	22.72	27.20	28.46	30.14	32.85	36.19	38.58	43.82
20	19.34	23.83	28.41	29.69	31.41	34.17	37.57	40.00	45.31
21	20.34	24.93	29.62	30.92	32.67	35.48	38.93	41.40	46.80
22	21.34	26.04	30.81	32.14	33.92	36.78	40.29	42.80	48.27
23	22.34	27.14	32.01	33.36	35.17	38.08	41.64	44.18	49.73
24	23.34	28.24	33.20	34.57	36.42	39.36	42.98	45.56	51.18
25	24.34	29.34	34.38	35.78	37.65	40.65	44.31	46.93	52.62
26	25.34	30.43	35.56	36.98	38.89	41.92	45.64	48.29	54.05
27	26.34	31.53	36.74	38.18	40.11	43.19	46.96	49.65	55.48
28	27.34	32.62	37.92	39.38	41.34	44.46	48.28	50.99	56.89
29	28.34	33.71	39.09	40.57	42.56	45.72	49.59	52.34	58.30
30	29.34	34.80	40.26	41.76	43.77	46.98	50.89	53.67	59.70

Answers to Selected Exercises

*The following are partial or complete answers to the exercises numbered in **bold** in the text.*

Chapter 1

1.2   a. .00043
1.5   a. 400
1.7   c. Randomized experiment.
     d. Observational study.
1.11   189/11,034, or about 17/1000, based on placebo group.
1.15   a. 150 mph.
     b. 55 mph.
     c. 80 mph.
     d. 1/2
     e. 51
1.19   No.
1.22   The base rate for that type of cancer.
1.26   a. 212/1525 = .139.
     b. $1/\sqrt{1525}$ = .026.
     c. .113 to .165.
1.28   a. Self-selected or volunteer.
     b. No. Readers with strong opinions will respond.

Chapter 2

2.1   a. Categorical.
     b. Quantitative.
     c. Quantitative.
     d. Categorical.
2.2   a. Categorical.
     b. Ordinal.
2.4   a. Not continuous.
2.5   a. Explanatory is amount of walking or running; response is lung function.
2.6   a. Population.
     b. Sample.
2.8   a. Support ban or not; categorical.
     b. Gain on verbal and math SATs after program; quantitative.

c. Smoker or not and Alzheimer sufferer or not; both categorical.
2.9   a. Question 1a.
2.11   Example: Letter grades (A, B, etc.) converted to GPA.
2.14   a. 1427/2530 = .564, or 56.4%.
     b. 100% − 56.4% = 43.6%.
2.15   a. 1700/2470 = .688, or 68.8%.
     b. 1056/1700 = .621, or 62.1%.
2.16   c. Overweight = 26.57%; about right = 69.23%; underweight = 4.20%.
2.18   a. Explanatory is smoked or not; response is developed Alzheimer's disease or not.
2.21   Either could be justified as being more informative.
2.23   a. Skewed to the right.
     b. 13 ear pierces may be an outlier.
     c. 2 ear pierces; about 45 women had this number.
     d. About 32 or so.
2.25   a. Roughly symmetric.
     b. Highest = 92.
     c. Lowest = 64.
     d. 5/20 = .25, or 25%
2.29   Example: The age of a person who is 80 years old would be an outlier at a traditional college, but not at a retirement home.
2.31   Whether it's the male author (then not an outlier) or the female author (then an outlier).
2.34   Yes. Values inconsistent with the bulk of the data will be obvious.
2.36   a. This is personal preference; some may prefer a very large family.
     b. An outlier (in the high direction).
2.37   a. Median = (72 + 76)/2 = 74; mean = 74.33.
     b. Median = 7; mean = 25.
2.39   a. Range = 225 − 123 = 102.
     b. IQR = 35.
     c. 50%

2.40 a Median = 12.

2.41 d. There are no outliers.

2.44 The median is 16.72 inches. The data values vary from 6.14 to 37.42 inches. The middle 1/2 of the data is between 12.05 and 25.37 inches, so "typical" annual rainfall covers quite a wide range.

2.48 0, 20, 55, 175, 450; data values are skewed to the right and there is an extreme outlier of 450 CDs. (Quartile values may differ slightly if using the computer.)

2.51 32, 45.5, 50, 57, 74.

2.54 a. Between 5.3 and 8.7 hours.
b. Between 3.6 and 10.4 hours.

2.56 a. $z = (200 - 170)/20 = 1.5$.

2.57 a. $\bar{x} = 20$; $s = 1.581$.

2.58 a. Range = $98 - 41 = 57$.
b. $s \approx 57/6 = 9.5$.

2.60 The Empirical Rule predicts about 68%, 95%, 99.7% within 1, 2, and 3 standard deviations of the mean; data show 72%, 97%, 98%, so the set of measurements fits well.

2.62 a. Population.
b. Population, 14.77.

2.66 a. Would hold without the two outliers; should still be close.
b. Yes, range is 10.75 cm, close to 6 standard deviations. Expect between 4 and 6 standard deviations.

2.68 Example: Male height of 73, $z = 1.00$.

2.70 a. $z = -0.5$; 0.3085.
b. $z = 2.5$; 0.9938.

2.72 50, 50, 50, 50, 50, 50, 50; no.

2.75 a. Mean = 51.47 years; standard deviation = 8.92 years (population standard deviation = 8.85).
b. Range is 42 years, 4.7 standard deviations, so it holds.
c. z for youngest CEO is -2.18, z for oldest CEO is 2.53; about as expected from the Empirical Rule.

2.78 a. Categorical.
b. Quantitative.

2.80 a. Yes; night light use, for example.
b. No.

2.84 a. Set 2—it covers a much wider range of heights.

2.86 a. Amount of beer consumed per unit of time (week, etc.) and systolic blood pressure.
b. Calories of protein consumed per day on average and whether or not they had colon cancer.

2.89 Question 5a: Do the students who studied the most last week tend to be the students with the highest grade point averages?

Chapter 3

3.4 c. Observational study.

3.6 a. Response variable is daughter's height.

3.8 a. Doesn't hold; professional basketball players do not represent any larger population.
b. Probably holds, students taking statistics most likely have representative pulse rates.

3.9 a. Probably not, because long-term meditation is a matter of choice, not easily randomly assigned.
b. Yes, volunteers could be randomly assigned to attend the program or not.

3.10 a. The explanatory variable is long-term practice of meditation; the response is blood pressure.

3.11 a. Amount of salt and sodium in the diet or other diet-related factors.

3.16 No. Although all songs have the same probability of being sampled, all *sets* of four songs do not. For instance, a sample consisting of the first four songs on the first CD is impossible.

3.19 a. No.
c. Yes.
e. No.

3.22 Completely randomized design.

3.24 a. Yes.
b. No.
c. No.

3.27 a. Single-blind; customers knew what plan they had but the technician did not.

3.28 a. Matched pairs.

3.29 a. The explanatory variable is the rate plan; the response is usage during peak hours.

3.30 a. A control group was used (matched with the treatment group); a placebo was not.

3.31 a. A random sample was used, so results can be extended to all customers.

3.37 a. Observational study.
b. The explanatory variable was whether or not the couple owned a pet, and the response variables were marriage satisfaction and stress levels.
c. An example is the amount of business travel they do. Couples who travel frequently may be less likely to own pets and also have more stress.

3.38 a. Interacting variable.

3.41 Explanatory variable is taking the new medication or not, response variable is weight gain or loss, interacting variable is gender.

3.45 a. Yes.
b. Yes.
c. No.

3.49 c. Confounding variables and the implication of causation (e.g., current diet). Relying on memory; it would be hard for people to remember how high in fat their childhood diet was.

3.52 Observational study.

3.56 Postmenopausal women who did not use hormone therapy at all.

3.60 Extending results inappropriately. If women today take lower levels of hormones than the women in the study, these results may not apply to women currently on hormone therapy.

3.63 Question 4a: Are the measurements similar across categories? Is self-esteem and self-judgment the same for the three groups, on average?

3.66 This is an example of an interacting variable, since the effect on self-esteem of thinking about their bad hair is different for men than for women.

3.67 a. Descriptive statistics.
b. Inferential statistics.

3.72 a. Individual unit is a tomato plant. Two variables were the number of tomatoes produced and whether the tomato plant was raised in full sunlight or partial shade.

3.74 a. Yes
b. Yes.
c. No.

Chapter 4

4.3 .014 or 1.4%.
4.6 c. Nonresponse bias.
4.9 b. 40% ± 3.15% or 36.85% to 43.15%.
4.12 For example, testing manufactured parts where the test damages the product.
4.15 a. .49 ± .03 = .46 to .52.
b. .47 ± .03 = .44 to .50.
c. No, the two intervals both cover 50% and overlap with each other.
4.17 a. Selection bias.
b. Nonresponse bias.
4.22 a. Use 01 to 49 as is and subtract 50 from 51 to 99 to get 15, 23, 20, 21, 29, 5.
b. It doesn't matter.
4.26 Any list that contains the two-digit strings in any order: 00, 07, 15, 19, 24, 33, 44, 51, 65, 99. For instance, 24190 03351 99076 54415.
4.30 Simple random sample.
4.34 a. Stratified sample: Use the three types of schools as strata. Create a list of all students for each of the three strata; draw a simple random sample from each of the three lists.
b. Cluster sample: Use individual schools or individual classes as clusters. Take a random sample of clusters; measure all students in those clusters.
c. Simple random sample: Obtain a list of all students in the classes at all schools; take a simple random sample from that combined list.
4.37 Cluster sample because a sample of exchanges is found and then only numbers within those exchanges are sampled.
4.39 a. All taxpayers.
b. Parents of all the school children.
4.45 Anonymous testing.
4.48 a. Dentists who subscribe to one of two dental magazines. Yes, because not all dentists subscribe to those two magazines.
b. Nonresponse.
c. Send a reminder or call those who didn't respond.
4.51 a. Put an ad in the local paper asking people to fill out the survey.
b. Ask "Don't you agree that there is too much trash in our streets and that more public trash containers are needed?"
4.53 Any two questions in which one changes the way respondents would think about the other. An example: "Are you aware that over 30% of homeless people in this city are mothers with children?" and "Do you think more public money should be used to help homeless people?"
4.56 Desire to please; confidentiality and anonymity.
4.59 a. Open-form question.
b. Yes, because of all the publicity he had just received.
c. Probably lower.
4.65 a. All students at the university.

b. All students enrolled in statistics classes.
c. The 500 students to whom the survey was mailed.
4.67 The sample sizes or margin of error. The 20% versus 25% may be within the margin of error for the surveys.
4.70 The quickie poll would probably be most representative.
4.72 Nonresponse.
4.73 b. Send the survey to a legitimate random sample but make the questions so outrageous that only those who support the position would respond; others would not take it seriously.
4.77 No, only those likely to vote should be used.
4.80 Larger; the sample size for Republicans only would be smaller.
4.82 a. .078; .66 ± .078 or 58.2% to 73.8%.
b. .066; .38 ± .066 or 31.4% to 44.6%.
c. Yes, intervals do not overlap.

Chapter 5

5.1 a. Negative association.
b. No association.
5.2 a. Negative association.
b. Roughly linear.
c. Highest average math SAT is about 600; less than 5% took the test.
d. Lowest average math SAT is about 475; about 60% took the test.
5.3 a. Yes, both variables are quantitative.
5.7 a. The speed of the car is the explanatory variable and the stopping distance is the response variable.
b. Positive association, somewhat linear but possibly curvilinear.
5.9 a. Average weight = −250 + 6 (70) = 170 lb.
b. On average, weight increases 6 pounds per each 1-inch increase in height.
5.11 b. Predicted average = 575 − 1.11 (8) = 566.12.
c. Residual = 573 − 566.12 = 6.88.
5.12 Deterministic relationship.
5.14 a. Height increases an average of 0.7 inches for each centimeter increase in handspan.
b. 65.1 in.
c. Residual = 66.5 − 65.1 = 1.4 in.
5.20 The two variables do not have a linear association.
5.21 a. Chest girth increases as length increases.
b. The value of r is high, so the relationship apparently is strong.
c. $r = +0.82$.
5.24 It would still be 0.95. Changing the units of measurement does not change the correlation.
5.27 a. Graph 2 is the strongest; Graph 3 is the weakest.
5.31 Both variables will have relatively high values in the winter and relatively low values in the summer.
5.34 Gender differences create the observed negative correlation. Females tend to be shorter, and have more ear pierces.
5.36 a. Estimated stopping distance when speed is 80 mph is −45 + 5.7(80) = 411 ft. No, this is probably not an accurate estimate. The relationship is probably not linear.

b. From the plot the relationship looks curvilinear. Estimated stopping distance is about 500 ft.

5.38 Men and women should not be combined. Women probably have higher scores but lower heights.

5.42 b. Positive, because larger cities have more of both.
c. Negative, because strength decreases with age.

5.44 $r^2 = 0.64$.

5.47 a. For a man, $53 + 0.7(140) = 151$ pounds. For a woman, $44 + 0.6(140) = 128$ pounds. The man wants to weigh 11 pounds more; the woman wants to weigh 12 pounds less.
b. No, there are no actual weights close to 0.
c. Yes. For each increase of 1 pound in actual weight, the slope provides the estimated average increase in ideal weight, which is 0.6 pounds for women and 0.7 pounds for men.

5.50 b. -0.312
c. The correlation for hardcover books only is 0.348. The correlation for softcover books only is 0.637.

5.54 a. Perhouse = $44.9 - (0.0212)$(Year). The estimated number of persons per household in 2010 is $44.9 - (0.0212)(2010) = 2.288$ persons.
b. Slope = -0.0212, indicating that the average number of people per household decreases by about 0.02 per year, or about 0.2 every 10 years.

5.56 a. For line 1, SSE = 10; for line 2, SSE = 4.
b. Line 2 is better because SSE is smaller.

5.61 b. Diff = $-52.5 + 0.312$ (Actual). For men who weigh 150 pounds the estimated average difference is $-52.5 + 0.312(150) = -5.7$ pounds. This is actual $-$ ideal, so they want to weigh more.
d. $r^2 = 0.353$ or 35.3%.

Chapter 6

6.3 a. Yes, both variables are categorical.
c. No, both variables are quantitative.

6.4 No. Totals are given for categories of each variable but counts for combinations are not provided.

6.6 a.

Anger?	Heart Disease	No Heart Disease	Total
No Anger	8	191	199
Most Anger	59	500	559
Total	67	691	758

b. $8/199 = 4.0\%$.
c. $59/559 = 10.6\%$.

6.10 a. 1.0

6.11 a. $10/100 = .1$.

6.12 a. $.1/.05 = 2$.

6.13 a. 2.11

6.17 1.4

6.20 a. Relative risk is $.106/.04 = 2.65$.
b. Percent increase in risk is $100\% \times (10.6 - 4.0)/4.0 = 165\%$.

6.23 a. Aged 18 to 29, 13.9%; aged 30 or over, 10.6%.
b. Relative risk is $.139/.106 = 1.3$. Adults under 30 are 1.3 times as likely to report seeing a ghost as adults 30 years old or older.

6.26 Probably not. The relationship between blood pressure and religious activity would need to be reversed when separate categories of a third variable are considered.

6.28 Occupation; men may be more likely to have jobs that require long-distance driving.

6.31 a. Response variable is outcome (successful or not).
c. Compare treatments separately within the two initial severity groups.

6.34 b. Do not reject the null hypothesis because the p-value is greater than .05.
f. Reject the null hypothesis because the chi-square statistic is greater than 3.84.

6.37 a. Yes
b. Yes.
c. No.

6.40 a. Null hypothesis is that there is no relationship between gender and opinions about capital punishment. Alternative hypothesis is that there is a relationship.
b. The relationship is not statistically significant because $0.19 > 0.05$.

6.41 a. Chi-square statistic = 1.714.

6.45 a. Yes, there appears to be a relationship based on the following percents:

Pierces	% with Tattoo
2 or less	7.2%
3 or 4	11.0%
5 or 6	26.0%
7 or more	54.5%

c. $(32 + 30)/107 = 57.9\%$ of women with a tattoo have five or more ear pierces. $(91 + 25)/571 = 20.3\%$ of women with no tattoo have five or more ear pierces.
d. $107/678 = 15.8\%$.
e. $245/678 = 36.1\%$.

6.49 a.

	Cancer	Control	Total
Bird	98	101	199
No Bird	141	328	469
Total	239	429	668

b. Bird owners: $98/199 = .4925$, or 49.25%. Nonowners: $141/469 = .301$, or 30.1%.
c. No. This was a case control study in which the researchers purposely sampled more lung cancer patients than would naturally occur in a group of 668 people. The sample percents falling into the cancer category are much higher than you would see in a random sample.
d. The relative risk for bird owners is $.4925/.301 = 1.64$.
e. The "baseline" risk of lung cancer for people like yourself.
f. No. This is an observational study and there may be confounding factors that explain the results.
g. Compare the smoking habits of the bird owners and nonowners and see whether or not the bird owners generally smoke more.

6.54 Chi-square = 4.817, p-value = 0.028; relationship is statistically significant.

6.58 Using the approximate odds given in Exercise 6.57, the odds ratio is $24/8.5 = 2.82$. The precise odds ratio is found by taking $(59 \times 191)/(8 \times 500) = 2.817$. The odds of remaining free of heart disease versus getting heart disease for men with no anger are about 2.8 times the odds of those events for men with the most anger.

6.61 a. For men, the odds for Program A are 400 to 250, or about 1.6 to 1. For women, the odds are 50 to 25, or 2 to 1. The odds ratio for women compared to men is therefore 2/1.6, or 1.25. The odds of being admitted versus being denied admittance for women are about 1.25 times what they are for men, so women have better odds.

b. For men, the odds for the combined programs are 450 to 550, or about .8 to 1. For women, the odds are 175 to 325, or about .54 to 1. The odds ratio (women to men) is therefore about .54/.8 = .675. The odds of being admitted versus being denied admittance for women are about 2/3 of what they are for men, so women have worse odds.

c. The last sentence in the solutions for parts (a) and (b) can be used for this purpose.

Chapter 7

7.2 1/125

7.4 c. Yes.
d. No.

7.7 a. Relative frequency; probably determined by observing the number injured by lightning for many years, divided by the average population in those years.
b. Personal; based on her assessment of the health of the plants and growing conditions.

7.9 Random circumstance: traffic light is green or not. Determine probability by observing the proportion of time it's green over many months of driving to work.

7.11 Play over and over again and record how many games he or she won out of the total number of games played.

7.13 Example: the probability that the United States President will be assassinated this year.

7.16 17/172.

7.19 a. Yes.
b. No.

7.22 a. Yes.
b. No.
c. Yes.

7.25 a. Yes, red die cannot be both 3 and 6.
b. No, red die can be 3 in the same toss that green die is 6.

7.26 a. No, knowing A occurred changes $P(B)$ to 0.
b. Yes, knowing the red die is a 3 does not change $P(B)$.

7.29 No, knowing the woman's age changes the probability that she is fertile.

7.32 a. Getting at least one tail.
b. 7/8

7.35 a. No.
b. .8

7.38 a. 11/12
b. 11/12
c. 121/144
d. 23/144

7.41 a. Long-run relative frequency of these flights arriving on time to get to the ship.
b. No, in Miami bad weather would increase the chances of being late for both.
c. .72
d. .26

7.43 a. .0625
b. .0577
c. .5625
d. .5577

7.44 b. 2/3
c. 1/2

7.47 .085

7.50 (1/13)(48/51)

7.53 P(calculus student) $= .4$; P(Junior|calculus student) $= .15/.4 = 0.375$.

7.56 45/10000

7.58 a. 8/50
b. 7/31
c. 1/50

7.60 Use only two digits, such as 8 and 9, to represent correct guesses.

7.63 a. .12
b. .4615

7.66 They are equally likely.

7.70 a. 1/365
b. .0136
c. No, that specific event has low probability but it is not unlikely that someone in the class would have a birthday matching a teammate's family member's birthday.

7.72 No, birth outcomes are independent.

7.76 Not surprising; $1 - P$(no match) $= .1829$.

7.79 a. .27
b. .86
c. .0225
d. .0064
e. .0145

7.82 a. 40%
b. 10%
c. 31%

7.86 a. .5 (problem stated that no responses equal median).
b. .0625
c. No, they would not be independent.

7.89 a. .1225
b. No, their statuses are not independent.
c. Relative frequency.
d. Statement must be about a separate randomly selected foreign-born person in each of the two years.

7.92 485/1669

7.95 238/612

7.97 .4125

Chapter 8

8.2 a. Discrete.
b. Continuous.
c. Discrete.
d. Discrete.

8.7 a. .80

8.8   c.

k	0	1	2
P(X = k)	42/90	42/90	6/90

8.12   Continuous example: amount of rainfall during the event; S includes the range from 0 to the maximum possible rain in one day. Discrete example: number of emergency vehicles with sirens driving by during event; $S = \{0,1,2, \ldots , k\}$, where k is the logical maximum for the situation.

8.15   a. $30
   b. $0 and $100 with probabilities .9 and .1.

8.17 $-$3.95

8.20 $E(X) = 1.5$.

8.23   Standard deviation = $29.67.

8.26 $E(X) = 1.69$; not a possible value, so meaningful only as a long-run average.

8.29   a. Yes, 200, .5.
   b. 100

8.31   a. $n = 30, p = 1/6$.

8.32   a. 5

8.35   a. The probability of success does not remain the same from one trial to the next.
   b. There are not a prespecified number of trials.
   c. Outcomes are not independent from one trial to the next; probability of success changes.

8.38   a. .2051
   b. .3504

8.39   b. The desired probability is $P(X \geq 110)$, n is 500, and p is .20.

8.43   b. $z = -1.0$.

8.44   b. .3632
   c. .6368

8.46   a. X is a uniform random variable.
   b. $f(x) = 1/100$ for any x between 0 and 100; $f(x) = 0$ for any x not between 0 and 100.
   c. $P(X \leq 15$ seconds$) = (15 - 0)/100 = .15$.

8.49   a. .0808

8.51   a. -1.96

8.54   a. .3665

8.55 $P(Z \leq -.675)$ is about .25, so about 25% of the women are shorter than $-.675 \times 2.7 + 65 = 63.2$ inches.

8.58   a. .0918 (for $z = -1.33$)
   b. .0228
   c. .3830
   d. .0023

8.61   a. .0571
   b. .0749

8.65   a. 400 seconds.
   b. 20 seconds.
   c. .0228

8.68   b. $X + Y$ is normal with mean = 150, variance = 325, standard deviation = 18.03.

8.71 $P(Z < -1.21) = .1131$.

8.73   a. $P(Z \leq -0.5) = .3085$.

8.75   .0918 (for $z = -1.33$)

8.77   Probably not. It's a mixture of two bell-shaped distributions with different means.

8.80   a. X is $-$2, or $70; $P(X = -$2$) = 37/38$, $P(X = 70) = 1/38$.

b. $E(X) = -$4/38 = -$.1053$, players lose about ten cents per $2 bet over the long run.

8.83   a. Probability of success is not the same for all trials.
   b. Outcomes are not independent from one trial to the next.

8.86   a. It makes sense to assume independence unless the children influence each others' choices, or some other common factor leads them both to pick a long or short story.
   b. $P(Z > 1.77) = .0384$.

Chapter 9

9.1   a. Statistic.
   b. Parameter.
   c. Statistic.
   d. Parameter.

9.3   a. $\hat{p}$
   b. p
   c. $\hat{p}$

9.5   a. $\bar{x}$
   b. μ

9.7   a. Parameters; they represent the population values.
   b. Yes, they know the means exactly, and the mean for males is $1500 higher.

9.10   a. 1/2
   b. Integers from 0 to 6.
   c. $(1/2)^6 = 1/64$.
   d. B is binomial with $n = 6$ and $p = 1/2$; probabilities for $B = 0, 1, 2, 3, 4, 5, 6$ are 1/64, 6/64, 15/64, 20/64, 15/64, 6/64, and 1/64, respectively.

9.12   a. Mean = .5; s.d. = s.d. = $\sqrt{\dfrac{(.5)(.5)}{400}} = .025$.
   b. Mean = .5; s.d. = .0125.

9.14   a. .55
   b. $\sqrt{\dfrac{(.55)(.45)}{100}} = .0497$, or about .05.
   c. $.55 \pm 3(.05)$, or between about .40 and .70.

9.16   a. $\hat{p} = 300/500 = .60$.
   b. s.e.$(\hat{p}) = .022$.

9.20 $.05 \pm 2\sqrt{\dfrac{(.05)(.95)}{400}}$, or .03 to .07.

9.23   a. .20
   b. .18
   c. .0384
   d. .20
   e. .04

9.26   a. Yes, this falls into Situation 1 described in Section 9.3.
   b. No. For skewed population, sample size is too small.
   c. Yes, this falls into Situation 2.

9.27   a. 6
   b. 3
   c. s.d.$(\bar{x})$ decreases when sample size is increased.

9.29   a. $\bar{x}$ and s.
   b. s.e.$(\bar{x}) = 47.7/\sqrt{28} = 9.014$.

9.32   b. Approximately normal with a mean of 210 pounds and a standard deviation of 3.95.
   d. $P(z > 2.53) = .0057$.

9.33   Standard deviation of the sampling distribution is $\sigma/\sqrt{n}$.

Standard error of the mean replaces population σ with the sample version s and is more commonly used because σ is rarely known.

9.35 a. The population of possible net gains is highly skewed (not bell-shaped) and the sample size is small.

  b. $-\$0.35$; $\$9.38$

9.40 Because of the symmetry of the situation, it should be a mirror image of the sampling distribution of H, the highest number, given in Figure 9.4, so it should be highly skewed to the right.

9.43 a. Binomial; $n = 10$, $p = .5$.

  b. 5

  c. 1.58

9.45 a. $z = 2$.

  b. $z = 2.828$.

9.47 a. $z = 1$.

  b. $z = -1$.

9.50 Mean μ and standard deviation σ.

9.53 a. -1.0 and $+1.6$.

9.55 a. $t = -1$; df $= 16 - 1 = 15$.

9.57 a. $t = 0.417$.

  b. $z = 0.5$.

  c. $t = -3$.

9.60 a. t-distribution.

  b. df $= 5$.

  c. Normal distribution.

9.61 a. $t = 2.5$.

9.65 Actual population exists (students at your college); fixed proportion are left-handed; $n = 200$ is large enough for number of left-handers to be sufficient.

9.68 b. Yes; the sample proportion is almost surely in the range .447 to .553, so .70 is unlikely.

9.72 Approximately normal with a mean of 105 and a standard deviation of 1.5.

9.75 a. Approximately normal, mean $= .20$, standard deviation $= \sqrt{.2 \times .8/2500} = .008$.

  b. Not likely. We expect virtually all samples to have between 17.6% and 22.4% of viewers.

9.78 a. No, sample size is too small.

  b. Yes.

  c. No, days in January and February only do not constitute a random sample.

  d. Yes.

9.82 225

Chapter 10

10.1 c. Sample proportion. $\hat{p} = .55$.

10.2 a. 36% ± 3.5%, or 32.5% to 39.5%. With 95% confidence, we can say that the percent of American adults who feel they don't get enough sleep is between 32.5% and 39.5%.

  b. If college students generally differ from others on this question, then this sample won't correctly estimate the percent for students.

10.5 a. 55% ± 3%, or 52% to 58%.

  b. Confidence interval is entirely above 50%, so it's reasonable to say more than 50% of population thinks their weight is about right.

10.8 About $0.95 \times 200 = 190$ intervals will capture the population proportion, so about 10 intervals will not.

10.9 The probability is .95 that the difference between sample and population proportions (or percents) is less than the margin of error, so the probability that the difference is greater than the margin of error (3% here) is $1 - .95 = .05$.

10.10 A confidence interval estimates a population value (not a sample value).

10.11 a. $2\sqrt{\dfrac{.56(1 - .56)}{100}} = .0993$.

  b. .0496

10.12 a. $\hat{p} = \dfrac{166}{200} = .83$.

  b. $.83 \pm 2\sqrt{\dfrac{(.83)(1 - .83)}{200}}$, or $.83 \pm .053$. With 95% confidence, we can say that in the population of patients with this disease, the proportion who would be cured with this treatment is in this interval.

10.16 a. $\dfrac{1}{\sqrt{501}} = .045$, or 4.5%. They apparently rounded up to 5%.

10.17 a. This can be calculated as $\sqrt{\dfrac{(.03)(.97)}{883}} = .006$.

  b. $2\sqrt{\dfrac{(.03)(.97)}{883}} = .011$.

10.18 a. .90

  b. .95

  c. .98

  d. .99

10.20 a. $\hat{p} = \dfrac{220}{400} = .55$.

  b. s.e.$(\hat{p}) = \sqrt{\dfrac{.55(1 - .55)}{400}} = .025$.

  c. $.55 \pm (1.96)(.025)$, or .501 to .599.

  d. $.55 \pm (2.33)(.025)$, or .492 to .608.

10.23 $.83 \pm 2.33\sqrt{\dfrac{(.83)(.17)}{200}}$, or $.83 \pm .062$. The 98% confidence interval is wider than the 95% interval.

10.25 $z^* = 1.28$. In a standard normal curve, the area under the curve between -1.28 and $+1.28$ is about .80.

10.26 a. 5% (or, .05).

  b. About $n = 2500$.

10.28 The 28% claim is not reasonable because it is not within the confidence interval.

10.32 At least $n = 400$.

10.34 95% confidence intervals for population percents in the two years do not overlap. It's reasonable to conclude that the population percent is higher in 2002.

10.35 a. $\hat{p} = \dfrac{45}{164} = .274$.

  b. $\sqrt{\dfrac{(.274)(1 - .274)}{164}} = .035$.

  c. $.274 \pm (2 \times .035)$, or $.274 \pm .07$.

d. The interval contains $1/4 = .25$, which is the proportion that would result from guessing. Based on this interval, we cannot rule out the possibility that the participants are merely guessing.

10.37 $.19 \pm 1.645 \sqrt{\dfrac{.19(1 - .19)}{300}}$.

10.40 a. Observational study. The researchers did not assign individuals to these groups.

b. There may be other differences between the two groups that could account for the difference. For instance, policeman may have worse diets or may smoke more.

f. The confidence intervals do not overlap so it is reasonable to conclude the two groups differ.

10.46 5%.

10.50 A confidence interval should not be computed because this is not a randomly selected sample, nor is it likely to be a representative sample. It is a self-selected sample, and this type of sample often is biased toward a particular trait or opinion.

Chapter 11

11.3 a. No.
b. Yes.
c. No.

11.5 a. There are five choices, so $p = 1/5$ or $.2$.
b. $H_0: p = .2$, $H_a: p < .2$.

11.7 Before they look at the data.

11.9 a. Population is all school-aged children in the state; p is the proportion of them living with one or more grandparents.

11.11 a. Do not reject the null hypothesis.
b. Reject the null hypothesis.

11.13 a. There is not sufficient evidence to conclude that smokers are more likely to get the disease.

11.17 a. Cannot reject the null hypothesis.
b. Reject the null hypothesis (or accept the alternative hypothesis).
c. Reject the null hypothesis (or accept the alternative hypothesis).

11.20 a. 0.0358
b. 0.0228
c. 0.1379
d. 0.00001

11.22 a. Yes.
b. No.
c. No.
d. No.

11.24 b. -2.68

11.25 b. 0.0037

11.26 Null standard error uses the null value p_0, and standard error uses the sample proportion.

11.28 a. No, don't know if it was a random sample.
b. Step 1: $H_0: p = 0.1$, $H_a: p > 0.1$.
Step 2: Sample size condition is met; see part (a) for other. Sample proportion is .12,

$$z = \frac{.12 - .1}{\sqrt{\dfrac{.1(.9)}{150}}} = 0.82$$

Step 3: Exact p-value = 0.242; p-value for z is 0.206 (or 0.207).
Step 4: Cannot reject the null hypothesis; result is not statistically significant.
Step 5: Cannot conclude that artists are more likely to be left-handed than the general population.

11.30 a. No, $0.07 > 0.05$.
b. Yes.

11.33 a. .0725
d. .0145

11.35 Very large.

11.37 a. 1000
b. .45
c. .45

11.41 Probably the finding was that there was no *significant* difference, in the statistical meaning, not the ordinary-language meaning. It could be because there really is no difference, or because the sample was too small to detect it and a type 2 error was made.

11.44 a. False.
b. True.
c. False.
d. False.

11.46 a. Correct decision.
b. Type 2 error.

11.49 a. Example: a diagnosis that leads to major surgery.
b. Example: an infection that could be cured by antibiotics but gets very serious if untreated.

11.52 c. Type 1 is probably more serious—voting for the bill that implements substantial change when actually a majority do not support it.
d. Type 1.

11.56 Step 1: $H_0: p = .1$, $H_a: p \neq .1$.
Step 2: Sample conditions are met. Sample proportion is $20/240 = .0833$,

$$z = \frac{.0833 - .1}{\sqrt{\dfrac{.1(.9)}{240}}} = -.86$$

Step 3: Exact p-value = 0.395; p-value for z is 0.3898.
Step 4: Cannot reject the null hypothesis; result is not statistically significant.
Step 5: Cannot conclude that UCD students' proportion who are left-handed differs from national proportion.

11.59 a. Population parameter = proportion of all people who suffer from this chronic pain that would experience temporary relief if taking the new medication. Hypotheses are $H_0: p = .7$, $H_a: p > .7$.

11.61 a. 0.1470
b. 0.0375

11.64 $z > 0$.

11.66 b. Population—only this class is of interest.

11.69 Assume students represent a random sample of a population of interest on this question and p = proportion of population who would choose first drink. $H_0: p = .5$, $H_a: p > .5$. Both 86 and 73 are large enough to meet conditions. $z = 1.03$, p-value = 0.1515, not statistically significant. Cannot conclude that the first drink presented is more likely to be selected.

Chapter 12

12.1 a. What proportion of parents in the school district support the new program?

b. Parameter is p = proportion of parents in school district who support the new program.

c. Sample estimate is $\hat{p} = 104/300 = .347$.

12.4 a. Is the mean number of CDs owned different for the populations of males and females?

b. Parameter $= \mu_1 - \mu_2$, where μ_1 = population mean for one gender (e.g., males) and μ_2 = population mean for the other (e.g., females).

c. $\bar{x}_1 - \bar{x}_1 = 110 - 90 = 20$.

12.5 a. Two independent samples.

12.8 One research question of interest would be, "What is the mean difference between hours spent studying and hours spent socializing?" The parameter is μ_d = mean difference for all students at the university. The sample estimate is $\bar{d}$ = mean difference for the 100 students in the sample.

12.9 Paired data. Two variables are measured on each individual.

12.13 a. s.e.$(\bar{x}) = \dfrac{2.7}{\sqrt{81}}$ 0.3; approximate 95% CI is 64.2 ± 0.6.

12.14 a. 2.2 cm.

b. $\sqrt{\dfrac{2.4^2}{36} + \dfrac{1.8^2}{36}} = 0.5$.

c. $2.2 \pm (2 \times 0.5)$, or 1.2 to 3.2 cm.

12.15 a. s.e. $(\bar{x}) = \dfrac{2}{\sqrt{64}} = 0.25$ cm. Roughly, for samples of this size, the average difference between the sample and population means is 0.25 cm.

12.18 Calculate the difference for each person. The standard error of the sample mean difference is calculated using the formula $s/\sqrt{n}$, where s = standard deviation of the sample of differences and $n = 50$.

12.21 d. The formula for a confidence interval for the difference between two independent means was used. The calculation is $(2.37 - 1.95) \pm 2\sqrt{\dfrac{1.87^2}{59} + \dfrac{1.51^2}{116}}$.

12.22 a. $(.63 - .29) \pm 2\sqrt{\dfrac{(.63)(1 - .63)}{300} + \dfrac{(.29)(1 - .29)}{300}}$.

12.24 a. $76 \pm 2.31\dfrac{6}{\sqrt{9}}$.

e. $100 \pm 2.13\dfrac{8}{\sqrt{16}}$.

12.26 a. $t^* = 2.52$ (df = 21).

b. $t^* = 2.13$ (df = 4).

12.31 a. Mean $\pm 2s$, which is $284.9 \pm 2 \times 96.1$, which is 284.9 ± 192.2. This interval describes expenditures for individuals and is not a confidence interval for a parameter.

12.34 a. Two-sample.

b. One–sample (for paired difference).

c. Two-sample.

12.35 a. The population means differ.

b. We cannot say the population means differ.

12.36 a. $(11.6 - 10.7) \pm 2.07\sqrt{\dfrac{3.39^2}{16} + \dfrac{2.59^2}{10}}$.

12.37 The variances (or, standard deviations) in the two populations are equal.

12.41 $s_p = 1.7$. The 95% confidence interval is $(8.1 - 4.5) \pm 2.01\sqrt{\dfrac{1.7^2}{23} + \dfrac{1.7^2}{25}}$.

12.43 a. Yes, there are two independent samples with sufficiently large sample sizes.

b. No, the sample sizes are too small.

c. No, methods of Section 12.5 should be used because the question is about mean values.

12.47 b. $(.466 - .380) \pm 2\sqrt{\dfrac{.466(1 - .466)}{4594} + \dfrac{.380(1 - .380)}{7076}}$.

12.48 This is a difference between the proportions in two different categories of a variable in the same sample. The method in Section 12.6 is for a difference between proportions with the same trait in two independent samples.

12.51 c. The interval is approximately $72.8 \pm 2\dfrac{72.2}{\sqrt{204}}$, or 72.8 ± 10.1.

12.53 Both intervals are entirely above 1, so it is reasonable to conclude that risk of death is greater in the abnormal heart rate recovery group than in the normal heart rate group.

12.59 a. The interval will be wider for 95% confidence.

b. The interval will become more narrow because the standard error will decrease.

c. Assuming the standard deviation stays the same, the interval width stays the same.

12.62 a. On each team, the ninth weight is the weight of the coxswain, who gives instructions about the rowing cadence but does not row. A coxswain's weight is much less than the weights of the rowers, so the distribution of weights is not symmetric.

12.66 b. This is paired data. The difference between the ages of a husband and wife can be determined for each married couple.

12.69 a. A "success" is an interval that captures the population mean.

b. .90

e. $.90^{100}$

Chapter 13

13.1 a. Parameter = proportion of adult Americans between the ages of 18 and 30 who think the use of marijuana should be legalized; null value = .5.

c. Parameter = difference in mean ages of death for left- and right-handed people; null value = 0.

13.4 a. 1.11

d. 4.00

13.6 a. Rejection region is $t \geq 1.72$; reject H_0.

c. Rejection region is $|t| \geq 2.09$; reject H_0.

g. Rejection region is $t \leq -1.83$; do not reject H_0.

13.9 a. H_0: $\mu = 72$ beats per minute, H_a: $\mu \neq 72$. Since $n = 57$ is large, the test can proceed. Test statistic is $t = -1.20$, df = 56, exact p-value = $2(0.1176) = 0.2352$. Cannot reject the null hypothesis; mean male pulse rate is not significantly different from 72.

13.10 $t = 3.4$; reject the null hypothesis.

13.13 b. H_a: $\mu_1 - \mu_2 < 0$.
    c. H_a: $\mu_1 - \mu_2 < 0$.

13.15 a. 1.414

13.17 Plots show necessary conditions are satisfied, $t = .43$; don't reject the null hypothesis.

13.20 a. Unpooled $t = 3.97$, pooled $t = 3.96$; reject the null hypothesis.
    b. Can use pooled because sample sizes and standard deviations are close.

13.23 a. 0.0401
    b. 0.9599

13.24 a. Reject the null hypothesis or accept the alternative hypothesis.
    b. Reject the null hypothesis or accept the alternative hypothesis.

13.25 c. $z < -1.645$; do not reject the null hypothesis.
    d. $z > 1.645$; do not reject the null hypothesis.

13.28 $z = 3.26$, $p = .001$; reject the null hypothesis and conclude a higher proportion of women would claim that they would return the money.

13.33 a. No.
    c. No.
    e. Yes.

13.36 a. $\mu_1 - \mu_2$ where μ_1 and μ_2 are the mean running times for the 50-yard dash for the populations of first-grade boys and girls, respectively.

13.38 a. Reject H_0 for $\alpha = 0.05$.
    b. Do not reject H_0 for $\alpha = 0.025$.

13.40 a. Can't tell. The entire interval could be greater than 10 or it could include 10.

13.42 a. Estimate the difference in proportions of college men and women who would answer yes, $p_1 - p_2$, and test whether $p_1 - p_2 = 0$. May also want to estimate p_1 and p_2 separately.

13.45 a. Type 1.
    c. Type 1.

13.47 a. 68% of the differences fall in the range -2 to $+6$.
    c. .695

13.49 Items 2, 4, and 5.

13.53 Larger samples should be used. Power for samples of 500 is only .2768.

13.55 a. .632

13.58 a. .23
    d. No.

13.61 a. All except item 5 rely on knowing the p-value.
    b. Items 2, 3, and 4.

13.63 b. $z = 3.19$; samples are large enough and independent, p-value = 0.001, conclude the population proportions were significantly different.

13.66 b. Type 1, people would take aspirin although it doesn't help; type 2, people would not take aspirin and heart attacks that could be prevented would not be; type 2 is more serious.

13.69 Reject the null hypothesis using $\alpha = 0.025$.

13.72 $z = 0.98$, p-value = 0.163 (exact p-value = 0.189); do not reject H_0; cannot guess at significantly better than chance level.

13.76 a. $z = 0.62$, p-value = 0.266; do not reject H_0 for $\alpha = 0.05$.
    b. $z = 4.59$; yes.

13.79 a. Type 1, patients are retested perhaps without cause; type 2, patients need to be retested but are not, so the cholesterol readings may be too high.

b. Type 2 seems more serious because a medical decision is not made using the most accurate data.

13.82 $z = 3.30$, p-value = 0.001; conclude proportions were significantly different.

Chapter 14

14.1 a. $\hat{y} = 119 - 1.64(40) = 53.4$.
    b. $y - \hat{y} = 50 - 53.4 = -3.4$.

14.3 a. $s = 2.837$ is roughly the average deviation of y-values from the sample regression line.
    b. Geographic latitude explains 91.7% of the variation in mean April temperatures.

14.4 a. The p-value is not less than 0.05 so result is not statistically significant.
    b. 0.552

14.7 a. $r^2 = \dfrac{\text{SSTO} - \text{SSE}}{\text{SSTO}} = \dfrac{500 - 300}{500} = 0.40$.

14.8 a. $\hat{y}_i = 513.79$, or about 514 ft.
    b. $e_i = 11.21$, or about 11 ft.
    c. Approximately 514 ± 100 ft.
    d. 650 ft is more than two standard deviations from the mean distance for drivers 21 years old, so it would be unusual.

14.10 a. $b_1 = -0.269$.
    b. $-0.26917 \pm (2 \times 0.06616)$, or about -0.401 to -0.137.
    c. H_0: $\beta_1 = 0$ versus H_a: $\beta_1 \neq 0$.
    d. $t = -4.07$, df = 114.
    e. The relationship is statistically significant (p-value = 0.000).

14.11 a. $s = 1.509$. This is roughly the average deviation of individual y-values from the regression line.

14.12 $b_0 = 7.56$. This is the mean hours of sleep for students who studied 0 hours.

14.13 d. The result may not have practical significance. The slope indicates only a small increase in television watching per one-year increase in age, and R^2 is small so the relationship is weak.

14.14 b. SSE = 18.
    c. $s = \sqrt{\dfrac{18}{6 - 2}} = 2.12$.

14.17 a. Prediction interval for a value of y.
    b. Confidence interval for the mean.

14.18 a. Grade point average of an individual who misses two classes per week.
    b. Mean grade point average of all who miss two classes per week.

14.21 a. $\hat{y}_i = 3.59 + (.967)(40)$.
    d. The width of the prediction interval reflects the variation among the individual ages of husbands married to women 40 years old. The confidence interval is narrow because, with $n = 170$, the mean age can be estimated with good precision.

14.22 $42.258 \pm (2 \times 0.313)$, which gives approximately the interval provided by Minitab.

14.23 $(3.746, 9.750)$. Of students in the population who studied 3 hours, about 95% slept between 3.746 and 9.75 hours. Or, the probability is .95 that the hours of sleep will be in the given interval for a student randomly selected from the population of students who studied 3 hours.

14.25 a. Conditions 2 and 4.
　　b. Conditions 1, 2, and 3.
　　c. Condition 5.
　　d. Conditions 1, 2, and 3.
14.27 There is an outlier; aside from the outlier the distribution might be normal.
14.28 Conditions 1, 2, and 3 hold. The form of the equation is probably correct, there are no outliers, and the standard deviation is about the same for all values of x.
14.29 Condition 4 seems to hold. It's not unreasonable to assume that the residuals have a normal distribution.
14.31 a. There appears to be an outlier so Condition 2 is violated.
14.32 c. The plot will have a U shape.
14.36 b. 0.165
14.37 b. $\hat{y} = 25.38 + 0.7069x$.
　　c. $\hat{y} = 78.4$.
　　d. 72.97 to 83.83.
14.39 a. $R^2 = 44.7\%$. Father's height explains about 44.7% of the variation in son's height.
　　b. $r = \sqrt{0.447} = 0.67$.
14.42 a. $H_0: \rho = 0$ versus Ha: $\rho \neq 0$. It may be that in reality the researchers' alternative hypothesis was $\rho > 0$, but computer software routinely provides a p-value for the two-sided alternative. The researchers have decided to reject the null hypotheses.
14.46 a. The two observations with the greatest chest girths could be outliers, or there might be a gently curving pattern.
　　b. $\hat{y} = -206.9 + 10.76x$.
　　c. $R^2 = 95.1\%$.
　　d. $\hat{y}_i = 223.5$.
　　e. 95% prediction interval is (180.47, 266.49).
　　f. 95% confidence interval is (211.41, 235.55).

Chapter 15

15.1 a. Appropriate. Both variables are categorical.
　　b. Not appropriate. Both variables are quantitative.
15.2 a. p-value = 0.05.
　　b. $0.10 < p$-value < 0.25 (0.153).
15.3 b. 16.81 (df = 6).
15.5 a. No, $\chi^2 < 6.63$.
　　b. No, $\chi^2 < 6.63$.
　　c Yes, $\chi^2 > 13.28$.
　　d. No, $\chi^2 < 30.58$.
15.7 a. H_0: Gender and type of class taken are not related for the population of students. H_a: Gender and type of class taken are related for the population of students.
　　b. df = 1.
　　c. 1.258
　　d. p-value = 0.262.
　　e. Do not reject the null hypothesis for $\alpha = 0.05$. The relationship between gender and type of class taken is not statistically significant.
15.10 c. H_0: Opinion about death penalty and opinion about marijuana legalization are not related; H_a: Opinion about death penalty and opinion about marijuana legalization are related.

　　d. $\chi^2 = 3.92$, df = 1, p-value < 0.05 (about 0.048).
　　e. $\chi^2 = 933(152 \times 159 - 561 \times 61)^2/(713 \times 220 \times 213 \times 720) = 3.92$.
15.13 For expected counts, the proportion with an ear infection is .22 within each treatment.
15.15 No. The response variable has 3 categories.
15.18 a. Given that 2 out of 10 participants have reduced pain, what is the probability that both of them would be in the magnet-treated group?
15.20 f. We're not telling; ask an engineer.
15.21 a. The researcher probably thought that short students would be more likely to be bullied. That is a one-sided hypothesis, so a z-test is appropriate.
　　c. $z = 3.02$, p-value = 0.001.
15.22 a. 100 for each of the 3 categories.
15.24 a. No.
　　b. Yes.
　　c. Yes.
　　d. No.
　　e. No.
　　f. No.
15.26 a. $H_0: p_1 = .3$, $p_2 = .6$, $p_3 = .1$ where p_1, p_2, and p_3 are the population proportions who drove, biked, and used another way that day, respectively.
　　c. 6.67
15.28 $\chi^2 = 1.766$, df = 2, $0.25 < p$-value < 0.50 (0.413).
15.30 a. HH, TH, HT, TT. $X = 0$, $p = 1/4$; $X = 1$, $p = 1/2$; $X = 2$, $p = 1/4$.
　　b. $H_0: p_0 = .25$, $p_1 = .50$, $p_2 = .25$.
　　c. $\chi^2 = 7.2$, df = 2, $0.025 < p$-value < 0.05 (0.027).
15.34 b. H_0: Number of ear pierces and likelihood of having a tattoo are not related, H_a: Number of ear pierces and likelihood of having a tattoo are related.
　　c. $\chi^2 = 90.544$, df = 3, p-value < 0.001. Decide H_a.
15.36 $\chi^2 = 15.8$, df = 5, $0.005 < p$-value < 0.01. Reject H_0: proportions are as stated at M&M website, conclude at least two proportions are not as stated.
15.38 a. $n = 10$, $p = .5$.

　　c.

Number Correct	Null Probabilities
≤2	.0547
3	.1172
4	.2051
5	.2461
6	.2051
7	.1172
≥8	.0547

15.40 a. The conditions are met.
　　b. $\chi^2 = 157$, df = 12.
15.44 a. $\chi^2 = 0.267$, df = 1, not significant.
　　b. $\chi^2 = 13.025$, df = 1, significant.
　　c. $\chi^2 = 6.748$, df = 1, significant.
　　d. $\chi^2 = 2.356$, df = 1, not significant.
15.47 Between 0 and 5.99.
15.49 b. $\chi^2 = 15.5$, df = 4, $0.001 < p$-value < 0.005. There is a statistically significant relationship.
　　c. The largest contributions occur in cells where student eye color and eye color student finds attractive are both blue or both brown.
15.51. a. Homogeneity.

Chapter 16

16.1 a. Appropriate.
d. Not appropriate. Not a comparison of independent groups.

16.2 a. Mean body weight is equal for the populations in the three age groups; $H_0: \mu_1 = \mu_2 = \mu_3$.
b. Mean body weight is not equal for the populations in the three age groups.

16.5. a. 1.018 to 4.362; interval does not cover 0 so it's reasonable to say means differ.

16.6 a. Reject the null; F-statistic is greater than 3.89, the critical value.

16.8 a. 60+ age group is significantly different from all other age groups.
b. 95% confident that all six intervals capture the corresponding population parameters.

16.9 b. Variation is about the same for each lab and there are no outliers.

16.10 a. $H_0: \mu_1 = \mu_2 = \mu_3 = \mu_4 = \mu_5$ versus H_a: Not all μ_i the same.
b. Yes. $F = 4.53$, p-value = 0.003.
c. p-value is area to the right of 4.53 under a skewed curve.
d. The mean for lab 4 is lower than the means for labs 2 and 5. Other comparisons are more difficult to judge.

16.12 Completed table is:

Source	df	SS	MS	F
Between groups	5	40	8	1.333
Error	10	60	6	
Total	15	100		

16.14 c. MSE = 1043/150 = 6.953; $s_p = \sqrt{6.953} = 2.637$.

16.17 a.

Source	df	SS	MS	F	p
Caffeine	2	61.40	30.70	6.18	0.006
Error	27	134.10	4.97		
Total	29	195.50			

b. $s_p = \sqrt{4.97} = 2.23$.

16.18 Each interval has the form $\bar{x}_i \pm 2.01\dfrac{0.4058}{\sqrt{11}}$, or $\bar{x}_i \pm 0.246$.

16.19 b. $F = 50/10 = 5$.
c. df = 3 and 20.

16.21 a. $\bar{x} = 8, \bar{x}_1 = 4, \bar{x}_2 = 11, \bar{x}_3 = 9$.
b. SS Groups = 78.
c. SS Total = 118.
d. SSE = 118 − 78 = 40.

16.23 a. Null is that median ratings would be equal for populations of students from the four types of hometowns; alternative is that median ratings would not be equal for the four types of hometowns.
b. p-value = 0.000; conclude that median ratings would not be equal.
c. Big city, median = 5.

16.25 a. Ordinal.

16.27 a. H_0: Medians are the same for the three populations of occupational groups; H_a: Population medians are not all the same.
b. Decide in favor of H_a (p-value = 0.002).

16.28 a. Interaction; amount of difference between members and nonmembers of Greek organizations depends upon gender.

16.30 a. Males.
b. Students from big cities rate rock music lower than students from other types of hometowns.
c. Pattern of relationship between rating and hometown is about the same for males and females so the interaction is weak or nonexistent.

16.33 a. The amount of difference between mean GPA of men and women would depend upon seat location.
b. The amount of difference between men's and women's mean GPAs would be the same in each seat location.

16.35 b. As the bacteria amount increases, the difference in mean rot for 10°C and 16°C increases.

16.38 a. $H_0: \mu_1 = \mu_2 = \mu_3 = \mu_4 = \mu_5$ versus H_a: Not all μ_i the same.
b. Population might be all public school children of this age in the United States.
c. Decide in favor of H_a. Not all the means are the same.
d. Perhaps children who engage in more violent behaviors also like to watch more television.

16.40 There may be interaction. The difference between boys and girls is smaller in the 6+ hours of television group than it is in other television watching groups.

16.43 SSE = 15,286, $s_p = \sqrt{46.18} = 6.796$.

16.44 Regression methods could be used. Both variables are quantitative.

Chapter 17

17.2 Yes, X is a hypergeometric random variable with $N = 250$, $n = 5$, $a = 100$.

17.5 $N = 100$, $n = 12$, $a = 40$.

17.6 X is an integer between 0 and the smaller of (a, n), so in this case it's an integer between 0 and 12.

17.9 X = number of winning numbers between 1 and 31; X is a hypergeometric random variable with $N = 51$, $n = 6$, and $a = 31$. $P(X = 6) = .0409$.

17.13 c. X is a binomial random variable; $\mu = np = 40$ and $\sigma = 5.66$.
d. X is a hypergeometric random variable; $\mu = np = 20$ and $\sigma = 2.836$.

17.16 0.0067

17.19 $P(X \geq 20) = 0.0035$.

17.22 a. $\mu = 30 \times 2 = 60$.

17.23 a. Mean = variance = 4, standard deviation = 2.

17.24 a. $e^{-4} = .0183$.

17.27 a. $n = 200, p = .01$.
b. X is approximately a Poisson random variable with $\mu = np = 2$, so the Poisson formula can be used to find probabilities related to X.
c. $P(X = 0) = (.99)^{200} = .1340$.
d. Approximate $P(X = 0) = e^{-2} = .1353$. The answers are very similar. There is about a 13.5% chance of no defective candies in a bag.

17.31 Mean = $np_i = 30(1/6) = 5$, variance = $np_i(1 - p_i) = 30(1/6)(5/6) = 4.167$.

17.34 Yes, this is a multinomial experiment. There are 1000 "trials" with 3 possible outcomes on each one and the outcomes are independent from one person to the next.

Chapter 18

18.1   a. Yes. H_0: η = median pulse = 72, H_a: η = median pulse > 72.

18.4   b. Sample median = $355.
    c. $S^+ = 4, S^- = 12, n_u = 16$.
    d. Find $P(X \le 4)$ in a binomial distribution with $n = 16$ and $p = .5$.
    e. Reject H_0, conclude that the median is less than $400. Exact p-value = 0.0384. With normal approximation but no continuity correction, $z = -2$, p-value = 0.0228. With continuity correction, $z = -1.75$, p-value = 0.0401.

18.7   a. H_0: Distribution of pulse rates is the same for populations of those who do and do not exercise.
    H_a: Pulse rates tend to be greater than for those who do not exercise (or lower for those who do).
    b. $W = 103.5$.
    c. $\mu_W = 8\dfrac{1 + 20}{2} = 84$,

$$\sigma_W = \sqrt{\frac{8(12)(8 + 12 + 1)}{12}} = 12.96.$$

    d. $z = \dfrac{103.5 - 84}{12.96} = 1.50$, p-value = $P(Z \ge 1.50) = .0668$.
    With continuity correction, $z = 1.466$, p-value = 0.0713.
    e. Cannot reject H_0; cannot say those who do not exercise tend to have higher pulse rates.

18.9   There's an obvious outlier in the diet plan group (value = 35).

18.14   a. $T^+ = 7.5 + 15 + 3 + 1 = 26.5$.
    b. $z = \dfrac{26.5 - 68}{19.339} = -2.12$.
    c. p-value = $P(Z < -2.12) = 0.017$. Reject H_0, conclude that in the population, the median amount spent is less than $400.

18.18   c. Median = 0 doesn't mean that the two measurements are equal for all people. Median is ≈ 0 if forearm is larger for half of all people and foot is longer for the other half. Also, failing to reject H_0 might be a type 2 error.

18.19   a. $\bar{R}_1 = 10, \bar{R}_2 = 12.8333, \bar{R}_3 = 5.6667$.
    b. $H = \dfrac{12}{18(18 + 1)}[6(10 - 9.5)^2 + 6(12.8333 - 9.5)^2 + 6(5.6667 - 9.5)^2] \approx 5.49$.

18.21   a. H_0: Distribution of weekly times of exercising is the same for each reason.
    H_a: Weekly times of exercising tends to be larger for at least one reason than for other reasons.

Chapter 19

19.4   Necessary conditions look to be supported. The plot shows a roughly horizontal band of points with no obvious pattern.

19.6   a. $\hat{y}_i = 20.8 - 0.415(49) + 1.24(56) = 69.905$.
    b. $e_i = y_i - \hat{y}_i = 64 - 69.905 = -5.905$.

19.8   Intercept estimates average height of males students whose mothers and fathers have height = 0. This is impossible.

19.9   0.539 inches (or 0.53897). This is the value of b_2.

19.10   a. 0.031. Reject H, conclude that the student height and mother's height are related.
    b. $t = \dfrac{0.19294}{0.08777} = 2.20$.

19.15   a. H_0: $\beta_1 = \beta_2 = 0$; H_a: At least one of β_1 and β_2 is $\ne 0$.
    b. Reject H_0 ($F = 15.19$, p-value = 0.000). Conclude that left arm length is related to at least one of left foot length and/or right foot length.

19.17   a. $s = 1.796$ cm. Roughly, the average absolute size of the deviations between actual and predicted arm lengths is about 1.796 cm.
    b. $R^2 = 36.9\%$. The foot length variables explain 36.9% of the variation among arm lengths.

19.18   a. $b_0 = 119.58, b_1 = 0.9636, b_2 = -1.0903, b_3 = 2.6295, b_4 = -1.2080, b_5 = -0.6413$.

19.21   a. $y_i = \beta_0 + \beta_1 x_{i1} + \beta_2 x_{i2} + \epsilon_i$, where we assume that ϵ_i are normally distributed with mean 0 and standard deviation σ.

Chapter 20

20.1   a. Response variable is weight loss after two months. Factor A is whether the person smokes or not, $a = 2$. Factor B is the exercise program, $b = 3$.

20.5   a. $\mu_{12} = 8.0$ hours.
    b. $\mu_{..} = 28.4/4 = 7.1$ hours.
    c. $\mu_{1.} = 7.0$ hours, $\mu_{2.} = 7.2$ hours.
    d. $\mu_{.1} = 6.5$ hours; $\mu_{.2} = 7.7$ hours.

20.8 $\mu_{11} = \mu_{..} + \alpha_1 + \beta_1 + \alpha\beta_{11} = 7.1 - 0.1 - 0.6 - 0.4 = 6.0$ hours.

20.10   a. Yes. Mean DDT appears to be much higher in the Arctic region than in the United States or Canada.

20.11   a. H_0: The population mean DDT levels in falcons are the same in the Arctic, the United States, and Canada.
    H_a: For at least one of the three sites the population mean DDT level is different.

20.12   a. p-value = 0.000; reject the null hypothesis; conclude that mean DDT levels differ across sites.

20.14   For Factor A, F = 8892.70/3.44 = 2585.1.

20.17   c. H_0: The difference in mean commute times for the three routes is the same across days.
    H_a: The difference in mean commute times for the three routes depends on the day.

20.18   c. $F = 8.40$, df = (4, 81). From Table A.4, p-value < 0.01 because 8.40 > 3.56. Reject H_0. Conclude that there is an interaction, so the difference in mean commute times for the routes depends on the day of the week category.

20.22   a. 1, based on subtracting the given df values from the total value of 53.
    b. 225, found as SS/df = 450/2.
    c. 840 because df = 1.

20.24   a. From knowing df = 2, $a = 3$.

20.25   c. Degrees of freedom for groups = Total df − Error df = 53 − 48 = 5. Also, there are 6 groups, so df = 6 − 1 = 5.
    d. Degrees of freedom for error is the same, 48.

Chapter 21

21.4   a. This would be a good idea.

21.7   No, the intervals overlap and there is no clear choice.

One method produces a slightly higher mean but also has more variability.

21.10 This condition was not met. The 9-year-old girl was in immediate proximity to the practitioners, and she knew which hand she was hovering over with her hand.

21.13 The main problem is that participants were deceived into thinking they were hurting others. But there would have been no way to conduct the experiment if they had been told the truth.

21.17 The problem is that they chose the confidence level to present the outcome they desired. They should have reported the standard 95% confidence interval.

21.19 The complaint raised by the letter is that the study was not measuring what it purported to measure.

Chapter 22

22.1 Study #1; explanatory variable is taking calcium or placebo and response variable is blood pressure.

22.4 Observational study; headline is not justified.

22.6 Drinking coffee or not. People who don't drink coffee may be more likely to drink tea and may also be more likely to conceive for other reasons.

22.9 Probably. We are not told who was in the sample, but there is no indication that it was an unrepresentative group with regard to diet and conception.

22.12 a. Statistical significance; the null hypothesis is that energy and fat intake are not related to tea consumption.
b. Confounding variables.

22.13 a. There was an association but it was not strong enough to be *statistically* significant for the sample size used.

22.15 a. Amount walked per day (less than 1 mile, between 1 and 2 miles, more than 2 miles) and whether the man died during the study (yes, no).

22.18 a. No, the entire population was already measured.
b. Yes, assuming the mechanism causing geyser eruptions stays the same.

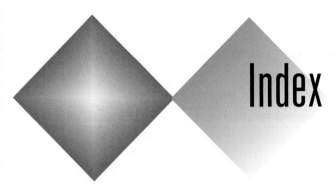

Index

Credits

This page constitutes an extension of the copyright page. We have made every effort to trace the ownership of all copyrighted material and to secure permission from copyright holders. In the event of any question arising as to the use of any material, we will be pleased to make the necessary corrections in future printings. Thanks are due to the following authors, publishers, and agents for permission to use the material indicated.

Photos

xxii © Tom Carroll/Phototake; **12** © Bob Daemmrich/Stock Boston/PictureQuest; **58** © Tom Stewart/CORBIS; **88** © Mark Scott/GettyImages; **130** © Stock Connection/PictureQuest; **170** © Joel Gordon; **202** © Will Hart/PhotoEdit/PictureQuest; **248** © Joseph Sohm, ChromoSohm Inc./CORBIS; **292** © Network Productions/Index Stock Imagery; **328** © Ron Chapple/GettyImages; **354** © Lester Lefkowitz/CORBIS; **390** © Ron Sherman/Stock Boston/PictureQuest; **436** © Stockbyte/PictureQuest; **492** © Bob Daemmrich/Stock Boston/PictureQuest; **526** © Chuck Savage/CORBIS; **560** © Mark Reinstein/Index Stock Imagery; **592** Andrew Lichtenstein/CORBIS; **608** Robert Glusic/PhotoDisc Red/GettyImages; **630** Randy Faris/CORBIS; **648** Ariel Skelley/CORBIS; **672** Peter Turnley/CORBIS; **692** © Keith Brofsky/PhotoDisc/GettyImages

Text

Chapter 1 **6** Case Study 1.7: From "Sad, Lonely World Discovered in Cyberspace," by A. Harmon, *New York Times*, August 30, 1998, p. A3. Copyright © 1998 by the New York Times Co. Reprinted by permission.

Chapter 2 **33** Example 2.9 excerpt: Amy Goldwitz, *The Davis Enterprise*, March 4, 1998, p. A1. Reprinted by permission of *The Davis Enterprise*.

Chapter 3 **66** Case Study 3.1: Sternberg, Steve, "Lead Exposure Linked to Bad Teeth in Children," *USA Today*. 23 June 1999. Internet. USA TODAY Copyright 6/23/1999. Reprinted with permission. **68** excerpt: *Sacramento Bee*, Stephanie McKinnon McDade, July 7, 1999, p. G5. Copyright, *The Sacramento Bee*, 2004. **78** Example 3.5 excerpt: "Study: Age doesn't sap memory," *Sacramento Bee*, Kathryn Dore Perkins, July 7, 1999, pp. A1 and A10. Copyright, *The Sacramento Bee*, 2004.

Chapter 4 **107** Example 4.8 excerpt: From "The Year of Believing in Prophecies," by T. Watanabe, *The Los Angeles Times*, March 31, 1999. Copyright © 1999 The Los Angeles Times Syndicate. Reprinted by permission. **117** Example 4.18 quote: Reprinted by permission of *The Davis Enterprise*.

Chapter 5 **162** table: Data source: *The World Almanac and Book of Facts 1999*. © World Almanac Education Group. All rights reserved.

Chapter 11 **377** quote: From "Sad, Lonely World Discovered in Cyberspace," by A. Harmon, *New York Times*, August 30, 1998, p. A3. Copyright © 1998 by the New York Times Co. Reprinted by permission. **381** excerpt: From "Probability Experts May Decide Pennsylvania Vote," by P. Passell, *New York Times*, April 11, 1994, A15. Copyright © 1994 by the New York Times Co. Reprinted by permission.

Chapter 22 **699** Example 22.3: "Fish Oil May Fight Psychiatric Disorders," 3 September 1998. Yahoo News Archive © 1998. **700–701** Example 22.4: "Visual Field Loss Ups Elderly Car Crashes," 7 April 1998. Yahoo News Archive © 1998. **701–702** Example 22.5: Reprinted by permission of the Associated Press. **703–704** Example 22.6: "Wine May Not Explain France's Lower Heart Disease Rate," 28 May 1999. Yahoo News Archive © 1999. **704** Example 22.7: "Women Remember Item Location Better," 26 March 1998. Yahoo News Archive © 1998.

CD Datasets and Text Examples

There are 46 datasets on the CD for this text; all are provided in MINITAB, Microsoft Excel, SPSS, JMP, and text formats. Full descriptions are in the files readme.html and printme.html within the dataset directory of the CD. When a dataset is either necessary or helpful for doing an exercise, the name of the relevant dataset is given within the exercise.

The connection between text Examples/Case Studies and the datasets is shown below. As an example, in Chapter 1, the dataset called **pennstate1** includes the data used for Case Study 1.1.

Chapter	Dataset	Where Used
1	**pennstate1**	Case Study 1.1
2	**pennstate1**	Examples 2.1, 2.3, 2.5
	MusicCDs	Example 2.8
	rainfall	Example 2.9
	pennstate1M	Example 2.10
	UCDavis1,pennstate1	Selected variables from each are used in the **Empirical** applet of Section 2.7
3	**ProfBooks**	Example 3.4
4	**chap4-heights**	Data used in **Sampling** applet of Section 4.2
5	**handheight**	Example 5.1
	signdist	Example 5.2
	pennstate1	Example 5.6
	sleepstudy	Example 5.9
	idealwt or **idealwtmen, idealwtwomen**	Case Study 5.1
6	**pennstate1**	Example 6.10
9	**Cash5**	Example 9.6
10	**pennstate3**	Case Study 10.3
11	**pennstate1**	Example 11.6
12	**pennstate5**	Example 12.2
	UCDavis1	Examples 12.6, 12.7, 12.9
	YouthRisk	Example 12.11
13	**UCDavis1**	Examples 13.4, 13.5
14	**handheight**	Example 14.1
	wthghtM	Example 14.2, Figures 14.8 and 14.9
	signdist	Example 14.3
	chugtime	Figure 14.11
15	**pennstate2**	Example 15.2
16	**UCDavis1,UCDavis2** (combined)	Examples 16.1 and 16.4 (GPAs > 4.0 deleted for Example 16.1 as was 78th observation of **UCDavis1**. They actually gave seat location as front/middle.)
	happyface	Example 16.6
18	**UCDavis2**	Examples 18.2, 18.4
19	**temperature**	Example 19.1
20	**UCDavis1**	Example 20.1
	happyface	Example 20.4